# AN INTRODUCTION TO MOLECULAR EVOLUTION AND PHYLOGENETICS

# AN INTRODUCTION TO MOLECULAR EVOLUTION AND PHYLOGENETICS

SECOND EDITION

**LINDELL BROMHAM**

Macroevolution and Macroecology,
Research School of Biology,
Australian National University

OXFORD
UNIVERSITY PRESS

Great Clarendon Street, Oxford, OX2 6DP,
United Kingdom

Oxford University Press is a department of the University of Oxford.
It furthers the University's objective of excellence in research, scholarship,
and education by publishing worldwide. Oxford is a registered trade mark of
Oxford University Press in the UK and in certain other countries



First edition 2008



Published in the United States of America by Oxford University Press
198 Madison Avenue, New York, NY 10016, United States of America

British Library Cataloguing in Publication Data
Data available

Library of Congress Control Number: 2015943760

ISBN 978–0–19–873636–3

Printed and bound by CPI Group (UK) Ltd, Croydon, CR0 4YY

For Arkady, Alexey, Gulliver and Asha,

wonderers who never cease.

# Preface

This book is designed to give a from-the-ground-up introduction to the use of molecular phylogenetic analysis in evolution and ecology. It describes how you can use analysis of DNA sequence data to reconstruct evolutionary history and processes. To do this, you need to have a basic understanding of molecular evolution: how mutations occur that change the DNA sequence, how some of these mutations become substitutions that create differences between populations or species, and how we can use the patterns of substitutions to reconstruct evolutionary history. The aims of this book, and the features it contains, are covered more fully in Chapter 1. Here is a brief summary with references to pages where each point is discussed in more detail.

## Who is this book for?

This book is aimed at biology students who need to learn how to use DNA sequence analysis to understand evolution and ecology. The book assumes no more than a basic grounding in biology and genetics, so should be suitable for entry-level undergraduate students (see Chapter 1, 'Who is this book for?', page 11). However, this book may also be useful to bioinformaticians with a background in maths, statistics, or computation, who need to learn evolutionary and genetic principles in order to interpret patterns in biological data.

## What is in this book?

This book is designed to be read in a variety of ways, depending on your interests or background knowledge (see Chapter 1, 'How to use this book', page 13). To facilitate you getting what you need from this book, each chapter is divided into several sections.

***The main text*** provides the background to key issues in molecular evolution and phylogenetics, and should be accessible to entry-level students with no previous knowledge of the subject.

- ***Points to remember*** provides a bullet-point summary of the key issues covered in the chapter, as a revision aid.
- ***Ideas for discussion*** draw on the subjects covered in the chapter, but asks you to step beyond the material covered to think more broadly about the concepts.
- ***Sequences used in this chapter*** gives the Genbank description and accession number for any sequence used for illustrative purposes in the chapter.
- ***Examples used in this chapter*** is not a comprehensive bibliography but the publication references of specific examples mentioned in the text.

There are three kinds of boxes, which provide optional additions to the main text, as follows.

TechBoxes go into more detail on specific methods or ideas. The language is more technical and they may be challenging to entry-level students, but more informative for students with some previous knowledge of molecular genetic analysis.

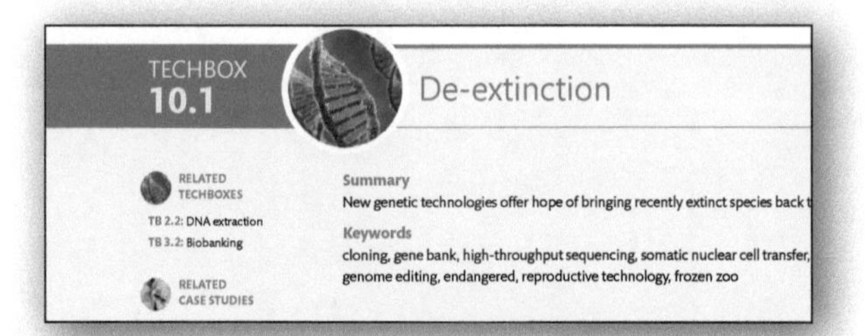


Case Studies illustrate recent scientific research related to the principles covered in the chapter, cross-referenced to TechBoxes that explain the techniques used. There will be many terms and concepts that will be unfamiliar to beginners, so they may be better suited to more advanced students. There are three ***'Check your understanding'*** questions which are based on material from the box. The ***'What do you think?'*** section aims to get you to think more deeply about the implications of the topics discussed in the box, and could be used for class discussions. ***'Delve deeper'*** encourages you to go beyond the material presented in the case study and do some independent research on a related topic. ***References*** cited in the box are listed at the end, as are cross-references to other useful boxes.

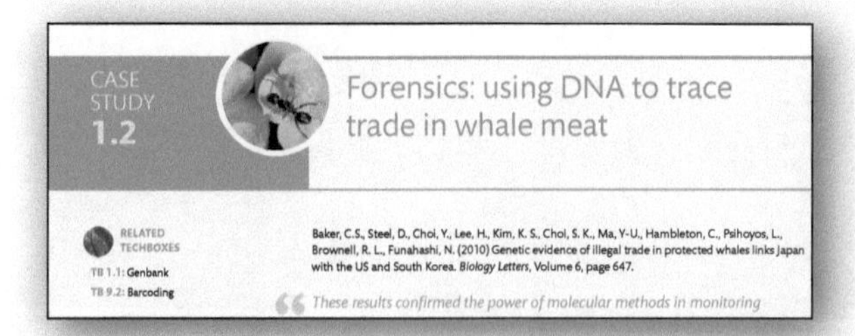

**Heroes of the Genetic Revolution** illustrate scientists who have worked in the area.

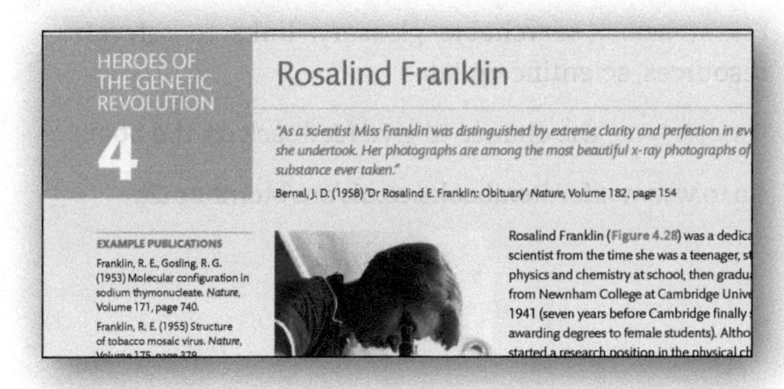
HEROES OF THE GENETIC REVOLUTION
4
Rosalind Franklin
*"As a scientist Miss Franklin was distinguished by extreme clarity and perfection in ev*
*she undertook. Her photographs are among the most beautiful x-ray photographs of*
*substance ever taken."*
Bernal, J. D. (1958) 'Dr Rosalind E. Franklin: Obituary' *Nature*, Volume 182, page 154

EXAMPLE PUBLICATIONS
Franklin, R. E., Gosling, R. G. (1953) Molecular configuration in sodium thymonucleate. *Nature*, Volume 171, page 740.
Franklin, R. E. (1955) Structure of tobacco mosaic virus. *Nature*,

Rosalind Franklin (Figure 4.28) was a dedica
scientist from the time she was a teenager, s
physics and chemistry at school, then gradu
from Newnham College at Cambridge Unive
1941 (seven years before Cambridge finally
awarding degrees to female students). Altho

## Other features of the text

**Cross-references** to information in other chapters, to help you navigate through the book and build up a knowledge base.

as unique physical characteristics are a key part of species taxonomy, unique sequence changes are exactly what we need if we want a diagnostic 'DNA barcode' to identify a species (TechBox 9.2). Secondly, autapomorphies can tell us about relative rates of change in particular biological lineages. For example, the marsupial mole has a relatively high number of autapomorphies, indicating either that it has a long, independent evolutionary history, or that it has a particularly rapid rate of change (or both).

*We look at rates of sequence change in Chapter 13*

But autapomorphies are no help in establishing relationships because, by definition, they are not shared by any of the other species in the alignment, so we can't use them to group related species together. To classify species, we need to group them on the basis of shared char-

reasonable than the alignment in Figure 10.11.

Manual sequence alignment has a number of advan
Humans are actually rather good at alignment, be
they can quickly learn to spot meaningful patter
they can interpret alternative alignments in ter
what they know about sequence evolution. For exa
a knowledgeable human will approach the alignm
a ribosomal RNA gene (with its stem and loop stru
very differently from a protein-coding gene (with its
structure) or an internal spacer or intron (which ha
of highly variable positions with the occasional isl
conserved sequence). They might spot particular pa
that help them make sense of the alignment, such a
ilarities between close relatives, or that some sequ
seem to have a faster rate of change than others.

Manual sequence alignment allows biologists

**Glossary** contains some basic biological terms and explains some tricky concepts in evolution and genetics.

**Phosphodiester bonds** join nucleotides to make a polynucleotide chain, through a covalent bond between 5' phosphate group of one nucleotide to the 3' hydroxyl group of the sugar of the next nucleotide, catalyzed by a polymerase enzyme (TechBox 2.1).

**Phylogeny** is a representation of the evolutionary history of biological lineages as a nested series of branching events, each marking the divergence of two or more lineages from a common ancestral lineage. Phylogenetics is the discipline of inferring phylogenies—there are many different phylogenetic techniques including those based on maximum likelihood and Bayesian inference. See Chapters 11 and 12.

**Phylum** (plural: phyla) is a taxonomic category: kingdoms are divided into phyla.

**Placental** mammals, more formally known as Eutheria, are one of three major lineages of mammals (the oth-

between populations or behavioural differences in ma
ing rituals), or may prevent the development or rep
duction of offspring resulting from hybrid crosses (su
as genetic incompatibility). Note that a population m
have 'fuzzy borders' if there are low levels of gene
exchange between connected populations, throu
occasional hybridization or rare migration events (s
Figure 9.10). Some populations maintain genetic d
tinctness even in the face of ongoing genetic exchan
with other populations (see Case Study 9.1).

**Positive selection**: see selection

**Posterior probability, Prior probability**: see Bayesi inference

**Prokaryotes** (bacteria and archaebacteria) are sing
celled organisms without a nucleus or mitochond
see eukaryote.

**Priors** are a special form of assumption used in Bayesi

**Figure legends** sometimes have interesting facts not directly related to the subject of the chapter, because that's what makes biology fun.

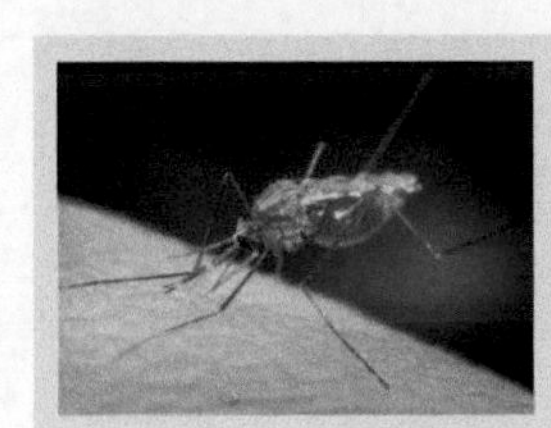

**Figure 5.9** Malaria derives its name from the early hypothe
was carried by bad air arising from swamps, but we now know i
by another factor that arises from swamps: mosquitoes. There a
of different species of *Anopheles* mosquitoes that can transmit t
different forms of *Plasmodium* parasite that cause malaria. *Anop*
*funestus*, shown here, and *Anopheles gambiae* are the main vect
malaria in Africa, so are indirectly responsible for the death of o
million people every year. Chromosomal races of these mosqui
show different habitat and climate preferences, complicating m
control programmes.
Photograph: James Gathany.

## What is not in this book?

***Instructions for computer programs:*** The aim of this book is to give you an understanding of the evolutionary principles which are the basis of molecular phylogenetic analysis, rather than to give you training in the use of particular programs. The reason that this book sticks to general principles and avoids specific methods is that new methods are constantly being developed. Any programs or methods I describe today are likely to be out of date within a few years. But by giving you the general principles underlying those methods, I hope to equip you with the intellectual tools you need to learn how to use particular methods, and to adapt your analysis when new methods become available.

***Equations:*** Unlike most molecular evolution textbooks, there is no maths in this book. This book is equation-free in an attempt to make the field of molecular evolution accessible to all biologists, including those who prefer explanations based on words or images rather than algebra (see Chapter 1, 'Where are the equations?', page 15). You might choose to read this book alongside a traditional molecular evolution text which will provide all the population genetic equations for the principles discussed here.

## What has changed since the first edition?

The first edition of this book was titled *Reading the Story in DNA: a beginner's guide to molecular evolution*. This new edition contains some of the same material, but has been considerably expanded, updated, and remodelled in response to feedback from readers. The book has increased from 8 chapters to 14, which allows the material to be divided more evenly into key concepts and may make the book better fit a semester teaching structure. There are more features designed to help students and teachers come to grips with the material, such as summary points at the end of chapters, and discussion questions in the chapters and case studies. There are now nearly twice as many case studies, and many of the previous case studies have been replaced with more recent examples. I have discovered that the problem with writing a book based on molecular genetics is that the field changes so fast that much of the first edition had been superseded only five years after its publication. For example, high-throughput (next-gen) sequencing, which took up only a paragraph in the first edition, is on its way to eclipsing Sanger sequencing as the method of choice in molecular phylogenetic analysis, so the chapters and case studies have been updated to reflect this fundamental change in data generation and analysis. It's inevitable that by the time this edition is in print, the field will have moved on even further. The best approach is to use this book as a jumping-off point for an examination of the most recent techniques in the scientific literature.

## What is in the Online Resource Centre?

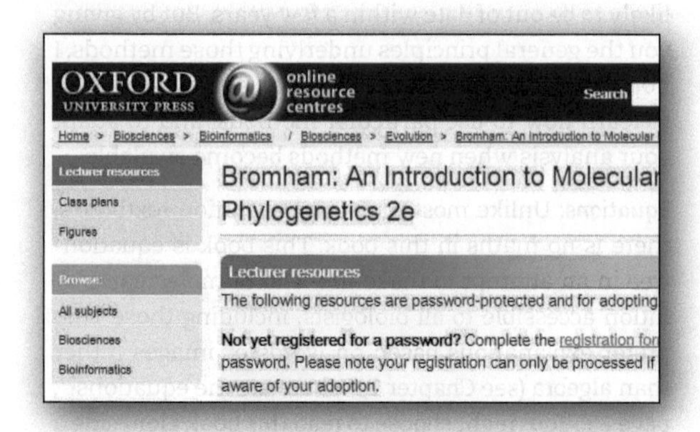


*For lecturers*: downloadable figures, tutorial exercises, practical projects.

*For students*: searchable glossary, links to relevant resources, scientific updates.

*For everyone*: the chance to give feedback on the book.

Go to **www.oxfordtextbooks.co.uk/orc/bromham2e**

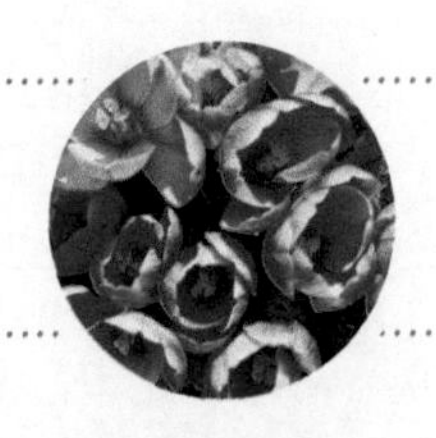

# Acknowledgements

Thanks to Jonathan Crowe for initiating this project, and for his optimism in supporting a second edition—it would never have crossed my mind to write a textbook had he not asked me to. Jess White has worked tirelessly to bring the whole project together, cheerfully dealing with everything from thorny devils to Tasmanian blobs to X-men. Thanks to Ainsley Seago for her charming drawings that bring the DNA to life; Tom Furby and the design team for the vibrant text design; Michael Protheroe and Heather Addison for their careful attention to detail in typesetting and proofreading the text. I am grateful to Alex Skeels, Caela Welsh, Emma Day, Conor Horgan, and Bill McAllister who bravely read and improved the entire book. Thanks to all those who have selflessly given their time to read and comment on parts of the text including Sally Adamowicz, Jeremy Austin, Scott Baker, Peter Bennett, Andrew Berry, Jason Bragg, Matt Brandley, Marcel Cardillo, Frederick Churchill, David Duchene, John Ewen, Rosemary Gillespie, Simon Greenhill, Eddie Holmes, Xia Hua, Ben Kerr, Marek Kohn, Rob Lanfear, Jo Meers, Fatima Mitterboeck, Camile Moray, Craig Moritz, Hélène Morlon, Matt Phillips, Alistair Potts, Andrew Rambaut, Dusty Rhoads, Ned Ruby, Haris Saslis-Lagoudakis, Jennifer Wernegreen, Jinliang Wang, and John Wiens. Many people generously allowed their images to be used in the book, including Raj Akinsanya, Ruby Akinsanya, Anne Aubusson-Fleury, Scott Baker, Barry Bromham, Beverley Bromham, James Bull, Alexey Cardillo, Arkady Cardillo, Asha Cardillo, Gulliver Cardillo, Marcel Cardillo, Tom Cardillo, Becky Chong, Ingrid Crozier, Erica del Castillo, Anita Dickinson, Ben Dickinson, Tom Dickison, Theo Evans, Simon Foale, Jenna Gallie, Bob Goldstein, Eric Harley, Carol Hartley, Xia Hua, Jo Kelly, Andras Keszei, Ben Kerr, Adam Lazarus, Haley Lindsey, Hélène Morlon, Bob Parsons, Rod Peakall, Ted Phelps, Andrew Rambaut, Hamish Rambaut, Paul Stothard, Freya Thomas, Christine Wakfer, Becky Wong, Meg Woolfit, and David Yeates.

My greatest appreciation (and admiration) is reserved for Marcel, because without his unwavering support, encouragement, and patience, this book could not have been written.

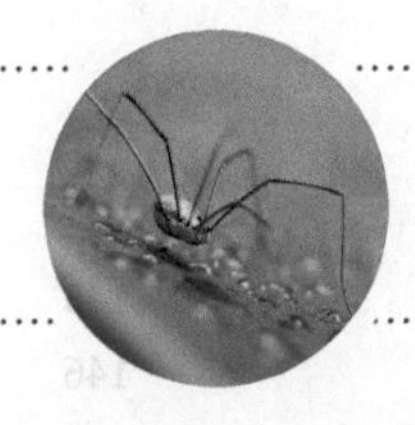

# Contents


*Dedication* v
*Preface* vi
*Acknowledgements* ix
*List of TechBoxes* xvi
*List of Case Studies* xvii

**1 Introduction**
*The story in DNA*
page 1

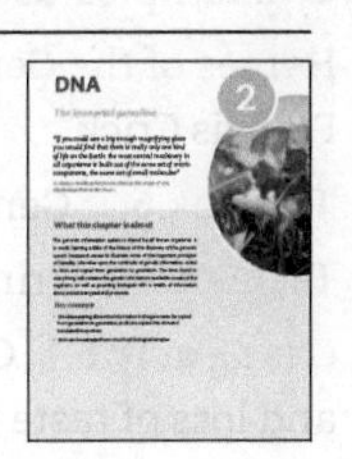

Reading the story in DNA 2
Individuals, families, and populations 3
Uncovering the evolutionary past 5
Tracing ancestors 5
About this book 10
What is this book for? 10
Who is this book for? 11
How the book is structured 12
How to use this book 13
Conclusion 15
Points to remember 16
**Ideas for discussion** 16
Examples used in this chapter 17
Heroes of the Genetic Revolution 1: Fred Sanger 18
TechBox 1.1: GenBank 19
TechBox 1.2: BLAST 23
Case Study 1.1: Identification: solving the mystery of the Chilean blob 25
Case Study 1.2: Forensics: using DNA to trace trade in whale meat 28

**2 DNA**
*The immortal germline*
page 31

Material basis of heredity 32
Principles of heredity 33
Discovery of the gene 36
Structure of DNA 38
The central dogma 41
DNA extraction 42
Ubiquity of DNA 42
Extracting DNA 42
Conclusion 43
Points to remember 44
**Ideas for discussion** 45
Sequences used in this chapter 45
Examples used in this chapter 45
Heroes of the Genetic Revolution 2: August Weismann 46
TechBox 2.1: DNA structure 47
TechBox 2.2: DNA extraction 52
Case Study 2.1: Environmental DNA: tracking changes in Arctic vegetation over time 54
Case Study 2.2: Ancient DNA: cataloguing diversity of extinct giant birds 57

**3 Mutation**
*We are all mutants*
page 61

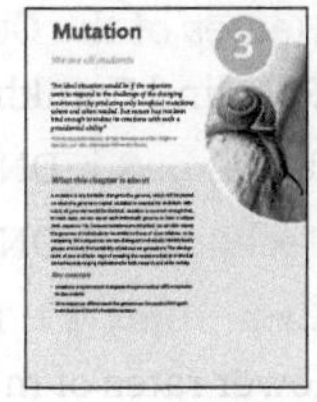

Mutation 62
Point mutations 62
Single nucleotide polymorphisms (SNPs) 65
Identifying individuals using SNPs 65
Using SNPs to track inherited traits 67
Estimating the mutation rate 68
Mutation rates are a compromise 72
Is mutation random? 73
Conclusion 75
Points to remember 75
**Ideas for discussion** 76
Examples used in this chapter 76
Heroes of the Genetic Revolution 3: James Crow 78
TechBox 3.1: Single nucleotide polymorphisms (SNPs) 79

TechBox 3.2: Biobanking 84
Case Study 3.1: SNPs: genetic resources for crop development in a non-model species 88
Case Study 3.2: Mutation rate: mutation accumulation in a laboratory population 91

**4 Replication**
*Endless copies*
page 95

DNA replication 96
Template reproduction 97
Replication 98
DNA amplification 102
Polymerase chain reaction (PCR) 103
Contamination 106
Copy errors create hierarchies 106
Sequence evolution 107
Hierarchies reveal history 110
Conclusion 113
Points to remember 114
Ideas for discussion 115
Sequences used in this chapter 116
Examples used in this chapter 116
Heroes of the Genetic Revolution 4: Rosalind Franklin 117
TechBox 4.1: DNA replication 118
TechBox 4.2: DNA amplification 121
Case Study 4.1: Replication: taller plants have lower rates of molecular evolution 124
Case Study 4.2: Hierarchies: genes and language track human settlement of Oceania 127

**5 Genome**
*Accident and design*
page 131

Genome size 132
The paradox of genome size 133
Doubling and dividing: genome replication 134
Recombination 136
Chromosomal rearrangement 138
Repeat sequences 139
Comparing repeat sequences 141
Transposable elements 145
LINEs, SINEs, and retroviruses 146
The genomic record of transposition 147
Conclusion 149
Points to remember 150
Ideas for discussion 150
Sequences used in this chapter 151
Examples used in this chapter 151
Heroes of the Genetic Revolution 5: Barbara McClintock 152
TechBox 5.1: Sanger sequencing 153
TechBox 5.2: High-throughput sequencing 157
Case Study 5.1: Duplication: gene copies and microsatellites help aphids tolerate nicotine 161
Case Study 5.2: Genome: a leukaemia virus becoming part of the koala genome 164

**6 Gene**
*Making an organism*
page 167

Genes 168
Gene structure and expression 170
Transcription and translation 171
Regulation of gene expression 173
Sequence evolution 175
Divided genes: exons and introns 175
Identifying genes 178
Regulatory sequences 181
Conclusion 182
Points to remember 183
Ideas for discussion 184
Sequences used in this chapter 184
Examples used in this chapter 184
Heroes of the Genetic Revolution 6: Francis Crick 185
TechBox 6.1: Genetic code 186
TechBox 6.2: Primers and probes 189
Case Study 6.1: Gene families: duplication and loss of taste receptor genes 191
Case Study 6.2: Genetic code: multiple proteins from a single human gene 194

**7 Selection**
*Descent with modification*
page 197

Variation 198
Substitution 200
Natural selection 201
The power of selection 202
Fitness 205
Genetic background 206
Linkage 208
Are humans still evolving? 209
Conclusion 214
Points to remember 214
Ideas for discussion 215
Sequences used in this chapter 215
Examples used in this chapter 215
Heroes of the Genetic Revolution 7: J. B. S. Haldane 216
TechBox 7.1: Fitness 217
TechBox 7.2: Detecting selection 220
Case Study 7.1: Selection: tracking insecticide resistance over time using museum specimens 222
Case Study 7.2: Variation: experiments on evolutionary rescue under environmental change 225

**8 Drift**
*Chance and necessity*
page 229

Neutral theory 230
Genetic load 230
Genetic drift 232
Neutrality 233
Population size 236
Inbreeding 238
Patterns of substitution 239
Chance and complexity 242
Conclusion 244
Points to remember 245
Ideas for discussion 246
Sequences used in this chapter 247
Examples used in this chapter 247
Heroes of the Genetic Revolution 8: Tomoko Ohta (太田朋子) 248
TechBox 8.1: Neutral theory 249
TechBox 8.2: Population size 252
Case Study 8.1: Substitution: mutation accumulation in asexual walking stick insects 255
Case Study 8.2: Population size: inbreeding depression in an endangered species 258

**9 Species**
*Origin of species*
page 262

Taxonomy 263
Classification 264
Relatedness 266
Divergence 267
Speciation 267
Hybrid incompatibility 268
Species 269
Defining species 272
DNA taxonomy 274
Conclusion 275
Points to remember 276
Ideas for discussion 277
Examples used in this chapter 277
Heroes of the Genetic Revolution 9: Rosemary Gillespie 278
TechBox 9.1: Phylogeography 279
TechBox 9.2: Barcoding 284
Case Study 9.1: Speciation: evolution of novel flower colour through reinforcement 288
Case Study 9.2: Barcoding: cataloguing weevil diversity in New Guinea 291

**10 Alignment**
*Same but different*
page 295

Homology 296
Systematic hierarchy 297
Shared by descent 300

DNA can reveal homology 301
Alignment establishes homology 302
Alignment 302
Alignment gaps 303
Homologous sites 304
Manual alignment 305
Automated alignment 306
How to make the best alignment 308
Conclusion 310
Points to remember 310
Ideas for discussion 311
Sequences used in this chapter 311
Examples used in this chapter 311
Heroes of the Genetic Revolution 10: Margaret Oakley Dayhoff 312
TechBox 10.1: De-extinction 313
TechBox 10.2: Multiple sequence alignment 317
Case Study 10.1: Alignment: identifying insertions and deletions in endosymbiont genomes 321
Case Study 10.2: Horizontal gene transfer: genes in parasitic plants derived from their hosts 325

**11 Phylogeny**
*Tree of life*
page 329

History 330
Life as a tree 332
Molecular phylogenetics 335
Phylogeny reconstruction 340
Splits 342
Similarity 345
Distance methods 345
Conflict 347
Networks 349
Conclusion 350
Points to remember 350
Ideas for discussion 351
Examples used in this chapter 352
Heroes of the Genetic Revolution 11: Joseph Felsenstein 353
TechBox 11.1: Distance methods 354
TechBox 11.2: Phylogenetic networks 358
Case Study 11.1: Networks: bobtail squid and their symbiotic bioluminescent bacteria 362
Case Study 11.2: Distance: tracing the origins and evolution of the domesticated apple 365

**12 Hypotheses**
*Seeing the wood for the trees*
page 369

Comparing trees 370
Possible histories 371
Evaluating alternative trees 372
Multiple hits 374
Statistical inference of phylogeny 376
Models of molecular evolution 377
Likelihood 378
Testing phylogenetic hypotheses 379
Phylogenetic support 380
Which phylogenetic method should I choose? 384
Conclusion 385
Points to remember 386
Ideas for discussion 387
Examples used in this chapter 387
Heroes of the Genetic Revolution 12: Hélène Morlon 388
TechBox 12.1: Maximum likelihood 389
TechBox 12.2: Bootstrap 392
Case Study 12.1: Epidemiology: using ancient DNA to identify the origins of a plague 396
Case Study 12.2: Prediction: relating ethnobotanical resources across different cultures 400

**13 Rates**
*Tempo and mode*
page 403

Rate of evolutionary change 404
Variation in the rate of evolution 405
Estimating branch lengths 406
Variation in rates across sites 408

Comparing rates 409
Mutation rate variation 409
Accounting for descent 411
Generation time effect 413
Lineage-specific rates 415
Comparing substitution rates 415
Testing for rate variation 418
Conclusion 419
Points to remember 420
Ideas for discussion 421
Sequences used in this chapter 421
Examples used in this chapter 422
Heroes of the Genetic Revolution 13: Xia Hua (華夏) 423
TechBox 13.1: Substitution models 424
TechBox 13.2: Bayesian phylogenetics 428
Case Study 13.1: Rates: flightless insects have faster substitution rates 432
Case Study 13.2: Diversification: mutation rates are linked to species richness in plants 434

**14 Dates**
*Telling the time*
page 438

Confidence 439
Fossilization: a chance at immortality 441
Alternative hypotheses 442
Divergence 443
Divergence dates 444
Estimating number of substitutions 447
Accuracy and precision 450
Dating with confidence 453
Molecular dating 455
Modelling rate change 456
Challenging assumptions 457
Conclusion 460
Points to remember 461
Ideas for discussion 461
Sequences used in this chapter 462
Examples used in this chapter 462
Heroes of the Genetic Revolution 14: Andrew Rambaut 464
TechBox 14.1: Calibration 465
TechBox 14.2: Molecular dating 471
Case Study 14.1: Calibration: did kauri survive the Oligocene drowning of New Zealand? 476
Case Study 14.2: Dates: using phylogenies to trace the source of disease outbreaks 480

**Coda: You are a scientist**
*What do I do now?*
page 484

Glossary 487
Index 505

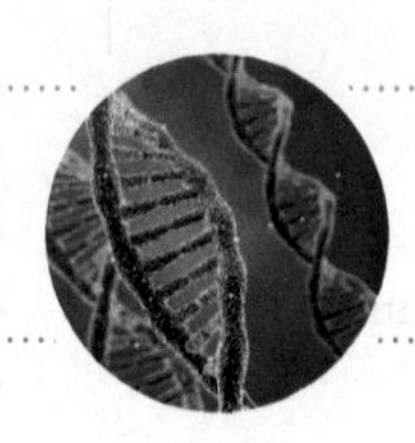

# List of TechBoxes


1.1 **GenBank** 19
1.2 **BLAST** 23
2.1 **DNA structure** 47
2.2 **DNA extraction** 52
3.1 **Single nucleotide polymorphisms (SNPs)** 79
3.2 **Biobanking** 84
4.1 **DNA replication** 118
4.2 **DNA amplification** 121
5.1 **Sanger sequencing** 153
5.2 **High-throughput sequencing** 157
6.1 **Genetic code** 186
6.2 **Primers and probes** 189
7.1 **Fitness** 217
7.2 **Detecting selection** 220
8.1 **Neutral theory** 249
8.2 **Population size** 252
9.1 **Phylogeography** 279
9.2 **Barcoding** 284
10.1 **De-extinction** 313
10.2 **Multiple sequence alignment** 317
11.1 **Distance methods** 354
11.2 **Phylogenetic networks** 358
12.1 **Maximum likelihood** 389
12.2 **Bootstrap** 392
13.1 **Substitution models** 424
13.2 **Bayesian phylogenetics** 428
14.1 **Calibration** 465
14.2 **Molecular dating** 471

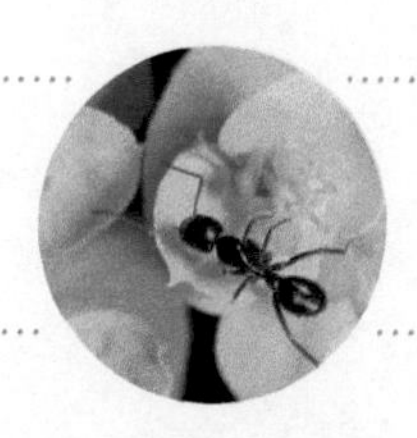

# List of Case Studies


| | | |
|---|---|---|
| 1.1 | **Identification:** solving the mystery of the Chilean blob | 25 |
| 1.2 | **Forensics:** using DNA to trace trade in whale meat | 28 |
| 2.1 | **Environmental DNA:** tracking changes in Arctic vegetation over time | 54 |
| 2.2 | **Ancient DNA:** cataloguing diversity of extinct giant birds | 57 |
| 3.1 | **SNPs: genetic** resources for crop development in a non-model species | 88 |
| 3.2 | **Mutation rate:** mutation accumulation in a laboratory population | 91 |
| 4.1 | **Replication:** taller plants have lower rates of molecular evolution | 124 |
| 4.2 | **Hierarchies:** genes and language track human settlement of Oceania | 127 |
| 5.1 | **Duplication:** gene copies and microsatellites help aphids tolerate nicotine | 161 |
| 5.2 | **Genome:** a leukaemia virus becoming part of the koala genome | 164 |
| 6.1 | **Gene families:** duplication and loss of taste receptor genes | 191 |
| 6.2 | **Genetic code:** multiple proteins from a single human gene | 194 |
| 7.1 | **Selection:** tracking insecticide resistance over time using museum specimens | 222 |
| 7.2 | **Variation:** experiments on evolutionary rescue under environmental change | 225 |
| 8.1 | **Substitution:** mutation accumulation in asexual walking stick insects | 255 |
| 8.2 | **Population size:** inbreeding depression in an endangered species | 258 |
| 9.1 | **Speciation:** evolution of novel flower colour through reinforcement | 288 |
| 9.2 | **Barcoding:** cataloguing weevil diversity in New Guinea | 291 |
| 10.1 | **Alignment:** identifying insertions and deletions in endosymbiont genomes | 321 |
| 10.2 | **Horizontal gene transfer:** genes in parasitic plants derived from their hosts | 325 |
| 11.1 | **Networks:** bobtail squid and their symbiotic bioluminescent bacteria | 362 |
| 11.2 | **Distance:** tracing the origins and evolution of the domesticated apple | 365 |
| 12.1 | **Epidemiology:** using ancient DNA to identify the origins of a plague | 396 |
| 12.2 | **Prediction:** relating ethnobotanical resources across different cultures | 400 |
| 13.1 | **Rates:** flightless insects have faster substitution rates | 432 |
| 13.2 | **Diversification:** mutation rates are linked to species richness in plants | 434 |
| 14.1 | **Calibration:** did kauri survive the Oligocene drowning of New Zealand? | 476 |
| 14.2 | **Dates:** using phylogenies to trace the source of disease outbreaks | 480 |

# Introduction

## The story in DNA

*"In nature's infinite book of secrecy*
*A little I can read"*

William Shakespeare (1623) *Antony and Cleopatra.*

1

## What this chapter is about

This chapter gives an overview of the kind of information that can be gained from analysing DNA sequences, such as identifying individuals, illuminating social interactions, understanding the evolution of major adaptations, tracing the evolutionary origins of lineages, and investigating the tempo and mode of evolution over all timescales. By taking a brief look at the use of DNA sequences in evolution and ecology, we will set the scene for topics covered in more detail in later chapters.

### Key concepts

- Evolutionary biology connects changes in individual genomes to population-level processes to the generation of biodiversity
- Information on all of these levels is available in the genome

# Reading the Story in DNA

*The mystery of the Chilean blob*

In July 2003, a 13 tonne blob washed up on a Chilean beach (**Figure 1.1**). With no bones, no skin, not even any cells, there was nothing obvious to identify the origin of the mystery blob. Theories abounded: it was a giant squid, a new species of octopus, an unknown monster from the deep, an alien. This was not the first appearance of a giant 'globster'. In 1896, an 18 metre blob washed up on St Augustine beach in Florida (USA). The identification of the blob varied from giant squid to whale blubber. However, when it was formally described in the scientific literature (albeit sight unseen), it was given the scientific name *Octopus giganteus*. There have been at least half a dozen other reported globsters, including the 1960 Tasmanian West Coast Monster, thirty feet long and eight feet tall, which, although badly decomposed, was described as being hairy.

The mystery of the Chilean blob was solved in 2004 when researchers sequenced DNA samples from the globster, and compared them to DNA sequences held in the gigantic public database GenBank (**TechBox 1.1**). The sequences matched that of a sperm whale (scientific name *Physeter catadon*: **Figure 1.2**). The globster was nothing but blubber (**Case Study 1.1**). The team also retrospectively solved previous sea monster mysteries. For example, DNA extracted from samples of the 'Nantucket Blob' of 1996 showed that it had been the remains of a fin whale (*Baleonoptera physalus*). As reported on Unexplained-Mysteries.com: 'One of the myths of the sea has been skewered by gene researchers'.

The DNA sample taken from the Chilean blob was enough to unambiguously identify it as the remains of a sperm whale. But that DNA sample could do far more. From that one tiny sample we could identify not only the species it came from but also, given enough data, which individual whale. We could use that DNA sample to predict where that individual whale was born and to understand its relationship to other whales within its immediate family and to members of its social group. That sample of DNA could help us to track whale movements across the globe, both in space and in time. We could use it to trace this whale's family history back through the ages, exploring how the whales responded to a changing world as ice ages came and went. The DNA sample could help us to reconstruct the evolution of important whale adaptations such as echolocation, and identify the nearest mammalian relatives of the whales, a question that had perplexed biologists for centuries. By comparing this whale DNA to the DNA of other mammals, we could ask whether the rise of modern mammals was contingent on the extinction of the dinosaurs. Deeper still, this DNA could help us look

**Figure 1.1** The blob that washed up on a Chilean beach had people guessing: squid? octopus? kraken? DNA analysis proved beyond doubt that the blob was the remains of a sperm whale. Similar blobs have been found on beaches all over the world. A prolonged period at sea may break down whale carcasses and turn their innards into unrecognizable blobs which eventually wash ashore.
Image courtesy of Elsa Cabrera/CCC.

**Figure 1.2** Sperm whales (*Physeter catadon*, also known as *Physeter macrocephalus*) derive their common name from spermaceti, a waxy, milky white substance that is found in abundance in sperm whales' heads. The exact function of spermaceti is not known: it might act as a sounding medium for echolocation or to aid buoyancy or possibly to give the head extra heft for headbutts in male-to-male combat. When whaling was a global industry, whale oil, derived from spermaceti, had a variety of industrial uses, including the production of candles.
Image courtesy of Gabriel Barathieu, licensed under the Creative Commons Attribution Share Alike 2.0 Generic license.

back into deep time, shedding light on one of the greatest biological mysteries, the explosive beginnings of the animals, and even back to the origin of the kingdoms. This chapter will use sperm whale DNA to briefly illustrate the kind of information we can get from analysing DNA sequences. The rest of the book will show you how.

## Individuals, families, and populations

The genome of a sperm whale contains over 3,000,000,000 nucleotides of DNA. Nucleotides are the basic units of DNA, and they come in four types, which are given the single-letter codes A, C, G and T. These four letters make up the DNA alphabet. All of the information needed to make the essential parts of a whale, such as the skin, the eyes, the blubber, and the blood, is coded in these four letters.

*The structure of DNA is covered in Chapter 2*

Most of the genome is exactly the same for all sperm whales, because all whales must be able to make functional skin, eyes, blubber and blood in order to survive. But some of the DNA letters can change without destroying the information needed to make a working whale. Because of this, some DNA sequences vary slightly between individual sperm whales. Most of these differences between genomes arise when DNA is copied from the parent's genome to make the eggs or sperm that will go on to form a new individual. DNA copying is astoundingly accurate, but it is not perfect. In fact, you could probably expect around 100 differences in the nucleotide sequence of the sperm whale genome between a parent whale and its calf, due to mutation. The upshot of this is that, although most DNA sequences are exactly the same in all sperm whales, every individual whale has some changes to the genome that make it unique. So given enough DNA sequence data it would be possible to tell not only which species the Chilean blob had come from, but also which individual whale (**TechBox 1.2**, **Case Study 1.2**). The same rationale applies to forensic DNA analysis to identify biological samples left at crime scenes: because each individual has a unique genome, if the DNA from the victim matches the blood on the accused's clothes, then the two must be linked.

*Chapter 3 explains how mutation makes individual genomes unique*

The sperm whale's genome is copied when it reproduces and any changes to its genome will also be inherited by the whale's offspring. By asking which individuals share particular sequence changes, biologists can use DNA sequences to reveal the relationships between individual whales. Since DNA is inherited from both father and mother with relatively few changes, it is possible to use DNA sequences to conclusively identify an individual's parents. The inheritance of specific DNA differences can also be traced back through a whale's family tree, to its parents, grandparents and great-grandparents, and so on. So, in addition to identifying specific individuals, if you take DNA sequences from a whole group of whales, you can tell who is whose mother or sister or cousin. More generally, you can start to understand how populations of sperm whales interact and interbreed.

*Chapter 4 explains how DNA replication results in related individuals being more genetically similar*

Sperm whales travel in social groups that co-operate to defend and protect each other and may even share suckling of calves. It is difficult to determine the membership of these groups from sightings alone because of the practical difficulties of observing whale behaviour, most of which happens underwater. To make things even more difficult, sperm whales can travel across entire oceans and can dive to a depth of a kilometre. Biologists who study whale behaviour have traditionally had to be content with hanging around in boats, waiting for their subjects to surface. But when they do surface, in addition to taking photos which allow individual whales to be identified, biologists can zip over in worryingly small boats and try to take a small biopsy or pick up the bits of skin that the whales leave behind on the surface when they submerge (**Figure 1.3**). The DNA extracted from these sloughed off bits of whale skin not only identifies the individuals in the group but also reveals their relationships to each other. This has allowed researchers to describe sperm whale social groups in detail.

Analysing DNA sequences from skin samples shows that sperm whale social groups are made up of 'matrilines', or female family groups (mothers, daughters, sisters and so on). Males leave the group before they mature. But not all the individuals in the group are related to each other—there are members of several different matrilines in each group. This suggests that, while adult males come and go and rarely stay with the group longer than it takes to father more offspring, female sperm whales form long-term, and possibly life-long, relationships. In this way, DNA sequencing is allowing biologists to gain insights into the private lives of animals that were previously hard to observe.

**Figure 1.3** DNA hunters: these researchers are trying to get close enough to a humpback whale to get a DNA sample.

Reproduced by permission of C. Scott Baker, Marine Mammal Institute, Oregon State University and School of Biological Sciences, University of Auckland.

### *Whale populations in space and time*

Sperm whales are a global species, found in every ocean. Male sperm whales leave the matrilineal social groups at a young age to join roaming 'bachelor schools', travel widely, going to cooler polar waters to feed, and returning to lower latitudes to mate. Female social groups inhabit relatively warm temperate and tropical waters and do not tend to move between oceans. So if a sperm whale calf is born in the Pacific Ocean, then we can be almost certain that its mother was also born in the Pacific and so was its grandmother.

The behavioural difference in distribution between males and females is reflected in sperm whale DNA. A sperm whale's nuclear DNA is carried on 44 chromosomes, located in the nucleus of each cell. Each individual whale inherits half of its chromosomes from its mother and half from its father. As males move around the world and mate, they spread their nuclear DNA around. But whale cells contain another source of DNA, in addition to the chromosomes in the nucleus. Mitochondria—energy-producing organelles found in the cellular cytoplasm—contain their own tiny genomes, less than 0.01 per cent the size of the nuclear genome. In sperm whales, as in most vertebrate species, mitochondria are passed from generation to generation in the egg cells supplied by the mother. Males do not pass their mitochondrial DNA to their offspring, because the mitochondria carried in sperm cells are jettisoned from the fertilized embryo. Therefore, mitochondrial DNA is inherited through the female line. Since females do not roam as widely as males, each whale will tend to have the mitochondrial sequence typical of the ocean it was born in. This means that if you were to give a sample of sperm whale skin (or blood, blubber, or tooth) to a biologist, by sequencing the mitochondrial DNA, they could probably tell you which region of the world that whale was born in (**Case Study 1.2**).

*Chapter 5 explores genome structure*

DNA sequences can reveal not only the current global distribution of sperm whales, but also how the population has changed over time. Changes accumulate in the mitochondrial genome as it is copied and passed from mother to daughter. Over time, these changes to the genome accumulate, increasing the number of differences between the genomes of individuals from different families. When measures of genetic similarity are taken for a whole population, it gives an indication of population size and mating structure. In a small population there is an increased chance of mating with a relative. This means that in a small, inbred population, the chances are that you would not have to trace two individuals' family trees back many generations before you found a shared ancestor. Since these two related individuals will have both inherited some of the same genetic variants from their common ancestor, you would expect their genomes to be more similar than when you compared two unrelated individuals. In a large population where mating between unrelated individuals is common, you would have to go much further back to find an ancestor shared by any two individuals, so two individuals sampled from a large population are likely to have less of their genome in common than two individuals sampled from a small population. Therefore, the average number of genetic differences between individuals can give an indication of population size.

*The effect of population size on DNA evolution is covered in Chapter 8*

Current estimates suggest that the size of the global sperm whale population is probably around 360,000 individuals. But the genetic diversity of this population is surprisingly low. The low number of differences between the mitochondrial DNA sequences of sperm whales from around the globe suggest that they are all descended from a small number of founding mothers who survived the last ice age. Since sperm whale females prefer warmer waters, their distribution may have shrunk towards the equator as the world's oceans cooled, reducing their population size. As the ice age ended and the climate warmed, the sperm whale population might have expanded and spread out around the globe once more.

*The nature and identification of species is discussed in Chapter 9*

## Uncovering the evolutionary past

Although the population may have reduced and expanded with the changing climate, sperm whales have swum in the oceans for millions of years. Sperm whales hunt using echolocation: they emit bursts of ultrasonic sound and use the sound reflected back from their surroundings to locate prey, such as giant squid. Echolocation is a characteristic that sperm whales share with other members of the Odontoceti, the group of predatory toothed whales that includes the dolphins and orcas. The other main group of cetaceans, the Mysticeti, have no need of echolocation. They are the baleen whales that use huge filter plates in their mouths (the baleen) to sieve plankton out of the water as they swim. The Mysticeti includes the gigantic blue whale, which, weighing up to 150 tonnes, is the largest animal that has ever lived.

Analysis of DNA sequences is revealing some of the changes that lead to the development of echolocation in toothed whales. Prestin is a protein that forms a molecular motor, attached to tiny mobile hairs in the mammalian ear, which acts to amplify sound signals. Not surprisingly, mutations in the *prestin* gene can cause hearing loss. But, rarely, fortuitous changes in the *prestin* gene sequence can also improve hearing. As it happens, the *prestin* gene has evolved rapidly in toothed whales, that rely on high-frequency hearing to hunt, compared to the baleen whales, who sieve their food from the water. *Prestin* has also undergone parallel evolution in some other echolocating lineages. Intriguingly, some of the changes in the echolocating whale Prestin protein are remarkably similar to the changes to Prestin in some lineages of echolocating bats. Could we be deceived into reclassifying sperm whales as a kind of aquatic bat on the basis of similarities in their Prestin proteins? It would take very little investigation to show this was not the case, because most of the rest of the genome would reflect the path of evolutionary history, being more similar between different kinds of whales than between echolocating whales and bats.

*Chapter 6 looks at gene structure, function, and evolution*

The fascinating case of Prestin evolution illustrates the way that evolutionary changes can sometimes overwrite the record of evolutionary history. But other parts of the genome will faithfully reflect the evolutionary history and relationships, even when peculiar adaptations to particular circumstances alter resemblances between relatives. This point is more clearly illustrated when we consider the place of the whales in the mammalian family tree.

*Detecting the footprint of natural selection in DNA sequences is discussed in Chapter 7*

## Tracing ancestors

Whales may look like giant fish on the outside, but inside they are typical mammals,with mammalian blood, bones, and organs. So although adaptations to life at sea have, in many ways, erased the signal of the whale's past, traces of their ancestry can be seen in the way that the mammalian finger bones have been modified into flippers. Some whale species even have the remains of a pelvis left over from an ancestor that walked on four legs. But what kind of animal was this four-legged ancestor? While fossil data is the ultimate source of information on the morphology of ancestral species, molecular data can provide important clues by revealing which living mammals are the whales' closest relatives.

*The importance of inherited similarities for uncovering evolutionary relationships is covered in Chapter 10*

The whale skeleton has been so highly modified by evolution that biologists have, in the past, argued over whether the whale is most closely related to artiodactyls (such as cows, camels, and pigs), perissodactyls (horses, rhinos, tapirs) or carnivorans (cats, dogs, bears). But even when morphology changes dramatically, the genome continues, by and large, to steadily accumulate changes. Just as changes to the genome every generation allow the relationships between individual whales to be traced, so the sum of these changes over longer time periods allows the evolutionary relationships between species to be uncovered. By comparing the similarities and differences in the DNA sequences of different species it is possible to reconstruct their history as an evolutionary tree, also known as a phylogeny.

*Chapter 11 explains how to estimate evolutionary trees from DNA sequence data*

When DNA sequences from whales are compared to other mammals, it is clear that they are most similar to artiodactyls, something that had been suspected for some time. But, more surprisingly, the DNA suggested

**Figure 1.4** Hippos (*Hippopotamus amphibius*), equally at home in the water and on the land, may provide clues to the origins of the fully aquatic Cetacea (whales and dolphins). Although hippos spend most of their time in shallow water and can dive for five minutes or more, on land they can run faster than humans.



that the whales' closest living relative is the hippo (Figure 1.4), and that whales and hippos share a more recent common ancestor than either do with the rest of the artiodactyls, such as pigs, goats, and camels. This initially startling idea came to be known as the Whippo Hypothesis. Although whales and hippos share some unusual characteristics, such as thick hairless skin insulated with layers of fat, these were previously considered to be convergent adaptations. That is, it was assumed that both whales and hippos independently evolved the same solutions to the shared problems of being warm-blooded mammals living in cold water. Molecular analyses suggest that, rather than being coincidental, these traits might have been inherited by both whales and hippos from a shared, semi-aquatic ancestor. The DNA evidence suggests that hippos might provide clues to the evolution of fish-like whales from their hairy, four-legged, land-dwelling relatives.

*Chapter 12 examines how we can test evolutionary hypotheses using DNA sequence analysis*

The fossil record of whales has improved dramatically in recent decades, with new finds providing more information on stages in the evolutionary series. The oldest whale fossils are 'walking whales' from the beginning of the Cenozoic period (which runs from 65 million years (Myr) ago to the present, and is sometimes given the romantic name 'Age of Mammals'). These four-legged mammals took to the water not long after the oceans had been vacated by the great marine reptiles, the fish-like ichthyosaurs, the serpent-like mosasaurs, and the Loch Ness monster-like plesiosaurs (Figure 1.5). Fossils of these great aquatic reptiles are known from the Mesozoic (from 250 to 65 Myr ago, known as the 'Age of Reptiles'). But, along with the non-avian dinosaurs, the icthyosaurs, mosasaurs and plesiosaurs all disappear from the fossil record by the beginning of the Cenozoic era.

The mammals that took over from the great sea-dwelling reptiles evolved similar adaptations to aquatic life, such as streamlined bodies and flippers rather than legs. That's why some odontocete whales, such as

**Figure 1.5** Not a whale: Mosasaurs were aquatic reptiles that disappeared from the fossil record along with the non-avian dinosaurs at the end of the Cretaceous (65 million years ago). They are more closely related to living reptiles than they are to whales and dolphins and the superficial resemblance is due to convergent adaptations to an aquatic way of life.



dolphins, look remarkably similar to the fish-like reptilian icthyosaurs. The independent acquisition of similar traits is called evolutionary convergence. The fossil record tells a similar story of convergence for other major groups of mammals which appear after the dinosaurs disappeared, such as the large-bodied hoofed mammals replaced browsing sauropods, swift sharp-teethed carnivorans replaced predatory dinosaurs.

This picture of an evolutionary scramble to fill a world vacated by the dinosaurs has strongly influenced biologists' views of the relationships between the major groups of mammals, such as primates, artiodactyls and bats. If, on being released from the tyranny of the giant reptiles, mammal groups all evolved simultaneously from a common ancestral stock, then their relationships might not be well-described by a serially branching evolutionary tree, but may more resemble a 'bush', with all branches arising at once from a common root. This conclusion was supported by morphological studies that often failed to resolve any clear relationships between the mammalian orders, suggesting a lack of hierarchical relationships between the major groups.

But as the amount of DNA sequence data increases, the evolutionary history of mammals is being resolved, and surprising new relationships have been suggested. For example, DNA sequences analysis has united an unlikely group of mammals, including aardvarks, elephants, and tenrecs, into a group now known as the Afrotheria (**Figure 1.6**). When the evolutionary tree of mammals is constructed using DNA sequence analysis, these Afrotherian lineages are amongst the earliest lineages to branch off from the other mammals. This has been used to suggest that many mammal lineages may have originated in Africa, then spread out from there in the Cenozoic. But the fossil record of mammals in Africa is currently poorly known, so it is difficult to trace this African genesis through fossils alone.

In an even greater challenge to our understanding of mammalian evolution, analysis of DNA sequences has suggested that the origins of the Afrotheria and other major branches of the mammalian evolutionary tree stretch back into the time of the dinosaurs. Because changes to the genome accumulate continuously, the longer two lineages have been evolving separately the more differences you expect to see between their genomes. For example, the genomes of two species of baleen whale are more similar to each other than either is to a sperm whale's genome because the baleen whales' genomes were copied from a more recent shared ancestor than either shares with the sperm whale. If changes to DNA accumulate at a predictable rate in all species, then we can use a measure of genetic difference to estimate when two species last shared a common ancestor. When DNA from mammals such as whales, cats, monkeys, and rats are compared, the results are surprising: there are far more DNA differences between the major mammal groups than you would expect if their common ancestor had lived in the Cenozoic. Instead, the molecular analyses suggest that major branches of mammalian evolutionary tree arose deep in the Mesozoic, long before the final extinction of the dinosaurs.

So analyses of DNA sequences paint a very different picture of mammal evolution, not just an explosive post-dinosaur radiation, but also an older and more gradual Mesozoic diversification. But these molecular date estimates are controversial. If changes to DNA happen when it is copied each generation, then the gigantic sperm whales that take a decade to mature might accumulate DNA differences at a slower rate than their diminutive artiodactyl ancestors. If that is the case, then assuming that DNA differences accumulated at the same rate in ancient walking whales as they do in their gigantic sperm whale descendants could lead to

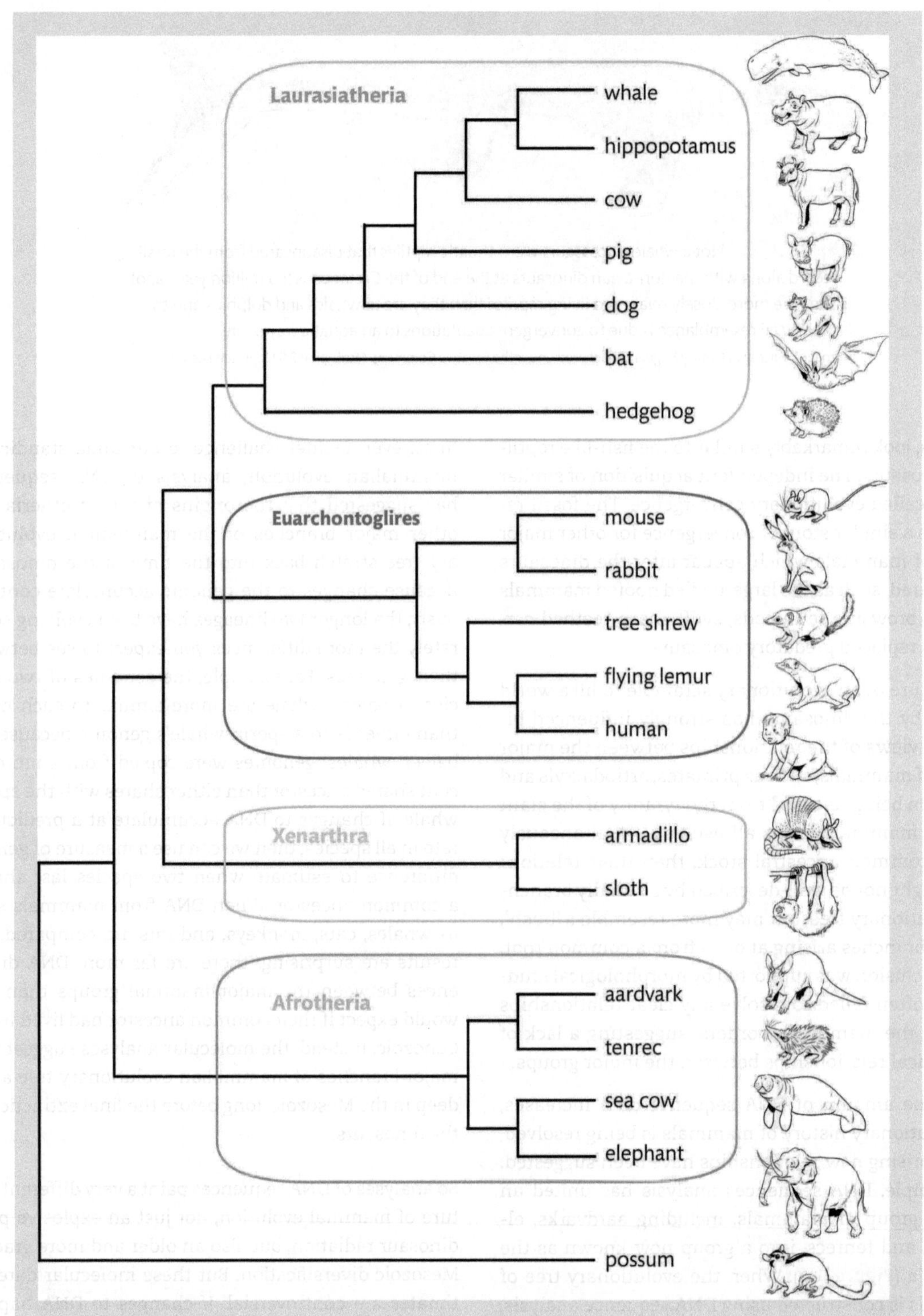


**Figure 1.6** The phylogeny of mammals has been revised in the last decade, reshaping ideas about mammalian evolution. In particular, molecular data suggests a new grouping of 'Afrotherians' as one of the early branching lineages of the mammalian radiation. This phylogeny shows the proposed relationships between some mammal lineages to illustrate the superordinal groupings. As with any phylogeny in this book, regard this as a hypothesis rather than a history: researchers disagree on the relationships between lineages and new data and new analyses lead to reevaluation of ideas on a regular basis.

**Figure 1.7** Barnacles growing on a whale. Just as whales don't look much like their mammalian relatives, due to their adaptation to the aquatic environment, so barnacles don't look much like their arthropod relatives, due to their adaptations to attaching to surfaces in the marine environment, such as this whale. Looking at a whale, you wouldn't guess that its close relatives have legs, ears and whiskers (**Figure 1.4**). Looking at a barnacle you wouldn't guess that its relatives have legs, eyes and feelers. But the genomes of these much-modified aquatic creatures reveal their affinities.
Image courtesy of Aleria Jensen, NOAA/NMFS/AKFSC.

incorrect estimates of the time of origin of the whale lineage. As our understanding of molecular evolution grows, and as analytical methods improve and more data become available, we will become better at deciphering the story in the DNA.

***Chapter 13 looks at the way that rates of molecular evolution can vary between species***

## *Evolution of animal body plans*

The DNA of a sperm whale can reach even further back into the whales' history, back to its vertebrate ancestors. A sperm whale's flipper looks nothing like a sheep's hoof, a monkey's hand or a bat's wing, and yet we can recognize the same underlying structure, not only in all mammals but also in a gecko's toes and a parrot's wing. These disparate animals also share the 'head and tail' body plan that unites the members of the phylum Chordata, which includes the mammals, birds, reptiles, and fish. The group gets its name from the central nervous cord which runs from the head to the tail. It is possible to piece together a more-or-less continuous series of forms that illustrate the evolution of the chordate body plan, linking gradual transitions along the evolutionary paths that connect the sperm whale to the monkey, the parrot, and the gecko.

But it is not so easy to follow the evolutionary paths that connect the chordates to other types of animals, for example linking the sperm whale to the barnacle stuck on its fin and the flatworm inside its guts (**Figure 1.7**). All animals are united by common features that reveal a shared ancestry, such as cell junctions that allow their multicellular bodies to be co-ordinated, absence of cell walls which permits movement and flexibility, and heterotrophy (consuming biological material as food, rather than producing their own energy from light or chemicals). But beyond basic shared features such as multicellularity, locomotion, and heterotrophy, the main groups of animals are strikingly different. The major divisions of the animal kingdom—the phyla—are often considered to represent different body plans, or basic ways of constructing animals. For example, the arthropod body plan consists of a segmented body with jointed appendages, all clothed in a jointed exoskeleton of protein and chitin (which is what makes bugs crunch underfoot). The echinoderm body plan, on the other hand, is characterized by pentameral symmetry (like a five-pointed star), an exoskeleton made of hard calcite, and a water vascular system that, in addition to transporting nutrients around the body, can power locomotion through hundreds of soft, hydraulically operated feet (turn a starfish over to see these little tube feet in action).

These animal body plans are so different from each other that there has been an intense (and often rather lively) debate about how they evolved. We can trace the whale and barnacle lineages back half a billion years, to Cambrian age rocks that contain fossilized animals with recognizable chordate and arthropod body plans (**Figure 1.8**: alas, flatworms leave few identifiable fossils). But there the trail ends rather abruptly. We don't have a continuous series of fossils showing the different animal body plans diverging gradually from each other, slowly modifying existing features to form the special characteristics of their phylum. Instead, a great diversity of different body plans appear almost simultaneously in the fossil record of the early Cambrian, complete with eyes, limbs, segments or armour, in an explosion of animal forms. While some interpret this pattern in the fossil record as the signature of a remarkable, rapid evolutionary radiation, others consider that earlier forms might have existed but have not been preserved, perhaps because they were small and squishy and lacked the skeletons, shells and spikes that would have granted them geological immortality. When genes shared by whales, barnacles, and flatworms are compared, there are more differences in the DNA sequences than expected from half a billion years of evolution, suggesting that these major branches of the evolutionary tree

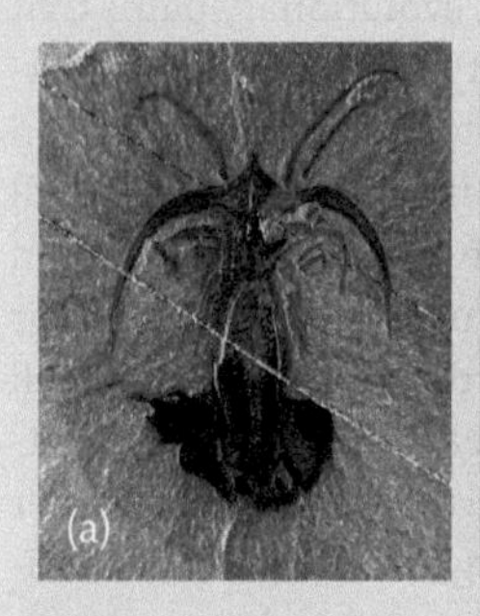


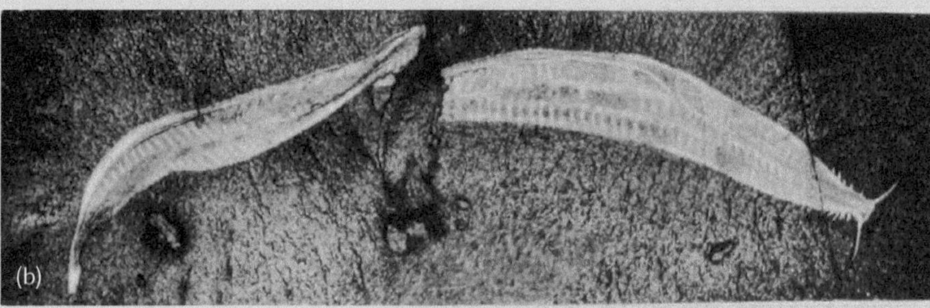


**Figure 1.8** The arthropod and chordate 'body plans' have a long history and can be found in Cambrian age rocks, over half a billion years old. Fossil beds such as the Burgess Shale in Canada and Chenjiang in China contain exquisitely preserved examples of these aquatic animals, such as *Marrella* (a, tiny arthropod 'lace crabs') and *Pikaia* (b, an early chordate, something like the modern amphioxus).



of animals stretch back beyond the record of readily recognizable fossils. If it is true that the beginnings of the animal kingdom lie in the deep past, but left little fossil evidence of its early history, then DNA evidence will play an important role in revealing its ancient history.

*Estimating evolutionary time from DNA sequences is covered in Chapter 14*

### *Back to the beginning*

Some sequences in the sperm whale genome code for such fundamental properties of life that they are shared not only with other animals but with all living things, including oak trees, mushrooms, seaweed, and bacteria. The DNA sequences of these basic genes reveal the deep history of all life. These shared parts of our genome demonstrate that all life on earth has a single common origin. The original genome, present in the last common ancestor of all plants, animals, fungi, algae, and bacteria, has been modified and expanded, but there is a part of that ancestral genome in all of us. The DNA from the globster was enough to prove that it was the remains of a sperm whale and not (as some hoped) an extra-terrestrial visitor. But if biologists ever do get their hands on a sample of alien life, the first thing they will do is check to see if it has DNA. If it does, it may well be our distant relative. Then, using the techniques described in this book, we will be able to use the alien DNA sequence to investigate the history and biology of our cousins from space.

A small piece of the mystery blob that washed up on a beach in Chile contains enough DNA to tell a long and wonderful story. The story begins with a sperm whale, born in the same ocean that its mother had been, and its grandmother, and great-grandmother, and so on back into history. Our sperm whale's ancestor was one of the founding mothers who survived the last ice age, although she may have had to retreat to tropical waters to do so. The story reaches back in time to tell of the origins of the sperm whales, whose predatory ancestors evolved echolocation to help them hunt aquatic prey, leaving their four-legged relatives wallowing in the mud. These ancestors were part of a radiation that exploded onto the world as the dinosaurs left the stage, and yet the roots of this radiation were planted firmly in the time of the reptiles. Mammals and reptiles were themselves products of the diversification of the chordate body plan, which appeared in the fossil record half a billion years ago, but may have more ancient beginnings. And, using the DNA from the blob, we can follow the story right back to the origins of the animal kingdom and, ultimately, back to the last living ancestor of all life. Not bad for a piece of blubber.

# About this book

## What is this book for?

The aim of this book is to provide a from-the-ground-up introduction to the use of DNA sequence data in evolutionary biology. Specifically, the idea is to give you the basic understanding of genome evolution and analysis that you need in order to use DNA sequences to trace evolutionary history and processes through phylogenetic analysis. Molecular phylogenies use patterns of relatedness between DNA sequences to reconstruct the history of the lineages they were sampled from. Phylogenies are now used in so many fields in biology that a basic understanding of how they are constructed and what they mean is a useful part of any biologist's intellectual toolkit. Given the wide reach of molecular phylogenies, people from a great diversity of backgrounds need to be able to learn how to create and interpret them. Let's be honest, not everyone wants to read a serious technical description of population genetic theory and phylogenetic analysis. So this book is designed to be friendly and easy to read. That means it might lack the detail that some readers are after, but I am hoping it will provide a starting point that will whet your appetite for further explorations.

In this book, we will cover the practical skills you need to collect DNA sequence data and analyse and interpret what it all means, but we will build those practical skills on the basis of the underlying evolutionary processes. Obviously, it is impossible to cover everything; there are many fascinating and useful ideas and techniques that are not covered here. And, in such a fast-moving field, it is inevitable that this book will be out-of-date as soon as it is printed. So rather than trying to give an exhaustive introduction to the field, the aim is to give you a basic grasp of the fundamental concepts and intellectual tools you need to understand the way DNA sequences are used in biology.

That is why you won't find instructions for specific programs or particular laboratory protocols in this book. A program that is all the rage today is likely to be superseded by improved methods next year. DNA sequencing methods are replaced on a regular basis by new techniques. The aim of this book is to give you the background knowledge you need to understand not only the techniques available today, but hopefully, to lay the groundwork for the new methods of the future. If you are taught to follow instructions to use particular software to produce a specific kind of result, you may not have the background knowledge you need to modify your approach if you get strange results, or to reshape your analysis plan in response to new ideas, or to expand what you know to undertake different analyses. But, if you have learned the general principles on which the programs are based, you will be better able to read the manual and make decisions for yourself about your own analysis. Therefore, the aim of this book is to introduce the basic principles of molecular evolution and phylogenetics in order to help you understand how the genome records information about evolutionary past and processes, how that information can be 'read', and what kinds of questions we can use that information to answer.

## Who is this book for?

This book should be suitable for university or college students who wish to gain a basic grounding in the application of molecular phylogenetic analysis to answering questions in evolutionary biology. The reader I have in mind has a basic background in biology (say, to high school level), but no specialist knowledge of genetics, evolution or ecology. Importantly, the focus of this book is on the 'whole-organism' biologist: someone who is primarily interested in how species persist and co-exist and evolve, whether they are interested in a particular group (say, chameleons) or a particular topic (such as sexual selection). Whole-organism biologists are increasingly using molecular techniques in their research: for example, using transcriptomics to uncover differences in gene expression between individuals in a social hierarchy, phylogenetics to judge the importance of biogeographic patterns of species richness, or molecular dating to understand the evolution of animal body plans. However, their interest remains, by and large, at the level of the organism, species and lineage. They use molecular techniques for the information they can gain about their organisms, not to illuminate the genetics or biochemistry of their subjects.

While there are a great number of bioinformatics textbooks available, most are focused on medical genetics or the human genome or gene expression studies. There are relatively few resources aimed at evolutionary biologists and ecologists who want to understand the application of DNA sequencing technology to their own fields. Many of these biologists do not find the somewhat abstract and computational approach found in many bioinformatics books very inviting or comfortable. Here, I have attempted to take a complementary approach: starting with evolutionary principles, and illustrated throughout with biological examples, I ask how we can make use of the information in the genome to answer the kinds of questions that a whole-organism biologist is interested in.

But I also hope that students coming from other backgrounds might benefit from this book. Many of the brightest minds in computational and statistical fields are attracted to the challenges posed by the analysis of biological data. But some bioinformatics texts are big on the informatics but less strong on the bio, presenting a simplified view of biological systems in order to emphasize the analytical techniques. While it may be convenient to treat DNA simply as computationally tractable strings of letters, the genome is actually a complex, fascinating and intricate adaptation. DNA, RNA and protein sequences are shaped by evolution. Therefore, a grounding in the principles of evolution is needed to underpin advances in bioinformatics.

Because this book could be used in a variety of different contexts, and be read by people from a range of scientific backgrounds, it's hard to get the level of biological and technical detail just right—too much detail and the introductory reader will be lost, too little and the more advanced reader will be bored. That is why the book is structured into main text and boxes, so that I can hopefully cater to different levels with the same book. The boxes are trickier to read than the main text, so you can mix and match parts of the book to suit your own personal level of understanding and areas of interest.

## How the book is structured

This book consists of four elements: the main text and three types of boxes. Any of these elements can be read without the others, though they will make most sense when read together. I have chosen this structure to make it easier for you to get what you want from this book. Here, I will explain how the way you use the book might differ depending on what you want to find out. First, though, I will briefly outline each of the elements of the book.

### *Main text*

The aim of each chapter is to introduce important concepts needed to build up a foundation for the application of molecular data to evolutionary biology. I have attempted to make the main text of each chapter as non-technical as possible, and to illustrate key concepts with examples from whole-organism biology. Wherever possible I have moved details of methods, procedures or biochemical information to the TechBoxes. The main text can stand alone without the details given in any of the boxes, but the boxes will give a fuller appreciation of the techniques and applications of molecular data.

You will see there are two kinds of figures in the main text. There are embedded figures that are there to bring a visual element to the explanation of basic concepts, for example through a simple drawing of DNA replication or to follow a mythical population of velvet worms as it evolves. These figures try to capture the way that we often explain things to each other, by making little sketches on paper, or diagrams on a blackboard, or scribbling on napkins in a cafe. Then there are images designed to brighten up the text and give some biological flesh to the explanatory bones of the chapter. Some of these have somewhat rambling figure legends to explain the story behind the image. You can skip these without losing the thread of the chapter, but they offer something of a diversion to break up the text.

At the end of the chapter there are the following sections:

- **Points to remember** boils down the ideas covered in the chapter to easy-to-digest bullet points.
- **Ideas for discussion** are not designed to test your memory or your knowledge, but to get your brain cells firing by thinking about issues arising from the topics covered in the chapter. You can discuss these with your friends, ponder them on your own, or start an online debate.
- **Sequences used in this chapter** gives you the details you need to find the sequences used in figures in GenBank. Wherever possible, I have used real DNA sequences in illustrations (though obviously only a tiny fragment of each sequence is shown), so you can download these and play with them if you like.
- **Examples used in this chapter** is not a comprehensive bibliography of the basic concepts covered, but a list of the specific examples of scientific research mentioned in the chapter. This is so that you can find that particular study I mentioned where they claimed, for example, that sperm whales underwent a genetic bottleneck in the ice age, or that the Prestin protein in whales is similar to that in bats. Actually, these days it isn't difficult to find exactly what you want to know and more by searching any large database of scientific publications. I haven't included citations in the text because it tends to break the flow of the text and make it a bit less comfortable to read. However, you will see that a scientific style of referencing is used in the boxes. While the precise format of scientific references varies between sources, the aim is to give you all the information you need to find the paper online or on a library shelf. In case this is the first time you are encountering scientific references, there is a quick guide to the information provided for each publication cited (**Figure 1.9**):

**Figure 1.9**

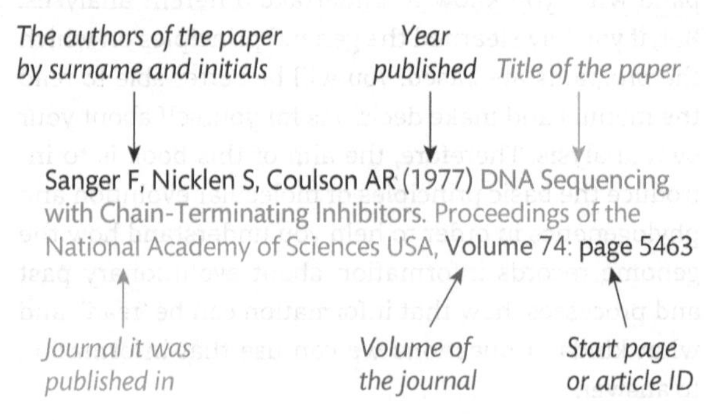

*Boxes*

- **TechBoxes** give more detailed information about methods, or take a more in-depth look at particular topics. They can be seen as 'optional extras', but each one contributes to an understanding of the topic as a whole. The main text and TechBoxes are complementary and can be read together to give a fuller account of the topic than the main text alone. However, readers may choose to ignore TechBoxes, or only read specific boxes on particular topics. TechBoxes are cross-referenced where appropriate, and linked to Case Studies where possible.
- **Case Studies** provide a summary of a scientific study that has relevance to the topics covered in that chapter. I have mainly used studies published in the last decade, rather than reporting 'classic' papers. The case studies are chosen for the diversity of topics and techniques used and do not represent the best possible studies in this field. Each study is set out with the same basic headings: background, aims, methods, results, conclusion, limitations, further work. The limitations section is not to disparage the study, but to emphasize that no scientific study is perfect, all have limits on the data that can be analysed or the conclusions that can be drawn. Note that the case studies will be technically harder to follow than the main chapter, so may be more suitable to readers with some experience in the field. Due to space limitations, case studies may contain terms or techniques that are not fully explained, so if you are interested, you should follow up the references given in the box. If you are new to the field, you might prefer to treat the case studies as a glimpse into the kind of work people are doing, without worrying too much about the technical details. To encourage you to use these case studies as springboards to further understanding, there are three kinds of questions at the end of each case study. **Check your understanding** asks you to think about the material you have just read in the box and make sure you understood why the researchers did what they did. **What do you think?** invites you to extend your understanding of the topic by considering the ideas or techniques in a broader context. **Delve deeper** allows you to move beyond the case study to do some independent investigation on a related topic.
- **Heroes of the Genetic Revolution:** The title is tongue-in-cheek, but the purpose is a noble one: to put some faces to the science (**Hero 1**). For each chapter, I have highlighted the work of one scientist who contributed to the area covered in that chapter (ideally, someone who has made a broad contribution to the field, rather than a single discovery). It hardly need be said that this is not an exhaustive list, nor should it be read as a 'best of' list, for there are a great many other scientists whose contribution has been as great or greater than those featured here. So please do not feel insulted if your favourite scientist (or yourself) has been left out. It is also an evidently biased list, including some of my friends and colleagues, but anybody's list of heroes would be similarly biased, so I make no apology. These boxes are intended merely as illustrations of the kind of work that has been done, and the type of people who have done it. The Heroes listed in this book range from illustrious scientists of the past, replete with fulsome beards, to younger scientists becoming established in the field today. I would be happy to receive nominations for Heroes to appear in subsequent editions (and yes, you are welcome to self-nominate).
- **Glossary:** I am expecting that most people reading this book have a basic knowledge of biology. Because of this, many basic concepts are assumed, such as a passing familiarity with cell division and Mendelian inheritance. However, I hope that the book will not be impossible to read even if you are unfamiliar with these basic concepts. To help with this, I have tried to include some useful biological terms in the glossary. If you can't remember what mitosis is, don't know what I mean when I say 'genotype', or have never heard of a moa, then you can turn to the glossary for some pointers. The glossary also reiterates terms introduced in the text, so it should also help if you are skipping through the book to relevant passages, rather than reading the book in order.

## How to use this book

The ideal way to read this book is, not surprisingly, taking each chapter in turn, reading all of the main text and the boxes, as each chapter is designed to build on the concepts introduced in the preceding chapters. The text is designed to help you to understand key aspects of molecular evolution, within the context of evolutionary biology and ecology, and see how these principles are applied in practice. For example, in Chapter 11, we discuss how in order to be able to estimate phylogenies (practice), you need to understand the process of genetic divergence (molecular evolutionary theory), which sheds light on diversification (evolutionary principles). Because each chapter has many roles, there are several different ways I could have ordered this book. I have tried to structure the material in the order of key concepts—DNA structure (Chapter 2), mutation (Chapter

3), replication (Chapter 4), genome organization (Chapter 5), gene structure and expression (Chapter 6), substitution by selection and drift (Chapters 7 and 8), homology (Chapters 9 and 10), descent and divergence (Chapters 11 to 14). I hope this works, but if it doesn't, you are free to take to the book with scissors and glue and rearrange it in a way that works best for you.

In any case, the structure of main text plus boxes is designed to allow you to approach the book in several different ways, depending on what you need to know:

- **If you want a general background in evolutionary biology,** then you can read through the main part of the text, without bothering with the case studies and the technical details. The main text is designed to stand alone from the boxes, so you need only dip into the TechBoxes, Case Studies or Heroes that interest you. But remember that this book is not designed as an introduction to evolution, but specifically to cover concepts you need to understand the molecular phylogenetics. So many topics of central importance in evolutionary biology will be skipped over briefly or missing entirely. If this is your first foray into evolutionary biology, then I hope you will find this brief introduction tantalizing enough to lead you to find out more.
- **If you want a basic introduction to molecular evolution,** you may find much of the main text useful, but there will be many digressions into whole-organism biology that you may find less interesting. You will find many of the TechBoxes useful, as they focus on specific aspects of theory (such as neutral theory: **TechBox 8.1**) and techniques (such as DNA amplification: **TechBox 4.2**). As you read the text, you can decide which ideas you would like to know more about, and follow the cross-references to the relevant boxes.
- **If you are after information on specific topics,** you might use the index to track down relevant TechBoxes. So, for example, if you are interested in designing a study to track cross-species transfers in a family of viruses, you might start by looking at how to use BLAST (**TechBox 1.2**) to get sequence data from GenBank (**TechBox 1.1**), then how to align those sequences (**TechBox 10.2**), how to estimate a phylogeny (**TechBoxes 11.1, 12.1,** and **13.1**) and molecular dates (**TechBox 14.2**), and test how well your data supports a particular hypothesis (**TechBox 12.2**). Cross-referencing between these boxes should make it easier for you to hop between relevant topics.
- **If you are after guidance for a particular research project:** If you know what kind of techniques you wish to use, then the TechBoxes may help you work out what you need to know to apply these tests yourself. If you are not sure what you need to do, the Case Studies may provide hints. Each Case Study has keywords that draw attention to the subjects, questions, and techniques used, so you can use these to find a case study relevant to your research. When you locate a TechBox or Case Study that's relevant to your work, you will find them in the chapters that give the background to those topics. For example, if you are interested in using single nucleotide polymorphisms (SNPs) to study population structure, you might go to **Case Study 3.1** (using SNPs to trace the evolution of switchgrass lineages). You will find it in a chapter on mutation, which is a foundation concept for understanding SNPs.

### *Where to go if you want more information*

This book is a very brief introduction to molecular evolution. This field moves so fast that any printed book cannot contain all relevant topics, and will be rapidly superseded by ongoing developments. So you should use your own initiative to follow the ideas introduced in this book and see where they lead. If you are interested in the examples mentioned in the chapter, a few key references are given at the end of the chapter, but this is not a full bibliography. Instead, I hope that the TechBoxes, Case Studies and examples given in the book will act as a springboard for interested readers to begin wider reading on the topic. Also remember that no book is perfect and I am pretty sure there will be mistakes in this one (please write and tell me if you spot any). But, since you are a scientist, you know that any one source of information should never be considered definitive and that you should use your own brain and your own investigations to explore any issue. Take heed of the motto of the Royal Society of London, one of the oldest scientific associations in the world: 'nullius in verba' (take no-one's word for it). A central tenet of science is that it is not authority that decides the answer, but the weight of evidence.

Not surprisingly, you will find publications available online more useful than books because books on molecular analyses date very quickly (which I have learned to my cost while endeavouring to update the first edition of this book). A quick search will reveal a great many freely available laboratory protocols, analysis software, and teaching resources where the latest techniques are explained. There are also several freely accessible databases of scientific papers—such as Google Scholar or PubMed central—where you can access original research. If you are a member of an institution that has subscriptions to scientific journals, then you should not find it too difficult to get access to primary research papers. Even if you don't have access to a library, many scientific studies are published in Open Access format, which anyone with an internet connection can read. However, if you are unable to obtain a particular scientific paper because you do not have access to the journal, try visiting the homepage of the author of the paper where there may be downloadable versions of their publications. If not, consider writing to the corresponding author: most scientists, particularly university academics, take their obligation to make their work accessible seriously and will be happy to send a copy of their papers to you.

### *Where are the equations?*

Many scientists in the field of molecular evolution will be shocked, and quite probably appalled, that there are no equations in this book. This is not because equations are bad. On the contrary, they are the most succinct and useful way of encapsulating many statements about molecular evolution. But equations are not everybody's cup of tea. If you are one of those whose eyes go blurry as they pass over equations, you may be disheartened by the statement by the great evolutionary biologist, John Maynard Smith, who said: 'if you can't stand algebra, stay away from evolutionary biology'. But consider for a moment how fortunate we are that Charles Darwin could not retrospectively heed this advice. Darwin never did get the hang of mathematics and produced his vast catalogue of work without recourse to any equations. Being handy with equations is a skill worth developing if you wish to understand molecular evolution and phylogenetics.; but it is, in my opinion, by no means an essential prerequisite. Far more important is the ability to think in terms of the evolutionary principles underlying molecular data, whether you phrase those principles in words or algebra.

There are plenty of texts on phylogenetics and molecular evolution containing equations but none, that I know of, with none. My aim in explaining molecular evolution without recourse to algebra is to ease the 'maths panic' felt by many biologists when they are learning to analyse molecular data. By describing the basic principles of molecular evolution in words and pictures rather than equations, I hope to give the mathematically nervous sufficient background, and courage, to read and understand more traditional texts and the primary literature. If you do feel comfortable with equations, you may wish to read a traditional molecular evolution text alongside this one, so you can see both verbal and algebraic statements of the basic principles.

## Conclusion

DNA can be extracted and sequenced from a wide range of biological samples, providing a wealth of information about evolution and ecology. The analysis of DNA sequences contributes to evolutionary biology at all levels, from dating the origin of the biological kingdoms to untangling family relationships. In this chapter, the information that can be gained from the analysis of DNA sequences has been illustrated by considering a single DNA sample and how it can shed light on the evolutionary history of individuals, families, social groups, populations, species, lineages, and kingdoms.

The aim of this book is not to provide you with protocols for DNA sequencing or instructions for software packages used in the production and analysis of DNA sequences. Instead, this book should provide you with the background knowledge you need to understand these techniques. The most important place to start is an understanding of the material basis of heredity. In Chapter 2, we will take a brief look at the history of the discovery that DNA carries genetic information. Outlining some of the important steps in the development of the field of genetics will serve to illustrate some of the key features that make DNA sequences so useful to evolutionary biologists.

## Points to remember

### Individuals, families, and populations

- Biological specimens can be identified by matching DNA sequences to a database of sequences from known sources.
- Most of the genome is identical between members of a species, but all individuals have some unique changes that can be used to identify them.
- Because changes to the DNA sequences are copied with the genome and passed on from parent to offspring, related individuals will share particular DNA differences, so patterns of similarity can reveal family relationships.
- Because individuals in a small population are likely to share a more recent common ancestor, population size can be estimated by considering the number of DNA differences between sampled individuals in a population.

### Uncovering the evolutionary past

- Selection can sometimes be detected by looking for genes that evolve more rapidly than the rest of the genome and that have changes at key functional sites.
- DNA analysis can reveal evolutionary relationships, even when adaptation and change has obscured physical similarities.
- DNA analysis is useful for tracing the evolutionary history of lineages that lack obvious similarities or informative fossils.
- Potentially, the entire evolutionary history of every lineage is contained within its genome.

### About this book

- A basic grounding in the principles of molecular evolution allows you to adapt your knowledge to new ideas and new techniques.
- This book is primarily aimed at whole-organism biologists who wish to use molecular phylogenies to understand ecology and evolution.
- This book may also be useful for researchers coming from outside biology to learn the molecular evolutionary principles that underlie bioinformatics.
- The main text gives the basic evolutionary principles and explains why they are important in understanding DNA analysis.
- TechBoxes give more detail on molecular and analytical methods.
- Case Studies present examples of studies that use these methods.
- Hero boxes illustrate the kind of people who have developed and applied these analyses.

## Ideas for discussion

**1.1** We discussed how DNA evidence can reveal the identity of mysterious globsters. What other mysteries could be solved with DNA evidence? What kind of information would you need to solve them?

**1.2** DNA evidence is increasingly being used in many ways in our society, from court cases to biosecurity to determining family relationships. Is this because DNA evidence is more reliable than other sources of information?

**1.3** How can we meet the challenge of teaching biology in a fast-moving field where today's big ideas are tomorrow's old news? Is it possible for science education to prepare you to work in a field that changes so rapidly, or to understand new discoveries as they are made?

# Examples used in this chapter


Alexander, A., Steel, D., Slikas, B., Hoekzema, K., Carraher, C., Parks, M., Cronn, R., Baker, C. S. (2013) Low diversity in the mitogenome of sperm whales revealed by next-generation sequencing. *Genome Biology and Evolution*, Volume 5, page 113.

dos Reis, M., Inoue, J., Hasegawa, M., Asher, R. J., Donoghue, P. C., Yang, Z. (2012) Phylogenomic datasets provide both precision and accuracy in estimating the timescale of placental mammal phylogeny. *Proceedings of the Royal Society B: Biological Sciences*, Volume 279, page 3491.

Gatesy, J., O'Leary, M. A. (2001) Deciperhing whale origins with molecules and fossils. *Trends in Ecology & Evolution*, Volume 16, page 562.

Liu, X. Z., Ouyang, X. M., Xia, X. J., Zheng, J., Pandya, A., Li, F., Du, L. L., Welch, K. O., Petit, C., Smith, R. J. H., Webb, B. T., Yan, D., Arnos, K. S., Corey, D., Dallos, .P, Nance, W. E., Chen, Z. Y. (2003) Prestin, a cochlear motor protein, is defective in non-syndromic hearing loss. *Human Molecular Genetics*, Volume 12, page 1155.

Liu, Y., Rossiter, S. J., Han, X., Cotton, J. A., Zhang, S. (2010) Cetaceans on a molecular fast track to ultrasonic hearing. *Current Biology*, Volume 20, page 1834.

Lyrholm, T., Leimar, O., Gyllensten, U. (1996) Low diversity and biased substitution patterns in the mitochondrial DNA control region of sperm whales: implications for estimates of time since common ancestry. *Molecular Biology and Evolution*, Volume 13, page 1318.

Lyrholm, T., Leimar, O., Johanneson, B., Gyllensten, U. (1999) Sex-biased dispersal in sperm whales: contrasting mitochondrial and nuclear genetic structure of global populations. *Proceedings of the Royal Society B*: Biological Sciences, Volume 266, page 347.

Ortega-Ortiz, J. G., Engelhaupt, D., Winsor, M., Mate, B. R., Rus Hoelzel, A. (2012) Kinship of long-term associates in the highly social sperm whale. *Molecular Ecology*, Volume 21, page 732.

Pierce, S. K., Massey, S. E., Curtis, N. E., Smith, G. N., Olavarri, C., Maugel, T. K. (2004) Microscopic, biochemical, and molecular characteristics of the Chilean Blob and a comparison with the remains of other sea monsters: nothing but whales. *Biological Bulletin*, Volume 206, page 125.

Poulakakis, N., Stamatakis, A. (2010) Recapitulating the evolution of Afrotheria: 57 genes and rare genomic changes (RGCs) consolidate their history. *Systematics and Biodiversity*, Volume 8, page 395.

Richard, K. R., Dillon, M. C., Whitehead, H., Wright, J. M. (1996) Patterns of kinship in groups of free-living sperm whales (*Physeter macrocephalus*) revealed by multiple molecular genetic analyses. *Proceedings of the National Academy of Sciences USA*, Volume 93, page 8792.

Zhou, X., Xu, S., Yang, Y., Zhou, K., Yang, G. (2011) Phylogenomic analyses and improved resolution of Cetartiodactyla. *Molecular Phylogenetics and Evolution*, Volume 61, page 255.

HEROES OF THE GENETIC REVOLUTION

1

# Fred Sanger

*"It is like a voyage of discovery into unknown lands, seeking not for new territory but for new knowledge. It should appeal to those with a good sense of adventure"*

Fred Sanger (Nobel Prize Banquet Speech, 10 December 1980) © The Nobel Foundation 1980

**EXAMPLE PUBLICATIONS**

Sanger, F., Nicklen, S., Coulson, A. R. (1977) DNA Sequencing with Chain-Terminating Inhibitors. *Proceedings of the National Academy of Sciences USA*, Volume 74, page 5463.

Sanger, F., Coulson, A. R., Friedmann, T., Air, G. M., Barrell, B. G., Brown, N. L., Fiddes, J. C., Hutchison, C. A. III, Slocombe, P. M., Smit, M. (1978) The nucleotide sequence of bacteriophage ϕX174. *Journal of Molecular Biology*, Volume 125, page 225.

**Figure 1.10** Fred Sanger (1918–2014) in front of the research institute that now bears his name, the Sanger Institute, near Cambridge, UK.

Credit: Wellcome Library, London.

Fred Sanger's interest in biology started at an early age, though he was not a high-achieving student. He chose science when he went to university, rather than his father's profession of medicine, because the focus of the scientific method appealed to him. He loved biochemistry and it was in this field that he first began to distinguish himself as a student. A conscientious objector during the Second World War, he studied for his PhD at Cambridge, on what he described as 'lysine metabolism and a more practical problem concerning the nitrogen of potatoes'. During his career he made almost countless advances in the techniques of protein sequencing and DNA sequencing. But perhaps more profoundly he put the idea of the sequence at the heart of biology. By showing that proteins consisted of a particular sequence of amino acids, he inferred that the gene, the unit of hereditary information, must have a similar sequential arrangement of units. He then provided a practical means for reading that sequence from the genes themselves.

In the early 1940s, there was a great deal of interest in protein structure. It was known that proteins contained characteristic proportions of the 20 amino acids, and it was assumed that this somehow gave proteins their specific functions and structures. But there was a debate between those who thought that the amino acids formed a particular sequence that gave the protein its function, and those who thought that different amino acids occurred at regular intervals throughout the protein, giving proteins a structural periodicity. Sanger and colleagues first determined that insulin was made up of four polypeptide chains, two each of two types. They then used a series of biochemical techniques, such as shearing particular chemical bonds then fractionating the portions of the chains using chromatography, to gradually deduce the amino acid sequence of each of the chains. The complete sequence demonstrated the 'classical peptide hypothesis' that the sequence of a protein was unique and consistent. Sanger also laid the groundwork for sequence comparisons, showing that most of the insulin sequence was identical between mammal species, except for three residues in one part of the A chain. He pre-empted the neutral theory by suggesting that the exact residue in these variable sites was relatively unimportant to the functioning of the protein (**TechBox 8.1**).

Sanger then moved to the Laboratory of Molecular Biology in Cambridge, where he became interested in gene sequencing. Although it was widely believed that DNA carried a linear code,

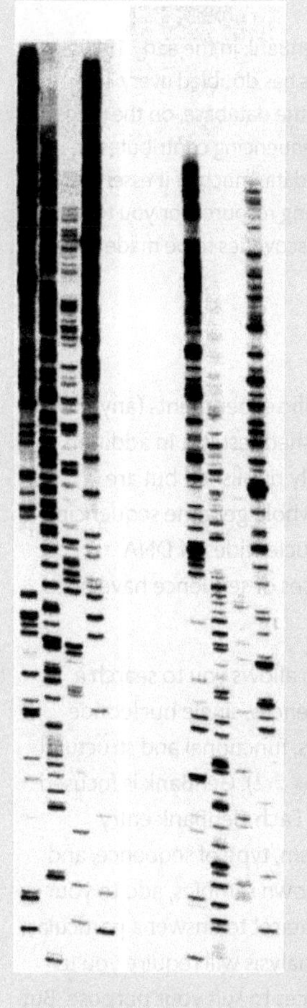


corresponding to the amino acid sequence in proteins, there was no method for reading the sequence of bases in a DNA molecule. Sanger developed methods for sequencing both RNA and DNA, and was the first to produce a whole-genome sequence, that of a bacteriophage (small viruses that parasitize bacteria: **Figure 2.5**). The DNA sequencing method developed by Sanger is known formally as the dideoxy method, or the chain termination method, but is more usually referred to as 'Sanger sequencing' (**Figure 1.11**: **TechBox 5.1**). Sanger and his colleagues went on to invent 'shotgun sequencing', where the genome is chopped up into fragments and the whole-genome sequence inferred by looking for regions of overlap between the fragments. In this way, they produced the first bacterial genome sequence and the first human mitochondrial genome sequence, and paved the way for the Human Genome Sequencing project (which is, in part, led by the research institute that bears Sanger's name: **Figure 1.10**). Fred Sanger's work has underpinned the whole field of molecular evolution. He is currently the only person to have gained two Nobel Prizes in Chemistry (there is no Nobel Prize for biology, so the study of biomolecules is considered chemistry and any other biological advance has to fit under the 'physiology or medicine' prize).

> "Often if one takes stock at the end of a day or a week or a month and asks oneself what have I actually accomplished during this period, the answer is often 'nothing' or very little and one is apt to be discouraged and wonder if it is really worth all the effort that one devotes to some small detail of science that may in fact never materialize. It is at times like the present that one knows that it is always worthwhile . . . "
>
> Fred Sanger, Nobel Prize acceptance speech, 1958. 

**Figure 1.11** An autoradiograph produced by the dideoxy method of sequencing, taken from Fred Sanger's 1980 Nobel Prize lecture. DNA sequencing is now done by automatic sequencing machines, making autoradiographs like this one virtually obsolete (but still of nostalgic value to biologists over a certain age).



# TECHBOX 1.1

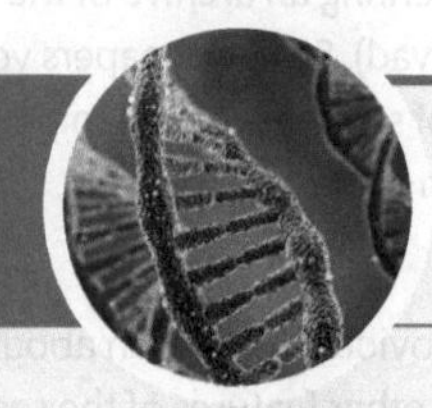

# GenBank

**RELATED TECHBOXES**

**TB 1.2:** BLAST

**TB 3.2:** Biobanking

**TB 10.2:** Multiple sequence alignment

**RELATED CASE STUDIES**

**CS 1.1:** Identification

**CS 1.2:** Forensics

**CS 2.2:** Ancient DNA

**Summary**

The gigantic online database of genomic information makes DNA sequences available to everyone.

**Keywords**

Entrez, DNA sequences, database, accession, annotation, NCBI, DataDryad

GenBank is a golden example of the international science community sharing data freely. It is based at the National Center for Biotechnology Information (NCBI) in the US, but it contains sequences submitted by scientists from all over the world. GenBank contains most of the DNA sequences that have ever been produced. Whenever a scientist sequences a section of DNA, they should submit the sequence to GenBank so that anyone else can access the sequence and use it in their own research. Submission to GenBank is usually a requirement of publication

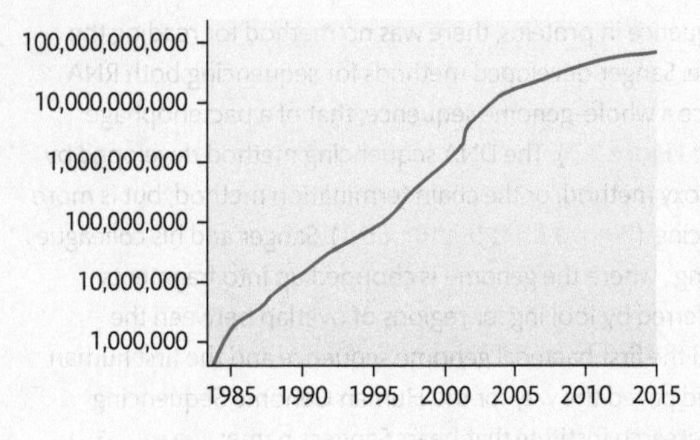


**Figure 1.12** Since the formation of GenBank in the early 1980s, the number of nucleotide bases it contains has doubled every 18 months (note that the number of bases in the database, on the y axis, is a log scale). Data from whole genome sequencing contributes an even greater amount of publicly available data, much of it essentially unanalysed. Genbank represents an amazing resource for you to explore and there are uncountable new discoveries to be made using the data it contains.

in the scientific press, in line with the ethos of repeatability of scientific experiments (anyone should have access to the data and materials needed to check published results). In addition, there are many sequences on GenBank that have never been formally published but are available for anyone to use, particularly sequences generated from whole genome sequencing projects. At the time of writing, GenBank contains over 180 billion nucleotides of DNA sequences, from over 300,000 species[1] (**Figure 1.12**). Even more bases of sequence have been contributed through the whole genome sequencing portal.

The easiest way to access GenBank is the Entrez search engine which allows you to search a large collection of databases at once, including whole genome sequences, single nucleotide polymorphism data (SNPs: **TechBox 3.1**), population-based datasets, functional and structural information on gene products, and taxonomic information (**TechBox 9.2**). GenBank is focused on cataloguing individual sequences, organized according to origin. Each GenBank entry contains a single sequence along with information about the organism, type of sequence, and so on. You can use these sequences to any purpose: to identify your own samples, add to your analysis of your own sequences or recombine to generate a novel dataset to answer a particular question. GenBank sequences are presented in a 'raw' format: any analysis will require you to identify (see **TechBox 1.2**) and align (see **TechBox 10.2**) the sequences to suit your purpose. But there are a growing number of other public databases that present sequence data organized by the project or publication it was derived from, representing an archive of the data in the form it was analysed in the publication (for example, DataDryad). So check papers you read for a link to a database where you can retrieve the alignment of sequences used to produce a particular phylogeny and the input files for a particular program and analysis.

### Sequence information and annotation

When a researcher submits a DNA sequence, they provide information about the organism it was sampled from, what kind of sequence it is, and other features of the sequence (this information is broadly known as sequence annotation). For example, if the sequence contains part of a protein-coding gene, the information given might include the location of the beginning and end of the coding sequence, the amino acid sequence of the protein product generated from it, and the likely function of the peptide. Genome sequencing projects often rely on automated annotation to identify and label features such as coding regions or gene regulation elements. Both manual and automated annotation can be misleading, though improvements in annotation techniques and methods to detect incorrect annotations should help limit errors[2].

Every GenBank submission is assigned a unique accession (identification) number that can be used to retrieve that sequence from the database (though NCBI has suggested that this may need to be relaxed in future given the vast numbers of sequences being submitted). Sequences can also be accessed by searching for an organism, gene, author name or keyword. For example, if you type 'Chilean blob' into the GenBank search engine, Entrez, you retrieve three sequences from the study described in **Case Study 1.1** with the accession numbers AY582746, AY582747, and AY582748. If you type these accession numbers into the query box on the Entrez search engine, that searches GenBank and related databases, it will locate the records of these sequences in the database so that you can access them. **Figure 1.13** shows the GenBank entry you will retrieve if you enter the accession number AY582746.

## Errors and contamination

Accepting submissions from any individual or laboratory is one of the strengths of GenBank, allowing it to rapidly expand to cover an ever-increasing breadth of species and genes. But it is also a weakness, as it is difficult to guarantee submission quality. All experienced GenBank users

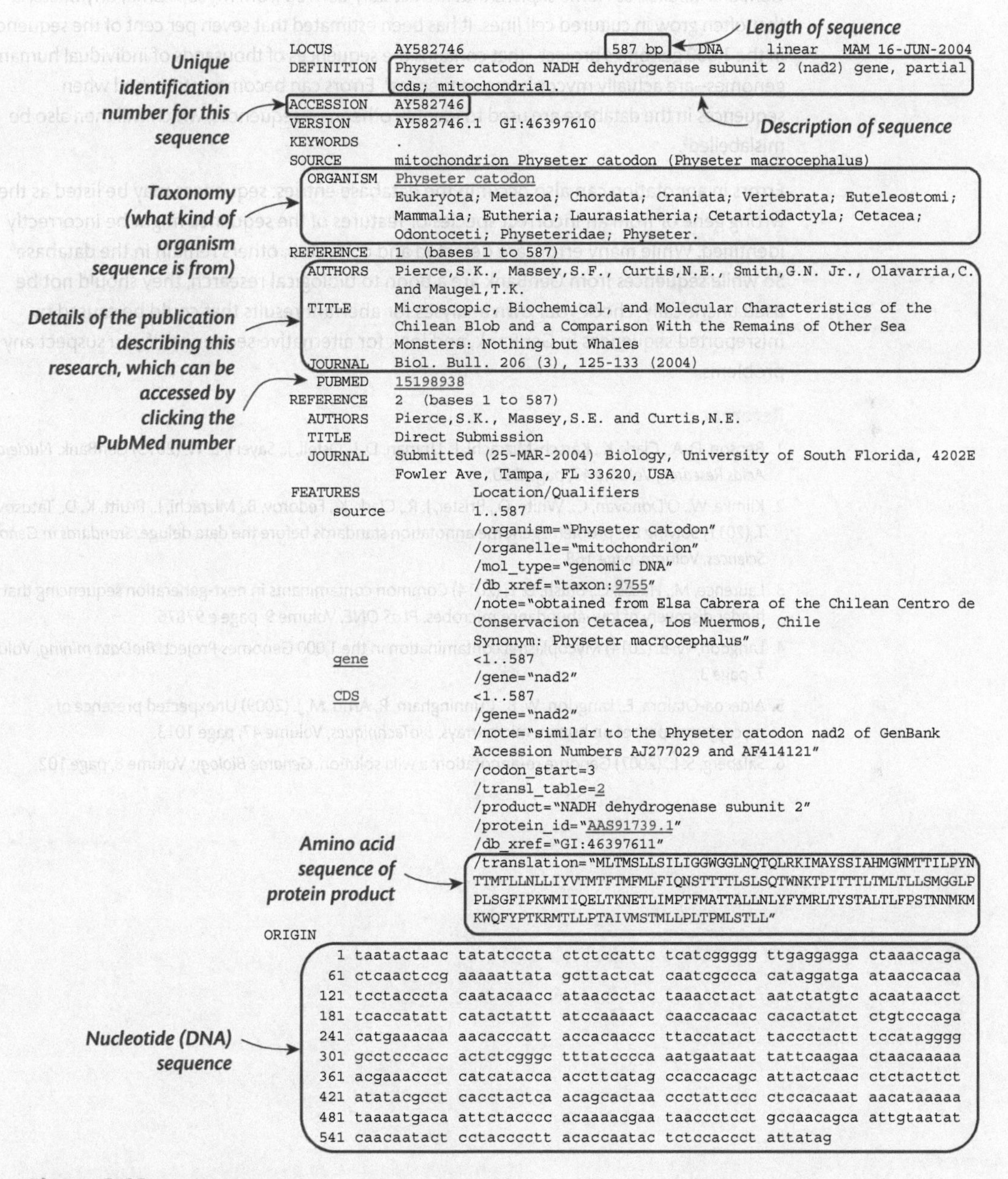

```
LOCUS       AY582746                 587 bp    DNA     linear   MAM 16-JUN-2004
DEFINITION  Physeter catodon NADH dehydrogenase subunit 2 (nad2) gene, partial
            cds; mitochondrial.
ACCESSION   AY582746
VERSION     AY582746.1  GI:46397610
KEYWORDS    .
SOURCE      mitochondrion Physeter catodon (Physeter macrocephalus)
  ORGANISM  Physeter catodon
            Eukaryota; Metazoa; Chordata; Craniata; Vertebrata; Euteleostomi;
            Mammalia; Eutheria; Laurasiatheria; Cetartiodactyla; Cetacea;
            Odontoceti; Physeteridae; Physeter.
REFERENCE   1  (bases 1 to 587)
  AUTHORS   Pierce,S.K., Massey,S.F., Curtis,N.E., Smith,G.N. Jr., Olavarria,C.
            and Maugel,T.K.
  TITLE     Microscopic, Biochemical, and Molecular Characteristics of the
            Chilean Blob and a Comparison With the Remains of Other Sea
            Monsters: Nothing but Whales
  JOURNAL   Biol. Bull. 206 (3), 125-133 (2004)
   PUBMED   15198938
REFERENCE   2  (bases 1 to 587)
  AUTHORS   Pierce,S.K., Massey,S.E. and Curtis,N.E.
  TITLE     Direct Submission
  JOURNAL   Submitted (25-MAR-2004) Biology, University of South Florida, 4202E
            Fowler Ave, Tampa, FL 33620, USA
FEATURES             Location/Qualifiers
     source          1..587
                     /organism="Physeter catodon"
                     /organelle="mitochondrion"
                     /mol_type="genomic DNA"
                     /db_xref="taxon:9755"
                     /note="obtained from Elsa Cabrera of the Chilean Centro de
                     Conservacion Cetacea, Los Muermos, Chile
                     Synonym: Physeter macrocephalus"
     gene            <1..587
                     /gene="nad2"
     CDS             <1..587
                     /gene="nad2"
                     /note="similar to the Physeter catodon nad2 of GenBank
                     Accession Numbers AJ277029 and AF414121"
                     /codon_start=3
                     /transl_table=2
                     /product="NADH dehydrogenase subunit 2"
                     /protein_id="AAS91739.1"
                     /db_xref="GI:46397611"
                     /translation="MLTMSLLSILIGGWGGLNQTQLRKIMAYSSIAHMGWMTTILPYN
                     TTMTLLNLLIYVTMTFTMFMLFIQNSTTTTLSLSQTWNKTPITTTLTMLTLLSMGGLP
                     PLSGFIPKWMIIQELTKNETLIMPTFMATTALLNLYFYMRLTYSTALTLFPSTNNMKM
                     KWQFYPTKRMTLLPTAIVMSTMLLPLTPMLSTLL"
ORIGIN
        1 taatactaac tatatcccta ctctccattc tcatcggggg ttgaggagga ctaaaccaga
       61 ctcaactccg aaaaattata gcttactcat caatcgccca cataggatga ataaccacaa
      121 tcctacccta caatacaacc ataaccctac taaacctact aatctatgtc acaataacct
      181 tcaccatatt catactattt atccaaaact caaccacaac cacactatct ctgtcccaga
      241 catgaaacaa aacacccatt accacaaccc ttaccatact taccctactt tccatagggg
      301 gcctcccacc actctcgggc tttatcccca aatgaataat tattcaagaa ctaacaaaaa
      361 acgaaaccct catcatacca accttcatag ccaccacagc attactcaac ctctacttct
      421 atatacgcct cacctactca acagcactaa ccctattccc ctccacaaat aacataaaaa
      481 taaaatgaca attctacccc acaaaacgaa taaccctcct gccaacagca attgtaatat
      541 caacaatact cctacccctt acaccaatac tctccaccct attatag
```



**Figure 1.13** The GenBank entry for a sequence from the Chilean Blob described in **Case Study 1.1**, with key features labelled in red.

have come across sequences that are strangely divergent from others and so they know to keep an eye out for entries that have unreliable base sequences or puzzling annotation. It is inevitable that some sequences contain mistakes made in the sequencing process. This is particularly true of raw data from high-throughput sequencing, which is error prone and requires verification (see **TechBox 5.2**).

Worse, some sequences may represent contaminants, rather than the target sequence reported. For example, it has been suggested that many metagenomic samples, which are used to describe the diversity of microbes in a sample, are contaminated by microbes from laboratory equipment, including 'ultrapure' water sources[3]. These contaminants can end up in GenBank identified as being from the target source. Embarrassingly, there are sequences in GenBank labelled as *Homo sapiens* that are actually derived from mycoplasma, tiny bacteria that often grow in cultured cell lines. It has been estimated that seven per cent of the sequences in the 1,000 genomes project—that contains the sequences of thousands of individual human genomes—are actually mycoplasma sequences[4]. Errors can become entrenched when sequences in the database are used to identify other new sequences, which will then also be mislabelled[5].

Errors in annotation can also occur in the database entries: sequences may be listed as the wrong gene or from an incorrect species or features of the sequence might be incorrectly identified. While many errors are detected and corrected, others remain in the database[6]. So while sequences from GenBank are a boon to biological research, they should not be used uncritically: check your own analyses for aberrant results that could be caused by misreported sequences in GenBank, and look for alternative sequences if you suspect any problems.

## References

1. Benson, D. A., Clark, K., Karsch-Mizrachi, I., Lipman, D. J., Ostell, J., Sayers, E. W. (2015) GenBank. *Nucleic Acids Research*, Volume 43, page D30.
2. Klimke, W., O'Donovan, C., White, O., Brister, J. R., Clark, K., Fedorov, B., Mizrachi, I., Pruitt, K. D., Tatusova, T. (2011) Solving the problem: genome annotation standards before the data deluge. *Standards in Genomic Sciences*, Volume, page 168.
3. Laurence, M., Hatzis, C., Brash, D. E. (2014) Common contaminants in next-generation sequencing that hinder discovery of low-abundance microbes. *PLoS ONE*, Volume 9, page e 97876.
4. Langdon, W. B. (2014) Mycoplasma contamination in the 1,000 Genomes Project. *BioData mining*, Volume 7, page 3.
5. Aldecoa-Otalora, E., Langdon, W. B., Cunningham, P., Arno, M. J. (2009) Unexpected presence of mycoplasma probes on human microarrays. *BioTechniques*, Volume 47, page 1013.
6. Salzberg, S. L. (2007) Genome re-annotation: a wiki solution. *Genome Biology*, Volume 8, page 102.

TECHBOX
1.2

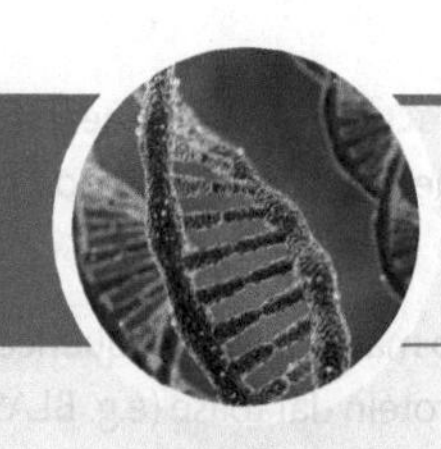

# BLAST

RELATED TECHBOXES

TB 1.1: GenBank

TB 10.2: Multiple sequence alignment

RELATED CASE STUDIES

CS 1.1: Identification

CS 1.2: Forensics

**Summary**

Unknown sequences can be identified by using an algorithm to match them to known sequences in a database.

**Keywords**

GenBank, alignment, database, homology, identity, E-value

One of the great advantages of massive DNA databases such as GenBank (**TechBox 1.1**) is that it can be used to identify an unknown sequence by matching a query sequence to a known sequence in the database (see **Case Studies 1.1** and **1.2**). This means that you can identify the source of a biological sample as long as you can obtain a DNA sequence from the sample and you can match it to another sequence in the database. You can also use BLAST to investigate: what kind of sequence you have sequenced, for example by finding similarities between your sequences and other gene families; or what kind of organism you have sampled, by seeing what kinds of organisms are the closest match.

Technically, BLAST stands for Basic Local Alignment Search Tool, though the name was originally intended as a play on the name of an earlier search method called FASTA[1] (so BLAST is an example of a not uncommon phenomenon in bioinformatics programming: think of acronym first then come up with a name that fits it). The scientific paper describing the BLAST algorithm was one of the most highly cited papers in the 1990s, now with over 50,000 citations in the scientific literature[2] (and a follow-up paper with improved methods has just as many[3]). BLAST has now been modified, extended, and diversified into a great variety of flavours by a large number of scientists and programmers.

BLAST is used when you wish to take a particular sequence and identify the closest possible match to it in a sequence database. Essentially, this is done by making a large number of pairwise alignments: the query sequence is aligned against a database of sequences and the highest scoring alignments are reported (see **TechBox 10.2**). The main problem with this approach is that the databases are huge: it would be impossible to provide an exhaustive alignment search against the millions of DNA sequences currently in GenBank in any reasonable timeframe. The breakthrough with BLAST was its speed: it is an heuristic search, which means it doesn't look at all possible alignments so it can't be guaranteed to find the absolute best match. But in practice, BLAST performs well and returns results almost immediately.

The sequence that you want to identify is usually referred to as the 'query sequence'. Rather than trying to align the whole query sequence to the database in one go, BLAST works by initially matching short segments of the query sequence against the database (hence the 'local alignment' in the acronym). Imagine that you lined up all of the sequences in GenBank, end to end. Then you cut your query sequence up into short fragments (say, 11 bases long) and you slid each fragment along the billions of nucleotides of sequence in GenBank until you found a match between the fragment and a sequence somewhere in the database. Then each time you got a match between your query fragment and a sequence in the database you tried to extend that match by seeing if the query sequence on either side of the 11-base fragment also matches to that sequence in the database. If you can find matches either side of the query fragment, then you have increased the chances that you have found an informative match, not just a chance similarity. You can then use an alignment algorithm to score the match of the whole query sequence to that particular database sequence (see **TechBox 10.2**). Repeat the process until you have made a thorough search of the database, then rank all of the good matches by their alignment scores.

When you blast you can choose whether you want to match a DNA sequence to the nucleotide database (BLASTn), an amino acid sequence to the protein database (e.g. BLASTp), or even to match a protein sequence to the nucleotide database (tBLASTn). There are also specialized blast searches that target particular kinds of sequences, such as conserved domains or regulatory sequences. In addition, there are different kinds of blast algorithms. For example, PSI-BLAST uses an iterative procedure to refine its search: after the first round of BLAST, positions in the matched pairs are scored according to how conserved the nucleotide sequence is across all pairs. These scores are used to generate a profile that informs a second round of BLAST searching. This iterative procedure that favours conserved positions makes PSI-BLAST good for finding divergent members of protein families.

BLAST hits are reported with E-values that reflect the probability that the observed match is due to chance alone, rather than reflecting a meaningful evolutionary relationship between the sequences. It is similar to the p-value reported for many statistical tests: the lower the E-value, the less likely it is that the similarity is due to chance. The probability of a random match decreases with increasing length of the query sequence. A four-base query sequence, such as AGTC, will perfectly match to very many sequences because there is a fairly high chance of those four bases occurring together just by chance. But any particular 25 base sequence has much less chance of occurring in unrelated sequences: for example, the nucleotide sequence CGTAGGGGTCAACATAATTTTCTTC matches exactly to only one sequence in the whole of GenBank (the cytochrome oxidase gene of a quagga, *Equus quagga*: **Figure 1.14**).

As with most bioinformatics programs, you can use the default parameters, pasting your sequence into the query box and hitting the 'Blast' button without worrying about the many different options on the page. But people running large numbers of searches using vast quantities of data can fine-tune BLAST (or other search algorithms) to optimize speed or

**Figure 1.14** The quagga is an extinct subspecies of zebra. Like other zebras, the quagga could produce occasional hybrids with horses (zebra hybrids are given the sci-fi-like name of zebroids). The quagga was the subject of one of the first ancient DNA studies (**TechBox 11.1**). A breeding program has been established to recreate the quagga by selectively breeding Plains Zebras for quagga-like coat patterning (see **Figure 10.2**). While this may create a simulacrum of the lost subspecies, it cannot retrieve the quagga genome.

performance for particular types of searches. If you take the time to learn about the options, and compare the output of searches using different parameters, you may get some interesting results (and possibly get hooked on blasting, which keeps some people amused for hours).

## References

1. Altschul, S. F. (2003) Foreword in BLAST eds.Korf, I., Yandell, M. and Bedell, J. (O'Reilly & Associates)
2. Altschul, S. F., Gish, W., Miller, W., Myers, E. W., Lipman, D. J. (1990) Basic local alignment search tool. *Journal of Molecular Biology*, Volume 215, page 403.
3. Altschul, S. F., Madden, T. L., Schaffer, A. A., Zhang, J., Zhang, Z., Miller, W., Lipman, D. J. (1997) Gapped BLAST and PSI-BLAST: a new generation of protein database search programs. *Nucleic Acids Research*, Volume 25, page 3389.

CASE STUDY 1.1

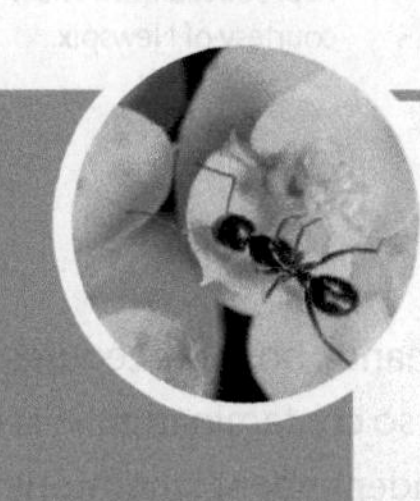

# Identification: solving the mystery of the Chilean blob

**RELATED TECHBOXES**

**TB 1.1:** Genbank
**TB 1.2:** BLAST
**TB 9.2:** Barcoding

**RELATED CASE STUDIES**

**CS 1.2:** Forensics
**CS 2.2:** Ancient DNA

Pierce, S. K., Massey, S. E., Curtis, N. E., Smith, G. N., Olavarri, C., Maugel, T. K. (2004) Microscopic, biochemical, and molecular characteristics of the Chilean Blob and a comparison with the remains of other sea monsters: nothing but whales. *Biological Bulletin*, Volume 206, page 125.

*"... to our disappointment, we have not found any evidence that any of the blobs are the remains of gigantic octopods, or sea monsters of unknown species"*

## Keywords

GenBank, sample preservation, ethanol, formaldehyde, contamination, mitochondrial, DNA barcoding, cryptozoology

## Background

Strange gelatinous material occasionally washes up on beaches around the world (**Figure 1.15**). The large mass that washed up on a beach in Florida in 1896 was too heavy to be removed using a team of horses, and so rubbery that axes bounced off it. The material generally lacks cells that would aid its identification and so these 'globsters' have variously been described as the remains of giant squid, unknown species of octopus, or whale blubber. The lack of a clear species identification has resulted in globsters being listed on websites that discuss unexplained phenomena. The blob that washed up on a beach in Chile in 2003 made international headlines and was discussed by biologists and conspiracy theorists all over the world.

## Aim

DNA sequences provide the means to unambiguously identify a biological sample, through comparison of the sequence to that from known species. Even if the sample is from a previously undescribed species, then comparing the sequence to a database of known sequences can show which species is the closest known relative. These researchers aimed to use DNA from the Chilean blob to work out what kind of organism the blob was derived from.

The Mercury

Established 1854

HOBART, FRIDAY, MARCH 9, 1962

NEARLY AS BIG AS A HOUSE!

IT WAS COVERED WITH FINE HAIR

ABOUT 20ft. long, 18ft. wide and about 4½ft. thick, with an estimated weight of between five and 10 tons.

"SEA MONSTER" FIND MAY BECOME WORLD TOPIC

AN expedition returned to Hobart from the West Coast yesterday with a story of a "sea monster" which could arouse world-wide interest.

They described the remains as conforming to no known animal, and examination may bring to light the fact that it is, indeed, unique.

Continued on Page 2

Tough going for party

**Figure 1.15** The first mystery beached blob to be referred to as a 'globster' was the enormous gelatinous lump, over five tons and covered in bristles, found on a Tasmanian beach in the 1960s. While zoologists identified it as whale blubber, it attracted rather more hysterical press coverage.

Reproduced from *The Mercury*, 9 March 1962 courtesy of Newspix.

### Methods

Small pieces of the blob were frozen or preserved in ethanol: both procedures protect the DNA in the sample from decaying. DNA is everywhere, so contamination in sequencing labs can be a problem. Analysing two different samples in independent laboratories provides a check against contamination (this can rule out incidental contamination from the lab, but not contamination at the source of the samples). One lab in the US analysed a frozen sample of the blob, and a lab in New Zealand analysed an ethanol-preserved sample. DNA was extracted from the samples and sequenced. The US team sequenced part of mitochondrial gene called *NADH2* which codes for the subunit of a fundamental metabolic enzyme. The NZ team sequenced the mitochondrial control region which does not code for any protein and is usually highly variable. These sequences were chosen because they tend to vary between species, and because they are relatively easy to sequence, so there are a lot of comparable sequences on the international sequence database, GenBank (**TechBox 1.1**).

### Results

To identify the origin of the blob, the DNA sequences were compared to existing sequences in GenBank (**Figure 1.16**). The sequences of *NADH2* from the blob were identical to two *NADH2* sequences in the database, which had both been taken from sperm whales (*Physeter catadon*, also known by the species name *Physeter macrocephalus*: **Figure 1.2**). Because this sequence normally varies between species, this is convincing proof that the blob was from a sperm whale. The mitochondrial control region sequence from the Chilean Blob was 99 per cent identical to *P.catadon* sequences in GenBank. The control region tends to evolve faster than *NADH2*, so may vary between individuals within a species, so it is often used for population-level studies. Although the control region sequence differed from the sperm whale sequence on GenBank, it was only as different as you would expect from two different individuals in the same species.

### Conclusions

The identical (or near-identical) match between the sequences from the blob and whale sequences in GenBank prove beyond doubt that the blob is the remains of a sperm whale.

### Limitations

Selection of an appropriate gene to sequence is critical—there has to be enough difference between species to allow discrimination, yet not so much difference that the relationship to other species is unclear. Sample quality and nature of preservation are important for DNA extraction. They could not extract DNA from earlier 'blob' samples that had been preserved in formaldehyde (including samples of the Tasmanian West Coast Monster: **Figure 1.15**), because formaldehyde destroys DNA. This is a shame, because many museum specimens are pickled in formaldehyde.

```
                  1                                                          60
Physeter catadon  TAATACTAACTATATCCCTACTCTCCATTCTCATCGGGGGTTGAGGAGGACTAAACCAGA
Chilean blob      TAATACTAACTATATCCCTACTGTCCATTCTCATCGGGGGTTGAGGAGGACTAAACCAGA
                  61                                                        120
Physeter catadon  CTCAACTCCGAAAAATTATAGCTTACTCATCAATCGCCCACATAGGATGAATAACCACAA
Chilean blob      CTCAACTCCGAAAAATTATAGCTTACTCATCAATCGCCCACATAGGATGAATAACCACAA

                  121                                                       180
Physeter catadon  TCCTACCCTACAATACAACCATAACCCTACTAAACCTACTAATCTATGTCACAATAACCT
Chilean blob      TCCTACCCTACAATACAACCATAACCCTACTAAACCTACTAATCTATGTCACAATAACCT

                  181                                                       240
Physeter catadon  TCACCATATTCATACTATTTATCCAAAACTCAACCACAACCACACTATCTCTGTCCCAGA
Chilean blob      TCACCATATTCATACTATTTATCCAAAACTCAACCACAACCACACTATCTCTGTCCCAGA

                  241                                                       300
Physeter catadon  CATGAAACAAAACACCCATTACCACAACCCTTACCATACTTACCCTACTTTCCATAGGGG
Chilean blob      CATGAAACAAAACACCCATTACCACAACCCTTACCATACTTACCCTACTTTCCATAGGGG

                  301                                                       360
Physeter catadon  GCCTCCCACCACTCTCGGGCTTTATCCCCAAATGAATAATTATTCAAGAACTAACAAAAA
Chilean blob      GCCTCCCACCACTCTCGGGCTTTATCCCCAAATGAATAATTATTCAAGAACTAACAAAAA

                  361                                                       420
Physeter catadon  ACGAAACCCTCATCATACCAACCTTCATAGCCACCACAGCATTACTCAACCTCTACTTCT
Chilean blob      ACGAAACCCTCATCATACCAACCTTCATAGCCACCACAGCATTACTCAACCTCTACTTCT

                  421                                                       480
Physeter catadon  ATATACGCCTCACCTACTCAACAGCACTAACCCTATTCCCCTCCACAAATAACATAAAAA
Chilean blob      ATATACGCCTCACCTACTCAACAGCACTAACCCTATTCCCCTCCACAAATAACATAAAAA

                  481                                                       540
Physeter catadon  TAAAATGACAATTCTACCCCACAAAACGAATAACCCTCCTGCCAACAGCAATTGTAATAT
Chilean blob      TAAAATGACAATTCTACCCCACAAAACGAATAACCCTCCTGCCAACAGCAATTGTAATAT

                  541                                            587
Physeter catadon  CAACAATACTCCTACCCCTTACACCAATACTCTCCACCCTATTATAG
Chilean blob      CAACAATACTCCTACCCCTTACACCAATACTCTCCACCCTATTATAG
```

**Figure 1.16** Alignment of a sequence from the mysterious Chilean blob with that of a sperm whale (*Physeter catadon*), showing that the sequences are identical for this part of their mitochondrial genome.

### Further work

Because DNA can be extracted from samples such as hair, faeces, or saliva, molecular analysis has the potential to solve many mysteries of cryptozoology (the field that seeks to establish whether apparently mythical creatures are in fact real species). For example, DNA from purported Yeti hair samples matched DNA sequences from bears[1]. The DNA barcoding movement seeks to catalogue DNA sequences of all species so that any unknown sample can be identified. (**TechBox 9.1**)

### Check your understanding

1. Do all biological samples contain DNA?
2. What results would you expect a BLAST search to have given if the blob had actually been the remains of a hitherto unknown species of giant octopus?
3. Why were the samples analysed in two different laboratories?

### What do you think?

One of the genes sequenced was identical between the blob and the sperm whale, and the other was 99 per cent similar. How similar do two DNA sequences need to be before you can confidently say they are from the same species?

### Delve deeper

What other cases in 'cryptozoology' have been addressed using DNA sampling? Will DNA evidence solve all outstanding cases of mysterious creatures? What, if any, are the limitations on this approach?[2]

### References

1. Sykes, B. C., Mullis, R. A., Hagenmuller, C., Melton, T. W., Sartori, M. (2014) Genetic analysis of hair samples attributed to yeti, bigfoot and other anomalous primates. *Proceedings of the Royal Society B: Biological Sciences*, Volume 281, page 20140161.
2. Edwards, C. J., Barnett, R. (2015) Himalayan 'yeti' DNA: polar bear or DNA degradation? A comment on 'Genetic analysis of hair samples attributed to yeti' by Sykes et al. (2014). *Proceedings of the Royal Society B: Biological Sciences*, Volume 282, page 20141712.

CASE STUDY 1.2

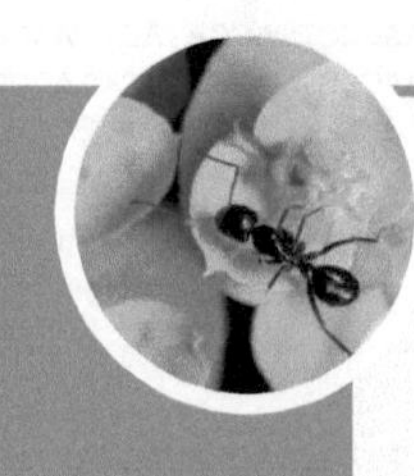

# Forensics: using DNA to trace trade in whale meat

RELATED TECHBOXES

**TB 1.1:** Genbank

**TB 9.2:** Barcoding

RELATED CASE STUDIES

**CS 1.1:** Identification

**CS 2.1:** Environmental DNA

Baker, C.S., Steel, D., Choi, Y., Lee, H., Kim, K. S., Choi, S. K., Ma, Y-U., Hambleton, C., Psihoyos, L., Brownell, R. L., Funahashi, N. (2010) Genetic evidence of illegal trade in protected whales links Japan with the US and South Korea. *Biology Letters*, Volume 6, page 647.

> *These results confirmed the power of molecular methods in monitoring retail markets and pointed to the inadequacy of the current moratorium for ensuring the recovery of protected species*[1]

### Keywords

PCR, conservation, wildlife forensics, CITES, geographic origin, identifying individuals, hunting, DNA barcoding, DNA surveillance

### Background

Following dramatic falls in global whale populations, commercial hunting of whales was banned by international treaty in 1986. However, regular whale hunts are still conducted in a number of countries and there are concerns that these exceptions to the moratorium threaten whale population recovery and persistence. Japan continues to hunt a number of whale species under a Special Permit programme commonly referred to as 'scientific whaling'. While meat from the Japanese whaling programme can be sold on the domestic market, it cannot be exported to other non-whaling countries such as the USA and South Korea.

### Aims

These researchers wanted to establish if whale meat sashimi purchased in 2009 at restaurants in Los Angeles and Seoul really did contain whale meat and, if it did, to determine where the whale meat had come from.

### Methods

Meat purchased from the Los Angeles restaurant was frozen, while the Seoul sample was stored in ethanol. Polymerase chain reaction (PCR: **TechBox 4.2**) was used to amplify two mitochondrial sequences (*cytochrome b* and *control region*), and seven microsatellite markers (see Chapter 5). They matched these sequences to the curated DNA Surveillance database[2] for cetaceans. A curated database contains sequences from samples of verified source and taxonomy (**TechBox 9.2**). Matches to this database were confirmed through a BLAST search of GenBank (**TechBox 1.1**).

### Results

The LA samples were identified as the endangered sei whale (*Balaenoptera borealis*). Furthermore, the sequences were identical to those from samples taken in Japanese markets in 2007 and 2008, suggesting that this whale meat was derived from the 'scientific whaling' programme. From 13 whalemeat products bought in Seoul, the researchers identified sei whale, minke whale, fin whale, and Risso's dophin. The fin whale sample was an exact match (for both the mitochondrial sequences and the microsatellite profile) to whalemeat purchased from Japanese markets in 2007, suggesting that the products were derived from the same individual whale. In contrast, sashimi sold as horse meat in the LA restaurant turned out to be from a cow.

### Conclusions

The findings support the contention that whale meat derived from the 'scientific whaling' programme is entering international trade despite international trading bans. Both cases of identified whale meat have been referred for criminal investigation.

### Limitations

Identifying species from market samples relies on having a comprehensive database of sequences from all possible target species, and finding sequences that differ in distinctive ways between species. Any database search will report the closest match between the sample DNA and the database, so caution is needed in interpreting results. Databases for molecular identification of individual whales, referred to as 'DNA registers,' have now been established in Norway[3] and Japan, though the Japanese database has not been released for independent review, and neither database is fully open access.

### Further work

This approach has been extended from identification of protected species to monitoring the demographic impact of hunting on species such as the North Pacific minke whale[4]. The researchers have also applied classic ecological techniques for estimating sample size to whale meat products, in order to estimate the number of individual whales ending up in the market[4]. Wildlife forensics is being increasingly applied to monitoring the illegal trade in endangered species[5].

### Check your understanding

1. Why did researchers freeze some sashimi samples, and put others in alcohol?
2. Why did they compare the sequences to a dedicated whale identification database, rather than just seeing what matches occurred on GenBank? Could these two searches give different results?
3. Is it possible to prove beyond doubt that the meat came from the 'scientific whaling' programme?

### What do you think?

The UN Convention on International Trade in Endangered Species (CITES) prohibits the import or export of products from endangered species across international borders. It has been argued that, while 'native DNA' extracted from biological tissues comes under CITES legislation, amplified DNA does not, because it is a synthetic copy of the DNA from the original sample, so it can be imported without permit[6, 7]. Do you agree that PCR product should be exempt from CITES rules, or should synthetic DNA from endangered species have the same protection as any other wildlife product (**Figure 1.17**)?

**Figure 1.17** Scott Baker using a 'portable laboratory', consisting of a PCR machine plus chemical reagents which were capable of being carried in a suitcase. Since international movement of products from endangered species is prohibited under CITES legislation, these researchers would buy whale meat products from the market, then surreptitiously extract and PCR the DNA in their hotel room. They took the PCR products back to universities in the US and New Zealand and sequenced them, and compared the whale meat sequences to sequences from known cetacean species. New DNA amplification methods that do not rely on cycles of heating and cooling may make mobile DNA testing much easier (see **TechBox 4.2**).

Reproduced by permission of C. Scott Baker, Marine Mammal Institute, Oregon State University and School of Biological Sciences, University of Auckland.

## Delve deeper

DNA wildlife forensics is being increasingly applied to understanding the trade in 'bush meat' derived from wild-caught animals in Africa. Why is it important to monitor this trade? What are the strengths and limitations of a DNA forensic approach?

## References


1. Baker, C.S., Lento, G. M., Cipriano, F., Palumbi, S. R. (2000) Predicted decline of protected whales based on molecular genetic monitoring of Japanese and Korean markets. *Proceedings of the Royal Society B: Biological Sciences*, Volume 267, page 1191.
2. Ross, H. A., Lento, G. M., Dalebout, M. L., Goode, M., Ewing, G., McLaren, P., Rodrigo, A. G., Lavery, S., Baker, C. S. (2003) DNA Surveillance: Web-based molecular identification of whales, dolphins, and porpoises. *Journal of Heredity*, Volume 94, page 111.
3. Glover, K. A., Haug, T., Øien, N., Walløe, L., Lindblom, L., Seliussen, B. B., Skaug, H. J. (2012) The Norwegian minke whale DNA register: a data base monitoring commercial harvest and trade of whale products. *Fish and Fisheries*, Volume 1, page 313.
4. Baker, C.S., Cooke, J. G., Lavery, S., Dalebout, M. L., Ma, Y., Funahashi, N., Carraher, C., Brownell, R. L. (2007) Estimating the number of whales entering trade using DNA profiling and capture-recapture analysis of market products. *Molecular Ecology*, Volume 16, page 2617.
5. Alacs, E., Georges, A., FitzSimmons, N., Robertson, J. (2010) DNA detective: a review of molecular approaches to wildlife forensics. *Forensic Science, Medicine, and Pathology*, Volume 6, page 180.
6. Bowen, B. W., Avise, J. C. (1994) Conservation research and the legal status of PCR products. *Science*, Volume 266, page 713.
7. Palumbi, A., Cipriano, F. (1998) Species identification using genetic tools: the value of nuclear and mitochondrial gene sequences in whale conservation. *Journal of Heredity*, Volume 89, page 459.

# DNA

## *The immortal germline*

*"If you could use a big enough magnifying glass you would find that there is really only one kind of life on the Earth: the most central machinery in all organisms is built out of the same set of micro-components, the same set of small molecules"*

**G. Cairns-Smith (1985) *Seven Clues to the Origin of Life*, Cambridge University Press.**

## What this chapter is about

The genomic information system is shared by all known organisms. It is worth learning a little of the history of the discovery of this genomic system because it serves to illustrate some of the important principles of heredity. Life relies upon the continuity of genetic information coded in DNA and copied from generation to generation. The DNA found in every living cell contains the genetic information needed to construct the organism, as well as providing biologists with a wealth of information about evolutionary past and processes.

### Key concepts

- DNA base pairing allows the information in the genome to be copied from generation to generation, and to be copied into RNA and translated into protein
- DNA can be extracted from most fresh biological samples

# Material basis of heredity

*Unity of life*

All life on earth has a common ancestor. If you trace your family tree back far enough, you will find you are related to the rats in your attic, the fly on your window, the mould on your bread, the rice in your cupboard, even the bacterium in your gut (Figure 2.1). We know this because every organism on earth uses the same basic system to carry the information that it needs to grow and reproduce. This genomic system consists of information stored in DNA, which is transcribed into RNA and translated into proteins. This system is so intricate, with so many complex interlocking parts, that we can be sure that it was not invented separately in different biological lineages. So we are all descended from an ancient simple life form that carried the same fundamental genomic system that we share today with all of the earth's biodiversity (TechBox 2.1).

Not only is the genomic system of DNA, RNA and proteins shared by life forms as different as rats, flies, mould, rice, and bacteria, but some of the actual information stored in the genome is also shared. In a sense this is not surprising because all living things require the instructions for constructing the genomic system itself. For example, all organisms must be able to convert the genomic information coded in DNA into RNA messages that provide the instructions needed to make a protein. One of the things they need in order to do this is a working copy of the enzyme RNA polymerase, which makes an RNA message from a DNA gene sequence. So all organisms must have a gene that codes for RNA polymerase.

In fact, eukaryotes (such as fungi, plants and animals) have several different types of RNA polymerase enzymes, and each one is constructed from many different subunits, in some cases combining a dozen or more different proteins. For simplicity, let's concentrate on a gene for a single subunit of one of these RNA polymerase enzymes, the beta-subunit of RNA polymerase II. Figure 2.2 shows part of the DNA sequence of this gene. Each horizontal line of letters represents the particular sequence of four nucleotide bases (A, C, T and G) in the version of the RNA polymerase gene found in a given species. In this alignment, we can compare the RNA polymerase II beta sequence from humans, rats, fruit flies, bread mould, rice, and bacteria. Although the exact DNA

**Figure 2.1** Rice (*Oryza sativa*) is one of the most important crop plants in the world. It is often said that humans are, genetically speaking, 99 per cent similar to chimpanzees. Perhaps more remarkably, it has been estimated that around half the genes in plants have recognizable homologs in humans. Remember that next time you eat a bowl of rice.

| | |
|---|---|
| **Homo sapiens** | CGTGATGGTGGCCTGCGTTTTGGAGAAATGGAACGAGATTGTCAGATTGC |
| **Rattus norvegicus** | CGTGATGGTGGCCTGCGCTTTGGAGAAATGGAGCGAGACTGTCAGATCGC |
| **Drosophila melanogaster** | CGTGATGGTGGCTTGCGTTTCGGTGAGATGGAGCGTGATTGCCAGATCTC |
| **Neurospora crassa** | AGAGACGGTGGTCTCCGTTTCGGTGAAATGGAACGTGACTGTATGATTGC |
| **Oryza sativa** | CGGTACGGCGGCGTCAAGTTCGGCGAGATGGAGCGCGACTGCCTCCTCGC |
| **Escherichia coli** | CAGTTCGGTGGTCAGCGTTTCGGGGAGATGGAAGTGTGGGCGCTGGAAGC |

**Figure 2.2** DNA sequences from the gene for the beta-subunit of the RNA polymerase II enzyme, from a human (*Homo sapiens*), a rat (*Rattus norvegicus*), a fly (*Drosophila melanogaster*), a mould (*Neurospora crassa*), rice (*Oryza sativa*), and a bacterium (*Escherichia coli*). This section represents approximately one per cent of the entire sequence of this gene.

sequence is slightly different in each of these organisms, each of the versions provides the necessary instructions for making a working copy of RNA polymerase.

*Gene transcription and translation are explained in Chapter 6*

This is just a small part of one gene. The genomes of most organisms contain thousands of genes. Every time the body manufactures a new molecule to help it grow or move or respond, particular genes must be located in the genome, unwound, transcribed and translated. Then the newly manufactured molecules must be folded, combined and transported to where they are needed. The expression of a single gene requires the coordinated action of dozens of enzymes, the manufacture of a great number of specialized molecular building blocks, and the coordination of a large number of tasks in time and space within the cell. Yet this complex process is being continuously performed by every single living cell at every moment to produce thousands of proteins and other molecules, all in the right place, at the right time, in the right amounts. The beauty and complexity of the genomic system never ceases to amaze me.

In order to truly appreciate the wonder of the natural world, you need to gain some insight into the workings of the molecular level of organization that underlie all of the functions of the living world. This is important for two reasons. Firstly, an appreciation of molecular biology is the best way to bring home the complexity and intricacy of organisms. Secondly, a grasp of the biochemical basis of heredity is essential to understanding evolution, as it is at this level that mutations occur and substitutions accumulate causing lineages to change and diverge over time. We are going to briefly consider the history of the discovery of the genomic system in order to review some of the key principles of heredity.

You may wonder why a textbook on modern DNA analysis would bother reviewing ideas from hundreds of years ago, many of which were subsequently shown to be incorrect. There are many reasons why a historical perspective is valuable. Science is cumulative: we build on the ideas of those who came before. What we think today is shaped by ideas of scientists who thought about these problems tens or hundreds of years ago. Considering biological explanations promulgated by scientists in the past, even those that turned out not to be correct, helps us understand our current ideas. More importantly, considering the flow of ideas over time reminds us that scientific ideas are hypotheses, proposed explanations that might need to be revised in future. It has been a major effort revising this edition, given all the changes in the field that have occurred in the seven years since the first edition. How many of the concepts explained in this book will still stand unchanged in twenty years' time? Or two hundred? By learning a little of the history of ideas in this field, how they were generated, tested, modified or rejected, we get a sense of science as a process, rather than as a growing body of unchanging facts. More fundamentally, these ideas are not just of historical interest, nor are they only relevant to those interested in genetics itself. These principles of heredity explain why we can use DNA as an information source in evolutionary biology and ecology.

## Principles of heredity

Our knowledge of the genomic system is surprisingly recent. The basic principle of heredity—that offspring tend to resemble their parents—has long been observed by human societies. But the exact mechanics of inheritance were subtle and unknown. It had always been recognized that animals and plants tend to arise from parents of the same species, but it was commonly believed that in certain circumstances living beings could arise spontaneously, such as flies being generated from rotting meat, wasps arising from galls on plants, or bacteria forming *de novo* in chicken broth. Spontaneous generation was finally put to rest 150 years ago with conclusive experiments in which potential parent organisms were carefully excluded from sterilized material. The lack of spontaneous appearance of living organisms where there were none before ultimately convinced scientists of the importance of genetic continuity. Only living organisms contained the necessary information to make another organism; life cannot arise without a copy of this genetic information. But what was the material basis of genetic continuity? Did reproductive cells contain tiny preformed creatures that grew into new adults? If so, then how were traits from both mother and father inherited? And how could an organism like a sponge reproduce by budding, where a small piece of its body could be induced to grow into a new individual? Somehow, cells must be able to transmit the information needed to create a new individual.

Of course, there were many different theories about the material basis of heredity. For example, in his natural history tract *Vénus Physique*, published in 1745, Pierre-Louis Moreau de Maupertuis proposed that particles corresponding to all parts of the body were provided by the parents and used to build the developing offspring (**Figure 2.3**). He suggested that these particles could be altered to give rise to new hereditary types, and might even undergo isolation in different parts of the world to produce new species. The idea that all parts of the parent's body contribute information to the offspring is referred to as pangenesis. Theories of pangenesis have a long history, going back to the philosophers of ancient Greece, but one of the most famous proponents of pangenesis was the father of evolutionary biology, Charles Darwin.

*Darwin's theory of evolution is discussed in more detail in Chapter 7*

Darwin recognized the central role of heredity in evolutionary theory, devoting whole volumes to recording and interpreting observations of inheritance in the natural world (particularly in domesticated animals and plants). Darwin knew that variations arose continually in natural populations, and that many variations could be inherited, but he could only guess at the mechanism. Critics of natural selection pointed out that if the characteristics of the parents were blended to create their offspring, then any favourable new variant would be diluted each generation and eventually lost.

**Figure 2.3** Pierre-Louis Moreau de Maupertuis (1698–1759) is variously described as a mathematician, an astronomer, a philosopher, and a literary wit. He is shown here in the clothes he wore on an expedition to Lapland to prove Newton's prediction that the earth is flattened at the poles. Maupertuis was also a keen observer of the natural world and developed a remarkably prescient theory of heredity. His works also contain a vague statement of the principle of natural selection. The broad concept of natural selection—that in each generation the better adapted individuals would tend to survive and reproduce—had been alluded to by many thinkers, but its power to explain biological diversity was not realized until Charles Darwin and Alfred Russel Wallace developed the idea of natural selection in much greater detail.

The lack of a clear mechanism for inheritance was, in some ways, a stumbling block for the development of evolutionary theory. To fill this gap, Darwin developed his theory of pangenesis, speculating that all the body's cells produced particles, called gemmules, which carried information. Gemmules collected in the reproductive cells prior to fertilization, ensuring the offspring inherited all the information needed to make a functioning individual. Because gemmules formed in the adult body, Darwin's theory of pangenesis specifically allowed for the inheritance of acquired characteristics: that is, modifications of the body during an individual's lifetime could be inherited by its offspring. He was at pains to point out that, while it was pure speculation, having a 'provisional hypothesis' of heredity was better than having no hypothesis at all. This theory of pangenesis was criticized by some of Darwin's contemporaries. One strong critic was Francis Galton, one of the founders of modern statistics and leader of the early eugenics movement. Galton showed that blood transfusions did not appear to move hereditary information from one individual to another, as would be expected if gemmules were carried in the blood. Even twenty years after the publication of *The Origin of Species*, the material basis of heredity was still unknown, despite being the focus of much study.

## The Weismannian barrier

*At the present time there is hardly any question in biology of more importance than this of the nature and causes of variability, and the reader will find in the present work an able discussion on the whole subject . . . Whoever compares the discussions in this volume with those published twenty years ago on any branch of Natural History, will see how wide and rich a field for study has been opened up through the principle of Evolution; and such fields, without the light shed on them by this principle, would for long or for ever have remained barren.*

Charles Darwin, Foreword to *The Study of Heredity* by August Weismann (1880).

August Weismann transformed evolutionary biology by arguing forcefully against the inheritance of acquired characteristics (**Hero 2**). He argued that the hereditary information (which he referred to as the 'germplasm') was isolated from the rest of the body, so was passed from one generation to the next essentially unchanged. Each new individual receives the same genetic instructions that their parents received. Weismann considered that the continuity of the germline meant that the genetic information was handed intact from one generation to the next, and was not rewritten by the trials and tribulations of the lives of the organisms that carry it.

His arguments were largely made on theoretical grounds, by considering the implications of patterns of inheritance for the process of evolution. For example, how could the non-reproductive castes of social insects evolve if there was no way that a sterile worker could pass its bodily modifications, having no offspring of its own? How could mutations arise in organs that must be fully formed before use, therefore had no opportunity for acquiring new characteristics by use and disuse? And more importantly, how could information about the state of adult organs be translated into a form of inheritable instructions?

Weismann also argued from observation: despite widespread belief, there was simply no evidence that acquired characteristics could be inherited. For example, it was clear that human societies that practised male circumcision over many generations did not give rise to offspring that no longer had foreskins. Weismann carried out experiments that demonstrated that bodily modifications acquired in an individual's lifetime were not passed to their offspring. One of his most famous experiments was to dock the tails of mice, then breed from the tailless individuals, and dock the tails of their offspring. He continued this process through 21 generations of mice, docking the tails each generation. But the tailless mice never produced tailless offspring. Although this experiment seems trivial, it is important to realize that this demonstration ran counter to the prevailing opinion of the times. Many contemporary animal breeders believed that it was possible to produce a tailless breed by cutting the tails of individuals then breeding from them.

Thus Weismann argued persuasively for 'hard inheritance': genetic information was not added to throughout life, as the body grew and changed, but was set immutable from conception. The reproductive cells (germline) were not influenced by changes in other cells of the body (soma) and so heritable information was passed from one generation to the next largely unchanged. Weismann defined two key principles of heredity—the continuity of the germline, and the isolation of the germline and soma—which combine to give us our modern view of heredity: that genetic information is passed from generation to generation, essentially unaffected by changes to the body. In this way, the genome represents an unbroken chain of information passed from parents to offspring, and so on down through the generations. Weismann's barrier denies the inheritance of acquired characteristics because it prevents the changes acquired during an individual's lifetime rewriting its genetic information.

Although he made detailed studies of developmental biology, Weismann's conceptual advances were primarily theoretical, made by considering the implications of various models of heritability for evolution. At the same time, largely unknown to those in the scientific community debating heredity and evolution, breeding experiments were being carried out that would shed light on the nature of the immortal germline. These experiments would eventually be used to counter Darwin's critics, by demonstrating that hereditary information did not blend and dilute down the generations, but was passed on in discrete units that could be carried over many generations.

## Discovery of the gene

*The 'gene' is nothing but a very applicable little word, easily combined with others... As to the nature of the 'genes' it is as yet of no value to propose any hypothesis; but that the notion 'gene' covers a reality is evident from Mendelism.*

W. Johannsen (1911) The Genotype Conception of Heredity. *The American Naturalist*, Volume 45, page 129.

Gregor Mendel was a researcher and teacher at a monastery in Moravia (now in the Czech Republic), which had a thriving research programme in many aspects of natural science. For nearly a decade in the mid 1800s, Mendel conducted large-scale experiments in plant breeding. He systematically crossed 34 different pure-bred strains of peas and recorded the characteristics of over tens of thousands of individual plants. Through the pioneering application of statistical analysis to the problem of heredity, he was able to show that heritable features of these pea strains—such as green peas versus yellow peas, or wrinkled peas versus smooth peas—were preserved down the generations. With this pioneering application of big data analysis to understanding inheritance, we should probably be thinking of Mendel as the first bioinformatician.

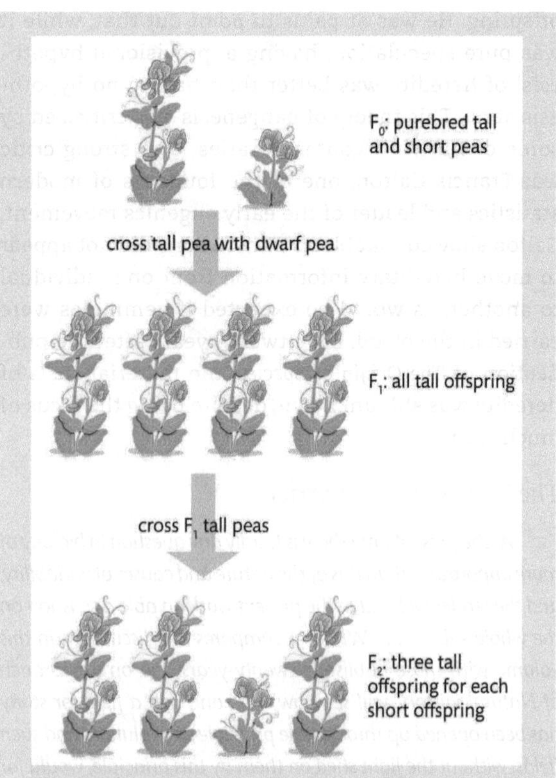


**Figure 2.4** Gregor Mendel crossed distinct pure-bred lines of peas and showed that, for certain traits, the first generation ($F_1$) offspring were not intermediate between the two parental types, but all resembled one parent. In the case illustrated here, crossing tall and dwarf varieties produced all tall offspring, no dwarfs. But the genetic information from the two parents was not lost. When Mendel crossed the first generation ($F_1$) offspring with each other, the second generation ($F_2$) offspring included both tall and dwarf plants. Mendel identified seven traits that varied discretely in this way, including wrinkly versus smooth seeds, yellow versus green peas, and purple versus white flowers. Incidentally, Mendel apparently suffered from severe exam anxiety, which is strangely comforting to know.

When Mendel crossed two varieties of peas, he found that offspring did not always have a simple blend of their parents' characteristics. Instead of being intermediate between the two types, the offspring resembled one parent or the other. For example, when he crossed a strain of tall peas with a strain of short peas, he got all tall offspring, not offspring of medium height (**Figure 2.4**). But when these tall offspring were crossed together, they produced both tall and short plants. The variation in the parents' generation was not lost, because it reappeared in subsequent generations.

Furthermore, the proportion of offspring of each parental type varied in predictable ratios. When tall and short plants were crossed, they produced all tall offspring. But when these tall offspring were crossed with each other, their offspring (the 'grandchildren' of the original tall by short cross) varied in height with three tall offspring to every one short. Mendel had discovered that heritability was governed by discrete factors that were copied and combined down the generations, and did not disappear through interbreeding. He was therefore the first person to describe the action of the hereditary units, which he called 'factors' or 'elements', but are now known as genes.

But what were these inherited factors? Early geneticists and evolutionary biologists studied the behaviour of genes in great detail. They described patterns of inheritance, how different genes combined to produce particular traits, and how these traits varied heritably within populations. Yet they did not actually know what genes were made of, or where in the cell they were located. Chromosomes seemed a good candidate for the genetic material. Chromosomes form dense bodies in cells that could be observed under a microscope (the 'chroma' comes from colour because they could be stained and observed), so their ordered behaviour at cell division had been studied since the late 1800s. At cell division, a copy of each chromosome went to each daughter cell, matching Mendel's description of the segregation and assortment of genetic factors. But chromosomes are made of both proteins and DNA. Which of these two types of molecules held the hereditary information?

DNA had been discovered in the 1860s by Friedrich Miescher. He collected cells, such as white blood cells taken from pus on bandages collected from a hospital, then used a number of protocols to lyse the cells and separate the cellular contents (**TechBox 2.2**). When he isolated the central nuclei of the cells he found them to be full of a phosphorus-rich material. He called this substance nuclein. Nuclein was found in every cell that Miescher tested, but it appeared to be inert: it was non-reactive and didn't appear to have any special metabolic role. Miescher initially concluded it might simply be a way of storing phosphorus in cells, though later he began to suspect it had some kind of role in fertilization.

As knowledge of nuclein was refined over the next sixty years, it was renamed deoxyribonucleic acid (DNA). It was shown that DNA was found in chromosomes, and that it was made up of phosphates, sugars, and bases linked together in long chains. But relatively few scientists were interested in DNA, since it did not seem to do anything exciting. DNA was always in the same inert form, did not appear to do anything other than lie around in chromosomes, and had only four different units—the nucleotide bases adenine (A), cytosine (C), guanine (G) and thymine (T) (**Techbox 2.1**). Proteins, on the other hand, were much more exciting: they existed in huge variety, did much of the important work as both structural elements and active enzymes, and were made up of over 20 different units (amino acids). Many scientists thought proteins were the obvious choice for storing the vast amount of information needed to make even the simplest cell. But the problem with the protein theory of heredity was not how information could be stored, but how it could be copied and passed to offspring. Could proteins be copied? And would a cell need to inherit a copy of every essential protein from its parent?

## DNA as genetic material

> *When asked what his idea of happiness would be, [Hershey] replied, 'to have an experiment that works, and do it over and over again.'*

J. Hodgkin (2001) Hershey and his heaven. *Nature Cell Biology*, Volume 3, E77.

In the 1940s, several experiments had suggested that it was DNA, not proteins, that carried genetic information. For example, Oswald Avery and colleagues showed that genetic information could be passed from one strain of the bacterium *Pnuemococcus* to another. They used a series of experiments to show that the 'transforming factor' (genetic information) was preserved even when enzymes were used to remove all proteins, sugars and RNAs. But if DNA was removed from the solution, then it could not transform cells. They concluded that it was the DNA that carried information from one cell to another. However, these experiments did not convince the majority of scientists working on the molecular basis of heredity, most of whom continued to concentrate on proteins.

Conclusive proof of the role of DNA in heredity was provided by Alfred Hershey and Martha Chase in 1952. Their elegant 'blender experiment' (named for their innovative use of kitchen equipment) showed that DNA was responsible for genetic continuity, by allowing them to follow the interaction between virus proteins and nucleic acids and living bacterial cells. They used bacteriophage to test whether it was proteins or nucleic acids that carried hereditary information. Bacteriophage (otherwise known as phage) are viruses that parasitize bacteria. Viruses lack the necessary equipment to replicate their own genomes, so to reproduce they must parasitize the molecular machinery of a living cell. A phage attaches to the host cell wall and injects its genome into the cell. The bacterial cell then makes copies of the virus genome, and the information in the virus genes is used to make the proteins that form the viral coat (i.e. the 'body' of the virus). The viral genomes are then packaged into the protein coats to form infectious virus particles.

Hershey and Chase labelled viral proteins with radioactive sulfur, and labelled viral DNA with radioactive phosphorous. They allowed these radioactive phages to infect bacteria, so that their genomes would be injected into the bacterial cells. Then they mixed the infected bacterial cells up in a Waring blender, an iconic 1950s domestic appliance (**Figure 2.5**). The blender separated the virus coats, which stayed outside the cell, from the genetic material which was injected into the bacteria. The radioactive labels allowed Hershey and Chase to show that viral protein did not enter the bacterial cells, but viral DNA did. DNA, not protein, was therefore the genetic material that carried the instructions for making new viral particles.

The effect of the Hershey–Chase experiment was to immediately convince both the leading scientists of the day (such as Linus Pauling) and less established researchers (like James Watson and Francis Crick: **Hero 6**) that DNA was the key to understanding heredity. So the race began to discover the structure of DNA.

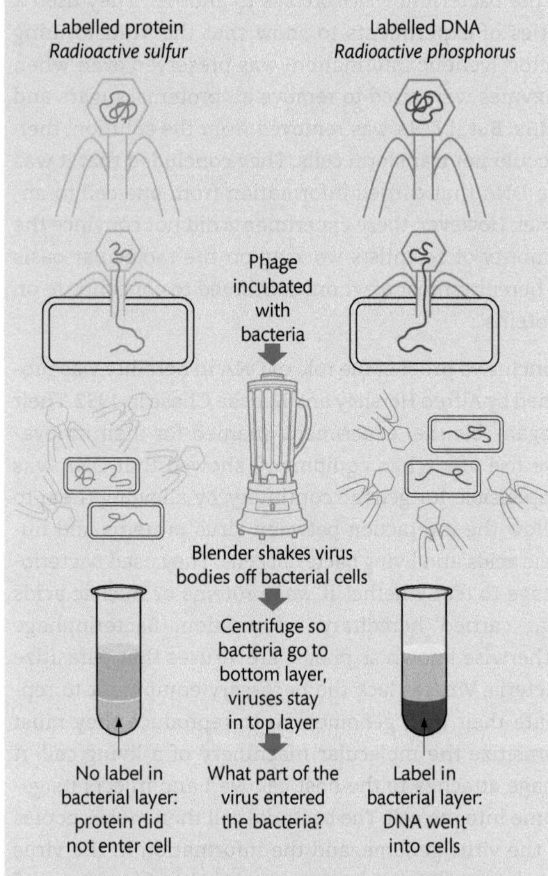


**Figure 2.5** The Hershey-Chase blender experiment: By attaching different radioactive labels to the protein and DNA in viruses, they could demonstrate that it was the DNA, not the protein, that transmitted the information needed to make a new virus.

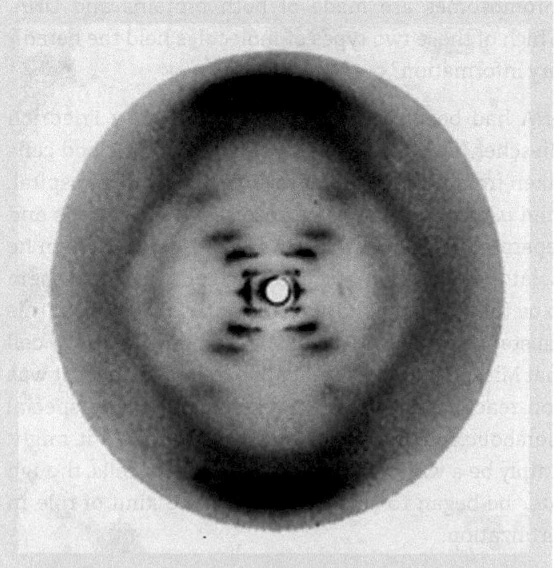

**Figure 2.6** The famous Photograph 51: Rosalind Franklin's X-ray diffraction images of DNA showed that DNA was a helix with the bases on the inside, an important advance that allowed Watson and Crick to correctly model DNA structure. That's apparently Linus Pauling's handwriting on the side. Pauling was one of the more established researchers trying to uncover the structure of DNA but he didn't have the benefit of seeing Franklin's striking X-ray images until after developing his own triple-helix model.



## Structure of DNA

It had been known since the 1920s that DNA was made up of regular patterns of three kinds of molecular subunits: phosphates, sugars and bases. But how were they connected together? Watson and Crick saw the outstanding X-ray diffraction pictures of DNA taken by Rosalind Franklin (**Hero 4**). These pictures suggested that DNA was a helix, with long chains of linked phosphate and sugar molecules twisted around each other in a regular pattern (**Figure 2.6**). Watson and Crick combined Franklin's discovery with an earlier observation made by Erwin Chargaff that the four types of bases of DNA were curiously evenly mixed. The number of adenine bases was always equal to the number of thymines, and the number of guanines was the same as the number of cytosines. Watson and Crick realized that Chargaff's pairing rule—A matches T, G matches C—was the key to the structure of DNA.

Watson and Crick constructed a large model, cutting shapes from tin plate to represent the four nucleotide bases (A, T, G and C). This model, with its the elegant spiral staircase of two intertwined strands of phosphates and sugars, connected by rungs of paired bases, is familiar to many biologists as the star of one of the most recognizable publicity photos in the history of biology (**Figure 2.7**). The complementary pairing of bases between the double strands of the helix, A with T and G with C, meant that each strand was an exact complement of the other. One strand could act as a template for the other, providing a means of copying information. The answer was so obviously right that Francis Crick is said to have announced in their local pub that evening: 'We have just uncovered the secret of life!'

The publication of their single, one-page scientific article in the journal *Nature* the following year put DNA at the heart of modern biology. This paper, in which James

**Figure 2.7** Francis Crick (left) and James Watson with the tin-plate model they built in 1953 (with the help of workshop technicians at Cambridge) to demonstrate their proposed helical DNA structure with complementary base-pairing.
Reproduced by permission of A. Barrington Brown/SPL.

Watson and Francis Crick described the molecular structure of DNA, concludes with a sentence of elegant understatement: 'It has not escaped our notice that the specific pairing we have postulated immediately suggests a possible copying mechanism for the genetic material.' This is the statement that launched the genetic revolution.

## *From DNA to RNA to protein*

The template copying mechanism identified by Watson and Crick is the key to understanding not only the replication of DNA, but also the way that the information in DNA is used to build and operate living cells. The genome, made of DNA, is often described as a blueprint. It holds the instructions for making a cell, but it is not directly involved in construction. Instead, the genetic information stored in the DNA is expressed through the actions of RNA and protein molecules. It is the RNA and protein molecules that build and operate the body of the cell. At the risk of oversimplification, DNA carries the genotype (hereditary information), RNA and protein create the phenotype (structure, development and behaviour).

To illustrate this process of the conversion of genetic information from one form to another, let's consider the gene shown in **Figure 2.2**. In humans, the gene that codes for the beta-subunit of RNA polymerase II (given the acronym RPB2) is found on the short arm of chromosome 4. It takes 3,525 bases of DNA to specify the amino sequence needed to make this protein. When the cell needs to make an RNA polymerase II enzyme, the *RPB2* gene must be located. Then the DNA containing the gene is unwound from the chromosome and the two strands of the double helix are unzipped to expose the base sequence of the gene. An existing RNA polymerase enzyme then uses the DNA template to make an RNA copy of the gene.

RNA is more or less the same as DNA but there are some differences. RNA is single-stranded (not double-stranded like DNA), uses a different sugar molecule in its backbone (ribose not deoxyribose), and one of the four bases is slightly different (it has uracil (U) instead of thymine (T)). The RNA copy of the gene is made by matching each base on the exposed DNA strand to its complementary RNA base. Where there is an A in the gene (DNA), it is matched by a U in the message (RNA), a T in the gene is matched by an A in the message, a G with a C, and a C with a G. So the DNA sequence of the human *RPB2* gene given in **Figure 2.2** begins "CGTGATGGT", but its complementary RNA would read "GCACUACCA" (reading 3' to 5': **TechBox 2.1**) (**Figure 2.8**).

*Complementary base pairing is covered in Chapter 4*

In eukaryotic cells, DNA is stuck in the nucleus. But RNA can move from the nucleus to the cytoplasm, the guts of the cell where most of the biochemical action takes place, including protein synthesis. When an RNA copy of a gene is made by complementary base pairing it contains the same information as the DNA strand it was copied from. This RNA strand is known as messenger RNA (mRNA) because it can take the information from the nucleus to the cytoplasm where it can be used to build useful things. When messenger RNA leaves the nucleus it is taken to a ribosome. Ribosomes are the workbenches of the genomic system, where information from the nucleus, transported in messenger RNA, is used to construct a protein. Here, complementary base pairing is used again to translate the information in the messenger RNA into the amino acid sequence of the protein product. The sequence of bases in the messenger RNA is matched to bases on transfer RNAs, each of which brings a specific amino acid to the ribosome.

There is a host of transfer RNAs (tRNAs) in the cytoplasm. Each tRNA has a particular 3-base recognition sequence and carries a specific amino acid. So

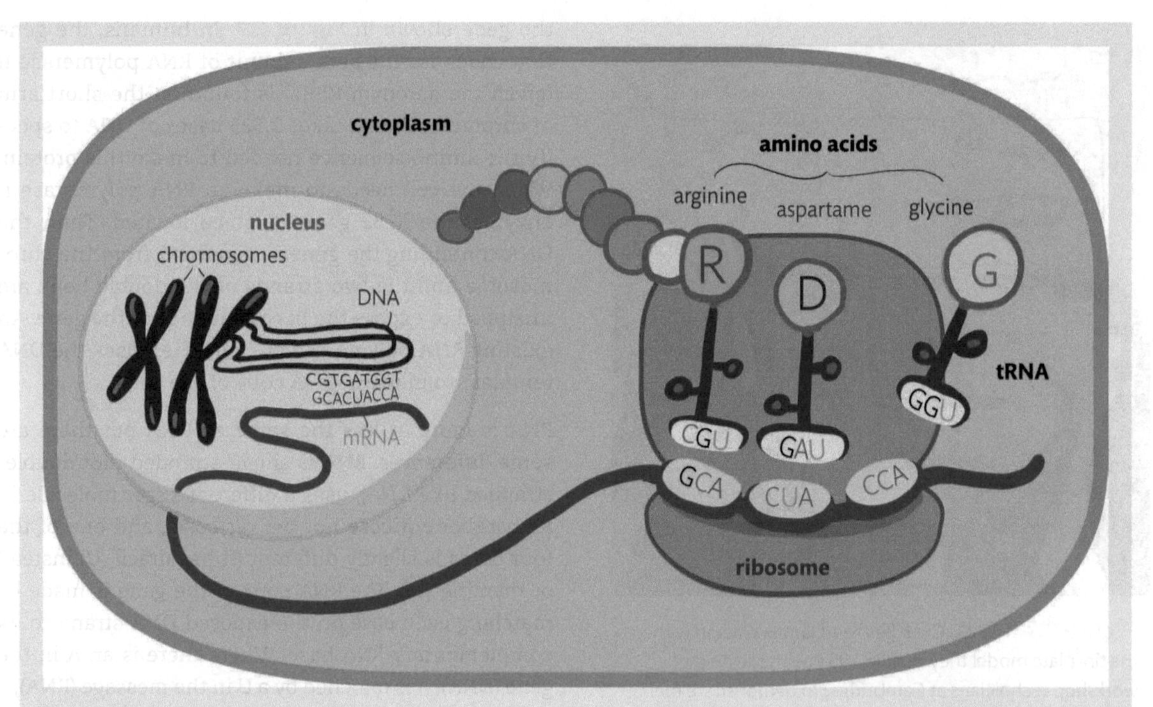


**Figure 2.8** Simplified diagram illustrating the way that complementary base pairing is used to transfer information from the DNA in the nucleus to the messenger RNA (mRNA), which then moves to the cytoplasm, where tRNAs with complementary 3-base sequences match the mRNA to bring the right sequence of amino acids to make the protein product.

the bases CGT in the *RPB2* gene are transcribed to GCA in the mRNA, which is matched by a tRNA with the recognition sequence CGU, which carries the amino acid arginine. Similarly, the next three letters in the gene are GAT, which matches CUA in the mRNA, which is matched by a tRNA with the sequence GAU, which carries the amino acid aspartane to join the protein. In this way the base sequence in the gene (DNA) determines the base sequence in the message (mRNA) that matches the recognition sequence of a particular tRNA which determines the sequence of amino acids in the protein. Figure 2.9 shows the DNA sequence given in Figure 2.1 translated into amino acids (one amino acid for every three bases of DNA: TechBox 6.1).

The amino acids are attached to the growing peptide chain in the order specified by the gene. When all of the bases in the message have been 'read', the ribosome falls away from the message, releasing the chain of amino acids. The forces of attraction and repulsion between the amino acids in the chain cause parts of the sequence to spontaneously fold into energetically stable helices and sheets, which then twist around each other to form a stable, three-dimensional structure (sometimes this folding process requires chaperone proteins to get the peptide into the right conformation). Finally, the completed protein might combine with other protein and RNA subunits to make the working enzyme. This complex series of biochemical events—transcription of genes into messenger RNA, translation of mRNA to protein by transfer RNAs—occurs continuously in every cell. The genomic system is the most remarkably complex yet wonderfully effective organization, and it is the bedrock of all living processes.

***Gene expression is discussed in Chapter 6***

| | |
|---|---|
| **Homo sapiens** | RDGGLRFGEMERDCQIA |
| **Rattus norvegicus** | RDGGLRFGEMERDCQIA |
| **Drosophila melanogaster** | RDGGLRFGEMERDCQIS |
| **Neurospora crassa** | RDGGLRFGEMERDCMIA |
| **Oryza sativa** | YGGGIRFGEMERDALLA |
| **Escherichia coli** | QFGGQRFGEMEVWALEA |

**Figure 2.9** Amino acid sequence of the beta-subunit of RNA polymerase II, translated from the DNA sequence shown in Figure 2.1, using a single letter per amino acid (see TechBox 6.1).

## The central dogma

> *The central dogma was put forward at a period when much of what we now know in molecular genetics was not established. All we had to work on were certain fragmentary experimental results, themselves often uncertain and confused, and a boundless optimism that the basic concepts involved were rather simple and probably much the same in all living things*

F. H. C. Crick (1970) Central dogma of molecular biology. *Nature*, Volume 227, pages 561–563.

The flow of information from DNA to RNA, and from mRNA to proteins, was described by Francis Crick (somewhat messianically) as the Central Dogma of Molecular Biology. You will sometimes see the central dogma stated like this: DNA copies DNA, DNA is transcribed into RNA, RNA is translated into protein (**Figure 2.10**).

**Figure 2.10**

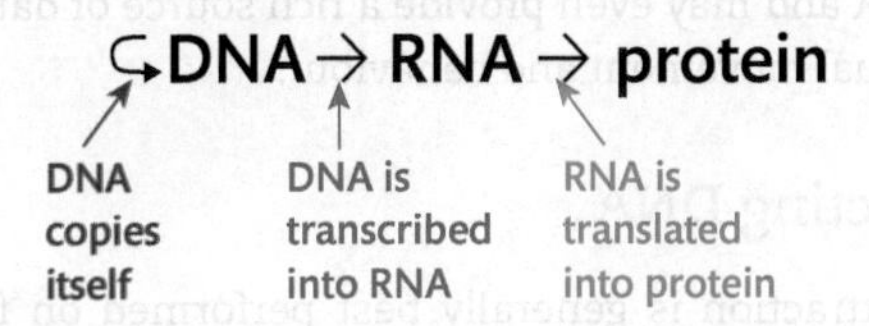


It is important to remember that the central dogma is more concerned with the flow of information than biochemical construction. The sequence of nucleotides in the DNA, if it is transcribed (not all DNA is), specifies the sequence of ribonucleotides in the RNA that is copied from it. If the DNA in question happens to code for a protein (not all DNA does), then we can add another step: the sequence of ribonucleotides in the messenger RNA specifies the sequence of amino acids in the protein.

***DNA that does not code for proteins is discussed in Chapter 5***

The label 'Central Dogma' is catchy, but also has an unfortunate whiff of unchallengeability about it, which was not, apparently, Crick's intention. Actually there are few unchallengeable dogmas in biology, which is one of the things that makes it so much fun. As with so many things in biology, the more we learn about the genome and cell function, the more complicated the picture gets. For example, there is an enzyme (reverse transcriptase) that can use an RNA template to make a complementary DNA molecule. We can modify the dogma to include this pathway (**Figure 2.11**).

**Figure 2.11**

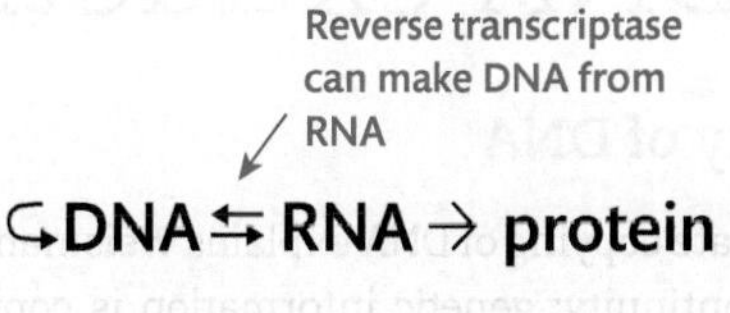

**Reverse transcriptase can make DNA from RNA**

**DNA ⇆ RNA → protein**

But the key message of the central dogma is this: the information in DNA and RNA is effectively interchangeable—DNA can be used to make a complementary RNA strand and vice versa (a fact exploited by many molecular technologies)—but the same is not true for the sequence of amino acids in a protein. Proteins don't have template copying. The sequence of amino acids in one protein cannot cause the formation of an identical copy of that sequence. Even prions, infectious proteins such as the one responsible for 'mad cow disease' (bovine spongiform encephalitis, or BSE), are thought to work by changing the conformation of existing prion proteins, rather than by creating new copies of themselves. There is no known biochemical mechanism for translating the sequence of amino acids in a protein into the nucleotide code of a gene.

So the central dogma is a molecular statement of Weismann's barrier: information flows from the germline to the soma but not back the other way (**Figure 2.12**). If the DNA sequence of a gene is changed, it may result in the formation of a protein with a novel sequence, and if that novel protein is advantageous, then carriers of that gene might have a fitness advantage and that version of the gene may rise in frequency by natural selection (see Chapter 7). But if a change is made directly to a protein, for example if the wrong amino acid is inserted as the protein is being constructed, then even if that change is advantageous, it cannot be coded back into the gene. Since it is the genes that carry hereditary information, a change to a protein cannot be stably inherited over many generations.

**Figure 2.12**

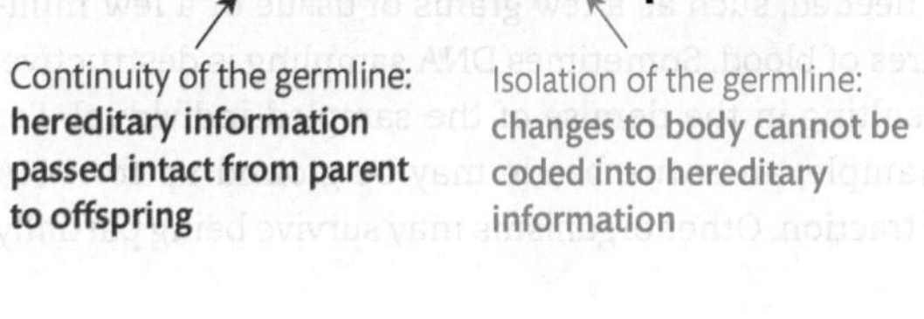

**DNA → RNA → protein**

Continuity of the germline: **hereditary information passed intact from parent to offspring**

Isolation of the germline: **changes to body cannot be coded into hereditary information**

# DNA extraction

## Ubiquity of DNA

The template copying of DNA explains Weismann's principle of continuity: genetic information is copied from generation to generation because DNA can be faithfully replicated by complementary base pairing. The central dogma of molecular biology explains Weismann's isolation of the germline: changes to the body cannot be written back into an individual's DNA sequence. We now know that the germline cells do not have to be physically isolated to preserve Weismann's barrier. In many plants, somatic (body) cells can give rise to reproductive tissue, so a mutation in a cell in a growing plant stem could be copied into cells that give rise to a flower, so ending up in pollen and ovules, and thus being passed on to the next generation. But still, in each of these somatic cells, any alteration directly to the products of gene expression (proteins) doesn't change the instructions for making those proteins (genes) that will be inherited by the cell's daughters, granddaughters and great-grand-daughters. Although changes in the body can influence gene expression, the underlying gene sequence remains the same until changed by chance mutation.

*Mutation is covered in Chapter 3*

### *How to get DNA samples*

So far we have concentrated on the germ cells that carry the genetic information from one generation to the next. But every somatic cell (with a few exceptions) also carries a full complement of the genetic information needed to make the whole organism, even though any particular tissue will only need a subset of that information. This means that we can collect nearly any biological tissue and extract DNA from it (**Case Study 2.1**). Incidentally, this fact also makes cloning possible, because each cell carries the instructions for the construction of a whole new organism (**TechBox 10.1**).

Most DNA sequences are derived from fresh biological samples, such as a leaf, a blood sample, or, for the more unfortunate study animal, a piece of liver. Some types of specimens are easier to extract DNA from than others (**TechBox 2.2**). Generally, the fresher it is, the easier it is to extract DNA. Usually only a small sample is needed, such as a few grams of tissue or a few millilitres of blood. Sometimes DNA sampling is destructive, resulting in the demise of the sampled individual. For example, an entire beetle may be ground up for DNA extraction. Other organisms may survive being partially sampled, such as taking a leaf, a flower or a fruit from a plant.

The harm done to study organisms by DNA sampling is an important ethical consideration. For example, it has been shown that the practice of toe-clipping amphibians—a common means of marking and taking tissue samples from captured individuals—can reduce their probability of survival. The distress caused to animals by being captured and handled to take a blood sample, or the destruction to habitat created by the search for elusive invertebrates, should not be underestimated. For many species, non-invasive methods of DNA sample collection are becoming widely advocated, such as using hair-traps, collecting faeces, or finding cast-off skin or feathers. Although these techniques may provide poorer quality samples than destructive sampling, they can sometimes provide a practical way of collecting DNA and may even provide a rich source of data on individual movement and behaviour.

## Extracting DNA

DNA extraction is generally best performed on fresh tissue, because DNA, like other biomolecules, degrades over time. DNA degradation can be reduced by preserving samples, particularly by freezing or immersion in ethanol. But although fresh material is the easiest to work with, DNA samples have been successfully taken from Egyptian mummies, frozen mammoths, dehydrated penguins, pickled thylacines, carved whale teeth, preserved food, ancient timber, and thousand-year-old marine sediments (**Case Study 2.2**). Even the last meal of Ötzi, the 'iceman' whose 5,000-year-old frozen body was found in the European Alps, has been determined through DNA analysis of his intestinal contents. Museum and herbarium specimens are proving particularly valuable for DNA analysis, although understandably curators might not always be terribly keen on having pieces taken out of rare specimens, so methods are now being developed for non-destructive DNA sampling of important artefacts. However, the inevitable decay of DNA means that there is little chance of recovering DNA from very old or poorly preserved biological specimens.

DNA extraction involves three basic steps. The first step is cell lysis, where the cells are broken open to release the DNA. This is typically done by gently heating the cells (not hot enough to destroy the DNA) with detergents that dissolve the cell membranes, and enzymes that break down the cellular proteins. Now that the

**Figure 2.13** The bright pigments that attract insects to carnivorous plants and the sticky goo that makes it so hard for the insects to leave, complicate DNA extraction from carnivorous plants such as (a) *Nepenthes*, (b) *Heliamphora*, and (c) *Drosera*.

*Heliamphora* image courtesy of Andreas Eils. *Drosera* image courtesy of Matthias Jauernig. Both images are licensed under the Creative Commons Attribution Share Alike 3.0 Unported license.

DNA has been released into a soup of cellular contents, the next step is to protect the DNA from enzymes that might destroy it, by adding chemicals that halt the action of enzymes. Finally, the DNA needs to be separated from the rest of the soup (**TechBox 2.2**).

While the basic processes of DNA extraction are common to most procedures, the actual details can vary, depending on the material being used and the kind of DNA fragments you want to end up with. Some samples contain many other biological molecules that get in the way of DNA extraction. For example, carnivorous plants often produce bright pigments to attract insects which are then trapped with sticky mucus (**Figure 2.13**). The pigments are made of phenolic compounds, which can bind to DNA and interfere with amplification, and the sticky mucilage is full of polysaccharides which can interfere with the action of enzymes used in DNA extraction and amplification (**TechBox 2.2**). So researchers found they got little or no DNA from carnivorous plants until they tweaked the extraction protocols. Different procedures require different kinds of DNA extraction; for example, procedures designed to extract large, intact DNA molecules for genomic libraries might minimize steps that risk breaking the DNA up into tiny fragments. If a serious amount of fiddling in the lab is needed to make a particular sample yield useable DNA, then DNA extraction can seem like more of an art than a science—but the degree of elation when DNA is successfully extracted is often proportional to the amount of time spent trying to get it to work.

# Conclusion

The mechanisms of inheritance have been revealed in an astonishingly short period of time.

While Darwin clearly illustrated the role of heredity in the process of evolution by descent with modification, he did not know how information about an organism's form or behaviour was transmitted to its offspring. Weismann reasoned that the information needed to make an organism was passed from one generation to the next, essentially unaffected by bodily changes acquired during an individual's lifetime. Mendel showed through breeding experiments that inheritance was controlled by genetic factors that were not blended but passed intact down the generations as discrete units of information. But the material basis of heredity was not uncovered until Franklin's images of the DNA helix allowed Watson and Crick to uncover how complementary base pairing provided a means to copy information endlessly down the generations.

DNA in the genome provides information needed to build, grow and operate organisms. It also carries that information from one generation to another. DNA is also incidentally a source of information about evolution, which is why you are reading this book. Genomes would not carry information if every individual had exactly the same DNA sequence. It is because DNA sequences differ between individuals, between species, and between evolutionary lineages that we can use DNA to understand evolutionary history and processes. In Chapter 3, we consider the process of mutation, whereby the genome of an individual is permanently changed. Mutation creates heritable variation, which is the raw material of evolution. It also makes each one of us unique, a fact that is increasingly exploited in biology and medicine.

# Points to remember

## Material basis of heredity

- All living things share the same genomic system in which information needed to construct an organism is stored in DNA, transcribed into RNA and translated into proteins.
- Even the simplest cell requires hundreds of genes and a complex system of regulation to express the right genes in the right place at the right time in the right amount.
- Some genes that code for fundamental biomolecules are similar between all living things.

## Principles of heredity

- The material basis of heredity was unknown until relatively recently.
- Weismann rejected inheritance of acquired characteristics (modifications of the adult body passed on to offspring) and argued for 'hard inheritance' (only traits already present in the germline can be inherited).
- Mendel conducted large-scale breeding experiments which demonstrated that traits could be determined by hereditary units that were passed unchanged down multiple generations.
- The physical form of these hereditary units (genes) was unknown until experiments showed that information could be transferred only by DNA, not proteins or RNA.

## Structure of DNA

- Two important experimental observations—X-ray diffraction images that showed DNA was a helix with bases on the inside, and biochemical analysis showing that pairs of nucleotide bases occurred in matched amounts—led to the double helix model of DNA.
- The double helix structure—two strands connected by paired bases—is the key to DNA replication, because each strand can act as a template for the formation of a new DNA strand.
- The nucleotide sequence in DNA can be used as a template to create a complementary RNA strand, which acts as a messenger to take the information to the ribosome where it can be used to build a protein.
- Protein sequences cannot be translated back into nucleic acid sequences.

## DNA extraction

- With few exceptions, every cell contains a copy of the whole genome, so most fresh biological samples can yield an informative DNA sample, as can some well-preserved older materials.
- DNA can be sampled from discarded biological material, such as hair or faeces, and provides a rich source of spatial and temporal data on species and individuals.
- DNA extraction involves three basic steps: breaking up the cells to release the DNA, halting the action of enzymes that would destroy the DNA, then separating the DNA from the other cellular components.

# Ideas for discussion

**2.1** Most cells in the body only use a fraction of the genes in the genome. So why do all cells carry a full copy of the genome, not just the subset of genes they need?

**2.2** Would inheritance of acquired characteristics (changes to the adult body) make evolution more efficient than having to wait for chance mutations?

**2.3** What kinds of biological samples might not contain DNA?

# Sequences used in this chapter

Table 2.1: RNA polymerase sequences used in Figure 2.2

| Description from GenBank entry | Accession |
|---|---|
| *Homo sapiens* polymerase (RNA) II (DNA directed) polypeptide B | NM_000938 |
| *Rattus norvegicus* polymerase (RNA) II (DNA directed) polypeptide B (Polr2b) | NM_001106002 |
| *Drosophila melanogaster* RNA polymerase II 140kD subunit (RpII140) | NM_057358 |
| *Neurospora crassa* OR74A DNA-directed RNA polymerase II 140 kDa polypeptide partial mRNA | XM_952013 |
| *Oryza sativa* (japonica cultivar-group) chromosome 10, section 62 of 77 of the complete sequence—partial sequence | AE017108 |
| *E.coli* RNA polymerase beta subunit (rpoB and rpoC) genes | M38303 |

# Examples used in this chapter

Fleischmann, A., Heubl, G. (2009) Overcoming DNA extraction problems from carnivorous plants. *Anales del Jardín Botánico de Madrid*, Volume 66, page 209.

Franklin, R. E., Gosling, R. G. (1953) Molecular configuration in sodium thymonucleate. *Nature*, Volume 171, page 740.

Hershey, A. D., Chase, M. (1952) Independent functions of viral protein and nucleic acid in growth of bacteriophage. *Journal of General Physiology*, Volume 36, page 39.

Rollo, F., Ubaldi, M., Ermini, L., Marota, I. (2002) Ötzi's last meals: DNA analysis of the intestinal content of the Neolithic glacier mummy from the Alps. *Proceedings of the National Academy of Sciences USA*, Volume 99, page 12594.

Watson, J. D., Crick, F. H. C. (1953) Molecular structure of nucleic acids. *Nature*, Volume 171, page 737.

HEROES OF THE GENETIC REVOLUTION

2

# August Weismann

*"I have gradually become aware, that, after Darwin, Weismann was the greatest evolutionary biologist of the nineteenth century. Further, the problems he was concerned with are often the same problems that concern us today"*

J Maynard Smith (1989) Weismann and modern biology. *Oxford Surveys in Evolutionary Biology,* Volume 6, pages 1–12

**EXAMPLE PUBLICATIONS**
(English translations)

Weismann, A. (1882) *Studies in the theory of descent*, translated by R. Meldola (Simpson Low, Marston, Searle, and Rivington).

Weismann, A. (1893) *The Germ-Plasm: a theory of heredity*, translated by W. N. Parker, H. Ronnfeldt (Charles Scribener's Sons).

**Figure 2.14** August Weismann (1834–1914) was a doctor and zoologist who, convinced by Darwin's theory of evolution, studied cell division and patterns of inheritance. Working before the widespread acceptance of Mendelism, Weismann described heritability in terms of units of information carried in cells. He inferred the existence of a replicating molecule that could grow and divide into two halves, both similar to the parent molecule. These information molecules could direct development, but were not themselves influenced by changes to an individual acquired during their lifetime. The lack of a means of coding phenotypic change into genotypic information is often referred to as the Weismannian barrier.

August Weismann (Figure 2.14) was one of the first to be called a 'neo-Darwinian' (not intended as a compliment then, and, regrettably, often used in the same vein today). Like Darwin, much of his work was prescient and it is surprising how many key ideas in modern evolutionary biology can be found in Weismann's work, such as the role of sexual reproduction in generating variation, and a discussion of the cellular causes of ageing. In particular, Weismann's careful observations of cell division contributed to understanding the role of chromosomes in heredity. Many of Weismann's books, like Darwin's, surprise modern readers with their vast catalogues of observations about the natural world. In Weismann's case, his special interest was in the colouration of caterpillars and butterflies. While these intimate studies of butterflies may seem whimsical, they provided Weismann with abundant raw material for understanding developmental biology and genetics.

His fondness of butterflies had an early genesis as he collected insects in the countryside as a boy. But, with few options for professional naturalists, he trained as a doctor, subsequently serving as an army doctor and a private physician to the aristocracy, though also conducting research in chemistry. In 1863, at the age of 29, he became an academic, joining the University of Freiburg's medical faculty, teaching zoology and comparative anatomy. He became the first director of the university's new zoological institute and natural history museum.

One of Weismann's most important contributions to biology was that he convincingly demonstrated that, counter to the prevailing viewpoint at the time, changes to the body during an individual's lifetime are not a source of heritable variation. Because he was working in the late 1800s when the molecular basis of heredity was unknown, Weismann's central theories are framed in terms of cell lines: the germline cells (which form sperm and eggs) are isolated from the somatic cells (all other cells in the body: see page 35). We can now interpret Weismann's principles in terms of DNA: the information in DNA is passed on exactly as it was inherited (bar the occasional mutation), because information about the state of the body is not recorded in an individual's DNA sequences during their lifetime. In other words, Weismann provided one of the first clear statements of the distinction between the genotype (hereditary information) and the phenotype (the result of expression of that information in a particular individual).

Weismann was an enthusiastic early adopter of Darwin's evolutionary theory. His work on inheritance was a key piece of evidence in support of Darwinian evolution. The following extract from a contemporary review of Weismann's work gives some sense of the impact of his ideas, and the controversy and excitement surrounding the problem of heredity:

> *"In spite of the difficulties involved in acceptance of Weismann's view, however, it has been enthusiastically accepted in England by the younger Darwinian school . . . . . . The old school of Lamarck seemed dead; even the ideas of Herbert Spencer and of Darwin himself as to 'use and disuse' began to be looked upon as antiquated and unphilosophical . . . At the present moment a reaction has set in; the battle is raging fiercely . . . Alike in Germany and in England, criticism and doubt as to Weismann's premises are beginning to take place of the paean of exultation . . . What is wanted now is some decisive experimental settlement of the question. Can it be shown that in any case a capacity or habit acquired beyond a doubt during the life-time of the individual is transmissible to the off-spring? If that can be proved, Weismannism falls at once to the ground, and we revert to the primitive Darwinian and Spencerian problem."*
>
> G. Allen (1890) The new theory of heredity. *Review of Reviews*, Volume 1, pages 537–538.

TECHBOX
**2.1**

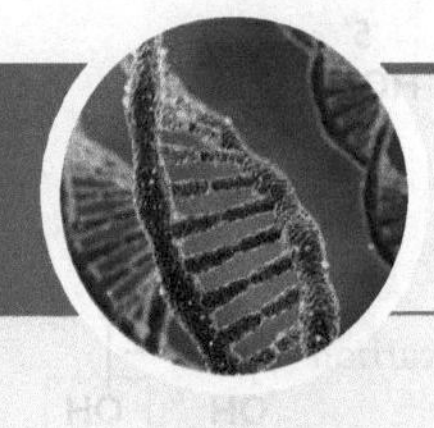

# DNA structure

**RELATED TECHBOXES**

**TB 4.1:** DNA replication

**TB 4.2:** DNA amplification

**RELATED CASE STUDIES**

**CS 3.1:** SNPs

**CS 3.2:** Mutation rate

**CS 4.2:** Hierarchies

**Summary**

The most beautiful molecule in the world is constructed of two intertwined strands of nucleotides that pair with each other.

**Keywords**

nucleotide, base, purine, pyrimidine, ribose, sugar, phosphate, 5', 3', double helix, complementary base pairing, phosphodiester bond

There are three basic subunits in DNA: bases, sugar and phosphate.

**Bases**

The ring-shaped bases are the most charismatic of the DNA subunits. They come in four types which fit together in pairs. This pairing forms the basis of the information carrying capacity of DNA. The bases are rings of oxygen, hydrogen, nitrogen and carbon molecules. Two of the bases, the pyrimidines, are single rings (**Figure 2.15**).

**Figure 2.15**

Cytosine (C)

Thymine (T)

The other two bases, the purines, are made of double rings (**Figure 2.16**).

**Figure 2.16**

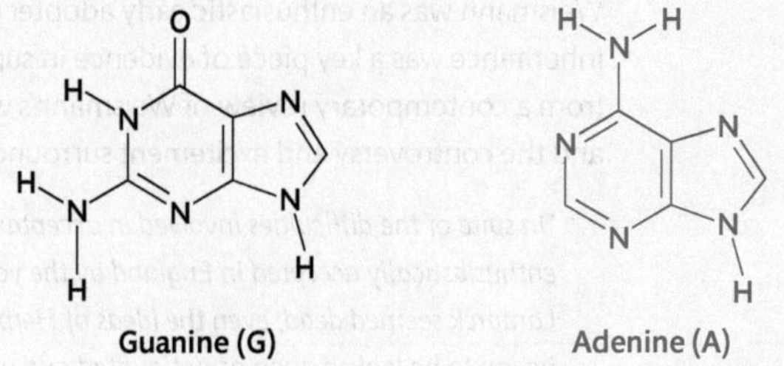

Guanine (G) Adenine (A)

## Sugars

Each base is joined to a sugar molecule, a 5-carbon (pentose) ring (**Figure 2.17**). In RNA, the sugar is ribose. In DNA, the sugar is a very similar molecule called deoxyribose (the 'deoxy' means that one hydroxyl (OH) group is missing from this form of ribose). Why do RNA and DNA have slightly different sugars? Deoxyribose makes DNA more chemically stable than RNA. Perhaps RNA represents an earlier form of information storage and DNA is the new, improved version.

**Figure 2.17**

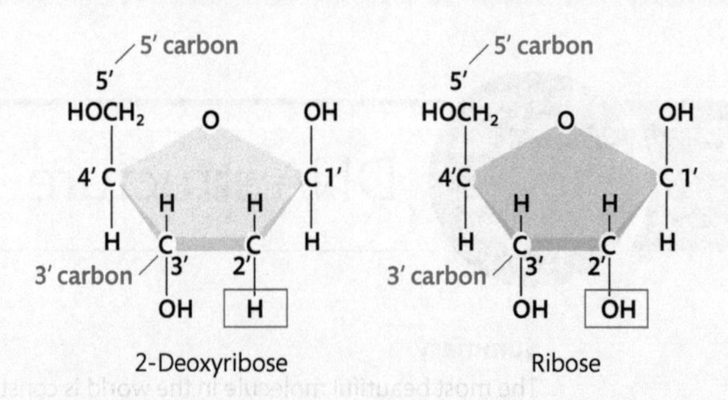


## Phosphate

Phosphate molecules provide the structural 'glue' that holds the DNA backbone together, because they form phosphodiester bonds (strong covalent bonds) which link the phosphate and sugar molecules (**Figure 2.18**).

**Figure 2.18**

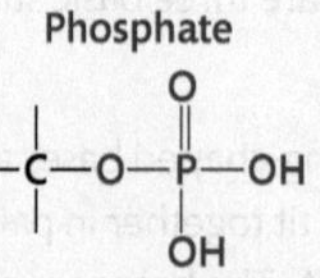

Phosphate

## Putting it all together

The combination of base + sugar + phosphate is a nucleotide, the basic structural unit of DNA (**Figure 2.19**).

**Figure 2.19**

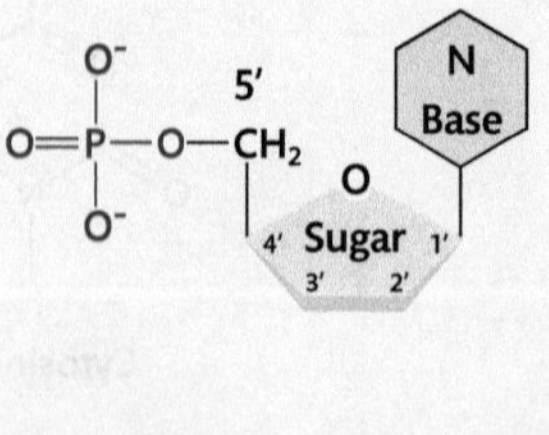

Base

Sugar

Figure 2.20

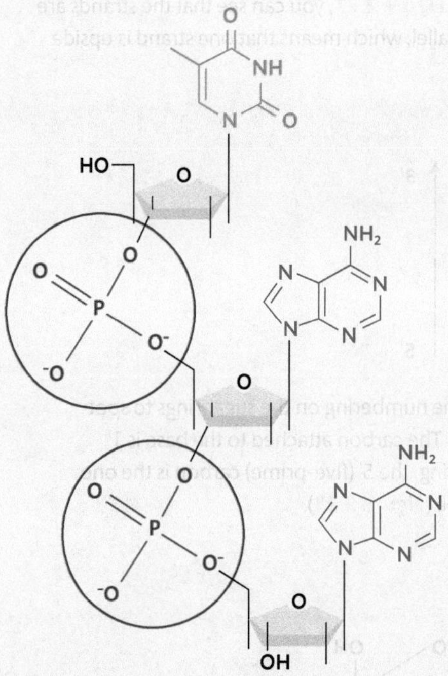

Nucleotides can form spontaneously under certain conditions, but linking them together into a DNA strand takes energy and specialized equipment (in the form of enzymes). A DNA polynucleotide strand is built by creating a phosphodiester bond that links the 3' carbon on the sugar of the growing chain with the phosphate attached to the 5' carbon of an incoming nucleotide. So the backbone of a polynucleotide strand is made of linked sugar-phosphate-sugar-phosphate, with one base joined to each sugar molecule (**Figure 2.20**).

### Base pairing

We have considered how the subunits of DNA—bases, sugars and phosphates—link together to form a linear polynucleotide. Now we can consider how two strands fit together to make the famous double helix. If two polynucleotide strands face each other, the sugar-phosphate backbone runs down each side, and the bases stick into the middle, like the steps of a spiral staircase. Complementary pairs of bases can spontaneously form hydrogen bonds—three bonds between a C and a G, two between an A and a T. Each pair consists of one double-ring purine and one single-ring pyrimidine, so the complementary base pairs maintain an even 'step' width between the two sugar-phosphate strands (**Figure 2.21**).

Figure 2.21

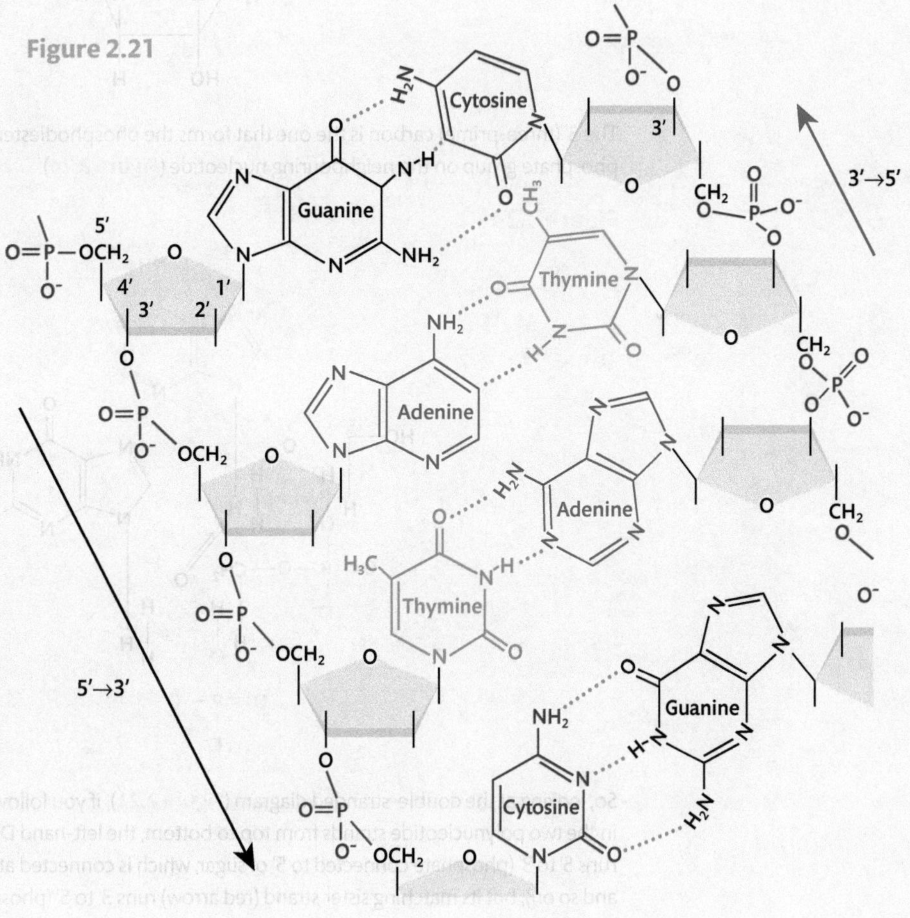

**What does 5' to 3' mean?**

If you look closely at the double strand of DNA in **Figure 2.21**, you can see that the strands are not mirror images of each other. They run antiparallel, which means that one strand is upside down with respect to the other (**Figure 2.22**).

**Figure 2.22**

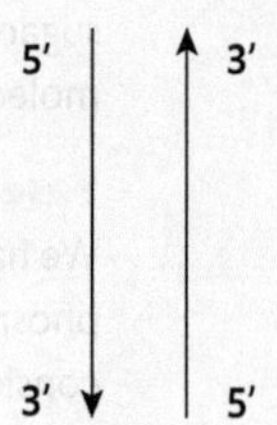


This may be hard to see at first, but you can use the numbering on the sugar rings to spot the difference. Each ribose has five carbon atoms. The carbon attached to the base is 1' (pronounced 'one-prime'). Counting around the ring, the 5' (five-prime) carbon is the one attached to the phosphate group of the nucleotide (**Figure 2.23**).

**Figure 2.23**

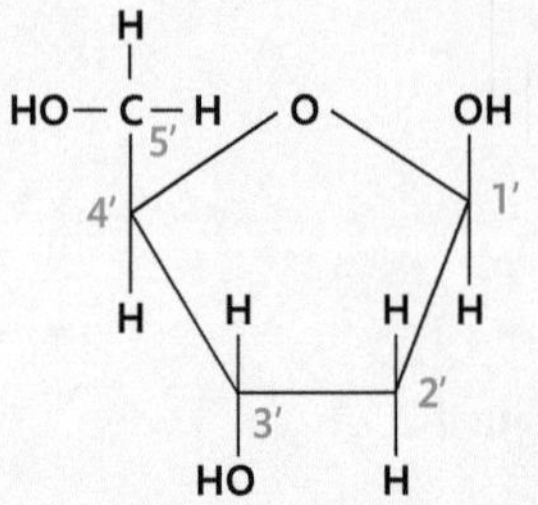

The 3' (three-prime) carbon is the one that forms the phosphodiester bond to link to the phosphate group on the neighbouring nucleotide (**Figure 2.24**).

**Figure 2.24**

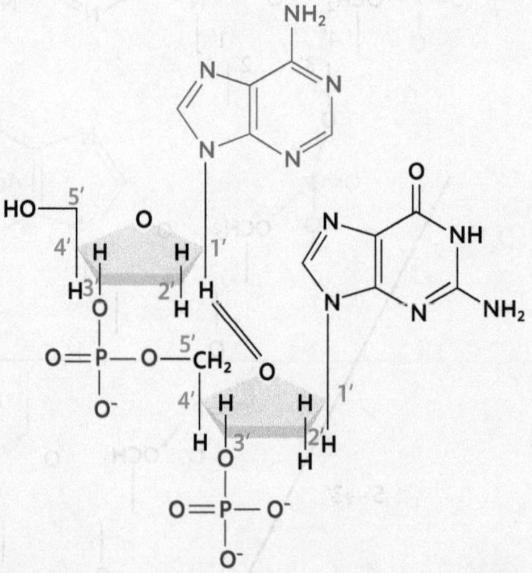

So, looking at the double-stranded diagram (**Figure 2.21**), if you follow the series of connections in the two polynucleotide strands from top to bottom, the left-hand DNA strand (black arrow) runs 5' to 3' (phosphate connected to 5' of sugar, which is connected at 3' to the next phosphate, and so on), but its matching sister strand (red arrow) runs 3' to 5' (phosphate connected to 3'

of sugar, which is connected at 5' to the next phosphate). By convention, when the sequence of bases in DNA is written down, it is usually given from the 5' end and moving to the 3' end (of course, if you have the base sequence on one strand you can work out the other strand using the base-pairing rules). So the sequence of bases in the short section of DNA in **Figure 2.21** would be written "GATC". 5' to 3' will also be important when we look at DNA replication (see Chapter 3).

### Why is DNA a helix?

Nucleotide bases attract each other along the edges, forming hydrogen bonds between the base on one strand and its matching base on the other strand. So why doesn't DNA form a simple ladder, with sugar-phosphate uprights and straight base-pair rungs? The flat faces of the bases are hydrophobic, so they repel water. If DNA was a simple ladder, the gaps between the 'rungs' would leave the bases exposed to water molecules, making the whole structure unstable. But if the bases are stacked not directly on top of each other but offset slightly and rotated, base pairs can fit snugly on top of each other and minimize the destabilizing effect of water molecules (**Figure 2.25**). Because each base-pair 'rung' turns at 32° from the previous pair, the double-stranded DNA molecule makes a complete turn every ten base pairs.

**Figure 2.25**

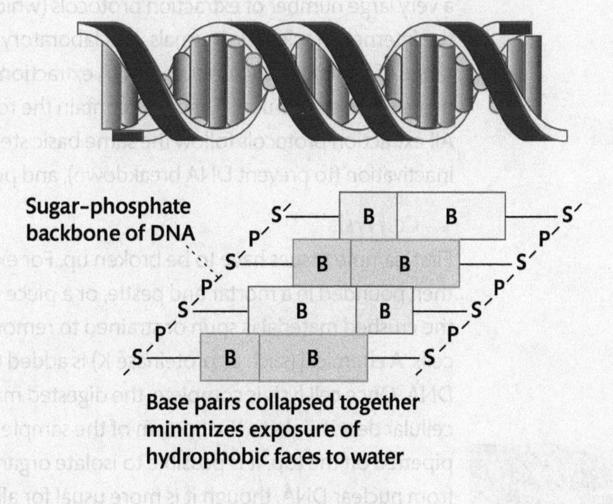


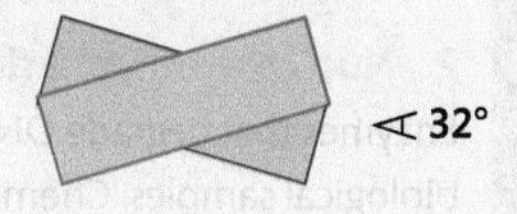

TECHBOX 2.2

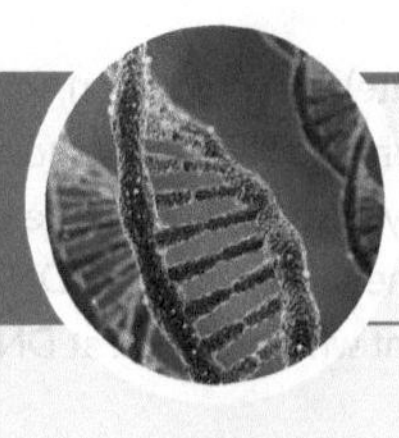

# DNA extraction

**RELATED TECHBOXES**

**TB 4.2:** DNA amplification

**TB 5.2:** High-throughput sequencing

**RELATED CASE STUDIES**

**CS 2.1:** Environmental DNA

**CS 2.2:** Ancient DNA


## Summary

The three basic steps of lysis, inactivation and purification are common to most DNA extraction methods.



## Keywords

lysis, purification, nuclease, proteinase, DNase, centrifuge, silica, column


DNA is present in most fresh biological samples. But before DNA can be sequenced, it must be isolated from the rest of the material in the sample. For some samples, DNA extraction is routine and reliable. For others, successful DNA extraction is an art that requires endless patience and some degree of tinkering in the laboratory. The details of DNA extraction techniques will vary from lab to lab, and different procedures will work best for particular samples. For this reason, there are a very large number of extraction protocols (which are essentially laboratory recipes) available on the internet, in scientific journals, and laboratory manuals. In addition, there is an ever-growing range of commercially produced DNA extraction kits, many of which are targeted at particular tissue types or end-uses. These kits contain the reagents needed to perform DNA extraction. All extraction protocols follow the same basic steps of cell lysis (to free the DNA), nuclease inactivation (to prevent DNA breakdown), and purification (to remove non-DNA molecules).

### 1. Cell lysis

First, sample tissues have to be broken up. For example, a leaf may be frozen in liquid nitrogen then pounded in a mortar and pestle, or a piece of liver might be pulverized in a blender. Then the crushed material is spun or strained to remove extraneous material, leaving just disassociated cells. A chemical (such as proteinase K) is added to the cells to burst the cell walls and release the DNA. Once cell lysis is complete, the digested material can be spun in a centrifuge, so that the cellular debris sinks to the bottom of the sample, permitting a liquid containing the DNA to be pipetted off the top. It is possible to isolate organelle DNA (from mitochondria or chloroplasts) from nuclear DNA, though it is more usual for all cellular DNA to be mixed together.

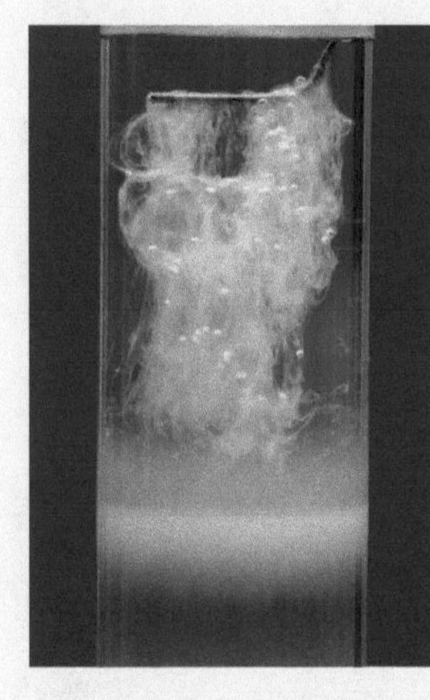

**Figure 2.26** It is possible to produce visible white strands of DNA from a biological sample (such as a salad) using chemicals available in supermarkets.



### 2. Nuclease inactivation

Enzymes that degrade DNA (deoxyribonucleases, or DNase for short) are present in most biological samples. Chemicals must be added to stop these enzymes from destroying the DNA in the sample, such as SDS (sodium dodecyl sulfate), EDTA (ethylenediaminetetraacetate), and proteinase K.

### 3. DNA purification

The DNA solution contains other biological molecules such as lipids, polysaccharides and proteins, which need to be removed. This is usually done by phenol-chloroform extraction or by running the solution through a column to separate the DNA from the other components. The DNA can then be precipitated out of solution using ethanol. The condensed DNA can then be isolated, for example by giving the sample a quick spin in the centrifuge which leaves the DNA in a pellet at the bottom of the test tube. Alternative methods for recovering DNA from the solution include binding the DNA to silica, using microfiltration, or column-based extraction.

In fact, it is possible to carry out these three steps in the home using commonly available materials, for example using household detergent to lyse cells, contact lens cleaning solution to inactivate nucleases, and rubbing alcohol to purify the DNA. With a bit of experimentation you can produce clearly visible strands of DNA (**Figure 2.26**).

**Figure 2.27** DNA analysis of amphoras (storage jars) from ancient Greek shipwrecks suggests they carried a wide variety of ingredients, not just wine.



## Ancient DNA

DNA extraction needs the most finessing when dealing with samples that are degraded. The field of 'ancient DNA' is booming, with DNA analysis being applied to samples as varied as giant ground sloth poo[1], mummified Nile crocodiles[2], Neolithic goatskin leggings[3], and amphora-stored foods on ancient Greek shipwrecks[4] (**Figure 2.27**). Since DNA degrades over time, samples such as these are challenging because they may contain relatively little DNA and it may be in quite a poor state, fragmented and with decay-induced changes to the structure and sequence of the DNA. Contamination is a serious problem with such samples, because recent DNA may be present in larger amounts and more easily extracted and amplified than the older, degraded DNA (for DNA amplification see **TechBox 4.2**)[5]. Because of the challenges presented by ancient DNA, protocols are constantly being revised and new techniques tried. New 'third generation sequencing' approaches may circumvent the need to amplify DNA by reading the sequence of single molecules of DNA[6]. For example, single-molecule sequencing was used to generate large sequences (over 100Mb) of DNA from a Pleistocene horse bone recovered from permafrost[7]. DNA capture techniques may also be helpful when there is a reference sequence available[8]: in this case 'bait' sequences, developed from the reference sequence, can be used to draw target sequences out of solution so that they hybridize onto a surface (say, onto tiny magnetic beads).

### References

1. Clack, A. A., MacPhee, R. D., Poinar, H. N. (2012) *Mylodon darwinii* DNA sequences from ancient fecal hair shafts. *Annals of Anatomy-Anatomischer Anzeiger*, Volume 194 (1), pages 26–30.
2. Hekkala, E., Shirley, M. H., Amato, G., Austin, J. D., Charter, S., Thorbjarnarson, J., et al. (2011) An ancient icon reveals new mysteries: mummy DNA resurrects a cryptic species within the Nile crocodile. *Molecular Ecology*, Volume 20, pages 4199–215.
3. Schlumbaum, A., Campos, P. F., Volken, S., Volken, M., Hafner, A., Schibler, Jr (2010) Ancient DNA, a Neolithic legging from the Swiss Alps and the early history of goat. *Journal of Archaeological Science*, Volume 37, pages 1247–51.

4. Foley, B. P., Hansson, M. C., Kourkoumelis, D. P., Theodoulou, T. A. (2012) Aspects of ancient Greek trade re-evaluated with amphora DNA evidence. *Journal of Archaeological Science*, Volume 39, page 389.
5. Rizzi, E., Lari, M., Gigli, E., De Bellis, G., Caramelli, D. (2012) Ancient DNA studies: new perspectives on old samples. *Genetics Selection Evolution*, Volume 44, page 21.
6. Schadt, E. E., Turner, S., Kasarskis, A. (2010) A window into third-generation sequencing. *Human Molecular Genetics*, Volume 19, page R227.
7. Orlando, L, Ginolhac, A., Raghavan, M., Vilstrup, J., Rasmussen, M., Magnussen, K., Steinmann, K. E., Kapranov, P., Thompson, J. F., Zazula, G., Froese, D., Moltke, I., Shapiro, B., Hofreiter, M., Al-Rasheid, K. A. S., Gilbert. M. T. P., Willerslev, E. (2011) True single-molecule DNA sequencing of a pleistocene horse bone. *Genome Research*, Volume 21, page 1705.
8. Carpenter, M. L., Buenrostro, J. D., Valdiosera, C., Schroeder, H., Allentoft, M. E., Sikora, M, et al. (2013) Pulling out the 1 percent: whole-genome capture for the targeted enrichment of ancient DNA sequencing libraries. *The American Journal of Human Genetics*, Volume 93, page 852.

CASE STUDY 2.1

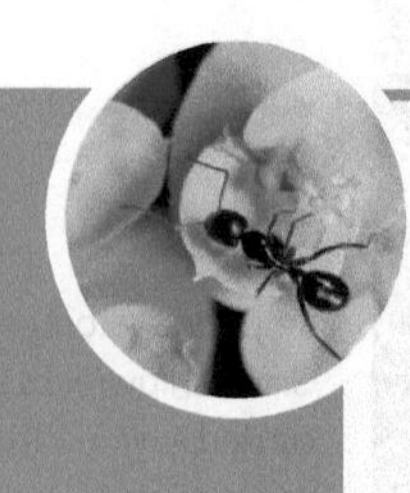

# Environmental DNA: tracking changes in Arctic vegetation over time

**RELATED TECHBOXES**

**TB 2.2:** DNA extraction

**TB 9.2:** Barcoding

**RELATED CASE STUDIES**

**CS 2.2:** Ancient DNA

**CS 9.2:** Barcoding

Willerslev, E., Davison, J., Moora, M., Zobel, M., Coissac, E., Edwards, M. E., Lorenzen, E. D., Vestergard, M., Gussarova, G., Haile, J., Craine, J., Gielly, L., Boessenkool, S., Epp, L. S., Pearman, P. B., Cheddadi, R., Murray, D., Brathen, K. A., Yoccoz, N., Binney, H., Cruaud, C., Wincker, P., Goslar, T., Alsos, I. G., Bellemain, E., Brysting, A. K., Elven, R., Sonstebo, J.H., Murton, J., Sher, A., Rasmussen, M., Ronn, R., Mourier, T., Cooper, A., Austin, J., Moller, P., Froese, D., Zazula, G., Pompanon, F., Rioux, D., Niderkorn, V., Tikhonov, A., Savvinov, G., Roberts, R. G., MacPhee, R. D. E., Gilbert, M. T. P., Kjaer, K. H., Orlando, L., Brochmann, C., Taberlet, P. (2014) Fifty thousand years of Arctic vegetation and megafaunal diet. *Nature*, Volume 50, page 47.

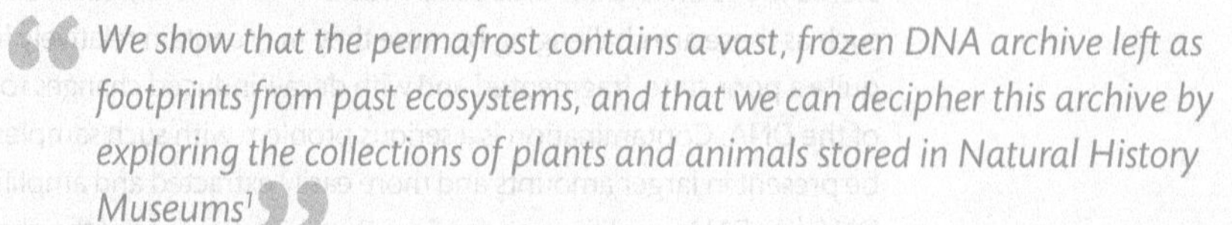
*“We show that the permafrost contains a vast, frozen DNA archive left as footprints from past ecosystems, and that we can decipher this archive by exploring the collections of plants and animals stored in Natural History Museums[1]”*

**Keywords**

nematode, megafauana, molecular operational taxonomic unit (MOTU), biodiversity, last glacial maximum (LGM)

**Background**

Changes in plant diversity over time can be studied through the changing composition of pollen in sediments, which, under certain conditions, can provide a continuous record (for example, lake sediments). But the palynological (fossil pollen) record might be skewed toward plants that produce the most pollen, particularly wind-pollinated taxa such as grasses. Environmental DNA analysis offers a different way of sampling the plants present in an area at a given time, but identifying traces of plant DNA left in the sediments. Species that may be unlikely to contribute large amounts of pollen to the palynological record might therefore be captured in environmental DNA. DNA can also potentially offer a finer taxonomic resolution than pollen morphology. Comparing DNA from different sources and taxa builds a picture of differences in biodiversity over time and space.

### Aim

The DNA in sediments may derive from many different sources, such as decomposed plant and animal matter, pollen and seeds, faeces and urine. As long as it has not moved across the landscape or down through the sediment layers, it should present a snapshot of the species found in that area at a given time. The Arctic permafrost provides an ideal environment for DNA preservation because the low temperature slows decomposition and limits DNA leaching through the sediment.

### Methods

In order to understand how the composition of the Arctic vegetation had changed over the last 50,000 years, a team of 50 researchers based in 12 countries collected 242 samples from 21 sites across the Arctic. They dated the samples with radiocarbon dating. They used generic plant primers to amplify *trnL* (an intron from the chloroplast) and *ITS* (an intergenic sequence from nuclear genome): both of these sequences are frequently used in DNA barcoding studies of plant diversity (see **TechBox 9.2**). They compared the sequences to reference libraries of sequences from Arctic plant taxa, based on identified herbarium specimens, and also to GenBank (**TechBox 1.1**). They grouped sequences into Molecular Operational Taxonomic Units (MOTU), sets of similar or identical sequences that are considered to represent a particular taxon (for example a species, or group of related genera, or a subfamily of plants). Then they counted the number of MOTUs found in the periods before, during and after the last glacial maximum (LGM), the most recent 'ice age' when the ice sheets were most extensive, approximately 26,500 to 20,000 years ago. They also identified nematode DNA from a subset of sites, and compared the relative proportions of sequences from two different families associated with different plant community types; one typically associated with tundra, and the other with steppe. In addition, they sequenced mammalian *16s* (a mitochondrial rRNA gene) from sites in Siberia, and used DNA analysis of ancient gut contents and coprolites (preserved faeces) to examine megafaunal diets (**Figure 2.28**).

### Results

Plant diversity, as measured by the number of identified MOTUs, was lowest during the last glacial maximum (LGM). The pre-LGM period was dominated by forbs (non-woody flowering plants), which persisted into the LGM, although at lower diversity. DNA from large grazing mammals was also found in these forb-dominated communities, and there was a high representation of forbs in the stomach contents and coprolites of megafauna such as mammoths and woolly rhinos. By contrast, the DNA samples from the most recent 10,000 year period was consistent with moist tundra and grassland. The nematodes associated with tundra were most common in the post-LGM period.

### Conclusions

These researchers challenge the classic picture of mammoths grazing on grasslands, suggesting instead that megafauna such as mammoths, woolly rhinos, bison and horses lived in an environment dominated by flowering herbs.

**Figure 2.28** A change in the weather. Were woolly mammoths hunted to death by humans or did they fail to keep up with a changing climate? Or both? Or neither?

### Limitations

Although plant diversity was lowest in the last glacial maximum, this is also the shortest period tested. Furthermore, the kind of samples available from the pre-LGM and LGM periods (mostly frozen deposits) differed from the post-LGM period (mostly soils) when it was warmer and wetter, so the preservational environment may have changed over time. As with any historical sampling, absence of a taxon from a sample does not necessarily prove it was not there at the time.

### Further work

This work suggests that the composition of the plant assemblage in the Arctic changed after the last glacial maximum, broadly coincident with the extinction of some charismatic megafauna such as mammoths. The link between these changes is a subject of ongoing debate: did the plants change because the large grazers disappeared, did the grazers respond to vegetation changes, or were both plants and animals responding to environmental change or human-induced habitat modification?

### Check your understanding:

1. How did these researchers measure changes in biodiversity of Arctic flora over time?
2. Why did the researchers include DNA analysis of coprolites in their study?
3. Why might the results of this study differ from studies based on pollen identification?

### What do you think?

Are MOTUs a fair assessment of biological diversity? The MOTUs in this study correspond to a range of recognized taxonomic levels, including species (e.g. *Hypochaeris maculata*, 27 occurences: **Figure 2.29**), genera (e.g. *Lathyrus*, 90), families (e.g. Ericaceae, 9539), and orders (e.g. Cornales, 10737) so counting the number of MOTUs does not give a simple tally of number of species present in an area. How might different ways of quantifying biodiversity influence patterns of change over time and space?

### Delve deeper

Permafrost provides ideal conditions for an environmental DNA study. Would this approach work for reconstructing the ecological history of other biomes? What, if any, sources of DNA samples could be use in non-permafrost environments to reconstruct biological assemblages over time?

### References

1. University of Copenhagen. 'A 'smoking gun' on Ice Age megafauna extinctions.' *ScienceDaily*, 5 February 2014.

**Figure 2.29** Flower power: new analysis questions the common picture of mammoths and other megafauna as predominantly grazing on grasslands and suggests that they may have browsed meadows of flowers such as *Hypochaeris maculata*, known as spotted cat's ear.

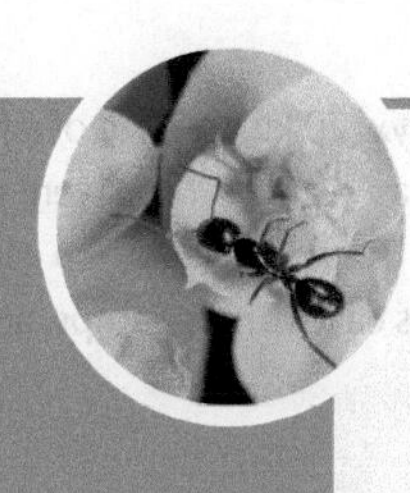

CASE STUDY 2.2

# Ancient DNA: cataloguing diversity of extinct giant birds

**RELATED TECHBOXES**

TB 2.2: DNA extraction

TB 13.2: Bayesian phylogenetics

**RELATED CASE STUDIES**

CS 2.1: Environmental DNA

CS 14.2: Dates

Baker, A. J., Huynen, L. J., Haddrath, O., Millar, C. D., Lambert, D. M. (2005) Reconstructing the tempo and mode of evolution in an extinct clade of birds with ancient DNA: The giant moas of New Zealand. *Proceedings of the National Academy of Sciences USA*, Volume 102, page 8257.

*. . . Ancient DNA methods provide powerful tools for inferring the number of lineages, as well as the tempo and mode of evolution of entire extinct groups of animals*

**Keywords**

museum specimens, genotyping, phylogeny, control region, molecular dating

## Background

Found only in New Zealand, moa were a morphologically diverse family of birds related to ostriches, emus and their allies (**Figure 2.30**). Currently six moa genera are recognized, with

**Figure 2.30** Moa—large flightless birds endemic to New Zealand—went extinct long before cameras were invented. This scene was posed in 1899 using a museum model in the botanic gardens being 'hunted' by medical students. One of the students (on the left) is Sir Peter Buck (Te Rangi Hiroa), who, amongst a great many other achievements, was the first Maori to qualify as a doctor.

Image courtesy of Alexander Turnbull Library, New Zealand. Kehoe, E. L. Mock Moa Hunt—Photograph taken by Guy. 1899. Ref no: PACol-1308.

species ranging in size from the Coastal Moa *Euryapteryx* (20 kg, the size of a largish turkey) to the Giant Moa *Dinornis* (up to 250 kg, twice the size of an ostrich). Moa went extinct not long after human settlement of New Zealand, less than one thousand years ago. However, moa bones are abundant both in natural collections, such as caves and swamps, and in middens (prehistoric rubbish heaps).

### Aim

Assignment of fossil remains to species is usually dependent on morphological similarity. Yet members of the same species may be morphologically very distinct, for example juveniles and adults, or males and females. Conversely, distinct non-interbreeding species may appear very similar from skeletal evidence alone. By sequencing DNA from a very large number of moa specimens, these biologists hoped to determine how many distinct lineages of moa had existed, and to explore reasons for the diversification of this endemic New Zealand group.

### Methods

Moa DNA samples were obtained by taking bone cores or shavings from 125 museum specimens. DNA was extracted from between 0.1 and 0.5 g of sampled bone using EDTA with proteinase K (**TechBox 2.1**). Samples were purified by phenol-chloroform extraction, then extracted DNA was amplified using polymerase chain reaction (PCR: see **TechBox 4.2**). They checked for the presence of multiple bands that might indicate nuclear copies of mitochondrial genes. DNA was sequenced along both strands. DNA was amplified in laboratories in Canada and New Zealand. Identical sequences were obtained from specimens sequenced in both laboratories. They used 658 nucleotides of the mitochondrial control region to genetically type their specimens, because it has a rapid rate of molecular evolution, so is expected to differ between different species or populations. For a subsample of specimens, they used a longer alignment of 2814 nucleotides of mitochondrial genes to estimate a phylogeny and molecular dates. They used a Bayesian phylogenetic method (**TechBox 13.1**) to estimate a phylogeny of the DNA samples, then used the genetic distance between samples (represented by branch lengths on the phylogeny in **Figure 2.31**), and the Bayesian probabilities on groupings, to judge which lineages represented distinct populations or species (**TechBox 13.1**).

### Results

Fourteen distinct lineages of moa were identified. Nine of these are currently recognized as species. The other five might be geographically separate populations or newly recognized species (or they may simply be capturing diversity within populations). The phylogeny showed that some sequences did not group with the rest of the sequences for that species. For example, a sequence from a sample which had been labelled as a Stout-Legged Moa *Euryapteryx geranoides* (yellow squares in **Figure 2.31**) is more similar to sequences from the Heavy-Footed Moa *Pachyornis elephantopus* (dark green triangles). The authors suggest that as many as a third of the museum specimens had been incorrectly assigned to species.

### Conclusions

While other ratite lineages are relatively species-poor, the interpretation of this study is that moa diversity was as high as other classic island endemic radiations, such as Darwin's finches. On the basis of molecular dates, they propose that much of the diversification of moa lineages was relatively recent, possibly driven by geographic reshaping of New Zealand, 4–10 million years ago. Populations may have become isolated and ecologically specialized as mountains rose and islands separated.

### Limitations

Genetic difference is a continuous scale, and species definitions are a matter of opinion—what one biologist considers a separate species, another may consider a regional sub-type, or a polymorphic population. Identification of mis-assigned specimens is a potentially valuable use of DNA data, but must be conducted with due recognition that DNA taxonomy is also prone to errors, through contamination, sequencing errors, or misleading phylogenetic inference.

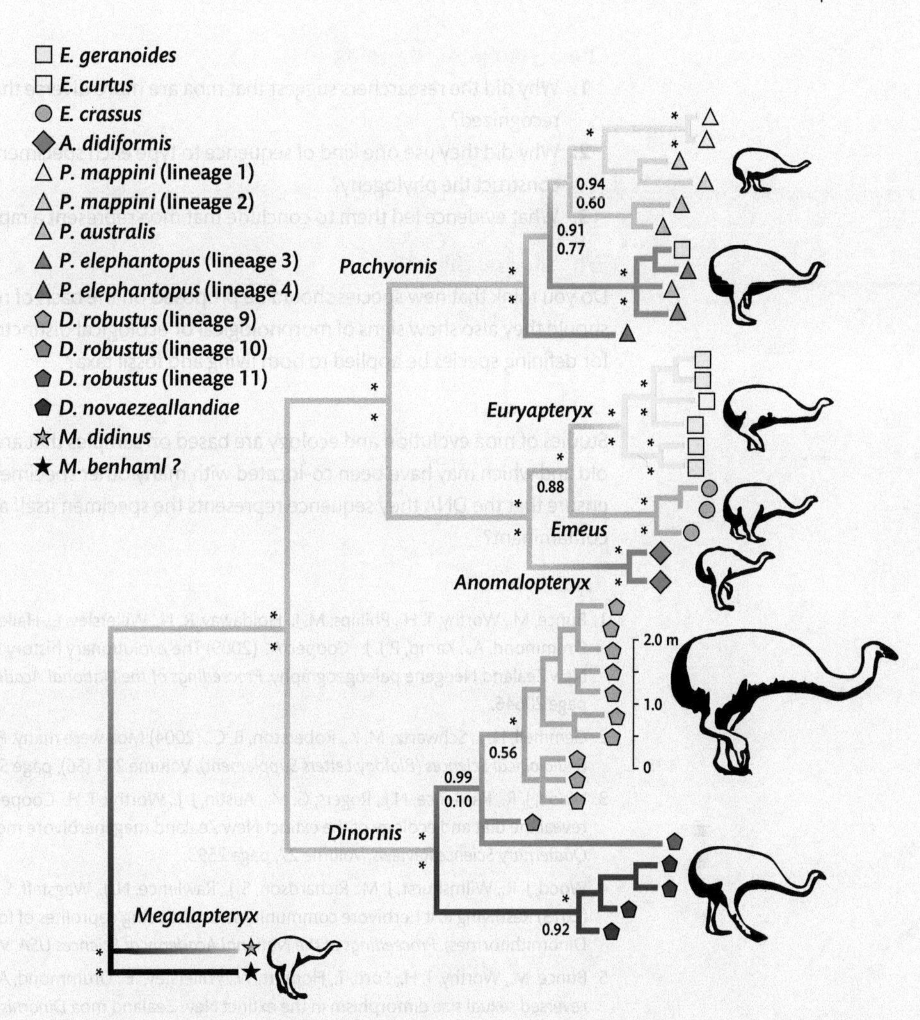


**Figure 2.31** Phylogeny (evolutionary tree) of moa, inferred from DNA extracted from moa bones. Moa bones are found in abundance in middens (prehistoric kitchen rubbish), suggesting moa were eaten by early colonists of New Zealand. A study that compared the species compositions of natural collections of bones in New Zealand (e.g. from animals that died from falling into caves) with those found in middens concluded that animals that were more often hunted and eaten were more likely to have gone extinct[7].



The molecular date estimates are dependent on the calibrations and methods used: other researchers obtained much younger date estimates from an alternative analysis[1].

### Further work

DNA sequence diversity has suggested that moa were present in much greater numbers than previously suspected[2]. Analysis of coprolites (faeces) suggest they were predominantly grazers, but also browsed from trees and shrubs[3, 4]. Bones previously attributed to two co-occurring species of moa were shown to be from sexually dimorphic males and females of the same species (with the females twice the size of the males)[5]. DNA from inside moa eggs was used to identify which species they belonged to and DNA from the outside of the surprisingly thin-shelled eggs was used to suggest that the smaller males must have incubated the eggs, though perhaps not by sitting on them[6]. These studies illustrate how DNA analysis may not only reconstruct evolutionary relationships between lineages but also shed light on ecology and behaviour of extinct species.

### Check your understanding

1. Why did the researchers suggest that moa are more diverse than has been previously recognized?
2. Why did they use one kind of sequence to type each specimen, but a different sequence to construct the phylogeny?
3. What evidence led them to conclude that moa represent a rapid, endemic radiation?

### What do you think?

Do you think that new species should be proposed on the basis of molecular data alone or should they also show signs of morphological or ecological distinctness? Can the same criteria for defining species be applied to both living and fossil taxa?

### Delve deeper

Studies of moa evolution and ecology are based on samples that are many hundreds of years old and which may have been co-located with many other specimens. How do the researchers ensure that the DNA they sequence represents the specimen itself and is not an incidental contaminant?

### References

1. Bunce, M., Worthy, T. H., Phillips, M. J., Holdaway, R. N., Willerslev, E., Haile, J., Shapiro, B., Scofield, R. P., Drummond, A. , Kamp, P. J. J. , Cooper, A. (2009) The evolutionary history of the extinct ratite moa and New Zealand Neogene paleogeography. *Proceedings of the National Academy of Sciences USA*, Volume 49, page 20646.
2. Gemmell, N. J., Schwartz, M. K., Roberston, B. C. (2004) Moa were many. *Proceedings of the Royal Society B: Biological Sciences (Biology Letters Supplement)*, Volume 271 (S6), page S430.
3. Wood, J. R., Rawlence, N.J., Rogers, G. M. , Austin, J. J., Worthy, T. H., Cooper, A. (2008) Coprolite deposits reveal the diet and ecology of the extinct New Zealand megaherbivore moa (Aves, Dinornithiformes). *Quaternary Science Reviews*, Volume 27, page 2593.
4. Wood, J. R., Wilmshurst, J. M., Richardson, S. J., Rawlence, N. J., Wagstaff, S. J., Worthy, T. H., Cooper, A. (2013) Resolving lost herbivore community structure using coprolites of four sympatric moa species (Aves: Dinornithiformes). *Proceedings of the National Academy of Sciences USA*, Volume 110, page 16910.
5. Bunce, M., Worthy, T. H., Ford, T., Hoppitt, W., Willerslev, .E , Drummond, A., Cooper, A. (2003) Extreme reversed sexual size dimorphism in the extinct New Zealand moa *Dinornis*. *Nature*, Volume 42, page 172.
6. Huynen, L., Gill, B. J., Millar, C. D., Lambert, D. M. (2010) Ancient DNA reveals extreme egg morphology and nesting behavior in New Zealand's extinct moa. *Proceedings of the National Academy of Sciences USA*, Volume 107, page 16201.
7. Duncan, R. P., Blackburn, T. M., Worthy, T. H. (2002) Prehistoric bird extinctions and human hunting. *Proceedings of the Royal Society B: Biological Sciences*, Volume 269, page 517.

# Mutation

3

## *We are all mutants*

*"An ideal situation would be if the organism were to respond to the challenge of the changing environment by producing only beneficial mutations where and when needed. But nature has not been kind enough to endow its creations with such a providential ability"*

**Theodosius Dobzhansky (1951) *Genetics and the Origin of Species*, 3rd edn, Columbia University Press.**

## What this chapter is about

A mutation is any heritable change to the genome, which will be passed on when the genome is copied. Mutation is essential for evolution: without it, all genomes would be identical. Mutation is common enough that, in most cases, we can expect each individual's genome to have a unique DNA sequence. Yet, because mutations are inherited, we can also expect the genomes of individuals to be similar to those of close relatives. So by comparing DNA sequences, we can distinguish individuals, identify family groups, and study the heritability of traits across generations. The development of new and faster ways of revealing the mutations that an individual carries has wide-ranging implications for both research and wider society.

### Key concepts

- Mutations are permanent changes to the genome that will be copied to its descendants
- DNA sequence differences in the genome can be used to distinguish individuals and identify heritable variation

# Mutation

*Mutants*

Would you like to be able to walk through walls, lift objects using only the power of your mind, or have the capacity to heal any wound instantly (and therefore be able to get into an awful lot of fights without risking lasting damage)? These are just some of the attributes of the X-Men, superhero-like characters first introduced in Marvel Comics in 1963, and popularized in a string of movies beginning in 2000 (**Figure 3.1**). The source of the X-Men's incredible powers were not alien birth (as for Superman), gamma radiation (the Incredible Hulk) or the bite from an irradiated spider (Spiderman). Instead, the X-men owed their supernatural abilities to mutation. Each of the X-Men carried a mutation that had the dramatic effect of conferring unnatural abilities. These mutants were vilified by the general public, yet considered to represent the evolutionary future of humanity (which is good news, if you want your descendants to be able to control the weather with their psychic powers).

**Figure 3.1** A mutant with a mission: does the evolutionary future of humanity hold the development of potentially useful new appendages, or a gradual decline into obesity and short-sightedness?
© Istockphoto.com/RichardD

In fact, we are all mutants. It is estimated that each human embryo begins life with as many as a hundred new mutations. Each of these mutations is a change that occurred in the genome of one of the parents during the production of sperm or eggs. If that egg or sperm forms an embryo, the mutation will be copied into every new cell added to the embryo as it grows into an adult. And if sperm or eggs from that adult go on to form another embryo, then the mutation may be perpetuated to a new generation. Alas, these mutations do not tend to confer super powers. Many mutations do not have any noticeable effect, especially those that occur in parts of the genome that do not seem to contribute to making a functioning body. Mutations that occur in functional sequences are often harmful, decreasing the chances of survival of that individual and its descendants. Only very rarely will a mutation produce any beneficial changes.

The word 'mutation' predates the genetic revolution, and can have different meanings in different contexts. For the purposes of this book, we will refer to a mutation as being any permanent, heritable change to the genetic material. Using this definition, we will call a change to the genome a mutation if it is irreversible and will be passed on to any copies made of that genome (usually when DNA is replicated when cells divide). In Chapters 4 and 5, we will see how mutations can arise from the rearrangement, duplication or deletion of strands of DNA. But as this book is primarily concerned with the analysis of DNA sequences, the mutations we will most commonly encounter are point mutations, which create single nucleotide changes to the DNA sequence. So that is where we are going to start.

## Point mutations

Point mutations are changes to a single base in the nucleotide sequence of a DNA molecule, where one of the four bases (A, C, T, G) is exchanged for another base, or a single base is inserted or deleted. Broadly speaking, changes to the nucleotide sequence of DNA come about through two processes, copy errors and damage. Actually, there is no clear division between these processes: copy error can arise through incorrectly repaired damage encountered during replication, and it seems possible that damage might occur more often when DNA is exposed during replication. In the next chapter, when we look at DNA replication, we will consider the mutations caused by copy errors, which form a substantial proportion of the mutations accumulated each generation. Here, we will focus on point mutations caused by DNA damage.

There are many physical and chemical factors, referred to as mutagens, that can damage DNA and cause change to the information in the genome. When people worry about mutagenic effects of radiation or chemicals, it is this incidental DNA damage they fear. But many mutagens are common features of the environment. So although damage to the genome may be increased by exposure to hazards such as radiation, DNA damage occurs all the time, in all cell types. In fact, there is a good chance that, in the time it takes you to read this paragraph, at least one of your cells will sustain damage to its genome.

Damage to DNA can be disastrous, because it can destroy the information needed to make and maintain living cells. The genome needs constant maintenance in order to repair this incidental damage to DNA, so there is a vast array of proteins and other biomolecules that work together to detect and repair any damage. Most damage will be corrected by the multitude of DNA repair pathways that monitor and repair the genome. But, however efficient they may be, no repair mechanisms are perfect. Sometimes, damage that changes the DNA sequence persists unrepaired, so that the altered sequence is copied along with the genome and passed to the cell's descendants. And sometimes the repair process itself causes a permanent change to the DNA sequence, if damaged DNA is replaced with a new, but not exactly identical, sequence. Once the sequence is altered so that the original sequence can no longer be recovered, that change is a mutation that will be included in any copies made from that genome. If the mutation happens to occur in a cell that will give rise to a new individual, then that mutation can be copied to a new generation.

*See Chapter 2 for the differing evolutionary fates of body and germline cells*

Because of the potentially dire consequences of DNA damage, cells have a variety of mechanisms for fixing damage. There are so many types of mutation, and such a diversity and complexity of repair systems, that we could not hope to cover them all. Instead, we will look at a single mutagen that causes a particular type of DNA damage, and give a brief summary of the strategies a cell employs to deal with this damage. The mutagen we will consider is ultraviolet light (a component of sunlight), which can cause the formation of thymine dimers (**Figure 3.2**). Thymine dimers are formed when two adjacent Ts become linked together, causing a distortion in the DNA helix. Formation of thymine dimers is potentially ruinous for a dividing cell, for two reasons. Firstly, when two Ts on one strand pair with each other, they can no longer pair with the opposite strand. Since DNA transmits information by complementary base pairing, any bases that don't pair represent a loss of transmissible information. If information is lost from a critical sequence, it may prevent the cell from making something important. Secondly, and more seriously, a thymine dimer can block the movement of a polymerase enzyme along the DNA helix, halting DNA replication. If a cell cannot replicate its DNA then it can't divide. And a cell that can't divide is a cell with no evolutionary future. So there has been strong selection pressure for cells to evolve mechanisms for detecting and repairing UV-induced thymine dimers. We will follow a typical bacterial repair pathway, but similar responses can be found in most organisms.

*DNA replication is described in Chapter 4*

## An example of a repair pathway: UV damage

There are several different repair pathways that deal with thymine dimers (**Figure 3.2**). There is a specific enzyme that detects thymine dimers, and cuts the bonds between the neighbouring Ts (as it happens, this enzyme gets its activation energy from visible light). This photoreactivation pathway reverses the damage caused by the thymine dimer, restoring the sequence to two normal Ts. If the dimer is not removed, it will attract the attention of enzymes that scan the genome for distortions in the helix, and trigger a general excision repair pathway. The damaged DNA strand is cut out, then a special DNA polymerase makes a new strand, and a ligase enzyme glues the newly made strand to the helix. If a thymine dimer is not detected and repaired by the excision repair pathway, then the next time that DNA is copied, the dimer will cause the polymerase enzyme to stall. To avoid halting DNA replication altogether, the cell may initiate a DNA repair pathway that bypasses the blockage caused by the thymine dimer, so it can continue replication after the block. This leaves a gap in the new DNA strand, which may be filled by a post-replication repair pathway that uses part of the other strand of the replication fork to replace the missing DNA sequence (recombination repair: **Figure 3.2**). If this strategy does not work, then some organisms have a last-ditch option: the SOS pathway. SOS is a complex series of responses to

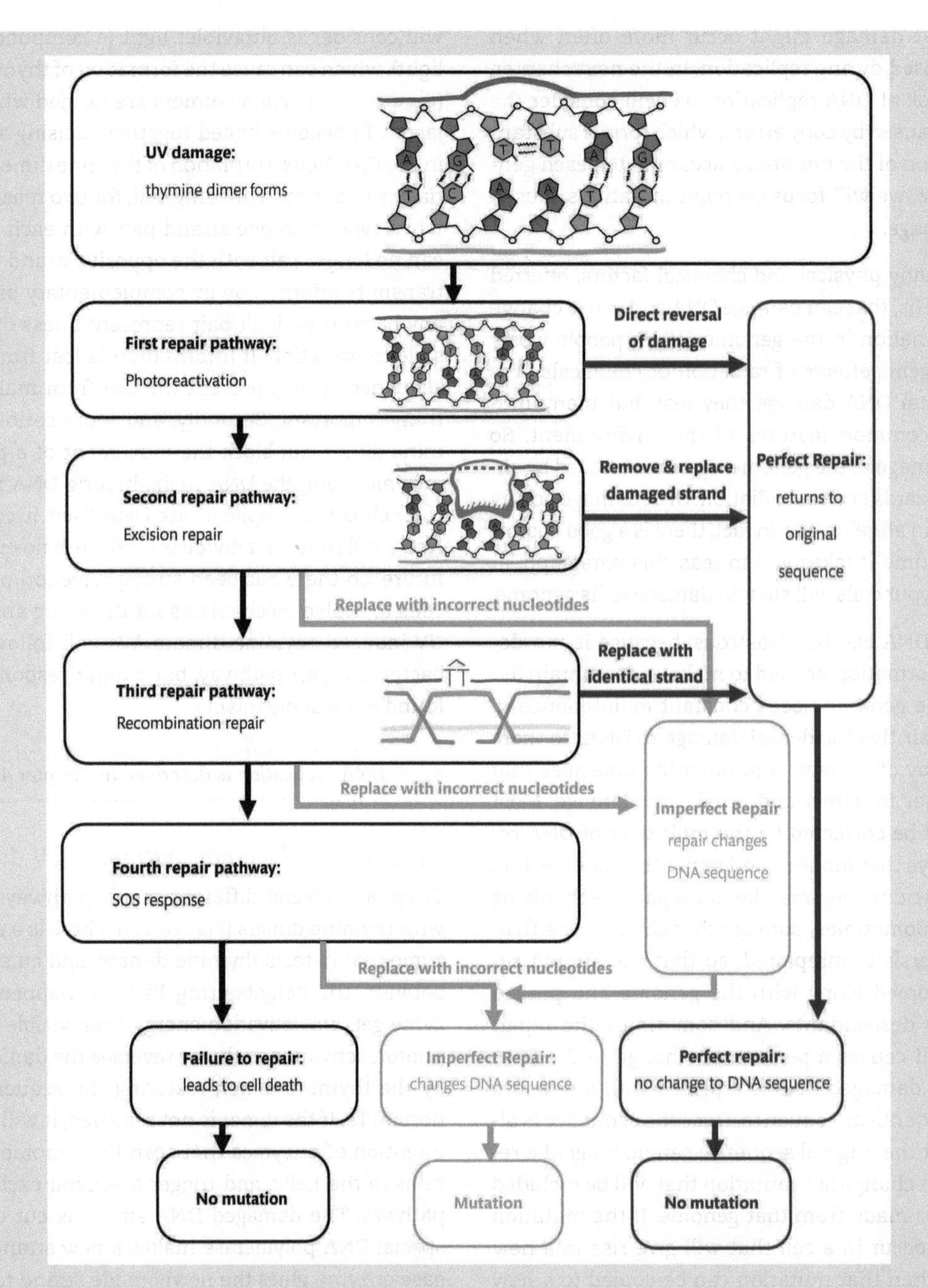


**Figure 3.2** Repair pathway for a thymine dimer. This flowchart describes four possible responses to UV damage, rather than representing the precise sequence of events in any given organism. Damage may be directly reversed, removed and replaced with excision repair, replaced with DNA from another strand by recombination repair or bypassed using the SOS response (this pathway has been described in bacteria, but it has been suggested that vertebrate genomes may have some of the elements of this pathway too). Imperfect repair creates a change in the base sequence (mutation), but perfect repair does not (sequence is returned to its original state) nor does failure to repair (damaged DNA can't be copied to the next generation).

an increasingly desperate situation: if the cell cannot fix the dimer, replication will cease and the cell will die. In a race against time, the cell tries a number of strategies to overcome the block in replication. The SOS response is error-prone: it may fix the damage, but in doing so it will probably change the DNA sequence. Inducing SOS repair is likely to lead to many mutations, each of which may be just as disastrous as the original damage itself. This is why the SOS response is the final attempt of a damaged cell to divide or die.

This is just one example of a specific type of DNA damage (thymine dimers) caused by a particular mutagen (UV light), to which the cell has a sophisticated response (enzymes that reverse damage, failing that excision repair, and failing that post-replication repair, and finally the last-chance SOS response). There are many other kinds of damage, brought about by a range of different mutagens. Some mutagens, like UV light, change the bases in the strand so that the information they carry is lost or changed. Some mutagens are base analogs: they fit into the DNA helix in place of a normal base but do not pair properly. The base analog 5-bromouracil (5BU) can be incorporated into a DNA strand instead of normal thymine, but unlike thymine (T) which pairs with adenine (A), 5BU will occasionally pair with a guanine (G). So at replication, the presence of a 5BU where a T should be can cause a change from A to G on the opposite strand. Other mutagens cause a physical change to the DNA helix. For example, intercalating agents, like ethidium bromide (commonly used in genetic labs), fit into the DNA helix where bases should go, often creating frameshift mutations, which make nonsense of the genetic code in the gene (see Chapter 6). Some mutagens can cause structural damage to DNA. Ionizing radiation can cause double-strand breaks in the DNA helix: the process of sticking the strands back together often gives rise to rearrangements and changes in the nucleotide sequence (see Chapter 4).

If you are interested, then you can learn about specific mutagens, types of DNA damage, and particular repair pathways in any general genetics or biochemistry textbook—and I recommend you do, if only to marvel at the richness and complexity of DNA maintenance. For the purposes of this book, it is enough to know that DNA damage occurs frequently, but thanks to a sophisticated system of repair pathways, it is usually repaired. But occasionally damage is imperfectly repaired, resulting in a change to the DNA sequence. From an evolutionary perspective, DNA damage only becomes a mutation when the changed base becomes a permanent part of the DNA sequence, so that it will be passed on to all copies made of that genome.

# Single nucleotide polymorphisms (SNPs)

Although most errors are corrected, and most damage repaired, the occasional change slips through and becomes a heritable change to the DNA sequence. Genomes contain a vast amount of DNA. The human genome is around 3 billion nucleotides, but many species have even larger genomes. Marbled lungfish, for example, have 133 billion nucleotides in their genomes (**Figure 3.3**). With billions of bases to copy every time a cell divides, even a phenomenally high rate of copy fidelity and a near-perfect DNA repair system will result in hundreds of new mutations occurring in each genome, every generation. The practical upshot of all this mutation is that every genome is unique. So if we can identify the sites in the genome where mutations have occurred, we can use DNA sequences to identify individuals.

Before we go on, we are going to have to clear up some potentially confusing terminology. In the previous chapter, we were introduced to the concept of the genotype, which we defined as the heritable genetic information that an individual can pass to its offspring—in contrast to the phenotype which is the particular anatomy, physiology and behaviour of an individual that arises from the interaction between the genotype, developmental processes and environmental influences. But when we start talking about the variation in genome sequence between individuals, we will come across the word 'genotype' used in a slightly different way. In this context, the genotype of an individual is the specific genome sequence that it carries, that distinguishes it from all other individuals in the population. And, to stretch the terminology even further, the word genotype is increasingly being used as a verb: to genotype an individual is to define the particular genetic variants it carries. Hopefully, it will usually be clear from context when people intend genotype to mean the genetic information as a whole, or the specific version of the genome an individual carries, or the process of determining which particular sequence variants a particular individual has.

## Identifying individuals using SNPs

Sites in the genome where the nucleotide sequence differs between some members of a population are often referred to as single nucleotide polymorphisms, or SNPs (pronounced 'snips': **TechBox 3.1**). 'Single nucleotide'

**Figure 3.3** Genome size champions. Genome size varies considerably between species. At the time of writing, the largest genome recorded is from the Marbled Lungfish (*Protopterus aethiopicus*), with around 130 billion bases of DNA. That's nearly forty-four times bigger than the human genome and more than six thousand times the genome size of *Pratylenchus coffeae*, a nematode that has the most compact animal genome yet recorded. *P. coffeae* is a major parasite of important crops such as bananas and (you guessed it) coffee.



refers to the fact that these are (usually) one-base changes to the DNA sequence that arise from point mutations (or single-base insertions or deletions). 'Polymorphism' means that these sequence variations are carried by some, but not all, members of a population: that is, that there are multiple (poly) different variants (morphs) of the sequence in the same population. Some formal definitions of polymorphism specify that a particular variant must be present in a minimum proportion of individuals in a given population to be considered a polymorphism (say, carried by at least one per cent of all members of a population). But, in practice, people usually refer to any locus that is not the same in all individuals in a population as polymorphic. It can be difficult to accurately judge the frequency of sequence variants where sampling is limited to a relatively small number of individuals, as is often the case in molecular ecology or medical genetics. For example, imagine you have sampled ten individuals, and one of them has a unique sequence variant found in no other individuals in your sample (such a variant is referred to as a singleton). In this case, you would be unable to tell whether a particular variant was present in only one individual or in ten per cent of the population or more or less. Actually you will find that the term 'SNP' is used in two different ways. One is to identify polymorphic sites at which individuals in a population can vary, and the other is to refer to the particular base found at a given polymorphic site in any individual.

***We consider population polymorphism in more detail in Chapter 7***

Methods for detecting and analysing SNPs are changing so fast that whatever I write in this book has a fair chance of being out of date by the time you are reading these words, but we can consider some basic approaches. One is to apply traditional targeted sequencing to a part of the genome known to vary between individuals. For example, researchers wanted to identify a set of SNPs that could be used to identify endangered white rhinos, so that they could monitor the population and detect the source of illegal rhino horn products (**Figure 3.4**). They sequenced non-protein-coding parts of the genome that are known to have a high rate of change, so tend to vary between individuals, then identified sites in these sequences where individuals did not all have the same base. By sequencing 30 individuals from a captive population they found ten sites in five different loci that varied between individuals. They attributed the relatively small number of variable sites to the small inbred populations in the endangered rhino. This is a bit sad, really: the kinds of populations in which we might most want to use DNA to identify individuals might also be the populations with the least genetic variation.

***The effect of inbreeding on molecular evolution is discussed in Chapter 8***

As new sequencing techniques make it quicker and cheaper to sequence large amounts of the genome (**TechBox 5.2**), SNP discovery is increasingly based on comparisons of genome sequences. Analysis of the sequences from the '1,000 genomes project' has identified more than 38 million SNPs in human genome data. Given the amount of data generated by these high-throughput sequencing platforms, computational analysis plays a crucial role in detection of SNPs. So the reliability of SNP genotyping rests not only on laboratory techniques and sequencing machines, but also on the analytical methods (Case Study 3.1). When looking for single base differences between individuals, there is always a risk that errors in sequencing could be counted as mutations, or that misalignment of sequences will influence the estimates of differences between two samples. Therefore, as with any scientific technique, genotyping must be reported with sensible confidence limits that reflect the possibility of errors or incorrect assignment (**TechBox 5.2**).

All this genetic diversity might lead you to wonder just what a genome sequence is meant to represent. If each individual carries around a hundred new mutations, and if approximately one in every thousand nucleotide positions in the genome is polymorphic, how is it possible to produce one official genome sequence for a species? The answer is that it's not possible. Published genome sequences are either a single individual's genome, or a consensus sequence for a sample of individuals, representing the most likely nucleotides to be found at any given position in the DNA sequence. It is extremely unlikely that any given individual would have the exact sequence reported by a genome sequencing project. But any given individual would have a high probability of having the same sequence at the majority of sites in the genome. Often in evolutionary biology, it is most helpful for us to use the consensus sequence representing the most common sequence variants for a population. However, for some genetic studies, it is the polymorphic sites we wish to study. In particular, SNPs are very useful for identifying genes that cause particular inherited traits, such as genetic diseases.

**Figure 3.4** White Rhinos aren't actually white. The species (*Ceratotherium simum*) was thought to be extinct until a population of around 100 individuals was discovered in 1895 in KwaZulu-Natal, South Africa. Careful management has resulted in a population of over 20,000 animals, so now White Rhinos are no longer listed as Endangered (unlike Black Rhinos (*Diceros bicornis*), which aren't particularly black, and are considered to be Critically Endangered). However, the northern subspecies (*Ceratotherium simum cottoni*) has been reduced to a handful of individuals, and is virtually extinct in the wild. A SNP database has been established to help monitor the population and to allow identification of the origin of illegally-traded rhino products. A set of established primers were used that attached to the coding regions (exons) of genes that also allowed them to sequence non-translated regions (introns) that tend to vary between individuals (**TechBox 6.2** looks at the use of primers and probes to target sequences).



## Using SNPs to track inherited traits

The ubiquity of mutations has the dual effect of making every genome unique, yet making related genomes similar. This is, of course, why children resemble their parents. This is good, when your children inherit your curly hair, keen eyesight or charming smile. But it is not so good when your children inherit your flat feet, attractiveness to mosquitoes, or propensity for heart attacks. We have already seen that we can use mutations to identify individuals, track family relationships, characterize populations and record movement of individuals in space and time (Chapter 1). We can also use the patterns of inheritance of mutations to try to determine why some sequence variants have undesirable effects on the individuals that carry them (**TechBox 3.2**).

Before we go on, we need to introduce a very useful word: allele. An allele is any heritable variant of a particular sequence in the genome. Recall Mendel's groundbreaking experiments in genetics (**Figure 2.4**): we can say that 'tall' and 'short' are both alleles of the 'height' gene. Every individual pea plant has the gene that specifies height, but they might inherit the tall variant or the short variant of that gene. Of course, if you have two copies of a particular sequence in your genome, you might be carrying two different alleles. Diploid organisms, like you, have two copies of every chromosome (with the exception of the sex chromosomes, where you might carry one X and one Y). So for every locus (place in the genome), you have two versions (alleles), one you inherited from your mother, and one you inherited from your father. These alleles might be the same as each other, or they might be different. Some alleles influence phenotype, like tall and short. But any heritable variant is an allele, whether or not it has any effect on phenotype: DNA fingerprinting is typically based on determining which alleles an individual carries at a number of non-coding loci that don't influence phenotype. Also note that there might be many different SNPs that occur close to each other in the genome, so that they are effectively linked together. All those closely linked variants will be inherited together, making them a single allele.

There are two ways that SNPs can be used to reveal the genetic basis of a trait. Firstly, the SNP may actually cause a particular trait by disrupting an important sequence. So, for example, two SNPs in the *human Period 2 (hPer2)* gene in humans have been found to be associated with being a 'morning person' (that is, being annoyingly chirpy early in the mornings, but socially challenged due to an inability to stay up late at parties). When researchers studied a particular family with a known pattern of inheritance for an extreme form of morningness called Advanced Sleep Phase Syndrome (ASPS), they found that all of the family members with ASPS had one particular SNP in the coding region of the gene, causing the amino acid glycine to be included in the protein where a serine should be. Since that SNP was not found in unaffected family members, it is a strong candidate for causing ASPS. But other cases of genetic causality are less clear cut. Another SNP found to be significantly associated with morningness occurs in the upstream regulatory regions of the *hPer2* gene, rather than in the protein-coding part of the gene. However, of the 210 people included in that particular study, 14 per cent of morning people had the upstream SNP, compared to only 3 percent of night-owls and 6 per cent of intermediate ('normal') people. So while this single nucleotide change seems to be associated with differences in daily rhythms, it would be problematic to say that it 'caused' morningness, as most morning people did not have the SNP, and some evening people did (**TechBox 3.2**).

Secondly, and far more commonly, a SNP may be located in a non-coding part of the genome, but close enough to the gene of interest that it is usually inherited along with a particular version (allele) of the gene. In this case, the SNP itself may have no effect on its carrier, but it can serve as a marker that indicates the presence of the allele of interest. SNPs are particularly useful as genomic markers because they are so abundant. On average, there is one SNP in every thousand bases of DNA in the human genome, so there is a high likelihood of finding one near each gene. It is because of the usefulness of SNPs that there are many international consortia developing databases of SNPs for a wide range of species, including crops, livestock, forestry trees and endangered species (**Case Study 3.1**).

# Estimating the mutation rate

Since the beginning of the genetic revolution, there has been intense interest in estimating the mutation rate (**Hero 3**). Mutation is one of the pillars of evolution, so an understanding of the rate at which new heritable traits arise is a fundamental part of evolutionary biology. As DNA sequence analysis is used in ever more fields of biology and medicine, estimates of the mutation rate also have a practical value. Many analyses rely on assumptions about the mutation rate, such as a uniform rate of mutation over time, across the genome, or between individuals and species (see Chapter 13). Many methods that draw on population genetic theory require a numerical value of mutations per generation in order to estimate population size, or evolutionary history, or the power of selection. So knowledge of the mutation rate is fundamental to biology. It's also surprisingly tricky to measure accurately.

The problem is that, because DNA repair systems are so devastatingly good at their job, the chance of any given nucleotide undergoing a mutation at any particular point in time is vanishingly small. For most organisms, we are looking for a relatively small number of mutations in a relatively huge genome. Take the human genome as an example: there may be 100 new mutations or so per generation, but they could have occurred in any of the three billion nucleotides in the genome. Finding those few sites that have undergone change makes looking for a needle in a haystack seem pretty easy. But there are several ways to increase the chances of accurately estimating the number of mutations that have occurred.

The only approach until recently was to track large numbers of individuals over enough generations that you would eventually observe enough mutations to be able to estimate the mutation rate across the genome. Traditionally, this was done by looking for detectable phenotypic change. This is why the tiny fruit fly *Drosophila melanogaster* played such a critical role in the early years of genetics. They could be bred in large numbers, with replicate populations housed in glass bottles and fed on yeast, and sorted under the microscope to spot any new variants in eye colour, wing shape or bristle length. But mutations that cause a visible change are relatively rare, so searching for mutant fruit flies takes a lot of time and patience. A study that screened nearly one million individual fruit flies found only 16 new mutations in eye colour.

Another classic approach to measuring mutation rate is to grow bacteria on a medium containing a food source that they cannot readily metabolize, then wait for a mutation to occur that allows its lucky carrier, and all its descendants, to utilize the new food source, giving rise to a growing bacterial colony. In fact, bacterial cultures are routinely used to judge whether particular chemicals are mutagenic (and therefore pose a danger to human health), by looking for an increase in the rate at which bacteria produce offspring that can metabolize something their parents can't (**Figure 3.5**). But this can only detect a very small fraction of mutations that occur in the genome. Only beneficial mutations that happened to endow the bacteria with the ability to thrive in the specific test environment can be detected.

So the classic fruit fly approach can only detect mutations that caused a visible difference, yet still permitted the fly to survive. And the bacterial approach only detected mutations that allowed the bacteria to do something useful it couldn't do before. These methods can't detect mutations that don't have a noticeable effect on phenotype, nor can they detect mutations that are sufficiently deleterious that they prevent the embryo from developing (though in breeding experiments the proportion of lethals can be estimated from the decline in fertility each generation).

The advent of DNA sequencing technology allowed scientists to look into the genome and see mutations that did not have obvious effects on phenotype. If you keep a laboratory population of something sufficiently short lived, say flatworms or flies or bacteria, then at regular intervals you sample the genomes present in the population, you should be able to develop a time series in which you can compare the sampled genomes to the ancestral genome they were all derived from (**Case Study 3.2**). *Drosophila melanogaster* can complete a generation, growing from an egg to an egg-laying female, in a week, and

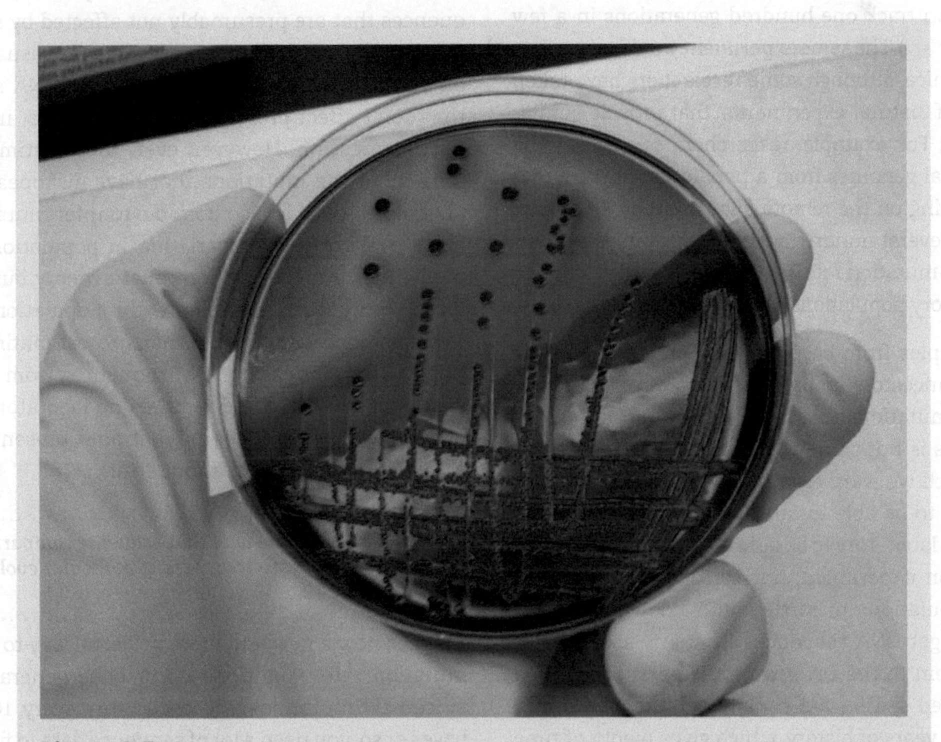

**Figure 3.5** The bacterium *Escherichia coli* is the workhorse of mutation studies. The potentially mutagenic effects of environmental agents, such as new commercially-produced chemicals, are sometimes investigated by measuring the rate of mutation reversal in *E. coli* colonies when exposed to that agent.

Reproduced by courtesy of Stephen Hill.

**Figure 3.6** Tiny heroes: *Drosophila* taught us much of what we know about mutation.

each female can lay hundreds of eggs. So, with sufficient patience, a mutation accumulation experiment using *Drosophila* can track one hundred generations in a few years (Figure 3.6). The same experiment would take a decade for lab mice, although some researchers have taken advantage of 'natural experiments' that provide similar information. For example, researchers sequenced the mitochondrial genomes from a population of mice that had been living on the remote sub-Antarctic Kerguelen Islands for several hundred years, and found that their estimates of mutation rate were similar to those derived from laboratory populations of mice.

These examples from bacteria, flies, and mice rely on comparing ancestors and descendants to estimate the number of mutations that have occurred in the germline. But this is not going to be possible for most species for which we want to know the mutation rate. We are unlikely to be able to raise enough generations of sloths in the laboratory to be able to perform a mutation accumulation experiment, nor can we expect to find enough populations of turtles with known histories. But we can get DNA sequences from any species and compare them to the DNA from related species. Even closely related species are usually separated by a million or more years of history, which gives plenty of time for a decent number of mutations to have occurred. It follows that if we know exactly how long it has been since two species last shared a common ancestor, then we can count the number of differences between their genomes, divide it by the amount of time they have been separated—and bingo—we have a rate of DNA sequence change.

But of course it's a bit more complicated than that (and a good rule of thumb in biology is that it almost always is more complicated than you first thought). Considering species divergences may technically give you millions of years' worth of mutations, but those mutations have been filtered through millions of years' worth of natural selection and drift. Every generation the genome has been subject to selection, which will have removed many of the deleterious mutations and, very occasionally, promoted advantageous mutations. Many other mutations were lost when their carriers, by unlucky chance, failed to leave descendants. Simply counting the differences between the sequences will only capture the proportion of mutations that persisted over enough generations to be counted, greatly underestimating the total number of mutations that actually happened.

*The actions of selection and drift are covered in Chapters 7 and 8*

One way to avoid the filtering effects of selection is to only consider changes to functionally unimportant sequences that are presumably not affected by selection. Cross-species mutation rate estimates are usually based on comparing changes in sites in the DNA sequence that don't affect protein function or do not impact on gene expression. However, even these estimates will miss all of the mutations that have disappeared simply by chance. It is also possible to infer mutation rate from levels of genetic variation in populations, using predictions from population genetic theory, but this will usually only work if you know the population size. So while the comparative approach to estimating mutation rate by comparing DNA sequences from different species is universal and requires no laboratory budget or field work (more's the pity), it does present its own challenges, and is not without error.

*We consider the way that sequence comparisons are used to estimate rates of molecular evolution in Chapter 13*

There is now a new and more powerful way to estimate mutation rate. Mutations occur each generation, but since new mutations will occur only every 10 million bases or so, you need a lot of sequence data to find them. High-throughput sequencing now makes it possible to sequence enough of an individual's genome to identify *de novo* mutations not found in their parents' genomes. For example, the '1,000 genomes' project has sequenced the genomes of several trios of mother, father and child

**Figure 3.7** Family relationships have traditionally been used to identify genes by tracking the inheritance of particular alleles. But now that whole genome sequencing is becoming more tractable, genome sequences of family members can be compared to detect how many mutations accumulate each generation.
Photograph by Lindell Bromham.

to estimate a mutation rate of $10^8$ per base pair per generation, or one in every hundred million bases. In the coming years, as whole genome sequencing becomes cheaper and more reliable, this technique may allow estimates of mutation rate in a wide range of species, and even the comparison of mutation rates between individuals (**Figure 3.7**).

## *Good, bad, and indifferent*

Mutations are the raw fuel of adaptation, but most of the preceding discussion has been about mutations that are harmful (such as those that wreck a particular gene product), or mutations that have little or no effect (such as those that occur in non-coding sequences). Why haven't we discussed the beneficial mutations that give you X-Men-like powers, or at least make you prettier, faster and stronger than your parents? Useful mutations that contribute to adaptation are relatively rare. There are several reasons for this.

Firstly, the useful bits of the genome are, for most complex organisms, in the minority. Most of the genome appears to have little direct impact on organism fitness. So any random point mutation has a fair chance of occurring in a nucleotide position in the genome where it doesn't much matter whether it's an A or a G or a C or a T. Most of the new mutations make no difference to their carrier at all.

***We take a closer look at DNA sequences that don't seem to affect fitness in Chapters 5 and 6***

Secondly, if a mutation does fall in a useful bit of the genome, it is more likely to do harm than good. We have discussed how organized and complex the genomic information system is (Chapter 2). So when considering the effect of mutation on a gene, we can think about what happens to an organized, complex, information-rich system when it is randomly altered. Here's an experiment to try at home: put on a blindfold, open up your computer and make a random change to its internal wiring. If I was to ask you to place a bet on one of three possible outcomes of this random rewiring—the computer will still work, the computer will no longer work, or the computer will run better than ever—which would you choose? I would guess that, unless you are particularly lucky (or well-practised at blind computer repairs), you stand a very good chance of ruining your computer with a random change.

The same basic principle applies to genetic information. Genes (and their associated regulatory elements) are highly organized and finely tuned to produce the right product in the right place at the right time. Just like a random change to a computer, a random change will almost certainly wreck a gene. So the majority of mutations in DNA sequences that hold important information will be deleterious—they will decrease the chance of that organism surviving. Mutations in sequences that don't affect the organism are unlikely to be deleterious, but by the same token they are also unlikely to be beneficial.

The disastrous nature of most mutations, and the sophistication of the DNA repair mechanisms in even the simplest cells, raises an important question: if natural selection against deleterious mutations drives the evolution of such a complex and finely tuned repair system, then why doesn't that selection pressure result in a perfect repair system that always returned a damaged sequence to its original state? After all, mutation is costly, because many mutations reduce that individual's chances of having great-great-grandchildren. Shouldn't all organisms have the lowest possible mutation rate? To explore this paradox, let's consider two possible strategies for perfecting the repair system: reduction of total mutation rate, and directing mutations towards useful changes.

***Chapter13 covers some of the factors that influence mutation rate in different species***

## Mutation rates are a compromise

Reduction in the mutation rate is costly in terms of time and energy. Any increase in repair efficiency must be paid for out of a cell's total metabolic budget: it takes energy and resources to make and operate repair enzymes. Furthermore, repair takes time: copying accurately is slower than copying with less regard for precision. No genome has unlimited resources. No genome can take an infinite amount of time to copy. At some point, the amount of resources used or time taken for repairs will outweigh the benefit gained from higher accuracy.

It may be helpful to consider the parable of the snail and the parasite. Snails have shells to protect their soft, vulnerable bodies from predator attack. A shell that is too thin will not save its owner from predation, so alleles for soft-shelledness are less likely to be passed on to the next generation. You may think that selection will tend to favour the thickest shell that the snail can carry. Yet this is clearly not the case, because snails infected with a certain parasite grow thicker shells than normal. Many parasites influence their host's morphology or behaviour in order to enhance their own chance of transmission (**Figure 3.8**), so it seems possible that the thicker shells are a results of the parasite's Machiavellian developmental manipulation. Why would the parasite make its snail host grow a thicker shell? Because if the snail gets killed, the parasite dies too. In this situation, the parasite benefits from more robust predation-protection just as much as the snail does. So if the snail is capable of growing a thicker shell when infected, then why does it not always do so? Surely natural selection should maximize protection from predators to increase the snail's chances of survival? A snail that spends all of its metabolic resources on growing a shell may have little left over for investing in reproduction. Natural selection cannot favour snails that survive predation but do not reproduce because, by definition, there will be no snail offspring to inherit the trait. The economy of nature must balance the costs and benefits of predation protection in the currency of reproductive output.

**Figure 3.8** My parasite made me do it. All species have parasites, and many parasites are not just passive passengers but active drivers of host behaviour. These darling little crabs (*Pachygrapsus crassipes*) may be playing host to several other species of crustaceans. The parasitic barnacle *Sacculina* has given up the rock-fixed, filter-feeding habits of its relatives. Instead, a female *Sacculina* larva swims until it finds a crab, then sneaks in through a joint in the crab's shell. Unlike its fellow barnacles, the adult *Sacculina* has no shell or appendages, instead it grows as a fungus-like mass inside the host crab's body. The microscopic male *Sacculina* larva finds an infected crab and inserts himself into the female *Sacculina*, becoming effectively a parasite on her. The growing female *Sacculina*, with her eggs fertilized by the parasitic males, releases a chemical that makes the crab behave like it is brooding its own egg sac, so that the possessed crab looks after the growing *Sacculina* as it if was its own offspring. If it happens to be inside a male crab, *Sacculina* 'feminizes' the crab so that it acts like a pregnant female. But some sacculinids carry their own parasitic crustaceans. *Liriopsis pygmaea* is a parasitic isopod that, unlike its pillbug relatives but like its host *Sacculina*, has no appendages or segments, but grows as a sac on the host. And just as *Sacculina* sterilizes the host crab, *Liriopsis* seems to sterilize the host *Sacculina*. So thanks to hyperparasitism, you can get a whole host of crustaceans riding around in one shell. That's biodiversity for you.



Economic considerations must also be applied to DNA repair. In the next chapter, we will look at the mechanisms employed to reduce mutations created by mistakes in DNA replication. One of these strategies is proofreading: the polymerase enzyme that copies DNA can check the accuracy of newly added bases and move 'backward' to delete them if they are incorrect. So any polymerase must balance 'forward' (replication) activity that adds nucleotides to the new DNA strand with 'backward' proofreading activity that removes incorrect bases. Mutations in the polymerase gene can change the balance between replication and proofreading, favouring copy accuracy at the expense of replication speed, or favouring speed over accuracy. For example, bacteria put in an experimental environment that favours fast reproduction can evolve a higher mutation rate, because individuals with a 'mutator' polymerase that copies more quickly by proofreading less can have a reproductive advantage over individuals with an 'antimutator' polymerase that copies more slowly and accurately by proofreading more. In this particular case, it appears that the benefit of fast replication outweighs the cost of acquiring more mutations. It is also possible that selection for rapid adaptation could alter the balance between the accuracy and speed of DNA replication. For example, it has been suggested that bacteria with a mutator polymerase are more likely to evolve antibiotic resistance. Because they copy fast with lots of errors, mutator bacteria have a greater chance of accidentally producing a mutation that confers resistance. However, although mutation is essential for adaptation, it carries a high cost. Mutators may, in certain circumstances, produce more 'winners' (advantageous mutations), but it will be at the expense of producing an awful lot of 'losers' (deleterious mutations).

Even if we accept that the mutation rate is shaped by natural selection, settling on a compromise between the competing costs of mutation and repair, we should not expect that the mutation rate is a perfect balance. Evidence is emerging, particularly from *Drosophila* studies, that individual members of a population can have different mutation rates depending on their personal condition. This makes sense if we view DNA repair as an investment: a starving individual will have less energy to invest in growth, reproduction and, possibly, basic processes like DNA replication fidelity. The power of selection to shape mutation rates, or any other trait, is limited by the availability of variation and the population size and structure. It has been suggested that species with small populations may experience an increase in mutation rates over time, because of accumulation of mutations that reduce the effectiveness of DNA repair. The next decade will be an exciting time for research into mutation rates as the power of high-throughput sequencing, and new methods of analysis, allow the dissection of the variation in mutation rates, and the way evolution shapes mutation rates over time.

*The role of population size in evolution is discussed in Chapter 8*

## Is mutation random?

Every textbook will tell you that mutation is random. It is important to think about exactly what this means. A random process is one that is not directed at a particular outcome. Another way to state this is that a random process is unpredictable: we may be able to describe the average result of a series of random events, but we cannot predict the exact result of any one event. The classic example is tossing a coin. Overall, we expect as many tails as heads, but on any given coin toss, we cannot predict whether we will get a head or a tail.

***The influence of chance events on molecular evolution is discussed again in Chapter 8***

The types of changes generated by mutation are not entirely random, because some mutations are more likely to occur than others. There are three important sources of bias (non-randomness) in mutations. Firstly, a genome can only step from where it is now. This arises as a logical consequence of the continuity of genetic information. The genome of the offspring is derived from that of its parents. Species evolve from existing species and are not created *de novo*. By definition, a mutation can create a genome that is one accidental change away from its parent genome. So the set of possible mutations for a genome is defined by the set of sequences that are one change (such as a chromosomal inversion, a genome duplication, a single nucleotide change) away from the genome as it is now.

Secondly, some genomic loci are more likely to mutate than others. In Chapter 5, we will see how repeat sequences are prone to increases in copy number. Point mutations are also more likely to occur at certain places in the genome, referred to as mutational 'hotspots'. When studies are made of the frequency of mutations along a particular gene, it is often found that a handful of sites account for most of the mutations. The positional bias in mutation can occur at a very fine scale. In many eukaryote genomes, cytosines (C) are far more likely to undergo mutation if they are next to a guanine (G). This is because cytosines tend to be methylated when they occur in a CG dinucleotide, and methylated Cs are prone to spontaneous deamination, where the amide group ($NH_2$) attached to the base is replaced with an oxygen molecule. When methylated cytosine is deaminated it turns into thymine (T). So C is more likely to mutate to T when it is sitting next to a G. Similarly, a thymine is more likely to undergo UV-induced mutation if it is next to another T than if it is next to a G, because the TT dinucleotide can form a thymine dimer (**Figure 3.2**).

Thirdly, some kinds of mutation are more likely to occur than others. For example, deletions are typically more common than insertions. Some base changes are more likely to occur than others. A pyrimidine (a one-ring base, T or C) is more likely to change to the other pyrimidine than it is to be exchanged for a purine (two-ringed base, A or G: **TechBox 2.1**). Similarly, a purine will more often be mutated to the other purine than to a pyrimidine. A change from one purine to the other purine, or from one pyrimidine to the other, is referred to as a transition. Transitions are usually far more frequent than transversions (changes from a purine to pyrimidine or vice versa). This may be partly due to the many mutation pathways that create specific transitions (e.g. deamination of C produces T).

So when we hear that mutation is random we then need to ask: 'random with respect to what?' The types of mutations that occur are not entirely random: some mutations are more likely to occur than others. And mutations are not random with respect to their occurrence in the genome: some parts of the genome are much more mutable than others. However, there is one very important sense in which mutations are considered to be random: mutations are random with respect to fitness. Another way of saying this is that mutation is not directed towards a particular outcome. As a general rule, mutations should be considered to be accidents. They arise without being designed for a purpose, in the same sense that accidentally breaking your coffee cup is not 'for' anything, it just happens.

If beneficial mutations contribute to adaptation, and deleterious mutations result in a reduced chance of leaving descendants, then surely there must be massive selection pressure to only produce good mutations? Wouldn't any genome that could decrease the number of deleterious mutations be selected for? Possibly so. In the late 1980s, there was a flurry of excitement when experiments were reported that appeared to show that *Escherichia coli* bacteria could preferentially generate mutations that would rescue them from a life-threatening situation. *Escherichia coli* can normally use lactose as an energy source using the enzyme lactase, but these experiments involved growing lactase-deficient *E. coli* (Lac$^-$) on a lactose-only medium. These Lac$^-$ bacteria would eventually starve if fed only on lactose that they couldn't metabolize. After some time, colonies of bacteria arose that could grow on lactose thanks to a reversal mutation in the lactase gene. The lactase-enabling mutations appeared to occur at a higher rate than expected by chance, leading to the hypothesis that the bacteria could somehow direct the process of genetic change to just the right genes that would rescue them from starvation. Because it challenged a long-held tenet of evolutionary biology—that mutation is random with respect to fitness—the argument over directed mutation was often heated. Indeed, the debate got so acrimonious that the mere mention of the phrase 'directed mutation' was enough to get blood boiling. One researcher suggested that, to allow the debate to proceed in a civilized fashion, the phrase 'directed mutation' should be replaced by something neutral, like 'Fred'.

But the debate has become much quieter and, as yet, no concrete evidence has been put forward that beneficial mutations are more likely to happen than expected by chance. Note that although random mutation is often considered a pillar of evolutionary theory it need not necessarily be true for natural selection to work. In fact, selection would be more efficient if mutation was biased toward useful changes. But, as Theodosius Dobzhansky noted in the quote at the beginning of this chapter, unfortunately for real organisms, it does appear to be the case that mutation is random with respect to fitness (**Figure 3.9**).

**Figure 3.9** One of my undergraduate lecturers considered being able to spell 'Theodosius Dobzhansky' a hurdle requirement of the evolutionary biology course. And fair enough too, because Dobzhansky played a key role in marrying Mendelian genetics to evolutionary theory to build the modern evolutionary synthesis. 'Synthesis' refers to the way that previously distinct fields of study, such as botany, zoology, genetics, systematics, and palaeontology were brought together to build one united conceptual framework for understanding evolutionary biology. Dobzhanksy was not only a leading researcher who did much to embed population genetics at the heart of modern biology, he also produced an impressive catalogue of books that sought to explain the new biology to a wide audience. So if you're a fan of the neo-Darwinian synthesis, then why not consider naming your first child Theodosius.

# Conclusion

On the one hand, mutation is rare. A barrage of DNA repair pathways ensures that most damage to DNA is fixed before the genome is replicated and that most errors made in DNA copying are corrected. On the other hand, mutation is very common; for although mutation rates are very low, genomes are constantly subject to damage (or copy error, as we shall see in the next chapter). The size of the genome, and its constant replication and activity, makes mutations inevitable and common. Not all of these mutations end up in successful offspring, but some do. In this way, mutation makes each one of us unique.

Without mutation there could be no evolution, because each genome would be exactly the same as all other genomes. Yet, although mutation is necessary for evolution, in most cases the results of mutation are either disastrous or unnoticeable. Very rarely does mutation produce a desirable feature. All your children will be mutants, but it is, alas, unlikely that these mutations will confer X-Men-like superpowers on them.

However, mutation does incidentally create something that is very useful to biologists: it creates a genome that it is a storehouse of information. Mutations make genomes unique, so that we can use genetic markers to identify individuals. Because these mutations are inherited, related individuals are more similar than random members of a population, so we can use DNA to describe family relationships and identify members of interbreeding populations. Conversely, we can use known family relationships to find genetic markers that are associated with important traits, such as inherited disease risk, and use those markers to track down the genes responsible. In order to understand how to track descent with DNA, we need to look at how mutations are copied from the genome in which they occur and then passed on to that genome's descendants. So in the next chapter we will cover one of the most important topics in the whole of biology–DNA replication.

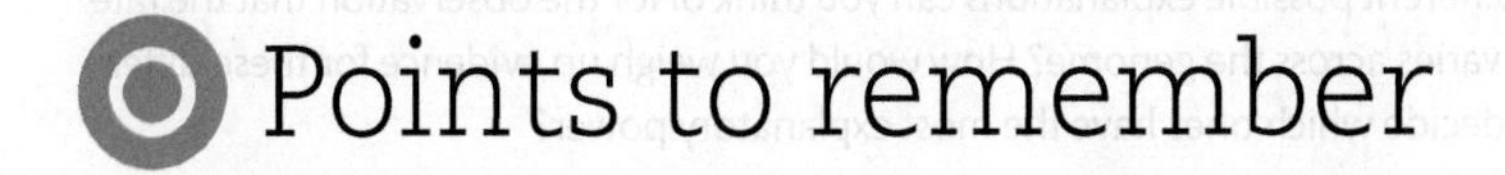

# Points to remember

### Mutation

- Genomes contain billions of nucleotides, so even with a very low mutation rate every individual will carry new mutations in their genomes.
- Point mutations are permanent changes to the base sequence that will be inherited by all copies made from that sequence.
- There are many different agents that cause mutation (mutagens), including UV light, radiation, and particular chemicals.
- There are many different repair pathways to repair this damage, for example excision repair where the damaged base is cut out and replaced.

### Single nucleotide polymorphisms (SNPs)

- Single nucleotide polymorphisms (SNPs) are places in the genome where the base sequence differs between individuals in a population.
- SNPs may be associated with a particular phenotype because they disrupt gene function or because the SNP occurs close to a locus that influences gene function, so that it is inherited with it.

### Estimating the mutation rate

- Mutations are rare so to count them you need to either:
  - screen a lot of individuals (e.g. *E. coli*)
  - track short-lived organisms over many generations (e.g. fruit flies)
  - sequence whole genomes of related individuals (e.g. humans)
  - estimate rates of change in DNA sequences unaffected by selection (any species).
- Most mutations are either harmful or without effect, very few improve their carrier's chances of survival and reproduction.
- Reducing the mutation rate takes time and energy, so organisms must balance the costs of increased repair against the costs of deleterious mutations.
- Some kinds of mutation are more likely to occur than others and some sites in the genome are more mutable than others.
- Mutation is random with respect to fitness: useful mutations are not more likely to happen than expected by chance alone.

# Ideas for discussion

**3.1** Can you think of a mechanism whereby useful mutations could occur at a greater frequency than random? Or a way of reducing the occurrence of harmful mutations? If you can't, why is it so hard to direct mutation to useful changes? If you can, why hasn't it been detected in nature yet?

**3.2** How many different possible explanations can you think of for the observation that the rate of mutation varies across the genome? How would you weigh up evidence for these different ideas to decide which ones have the most explanatory power?

**3.3** Given that the mutation rate is dependent on the operation of a large number of gene products such as repair enzymes, what effect might a high mutation rate across the genome have on the efficiency of repair apparatus? Could an 'error catastrophe' (mutations in DNA repair that cause more mutations) lead to the death of an individual or the extinction of a population?

# Examples used in this chapter

Agrawal, A. F, Wang, A. D. (2008) Increased transmission of mutations by low-condition females: evidence for condition-dependent DNA repair. *PloS Biology*, Volume 6, page e30.

Altshuler, D. M., Lander, E. S., Ambrogio, L., Bloom, T., Cibulskis, K., Fennell, T. J., Gabriel, S. B., Jaffe, D. B., Shefler, E., Sougnez, C. L. (2010) A map of human genome variation from population scale sequencing. *Nature*, Volume 467, page 1061.

Bjorkholm, B., Sjolund, M., Falk, P. G., Berg, O. G., Engstrand, L., Andersson, D. I. (2001) Mutation frequency and biological cost of antibiotic resistance in *Helicobacter pylori. Proceedings of the National Academy of Sciences USA*, Volume 98, page 14607.

Carpen, J. D., Archer, S. N., Skene, D. J., Smits, M., Schantz, M. (2005) A single-nucleotide polymorphism in the 5'-untranslated region of the hPER2 gene is associated with diurnal preference. *Journal of Sleep Research*, Volume 14, page 293.

Chao, L., Cox, E. C. (1983) Competition between high and low mutating strains of *Escherichia coli. Evolution*, Volume 37, page 125.

Hardouin, E. A. , Tautz, D. (2013) Increased mitochondrial mutation frequency after an island colonization: positive selection or accumulation of slightly deleterious mutations? *Biology Letters*, Volume 9, page 20121123.

Labuschagne, C., Kotzé, A., Grobler, J. P., Dalton, D. L. (2013) A targeted gene approach to SNP discovery in the White Rhino (*Ceratotherium simum*). *Conservation Genetics Resources*, Volume 5, page 265.

Lovrich, G. A., Roccatagliata, D., Peresan, L. (2004) Hyperparasitism of the cryptoniscid isopod *Liriopsis pygmaea* on the lithodid *Paralomis granulosa* from the Beagle Channel, *Argentina. Diseases of Aquatic Organisms*, Volume 58, page 71.

The 1,000 Genomes Project Consortium (2012) An integrated map of genetic variation from 1,092 human genomes. *Nature*, Volume 491, page 56.

Toh, K. L., Jones, C. R., He, Y., Eide, E. J., Hinz, W. A., Virshup, D. M. et al. (2001) An hPer2 phosphorylation site mutation in familial Advanced Sleep Phase Syndrome. *Science*, Volume 291, page 1040.

Yang, H-P., Tanikawa, A. Y., Kondrashov, A. S. (2001) Molecular nature of 11 spontaneous *de novo* mutations in *Drosophila melanogaster. Genetics*, Volume 157, page 1285.

HEROES OF THE GENETIC REVOLUTION

3

# James Crow

*"I think the best way to describe what I have done is that there hasn't been any central theme. I have jumped from one subject to another without very much of a coherent pattern to it. I don't think that is the best way to get ahead. But for me it worked fine and it's made for a much richer life than if I'd stuck with one problem and worked at it continuously"*

'Interview with Professor James Crow' (2006) *BioEssays*, Volume 28, page 660.

**EXAMPLE PUBLICATIONS**

Crow, J. F., Kimura, M. (1970) *Introduction to Population Genetics Theory*. (Harper and Row).

Crow, J. F. (2000) The origins, patterns and implications of human spontaneous mutation. *Nature Reviews Genetic*, Volume 1, page 40.

**Figure 3.10** James Crow (1916–2012). Music was an important part of Jim's life (indeed he met his wife Ann when they both played in the student orchestra at the University of Texas). He was a member of the Madison Symphony and gave a viola recital to the Genetics Department on his 90th birthday.

Credit: Craig Schreiner–Wisconsin State Journal.

When I was a graduate student, I had the remarkably good fortune to have dinner with Jim Crow and John Maynard Smith (in a chalet in the Swiss Alps, as it happens) and the subject of conversation turned, naturally, to fitness. Both Jim and John admitted that, if you counted grandchildren, neither was particularly impressive in terms of evolutionary fitness. But then we defined 'academic fitness' as the number of full professors that were descended from an academic supervisor—and, by this measure, it would be hard to find someone with greater fitness than Jim Crow (**Figure 3.10**). I was amazed by how many influential evolutionary biologists we could mention and Jim would say 'yes, he was one of mine too', including many scientists mentioned in this book such as Motoo Kimura (Chapter 8) and Joe Felsenstein (**Hero 11**). Through his memorable undergraduate lectures, textbooks and research supervision, Crow's contribution to teaching evolutionary biology was almost as phenomenal as his contribution to research over seven decades.

Born in 1916, Crow 'participated in the 1918 flu epidemic but survived it'. He went to the University of Texas to study with the pioneering geneticist H. J. Muller (after whom 'Muller's ratchet' is named). But Muller had gone to Russia, so Crow undertook a *Drosophila* genetics project focusing on mechanisms that caused reproductive isolation, and took graduate courses in maths and statistics. He wanted to move on to postdoctoral research with Sewall Wright (**Figure 7.5**), but the Second World War got in the way, so he went to Dartmouth College in New Hampshire where, as the faculty was thinned out by military service, he taught an ever increasing range of teaching subjects, including parasitology, haematology, and even navigation—subjects considered sufficiently critical to the war effort that he was not called for military service. His lectures at the University of Wisconsin-Madison, where he moved in 1948, were legendary and fondly remembered for decades afterwards by generations of students. His printed lecture notes took on a life of their own as 'Crow's Notes', and his pioneering textbook *An Introduction to Population Genetics Theory* (1970), written with Motoo Kimura, made a lasting impression on a whole generation of biologists (see **Hero 11**).

Crow's research career spanned many of the most important developments in genetics. He knew the great men who forged population genetics, having bonded with R. A. Fisher (**Figure 7.16**) over a late-night bottle of champagne, acted as conduit of ideas between Fisher and Wright, and discussed genetic load with J. B. S. Haldane (**Hero 7**). Crow helped shape the growing field of theoretical population genetics, examining how the process of substitution is influenced by the mutation rate, sexual reproduction and effective population size. He developed the concept of genetic load, not only as a theoretical tool but also as a practical concern, writing on the costs of mutation and inbreeding to human population health, sitting on national committees that reviewed the safety of radiation, and applying his knowledge to the practical problem of insecticide (DDT) resistance. He also made important contributions to recording the history of the exciting phase of the development of genetics that he had lived through, for example by commissioning a fascinating series of 'perspectives' articles in the journal *Genetics*.

TECHBOX **3.1**

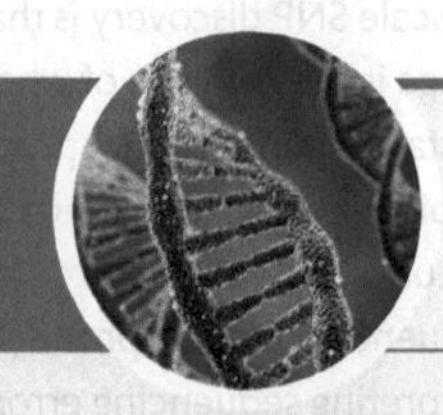

# Single nucleotide polymorphisms (SNPs)

**RELATED TECHBOXES**

**TB 3.2:** Biobanking

**TB 5.2:** High-throughput sequencing

**RELATED CASE STUDIES**

**CS 3.1:** SNPs

**CS 3.2:** Mutation rate

**Summary**

Sites in the genome that differ in base sequence between individuals can be detected by a growing number of techniques.

**Keywords**

genotype, HapMap, disease genes, genetic diversity, heteroduplex, microarray, genotype, reduced representation (RR), RADseq, marker-based breeding, pipeline, genotypings

Any two individuals from a given species will have identical DNA sequences across most of the genome. But, with the possible exception of clones (potentially including some identical twins: **Figure 10.18**), any two individuals will have occasional differences in their DNA sequences. In Chapter 2, we saw that the word 'genotype' was invented to describe the genetic information that is passed from parent to offspring. Here, just to confuse you, we are going to meet a slightly different use of the word: the genotype of an individual is the particular set of genetic variants that they have in their genome that distinguishes them from other members of the population. Genotype is also increasing used as a verb to describe any process that leads to the identification of the genetic differences between individuals. Identifying the sites that differ between individuals is not only used for genotyping, but is also important for investigating genetically determined traits, and studying population dynamics. The differences in the DNA sequence between individuals in a population are often referred to as single nucleotide polymorphisms (SNPs).

**Detecting SNPs**

There are a large number of methods for detecting positions in the genome where the sequence can differ between individuals (SNPs), and new methods are being constantly devised. It would be impractical and pointless to attempt to list the methods here, but it is possible to consider a few of the general approaches to finding SNPs and gauging their diversity. It will be impossible to discuss SNPs without mentioning ideas and techniques that we won't discuss in detail until later chapters; you can choose to skip ahead and read the relevant sections, or just skate over these concepts knowing you will meet them again soon.

### Comparing aligned nucleotide sequences

The most obvious way to detect SNPs is to sequence the same parts of the genome in different individuals and then compare the sequences and see where they differ. The problem is that you need to find parts of the genome that are similar enough that you know you are comparing homologous sequences (all descendants of the same ancestral sequence), yet variable enough that you can find sufficient sites that vary between individuals.

One approach is to design a useful set of primers that match conserved gene sequences that adjoin variable regions: for example, the primer matches a conserved exon (protein-coding sequence) which is the same in all individuals, but the DNA amplification product includes the adjacent SNP-containing intron (non-protein-coding sequence: **TechBox 6.2**). Sequencing only part of the genome can be a hit-and-miss strategy, as you may fail to find informative polymorphic sites, so primers have to be tested on each population of interest to establish that they can reveal useful variation (**Figure 3.4**).

Now that it is feasible to sequence individual genome sequences, SNP discovery does not need to be tied to specific loci. The advantage of genome-scale SNP discovery is that it can provide thousands of SNPs from a single analysis. The challenge is that analysis of whole genome sequences is dependent on the reliability of a multi-stage bioinformatics processing of the raw sequence data. This process is often referred to as a 'pipeline' because the data must be passed through sequential filtering stages to remove false or unreliable SNPs (**Case Study 3.1**). This is very important to remember: high-throughput sequencing is error-prone (**TechBox 5.2**), and you don't want to waste your precious time interpreting sequencing errors as biologically meaningful variation.

Detecting SNPs by sequencing could follow a procedure something like this: first, DNA samples are sequenced using a high-throughput method (**Techbox 5.1**), creating a set of 'reads', or short nucleotide sequences. Read length varies between sequencing platforms but is typically a few hundred bases. Sequencing platforms also differ in the rate and type of errors produced, which must be taken into account when assessing which sites have meaningful sequence differences. Determining the sequences of these fragments is sometimes referred to as 'base calling'. There will be a lot of sequencing errors in the initial reads, so the data must be cleaned up to avoid these errors being interpreted as SNPs. For example, the ends of reads may be trimmed if they have a high level of errors. The best way to reduce error rates is to have deep coverage, so that each part of the genome is sequenced many times. But greater coverage takes more time and money, two things that are often in limited supply.

To detect SNPs, you need to compare the bases in one sequence to the bases in another homologous sequence. In other words, you need to know that you are comparing sequences that are nearly identical because they were both copied from the same ancestral sequence, so any differences between them were acquired since they last shared a common ancestor. Homology is established by alignment (see **TechBox 10.2**). For species with a fully assembled genome sequence, reads will usually be aligned against a reference sequence (or against an assembled genome from a related species: **Case Study 3.1**). Alternatively, matching fragments of the genome can be aligned together. This approach can be made more tractable by focusing on a subset of genome, referred to as reduced-representation sequencing (see **TechBox 5.2**).

Once you have aligned sequences you can identify the sites at which the sequences vary in their base sequence, a procedure known as SNP calling. This step requires some quality control techniques to prevent errors or misalignments being called as polymorphic sites. Genotype calling is the process of inferring the specific set of alleles a particular individual has at those variable loci. This may be in the form of a statistical statement that reflects the confidence that we can assign a particular genotype based on the data that we have, taking into account the likely error rates[4]. Since closely linked alleles will typically be inherited together, patterns of linkage disequilibrium (a measure of the extent to which alleles co-occur rather than being inherited independently) can be used to improve genotype calling.

It is important to recognize that all SNP detection methods are error-prone, and that every stage of detection and analysis can produce both false positives (calling an error a SNP) and false negatives (failing to recognize a real SNP). Error rates for some data or methods can be very high, so it is important to consider SNP identification and SNP-based genotyping in light of these uncertainties, and not treat your results as error-free facts.

### *Hybridization to sequence arrays*

If you know what your target sequence is you can design an array of single-stranded DNA sequences that differ by just one nucleotide. When a DNA fragment in your sample pairs with a particular sequence in the array you will know what the base sequence of your sample is without actually having to sequence it. So you take the PCR product from your DNA sample, label it with some kind of reporter, and hybridize it to the array of sequences so that your sample sticks to the matching sequence on the array. One way to do this is to position the sequences on a chip (referred to as a microarray or 'SNP chip') which can carry tens of thousands of alternative versions of the sequence, corresponding to all possible variants for part of the genome. By labelling your PCR product with a fluorescent marker you will be able to tell which sequence on the array it hybridizes to. Alternatively, the array can be attached to tiny beads only micrometres in diameter, each of which carries one specific sequence and is coded using varying amounts of two coloured dyes. After the sample DNA is hybridized to the beads, they are passed through a flow cytometer, which is able to separate the beads by colour and indicate the amount of DNA hybridized to each kind.

### *Detecting mismatches in mixed samples*

An alternative approach takes advantage of the fact that mismatched DNA causes a distortion in the helix. DNA from different individuals can be hybridized together so that a sample contains DNA molecules that have one strand from one individual and one from another (termed heteroduplex DNA). Where the sequences are the same, the two strands will fit together snugly. But where the sequences from the two individuals differ, the two strands pair, so there will be a 'bubble' between the strands at that point, which may destabilize the surrounding helix (see Chapter 4). There are several ways of detecting these mismatches in heteroduplex DNA most of which rely on mismatched DNA having a lower melting temperature (takes less heat to break the mismatched strands apart) or slower movement down a gradient (such as a gel or a chromatography column) due to the distorted shape of the molecule[1]. Alternatively, heteroduplex methods can provide a practical way to generate SNPs by using enzymes that cut DNA at heteroduplexes[2] (**Figure 3.11**, **Figure 3.12**). Or, you can take advantage of the

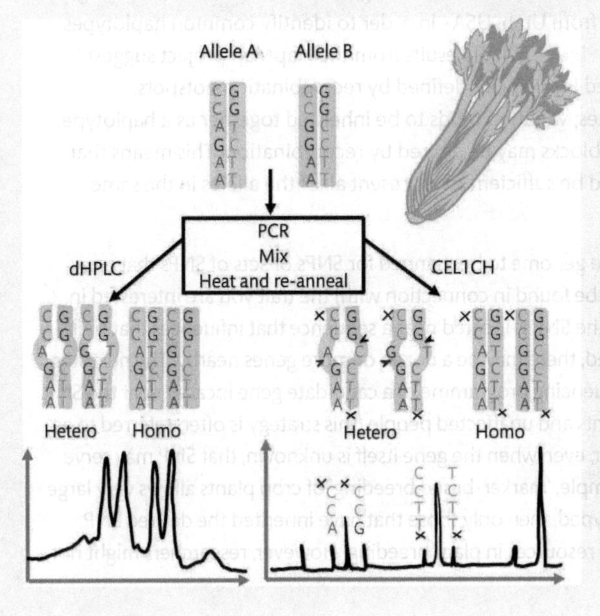


**Figure 3.11** SNP detection from heteroduplex DNA using TILLING (Targeting Induced Local Lesions IN Genomes). Amplified DNA from two sources is mixed together in equal proportions, then melted and reannealed so that both homoduplex (two identical strands joined together in a helix) and heteroduplex (joining two strands that differ in sequence) helices form (**TechBox 11.1**). The heteroduplex DNA, which contain an allele at which the sequence differs between strands, can be separated by dHPLC (Denaturing High Pressure Liquid Chromatography) because the melting temperature of the DNA fragments determines the movement through the column. Alternatively, the SNPs can be detected using enzymes that recognize mismatches and cut the DNA, creating strands of different lengths (in this case, using the enzyme CEL1 which can be obtained from celery stalks).

Redrawn from Fusari, C. M. et al. (2011) *Molecular Breeding*, Volume 28, pages 73–89, with kind permission from Springer Science and Business Media.

**Figure 3.12** EcoTILLING: Enzymes that cut DNA at hereroduplex DNA have been used as a cost-effective way to survey SNPs in sunflowers (*Helianthus annuus*), with the aim of extending the reach of genetic analysis to a wider range of populations and land races.



fact that DNA with mismatched bases will have a different melting temperature to separate out the heteroduplex DNA on a chromatography gradient.

## Analysing SNPs

### *Identifying genes of interest*

Some SNPs will have a direct effect on a particular trait of interest, because the nucleotide sequence changes a gene product or expression pattern. But since closely linked SNPs tend to be inherited together in haplotypes, a SNP does not have to play a causal role in a disease or trait to be informative. If a particular SNP is located near the allele of interest, then it will tend to be inherited along with that allele. If we are lucky, then a causal difference in the gene (e.g. an allele associated with a particular disease) will just happen to be connected to an identifiable SNP nearby. Then if we find someone has that SNP, there is a good chance they also have the allele we want to study. It is this incidental association between the SNP and the trait that allows the position of the gene to be deduced. For this reason, one of the earliest SNP databases, the human HapMap, aimed to identify a comprehensive and useful set of genomic markers that any researcher could use to screen human genetic data (**TechBox 3.2**).

Human populations differ in the presence or frequency of many haplotypes, so while some SNPs are found in many ethnic groups, others are specific to particular populations. The HapMap international consortium analysed DNA from people from distinct populations—including Japanese from Tokyo, Yoruba from Ibadan in Nigeria, Han Chinese from Beijing, and people of European descent from Utah, USA—in order to identify common haplotypes found in all human populations (**Figure 3.13**). Results from the HapMap project suggest that the human genome is divided into blocks, defined by recombination hotspots. Between these recombination sites, variation tends to be inherited together as a haplotype, but the association between the blocks may be altered by recombination. This means that one good informative SNP should be sufficient to represent all of the alleles in the same block.

Genome-wide coverage allows the genome to be scanned for SNPs or sets of SNPs that segregate with (i.e. tend always to be found in connection with) the trait you are interested in. There is then a good chance that the SNP is located near a sequence that influences that trait. Once an informative SNP is located, there may be a dozen or more genes nearby. The next step can be to do a more targeted sequencing programme of a candidate gene located near the SNP, from a large sample of both patients and unaffected people (this strategy is often referred to as 'medical resequencing')[3]. However, even when the gene itself is unknown, that SNP may serve as an informative marker. For example, 'marker-based breeding' of crop plants allows very large numbers of seedlings to be genotyped, then only those that have inherited the desired SNP grown up, saving a lot of time and resources in plant breeding. However, researchers might not

**Figure 3.13** A Yoruba man and his daughter. Yoruba people from Ibadan in Nigeria are one of four source populations for the human HapMap database which aims to develop a set of genome markers that can be used in genomic research. Unlike biobanks (**Techbox 3.2**), no data on individuals was collected with the samples, so the data from HapMap cannot in itself be used to identify disease genes[4].

Photograph by Marcel Cardillo.

find a perfect link, such that having the SNP guarantees you have the trait. Instead, most studies report a statistical association between a SNP haplotype and a particular outcome.

### *Population genetics with SNPs*

Interbreeding populations share a common pool of alleles. So in some cases SNPs can be identified that are characteristic of a particular population. So SNPs are increasingly being used to identify the population of origin of biological material, for example, in wildlife forensics (**Figure 3.4, TechBox 1.2**). This approach relies on accurate genotyping, and having an informative reference collection against which to compare your sample. But the use of SNPs is moving beyond sample identification and being used to evaluate population structure or effective population size, assess levels of genetic variation, and detecting alleles under selection. Studies of population variation have traditionally relied on allozymes or microsatellite markers, because they tend to be highly variable within populations and can be relatively simple to measure (Chapter 5). SNPs may provide many more polymorphic sites (potentially tens of thousands), but each variable locus has fewer variants (most SNPs are biallelic, meaning only two alternative sequences are found in the population). Genome sequencing has typically been used to sequence large amounts of sequence from one or few individuals, but informative population genetic analysis requires representative samples from as many individuals as possible. It is important to avoid ascertainment bias, where a biased sample of individuals gives you an inaccurate picture of the variation present in the population. Don't be blown away by the wonderfully large numbers involved in SNP studies—the power of any given analysis to detect the processes of interest should be formally assessed, as for any data, to make sure that your study design will allow you to see what you are looking for and not lead you astray into misleading patterns.

### References

1. Morin, P. A., Luikart, G., Wayne, R. K. (2004) SNPs in ecology, evolution and conservation. *Trends in Ecology & Evolution*, Volume 19, page 208.
2. Fusari, C. M., Lia, V. V., Nishinakamasu V., Zubrzycki, J. E., Puebla, A. F, Maligne, A. E, et al. (2011) Single nucleotide polymorphism genotyping by heteroduplex analysis in sunflower (*Helianthus annuus* L.). *Molecular Breeding*, Volume 28, page 73.
3. Ogden, R. O. B. (2011) Unlocking the potential of genomic technologies for wildlife forensics. *Molecular Ecology Resources*, Volume 11, page 109.
4. Nielsen, R., Paul, J. S., Albrechtsen, A., Song, Y. S. (2011) Genotype and SNP calling from next-generation sequencing data. *Nature Reviews Genetics*, Volume 12, page 443.

TECHBOX 3.2

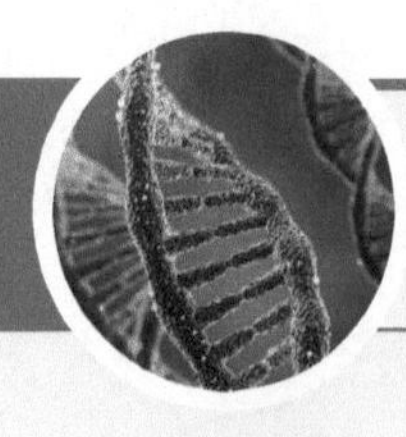

# Biobanking

RELATED TECHBOXES

**TB 1.1:** GenBank

**TB 3.1:** Single nucleotide polymorphisms (SNPs)

RELATED CASE STUDIES

**CS 3.1:** SNPs

**CS 4.2:** Hierarchies


## Summary

Databases of genetic variation and individual information give unprecedented power to detect heritable traits.



## Keywords

haplotype, association studies, disease, ethics, diabetes, informative pedigree, informed consent, common allele


A biobank is any large collection of biological samples, usually linked to information about the individuals from which the samples were taken. The largest collections of human biological samples are those used in forensic identification of individuals (see Chapter 5), but the term 'biobank' is most commonly applied to samples collected for research into human health. These biobanks essentially act as large-scale extensions of traditional family association studies.

The classic way to identify genes that cause human diseases is to find an 'informative pedigree' (a family that shows a clear pattern of inheritance of the disease), then try to establish which parts of the genome are shared by affected family members, but not those unaffected. The gene for Huntington's disease, an early success story of genetic mapping, is a good example of a disease suited to association studies: it is caused by a single gene, has high penetrance (all people with the disease allele will develop the disease), is genetically dominant (a single copy of the disease allele will give rise to the disease, so half of the offspring of an affected person are also expected to be affected by the disease), and with large and well-described pedigrees (particularly because Huntington's disease is usually expressed later in life, so many carriers have already had children: **Figure 5.16**). Association studies have revealed the genetic basis of many single-gene disorders with clearly defined pedigrees. But they are limited in power for complex diseases that may be influenced by many factors, both genetic and environmental. For example, heart disease has a strong genetic component (it runs in families) but it is clearly influenced by many different genes and environmental factors, so has incomplete penetrance (e.g. even in the presence of inherited risk factors, the chance of developing heart disease is influenced by diet and exercise).

Instead of identifying informative pedigrees from particular families, biobanks contain a large number of samples from a human population, so statistical analyses can be used to identify genomic markers that are significantly associated with specific diseases. This approach does not work well for rare diseases: even in a large sample, there will be too few affected people to allow association with genetic markers to be detected. Instead, biobanks focus on 'common alleles', typically those present in five per cent or more of the population. Therefore, biobanks work best for identifying common diseases that are influenced by variation in many genes, rather than for searching for single mutations that cause rare diseases. These complex diseases are often described by the phrase 'common disease, common variant': it may be the combination of many different commonly occurring factors, both genetic and environmental, that leads some people to be affected by a certain disease.

Biobanks target multi-factorial diseases by connecting individual genetic information with health data, then looking for a statistically significant association between certain genotypes and particular health outcomes. For example, one of the pioneering biobanks, Iceland's deCODE database, published a study identifying a genetic risk factor for Type 2 diabetes (**Figure 3.14**). They compared the frequency of genetic markers in 1185 people affected by Type 2 diabetes and 931 unaffected people. They found that a particular haplotype was over-represented in

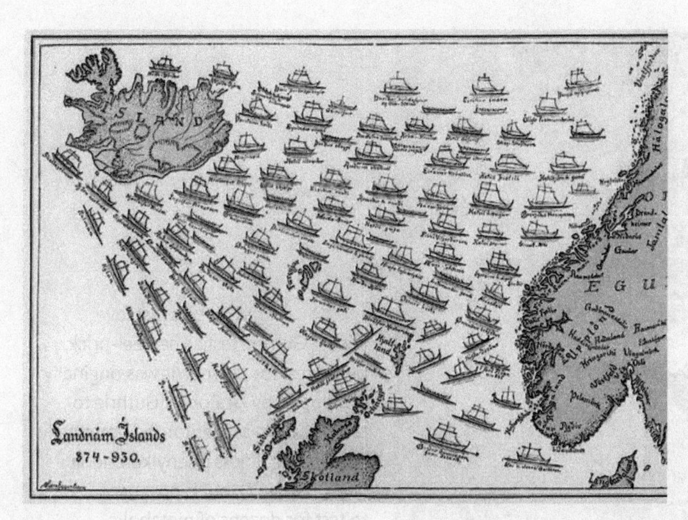


**Figure 3.14** The well-documented history of the Icelanders and their relatively genetically homogeneous population has made them an excellent population for the hunt for genes involved in common human diseases.

Reproduced with kind permission of the National Library and University of Iceland, Reykjavik.

the affected sample (and a different haplotype that was under-represented in people with diabetes)[1]. However, the frequencies illustrate how these alleles play only a small part of the risk of developing diabetes: only one third of the affected patients carried the at-risk haplotype, and one quarter of the unaffected people also carried the at-risk haplotype. There are many common alleles associated with increased risk of Type 2 diabetes, but the disease also has strong environmental risk factors, particularly related to diet and exercise. Type 2 diabetes has increased rapidly in many countries within a single generation, so this increase cannot be due to genetic factors alone.

Many countries have established biobanks as a valuable research tool in medical genetics. For example, the UK Biobank, funded by government and medical research funding bodies, recruited half a million people to provide samples and health information. Most of the current research infers heritability from health and lifestyle information provided with the samples: for example, analysis of the database records of nearly a quarter of a million people suggests a link between father's Type II diabetes and low birthweight in their children[2]. But genetic material stored in the UK Biobank may also be used in research. Not all biobanks are formed for medical research. For example, the privately funded Genographic project is building a global database of human genetic variation, partly by selling kits that allow individuals to take a cheek swab and post it to a laboratory for sequencing. Many private 'ancestry' companies use DNA samples to provide individuals with a statement linking their genotype to different populations around the world and identifying possible genetic risk factors for common diseases.

The establishment of biobanks has not been without controversy. For example, a proposed national biobank in the Kingdom of Tonga was abandoned due to strong opposition to the government granting exclusive rights to residents' DNA to a private company. Some people from indigenous populations fear that, by donating their DNA to a biobank, people may be giving away a valuable resource: the genetic material of their particular population, which may contain medically valuable variations. To some extent, this issue of shared genetic material affects all donors to biobanks, because any individual is likely to share genomic variants with their relatives, who may not have consented to the donation.

Some of the ethical issues raised by large-scale biobanks are[3]:

- **Consent:** The rapid pace of advance in the field of genomics poses problems for the informed consent of donors, when the use of the sample may change as technology and scientific knowledge develop. Therefore, it is not possible to explain to a donor the uses to which their sample will be put in the future. The problem of informed consent for future use

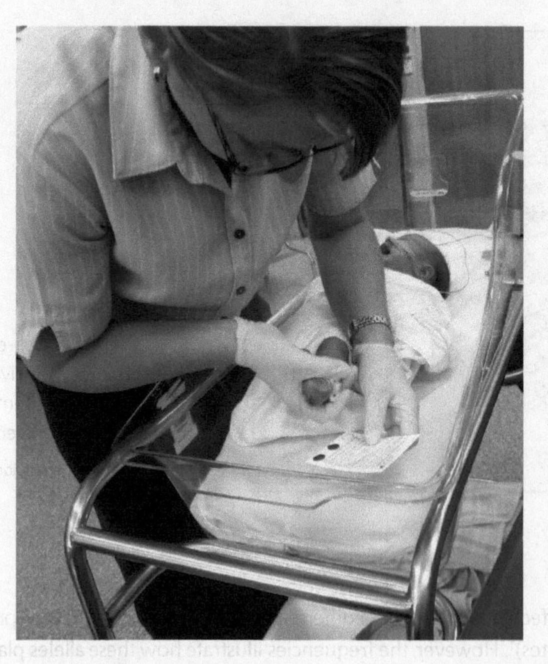

**Figure 3.15** Babies just love having Guthrie tests. The heel-prick blood test for newborns was originally developed by Dr Robert Guthrie to test newborn babies for the potentially debilitating illness phenylketonuria (see Chapter 7), but is now used to test for dozens of metabolic illnesses, allowing early detection and treatment. DNA can also be extracted from the dried blood spots on the cards.
Photograph by Lindell Bromham.

is also raised by existing collections of biological samples. For example, most people born in Australia in the last three decades have had a blood sample taken and stored (potentially indefinitely) on Guthrie cards (**Figure 3.15**). These cards have been accessed by medical researchers[4] and the police[5] without the knowledge or consent of the children or their parents. Should a donor have the right, at any point in the future, to request information on the use of their sample or to demand the samples be destroyed? Or should 'informed consent' be assumed to cover all unknown future uses? Is it reasonable for parents to give access to their children's genetic information?

- **Information:** If an individual has their genome sequenced for any reason, say to establish ancestry or identify particular alleles, should they be told if it also contains alleles that are associated with other diseases? Reporting of such 'incidental' findings could have both positive and negative outcomes. For example, if a patient who gets their genome sequenced to assess risk factors for diabetes is told they also carry an allele of the BRCA1 gene that is associated with a high risk of breast cancer, they might reduce their cancer risk by regular screening, but they might also suffer significant anxiety and face ethical issues about informing family members who might have also inherited that allele.
- **Relatives:** By definition, you share many of your alleles with your biological relatives. If you provide your genetic information to a biobank, then you are indirectly contributing information about your relatives too. Some commercial 'ancestry' genomics companies now allow you to search for closely related genetic profiles, and this has led to people discovering relatives they didn't know they had. Biobanks are now being used by adopted people to trace their biological relatives. But many people using the commercial services will be unaware of the consequences of providing their genetic information, which may reveal family history that had been hidden[6]. Forensic databases have even been used to search for the relatives of suspects: for example, police in Canberra, Australia, reported that samples found at the scene of a minor property crime were likely to be from the son of a suspect in a murder case ten years before. They reopened the murder case by looking for the teenage perpetrators of the property crime, in the hope they may lead them to the murder suspect[7], though further analysis ruled out a link between the two crimes[8].

- **Profit:** Some biobanks are either wholly or partly owned by pharmacogenomics companies, who are investing in the future rights to marketable medicines arising from research on the biobank. There may be tension between the potential public health benefits of biobank discoveries and private-company profits. A compromise strategy might be profit-sharing, but who should benefit: the whole community or only those who donated to the biobank? The ownership and accessibility of the data impacts on this debate, as public benefit is best served by researchers having open access to biobank data, but commercial gain relies on at least some degree of non-disclosure or data protection.
- **Anonymity:** Biobanks differ in the degree to which samples and information can be identified to specific individuals. Some are encrypted, so that only a limited number of people can connect individuals to their samples and data. Others are anonymized, so that individual identity is stripped from the data and samples (in which case, it would be no longer possible for the individual to exert rights over their biobank entry). Studies have shown that analysis of published sequence data can re-identify 'de-identified' samples[9], which might then make aspects of medical records or genetic risk factors available to employers or health insurance companies. For example, even though no information about donors is given in HapMap, it seems possible that relatively small amounts of genetic information can be used to identify an individual on a database, then gain the entire publicly available genomic information for that individual[10]. Because of the complex issues surrounding public use of genetic data, the HapMap project (**Figure 3.13**) involved substantial input from ethicists at every stage of its planning and enactment[11].

## References


1. Grant, S. F., Thorleifsson, G., Reynisdottir, I., Benediktsson, R., Manolescu, A., Sainz, J., Helgason, A., Stefansson, H. , Emilsson, V. , Helgadottir, A. (2006) Variant of transcription factor 7-like 2 (TCF7L2) gene confers risk of type 2 diabetes. *Nature Genetics*, Volume 38, page 320.
2. Tyrrell, J.S., Yaghootkar, H., Freathy, R. M, Hattersley, A. T., Frayling, T. M. (2013) Parental diabetes and birthweight in 236 030 individuals in the UK Biobank Study. *International Journal of Epidemiology*, Volume 42, page 1714.
3. Cambon-Thomsen, A. (2004) The social and ethical issues of post-genomic human biobanks. *Nature Reviews Genetics*, Volume 8, page 866.
4. Barnes, G., Srivastava, A., Carlin, J., Francis, I. (2003) Delta-F508 cystic fibrosis mutation is not linked to intussusception: Implications for rotavirus vaccine. *Journal of Paediatrics and Child Health*, Volume 39, page 516.
5. Phillips, G. (2003) Guthrie cards. Transcript of Catalyst, ABC television http://www.abc.net.au/catalyst/stories/s867619.htm.
6. Belluz, J. (2014) Genetic testing brings families together–And sometimes tears them apart. *Vox* 18 December 2014: http://www.vox.com/2014/9/9/6107039/23andme-ancestry-dna-testing.
7. Pianegonda, E., Osborne, T. (2014) 'Father-son DNA link' between 1999 Irma Palasics killing and 2010 Woden break-in. *ABC News Online*, 3rd November 2014.
8. Gorrey, M. (2015) Police rule out family DNA link to Irma Palasics' cold case murder. *Canberra Times*, July 14, 2015.
9. Gymrek, M., McGuire, A. L., Golan, D., Halperin, E., Erlich, Y. (2013) Identifying personal genomes by surname inference. *Science*, Volume 339, page 321.
10. Lin, Z., Owen, A. B., Altman, R. B. (2004) Genomic Research and Human Subject Privacy. *Science*, Volume 305, page 183.
11. Foster, M. W. (2004) Integrating ethics and science in the International HapMap Project. *Nature Reviews Genetics*, Volume 5, page 467.

CASE STUDY 3.1

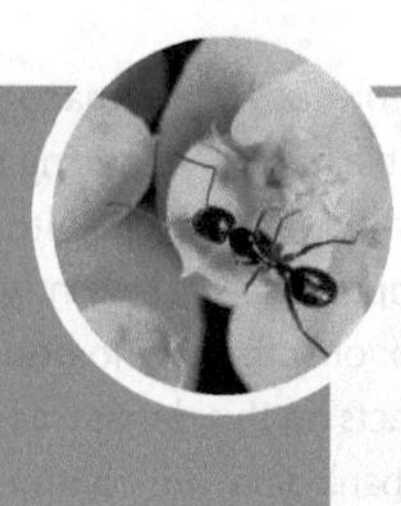

# SNPs: genetic resources for crop development in a non-model species

RELATED TECHBOXES

TB 3.1: Single nucleotide polymorphisms (SNPs)

TB 5.2: High-throughput sequencing

RELATED CASE STUDIES

CS 1.2: Forensics

CS 12.2: Prediction

Lu, F., Lipka, A.E., Glaubitz, J, Elshire, R., Cherney, J. H., Casler, M. D., Buckler, E. S., Costich, D. E. (2013) Switchgrass genomic diversity, ploidy, and evolution: novel insights from a network-based SNP discovery protocol. *PLoS Genetics*, Volume 9, page e1003215.

> *This is only the beginning: we believe [this approach] offers the key to the exploration and exploitation of the genetic diversity of thousands of non-model species*

### Keywords

high-throughput sequencing, restriction enzyme, reduced representation library (RRL), genome-wide association studies (GWAS)

### Background

The world needs a greater diversity of crop species, not only to feed a growing population but also to provide biofuels without reducing edible harvests. Switchgrass (*Panicum virgatum*) is a North American grass species being developed in agriculture, including as a biofuel crop, because it can provide high biomass yield on marginal land that would be less valuable for food plants (**Figure 3.16**). But, as a 'non-model' species, it lacks the genetic and breeding resources of the major crop plants. There is no reference sequence for its large and complex genome and it is highly genetically variable. This research team set out to create a reference set of SNPs to aid the development of switchgrass as a biofuel crop.

### Aim

One of the challenges for SNP calling without a reference genome is filtering out paralogs: if a read is paired not to a homologous sequence but to a similar duplicated sequence elsewhere in

**Figure 3.16** Switchgrass (*Panicum virgatum*), a common prairie grass in North America, has a number of practical uses including forage, soil conservation, carbon dioxide sequestration and biofuel production.

Photograph by Lynn Betts, USDA Natural Resources Conservation Service.

the genome, then any base mismatches are 'false SNPs' because they don't represent alternative alleles of the same gene. The 'genotyping by sequencing' approach used here is a reduced-representation approach to high-throughput sequencing which aims to target a tractable and informative proportion of the genome, avoiding highly repetitive DNA and preferentially sequencing low-copy number sequences[1].

### Methods

The genome was sheared with a restriction enzyme that cuts DNA at a specific recognition sequence: in this case, the *ApeKI* enzyme which cuts at the sequence CGWGC, where W could be an A or a T. This short sequence will occur throughout the genome, so this approach cuts the genome at a large number of sites. Barcode sequences were added to the 'sticky ends' of DNA created by the cuts to allow fragments close to the restriction site to be isolated and sequenced (the same approach used in RADseq[2]). High-throughput sequencing of this reduced set of genomic fragments produced reads 64 bases long (after trimming the ends). Matched fragments with the same sequence are referred to as 'tags'. Pairs of tags that differ only in one base were used to identify potential SNPs, but any rare pairs that occur below a threshold frequency were rejected as potential sequencing errors. Network analysis was used to identify sets of reciprocal tag pairs, which should bias the dataset towards SNPs with high sequencing coverage, reducing the likelihood of sequencing errors being mistaken for SNPs. Expected allele frequencies were used as an additional filter, because true SNPs should typically conform to the frequencies expected from a segregating Mendelian marker, so departures from these ratios might indicate paralogs or sequencing errors.

### Results

Over one million putative SNPs were identified in switchgrass. Sequencing coverage, and therefore SNP reliability, varied. Average sequencing coverage was low (around 0.5×), but there were some SNPs with coverage greater than 6×. Computational approaches to producing a linkage maps of SNPs failed, so the researchers used the assembled genome of foxtail millet (*Setaria italica*) to align the SNPs into linkage groups, which were used to map 7,245 SNPs to chromosomes. These linkage groups were then used to place 88,217 SNPs on a high-density map.

### Conclusions

The analysis pipeline developed by this team aligns genome-sequencing reads against each other, rather than against a reference genome, so can technically be applied to any species. There are several filtering stages which are designed to reduce the number of false SNPs. This platform (which they call Universal Network-Enabled Analysis Kit, or UNEAK) has been adopted for other model organisms, for example to characterize diversity in hops plants[3] (**Figure 3.17**).

**Figure 3.17** Hops are the female flowers of the plant *Humulus lupulus* (a), used primarily to make beer. Hops are traditionally grown up long poles, as seen in this picture of a hops harvest in Slovenia in the 1960s (b). Hops are also a potent source of phytoestrogens, and women who harvested hops often experienced changed menstrual cycles as a result. Of course, what these people really need is SNP technology.

### Limitations

Although this method was designed to be applied to non-model organisms, they used the foxtail millet genome, a reasonably close relative, to produce a linkage map. It may be more difficult to generate a linkage map of SNPs for species that don't have a closely related reference genome. By using a targeted portion of the genome, this approach reduces sequencing costs, but the average coverage was still quite low: increasing the average coverage could increase the accuracy of SNP calling.

### Further work

Like the HapMap database (**Techbox 3.1**), the aim of this study was to generate a useful set of genetic markers. These markers could now be used in a targeted breeding programme, by identifying SNPs that segregate with desirable traits such as biomass production.

### Check your understanding

1. What effect did the restriction enzyme have on the genetic material? Would they have got a different result if they had used a different enzyme that cuts at a different recognition sequence?
2. What approaches were taken in this study to filter out sequencing errors? Have these filters eliminated all sequencing errors?
3. Why did they compare sequences to the foxtail millet genome?

### What do you think?

What advantages do SNP-based technologies have in breeding new crop plants?

### Delve deeper

This study used restriction enzymes to reduce the amount of the genome sequenced. What other approaches could be used? For marker assisted breeding, how could you ensure that your sequences cover the areas of the genome that are most likely to contain traits of interest?

### References


1. Elshire, R. J., Glaubitz, J. C. , Sun, Q., Poland, J.A., Kawamoto, K., Buckler, E. S., Mitchell, S. E. (2011) A Robust, Simple Genotyping-by-Sequencing (GBS) Approach for High Diversity Species. *PLoS ONE*, Volume 6: e19379.
2. Baird, N. A., Etter, P. D., Atwood, T. S., Currey, M. C., Shiver, A. L., Lewis, Z. A., Selker, E. U., Cresko, W. A., Johnson, E. A. (2008) Rapid SNP discovery and genetic mapping using sequenced RAD markers. *PLoS ONE*, Volume 3: e3376.
3. Matthews, P., Coles, M., Pitra, N. (2013) Next Generation Sequencing for a plant of great tradition: application of NGS to SNP detection and validation in hops (*Humulus lupulus* L.). *BrewingScience (Monatsschrift für Brauwissenschaft)*, Volume 66, page 186.

CASE STUDY 3.2

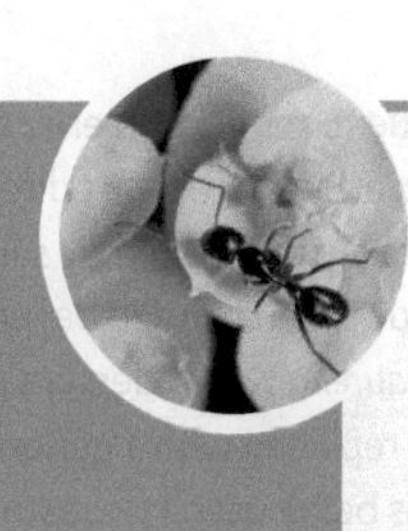

# Mutation rate: mutation accumulation in a laboratory population

**RELATED TECHBOXES**

TB 3.1: Single nucleotide polymorphisms (SNPs)

TB 4.1: DNA replication

**RELATED CASE STUDIES**

CS 4.1: Replication

CS 5.1: Duplication

Sung, W., Tucker, A. E., Doak, T. G., Choi, E., Thomas, W. K., Lynch, M. (2012) Extraordinary genome stability in the ciliate *Paramecium tetraurelia*. *Proceedings of the National Academy of Sciences USA*, Volume 109, page 19339.

*Should future work with additional key taxa uphold the inverse scaling between μ [mutation rate] and $N_eP$ [effective population size and proteome size], we will have achieved a fairly general theory for one of the key evolutionary-genetic features of species*[1]

**Keywords**

mutation accumulation, germline, *Paramecium*, SNP, copy fidelity, effective population size, selection

**Background**

DNA is copied with a high degree of fidelity in all organisms, but the number of mistakes made for every base copied varies a lot between species—for example the per-base error rate in mammals is a million times lower than the per-base error rate in HIV. It has been suggested that genome size drives these differences in error rate: organisms with large genomes need lower per-base error rates to achieve the same per-genome error rate as an organism with a small genome, so that the mutation rate per genome replication is much the same for many very different microbes (a pattern known as Drake's rule[2]). It's tricky to evaluate this pattern in multicellular organisms: because the genome goes through many copies per generation as the cells in the germline divide, the mutation rate per generation is much higher in animals than in unicellular microbes, but the per-replication mutation rate is relatively low.

**Aim**

*Paramecium tetraurelia* provides an interesting case study in the evolution of mutation rates because, in addition to the working genome held in a 'macronucleus', a quiescent, non-transcribed copy of the genome is stored in a 'micronucleus'[3] (**Figure 3.18**). Both genomes are

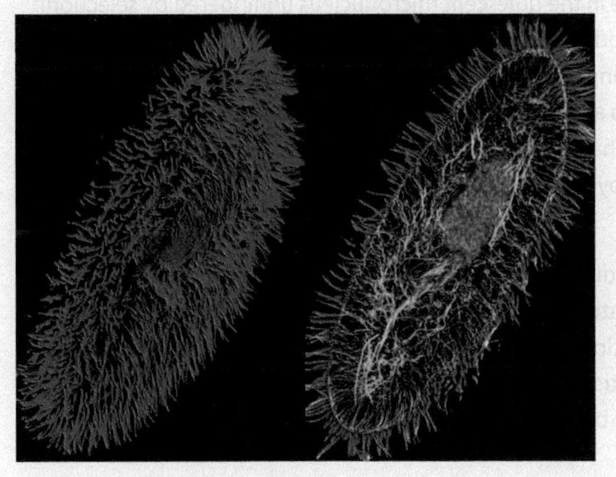

**Figure 3.18** Bat my lashes. Ciliates are single-celled organisms that are covered in a carpet of cilia, tiny projections used for swimming, sensing and feeding. *Paramecium* was one of the first microscopic organisms to be described using newly-invented microscopes in the 1600s and has long been a favourite for microbiological experiments. Most *Paramecium* eat other microscopic organisms—in this experiment, they were fed on a wheat grass medium containing *Klebsiella* bacteria—but some *Paramecium* species harbour endosymbiotic green algae. These cells have been stained by immunofluorescence methods to decorate the cilia, the cytoskeleton and the nuclei.

Biphotonic microscopy courtesy of A. Aubusson-Fleury.

copied every cell division, so both accrue mutations due to replication errors. But, because the genome in the micronucleus is not used to make products, it doesn't directly affect the phenotype. So mutations in the micronucleus are normally unaffected by selection: if genes aren't actually expressed, then wrecking them doesn't do any harm. However, after around 75 generations of cell division (fewer if starving), *P. tetraurelia* will undergo a process of 'self-fertilization', where the macronucleus is dissolved and replaced with a genome copied from the micronucleus. Now, the genome in the micronucleus is being used to construct the phenotype, so any mutations that accumulated in the micronucleus will be exposed to selection. This mutation accumulation experiment was designed to test whether this genome sequestration influences the evolution of mutation rates.

### Methods

The researchers kept 100 independent lines (separate populations) of *Paramecium tetraurelia* in the lab. Every day, for each of these 100 replicates, they would transfer a single cell to a fresh medium, where the cell would start dividing again and the next day one of the progeny cells would be sampled and so on for four years. As long as the cells had fresh medium every day, they would keep dividing asexually. Every three weeks cells from each line were transferred to a well (tiny tissue culture vessel), where, after several days, they began to starve, triggering the self-fertilization process whereby the quiescent micronucleus replaces the macronucleus[4]. Aceto-Carmine, which stains chromosomes, was used to detect the break-up of the old macronucleus, then a single cell was sampled and cultured as before. Half of these culture lines died out over the four-year experiment. Eight of the remaining lines were selected for genome sequencing. Sequences were aligned to the reference genome, and compared to a consensus sequence from the other experimental lines to identify 111 putative new mutations. After filtering for paralogy and sequencing errors, 29 mutations remained. To check how reliable the mutation calling procedure was, random samples of both identified mutations and filtered (false) mutations were confirmed by resequencing. They then calculated the mutation rate by dividing the observed number of mutations by the number of nucleotide sites analysed, multiplied by the estimated number of cell generations in the experiment.

### Results

Based on models developed in other unicellular organisms, the researchers predicted that, given its genome size, *Paramecium tetraurelia* would have around $1.5 \times 10^{-9}$ mutations per site per generation, in which case they should have seen hundreds of mutations over the course of the experiment. Instead, they detected relatively few mutations, and the estimated mutation rate was orders of magnitude lower than predicted, at around $2 \times 10^{-11}$ mutations per site per generation, one of the lowest mutation rates ever recorded (**Figure 3.19**).

### Conclusions

The authors interpret their surprisingly low mutation rate estimates in light of selection for lower mutation rates. Mutations accumulate in the quiescent micronucleus unhindered by selection. But when that micronucleus is activated at self-fertilization, all of those accumulated mutations can be expressed, so individuals carrying harmful mutations in their 'germline' genome will be less likely to reproduce and survive. To avoid being overwhelmed by mutations accumulated in the micronucleus, *Paramecium* might have evolved a greatly reduced mutation rate, for example through selection for greater DNA copy fidelity in polymerase enzymes.

### Limitations

As with all mutation accumulation studies, this experiment cannot detect all mutations that have occurred—for example, any mutation that was lethal could not have been passed on through multiple generations. In addition, the sequencing process is highly error prone, hence the need for filtering of putative mutations. If the filtering stage is too rigorous or too lax, the number of estimated mutations could be affected.

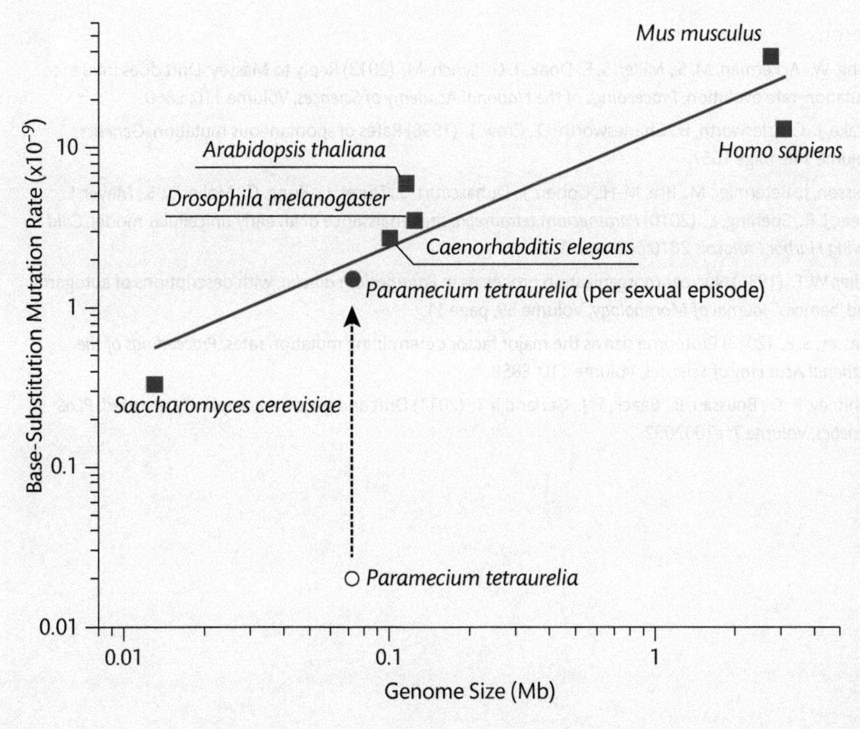


**Figure 3.19 Estimated mutation rate per generation** is lower for organisms with small genomes. The estimated mutation rate per cell generation for the unicellular *Paramecium tetraurelia* is astoundingly low, but when the mutation rate is calculated per sexual generation (from one autogamous reproduction to the next), the rate is not dramatically different to that of multicellular animals and plants.

Figure from Sung, W. et al. (2012) *Proceedings of the National Academy of Sciences*, Volume 109(47), pages19339–19344.

### Further work

Mutation rate estimates are critical to estimation of population size, so previous estimates of population size in ciliates might need to be re-evaluated in light of this new mutation rate estimate. More broadly, the authors interpret their results in light of a driving role for effective population size in mutation rate evolution. Species with smaller genomes tend to have larger effective population sizes, and therefore selection has more power to remove slightly deleterious mutations that increase mutation rates, and to promote mutations that cause slight improvements to DNA copy fidelity. This hypothesis has been challenged: for example, some researchers consider that the size of the 'proteome' (which determines the size of the mutational target) provides a better predictor of mutation rate[5], while others point out that analysis of mutation rates across species can be misleading if phylogenetic history is not taken into account[6].

### Check your understanding

1. Why do mutations accumulate more readily in the micronucleus than the macronucleus?
2. Why did they resequence some of the identified SNPs? Why did they include mutations that had been filtered out as potential errors in their resequencing procedure?
3. From an evolutionary point of view, in what way is the *Paramecium* micronucleus like the germline (sperm and egg) cells of an animal?

### What do you think?

Do you think the mutation identification procedure used will have overestimated, underestimated, or accurately estimated the mutation rate?

### Delve deeper

Can you extrapolate from these results to infer anything about evolution of mutation rates in multicellular organisms that maintain a separate germline? Would you expect the mutational dynamics in plants (that do not have a dedicated germline, since gametes can form on many parts of the body) to differ from animals (where the germline is sequestered such that only special cells give rise to gametes)?

## References

1. Sung, W., Ackerman, M. S., Miller, S. F., Doak, T. G., Lynch, M. (2013) Reply to Massey: Drift does influence mutation-rate evolution. *Proceedings of the National Academy of Sciences*, Volume 110: E860.
2. Drake, J., Charlesworth, B., Charlesworth, D., Crow, J. (1998) Rates of spontaneous mutation. *Genetics*, Volume 148, page 1667.
3. Beisson, J., Betermier, M., Bre, M-H., Cohen, J., Duharcourt, S., Duret, L., Kung, C., Malinsky, S., Meyer, E., Preer, J. R., Sperling, L. (2010) *Paramecium tetraurelia:* the renaissance of an early unicellular model. *Cold Spring Harbor Protocols 2010*:pdb.emo 140.
4. Diller, W. F. (1936) Nuclear reorganization processes in *Paramecium aurelia*, with descriptions of autogamy and 'hemixis'. *Journal of Morphology*, Volume 59, page 11.
5. Massey, S. E. (2013) Proteome size as the major factor determining mutation rates. *Proceedings of the National Academy of Sciences*, Volume 110: E858.
6. Whitney, K. D., Boussau, B., Baack, E. J., Garland,Jr T. (2011) Drift and genome complexity revisited. *PLoS genetics*, Volume 7: e1002092.

# Replication

## *Endless copies*

*"If the account of heredity… is correct, it follows that the whole pageant of evolution since pre-Cambrian times—ammonites, dinosaurs, pterodactyls, mammoths, and man himself—is merely a reflection of changed sequences of bases in nucleic acid molecules. What is transmitted from one generation to another is not the form and substance of a pterodactyl or a mammoth, but primarily the capacity to synthesize particular proteins. The development of specific form is a consequence of this capacity, and the capacity itself depends on the self-replication properties of DNA"*

Maynard Smith J. (1958) *The Theory of Evolution*. Penguin.

## What this chapter is about

The evolution of life depends on hereditary information being copied from one generation to the next. A basic grasp of DNA replication is essential for anyone wishing to understand evolution. Furthermore, familiarity with the processes of DNA replication is the key to understanding many molecular techniques. DNA amplification (making millions of copies of a DNA sequence in the laboratory) relies upon the domestication of the DNA copying processes that occur in living cells. Understanding DNA replication is also central to appreciating the nature of biological information stored in DNA. DNA replication creates a nested hierarchy of differences between genomes that reveals the relationships between organisms and the processes of evolution.

### Key concepts

- DNA replication provides a near-perfect copy of the DNA sequence of the genome
- DNA amplification uses DNA replication to create large numbers of copies of DNA sequences
- DNA replication generates a hierarchy of similarities that allows us to trace the history of descent

# DNA replication

## *Blame your father*

In the previous chapter, we learned that we are all mutants. Each of us begins life with dozens of new mutations, most of which occurred in our parents' bodies, in the cells that divided again and again to make sperm or eggs. You might assume that, as you inherit half your DNA from your mother and half from your father, that your new mutations had an equal chance of coming from either parent. But this is not the case. Most of the new mutations in your genome came from your father, not your mother.

As we saw in Chapter 3, some mutations are caused by damage to the DNA, biochemical accidents that muddle the nucleotide sequence, causing a permanent change to the information in the genome. But a large fraction of the mutations that occur every generation arise not from incidental damage due to mutagens, but to mistakes made when copying DNA. DNA replication is astoundingly accurate, but it is not perfect. The error rate varies between species (and even between different kinds of polymerase enzymes within a cell). In mammals, the DNA copy error rate is typically around one in a billion—that is, for every billion bases of DNA copied, there will usually only be one mistake. But, given that the human genome is more than three billion nucleotides long, this is still enough to create several new mutations every time the genome is copied. And the human genome is copied many times per individual lifetime, as cells divide again and again to form the germ cells that will carry a copy of the genome to a new generation.

DNA is copied almost (but not entirely) without error, and most (but not all) DNA damage is repaired. So your DNA is almost, but not exactly, the same as that of your parents. When you copy your DNA to give to your own offspring, new mutations will be added to the ones you collected, and more still when your children copy their genomes to produce your grandchildren, and so on. Every time DNA is copied, there is a chance for copy errors to occur. So the more times the genome is copied, the more mutations should accumulate. And this is why you can blame your father for most of the new mutations you have (since advantageous mutations are rare, I am assuming most of these mutations have sadly not resulted in improvements for which you wish to thank your father).

*For an explanation of why most mutations are deleterious, see Chapter 3*

'Male-driven evolution' arises from the different ways that sperm and eggs are produced. In mammals, like yourself, eggs are formed by a process of symmetrical cell divisions. A germ cell divides, then each of the daughter cells divides again, and so on until all the eggs (and their accompanying cells) are produced. Each egg is formed from the same number of cell divisions. But sperm cells are produced by an asymmetrical pattern of cell divisions (**Figure 4.1**). When each germ cell divides in two, one daughter cell goes on to produce sperm cells, and one goes back into the pool of germ cells, when once again it will divide to produce one cell fated for sperm production and another to go back into the pool of germ cells. Unlike eggs, sperm cells are continuously produced in adult males, so the pool of germ cells is constantly dividing and re-dividing.

**Figure 4.1**

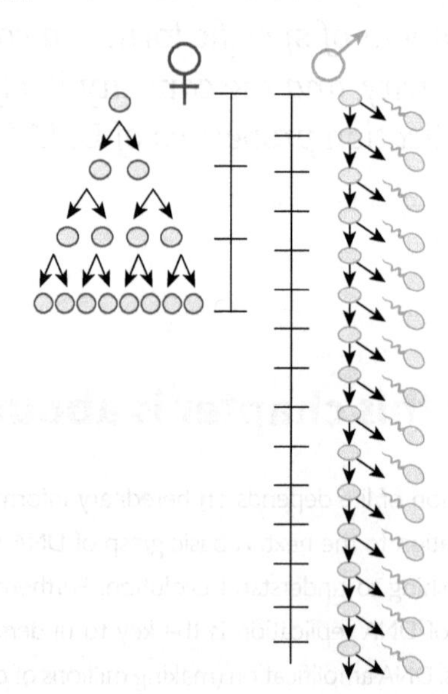

The pattern of sperm cell formation has two important consequences. Firstly, it takes far more cell divisions to produce sperm than eggs. This means that there are more opportunities for copy errors to occur in the manufacture of sperm, simply because the genome is copied more times to produce each sperm cell than it is to produce each egg. Secondly, the more sperm that are produced, the more cell divisions the germline undergoes. This means that, as a male ages, he accumulates ever more mutations in his germline, and these mutations are copied to the newly formed sperm cells. The older the father, the more mutations his sperm are likely to carry (**Figure 4.2**).

Figure 4.2

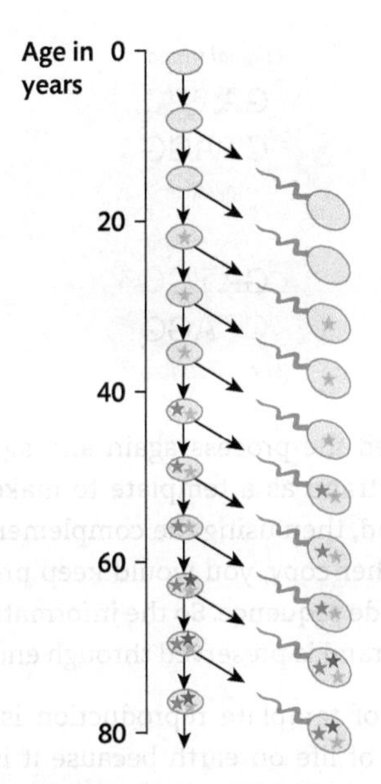


The example of male-driven evolution is not included to make older fathers feel nervous, but to illustrate an important point (Figure 4.3). The fact that the mutation rate is measurably higher in males than in females demonstrates that DNA copy errors are a major component of the mutations that accumulate every generation. The more the genome is copied, the more chance there are for copy errors to accumulate (Case Study 4.1). And the nature of DNA replication is such that once a mutation occurs, through damage or copy errors, it will then be faithfully copied and distributed to offspring produced from that genome.

## Template reproduction

Understanding the template reproduction of DNA is central to several key concepts at the heart of molecular phylogenetics. Firstly, many molecular genetic techniques are based upon the ability to make myriad copies of DNA sequences in the lab, which is achieved through the 'domestication' of the processes of DNA replication. Secondly, template copying is the key to understanding the evolutionary information stored in DNA. As DNA is copied from parent to offspring to grandchildren and great-grandchildren, it accumulates sequence changes that are inherited and copied down the generations. In this way, DNA replication creates a hierarchy of genetic differences—more closely related individuals have

**Figure 4.3** The Common Eider (*Somateria mollissima*) breeds in the Arctic. The female (right) keeps her chicks warm by lining the nest with soft down (which can also be used to keep humans warm when collected and made into eiderdown quilts). Males, on the other hand, make little material contribution to the chicks' wellbeing, but they do make a disproportionate contribution to their offspring's mutation load. In birds, as in mammals, most heritable mutations occur in males, because it takes far more cell generations to make sperm than to make eggs. The detection of male-driven evolution in birds proves that it is not simply an artefact of slower rates of molecular evolution in the mammalian X-chromosome. Male-driven evolution has been demonstrated in mammals by showing that sequences on Y-chromosomes (that spend all of their time in males) have faster rates of molecular evolution than sequences on autosomes (that spend equal time in males and females) which have faster rates than sequences on the X-chromosome (that spend two thirds of their time in females). However, since mammalian females have two copies of the X-chromosome (XX), but males have only one (XY), sequences on the X-chromosomes may be slow due to the removal of recessive mutations expressed in hemizygous males. But in birds, the situation is reversed, because females are hemizygous (WZ) but males are homozygous (ZZ) for the sex chromosomes. Sure enough, Z-chromosomes that spend more time in males have a faster rate of molecular evolution than autosomes, which are faster than W-chromosomes which spend all their time in females.

Reproduced by permission of Andreas Trepte, Marburg.

more similar genomes, more distantly related individuals have more differences between their genomes. Many molecular techniques in biology exploit this simple fact.

In this chapter, we are going to start with an overview of DNA replication, and consider some of the mechanisms that make DNA replication so phenomenally accurate, in order to understand how the occasional copy error slips through the net to create a new mutation. Then we will see how the process of DNA replication has been domesticated to provide a means of copying DNA in the laboratory. Finally, we are going to step back and see the bigger picture: how the process of copying DNA, generation after generation, leads to the accumulation of a hierarchy of genetic differences between individuals, populations and species. It is this hierarchy of changes that allows us to reconstruct evolutionary history using DNA sequences.

### *Copying by base pairing*

The discovery that DNA is the hereditary material came about when scientists demonstrated that it was DNA, not proteins, that carried genetic information from one individual to another. Proteins had been considered the natural candidate for the hereditary material because they were made of 20 different amino acids, giving a generous 'alphabet' for information storage. But information storage alone cannot provide a basis for heredity. The information must be able to be copied from one generation to the next, and there was no obvious way to copy the information in proteins. The discovery of the double helix structure of DNA, with the phosphate backbones on the outside and the bases facing each other on the inside (**Hero 4**), revealed the answer. The base pairing rules of DNA (C ≡ G, A = T) provide a mechanism for the transfer of genetic information through template copying.

***The discovery of the structure of DNA is covered in Chapter 2***

The two strands of the DNA helix run antiparallel, which means that one runs 5′ to 3′, but its complementary strand matches it in the 3′ to 5′ direction (**Figure 4.4**: see **TechBox 2.1** for an explanation of 5′ and 3′).

**Figure 4.4**

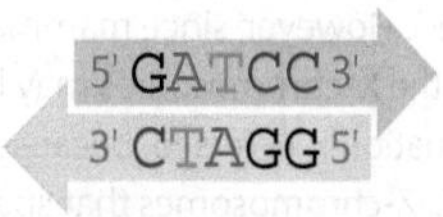

5′ GATCC 3′
3′ CTAGG 5′

Any strand of nucleic acid can be used as a template to make a complementary strand, by matching the base pairs (**Figure 4.5**).

**Figure 4.5**

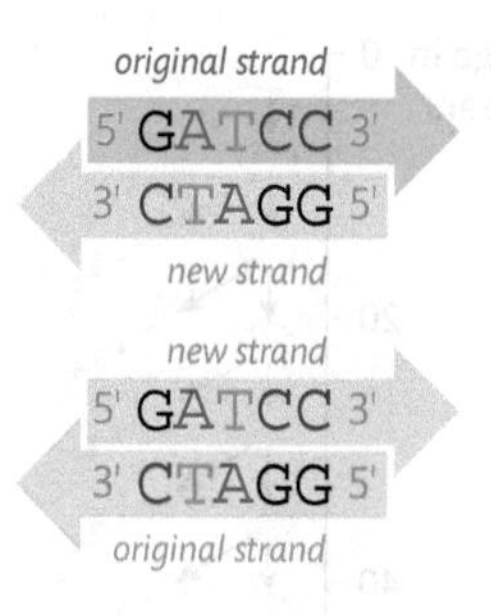


If you repeated the process again and again, using a nucleic acid strand as a template to make a complementary strand, then using the complementary strand to make another copy, you would keep producing the same nucleotide sequence. So the information coded in the original strand is preserved through endless copies.

This process of template reproduction is the key to the evolution of life on earth because it is the single known means of reliable self-replication, the fundamental property of living systems. All life on earth uses the same nucleic acid replication system: there are no known alternatives. If you want to understand life, and its evolution, the best place to start is to become familiar with the wonders of DNA replication. There is no better system for understanding why life exists and evolves.

## Replication

DNA replication involves the coordinated action of a vast number of enzymes, regulatory factors and molecular building blocks. The exact details differ between species, but the basic processes of DNA replication are common to all organisms. We will briefly look at some of the key steps in DNA replication here, but you can find more detailed explanation of the replication machinery and processes in **TechBox 4.1**.

The entire genome must be copied every time a cell divides. So, prior to cell division, complexes of replication proteins form, which then bind to specific sites in the genome. Some small, circular genomes, like bacterial or mitochondrial genomes, have a single origin of replication. But the much larger eukaryote genomes, with multiple linear chromosomes, must initiate replication at numerous sites throughout the genome, to allow the entire genome to be copied in a reasonable time (which, depending on how rapidly the cells are dividing, may range from a few minutes to a few hours). There can be many thousands of replication origins in a large genome. The co-ordination of replication origins is very important, to make sure the entire genome is copied before cell division, and no region is copied more than once.

At the site of replication initiation, the two strands of the DNA double helix are separated by breaking the hydrogen bonds between the base pairs. These separated strands form a replication bubble (**Figure 4.6**).

**Figure 4.6**

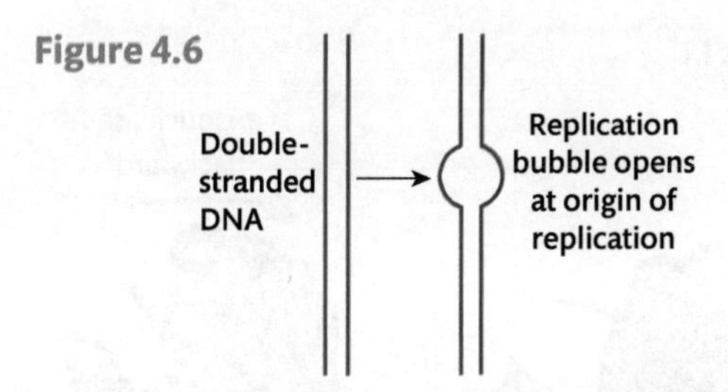


At the site of the replication bubble, the DNA helix is unwound by a suite of enzymes, separating the two complementary strands of the DNA helix to expose the sequence of nucleotide bases (**Figure 4.7**). This unzipped section of the DNA helix is referred to as the replication fork.

**Figure 4.7**

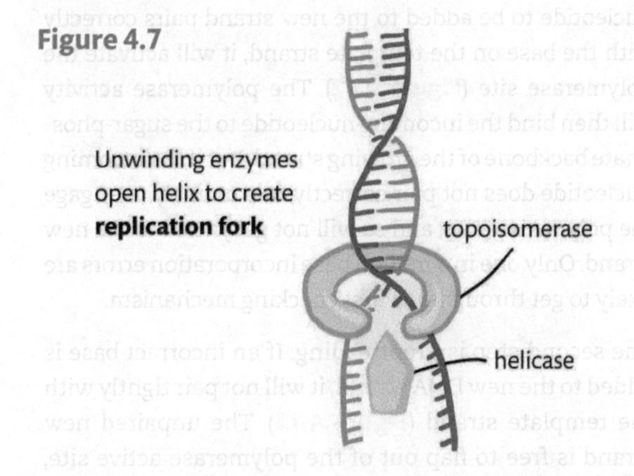


Each of the exposed strands forms a template for the construction of a new complementary strand. A primase enzyme makes short RNA polynucleotides along the exposed single-stranded DNA. These RNA primers provide a starting block for DNA polymerase. Starting at the end of the primer, DNA polymerase 'grows' a new complementary strand by adding nucleotides that pair with those on the existing strand, and binding them together along their phosphate-sugar backbone.

DNA polymerase works in the 5' to 3' direction. Remember that the two DNA strands in the double helix run antiparallel (**Figure 4.4**). This means that on one of the exposed DNA strands, the polymerase can work toward the replication fork, adding new nucleotides to the growing 3' end of the strand (which is called the leading strand). But on the other strand of the open helix, polymerase must follow the 5' to 3' direction and move away from the replication fork to create the lagging strand (**Figure 4.8**).

**Figure 4.8**

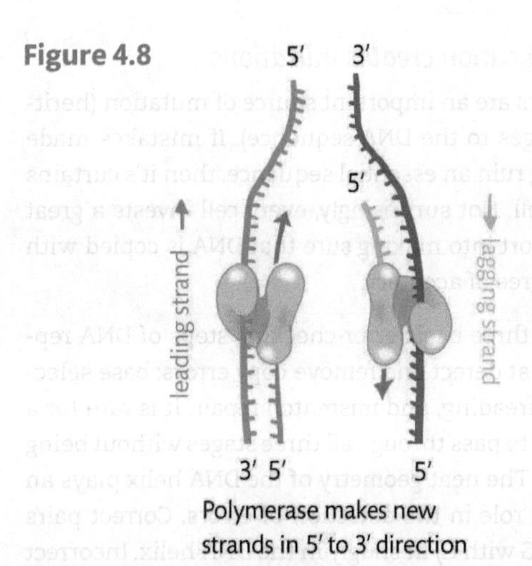


How can the lagging strand be made if the polymerase can't run toward the replication fork as it moves along the helix? The polymerase makes a series of short polynucleotides, each one from 5' to 3', then starts a new fragment closer to the replication fork, and so on. The fragments are then glued together to make a continuous strand (**TechBox 4.1**).

So the replication fork moves along the helix. Unwinding enzymes continue to untangle the DNA ahead of the replication fork, and the replication complex extends the newly created strands to match the unzipped DNA. The end result is two identical double-stranded DNA molecules, each consisting of one strand of the original molecule and one newly synthesized strand (**Figure 4.9**).

**Figure 4.9**

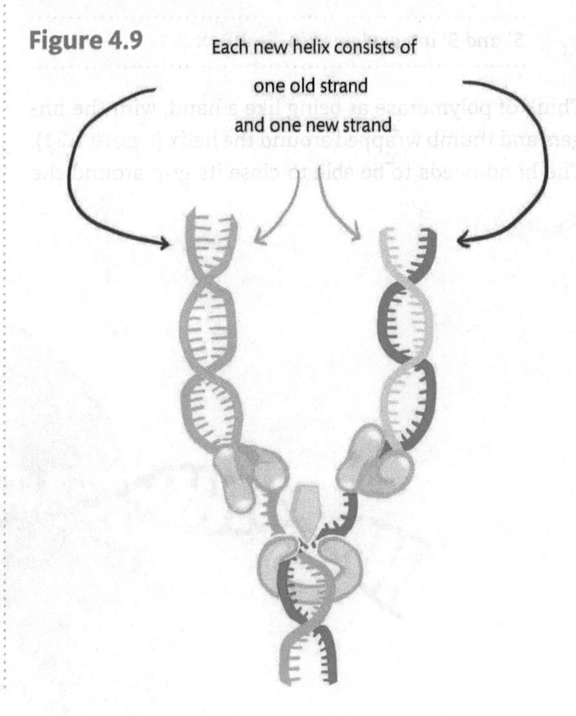

## *DNA replication creates mutations*

Copy errors are an important source of mutation (heritable changes to the DNA sequence). If mistakes made in copying ruin an essential sequence, then it's curtains for that cell. Not surprisingly, every cell invests a great deal of effort into making sure that DNA is copied with a high degree of accuracy.

There are three basic error-checking steps of DNA replication that detect and remove copy errors: base selection, proofreading, and mismatch repair. It is rare for a copy error to pass through all three stages without being corrected. The neat geometry of the DNA helix plays an important role in the detection of errors. Correct pairs (A with T, G with C) fit snugly in the DNA helix. Incorrect pairs do not. Distortions to the helix caused by damage or mis-pairing underlies each of the three error checking stages of replication.

*Complementary base pairing is discussed in* **TechBox 2.2**

This is particularly important when considering the interaction between the polymerase enzyme and the DNA helix. Most DNA polymerase enzymes have two main active sites, each of which does a different job (**Figure 4.10**). The polymerase site takes DNA synthesis 'forward' (5' → 3'), adding nucleotides to the end of a growing strand. The exonuclease site takes the enzyme 'backwards' (3' → 5'), removing nucleotides from the end of the growing strand.

*5' and 3' are explained in* **TechBox 2.1**

Think of polymerase as being like a hand, with the fingers and thumb wrapped around the helix (**Figure 4.11**). The hand needs to be able to close its grip around the helix in order to engage the polymerase active site. A distorted helix prevents the hand closing, so that the polymerase function cannot be employed, but the exonuclease site may be engaged.

**Figure 4.10**

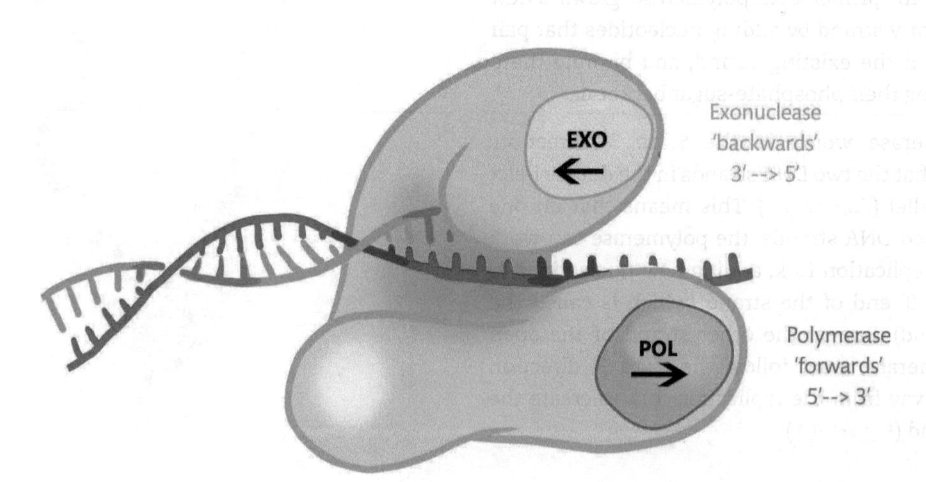


**Figure 4.11**

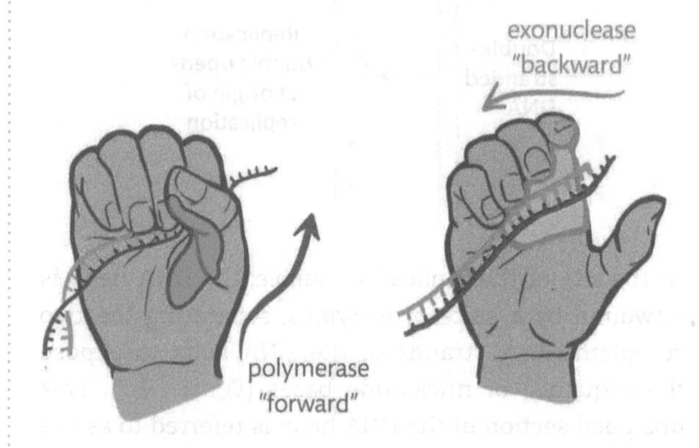


The first step of error correction is base selection. If a nucleotide to be added to the new strand pairs correctly with the base on the template strand, it will activate the polymerase site (**Figure 4.12**). The polymerase activity will then bind the incoming nucleotide to the sugar-phosphate backbone of the growing strand. But if the incoming nucleotide does not pair correctly, it is unlikely to engage the polymerase site, and so will not get joined to the new strand. Only one in a million base incorporation errors are likely to get through this first checking mechanism.

The second step is proofreading. If an incorrect base is added to the new DNA strand, it will not pair tightly with the template strand (**Figure 4.13**). The unpaired new strand is free to flap out of the polymerase active site, and into the exonuclease site. The exonuclease activity chews up the most recently added nucleotides, erasing the incorrect base. Efficiency of proofreading varies between different versions of the polymerase enzyme,

Figure 4.12

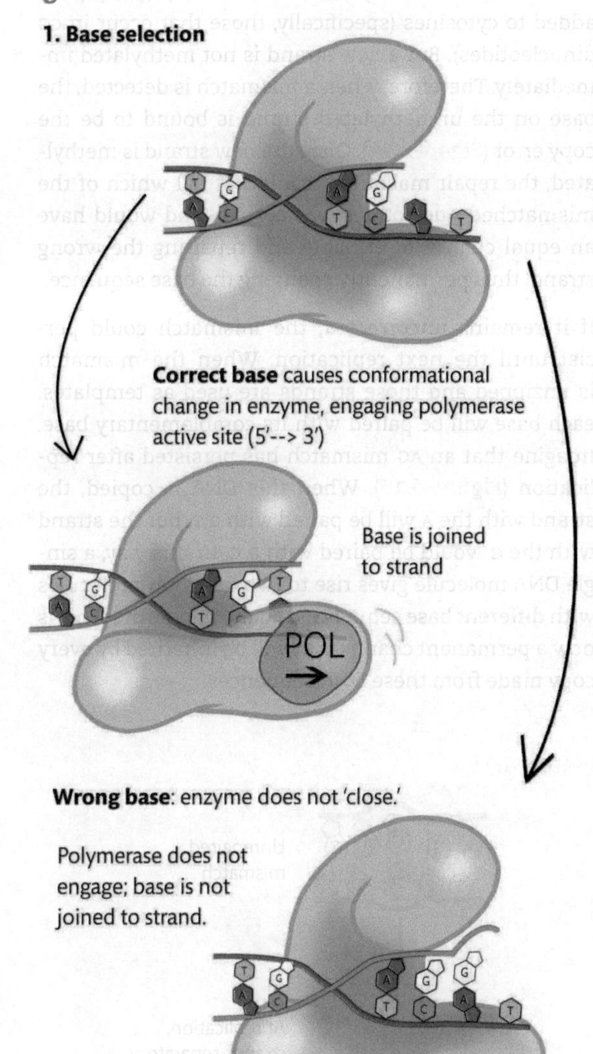


Figure 4.13

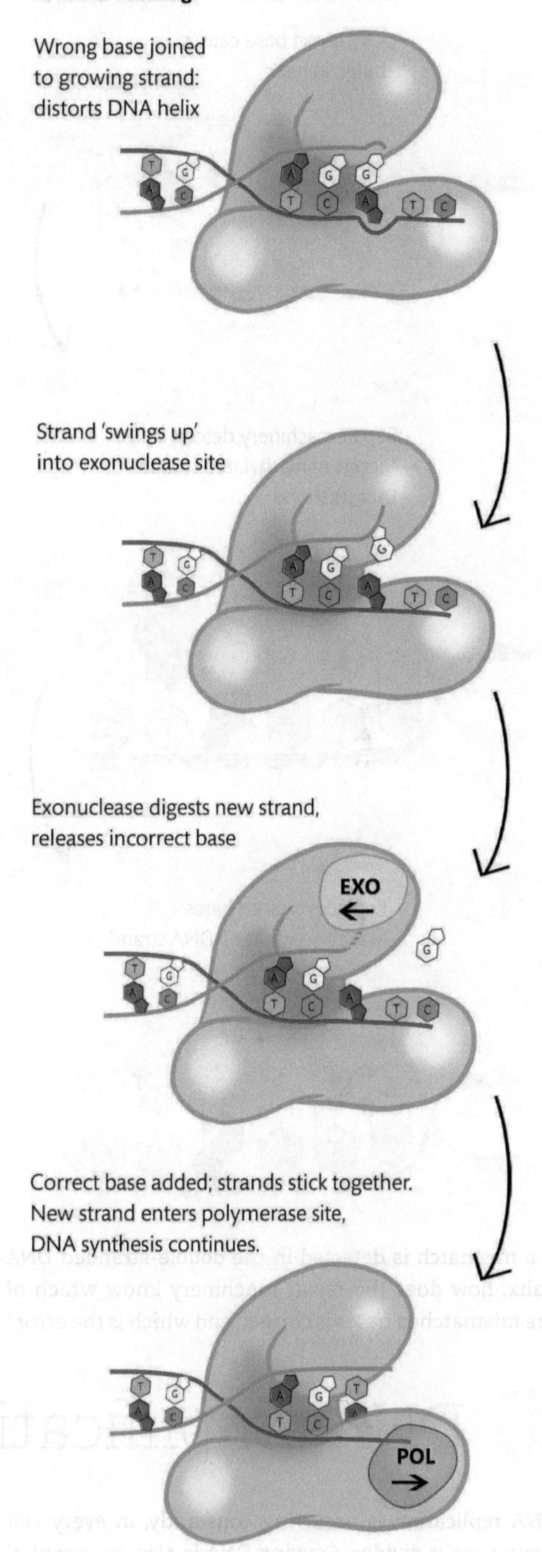


ranging from no proofreading to 99.99 per cent of errors corrected. Even with high fidelity, the occasional error will make it past the proofreading stage.

The third stage of copy error correction is post-replication mismatch repair. If an incorrect nucleotide has managed to avoid correction by base selection and proofreading, it may be incorporated into the new DNA strand. But the wrong nucleotide will not pair correctly with the mismatched base on the template strand. This causes the DNA helix to bulge out. Mismatch repair enzymes scan the DNA for bulges that indicate an incorrect base. Once found, the base can be chopped out using an endonuclease, an enzyme which removes nucleotides from the middle of a DNA strand. The excised strand is then replaced with the correctly matching bases (**Figure 4.14**).

**Figure 4.14**

**3. Postreplication repair**

- Incorrect base causes bulge in helix

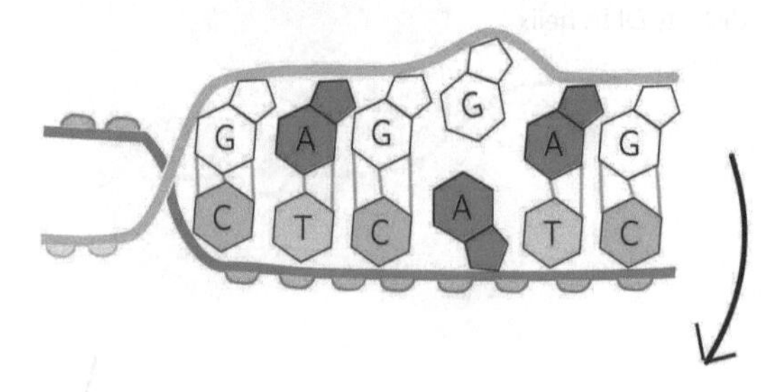


- Repair machinery detects bulge
- Targets unmethylated strand
- Digests strand

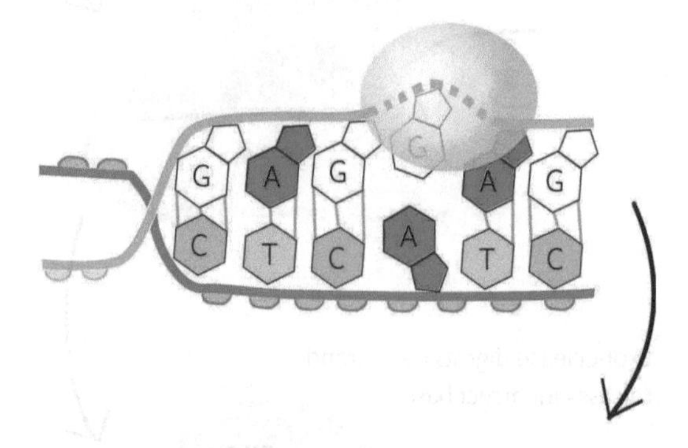


- Replaces excised bases
- Makes new joining DNA strand

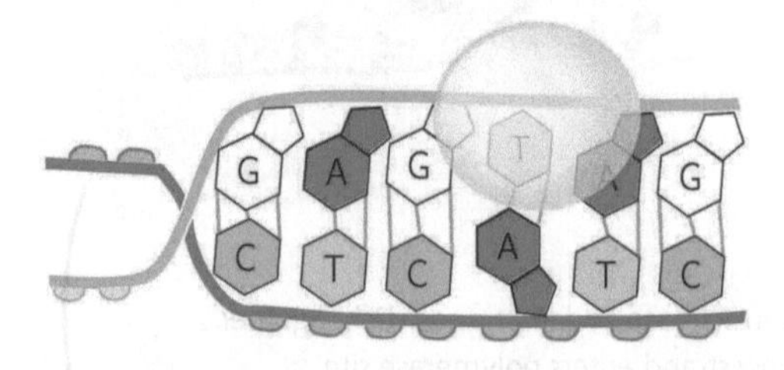


If a mismatch is detected in the double-stranded DNA helix, how does the repair machinery know which of the mismatched bases is correct, and which is the error?

DNA strands are methylated, with a methyl group ($CH_3$) added to cytosines (specifically, those that occur in CG dinucleotides). But a new strand is not methylated immediately. Therefore, when a mismatch is detected, the base on the unmethylated strand is bound to be the copy error (**Figure 4.14**). Once the new strand is methylated, the repair machinery could not tell which of the mismatched nucleotides was correct, and would have an equal chance of excising and repairing the wrong strand, thus permanently changing the base sequence.

If it remains uncorrected, the mismatch could persist until the next replication. When the mismatch is unzipped and those strands are used as templates, each base will be paired with its complementary base. Imagine that an AG mismatch has persisted after replication (**Figure 4.15**). When this DNA is copied, the strand with the A will be paired with a T, but the strand with the G would be paired with a C. In this way, a single DNA molecule gives rise to two daughter molecules with different base sequences: AGGAG and AGTAG. This is now a permanent change that will be inherited by every copy made from these DNA sequences.

**Figure 4.15**

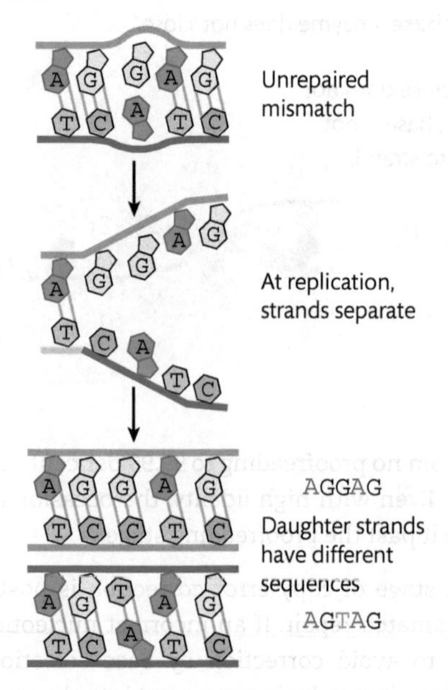

# DNA amplification

DNA replication is occurring constantly, in every cell every time it divides. Copying DNA is also an essential step in many molecular genetic techniques, which often rely on the ability to successfully amplify DNA from natural samples. Sometimes, scientists borrow the replication abilities of a living organism to copy DNA for them, for example by inserting target DNA into bacteria who then helpfully use their own DNA replication machinery to make myriad copies of the inserted sequence (this approach is referred to as cloning). But most of the time, scientists use an artificial process to amplify DNA, mixing the DNA with enzymes and reagents and 'cooking' it to a particular recipe. However, even these lab-based amplification procedures are not strictly speaking, entirely a human invention (despite the fact that one of the early techniques resulted in a particularly lucrative patent). Instead, DNA amplification relies on the domestication of natural processes of DNA replication.

### *Domesticating DNA replication*

DNA replication *in vivo* (in living cells) is a very complex process. The replication machinery, consisting of dozens of enzymes all working together, must be formed at the right time in the right places to allow the entire genome to be copied in a coordinated fashion before cell division. Enzymes and cofactors must come together at the sites of replication origin to form the machinery that will open the double-stranded DNA helix to form replication bubbles, unwind the DNA helix to expose replication forks, make short DNA primers along the exposed strands, then grow new DNA strands to match both the leading and lagging strands, creating two double helices from one. This is far too complex a process to replicate in the laboratory.

DNA replication *in vitro* (in the laboratory) must deal with several important differences from the well-organized living cell. Firstly, rather than dealing with DNA that is packaged into chromosomes, the process of extracting DNA from biological samples usually results in the genome being chopped into pieces and mixed up. Secondly, it would be extremely tricky to get all of the correct enzymes working together *in vitro* with the right conditions and cofactors to recreate the complex replication machinery found in the living cell. So the DNA replication method adopted by a living cell—starting at defined origins of replication and constructing replication machinery which moves with the replication fork along the double helix until the whole genome is copied—is not practical in a test tube.

Lab-based DNA amplification overcomes these problems by being much, much simpler. Instead of using dozens of enzymes working in concert, the most common techniques rely on essentially three factors to copy DNA: heat, primers, and polymerase. Instead of enzyme complexes creating replication forks at defined sites of replication, many lab-based methods take the brutally effective approach of heating the sample to break the hydrogen bonds between base pairs, melting the entire helix so that all of the DNA in the sample is converted to single strands. This bypasses the need to have enzyme complexes to initiate and maintain replication forks: instead, all DNA in the sample is instantly exposed as a template (some amplification methods don't use heat, see **TechBox 4.2**). Living cells begin DNA synthesis by using primase to create short RNA polynucleotides along the exposed replication fork. The need for primase is circumvented *in vitro* by the addition of short DNA sequences (primers: **TechBox 6.2**) that bind to specific sequences on the exposed single-stranded DNA, providing a starting point for DNA polymerase. And instead of the co-ordinated set of polymerase enzymes and cofactors found in most living cells, typically only one robust DNA polymerase is used in the lab to do all of the DNA synthesis.

## Polymerase chain reaction (PCR)

> "*the truly astonishing thing about PCR is precisely that it wasn't designed to solve a problem; once it existed, problems began to emerge to which it could be applied*"
>
> Rabinow, P. (1996) *Making PCR: A story of biotechnology*. University of Chicago Press.

Rarely has an invention had such a rapid impact on scientific practice and the wider culture as the development of the polymerase chain reaction (PCR). This laboratory procedure allowed the amplification of even small samples of DNA to produce enough copies of just the right sequence to allow a range of analyses. Within a few years of the invention of PCR, it had been used to detect infectious agents in samples of blood and water, to identify missing persons, to settle paternity disputes, and as a forensic tool that resulted in both convictions and acquittals in criminal courts.

The PCR protocol is essentially a recipe describing the chemicals and conditions that can be employed to make multiple copies of a DNA sequence in a test-tube (**TechBox 4.2**). It works by cycling the DNA sample through a series of heating and cooling cycles, with raw materials and enzymes needed for DNA synthesis. Through the process of template reproduction, the number of copies of the target DNA sequence increases exponentially. In the first round, complementary copies are made of the original DNA, then in the next round copies are made of the original DNA and the first set of copies, and so on. Three main technical breakthroughs were needed to make PCR work: primers, polymerase, and temperature cycling.

### *Primers: starting blocks for DNA synthesis*

One of the first technical breakthroughs that allowed the development of DNA amplification was the ability to make short nucleotide chains with a specific base sequence. DNA polymerase cannot start from scratch, but can only add nucleotides to the end of an existing polynucleotide chain. In living systems, an enzyme with primase activity manufactures short RNA polynucleotides that match the sequence of bases on the exposed DNA of the replication fork. But in the lab, researchers circumvent the need for primase by adding ready-made primers. PCR primers are typically short sequences of DNA that stick to complementary DNA sequences in the sample, providing starting blocks for polymerase to work from. Vast numbers of copies of the primer are added to the extracted DNA. These primers sticks to the DNA in the sample, providing a free 3' OH to attach new bases to.

In addition to providing a starting block for DNA polymerase to act upon, primers also solve the problem of where to begin DNA synthesis. *In vivo*, the replication enzyme complex recognizes specific sequences in the genome that act as replication initiation sites. But *in vitro*, this approach won't work. Since the genome in the sample is chopped up into pieces, replication initiation sequences will be found on a number of disassociated fragments. If polymerase began at these recognition sequences it may not get very far before it fell off the end of the fragment, and fragments that did not contain any initiation sequences would not get copied at all. In particular, the sequences you want to amplify might not be near a replication initiation site. Happily, this problem can be solved by adding primer sequences that stick to the sequences of DNA you wish to copy (**TechBox 6.2**).

Because complementary base pairing is a universal feature of DNA, if you make a short polynucleotide chain of a particular sequence and add it to a sample of single-stranded DNA, then it will stick to any complementary sequence that occurs anywhere in the DNA sample. So if you know (or can guess) a short base sequence from the region of DNA you wish to amplify, you can design a primer that will stick to that sequence (and to no other sequence in the genome). Adding large numbers of copies of this primer sequence to your sample will make the polymerase enzyme start exactly where you want it to, and copy only your target DNA, even if that sequence represents less than a ten-thousandth of a per cent of the DNA in the sample. Alternatively, if you want to copy the whole genome, you can throw in a whole heap of random primers that will stick to places all over the genome, providing multiple starting blocks for synthesis.

## *Temperature cycling: from double-stranded to single-stranded and back again*

Primers work by complementary base pairing, so they can only stick to exposed nucleotides on single-stranded DNA. In living systems, the double-stranded DNA helix is gently prised apart by complexes of enzymes which create and maintain the replication fork. But most lab techniques employ a much less sophisticated means of exposing single-stranded DNA templates. The entire sample is heated to the temperature at which the hydrogen bonds holding the nucleotide 'rungs' of the helix melt away (**TechBox 4.2**). Heating is a pretty brutal way to treat a complex biomolecule. DNA loses its higher-order structure when heated—the supercoils unravel, the helix unwinds, and the base pairs separate. All that is left are single strands of polynucleotides. This loss of structure is referred to as 'denaturing', and it's an appropriate term because it's exactly what should not happen in a natural system.

When the sample is cooled, the single-stranded DNA begins to anneal, sticking back together along the exposed nucleotide sequences to form double helices once more. The degree of matching between two strands of DNA will determine how strongly they stick together, because the hydrogen bonds linking the bases form between matched base pairs. The more bases match, the more bonds form between the two strands, so two perfectly matched strands will be more strongly stuck together than an imperfect match.

The universality of complementary base pairing means that, as the single-stranded DNA is cooled, any complementary DNA sequences will anneal together when the sample is cooled again. As primers are added in excess and outnumber the source DNA, it's more likely that a target sequence will be matched up with a primer than with a complementary strand of source DNA. So when the DNA is cooled, most copies of your target sequence will be bound to a primer. If a primer sticks to its exact complement, it is more likely to stay stuck as the sample is heated again to raise it to the temperature needed for polymerase to extend the sequences. DNA polymerase can now start at the primer sequences and make the rest of the matching strand. Now you repeat the cycle of heating and cooling to create more and more copies of the sequence that the primers stick to (**Figure 4.33**).

This cycle of heating and cooling can be done manually by anybody with several water baths (one at each temperature), a reliable stopwatch and sufficient patience. However, PCR has been made more efficient, and rather less tedious to perform, by the development of thermal cyclers (sometimes referred to as 'PCR machines') which do all the heating and cooling for you. But the real breakthrough for automating DNA amplification was in the taming of the polymerase enzyme.

**Figure 4.16** Some like it hot: *Thermophilus aquaticus* is an 'extremophile' bacterium, first found in hot springs at Yellowstone National Park, USA. *T. aquaticus* is the original source of the *Taq* polymerase now used to amplify DNA in the lab. Because of its key role in the development of molecular genetics, *Taq* was awarded the inaugural 'Molecule of the Year' award in 1989. *Taq* earned the patent holders hundreds of millions of dollars, but the patent was the subject of lengthy legal battles.

## *Taq polymerase: Molecule of the Year 1989*

Denaturing DNA would spell disaster for a living cell, because without its higher order structure DNA could no longer be reliably maintained, copied and transcribed. Heating also denatures most proteins, including those involved in DNA synthesis. When PCR was first developed in the mid 1980s, back when samples had to be manually heated and cooled, fresh polymerase had to be added each cycle to replace the denatured enzymes ruined by the last round of heating. Discovery of a DNA polymerase enzyme that could survive thermal cycling was an important breakthrough in the development of biotechnology.

There are a great many variants of the DNA polymerase enzyme in the living world: not only does the enzyme vary slightly between species, but some species have many different kinds of polymerase, that perform different DNA replication tasks in the cell. For example, mammalian cells have a dozen or more different kinds of polymerase, functioning in both DNA synthesis and repair. Lab-based DNA amplification requires an enzyme that can work in unusual conditions. The logical place to look for a polymerase enzyme that could survive repeated heating was in an 'extremophile' organism that lives at temperatures that would cook ordinary organisms.

*Thermophilus aquaticus* lives in hot springs, at temperatures that denature most genomes (**Figure 4.16**). Not surprisingly, its version of DNA polymerase is able to operate at high temperatures. *T. aquaticus*' DNA polymerase, now referred to as *Taq*, can persist in the PCR sample during the heating phase, then begin copying when the sample is cooled. *Taq* is not the only polymerase enzyme that can be used in PCR, but it is one of the most common. DNA replication *in vitro* using *Taq* differs from DNA replication in living systems in a number of important ways. Firstly, *Taq* can only copy relatively short DNA sequences, typically less than 2,000 bases. Secondly, unlike most polymerase enzymes, *Taq* has no proofreading activity, so it has an inherent error rate of around one mistake for every ten- to a hundred-thousand bases copied. Given that a PCR reaction involves making millions of copies of a DNA sequence, this error rate is not trivial. Mistakes can be identified through multiple coverage, so most genome sequencing projects sequence each part of the genome many times over. Using alternative thermostable polymerases that can proofread could reduce the PCR error rate (TechBox 4.2).

## Contamination

The great strength of DNA amplification—that it can take a tiny amount of target DNA and make millions of copies of it—is also one of its dangers. There is DNA everywhere—on your hands, in your lunch, on your lab bench. So your sample may contain DNA from more than one source: non-target DNA is usually referred to as contamination. DNA from all sources is structurally the same, and it is all copied by the same mechanism. Therefore, a PCR reaction will amplify any sequence that your primer sticks to, whether or not it was the sequence you wanted. Contamination is a particularly serious problem when dealing with very small samples, or samples with degraded DNA, such as museum specimens. In these cases, contaminating DNA can overwhelm the target DNA, so the PCR product can consist entirely of non-target sequences. The problem of contamination can be countered by scrupulous attention to lab hygiene, by careful primer design, and through careful consideration of the resulting DNA sequences.

In some cases, contamination can be detected by the surprising similarity of your amplified DNA to something other than your intended sample. You may not know the exact DNA sequence of a certain gene in your target species, but you will have some idea of which other species you expect it to be most similar to. For example, in the mid 1990s two separate laboratories made the surprising claim that they had sequenced DNA isolated from dinosaur bones and eggs. Two lines of argument were used to refute these claims. Firstly, there is currently doubt that DNA can survive in a sample longer than a million years without being totally degraded, so there is virtually no hope of finding 80 million-year-old dinosaur DNA. Secondly, comparing the dinosaur-derived sequences to the DNA of living species revealed some unexpected relationships. Dinosaur origins and evolution may be debated, but most people would expect dinosaur sequences to be most similar to birds and other reptiles. But careful analysis of the supposed dinosaur sequences showed that, in one case, they were most similar to sequences from plants and fungi, and in the other case, most similar to human sequences. It is easier to believe that these sequences were actually contaminants than it is to rethink dinosaurs as giant mushrooms, or to imagine that the dinosaur genome underwent an improbable degree of convergence with the human genome.

Contamination will be less obvious when the target DNA is closely related to the contaminant. In fact, a pervasive source of contamination is from the PCR product of a previous reaction. A drop of liquid from a PCR reaction can contain many thousands of copies of a sequence, overwhelming the target DNA in your sample. Good lab practice will reduce the risk of contamination, such as conducting DNA extraction, pre- and post-PCR preparation in different parts of the lab, not sharing lab equipment, and generally avoiding activity that will spread aerosols of PCR product. In addition, a number of checks for contamination can be performed. Blank controls, with no sample material, should be included in PCR procedures: a positive PCR result for a blank control demonstrates that the PCR product came from contamination.

We have now considered how DNA is replicated by using the two strands of the helix as templates to make complementary strands to form two new helices. We have also seen how the process of DNA replication has been dramatically simplified to provide innumerable copies of a DNA sequence for molecular genetic analysis. Now we are going to have a more detailed look at the process of DNA replication in order to understand how the combination of replication and mutation turns the genome into a phenomenally useful source of historical information.

# Copy errors create hierarchies

### *History in the genome*

DNA replication is the basis of life on earth. It is also the basis of molecular phylogenetics, because the process of copying creates an array of genomes, each of which is more similar to close relatives, and increasingly dissimilar to ever more distant relatives. In other words, copying creates a hierarchy of similarities between genomes. Due to the process of descent with modification, this hierarchy spans all levels of biological organization, from individuals to populations to lineages to distantly related kingdoms. We saw this in Chapter 1: DNA from the mystery blob could track back through time, from the relationship of a single whale to its parents and grandparents, to the past whale populations, to the diversification of the whales, the radiation of mammals, the origin of the animal kingdom, and potentially way back further still. In order to look more carefully at how DNA replication leaves a hierarchy of changes that we can read at all levels of biological organization, we are going to revisit an example we introduced in Chapter 2: the gene that codes for a subunit of the transcription machinery, RNA polymerase II beta.

## Sequence evolution

All organisms must have the information needed to make all the essential components of the genomic information system, including the enzyme RNA polymerase. And yet we saw in **Figure 2.2** that the DNA sequence for the RNA polymerase II beta subunit gene is not identical in all organisms. If they all require the same enzyme to do the same job, then why not have exactly the same gene?

To picture how such an important gene could change over evolutionary time, we can invent a fable about a recipe which is handed down through the generations of a family. Imagine that someone, let's call her Mrs Smith, was famed for her excellent sponge pudding. The pudding was so much admired by all who partook of it that each of Mrs Smith's sons and daughters took a copy of the recipe with them when they left home, so they too could create the much-loved dessert. Their children too would copy the recipe upon leaving home, and so would their grandchildren, and great-grandchildren, and so on down the generations. Occasional mistakes were made in copying, some of which ruined the recipe so that it was no longer used and copied. But if the copy mistake did not prevent the cooking of a delicious sponge pudding, it was not noticed, and so some little mistakes in the recipe would be copied again by the next generation of pudding-makers. Eventually, after many generations, all 75 of Mrs Smith's great-great-great-grandchildren have their own copy of the recipe, yet their recipes are all slightly different—a spelling mistake here, a change of ingredients there. But every single recipe produces a fine sponge pudding.

For any process of sequential copying—where each copy is used as a template for the next copy—comparison of shared copy errors can be used to reconstruct the copying process. Much of this book concerns the application of this line of thinking to DNA, given that each organism's DNA was copied from its parents, which was copied from their parents, and so on. But the same idea can be applied to reconstructing the history of any copied information. Techniques developed for the analysis of DNA data are now being applied to understanding language evolution, reconstructing the history of ancient texts, and studying the development of cultural artefacts (**Case Study 4.2**).

***In Chapter 10 we see how the hierarchy of similarities is revealed by sequence alignment***

Before the advent of printing technology, books were copied by scribes, each new copy being made from an existing manuscript. A mistake made in copying (or an 'improvement' added by the scribe) would be perpetuated in any copies made from the new manuscript. For example, Geoffrey Chaucer's *Canterbury Tales*, a series of stories told by pilgrims on their way to Canterbury, was written in the late fourteenth century (**Figure 4.17**). The original version of the *Canterbury Tales* has disappeared, but there are around eighty surviving copies. By comparing the similarities and differences of the surviving copies, it is possible to group the manuscripts into a nested hierarchy. A research group including specialists in both literature and evolutionary biology

**Figure 4.17** Geoffrey Chaucer himself is depicted as a pilgrim in the Ellesmere Manuscript of *The Canterbury Tales*. The Ellesmere Manuscript is one of the oldest surviving copies of the *Tales*, and may have been created not long after Chaucer's death.

compared 850 lines from one of the tales, 'The Wife of Bath's Prologue', from 58 surviving fifteenth-century manuscripts. They constructed nested hierarchies of shared differences in spelling, word usage, and punctuation and used this information to draw a diagram indicating the relationships between the manuscripts. They found five distinct groups of related manuscripts, each representing a chain of copies. Importantly, their research also identified a group of key texts that appear to be much closer to the ancestral text, suggesting that these are the manuscripts that will give the most clues to Chaucer's original version.

In fact, any culturally learned behaviour can be interpreted as the result of a copying process with occasional errors. So it is sometimes possible to reconstruct the history of technologies and crafts by reconstructing hierarchies of similarity. For example, the nomadic Turkmen people (from Turkmenistan and neighbouring regions of Iran and Afghanistan) use woven textiles to make both practical and ceremonial artefacts, such as carpets, saddlebags, and wedding decorations (**Figure 4.18**). Carpet weaving is a difficult skill to learn, particularly as Turkmen carpets incorporate intricate designs, so Turkmen girls traditionally learned to weave from their mothers over a long period of time. Particular carpet designs were passed from mother to daughter, with the occasional modification. An analysis of Turkmen textiles produced from the eighteenth to twentieth centuries reveals a hierarchy of similarities that result from a pattern of descent with modification, with particular designs copied down generations, just as genes are.

Much the same process of copying with occasional errors has occurred in the evolution of the RNA polymerase II beta gene. All organisms must have a functioning copy of this gene, which they inherit from their parents. Very rarely, mistakes are made in copying this gene—an A inserted instead of a T, or a G where there should be a C. Most mistakes will simply ruin the gene, which can then only make a faulty copy of the enzyme, reducing the chance that an organism carrying that mutation will reproduce and pass the mutation on to the next generation. But occasionally, a copy error is made that does not

**Figure 4.18** Cultural evolution: Because traditional Turkmen weaving techniques and patterns are passed from mother to daughter, with occasional changes, the history of textile designs can be viewed as a process of descent with modification. Phylogenies of textiles have been produced using the same techniques used to determine the evolution of species.

ruin the enzyme, which can still function even though it is slightly different from its parent copy. If the individual with this mutation survives and reproduces, that gene may be copied to their offspring, mutation and all. The mutation will then become an inherited difference, present in descendants of that lineage. Note that this mutation did not have to make the enzyme better, it simply didn't make it much worse, so that it didn't reduce its chances of ending up in successful progeny.

## *Sequences show a nested hierarchy of similarities*

In Chapter 2, we saw that the DNA sequences that code for the RNA polymerase II beta are recognizably similar, but not identical, between very distantly related species (**Figure 2.2**). Let's look at that example again now. You will notice that, although they are all different, some sequences are more different than others. The two most similar sequences are from the two most closely related species—the two mammals, rat (*Rattus*) and human (*Homo*). Their sequences are almost identical, differing only at four places in the sequence (the rat has a C instead of a T at position 3531, a G instead of an A at 3546, and a C instead of a T at positions 3552 and 3561: **Figure 4.19**).

The two mammal sequences are also similar to the other animal sequence, from the fruit fly *Drosophila*. Here we can see that the human and fly sequences differ at nine places in the sequence (**Figure 4.20**).

When we compare sequences from the animal and fungi kingdoms, we see more sequence differences. The human and bread mould (*Neurospora*) sequences have 11 differences between them. The plant sequence is even more different from the fungus and both the animals: there are 20 differences between human and rice. The most distantly related organism to us, the bacterium, is also the most different, with 23 sequence differences between human and *E. coli*. Yet each of these sequences contributes to the manufacture of a functional RNA polymerase II beta subunit, a necessary part of a functioning cell.

This hierarchy of similarity can also been seen in the amino acid sequence for this section of the RNA polymerase II beta protein subunit (**Figure 2.9**). When we translate the base sequence into amino acids (see **TechBox 6.1**), we can see that (for this small section of the gene) although the three animals (human, rat and fly) have different DNA sequences, they all translate to the same amino acid sequence. This is because not all changes in the DNA sequence cause a change in the amino acid sequence, due to the redundancy of the genetic code (see **TechBox 6.1**).

*Chapter 6 explains the translation of DNA sequences of genes into the amino acid sequences of proteins*

Looking at the nucleotide alignment (**Figure 4.19**), read along the scale bar at the top to locate position 3531. Where the human has CGT, the rat has CGC. But both of these codons specify the amino acid arginine (R). So although the DNA sequence of the gene changed, the amino acid sequence remained the same: both versions of the sequence make exactly the same protein (**Figure 4.21**).

Since this change from T to C in the rat lineage didn't make any difference to the functioning of the RNA polymerase enzyme, it continued to be copied down the rat generations.

**Figure 4.19**

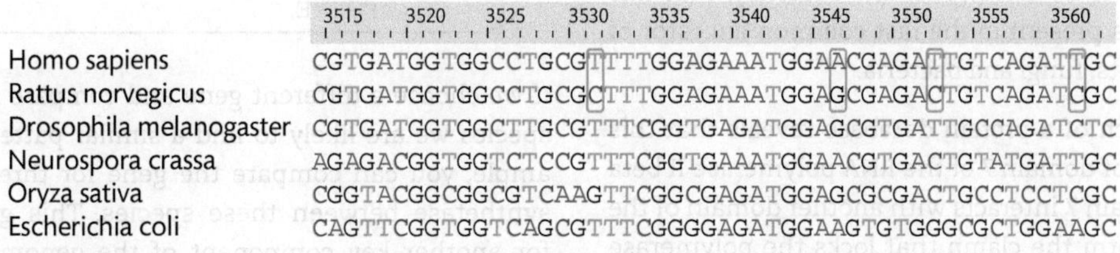

```
                          3515 3520 3525 3530 3535 3540 3545 3550 3555 3560
Homo sapiens              CGTGATGGTGGCCTGCGTTTTGGAGAAATGGAACGAGATTGTCAGATTGC
Rattus norvegicus         CGTGATGGTGGCCTGCGCTTTGGAGAAATGGAGCGAGACTGTCAGATCGC
Drosophila melanogaster   CGTGATGGTGGCTTGCGTTTCGGTGAGATGGAGCGTGATTGCCAGATCTC
Neurospora crassa         AGAGACGGTGGTCTCCGTTTCGGTGAAATGGAACGTGACTGTATGATTGC
Oryza sativa              CGGTACGGCGGCGTCAAGTTCGGCGAGATGGAGCGCGACTGCCTCCTCGC
Escherichia coli          CAGTTCGGTGGTCAGCGTTTCGGGGAGATGGAAGTGTGGGCGCTGGAAGC
```

**Figure 4.20**

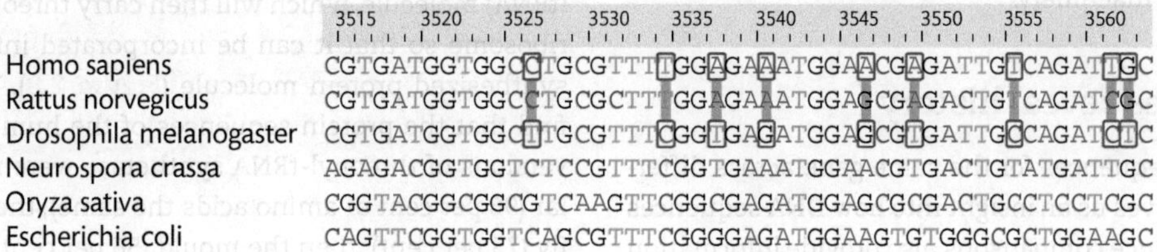

```
                          3515 3520 3525 3530 3535 3540 3545 3550 3555 3560
Homo sapiens              CGTGATGGTGGCCTGCGTTTTGGAGAAATGGAACGAGATTGTCAGATTGC
Rattus norvegicus         CGTGATGGTGGCCTGCGCTTTGGAGAAATGGAGCGAGACTGTCAGATCGC
Drosophila melanogaster   CGTGATGGTGGCTTGCGTTTCGGTGAGATGGAGCGTGATTGCCAGATCTC
Neurospora crassa         AGAGACGGTGGTCTCCGTTTCGGTGAAATGGAACGTGACTGTATGATTGC
Oryza sativa              CGGTACGGCGGCGTCAAGTTCGGCGAGATGGAGCGCGACTGCCTCCTCGC
Escherichia coli          CAGTTCGGTGGTCAGCGTTTCGGGGAGATGGAAGTGTGGGCGCTGGAAGC
```

**Figure 4.21**

```
                    3515      3520      3525      3530      3535      3540      3545      3550      3555      3560
Homo sapiens        CGTGATGGTGGCCTGCGTTTTGGAGAAATGGAACGAGATTGTCAGATTGC
                     R  D  G  G  L  R  F  G  E  M  E  R  D  C  Q  I  A
Rattus norvegicus   CGTGATGGTGGCCTGCGCTTTGGAGAAATGGAGCGAGACTGTCAGATCGC
```

Similarly, in **Figure 4.17**, you can see at position 3532, TTT (human and rat) and TTC (fly) both code for phenylalanine (F), so when the C mutated to T in a distant mammalian ancestor, it continued to be copied and was eventually inherited by both the rat and human lineages. You can also see that the fungus sequence differs from the human sequence by 11 nucleotides, but differs in only one amino acid (M instead of Q: **Figure 4.22**). Rice differs from the animal sequences at six amino acid positions, and the bacterium differs from the animal sequences at eight positions.

*Silent (non-amino-acid changing) mutations are considered in more detail in Chapter 11*

**Figure 4.22**

| | |
|---|---|
| Homo sapiens | RDGGLRFGEMERDCQIA |
| Rattus norvegicus | RDGGLRFGEMERDCQIA |
| Drosophila melanogaster | RDGGLRFGEMERDCQIS |
| Neurospora crassa | RDGGLRFGEMERDCMIA |
| Oryza sativa | YGGGIRFGEMERDALLA |
| Escherichia coli | QFGGQRFGEMEVWALEA |

All of the sequences, from bacteria to humans, have the central amino acid motif 'RFGEME' (arginine-phenylalanine-glycine-glutamic acid-methionine-glutamic acid: **Figure 4.22**). The universality of this motif suggests two things—firstly, that this sequence is so important to the function of the enzyme that any change to it is likely to be disastrous, and secondly, that this sequence was present in the last common ancestor of animals, plants, fungi and bacteria.

A quick search of a protein database reveals that this motif is part of domain 7 of the RNA polymerase II beta protein. Domain 7 interacts with another domain of the enzyme to form the clamp that locks the polymerase enzyme onto the DNA strand. So this amino acid sequence has a critical role to play in the formation of the transcription machinery.

## Hierarchies reveal history

Comparing sequences for the same gene across different species gives us an insight into how DNA sequences evolve. Sequence comparisons also provide information about the relationships between species. The action of copying DNA every generation, occasionally passing on a mutation, leaves a historical record in the genome. Shared inherited changes reveal lines of descent. We can draw the shared sites of this short amino acid sequence as a nested hierarchy (**Figure 4.23**). All of the animals have the same sequence, which differs at one position from fungi. Plants share ten of these amino acids with fungi and animals, and the bacterium shares only eight amino acids with the other species. This pattern is unsurprising, given that human, rat and fly are all more closely related to each other than each is to fungi, and that animals and fungi are more closely related to each other than either is to plants. Bacteria are the most distantly related. So this hierarchy of shared changes reflects the evolutionary relationships between the species. Indeed, we should expect that most heritable characters will show a comparable pattern of shared similarities (see **TechBox 11.1**).

**Figure 4.23**

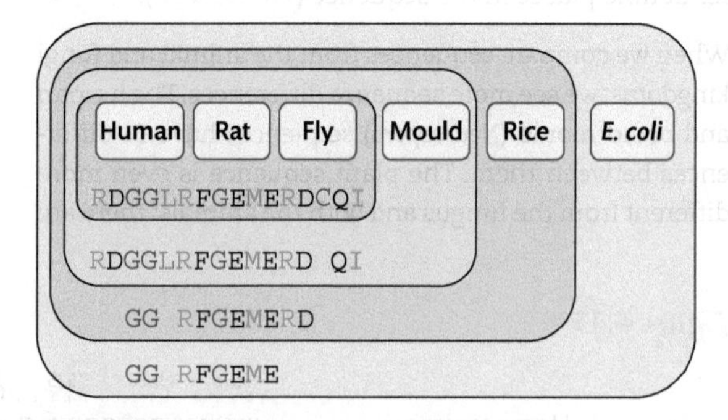


If we choose a different gene and compare it between species we are likely to find a similar pattern. For example, you can compare the gene for threonyl-tRNA synthetase between these species. This gene codes for another key component of the genomic system. Threonyl-tRNA synthetase is an enzyme that binds the amino acid threonine to the appropriate transfer RNA (tRNA) molecule, which will then carry threonine to the ribosome so that it can be incorporated into a newly synthesized protein molecule (**Figure 2.8**). You would find that the protein sequences of the human and rat versions of threonyl-tRNA synthetase were most similar (96 per cent of amino acids the same), then the fruit fly (75 per cent), then the mould (56 per cent), then rice

| | |
|---|---|
| *Bacillus anthracis GJ-2* | CGTCTGCGTTCTGTTGGAGAACTATTACAAAATCAATTCCGTATCGGTCTTTCTCGTAT |
| *Bacillus anthracis GJ-1* | CGTCTGCGTTCTGTTGGAGAACTATTACAAAATCAATTCCGTATCGGTCTTTCTCGTAT |
| *Bacillus anthracis Army* | CGTCTGCGTTCTGTTGGAGAACTATTACAAAATCAATTCCGTATCGGTCTTTCTCGTAT |
| *Bacillus anthracis ATCC* | CGTCTGCGTTCTGTTGGAGAACTATTACAAAATCAATTCCGTATCGGTCTTTCTCGTAT |
| *Bacillus anthracis Sterne* | CGTCTGCGTTCTGTTGGAGAACTATTACAAAATCAATTCCGTATCGGTCTTTCTCGTAT |
| *Bacillus thuringiensis* | CGTCTGCGTTCTGTTGGAGAACTATTACAAAACCAATTCCGTATCGGCCTTTCTCGTAT |
| *Bacillus mycoides* | CGTCTGCGTTCTGTAAGAGAGTTACTACAAAACCAATTCCGTATCGGTCTTTCTCGTAT |
| *Bacillus cereus* | CGTCTGCGTTCTGTAAGTGAATTGTTACAAAACCAATTCCGTATCGGTTTATCTCGTAT |

**Figure 4.24** Part of the RNA polymerase II beta gene from a number of different types of *Bacillus* bacteria, including several different strains of the anthrax-causing *Bacillus anthracis. Bacillus thuringiensis* is the source of the Bt toxin widely used as an insecticide. Bt insecticide can be sprayed on plants, or produced by crops genetically modified to carry the Bt gene (interestingly, Bt spray is allowed under organic farming regulations but GM crops that produce their own Bt are not). *Bacillus cereus* is a soil bacterium that is usually harmless to humans but can cause illness, for example as a source of 'fried rice illness' which can be picked up from eating cooked rice that has been sitting at room temperature for many hours. *Bacillus mycoides* is a very interesting little beast because, apart from an apparent ability to eat the explosive TNT, it forms rather beautiful spiral patterns when growing in culture and the direction of the spiral is heritable.

(55 per cent). Less than 50 per cent of this sequence is identical between all of the species including the bacterium *E. coli*.

In fact, we would expect to see a similar pattern of a nested hierarchy of similarities for other heritable traits. The two mammals are very similar in body organization, in which they differ from the fly. But all three animals share a great many features of cell structure and organization, which are not shared with fungi or plants. Plants, fungi and animals all share features of the eukaryotic cell—such as the cell nucleus—that differ from bacterial cells. All six species—human, rat, fly, mould, rice and *E.coli*—share features of the DNA–RNA-protein genomic system, common to all life on earth.

## *Hierarchies of similarity in* Bacillus

This hierarchy of similarities can be seen at all levels of biological relationships. We have seen how the RNA polymerase II beta gene can be compared between the different kingdoms of life. Now let us use a different part of the gene sequence to look at the way the same gene can reveal relationships between closely related populations and species. **Figure 4.24** shows a section of RNA polymerase II beta gene from a number of different types of *Bacillus* bacteria, a group of spore-forming bacteria found in soil, water and dust.

One particular species of *Bacillus* is deadly to humans (**Figure 4.25**). If *Bacillus anthracis* enters the human body through a cut on the skin, it can cause nasty black ulcers to

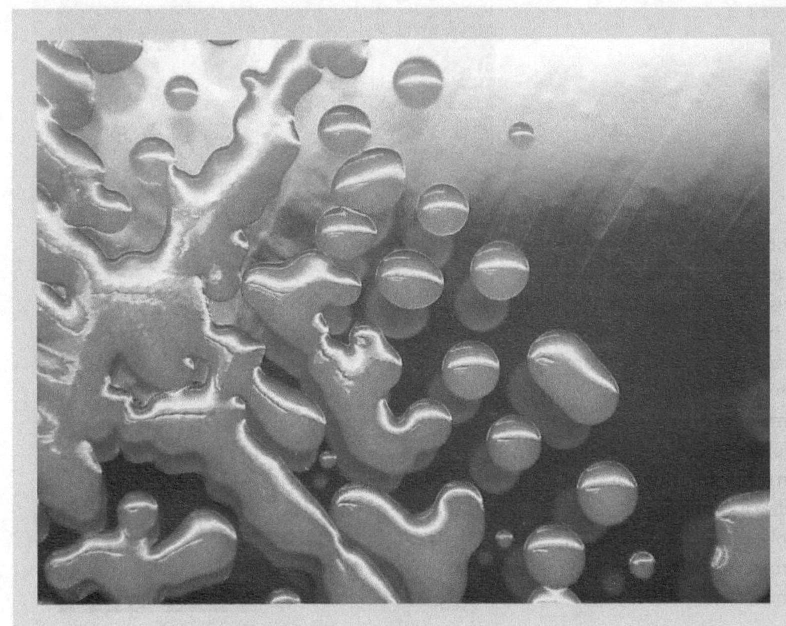

**Figure 4.25** The beauty of bacteria: colonies of *Bacillus anthracis*.

CDC/Courtesy of Larry Stauffer, Oregon State Public Health Laboratory.

develop (cutaneous anthrax). If *Bacillus anthracis* enters the lungs or digestive tract, it can cause a potentially fatal illness (inhalation or gastrointestinal anthrax). Anthrax can be passed from cows and sheep to humans, but cannot be passed from human to human. Its ability to form tough spores makes anthrax an ideal biological warfare agent, because it can be dispersed in powder form and can survive a wide variety of conditions, even low-level radiation.

The importance of anthrax to human health and agriculture, and its potential use in bioterrorism, makes reliable identification of suspected anthrax samples essential (**Figure 4.26**). In **Figure 4.24**, we can see that, for this section of the RNA polymerase II beta gene, the nucleotide sequence is identical between different strains of *Bacillus anthracis*. But this sequence differs from other *Bacillus* species. This observation suggests that DNA sequencing could be used to rapidly identify the presence of anthrax in a suspected bioterrorist sample. You could extract DNA from the sample, and use PCR to amplify the RNA polymerase II beta gene, then sequence it. If the sequence generated was identical to the *B. anthracis* sequence, you could be fairly confident that the sample contained anthrax.

Hierarchies of similarities exist within species as well. Your genome will be more similar to your cousin's than it will be to an unrelated individual because you both have DNA copied from the same grandparents' genomes. The same principle is true for *Bacillus*. Strains that are recently derived from a common stock will have more similar genomes than more distantly related strains. So although the RNA polymerase II beta gene may be identical in all *Bacillus anthracis*, there are other genomic differences that reveal the history of particular anthrax samples.

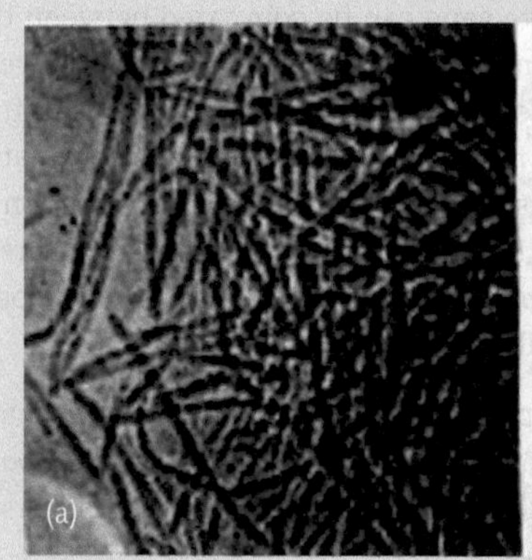
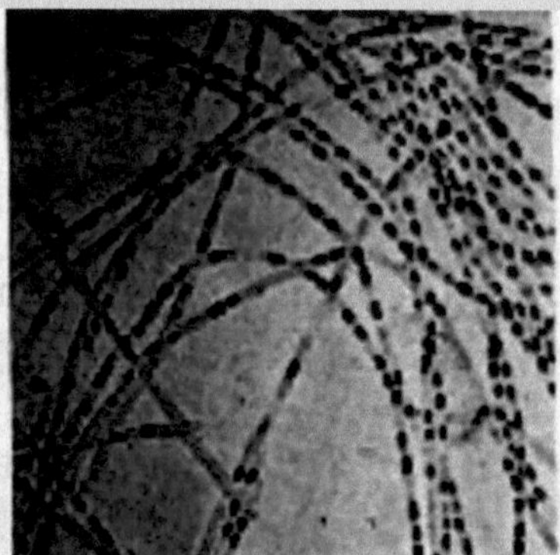
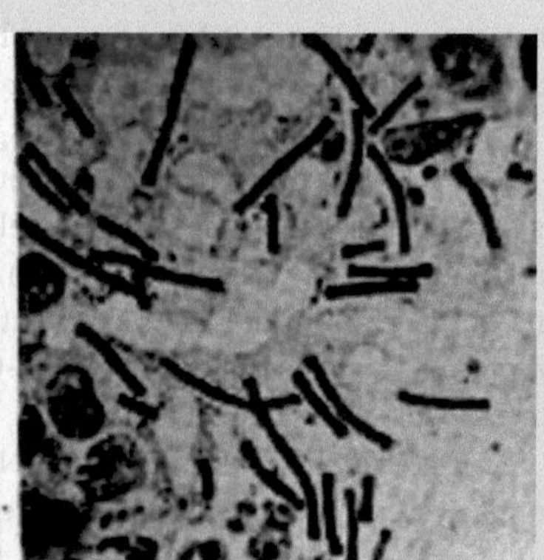

**Figure 4.26** Infection detection: (a) Robert Koch's original photomicrographs of *Bacillus anthracis* taken in 1877. Koch's work on *Bacillus* was pioneering not only in his use of micrographs to make images of bacteria, but also because his study of anthrax was, arguably, the first case where an infectious agent was proved beyond doubt to be the cause of a disease. (b) Koch's work on this and other infectious diseases earned him a Nobel Prize in 1905 and his approach continues to be used today by those who follow 'Koch's postulates' to demonstrate disease causation: association (agent found in affected but not unaffected individuals), isolation (disease-causing organism is isolated from affected individuals), inoculation (isolated organism causes disease in healthy individuals) and re-isolation (to prove that the original organism can be recovered from inoculated individuals).

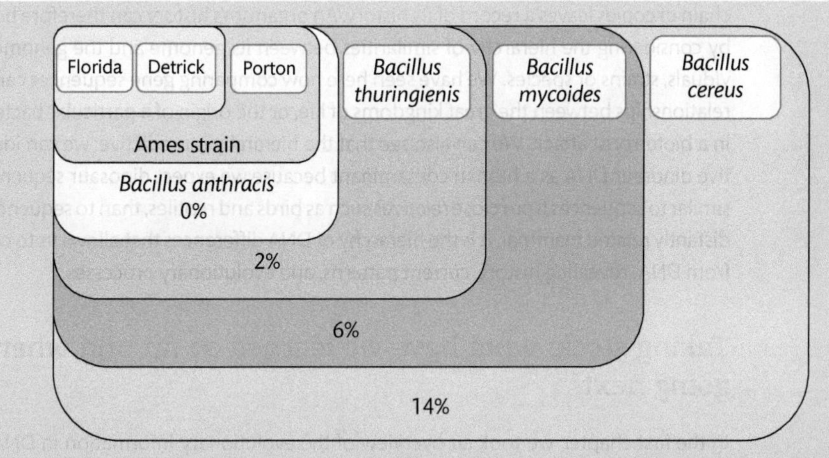


**Figure 4.27** Approximate percentage similarity between RNA polymerase II beta genes between various types of *Bacillus* bacteria. Although the *Bacillus anthracis* strains all have the same sequence for this particular gene, genomic analysis has identified other heritable variants that allow the relationships between strains to be uncovered and the source of anthrax samples to be traced.

In 2001 when America was reeling from the shock of the attacks on the World Trade Center and Pentagon, five people died and 17 more were infected by anthrax spores that had been sent through the post. Can sequence analysis identify the source of the anthrax? Previous research suggested that anthrax genomes are nearly identical, making identification of different strains difficult. But by using whole or nearly complete genome sequences, researchers were able to identify a number of polymorphic sites that differed between strains, including single nucleotide differences (SNPs), variable number tandem repeat regions (VNTRs), and insertions and deletions (indels). Some of these changes may have no noticeable effect on phenotype but several were predicted to change important genes, thus might have an effect on the activity or infectivity. Furthermore, genetic analysis of *Bacillus anthracis* strains demonstrated that these genetic differences can arise very rapidly, even in as little as a few years, so can arise within particular lab cultures of anthrax. In this case, the investigation demonstrated that the Florida bioterrorist anthrax was most similar to the strain maintained and distributed by the US military research facility Fort Detrick. So we could draw a nested hierarchy of genetic differences between different *Bacillus* strains, just as we did between members of different kingdoms (**Figure 4.27**). This suggests that whole genome sequences may be used in future to identify the sources of outbreaks, even for a bacterial species previously considered to show little genetic variation.

# Conclusion

Understanding DNA replication is important for all biologists, not just biochemists or geneticists. Everyone who wants to understand the basis of life on earth should familiarize themselves with the way that the information in DNA is copied by template reproduction. The astounding complexity of the DNA replication machinery ensures that the genome is copied accurately and efficiently at cell division. This process is considerably simplified in the laboratory, where the entire DNA sample is melted to expose single strands, primers are added to provide a starting point for a domesticated polymerase enzyme to work from, and the sample put through repeated cycles of melting and copying to produce millions of copies of the same DNA sequence.

No copying process is perfect. Errors in laboratory DNA amplification are to be avoided at all costs. But ironically it is errors in natural DNA replication that provide us with one of the most useful tools in biology. Errors in DNA replication can be passed on to the next generation, thus the chain of copies leaves a record of its history. An organism's history can therefore be reconstructed by considering the hierarchy of similarities between its genome and the genome of other individuals, strains or species. We have seen here how comparing gene sequences can illuminate the relationships between the great kingdoms of life, or the origin of a particular bacterial strain used in a bioterrorist attack. We can also see that the hierarchy is predictive: we can identify the putative dinosaur DNA as a human contaminant because we expect dinosaur sequences to be more similar to sequences from close relatives such as birds and reptiles, than to sequences from a more distantly related mammal. It is the hierarchy of DNA differences that allows us to construct stories from DNA, revealing history, current patterns, and evolutionary processes.

## Taking stock: what have we learned so far and where are we going next?

In the first chapter, we took an overview of the evolutionary information in DNA. We saw how it can be used to shed light on individuals, populations, and lineages. To understand how DNA stores information, we took a brief look at DNA structure in Chapter 2. We saw how the complementary base pairing of nucleotides provides a means of transferring information between nucleic acids, providing a pathway for the genetic instructions to move from the genome to the sites of production of the working parts of a cell, as well as a mechanism for the transfer of information between generations. In Chapter 3 we saw how mutation changes the information in the genome, by altering the way genes function, and by creating inherited differences that will be passed on to descendants. In Chapter 4 we learned how DNA is copied almost, but not entirely, without error. The process of copying with occasional errors creates the historical information in DNA by creating a chain of increasingly dissimilar genomes.

So now we know how mutations occur and we have seen how DNA is copied, it's time to think about the evolutionary consequences of it all. In Chapter 5, we look at how the machinery for DNA copying leads to expansion of genomes into enormous, complex collections of interacting DNA sequences. In Chapter 6, we consider how the information in the genome is used to build and operate a complex organism, by focusing on gene expression and evolution. Then in Chapters 7 and 8 we get to grips with the great evolutionary processes that drive the evolution of genes and genomes, natural selection and drift, the engines of the diversification of life.

### DNA replication

- DNA replication is incredibly accurate, but rare errors occur.
- Each nucleotide strand forms the template for the construction of a new strand, formed by complementary base pairing.
- Replication begins with opening and unwinding the double helix to expose single strands to act as templates.
- Short primer sequences that pair to the exposed single strands act as starting points for DNA synthesis.
- DNA polymerase adds nucleotides to the 3' end of a new strand, moving toward the replication fork on the leading strand.
- On the lagging strand, DNA polymerase moves away from the replication fork, making a series of nucleotide strands that are then joined together.

- Accuracy of DNA replication is maintained by:
  - base selection (making sure the correct complementary base is added to the growing strand)
  - proofreading (removing incorrect bases before continuing to grow the new strand)
  - post-replication repair (detecting and replacing incorrect base pairs in a new DNA helix).

### DNA amplification

- Many molecular techniques require large numbers of copies of the target DNA sequence.
- DNA amplification is a greatly simplified form of DNA replication, achieved by incubating DNA samples with enzymes and other chemicals.
- Heat or enzymes are used to denature DNA into single strands to expose the base sequences for template copying.
- Primers, short sequences of DNA, are added to the DNA sample to stick to the exposed sequences and act as starting points for DNA synthesis.
- Polymerase enzymes start at the primers bound to DNA sequences in the sample and make new complementary strands.
- Newly synthesized DNA acts as templates for copying in subsequent cycles of heating and cooling, leading to exponential increase in number of copies of the target sequences.

### Copy errors create hierarchies

- Replication is amazingly accurate but never error-free, so the more times DNA is copied, the more errors will accumulate.
- Copy errors can be perpetuated down generations if they do not prevent the cell from functioning and reproducing.
- Mutations copied with the genome will be added to by more copy errors, creating a hierarchy of differences between DNA sequences.
- The hierarchy of differences between DNA sequences reveal their history, allowing reconstruction of the chain of copies that gave rise to the nested pattern of similarities.

## Ideas for discussion

**4.1** 'Male-driven evolution' results from a difference in the number of DNA copies in the male and female germlines every generation. Can you think of other examples of this copy-frequency effect, where some genomes are copied more often than others so change at a faster rate?

**4.2** Why is DNA replication in living cells so much more complicated than DNA amplification in the lab? Could organisms evolve a system as simple as PCR?

**4.3** If you take DNA from two sources, melt it to single strands, cool it to reanneal, then reheat, the temperature at which the hybrid strands melt is a sign of how similar they are. Why? What could you find out by comparing sequences in this way?

# Sequences used in this chapter

| Table 4.1: RNA polymerase sequences used in Figure 4.19 | |
|---|---|
| Description from GenBank entry | Accession |
| *Homo sapiens* polymerase (RNA) II (DNA directed) polypeptide B | NM_000938 |
| *Rattus norvegicus* polymerase (RNA) II (DNA directed) polypeptide B (Polr2b) | NM_001106002 |
| *Drosophila melanogaster* RNA polymerase II 140kD subunit (RpII140) | NM_057358 |
| *Neurospora crassa* OR74A DNA-directed RNA polymerase II 140 kDa polypeptide partial mRNA | XM_952013 |
| *Oryza sativa* (japonica cultivar-group) chromosome 10, section 62 of 77 of the complete sequence—partial sequence | AE017108 |
| *E.coli* RNA polymerase beta subunit (rpoB and rpoC) genes | M38303 |

| Table 4.2: *Bacillus rpoB* sequences used in Figure 4.24 | |
|---|---|
| Description from GenBank entry | Accession |
| *Bacillus anthracis* strain GJ-2 RNA polymerase beta subunit (rpoB) | AY169514 |
| *Bacillus anthracis* strain GJ-1 RNA polymerase beta subunit (rpoB) | AY169513 |
| *Bacillus anthracis* strain Army RNA polymerase beta subunit (rpoB) | AY169512 |
| *Bacillus anthracis* strain ATCC 14185 RNA polymerase beta subunit (rpoB) | AY169511 |
| *Bacillus anthracis* strain Sterne RNA polymerase beta subunit (rpoB) | AY169510 |
| *Bacillus thuringiensis strain IMSNU* 10051 RNA polymerase beta | AY169538 |
| *Bacillus mycoides* strain KCCM 40260 RNA polymerase beta subunit (rpoB) | AY169540 |
| *Bacillus cereus* strain ATCC 9634 RNA polymerase beta subunit (rpoB) | AY169515 |

# Examples used in this chapter


Barbrook, A. C., Howe, C. J., Blake, N., Robinson, P. (1998) The phylogeny of the *Canterbury Tales. Nature*, Volume 394, page 839.

Di Franco, C., Beccari, E., Santini, T., Pisaneschi, G., Tecce, G. (2002) Colony shape as a genetic trait in the pattern-forming *Bacillus mycoides. BMC Microbiology*, Volume 2, page 33.

Lin, H-y., Yu, C-P., Chen, Z-l. (2013) Aerobic and anaerobic biodegradation of TNT by newly isolated *Bacillus mycoides. Ecological Engineering*, Volume 52, page 270.

Read, T. D., Salzberg, S. L., Pop, M., Shumway, M., Umayam, L., Jiang, L., Holtzapple, E., Busch, J. D., Smith, K. L., Schupp, J. M. (2002) Comparative genome sequencing for discovery of novel polymorphisms in *Bacillus anthracis. Science*, Volume 296, page 2028.

Tehrani, J., Collard, M. (2002). Investigating cultural evolution through biological phylogenetic analyses of Turkmen textiles. *Journal of Anthropological Archaeology*, Volume 21, page 443.

HEROES OF THE GENETIC REVOLUTION

4

# Rosalind Franklin

*"As a scientist Miss Franklin was distinguished by extreme clarity and perfection in everything she undertook. Her photographs are among the most beautiful x-ray photographs of any substance ever taken."*

Bernal, J. D. (1958) 'Dr Rosalind E. Franklin: Obituary' *Nature*, Volume 182, page 154

**EXAMPLE PUBLICATIONS**

Franklin, R. E., Gosling, R. G. (1953) Molecular configuration in sodium thymonucleate. *Nature*, Volume 171, page 740.

Franklin, R. E. (1955) Structure of tobacco mosaic virus. *Nature*, Volume 175, page 379.

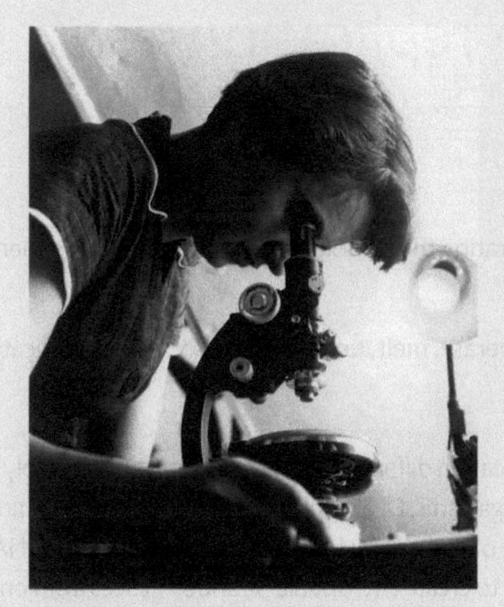

**Figure 4.28** Rosalind Franklin (1920–1958): Although her structural detective work unlocked the secret of the double helix, she was not named on the 1962 Nobel prize for the structure of DNA. Nobel prizes are never awarded posthumously, nor can they be shared by more than three people, so the Nobel prize for DNA went to James Watson, Francis Crick and Franklin's colleague Maurice Wilkins. The unfair portrayal of Franklin in Watson's memoir sparked outrage, which serves as a reminder that no one account of a scientific discovery should be taken as an unbiased history of events.

Reproduced from Glynn, J. (2012) My Sister Rosalind Franklin, by permission of Oxford University Press.

Rosalind Franklin (Figure 4.28) was a dedicated scientist from the time she was a teenager, studying physics and chemistry at school, then graduating from Newnham College at Cambridge University in 1941 (seven years before Cambridge finally started awarding degrees to female students). Although she started a research position in the physical chemistry lab at Cambridge, she was unhappy with her supervisor and resigned, instead moving to a coal research institute to fulfil her national service. Her work on structural chemistry—which she described as focusing on the 'holes in coal'—contributed to assessing the use of coal for wartime purposes such as gasmasks and earned her a doctorate from Cambridge, awarded in 1945. She learnt X-ray diffraction techniques on a productive fellowship at the Centre National de la Recherche Scientifique (CNRS) in Paris.

In 1951, she returned to England to work at King's College London, and used her skills in X-ray diffraction to tease apart the structure of DNA. By studying the X-ray patterns, and comparing two alternative conformations of DNA under different conditions, she demonstrated that DNA was a helix with the phosphate molecules on the outside (contrary to some other proposals that had the bases facing out). It was Franklin's X-ray diffraction images that revealed the helical structure of DNA to Watson and Crick, who then used her careful measurements of the structure of the helix to build their three-dimensional model. Franklin's paper describing the structure of the A-form of DNA was submitted to the journal *Acta Crystallographica* in March 1953, but it was not published until September, several months after the paper on DNA structure by Watson and Crick was published in *Nature* in April. She was preparing to publish work on the helical structure of the B-form when she received news of Watson and Crick's model. She added a sentence to her draft manuscript stating that it supported their structure and the paper was published alongside Watson and Crick's paper in *Nature* in April 1953.

She published several more papers on DNA before moving her focus to the structure of viruses. It is this work on viral structure that is referred to in the inscription on her tombstone, which says 'Her research and discoveries on viruses remain of lasting benefit to mankind'. Rosalind Franklin died tragically young in 1958, only 37 years old.

"...my method of thought and reasoning is influenced by a scientific training – if that were not so my scientific training will have been a waste and a failure. But you look at science (or at least talk of it) as some sort of demoralizing invention of man, something apart from real life, and which must be cautiously guarded and kept separate from everyday existence. But science and everyday life cannot and should not be separated."

Rosalind Franklin in a letter to her father, Ellis Franklin, 1940

TECHBOX
**4.1**

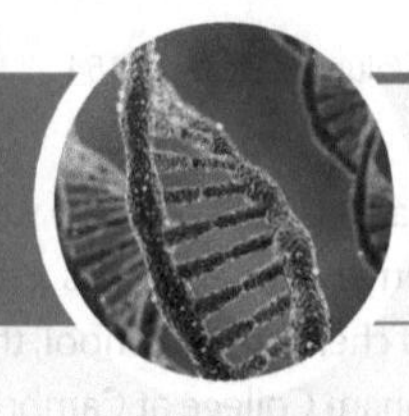

# DNA replication

**RELATED TECHBOXES**

TB 2.1: DNA structure

TB 4.2: DNA amplification

**RELATED CASE STUDIES**

CS 3.2: Mutation rate

CS 4.1: Replication

**Summary**

DNA is copied by separating the double helix and making a complementary copy of each strand.

**Keywords**

primer, helicase, polymerase, melt, unwind, topoisomerase, replication fork, Okazaki fragment, 5′→3′, primase

The details of DNA replication differ between organisms (most notably between prokaryotes and eukaryotes). But in all organisms, DNA replication involves the formation of a replication complex made of a large number of enzymes, which split the two strands of DNA and construct two new complementary strands to create two double-stranded helices from one. The basic series of events is:

1. Melt: strands are separated at the origin of replication.
2. Unwind: replication forks expose single strands to act as templates.
3. Prime: synthesis of a new strand begins with a short RNA strand.
4. Grow: DNA polymerase adds nucleotides to form a complementary strand.

Since different species have different enzymes, here we will use enzyme names ending in -ase, like polymerase or helicase, to represent particular activities, which might be performed by different proteins in different organisms.

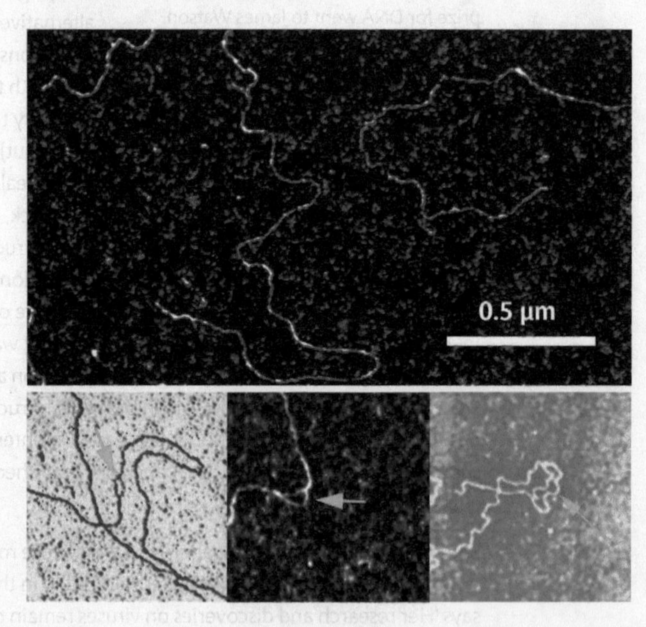


**Figure 4.29** Electron micrographs of replication bubbles (marked with arrows) in DNA from budding yeast.

Reproduced courtesy of Professor Zhifeng Shao, University of Virginia.

**Melt**

DNA replication occurs by template reproduction: the base sequence on one strand is used to create a new complementary strand (**Figure 4.5**). So to replicate DNA, the first step is to separate the two strands of the double helix so that each can act as a template for the production of a new strand. Replication origins are sequences in the genome (typically less than 250 bases in bacteria, but often much longer in eukaryotes) that are recognized by enzymes that can locally destabilize the DNA helix to break the hydrogen bonds between base-pairs. Then helicase 'melts' the double helix, separating base pairs to open up the two DNA strands to form a replication bubble (**Figure 4.29**).

**Unwind**

If you take a two-stranded rope or string, fixed at one end, and pull the strands apart, you will soon find that the rope snarls up and the strands cannot be separated. The same thing would happen to the DNA double helix if it were not for the unwinding enzymes (topoisomerases) that prevent the formation of positive supercoils as the strands are separated. Topoisomerases cut one or both DNA strands ahead of the replication fork, allow the strands to unwind, then rejoin the strands.

In circular chromosomes, such as those found in mitochondria or bacteria, two replication forks move in opposite directions around the chromosome, until they meet on the other side of the chromosome, having made an entire copy of the genome (**Figure 4.30**).

**Figure 4.30**

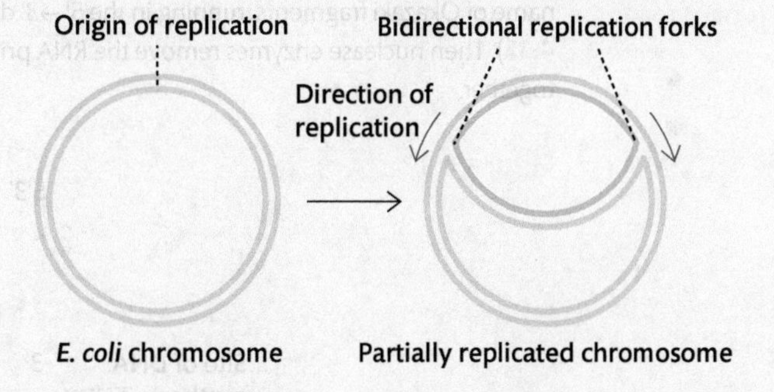


Linear eukaryotic chromosomes split at numerous locations, then the replication forks move in both directions along the chromosome until they meet up (**Figure 4.31**).

**Figure 4.31**

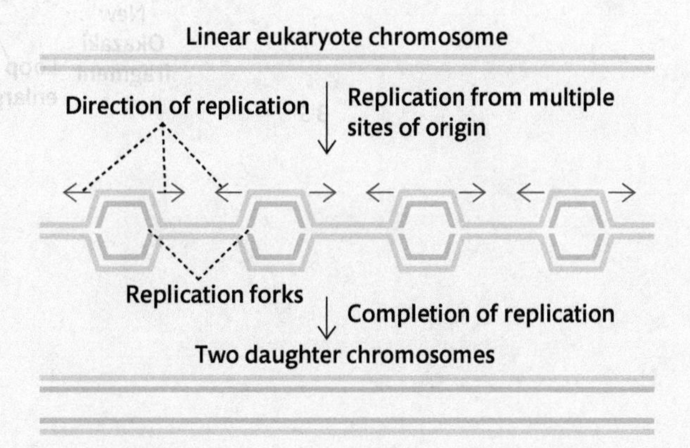


**Prime and grow**

DNA polymerase, the enzyme responsible for making new DNA strands, works by adding nucleotides to an existing polynucleotide chain: it forms a phosphodiester bond between the OH group attached to the 3' carbon of the sugar of the last nucleotide in the chain to the phosphate of the incoming nucleotide (**TechBox 2.1**). This mode of chain extension introduces two important limitations. If DNA polymerase adds nucleotides to the 3' end of an existing nucleotide chain, then how does it get started? And how does it copy both the 5'→3' and 3'→5' strands?

Unlike DNA polymerase, RNA polymerase is able to start from scratch, joining two nucleotides together to start a chain. So primase begins the DNA replication process by making a short RNA molecule (approximately ten nucleotides long), matching nucleotides to the open replication fork. This short RNA strand acts as a primer that DNA polymerase can then add more nucleotides to. The RNA primer is later excised from the newly synthesized DNA strand.

Primase also plays a role in solving the problem of the directionality of DNA polymerase. Each DNA molecule consists of two antiparallel strands, one that runs 3'→5', and one that runs 5'→3' (**Figure 4.5**). DNA polymerase can only make a new strand in the 5'→3' direction, by adding nucleotides to the 3' carbon atom of the sugar of the previous base. So, for the leading strand, polymerase can move along the exposed strand toward the replication fork, making a 5'→3' strand to match the 3'→5' template strand. But, for the lagging strand, the template runs 5'→3'. DNA polymerase can't synthesize in the 3'→5' direction because it can't add nucleotides to the 5' end of the new chain. So the polymerase enzyme has to travel along the strand in the opposite direction to the replication fork. So, as the DNA unwinds, primase repeatedly places short RNA primers along the template strand, between 1,000 and 2,000 bases apart. DNA polymerase then uses these primers to make a series of short fragments, given the charming name of Okazaki fragments, running in the 5'→3' direction along the exposed strand (**Figure 4.32**). Then nuclease enzymes remove the RNA primers and ligase joins the Okazaki fragments together.

**Figure 4.32**

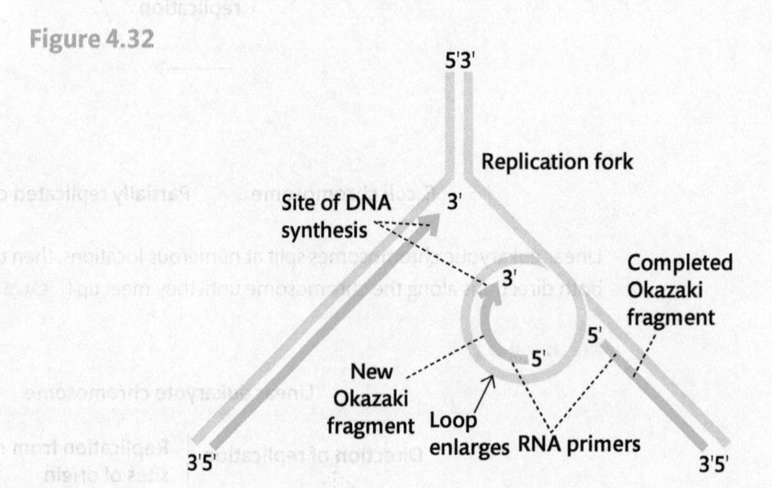

TECHBOX
4.2

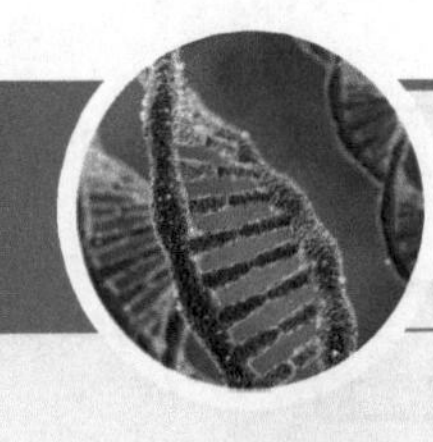

# DNA amplification

RELATED
TECHBOXES

**TB 4.1:** DNA replication

**TB 5.1:** Sanger sequencing

**TB 5.2:** High-throughput sequencing

RELATED
CASE STUDIES

**CS 1.2:** Forensics

**CS 2.1:** Environmental DNA

**Summary**

To make multiple copies of a DNA sequence, you need to expose single strands and attach primers as starting points for polymerase to make complementary copies.

**Keywords**

primer, nucleotides, dNTP, polymerase, PCR, thermal cycling, rolling-circle, strand displacement amplification, helicase-dependent amplification, isothermal, primer walking

Because DNA amplification is a domesticated version of DNA replication, it is important to read through the DNA replication box first (**TechBox 4.1**). DNA amplification will be explained with respect to the processes that occur in normal DNA replication, because it follows the same basic pattern of melt/unwind, prime, and grow.

## Ingredients

DNA amplification, like many lab procedures, is like following a recipe, with a set of ingredients and a series of 'cooking' instructions. The ingredients for a DNA amplification protocol might include:

- ***Primer:*** oligonucleotides (short strings of nucleotides) that are complementary to a part of the target sequence in the sample DNA, so they can provide a starting point for DNA polymerase (**TechBox 6.2**).
- ***Nucleotides:*** the raw material for DNA synthesis is provided in the form of deoxyribonucleotides (dNTP) each consisting of a phosphate plus sugar plus base (**TechBox 2.1**). Four dNTPs need to be provided, one for each of the four 'letters' of DNA (i.e. dATP, dCTP, dGTP and dTTP).
- ***Polymerase:*** the enzyme that builds new DNA strands by matching nucleotides to the single-stranded DNA in solution (**TechBox 4.1**).
- **$Mg^{2+}$:** magnesium is an essential cofactor for polymerase function.
- ***Buffer:*** provides the correct chemical environment for the sequencing reactions.

## Recipe

The standard approach to amplifying DNA is called polymerase chain reaction (PCR). PCR builds copies of a DNA sequence through multiple rounds of the following cycle (often up to thirty cycles: **Figure 4.33**):

1. Melt (denature): heat to separate double helices into single strands.
2. Prime (anneal): cool so that the primers bind to the target sequences in the DNA.
3. Grow (elongate): polymerase adds nucleotides to make a complementary strand of DNA.
4. Melt again (denature): so that all of the DNA in the sample, both original and newly-copied, can be used as templates for the next round of prime and grow.

If you compare this to the DNA replication box (**TechBox 4.1**) you will see that 'unwind' is missing. That's because, in PCR, heat is used to denature all DNA in the sample into single strands, so there are no helices to be unwound. However, there are DNA amplification methods that use unwinding enzymes rather than heat to separate the strands.

1. ***Denature (melt):*** Many DNA amplification methods use heat to denature all the DNA in the sample. Because melting temperature depends on the strength of bonds between the DNA helices, it may vary between samples. For example, DNA with a high proportion of Gs and Cs generally requires higher melting temperatures, because there are three hydrogen bonds between each GC pair but only two for each AT pair. It may be necessary to adjust the temperatures to get a PCR reaction to work well for your samples.

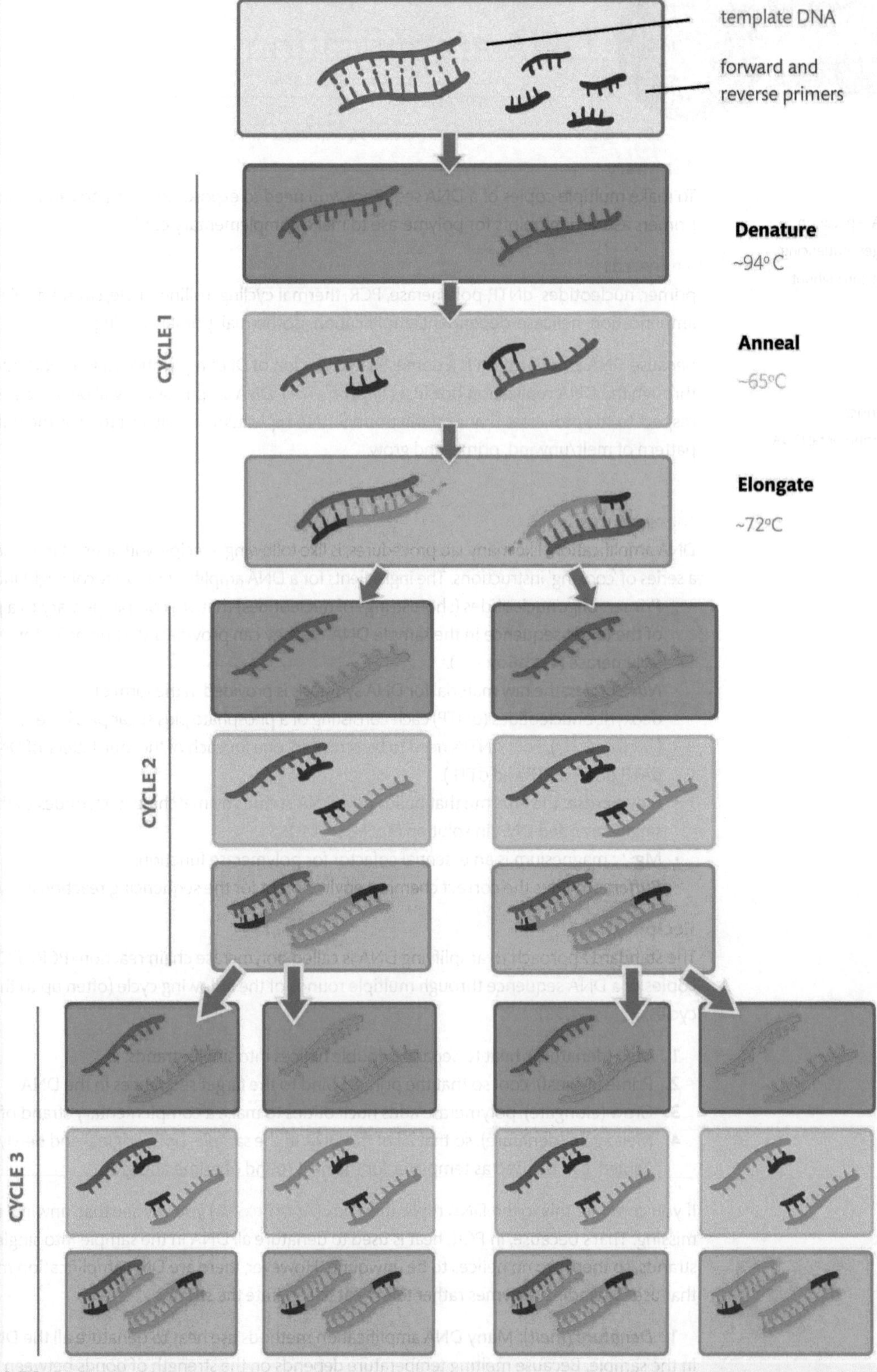


**Figure 4.33** Schematic representation of the amplification of DNA by the polymerase chain reaction (PCR). The template DNA is heated so it separates (denature), then cooled so that primers will stick to the templates (anneal), then warmed again to allow copying (elongate). You can see that the more cycles of denature–anneal–elongate, the more synthetic DNA copies there are in the solution, and these synthetic DNA strands act as templates in the next round of heating and cooling. So the target DNA is rapidly overtaken by the newly copied DNA, and the amount of DNA increases exponentially with each cycle.

Not all DNA amplification techniques require temperature cycling to expose single-stranded DNA templates. For example, strand displacement amplification uses a single round of heating with enzymes and primers to begin a reaction that then continues at low temperature: endonucleases cut completed strands which are then displaced by the polymerization of a new strand. Helicase-dependent amplification avoids heating altogether by using a helicase enzyme to unwind DNA at low temperature. By circumventing the need for a thermal cycling machine, single-temperature (isothermal) amplification techniques might allow the development of mobile DNA amplification kits that could be taken out of the laboratory.

2. ***Anneal (prime)*:** There are two roles to priming. One is to provide a starting block for polynucleotide synthesis, because DNA polymerase cannot begin a new chain from scratch (**TechBox 4.1**). The second is to specify the DNA sequence you wish to amplify. The success of a PCR reaction depends on designing short DNA sequences (each around 18–30 nucleotides long) that will stick to the right sequence at the right time with the right strength (**Techbox 6.2**). In fact, PCR usually uses two primers, one which defines the beginning of the sequence to be copied, and one which defines the end. This requires some knowledge of the sequence you wish to amplify, however primer walking can be used to extend amplification into unknown territory: the PCR product from one reaction can be used to design primers to sequence an adjacent unknown sequence. Alternatively, random primers can be used to amplify large amounts of the genome by providing lots of different starting points for DNA synthesis. Not all DNA amplification techniques use pairs of linear primers. For example, padlock probes have their recognition sequence split into two connected by a linking sequence, so when both ends hybridize to the target sequence, the probe forms a circle. These circularized probes can be labelled for gene detection, or used to prime rolling circle amplification.

3. ***Elongate (grow)*:** There are many different 'domesticated' polymerases which are used to copy DNA sequences *in vitro*. All of them work by adding nucleotides to the 3' end of a nucleotide chain (hence the need for a primer), matching the sequence on a template strand (hence the need for heat or enzymes to expose single-stranded DNA). The most common is the *Taq* DNA polymerase (**Figure 4.9**). But there are other thermostable polymerases which can survive thermal cycling, such as *Pfu* and *Pwo* (from the archaebacteria *Pyrococcus furiosus* and *Pyrococcus woesei*). Some amplification techniques use multiple polymerases. Long-range PCR techniques amplify much longer sequences than standard PCR by combining the strengths of two polymerases, such as *Taq* (efficient 5'→3' polymerase activity) and *Pwo* (3'→5' proofreading abilities). Isothermal DNA amplification techniques can make use of a wider range of polymerases, because they do not need to survive temperature cycling.

4. ***Repeat*:** In PCR, the cycle of melt, prime and grow is repeated again and again, often around 30 times. The number of copies of the target sequence increases exponentially, as each round copies the original DNA plus the copies made previously. Temperature cycling is usually done by placing test tubes of DNA plus reagents in a thermocycler, a machine that is programmed to move between temperatures at defined times. In both PCR and isothermal amplification, enzymes continue to copy both the original sequences and the copies previously produced, but eventually the copies overwhelm the original DNA. This means that any bias in amplification will be propagated and increased as the reaction proceeds. This is a problem if you are using DNA amplification to characterize the whole genome, because you might end up with some sequences overrepresented and some underrepresented.

CASE STUDY 4.1

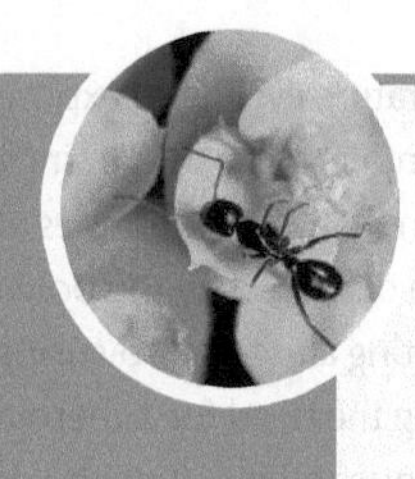

# Replication: taller plants have lower rates of molecular evolution

**RELATED TECHBOXES**

**TB 1.1:** GenBank

**TB 4.1:** DNA replication

**RELATED CASE STUDIES**

**CS 3.2:** Mutation rate

**CS 13.2:** Diversification

Lanfear, R., Ho, S. Y. W., Davies, T. J., Moles, A. T. A., Aarssen, L., Swenson, N. G., Warman, L., Zanne, A. E., Allen, A. P. (2013) Taller plants have lower rates of molecular evolution: the rate of mitosis hypothesis. *Nature Communications*, Volume 4, page 1879.

> *From an evolutionary standpoint, trees have several intriguing and apparently paradoxical features. In particular, they often have high levels of genetic diversity but experience low nucleotide substitution rates and low speciation rates... why [do] trees have such a low pace of evolution at longer timescales, both in terms of DNA sequence and character change within lineage and in terms of diversification rate.*[1]

**Keywords**

generation time effect, sister-pair comparisons, species richness, DNA replication, germline, mutation

## Background

DNA replication is never entirely error-free, so the more times DNA is copied, the more copy errors it will accumulate. This copy-frequency effect might be the basis of the generation time effect in animals, because species with short generations copy their germline DNA more often per unit time[2]. But, unlike most animals, flowering plants don't have a dedicated germline: reproductive tissues form on the periphery of the body, as flowers grow on the tips of stems and trunks (**Figure 4.34**). This means that a tall plant, which must grow a large stem before

**Figure 4.34** Somatic mutation: Researchers used fruit set from controlled pollination in blueberry bushes to examine the accumulation of mutations as cells divide along growing branches[3]. Flowers pollinated by pollen taken from the same branch set more fruit than flowers pollinated by pollen from a more distant branch on the same bush. The researchers concluded that the more cell divisions have occurred between two flowers, the more different their genomes will be due to the accumulation of replication errors, so reducing the compatibility of genomes taken from different parts of the plant.

flowering, might have more cell divisions per generation than a short plant. On the other hand, short plants typically have faster rates of growth, so might have more cell divisions per unit time than taller plants. So which should have faster rates of molecular evolution, taller or shorter plants?

### Aims

Large plants often have lower absolute growth rates[1], so they would be expected to undergo fewer cell divisions in their growing stems per unit time. Fewer cell divisions means fewer opportunities for replication errors to occur, so a taller plant might collect fewer DNA replication errors per unit time than a shorter plant. These researchers asked whether the relative rate of molecular evolution is influenced by plant height.

### Methods

This study is based on sister group analysis, which means that pairs of closely related plant families were compared in order to see if the family with the shorter average height tended to also have the fastest rate of molecular evolution. Each of the two sister families have had the same amount of time to accumulate DNA sequence changes since their shared common ancestor, so differences in the number of changes accumulated in each family may reflect variation in the rate of molecular evolution. Each sister comparison becomes one datapoint in a statisical analysis. Sixty-nine pairs of flowering plant families were compared. For each comparison, the average height of species was compared to the rate of molecular evolution estimated from chloroplast and nuclear sequences representing each family. Environmental factors were also included, using family-level estimates of species richness, temperature, levels of UV radiation and latitude. They compared the number of pairs in which the family with the taller average height had the slower rate of molecular evolution using a signs test, which compares the observed direction of rate changes across pairs to that expected if rate variation was random with respect to average height.

### Results

More often than expected by chance, the plant family with the shorter average height also had the faster rate of molecular evolution. Regression analyses showed that, in addition to a significant influence of height on rates of change of both chloroplast and nuclear sequences, rates of molecular evolution in the chloroplast sequences were associated with differences in species richness between families. There was no significant effect of the environmental variables on rate of molecular evolution.

### Conclusions

Average height for a family was consistently and significantly associated with rates of molecular evolution in this analysis (**Figure 4.35**). Some plant families contain species of a fairly similar height, but many will have very disparate heights: for example the Proteaceae includes both ground hugging plants and tall trees. So the significant signal for height in this dataset either suggests that the trend is strong enough that it can be detected even with average differences in height, or that height co-varies with some other property that varies between families and has a strong influence on rates of molecular evolution.

### Limitations

The interpretation of these results in terms of a copy number effect rests on the assumption that taller plants will have slower rates of cell division in the cell lines leading to reproductive tissues, so it doesn't provide a direct test of the hypothesis that frequency of genome copies drives rate of molecular evolution. Height only accounts for a fifth or less of the variation in rates, so as with most things in biology, this study does not provide a simple explanation for differences in substitution rates between plant families and there are bound to be many other factors that influence patterns and rates of molecular evolution. The lack of a relationship between the environmental variables and rates of molecular evolution may reflect low power, so should not be considered a demonstration of a lack of influence of UV or temperature on rates.

**Figure 4.35** Tall timber: Mountain ash trees (*Eucalyptus regnans*) with tree fern understory in Sherbrooke Forest in the Dandenong Ranges, Victoria, Australia.

Photograph by Lindell Bromham.

### Further work

Teasing apart the factors governing rates of genome copying and molecular evolution in plants will require studies where species that differ consistently in traits of interest can be compared. In particular, finer scale analyses may be needed to examine environmental factors that could influence growth rates, and thus cell division rates. Since plants do not have a dedicated germline, the rate of accumulation of mutations can be compared across tips of long-lived plants to investigate the accumulation of mutations[3].

### Check your understanding

1. Why do tall plants have more genome replications per generation than their shorter relatives?
2. Why did these researchers compare pairs of plant families, rather than treating each family as one datapoint in an analysis of the relationship between rates and height across all families?
3. If taller plants have more cell divisions per generation, why don't they collect more replication errors per unit time and so have a faster rate of molecular evolution?

### What do you think?

What effect might these patterns of rate variation have on the use of DNA sequence analysis to reconstruct the evolutionary history of plant groups?

### Delve deeper

If growing taller involves more cell divisions, and each cell division carries a risk of copy error, might larger plants evolve better DNA repair in order to compensate and reduce their mutation rates?

### References


1. Petit, R.J., Hampe, A. (2006) Some evolutionary consequences of being a tree. *Annual Review of Ecology, Evolution, and Systematics*, Volume 2006, page 187.
2. Bromham, L. (2009) Why do species vary in their rate of molecular evolution? *Biology Letters*, Volume 5, page 401.

3. Bobiwash, K., Schultz, S., Schoen, D. (2013) Somatic deleterious mutation rate in a woody plant: estimation from phenotypic data. *Heredity*, Volume 111, page 338.

CASE STUDY 4.2

# Hierarchies: genes and language track human settlement of Oceania

**RELATED TECHBOXES**

**TB 9.1:** Phylogeography

**TB 14.2:** Molecular dating

**RELATED CASE STUDIES**

**CS 2.2:** Ancient DNA

**CS 12.1:** Epidemiology

Hurles, M. E., Matisoo-Smith, E., Gray, R. D., Penny, D. (2003) Untangling Oceanic settlement: the edge of the knowable. *Trends in Ecology and Evolution*, Volume 18, pages 531–539.

*During major migrations to previously uninhabited lands, we might expect the simultaneous transmission of genes, language and culture. By contrast, we might expect the decoupling of biology, culture and language during other periods. Thus, correspondence between the different forms of data needs to be evaluated rather than assumed*

**Keywords**

linguistics, migration, Polynesia, tree-like data, mitochondria, Y-chromosome, cultural evolution

## Background

The people of Oceania arose from one of the world's greatest ocean-travelling civilizations. Melanesian people have inhabited the islands of Near Oceania (such as Papua New Guinea) for at least 50,000 years. Then, around 3,300 years ago, the first Polynesian people (with the distinctive Lapita culture and artefacts), settled the coastal areas of these islands (**Figure 4.36**). Soon there were Lapita settlements throughout the Pacific, on previously uninhabited islands such as Vanuatu, Fiji, and Samoa. Migrating in ocean-going canoes, Polynesian people reached even the remotest islands of Oceania, such as Hawaii, Rapanui (Easter Island) and Aotearoa (New Zealand). But the precise pattern of Oceanic settlement has been debated: was there a fast, eastward migration of Polynesian people, creating a distinct series of related populations (the 'Express Train' model), or has Pacific history involved much more movement and exchange between established populations (the 'Entangled Bank' model)? Colonists arriving in a new land bring with them their genes, culture, and language, so each of these data sources can show the historical signal of descent with modification. However, it is also possible for the historical signal to be erased, for example when genes mix by migration and intermarriage, languages mix through trade or conquest, and cultures mix with the borrowing of ideas and technology.

## Aim

The researchers wished to compare the Express Train and Entangled Bank models of oceanic settlement, by testing the predictions they make about genes, cultures, and languages. If the

**Figure 4.36** The island of Tikopia is only five square kilometres in area yet is home to a distinct Polynesian culture with its own language. The Polynesians, who travelled between the distant islands of Oceania, are one of history's greatest sea-faring cultures. Some early anthropologists believed that the colonization of the Pacific had been primarily due to chance, with the occasional lucky canoe chancing upon an uninhabited island. But DNA analysis is concordant with the oral history preserved by many Polynesian cultures which tells of accomplished navigators able to cross vast distances between Oceanic islands in a clearly directed migration, and to conduct return voyages using navigational aids such as star maps.

Reproduced courtesy of Simon Foale.

Express Train model is the best description of Oceanic settlement, then populations might show a nested hierarchy of similarities, each founding population receiving 'copies' of the genes, language and culture of its parent population. Because a nested hierarchy can be drawn as a phylogenetic tree, such data is described as 'tree-like' (see Chapter 11). If the Entangled Bank model is a better description of oceanic settlement, then there should be no clear hierarchical pattern, so the data would not be 'tree-like'. The researchers aimed to trace migration through the genetic history of the people, their language, and the commensal animals that travelled in their canoes with them.

### Methods

A cross-disciplinary team reviewed analyses based on DNA sequences, linguistic information, and archaeological studies of Polynesian cultures. Over 5,000 shared words for over 200 Austronesian languages were coded as a matrix, scoring the presence of cognate (homologous) words in different languages. For example, the ocean-going canoes used by Polynesian people are known as *waqa* in Fiji, *wa'a* in Hawaii, and *waka* in Aotearoa. DNA sequences from Polynesian people from across Oceania were analysed, including mitochondrial DNA (passed on by females) and Y-chromosome sequences (passed on by males). They also analysed DNA sequences from the Pacific rat (*Rattus exulans*), which were often carried as a food source by Polynesians as they migrated across the Pacific in ocean-going canoes.

### Results

Analysis of the language data shows a clearly structured hierarchy, suggesting an eastward chain of colonization, from continental Asia to Remote Oceania. The pattern closely matches the archaeological evidence of the spread of Polynesian people. The mitochondrial data also suggests eastward expansion, with a hierarchy of changes from Asia to Micronesia and Melanesia, to the islands of Remote Oceania. Genetic data from both contemporary and archaeological specimens of the Pacific rat suggests that there was ongoing contact between remote oceanic settlements.

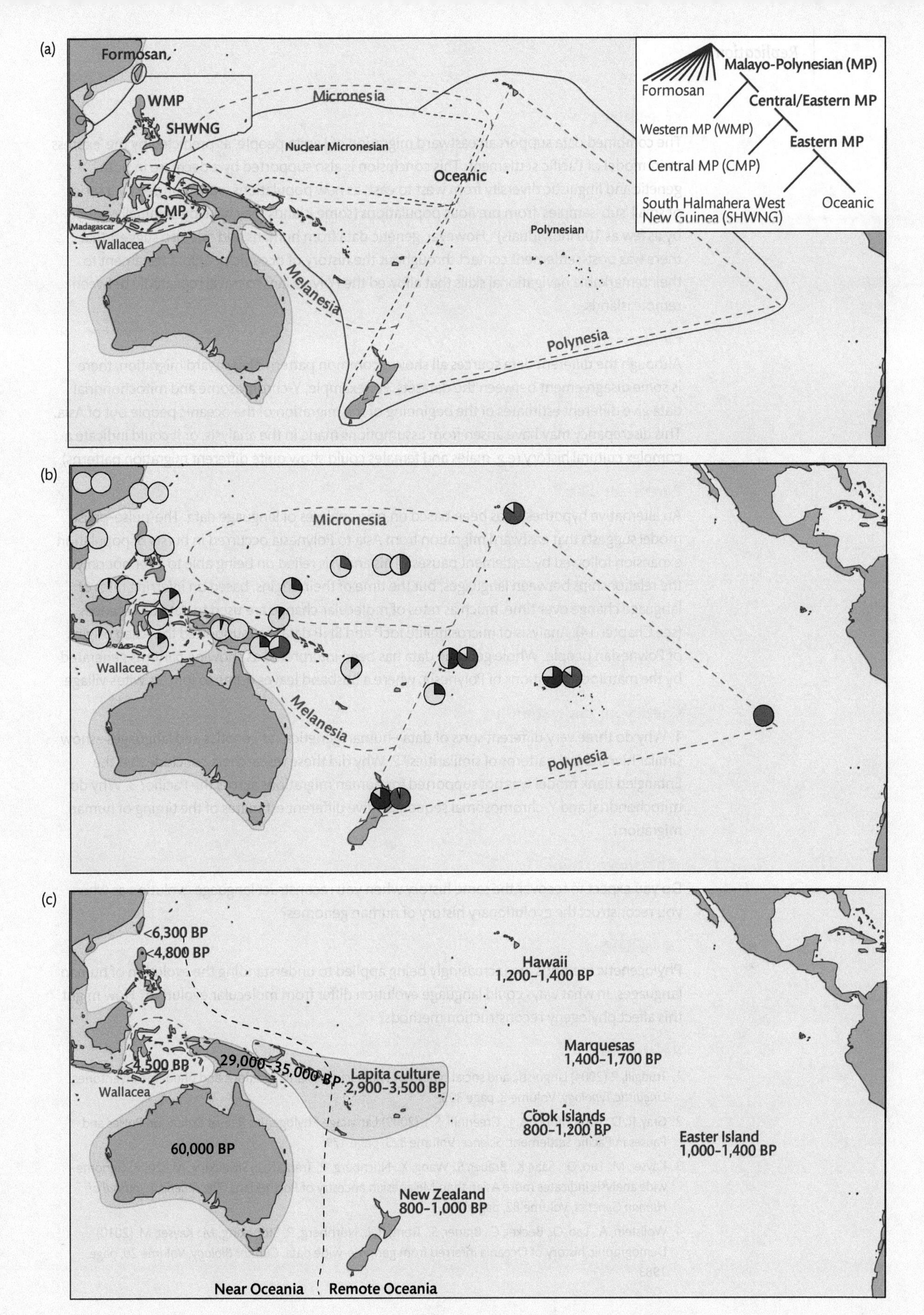


**Figure 4.37** Map of Oceania showing similarity between languages (a) and genes (b), and the dates of colonization estimated from the archaeological record (c).

Reprinted from Hurles, M. E. et al. (2003) *Trends in Ecology and Evolution*, Volume18, pages 531–539, copyright © 2003, with permission from Elsevier.

### Conclusions

The combined data support an eastward migration of Lapita people, as predicted by the 'express train' model of Pacific settlement. This conclusion is also supported by a decrease in both genetic and linguistic diversity from west to east, as new populations were founded by a series of small 'sub-samples' from previous populations (some islands may have been initially settled by as few as 100 individuals)[1]. However, genetic data from humans and rats also suggest that there was post-settlement contact throughout the history of oceanic people, a testament to their remarkable navigational skills that allowed the Polynesians to travel repeatedly between remote islands.

### Limitations

Although the different data sources all show a common pattern of eastward migration, there is some disagreement between the datasets. For example, Y-chromosome and mitochondrial data give different estimates of the beginning of the migration of the oceanic people out of Asia. This discrepancy may have arisen from assumptions made in the analysis, or it could indicate a complex cultural history (e.g. males and females could show quite different migration patterns).

### Further studies

An alternative hypothesis has been based on new analyses of language data. The 'pulse-pause' model suggests that eastward migration from Asia to Polynesia occurred in bursts of population expansion followed by settlement pauses[2]. This analysis relied on being able to infer not only the relationships between languages, but the time of their origins, based on inferring rates of language change over time, much as rates of molecular change are used to date divergences (see Chapter 14). Analysis of microsatellite loci[3] and SNP data[4] has supported the Asian origins of Polynesian people. Whole genome data has been interpreted as showing patterns generated by the matrilocal traditions of Polynesia, where a husband leaves home to join his wife's village.

### Check your understanding

1. Why do three very different sorts of data—human genetics, rat genetics and languages—show similar hierarchical patterns of similarities? 2. Why did these researchers conclude that the Entangled Bank model was not supported for human migrations across the Pacific? 3. Why do mitochondrial and Y-chromosomal sequences give different estimates of the timing of human migration?

### What do you think?

Do you expect to recover the same history when you reconstruct language evolution as when you reconstruct the evolutionary history of human genomes?

### Delve deeper

Phylogenetic methods are increasingly being applied to understanding the evolution of human languages. In what ways could language evolution differ from molecular evolution? How might this affect phylogeny reconstruction methods?

### References

1. Trudgill, P. (2004) Linguistic and social typology: The Austronesian migrations and phoneme inventories. *Linguistic Typology*, Volume 8, page 305.
2. Gray, R. D., Drummond, A. J., Greenhill, S. J. (2009) Language Phylogenies Reveal Expansion Pulses and Pauses in Pacific Settlement. *Science*, Volume 323, page 479.
3. Kayser, M., Lao, O. , Saar, K., Brauer, S., Wang, X., Nürnberg, P., Trent, R. J., Stoneking, M. (2008) Genome-wide analysis indicates more Asian than Melanesian ancestry of Polynesians. *The American Journal of Human Genetics*, Volume 82, page 194.
4. Wollstein, A., Lao, O., Becker, C., Brauer, S., Trent, R. J., Nürnberg, P., Stoneking, M., Kayser, M. (2010) Demographic history of Oceania inferred from genome-wide data. *Current Biology*, Volume 20, page 1983.

# Genome

## Accident and design

*"Every scientific problem solved gives us a number of new ones to puzzle on"*

J. B. S. Haldane (2009) *What I require from life: writings on science and life from J. B. S. Haldane*, ed. Krishna Dronamraju. Oxford University Press.

## What this chapter is about

Genome size varies dramatically between species. Much of the variation is due to repeated sequences, rather than differences in number of different genes. Whole genome duplications can provide the raw material for establishing new lineages. On a finer scale, shorter repeat sequences are prone to accidents in replication that increase or decrease the number of copies. Because of this high rate of change, short repeats are likely to vary in number between individuals. Other repeat sequences play a more active role in their own duplication, containing regulatory signals and genes that cause them to be copied and inserted throughout the genome. Transposable elements not only fill the genome with copies of themselves, they also remodel their host's genome. The changes caused by these 'jumping genes' have profound effects on genome structure and evolution.

### Key concepts

- Changes in the number of copies of genomes, genes or sequences can generate new heritable material, providing fuel for evolutionary change
- Repeat sequences often have a high rate of change in copy number, so a 'DNA fingerprint' that identifies individuals can be generated by counting the number of repeats at different loci
- Transposable elements that promote their own duplication within the genome can increase the size of the genome, cause mutation through insertion in or near genes, and provide novel genetic material that is incorporated into genome function

# Genome size

## *The great chain of being*

*And since the law of continuity requires that when the essential attributes of one being approximate those of another all the properties of the one must likewise gradually approximate those of the other, it is necessary that all the orders of natural beings form but a single chain, in which the various classes, like so many rings, are so closely linked one to another that it is impossible for the senses or the imagination to determine precisely the point at which one ends and the next begins*

Gottfried Wilhelm Leibniz (1753) Lettre Prétendue de M. De Leibnitz, à M. Hermann dont M. Koenig a Cité le Fragment (1753), cxi-cxii, trans. in A. O. Lovejoy, *Great Chain of Being: A Study of the History of an Idea* (1936), 144–5.

Seeking order in the natural world, people have arranged species in a wide variety of classification systems that reflect use, similarity, importance or place in the economy of nature. One of the classic schemes of organization is the *scala naturae*, or Great Chain of Being. This idea long predates evolutionary theory, and yet the basic structure of organizing things in a simple-to-complex chain is still often reflected in the way we present biological diversity. Humans are commonly presented as the end point of a chain of species, even when the comparisons are made with our contemporaries, not our ancestors (**Figure 5.1**). Many biology textbooks resemble this scheme, with the earliest chapters on bacteria, working through the plant and animal kingdoms before arriving at the final chapters on human biology, as if strange social hairless apes were the end point that all of those millions of years of evolution was working towards. Even more curiously, the tips of a phylogeny are often arranged with species perceived as simpler or older at the left and those considered more complex and recent lineages at the right, implying a progression. These tendencies echo earlier notions of orthogenesis, the idea that evolution is somehow driven on a directional trend to producing bigger and better organisms.

**Figure 5.1** The Great Chain of Being, or *scala naturae*, is an idea at least as old as Aristotle, and has had a great impact on the way that European philosophers think about the natural world. Even though evolutionary theory has largely superseded the idea of an ordered natural world, species are still often presented in a hierarchy from those perceived to be simple and ancient to those considered complex and advanced. This example was published by Ernst Haeckel in 1874, in his book on human evolution, *Anthropogenie; oder Entwicklungsgeschichte des Menschen* (Anthropogeny: Or, the Evolutionary History of Man). Haeckel, a prolific zoologist and taxonomist, was enthusiastically committed to evolutionary explanations of biodiversity. He promoted the idea that 'ontogeny recapitulates phylogeny': that animal species can be ordered on a scale of developmental complexity, with simpler organisms representing truncation of development at earlier stages. He was convinced by the similarity between humans and apes that fossils of ape-men would be found in South East Asia, and dispatched his students to find them. His predictions proved correct: one of his students found the first evidence of an extinct hominid, Java man (*Homo erectus*).

So it is not surprising that when human gaze turned to the genome, they expected to see the pinnacle of the natural world, a genome bigger, better, and generally more wonderful than all other species. The results came as a bit of a shock. While human genomes are many hundreds of times larger than bacterial genomes, they are also forty times smaller than genomes of the lungfish (**Figure 3.3**) or fritillary (**Figure 5.2**). When it was found that much of the variation in genome size was due to highly repetitive DNA that did not seem to be needed to build the organism, humans breathed a collective sigh of relief. It was assumed that, even if human genomes weren't the largest, they would have the greatest number of working genes. In the year 2000, a scientist working on the human genome project started an international wager on the likely number of genes. Not a single scientist placed a bet as low as the real figure, which is currently thought to be

**Figure 5.2** My genome's bigger than yours. (a) The fritillary (130 billion bases), (b) onion (17 billion bases) and (c) *Paris japonica* (150 billion bases) all have way more DNA than your measly 3 billion bases.

around 20,000 genes. This is marginally more genes than it takes to build *Caenorhabditis elegans*, a millimetre-long transparent flatworm studied for its simplicity. So even in number of genes, humans are outclassed by other species, such as the tiny pond-dwelling water flea, *Daphnia*, which has an estimated 31,000 genes (**Figure 5.3**).

In the next chapter, we will consider genes and regulatory elements. But for many organisms, genes make up a relatively small fraction of the DNA in the genome. For example, in humans, genes are only a few per cent of the three billion bases of DNA in the nuclear genome.

**Figure 5.3** I've got more genes than you. Water fleas (*Daphnia pulex*) have 50 per cent more genes than humans. Not bad, given that they are see-through.



So what is the rest? Some of it is additional sequences needed to run the genome. After all, with 20,000 genes embedded in a massive, multi-chromosomal genome, there is going to need to be a lot of regulatory machinery to get the right genes expressed at the right times in the right places. But a lot of the DNA in the genome consists of repetitive DNA that does not seem to be directly connected to making and operating the owner of the genome. Some of these repeats appear to be functionless DNA that arises through errors in replication. But some repeat sequences result from the active propagation of elements that have features that promote copying, allowing them to proliferate in the genome.

In order to understand how genomes evolve, we need to become familiar with the way genomes can be changed. In this chapter, we are going to focus on the repetitive sequences that bulk out the genome. These sequences help us to understand the mechanics of DNA copying and genome maintenance. They also provide a valuable insight into the selective forces that shape the genome, which turn out to be a lot more complex than might first be supposed. And, like all of the processes we discuss in this book, the evolution of repetitive DNA leaves a useful record in the genome that allows us to identify individuals, trace family relationships, and decipher population dynamics.

## The paradox of genome size

When DNA is 'melted' to separate the two strands, the rate at which the strands come back together to form a double helix (reanneal) is a measure of sequence similarity. A DNA sample that contains entirely single copies of unique sequences will take longer to anneal because it will take time for each sequence to find its one and only partner. But a DNA sample containing multiple copies of the same nucleotide sequence will anneal more quickly as there is more chance of each strand bumping into one of the many matching strands in the solution. So when denatured (single-stranded) DNA is cooled, the rate of re-association into double strands will be partly determined by how similar the DNA strands in the sample are. When a mixture of DNA from difference sources is heated and cooled, the re-association rate is a measure of the similarity of the two samples, a technique known as DNA hybridization (**TechBox 11.1**). Reannealing the sequences from a single genome reveals how much repetitive DNA it contains, because self-similar repeat sequences will anneal more rapidly than unique sequences. The re-association can be plotted as percentage re-association against concentration ($C_o$) by time (t), known as a CoT curve.

In the 1960s, scientists found that when they melted genomic DNA from a mouse then reannealed it, they got distinct patterns of re-association, with a band in the CoT curve that annealed far more quickly than would be expected from a sample of unique DNA sequences. These peaks of rapidly annealing DNA were referred to as satellite bands (a satellite is something peripheral to a larger centre, like a satellite town near a major urban centre, or an astronomical satellite orbiting a planet). They demonstrated that this rapidly reannealing fraction did not contain genes by showing that messenger RNA (gene transcripts) matched only to the more slowly reannealing fraction of the genome. They concluded that this satellite DNA consisted of short DNA sequences present in a million or more near-identical copies.

This observation went some way to explaining the 'C-value paradox'. C-value is the total size of the nuclear DNA content, often measured in picograms (a trillionth of a gram). The paradox was that there didn't seem to be any obvious relationship between perceived complexity of a species and its genome size. Many people assumed that simpler organisms wouldn't need as many genes, yet they often had larger genomes than fancier species. Incidentally, the 'C' in 'C-value' does not stand for complexity, or complementary, or CoT curve, but instead refers prosaically to 'constant', a legacy of the observation that the amount of DNA is the same in all body cells of the same organism. Of course, we now know that this constancy of amount of DNA is because each cell carries an entire copy of the genome, even though it will only use a subset of the information to build each different cell type. The specificity of information needed in different parts of the body at different points in time is generated by regulating which genes are switched on or off (see Chapter 6).

These repetitive sequences in the genome can consist of duplicated functional genes, transposable elements or short repeated base sequences. How did they get there? To understand the origins of repeat sequences, we need to look at the process of duplication of DNA sequences. We will start by considering the duplication of whole genomes, entire chromosomes, or parts of chromosomes, events that can have severe effects on individual development, but may also play an important role in speciation in some lineages. Then we will focus down on the duplication and deletion of shorter segments of DNA, from whole genes to sequences of only a few base pairs. These insertions and deletions can be considered accidents, hiccups of genome copying and maintenance. But some sequences contain features that actively promote their replication, so they tend to increase in number in the genome. In addition to increasing the amount of DNA in the genome, these transposable elements can jump from one place to another, moving sequences around the genome and disrupting normal gene function. All of these types of genome rearrangements leave a mosaic of changes that allow us to distinguish individual genomes, trace relationships, and uncover the evolutionary history of species.

## Doubling and dividing: genome replication

To understand how changes to genome structure contribute to evolution, we first need to refresh a few basic concepts on how the genome is copied, divided, and combined at reproduction. However, this is not the place to provide a primer on cell division. If you are unfamiliar with the way cells copy and divide their chromosomes, it's a good idea to pick up an introductory biology text at this point and have a look at meiosis and mitosis. Understanding cell division is an important foundation to studying genome evolution.

Organisms with relatively small genomes may have all their DNA contained in a single molecule. The mitochondria in your cells, like their free-living bacterial ancestors, have single-molecule circular genomes, as do chloroplasts in plants and algae (**Figure 5.4**). But there is a limit to how large a genome can get and still be maintained and copied as a single strand. So most organisms with larger genomes organize their DNA into discrete parcels called chromosomes. If a cell carries its genome in multiple chromosomes, it needs a reliable way of organizing, copying, and dividing DNA at cell division, so that each daughter cell ends up with one complete copy of each chromosome. We will briefly review the behaviour of chromosomes at cell division, so that we can understand how some different kinds of chromosomal rearrangements can occur.

The word 'ploidy' refers to the number of copies of each chromosome, so cells (or individuals) with one copy of the entire genome are haploid, those with two copies of the genome are diploid, those with three copies triploid and so on (**Figure 5.5**). More generally, cells with more than two copies of the genome are usually referred to as

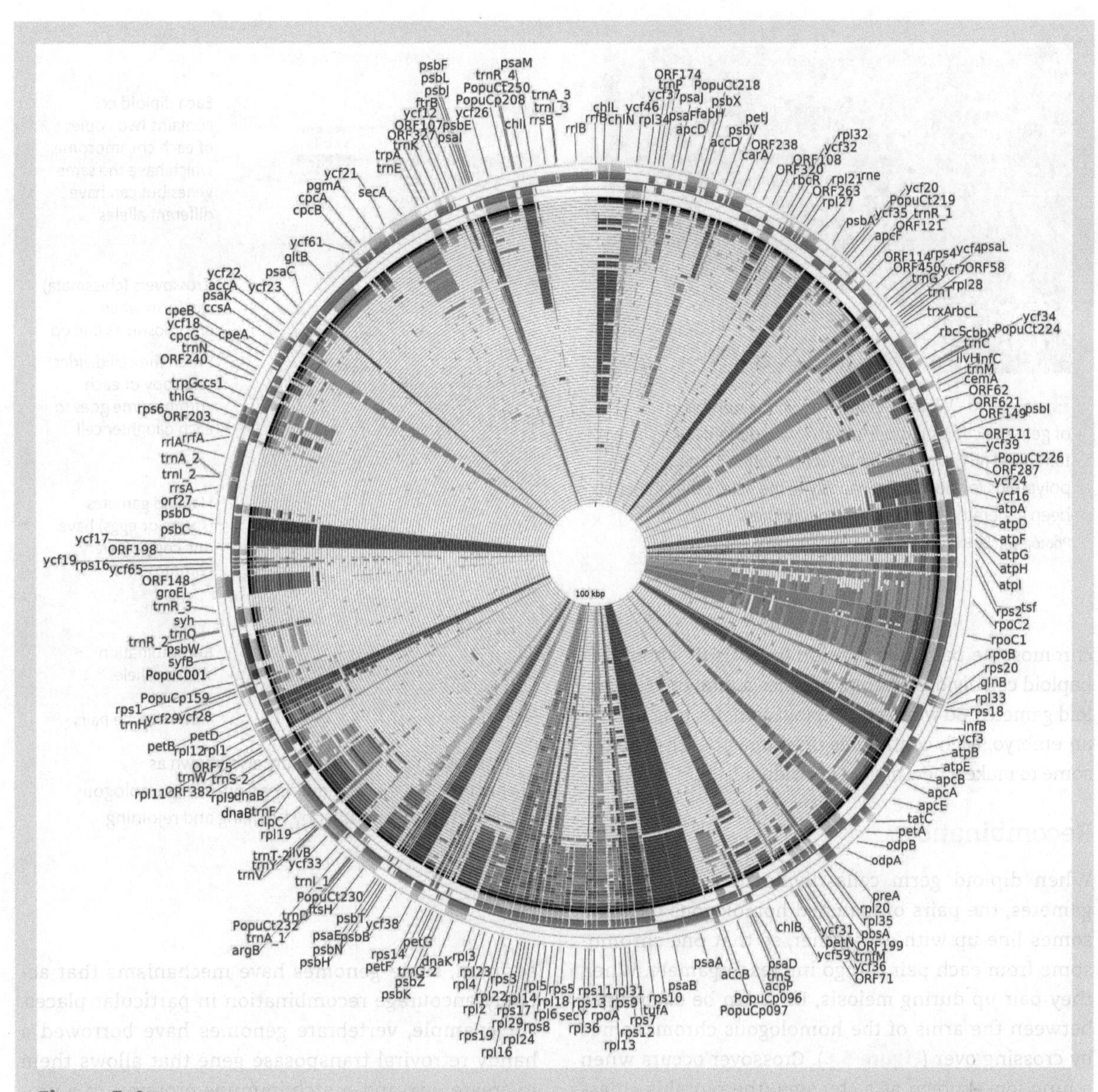


**Figure 5.4** The chloroplast genome of *Porphyra purpurea*, a red algae widely enjoyed as nori (the paper-like seaweed sheets used in the production of sushi). It has been suggested that enzymes for digesting porphyran (a polysaccharide found in *Porphyra*) have been transferred from marine bacteria to the gut bacteria of people who regularly eat seaweed.
Image courtesy of Paul Stothard.

polyploid. Your body cells are diploid (just like the body cells of most animals, many plants and some fungi), so every cell contains two copies of each chromosome. These pairs of matching (homologous) chromosomes contain copies of the same genes, though they may differ slightly in the particular nucleotide sequence of each gene. Alternative sequence variants are referred to as alleles (see Chapter 3). A diploid cell has two copies of each chromosome, so it can carry two alternative versions (alleles) of each homologous sequence. In most diploid organisms, one copy of each chromosome came from each parent through the process of sexual reproduction, where genomes from two individuals are fused together to make a new individual.

Clearly, sexual organisms can't keep doubling their genomes every time they fuse cells from two different parents to make a single offspring. They avoid an ever-increasing number of genome copies by undergoing reduction/division (meiosis). The diploid germ cells that will give rise to gametes (sperm and eggs) go through two rounds of cell division: reduction (meiosis I) where the two copies of the genome are split to make two haploid daughter cells, each with only one copy of each

Figure 5.5 Wheat varieties vary in the number of genome copies per cell (ploidy), with diploid (2N), tetraploid (4N) and hexaploid (6N) varieties. Some polyploids formed spontaneously in the field, others have been generated by breeding programmes.
Photograph by Barry Bromham.

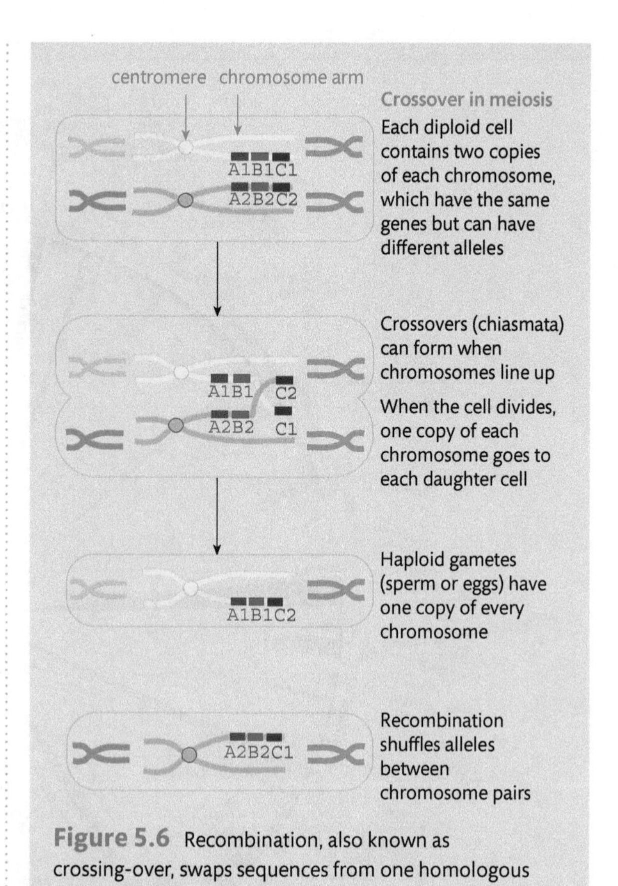


Figure 5.6 Recombination, also known as crossing-over, swaps sequences from one homologous chromosome to another by breaking and rejoining DNA strands.

chromosome pair, then division (meiosis II) when each haploid cell divides again to produce a total of four haploid gametes. So when two haploid gametes fuse to form an embryo, each contributes one copy of each chromosome to make a new diploid individual.

## Recombination

When diploid germ cells divide to make haploid gametes, the pairs of matched homologous chromosomes line up with each other, so that one chromosome from each pair can go into each gamete. When they pair up during meiosis, DNA can be exchanged between the arms of the homologous chromosomes by crossing over (Figure 5.6). Crossover occurs when a junction—known as a chiasma (the plural is chiasmata)—connects two strands of DNA from different chromosomes. Formation of junctions between chromosomes is an essential part of DNA repair and genome maintenance. Double-strand break repair uses junctions to patch missing or damaged sequences using the matching sequence from the homologous chromosome.

Crossing over is a reasonably common occurrence, resulting in the reshuffling of sequences between homologous chromosomes. This shuffling is an important source of genetic variation in sexually reproducing populations. It can also result in 'gene conversion' where the allele from one chromosome is replaced by a copy of the allele from the other chromosome. Chiasmata are more likely to form in some locations in the genome than others. Recombination hotspots are scattered throughout the genome. In addition, some genomes have mechanisms that actively encourage recombination in particular places. For example, vertebrate genomes have borrowed a handy retroviral transposase gene that allows them to create mix-and-match immune proteins, in a process known as V(D)J recombination (Variable, Diverse, Joining). DNA strands are broken at specific sequences between the protein-coding segments, which are shuffled and glued back together, to create different combinations of antigen-specific receptors. This results in the rapid production of diverse sets of novel antibodies to be used in immune response.

Sites that have a high frequency of crossing-over are called recombination hotspots. By concentrating crossover at particular points along the genome, recombination hotspots create blocks of sequence defined by two adjacent crossover points. Recombination may shuffle the connections between the blocks, but is less likely to break up combinations of mutations within the blocks, so all mutations occurring in that block will tend to be inherited together as a set. The presence of these haplotype blocks in the genome has practical consequences for gene mapping and studies of heredity, because one informative marker can act as a 'tag' that signals the presence of the whole haplotype block. These markers do not actually have to be in the gene of interest, as long as they are reliably inherited in the same haplotype block (**TechBox 3.1**). But it is important to remember that although recombination is more common at the hotspots between haplotype blocks, it might also occasionally occur within blocks, breaking up the linkage between the marker and the trait of interest.

### *Genome duplication in evolution*

Now that we have refreshed ourselves on the basics of chromosome behaviour at cell division, we can consider some of the ways in which hiccups in this process can lead to changes in genome structure and content. Let's start with the biggest possible change to genome size: increase in the number of whole genome copies. Genome doubling can occur if a cell that has not undergone chromosome reduction fuses with another cell. Alternatively a cell might copy all its chromosomes then fail to divide, leaving it with double the number of chromosomes in its nucleus. While increases in ploidy can be fatal to many species (including humans), some organisms seem to be able to tolerate whole genome doubling. In particular, polyploid plants can often function perfectly well (**Figure 5.5**). But it can be difficult for a polyploid individual to mate sucessfully with another member of the population, because mixing gametes with different ploidy levels can result in offspring with odd numbers of chromosomes. For example, if a newly minted tetraploid, with four copies of each chromosome, mates with a standard diploid individual, the offspring will tend to be triploid (two copies of each chromosome from the tetraploid parent, one copy from the diploid parent). Because it's difficult to reliably pair three copies of each chromosome at cell division, triploid individuals tend to produce unbalanced gametes with assorted numbers of chromosomes, and are therefore often infertile.

Imagine that an accident of cell division, or a quirk of fertilization, results in the formation of a triploid individual in a population of diploids. This triploid individual may be perfectly healthy, and it may even have traits that give it certain advantages over others in the population, due to changes in genome expression or by virtue of having extra gene copies. So it thrives and survives. But if it can't interbreed successfully with the diploids around it, then either it dies without issue or it exchanges gametes with another lucky triploid or it starts its own triploid line by uniparental reproduction. Now, any alleles carried in the triploid line won't be shared with the surrounding diploid population, because they produce no (or very few) hybrid offspring. Because polyploidy can create a barrier to interbreeding with the parent stock, some people have viewed polyploidy as a means to 'instant speciation'. Of course, this will only work if the polyploid individual can produce offspring on its own, being either self-fertile or capable of clonal reproduction, or if there are other polyploids it can mate with. Not surprisingly, polyploid speciation appears to be more common in lineages capable of uniparental reproduction. Polyploidy does not always create a complete barrier to gene flow, but there is a growing appreciation that it has been an important contributing factor in speciation and diversification, particularly in plants. Debate continues over the role of genome doubling in the formation of new lineages in animals (**Figure 5.7**).

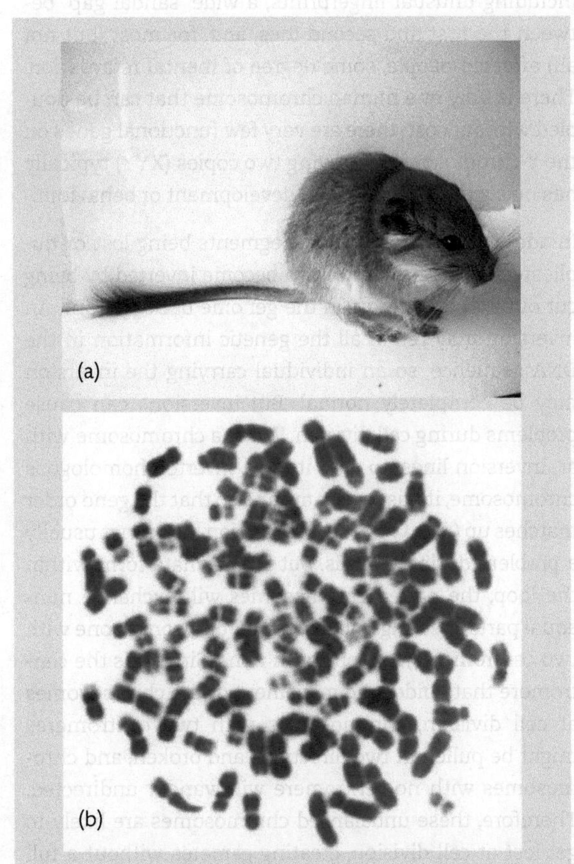


**Figure 5.7** I've got more chromosomes than you. (a) The red viscacha rat (*Tympanoctomys barrerae*) has 102 chromosomes (b), more than double the number in humans. It's functionally a diploid (its chromosomes come in pairs), but it has been suggested that the species arose either from a tetraploid ancestor, or through hybridization between two diploid species.

## Chromosomal rearrangement

Sometimes only part of the genome is affected by a change in copy number. Whole chromosomes may be duplicated, and parts of chromosomes can be copied, rearranged or deleted. Loss of all or part of a chromosome is usually lethal, because it's likely to result in loss of information essential to an organism's development and operation. Less obviously, creating an extra copy of all or part of a chromosome can be just as damaging, probably due to the disturbing effect of having extra doses of the duplicated genes' products or disruption of normal regulation. There are some chromosomal losses and gains that can be tolerated, though with potentially severe consequences. People with Down's syndrome have an additional copy of part or all of chromosome 21, a condition that gives rise to a range of characteristics including unusual fingerprints, a wide 'sandal gap' between the first and second toes, and, for most (but not all) affected people, some degree of mental retardation. There is only one human chromosome that can be doubled without cost: there are very few functional genes on the Y-chromosome, so having two copies (XYY) typically has no significant impact on development or behaviour.

In addition to chromosomal segments being lost or duplicated, a DNA sequence can become inverted by being cut out and re-inserted in the genome back-to-front. An inversion may retain all the genetic information in the DNA sequence, so an individual carrying the inversion may be completely normal. But inversions can cause problems during cell division. When a chromosome with an inversion lines up with its non-inverted homologous chromosome, it has to loop around so that the gene order matches up (**Figure 5.8**). This inversion loop is not usually a problem during meiosis. But if chiasmata form within the loop, the paired chromosomes will exchange non-equal parts, creating unbalanced chromosomes, one with two centromeres, and one with none. Since it is the centromere that guides the movement of the chromosomes at cell division, chromosomes with two centromeres might be pulled in two directions and broken, and chromosomes with no centromere will wander undirected. Therefore, these unbalanced chromosomes are likely to get lost at cell division, creating gametes without a full chromosome complement. Such gametes are unlikely to be able to form working offspring. The practical result of all this is that individuals that carry one inverted and one normal chromosome are likely to have reduced fertility.

Because of the reduced reproductive success of individuals with mixed chromosomes, those that only mate with others with the same chromosome type will have a greater chance of producing viable offspring. Thus

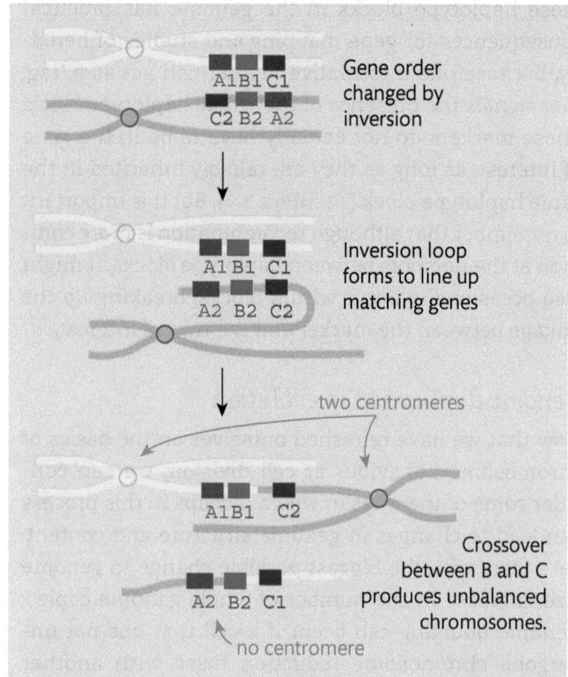


**Figure 5.8** Chromosomal inversions retain genetic information but can cause problems at meiosis when crossing over results in unbalanced chromosomes that are unlikely to result in well-balanced offspring.

inversions can lead to the separation of a population into two non-interbreeding species. For example, researchers at the University of Ouagadougou examined the karyotype (chromosome complement) of over four thousand *Anopheles funestus* mosquitoes from villages around a swamp in Burkina Faso. *Anopheles funestus* mosquitoes are vectors for malaria, a disease that kills over half a million people every year (**Figure 5.9**). The researchers found two distinct types of *Anopheles funestus* mosquitoes, physically identical but each with a characteristic pattern of chromosomal inversions. Importantly, they found relatively few hybrids that carried both types of inversions. From this observation they inferred that, although the two types of *Anopheles funestus* occur in the same places and are indistinguishable to humans, they don't tend to interbreed successfully. This implies that the two chromosomal strains of *Anopheles funestus* are in the process of evolutionary divergence, perhaps ultimately leading to the formation of two distinct species. Further research has provided evidence of divergence, showing that the chromosomal races have different seasonal abundances, possibly connected to evolution of different habitat and environmental preferences.

*Chapter 9 includes a discussion of speciation*

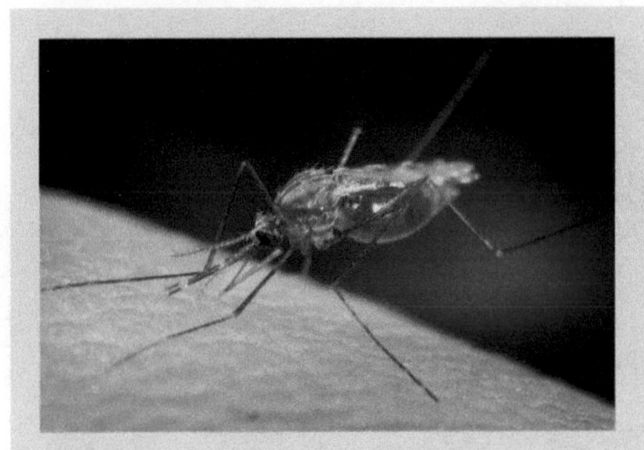

**Figure 5.9** Malaria derives its name from the early hypothesis that it was carried by bad air arising from swamps, but we now know it is carried by another factor that arises from swamps: mosquitoes. There are dozens of different species of *Anopheles* mosquitoes that can transmit twhe four different forms of *Plasmodium* parasite that cause malaria. *Anopheles funestus*, shown here, and *Anopheles gambiae* are the main vectors for malaria in Africa, so are indirectly responsible for the death of over half a million people every year. Chromosomal races of these mosquitoes can show different habitat and climate preferences, complicating mosquito control programmes.
Photograph: James Gathany.

# Repeat sequences

We have seen how large regions of the genome can be lost, doubled, or turned around by chromosomal rearrangements. Now we will consider finer-scale rearrangements that act on smaller sequences. Duplicate sequences can be created when a copy of a DNA sequence is inserted elsewhere in the genome. If the new copy is inserted next to the original version, they are said to be tandem (side-by-side) repeats. Once created, tandem repeats are prone to increase in number through a process called unequal crossing over. When chromosomes pair at meiosis, tandem repeat sequences may be sufficiently similar that they can pair with neighbouring sequences. If recombination occurs between these misaligned repeats, one chromosome may end up with more repeats, and the other with fewer (**Figure 5.10**). Repeat sequences play a very important role in genome evolution and maintenance. For example, both the middles (centromeres) and ends (telomeres) of chromosomes have substantial numbers of repeat sequences. But in this chapter we are going to focus solely on repeat sequences that are commonly used in phylogenetics, particularly microsatellites.

Tandem gene duplication can generate gene families: sets of related sequences with similar functions (**Case Study 6.1**). Gene duplication can provide the raw material for evolutionary change by creating a 'spare copy' of essential sequences. If one copy of the gene can keep producing the original gene product, the spare gene copy may be able to change without jeopardizing the function of the original gene. Many duplicate genes simply decay, acquiring changes that destroy the sense of the coding information (these non-functional copies are usually referred to as pseudogenes). But in some cases duplicate genes can evolve new functions, or become specialized to perform just one of the functions of the original gene.

Not all duplications and deletions are of whole genes. In fact, some short sequences of DNA are particularly prone to duplication and deletion (**Case Study 5.1**). Most eukaryotic genomes contain large numbers of short tandem repeats, where the same short sequence of nucleotides is repeated over and over again, such as CAGCAGCAGCAGCAG. The runs of tandem repeats that make up satellite DNA are thought to be generated primarily by slippage during replication, though unequal

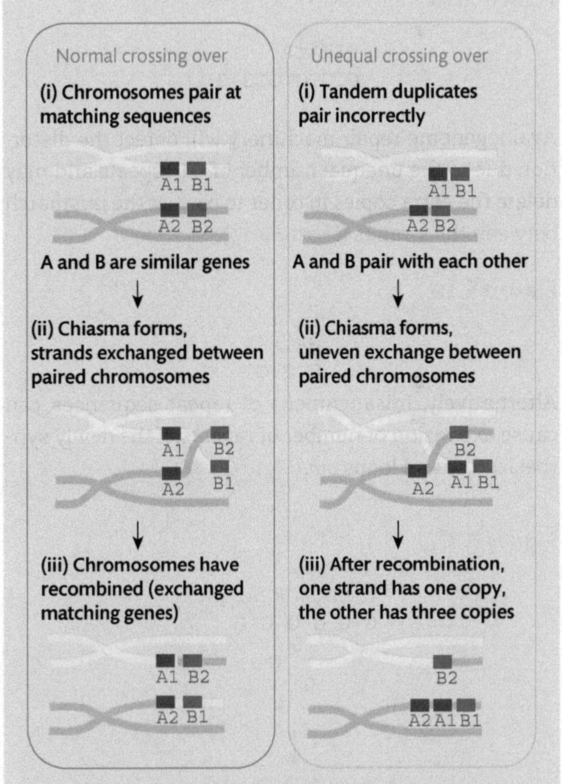


**Figure 5.10** Unequal crossing-over results when side-by-side (tandem) repeats pair asymmetrically, then recombination results in one chromosome getting more of the repeats, and one getting fewer.

crossing over may also generate variable numbers of repeats. Slippage is a by-product of the universality of complementary base pairing (Chapter 2). For example, the repeat run CACACACACACA pairs with the complementary sequence GTGTGTGTGTGT (**Figure 5.11**).

**Figure 5.11**

```
CACACACACACA
GTGTGTGTGTGT
```

But any CA repeat can pair with any GT. If the sixth CA was to pair with the third GT, the intervening repeats would loop out (**Figure 5.12**).

**Figure 5.12**

```
      C A
     A   C
      C A
  CACA   CA
  GTGTGTGTGTGT
```

This loop can be removed by the genome's repair machinery (which can detect such distortions of the DNA helix), causing reduction in the number of repeats (**Figure 5.13**).

**Figure 5.13**

```
CACACA
GTGTGTGTGTGT
```

Again, genome repair machinery will detect the distortion due to the unequal number of GT repeats and may delete the extra copies in order to correct the mismatch between the strands (**Figure 5.14**).

**Figure 5.14**

```
CACACA
GTGTGT
```

Alternatively, misalignment of repeat sequences can cause expansion of number of repeats, if the newly synthesized strand loops out (**Figure 5.15**).

**Figure 5.15**

```
  CACACACACACA
 GTGTGTGTGTGT
        G T
       T   G
       G   T
        T G
         ↓
CACACACACACA
GTGTGTGTGTGTGTGTGTGT
         ↓
CACACACACACACACACACA
GTGTGTGTGTGTGTGTGTGT
```

**Figure 5.16** A woman living in a stilt village on Lake Maracaibo in the early nineteenth century unwittingly gave rise to one of the most informative human genetic studies in the world. This woman (who died of Huntington's disease) gave rise to ten generations of descendants, encompassing around 15,000 people, many of whom carry disease-causing alleles of the Huntington's disease (*HD*) gene. The *HD* gene was one of the first human disease genes to be mapped (in 1983). It took a decade to be sequenced, a heroic undertaking in the pre-genomic era.
Reproduced courtesy of Randy Trahan; Photograph: Geoffery Charles.

The more repeats there are, the more chance for slippage to occur, and the greater the likelihood of increasing or decreasing the number of repeats. So parts of the genome with many repeat sequences are unstable, prone to changes in the repeat copy number.

This instability can cause havoc in the genome. For example, Huntington's disease is a neurodegenerative disorder that causes movement abnormalities and loss of cognitive functions (**Figure 5.16**). Huntington's disease is a dominant genetic disorder, which means a single copy of the disease-causing allele is sufficient to cause the disease. A person carrying the Huntington's disease allele has a 50 per cent chance off passing the disease allele to each offspring. Any child that receives the disease allele will definitely get the disease, though they may not fall sick until after they have had their own children, each of which also has a 50 per cent chance of inheriting the Huntington's disease allele. But Huntington's disease can show patterns of inheritance not expected from a normal Mendelian trait. Although most carriers don't develop symptoms until after the age of 40, some individuals develop the disease earlier, but only if they inherited the disease from their father, not their mother. Furthermore, in some cases the age of onset gets earlier with each subsequent generation, so that the son of a man with early onset Huntington's disease may have even earlier disease onset.

These mysteries of inheritance were solved when the Huntington's disease gene (*HD*) was sequenced in 1993 (**TechBox 5.1**). The beginning of the gene had multiple repeats of CAG, which codes for the amino acid glutamate. But the number of CAG repeats varied between individuals. Most people have between 9 and 35 CAG repeats. People with more than 40 CAG repeats in the *HD* gene develop Huntington's disease. The age of onset is correlated to the number of repeats: the more repeats, the earlier the age of onset. Furthermore, once the number of repeats goes above 40, the frequency of changes to the number of repeats also increases, a process known as 'dynamic mutation'. Expanding repeats are mutations that increase their own rate of mutation.

There are at least nine human diseases that, like Huntington's disease, are known to be caused by unstable repeats. For example, Fragile X syndrome, the most common heritable form of moderate mental retardation, is caused by a massive increase in the number of copies of a CGG repeat in the upstream regulatory region of a gene called *FMR1*. But many tandem repeats, especially those occurring in non-coding regions of the genome, do not have noticeable effects on their carriers.

## Comparing repeat sequences

Before we go on, we need some terminology to describe the different regions of tandem repeats in the genome. A particular place in the genome is usually referred to as a 'locus', so run of repeats occurring in a given place in the genome is termed a 'repeat locus' (being a Latin word, the plural of locus is 'loci'). Repeat loci are known by many different names, including variable number of tandem repeats (VNTRs), simple sequence repeats (SSRs), and short tandem repeats (STRs). Remember that repeat DNA was originally discovered from satellite bands in the DNA re-association studies (CoT curves). If the sequence that is repeated is between 10 and 100 nucleotides, it is sometimes called a minisatellite. If the sequence that is repeated is only a few nucleotides long, such as CACACACACA, the repeat region is usually referred to as a microsatellite.

Thanks to dynamic mutation, these repeat regions provide an extremely useful tool for distinguishing individuals in a population. Because the mutation rate of tandem repeats may be orders of magnitude faster than in the rest of the genome, these sequences are so prone to change that two individuals are unlikely to share exactly the same number of repeats in all of their repeat loci. Counting the number of repeats at many different loci can often provide a unique identifier for individuals, often referred to as a DNA fingerprint.

If you wanted to compare repeat sequences between individuals, you could sequence the entire repeat locus then simply count the number of repeats (although sequencing of repeat regions is error-prone due to polymerase slippage). But there is a simpler way of comparing repeat number, which takes advantage of the fact that increasing or decreasing the number of repeats at a locus changes the length of the repeat region. To compare the size of repeat regions between individuals, you need to identify non-repeated DNA sequences (flanking regions) on either side of the run of repeats (**Figure 5.17**).

**Figure 5.17**

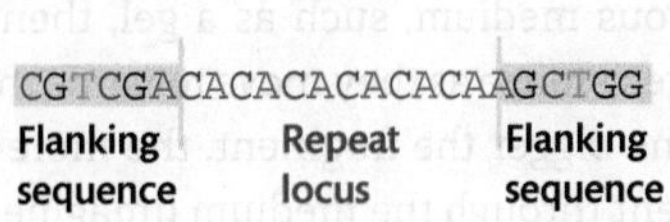

If the flanking regions have a lower rate of change than the repeat region, they can remain much the same even when the number of repeats between the flanking regions changes. You can then use these conserved flanking sequences to produce fragments of DNA that start and end with the same sequence, but have variable numbers of repeats in between (**Figure 5.18**).

**Figure 5.18**

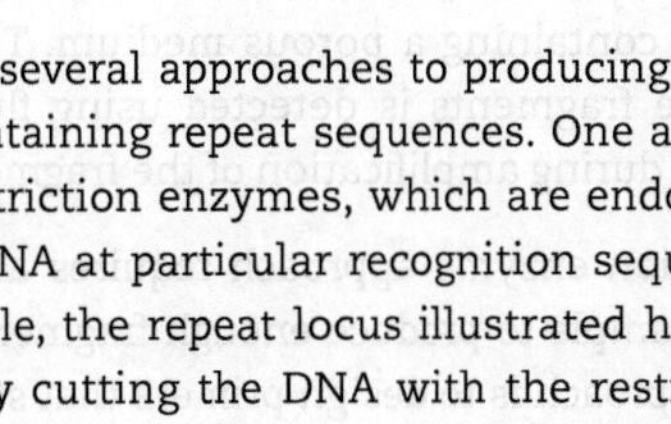

There are several approaches to producing DNA fragments containing repeat sequences. One approach is to use restriction enzymes, which are endonucleases that cut DNA at particular recognition sequences. So, for example, the repeat locus illustrated here can be isolated by cutting the DNA with the restriction enzyme AluI, which cuts the sequence AGCT between the G and the C, thus dividing the DNA into different size fragments that all begin and end with the recognition sequence (Figure 5.19).

Figure 5.19

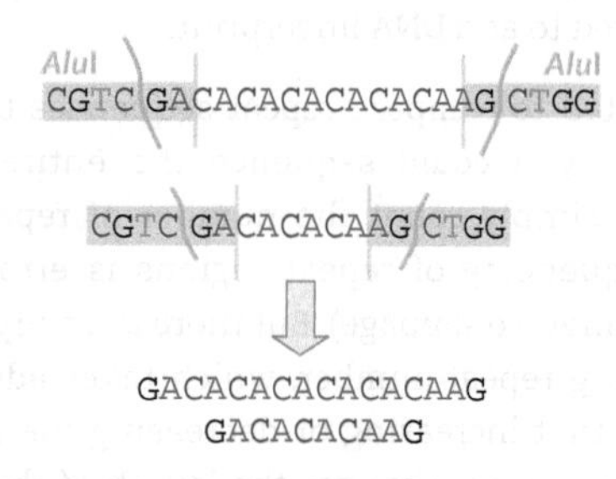


Once you have your fragments, you can separate them by size using electrophoresis, which is the movement of a charged substance in an electric field. DNA has a negative charge, thanks to the phosphate ions in the backbone of the helix (TechBox 2.1). So if you subject DNA molecules to an electric current, they will tend to move toward the positive electrodes. But if your DNA samples are in a porous medium, such as a gel, then they will meet some resistance as they move towards the positive electrode. The bigger the fragment, the more it will resist movement through the medium (imagine using the same force to drag a basketball and a golf ball through a bog—the larger ball will move more slowly as it encounters more resistance). So the shorter DNA fragments will move more rapidly than the longer fragments, and therefore get further down the gel (Figure 5.20). The gel can then be stained to reveal the presence of bands of different length fragments of DNA.

Figure 5.20

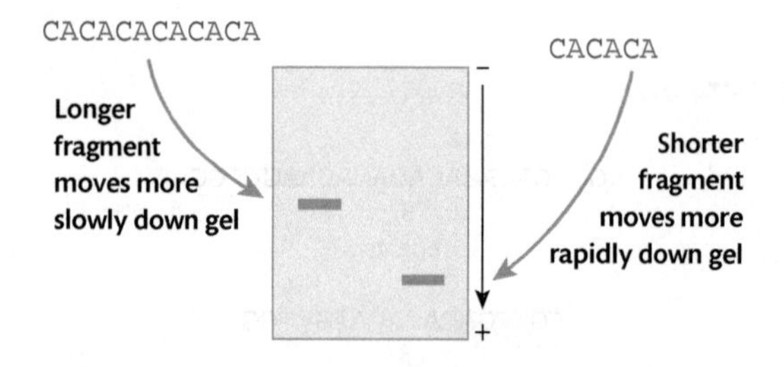


Capillary electrophoresis relies on the same basic principle, but the DNA fragments are drawn through thin glass tubes containing a porous medium. The movement of the fragments is detected using fluorescent dyes, added during amplification of the fragments.

The restriction enzyme approach requires a relatively large DNA sample to produce enough fragments. An alternative approach is to design primers that stick to the flanking regions, then use DNA amplification to produce vast numbers of copies of the fragment containing the repeat (TechBox 4.2). This approach can be used on a wide variety of DNA samples, because it does not require large amounts of DNA to start with. Because of the rapid rate of change of repeat number, microsatellites are typically used in population genetic studies, or to identify individuals, or trace family relationships. Researchers are increasingly using analysis of single nucleotide polymorphisms (SNPs) for genotyping, because they occur in greater numbers in the genome than microsatellite sequences (TechBox 3.1). However, microsatellites have a number of advantages that promote their continued use. Because they target the conserved flanking regions, the primers developed for one species can often be used on related species. Conversely, the repeat region will often be highly variable, with a much higher mutation rate than most nucleotide sequences. Whereas most SNPs have only two variants in a population, large microsatellite loci (with a decent number of repeats) can have many different alleles. This makes microsatellites handy for detecting recent population changes. However, the downside of all this variation is that it takes a lot of samples to accurately gauge allele frequencies in a populations.

## *Using microsatellites to identify individuals*

> *I took one look, thought 'what a complicated mess', then suddenly realised we had patterns. There was a level of individual specificity that was light years beyond anything that had been seen before. It was a 'eureka!' moment. Standing in front of this picture in the darkroom, my life took a complete turn. We could immediately see the potential for forensic investigations and paternity, and my wife pointed out that very evening that it could be used to resolve immigration disputes by clarifying family relationships.*

Alec Jeffreys (originator of DNA fingerprinting) 2004, http://www.genome.wellcome.ac.uk

The mutational instability of short tandem repeats, and the relative simplicity of comparing repeat regions, makes them ideal for identifying individuals. The number of repeats changes so frequently that even closely related individuals are likely to have different numbers of repeats in at least some places in the genome. A particular number of repeats at a given locus is considered to be an allele, a discrete heritable variant. Individuals may be characterized by a specific combination of alleles across many repeat loci: the set of alleles that define an individual is often referred to as their genotype. If you score enough repeat loci, you should be able to produce a 'DNA fingerprint': a specific combination of alleles that is unlikely to be found in more than one individual.

The first step is to identify repeat loci that tend to vary in number of repeats between individuals. For example, researchers wanted a reliable way to measure population size and diversity in the endangered Australian lungfish (*Neoceratodus forsteri*: **Figure 13.6**). Lungfish are particularly special animals, and not just because of their hefty genomes (**Figure 3.3**). The Australian lungfish is the only living representative of an ancient family of air-breathing fish. You would think this would accord them respect and protection as a living treasure, yet the persistence of the remaining populations of the Australian lungfish is continually threatened by development, including dams. Direct estimation of population size is tricky for such a shy fish, and molecular monitoring is hampered by low levels of genetic variation. So researchers set out to find useful genetic markers that would vary between lungfish. They extracted genomic DNA from some fin clips then used high-throughput sequencing to generate a large number of sequencing reads (**TechBox 5.2**). Then they used a bioinformatics pipeline (which defined a series of automated searching strategies) to identify sequence reads containing microsatellite sequences. Now they needed to know which of these microsats would be handy at telling lungfish apart, so they designed primers that allowed them to amplify some of these microsatellite loci for a bunch of different individual lungfish (**TechBox 6.2**), and identified which loci were polymorphic, meaning that the number of repeats at these loci differed between lungfish from these populations. Now that they have these variable loci, they can use them to estimate population size for the Australian lungfish in the five river catchments in which it is found.

*We look at the relationship between population size and genetic diversity in Chapter 8*

In addition to being a handy tool in population biology, DNA profiling has been embraced by the legal system as a way of determining whether a particular individual is associated with a biological sample found at a crime scene (such as blood or hair). The United States Federal Bureau of Investigation keeps a database called CODIS (Combined DNA Index System) of DNA samples recovered from crime scenes, from convicted offenders, and from people who have been arrested. A CODIS profile scores the alleles present at 13 particular repeat loci, plus a marker for the Y chromosome (**Figure 5.21**). CODIS is currently the largest database of DNA fingerprints in the world, containing over 10 million individual profiles. But the United Kingdom's National DNA Database contains a larger proportion of its source population, with DNA profiles from around 10 per cent of the UK population (around 6 million people). Anyone arrested in the UK can have their DNA profile stored in the database, whether they are subsequently convicted or not. However, new laws enacted in 2013 should allow for DNA profiles from innocent people to be deleted from the database.

DNA databases such as these have been used to identify suspects in crimes that would otherwise have been unsolved, and DNA profiling has also helped to exonerate people who have been falsely convicted (including a number of people on death row in the USA). But the rise of DNA forensic databases has also been criticized. Retaining DNA fingerprints on a database raises important ethical issues (**TechBox 3.2**). For example, concerns have been raised over the strong racial skew of the UK National DNA Database and the large number of samples from children. The interpretation of

| Locus | D3S1358 | vWA | FGA | D8S1179 | D21S11 | D18S51 | D5S818 |
|---|---|---|---|---|---|---|---|
| Genotype | 15, 18 | 16, 16 | 19, 24 | 12, 13 | 29, 31 | 12, 13 | 11, 13 |
| Frequency | 8.2% | 4.4% | 1.7% | 9.9% | 2.3% | 4.3% | 13% |

| Locus | D13S317 | D7S820 | D16S539 | THO1 | TPOX | CSF1PO | AMEL |
|---|---|---|---|---|---|---|---|
| Genotype | 11, 11 | 10, 10 | 11, 11 | 9, 9.3 | 8, 8 | 11, 11 | X Y |
| Frequency | 1.2% | 6.3% | 9.5% | 9.6% | 3.52% | 7.2% | (Male) |

**Figure 5.21** This is the CODIS profile of Bob Blackett, a forensic DNA analyst. At each of 13 loci, Mr Blackett has two alleles, one on each chromosome. The 14th box is a marker found on the Y chromosome. Each allele is given a number, and because Mr Blackett is diploid, he has two alleles listed for each locus (in some cases the same allele—homozygous—and in some cases two different alleles—heterozygous). The frequency of the combination of alleles in the reference population is given below. These frequencies are sometimes used to calculate how likely a random match would be—that is, the probability that two people could have exactly the same pattern of alleles. However, these probability calculations depend on identifying an appropriate reference population, since the frequencies vary between countries, and within ethnic groups (**TechBox 3.2**).

Reproduced by permission of Bob Blackett/Rick Hallick, University of Arizona.

matches to the database is also a subject of ongoing debate. When a new sample is entered in the database, it can be searched against records of unsolved crimes for possible matches. The chance of false matches produced by these searches needs to be evaluated. After all, even a one-in-a-billion chance of a match means that there might be half a dozen people in the world with exactly the same profile as you. Most of the statistics used to interpret the likelihood of a chance match assume that DNA profiles represent random independent draws from a population of alleles. But microsatellite alleles are not randomly drawn from the population, they are inherited (with some mutation) from parents. So not only will alleles be more similar among close relatives than a random draw from the population (hence their use in paternity tests), they will also tend to be more similar within interbreeding groups, defined by social networks, geographic regions, races or cultures.

More problems arise when profiles generated in one jurisdiction are tested against different countries' databases, because only the shared loci used by both systems can be compared. If the number of loci compared is reduced to those matching between databases, then the chance of random matches between a sample and an unrelated profile in the database (false hits) will increase. A data-sharing agreement between European countries requires comparison of only a minimum of six microsatellite loci to be used as the basis of a search, which, given the large number of samples in the combined databases, greatly increases the chance of a false hit. For example, in a test of accuracy using dozens of samples, over half of the supposed identifications based on six loci were false positives, incorrectly matching an individual to a profile in the database that wasn't actually theirs. So while DNA profiling has been a boon to the justice system, it should not be regarded as foolproof, but as one possible line of evidence that must be considered in light of possible errors.

### *Using microsatellites to uncover relationships*

Although tandem repeats have a very high rate of change, they still carry historical signal. The repeat loci in your parents' genomes were copied when they made the gametes that came together to make you. While some of these repeat sequences might have increased or decreased in length, on the whole your DNA fingerprint will be more similar to your parents than it is to unrelated members of the population. Therefore, analysis of repeats can reveal family relationships. Indeed, one of the first applications of DNA fingerprinting was in an immigration dispute in 1985, where genetic data was used to prove that a child that had been refused British residency was indeed the son of a UK citizen.

**Figure 5.22** Alpine marmots (*Marmota marmota*): not as innocent as they look.

Image courtesy of François Trazzi, licensed under the Creative Commons Attribution Share Alike 3.0 Unported license.

More broadly, DNA fingerprinting can be used to understand social dynamics and population structure. In particular, analysis of repeat regions has provided a way of assessing mate choice and paternity in wild populations, often revealing relationships that would not have been evident from behavioural studies alone. For example, the alpine marmot (*Marmota marmota*: **Figure 5.22**) is one of the few mammal species considered to be monogamous (forming exclusive breeding relationships between one male and one female). Alpine marmots live communally in family groups, consisting of a breeding pair, plus non-breeding subordinates and juveniles. But, using DNA extracted from hair samples, analysis of six microsatellite loci revealed that one fifth of the marmot juveniles were not fathered by the resident male in the family group, and a third of the litters contained the offspring of two different fathers.

Because repeat alleles are inherited, close relatives will tend to have similar alleles. This means that a small, interbreeding population will tend to share a relatively small number of alleles, which will be distinct from other such populations. So repeat sequences such as microsatellites can, in many cases, be used to tell which population an individual is from. This is becoming increasingly useful as a way of monitoring the trade in endangered species. For example, DNA can be extracted from timber to check not only the species of tree it was taken from, but also the geographic region of origin. Genetic analysis will play an increasingly important role in policing the illegal logging of tropical hardwoods, to make sure that commercially available timber has not been harvested from endangered populations or protected areas. DNA fingerprinting can also improve traceability and quality control of timber. The global increase in wine sales has led to a growth of the cooperage (barrel-making) industry, and subsequent shortages of suitable oak timber. DNA fingerprinting can be used to check the provenance of oak timber, to ensure the barrels are made from the optimum type of timber for maturing wines.

# Transposable elements

*So, naturalists observe, a flea*
*Has smaller fleas that on him prey;*
*And these have smaller still to bite 'em,*
*And so proceed ad infinitum.*

Jonathan Swift, 1733, from *On poetry: a rhapsody*

We have seen that, in addition to the genes and regulatory sequences needed to build the organism, much of the DNA in the genome consists of repetitive sequences that don't appear to be directly connected to building and running the organism. Some of this extra DNA in the genome has resulted from accidents of chromosomal management or DNA copying. But not all duplications are the result of passive processes. Some sequences have features that actively encourage their own duplication, leading to a proliferation of these sequences in the genome. The more copies there are, the more extra copies will be made, which can then be inserted at different locations across the genome.

Transposable elements, also known as transposons, were originally described by Barbara McClintock (**Hero 5**) as 'jumping genes'. They are sequences that can move location in the genome. Transposon sequences will be copied along with all the rest of the genome every time the cell divides, so they will be passed on to offspring just like any other genes. But in addition to being replicated at cell division, transposable elements can increase in number within the genome. Many of these sequences bear a close resemblance to viral genomes (**Case Study 5.2**). Like viruses, transposable elements rely on their host's cellular equipment to propagate their genomes. And, like parasites, some transposable elements have cunning ways of encouraging their hosts to act in the element's own best interest (see **Figure 3.8**). Transposable elements often contain strong promoter and enhancer signals that attract the host replication machinery, causing them to make many copies of the transposable element sequence. It's essentially like sitting in the genome screaming 'Copy me! Copy me!'. The evolutionary success of transposable elements is evident by their sheer numbers. Based on rough estimates from genome sequences, it seems that DNA derived from transposable elements makes up a much larger proportion of your genome than the genes needed to construct a human being (you may feel this raises an interesting philosophical question about exactly who you are, given that your genome could be considered to be more virus than human).

Many transposable elements are characterized by terminal repeats: that is, they begin and end with copies of the same sequence (**Figure 5.23**).

**Figure 5.23**

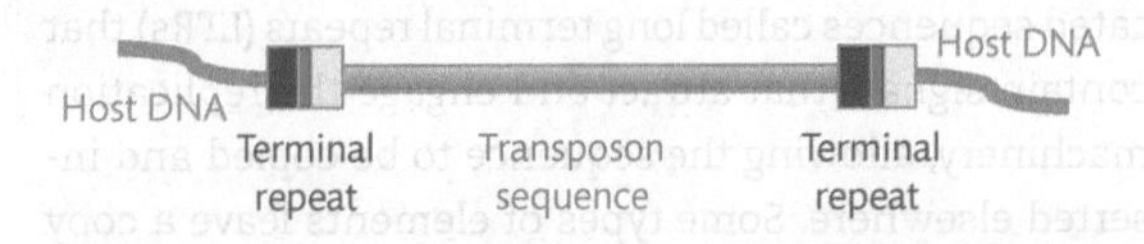


These paired sequences can come together by complementary base pairing, causing the intervening sequence to loop out (**Figure 5.24**). The loop can then be excised, freeing the element to insert into a different location in the genome.

**Figure 5.24**

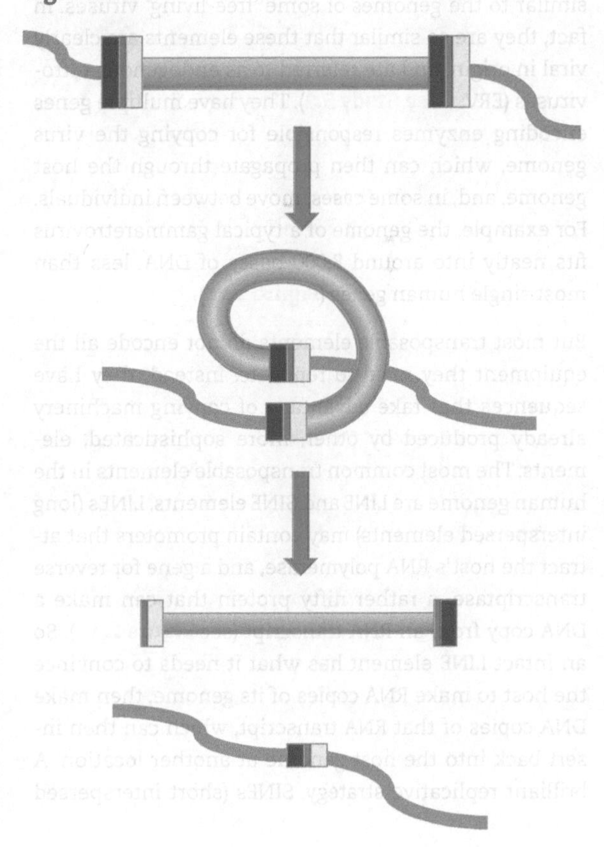

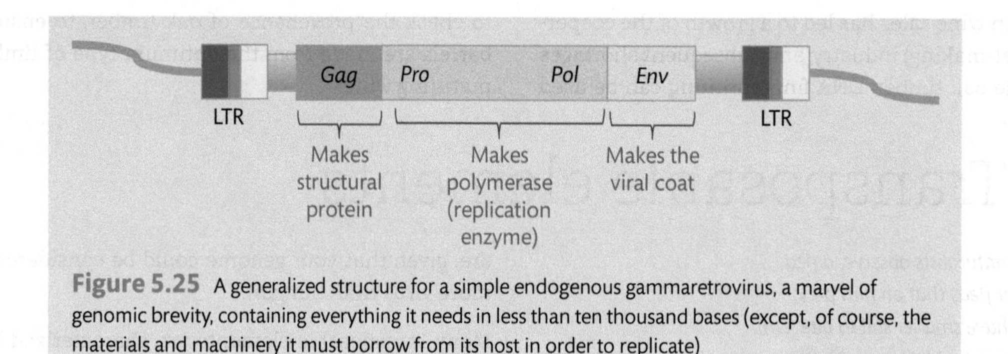


**Figure 5.25** A generalized structure for a simple endogenous gammaretrovirus, a marvel of genomic brevity, containing everything it needs in less than ten thousand bases (except, of course, the materials and machinery it must borrow from its host in order to replicate)

Some transposable elements are flanked by sophisticated sequences called long terminal repeats (LTRs) that contain signals that attract and engage the replication machinery, allowing the sequence to be copied and inserted elsewhere. Some types of elements leave a copy in the original location, so that each transposition event increases the number of copies in the genome. Others just leave a copy of one LTR. If any part of the sequence is left behind, it leaves the footprint of transposon activity.

## LINEs, SINEs, and retroviruses

The most sophisticated transposable elements are very similar to the genomes of some 'free-living' viruses. In fact, they are so similar that these elements are clearly viral in origin, and are referred to as endogenous retroviruses (ERV: Case Study 5.2). They have multiple genes encoding enzymes responsible for copying the virus genome, which can then propagate through the host genome, and, in some cases, move between individuals. For example, the genome of a typical gammaretrovirus fits neatly into around 8,000 bases of DNA, less than most single human genes (Figure 5.25).

But most transposable elements do not encode all the equipment they need to replicate. Instead, they have sequences that take advantage of copying machinery already produced by other, more sophisticated, elements. The most common transposable elements in the human genome are LINE and SINE elements. LINEs (long interspersed elements) may contain promoters that attract the host's RNA polymerase, and a gene for reverse transcriptase, a rather nifty protein that can make a DNA copy from an RNA transcript (see Figure 2.12). So an intact LINE element has what it needs to convince the host to make RNA copies of its genome, then make DNA copies of that RNA transcript, which can then insert back into the host genome at another location. A brilliant replicative strategy. SINEs (short interspersed elements) don't have a reverse transcriptase gene, so they wouldn't be able to replicate on their own. But a genome containing intact LINE elements is a genome that produces reverse transcriptase. SINEs borrow the reverse transcriptase produced by other elements—another brilliant strategy. If transposable elements are genomic parasites, then SINE elements are parasitic on other parasites, just as in Swift's pithy homage to the economy of nature.

As loose cannons in the genome, transposable elements make a substantial contribution to the mutation rate in many species. There are several ways that transposition causes heritable changes in the genome. Firstly, if a transposable element inserts into a working gene, it will almost certainly destroy gene function, causing a 'knock-out' mutation. Secondly, transposition can disrupt normal gene expression. Many transposable elements contain strong regulatory signals that attract the cellular machinery that makes RNA transcripts. The presence of these strong signals can cause overexpression of genes in the vicinity. This may be how some genomic viruses cause disease (Case Study 5.2). Thirdly, transposable elements can move DNA around the genome: they might pick up pieces of the host genome when they are copied or excised and then translocate them to another part of the genome (or even to the genome of a different individual). Fourthly, transposable elements can be a source of novel DNA which can be co-opted into normal host development. One fascinating example is syncytin, which is active in the formation of the placenta in mammals. The syncytin gene is clearly made out of a retrovirus genome: though mutations have rendered the *gag* and *pol* genes non-functional, the *env* gene is intact and triggers cell fusion in the developing placenta (Figure 5.26). The V(D)J recombination system used in the mammalian immune system also relies on a tamed transposon-derived sequence.

**Figure 5.26** We wouldn't be here without our tame retroviruses. The placenta does one of the trickiest jobs in biology, acting as a link between two genetically different individuals (mother and child), without triggering an immune reaction in either of them. Retroviral sequences play a key role in building this multigenerational bridge. The envelope gene gives a virus the capacity to fuse with a host cell membrane, thus allowing the viral genome entry into the cell. On the other hand, when *env* is expressed by an infected host cell, it can saturate the surface receptors on the cell, making it less vulnerable to entry by another virus, thus endogenous expression of *env* can confer a degree of host resistance. *Env* has been co-opted to placental formation in mammals. Two *syncytin* genes, both derived from retrovirus *env* genes (one of which is still embedded in an ERV proviral genome), drive the fusion of cells in the placenta to form a syncytial (multinucleate) cell layer which invades maternal tissue, rupturing capillaries (tiny blood vessels) to promote exchange of nutrients from mother to embryo. Amazingly, it seems that different retroviral sequences have been independently co-opted into placenta formation in several different mammalian lineages.
Photograph by Lindell Bromham.

## The genomic record of transposition

Transposition can leave a historical record in the genome. For example, when an LTR-transposon inserts into the genome, it creates two copies of the terminal repeat regions from a single template. This means that each inserted element starts with two identical copies of the LTR sequences. The whole element is copied by the host every time the genome is replicated, and, since no copy process is error-free, occasional replication errors introduce mutations into the sequence (Chapter 4). But, because most transposable elements play no useful role in growth and development, a mutation that renders the element inactive will come at no cost to the owner of the genome, who will continue to live and reproduce as if nothing happened. Since these mutations in inserted transposons will not be cleared away by natural selection, they will accumulate over time, slowly altering the transposon sequence. So although the two LTR sequences start out identical to each other, they get progressively more different from each other over time, as each of the LTR sequences independently acquires random mutations. The difference between the paired LTRs therefore gives an indication of how long it has been since the element was inserted. For example, by comparing the differences between the paired LTRs of various transposable elements, researchers estimated that much of the increase in the rather bloated maize genome was due to an explosive burst of transposable element activity within the last few million years (Figure 5.27). More generally, patterns of insertion ages

**Figure 5.27** Maize was first domesticated in Mesoamerica over 6,000 years ago, where many varieties are still grown. (a) Some of the colour variants in maize varieties are due to the action of transposable elements. (b) The most common agricultural variants of maize have now spread over the globe, grown not only in large-scale agribusiness but also small holdings, such as these farmers on a smallholding in Kenya.
(a) Courtesy of Jenny Mealing. (b) Courtesy of McKay Savage. Both images are licensed under the Creative Commons Attribution 2.0 Generic license.

have been used to suggest that, rather than being the result of continuous, ongoing transposition, most new insertions occur in bursts of activity. This boom-and-bust pattern has been interpreted as evidence that the initial unchecked proliferation of novel elements is eventually brought under control by host genome mechanisms. For example, although the maize genome is puffed out with transposable elements, few if any appear to be active, suggesting that they have been 'silenced'.

The pattern of insertions across the genome can be exploited as a source of historical information. Like any mutation, a novel transposon insertion will be inherited by any descendants that receive a copy of that genome. On the assumption that transposable element insertion is more-or-less random across the genome, if two sampled genomes have a transposon at the same locus then they probably inherited it from a common ancestor. Over time, populations become characterized by different patterns of transposon insertions. This can be exploited to reveal population structure and history. For example, foxtail millet is one of the earliest grain crops grown in Eurasia (**Figure 5.28**). The foxtail millet genome contains several different families of active transposable elements, so researchers used restriction enzymes to identify parts of the genomes that showed insertion polymorphisms (that is, where some but not all members of the species have a transposon inserted at a particular locus). By grouping populations on the basis of shared

**Figure 5.28** Foxtail millet is one of the oldest domesticated grains, grown in China for at least eight thousand years. Like many cultivated grains, a major change was in retention of seed heads: wild millets shatter the seed head so that they can disperse, but domesticated grains retain the seed head on the plant so it can be harvested.



insertions, they showed strong geographic structuring in foxtail millet, suggesting that 'land races' have become distinct during the long history of cultivation in each separate area (see also **Case Study 3.1**).

Transposon insertions can also be used to infer deeper history. Because descendants inherit copies of all the transposable elements in the genome, SINEs have been used as markers for mammalian phylogeny, grouping together lineages that share particular SINE element insertions. Since SINE elements are present in hundreds of thousands of copies in mammal genomes, they provide a potentially rich source of historical information. However, like any other source of historical information in biology, transposable elements must be interpreted with caution. For example, a surprising new grouping of mammals was suggested on the basis of LINE element insertions shared by bats and horses. This novel group, named Pegasoferae after the winged horse of Greek myth, has not been supported by subsequent phylogenetic analyses of more extensive genomic sequences.

## Selfish DNA or helpful symbionts?

> *In the future attention undoubtedly will be centered on the genome, and with greater appreciation of its significance as a highly sensitive organ of the cell, monitoring genomic activities and correcting common errors, sensing the unusual and unexpected events, and responding to them, often by restructuring the genome.*

Barbara McClintock, 1983, Nobel Prize acceptance speech.

Transposable elements of one kind or another make up a substantial proportion of the genome in most animals and plants. Maize, from which jumping genes were first identified, is a classic example: over 85 per cent of the maize genome is made up of transposable elements (**Figure 5.26**). In one way, the revelation that much of the genome is made up of transposons solves the C-value paradox. Why is the size of the genome apparently unrelated to organismal complexity? Because a lot of it is transposable elements which are not directly involved in constructing and operating the organism's body. But this raises even more intriguing questions: why do organisms maintain and copy vast amounts of DNA not directly relevant to survival and reproduction? And why does the proportion of the genome taken up by transposable elements differ so much between species?

There are two broad ways to interpret the presence of large numbers of transposable elements in eukaryote genomes. One is to view transposons as genomic parasites, free-riders that use the host's genomic machinery to replicate themselves but give nothing in return. After all, all organisms have parasites, whether it's fleas on their outsides, tapeworms on their insides, or cuckoo eggs in their nests. Are most species locked in a constant battle with their genomic parasites, forever working to remove or disable the transposable elements? Evidence cited in favour of this view includes the tendency of transposable elements to proliferate at the greatest rate when newly introduced to a genome or lineage, when an individual is under great stress, or when the immune system is compromised. These patterns could be the result of the failure of mechanisms that, in better times, act to recognize and inhibit replication of transposons. Species with large numbers of transposable elements might be those that have essentially lost the battle, albeit temporarily. Or it could be that for species with large numbers of genomic parasites, the cost of carrying transposons in the genome is insufficient to provide strong enough selection pressure for their removal. It has been suggested that species with small effective population sizes tend to have larger genomes. In small populations, selection is less effective at removing mutations of relatively small fitness cost (**TechBox 8.2**), so extra copies of transposons might accumulate unchecked as long as none causes serious harm.

An alternative view is that the benefits of transposable elements outweigh the costs of extra DNA copying and transpositional mutagenesis. Transposition sculpts the genome, moving sequences, adding regulatory elements, switching genes on or off. The tendency of transposition activity to increase in plants under stress has been interpreted as a desperate attempt to bring about changes in gene activity that just might provide a solution to the crisis. Barbara McClintock, the discoverer of transposable elements, considered that transposition allowed the genome to be rapidly reconfigured, providing a quick response in times of environmental stress or 'genome shock', potentially providing the foundation for adaptation and speciation. Is the significant role of transposition in shaping the eukaryote genome an accidental by-product of the presence of selfish elements, or a valuable engine of evolvability? Could it be both at once?

# Conclusion

> *I would be quite proud to have served on the committee that designed the E. coli genome. There is, however, no way that I would admit to serving on a committee that designed the human genome. Not even a university committee could botch something that badly.*

David Penny, quoted in Graur et al. 2013 *Genome Biology and Evolution, Volume* 5, pages 578–590.

Much of the delight in evolutionary biology comes in discovering the myriad ways that organisms have found to survive and thrive. In some cases, this is achieved through dazzling simplicity, in other cases with a bewildering complexity. Yet the ghost of the Great Chain of Being continues to haunt biology. You can catch a glimpse of it whenever you see species arranged in a way that suggests progress in evolution, climbing a ladder of ever-increasing complexity, from pond slime up to people. But, although we have a natural tendency to see our own species as the pinnacle of the natural world, the view from the genome gives us pause for thought. The human genome is not the largest, nor the most gene rich, it is neither the most efficient nor the most complicated. Like the genome of any other species, the human genome has been shaped by selection and chance, being expanded and contracted by accidental duplications and deletions, rearranged by recombination and transposition, and remodelled by transposable elements. The human genome is not pretty or efficient, but the fact that you are reading this now proves that it works.

In some ways the C-value paradox has been solved: the genome is stuffed full of repetitive sequences of one kind or another, and species with huge genomes tend to have more repetitive sequences, not more genes. But in other ways, the paradox has got even more puzzling: why on earth is so much of the genome made up of sequences that have little direct role in making the organism? We are beginning to appreciate that the genome is far from a simple recipe, read gene-by-gene, for building an organism. In the next chapter, we will explore the tangled relationship between genotype and phenotype.

# Points to remember

## Genome size

- For many species, repeated sequences form a greater proportion of the genome than genes do.
- The observation that genome size does not have an obvious relationship with perceived complexity of the organism is referred to as the C-value paradox.
- At meiosis the genome is duplicated and divided to produce haploid gametes with only one copy of each chromosome, so that fusing two gametes creates a new diploid cell.
- Whole or part of the genome can be doubled if chromosomes fail to separate at meiosis, or if cells fuse without first reducing the number of chromosomes.
- Changes in ploidy (number of genome copies) are often fatal, but can lead to the formation of new polyploid lineages.

## Duplications and deletions

- Matching copies of chromosomes (homologs) must pair at meiosis so that one copy can go in each daughter cell.
- Pairing of homologous chromosomes provides an opportunity for recombination (exchange of DNA between chromosomes).
- Inversion of part of a chromosome may preserve the DNA sequence, but can lead to reproductive incompatibility between organisms with different gene orders.
- Tandem (side-by-side) repeat sequences are prone to changes in copy number due to unequal crossing over when chromosomes are paired at meiosis, or due to replication slippage when DNA is being copied.
- Short tandem repeats (microsatellites) provide fast-changing loci that can be used to genotype individuals, identify relationships or characterize populations.

## Transposable elements

- Transposable elements have features that encourage the sequence to be copied and reinserted elsewhere in the genome.
- Transposable elements are a large proportion of many species' genomes.
- Transposition (copying and insertion of elements) can cause mutations.
- The dynamics of transposable element activity can be inferred from the pattern of insertion sequences across the genome and the sequence divergence between elements.
- The position of transposable elements can be used as informative markers of shared evolutionary history.

# Ideas for discussion

**5.1** Do genomes have a tendency to keep growing in size? What, if anything, stops them from getting bigger?

**5.2** How could you assess the relative importance of polyploidy in the formation of new species?

**5.3** Can you devise a test that could compare the two alternative hypotheses for the relationships between transposable elements and their host genomes, either that they are harmful genomic parasites or helpful genomic architects?

# Sequences used in this chapter

| Table 5.1: Sequence of *Porphyra purpurea* chloroplast genome from Figure 5.4 | |
|---|---|
| Description from GenBank entry | Accession |
| *Porphyra purpurea* chloroplast, complete genome: genome map produced using CGView Comparison Tool | NC_000925 |

# Examples used in this chapter


Deguilloux, M-F., Pemonge, N-H., Petit, R. J. (2004) DNA-based control of oak wood geographic origin in the context of the cooperage industry. *Annals of Forest Science*, Volume 61, page 97.

Geddes, L. (2011) DNA super-network increases risk of mix-ups. *New Scientist*. 5 September 2011.

Goossens, B., Graziani, L., Waits, L. P., Farand, E., Magnolon, S., Coulon, J., Bel, M-C., Taberlet, P., Allaine, D. (1998) Extra-pair paternity in the monogamous Alpine marmot revealed by nuclear DNA microsatellite analysis. *Behavioural Ecology and Sociobiology*, Volume 43, page 281.

Guelbeogo, W. M., Sagnon, N. F., Grushko, O., Yameogo, M. A., Boccolini, D., Besansky, N. J., Costantini, C. (2009) Seasonal distribution of *Anopheles funestus* chromosomal forms from Burkina Faso. *Malaria Journal*, Volume 8, page 239.

Hehemann, J-H., Correc, G. I., Barbeyron, T., Helbert, W., Czjzek, M., Michel, G. (2010) Transfer of carbohydrate-active enzymes from marine bacteria to Japanese gut microbiota. *Nature*, Volume 464, page 908.

Hirano, R., Naito, K., Fukunaga, K., Watanabe, K. N., Ohsawa, R., Kawase, M., Belzile, F. (2011) Genetic structure of landraces in foxtail millet (*Setaria italica* (L.) P. Beauv.) revealed with transposon display and interpretation to crop evolution of foxtail millet. *Genome*, Volume 54, page 498.

Huey, J. A., Real, K. M., Mather, P. B., Chand, V., Roberts, D. T., Espinoza, T., McDougall, A., Kind, P. K., Brooks, S., Hughes, J. M. (2013) Isolation and characterization of 21 polymorphic microsatellite loci in the iconic Australian lungfish, *Neoceratodus forsteri*, using the Ion Torrent next-generation sequencing platform. *Conservation Genetics Resources*, Volume 5, page 737.

Suárez-Villota, E., Vargas, R., Marchant, C., Torres, J., Köhler, N., Núñez, J., de la Fuente, R., Page, J., Gallardo, M., Jenkins, G. (2012) Distribution of repetitive DNAs and the hybrid origin of the red vizcacha rat (Octodontidae). *Genome*, Volume 5, page 105.

Wexler, N. S., Lorimer, J., Porter, J., Gomez, F., Moskowitz, C., Shackell, E., Marder, K., Penchaszadeh, G., Roberts, S. A., Gayan, J., Brocklebank, D., Cherny, S. S., Cardon, L. R., Gray, J., Dlouhy, S. R., Wiktorski, S., Hodes, M. E., Conneally, P. M., Penney, J. B., Gusella, J., Cha, J. H., Irizarry, M., Rosas, D., Hersch, S., Hollingsworth, Z., MacDonald, M., Young, A. B., Andresen, J. M., Housman, D. E., De Young, M. M., Bonilla, E., Stillings, T., Negrette, A., Snodgrass, S. R., Martinez-Jaurrieta, M. D., Ramos-Arroyo, M. A., Bickham, J., Ramos, J. S., Marshall, F., Shoulson, I., Rey, G. J., Feigin, A., Arnheim, N., Acevedo-Cruz, A., Acosta, L., Alvir, J., Fischbeck, K., Thompson, L. M., Young, A., Dure, L., O'Brien, C. J., Paulsen, J., Brickman, A., Krch, D., Peery, S., Hogarth, P., Higgins, D. S., Jr, Landwehrmeyer, B. (2004) Venezuelan kindreds reveal that genetic and environmental factors modulate Huntington's disease age of onset. *Proceedings of the National Academy of Sciences USA*, Volume 101, page 3498.

HEROES OF THE GENETIC REVOLUTION

5

# Barbara McClintock

*"I was entranced at the very first lecture I went to. It was zoology, and I was just completely entranced. I was doing now what I really wanted to do, and I never lost that joy all the way through college"*

B. McClintock, quoted in Keller, E. F. (1986) *A feeling for the organism*. W. H. Freeman and Company.

**EXAMPLE PUBLICATIONS**

Creighton, H. B., McClintock, B. (1931) A correlation of cytological and genetical crossing-over in *Zea mays*. *Proceedings of the National Academy of Sciences USA*, volume 17, page 492.

McClintock, B. (1953) Induction of instability at selected loci in maize. *Genetics*, page 38, page 579.

**Figure 5.29** Barbara McClintock (1902–1992) was one of the pioneering geneticists who linked genes as observable phenomena (patterns of inheritance) to physical locations on chromosomes, long before the biochemical basis of genes was discovered (as nucleotide sequences in DNA molecules).

Credit: Smithsonian Institution Archives. Image SIA2008-5609.

Barbara McClintock's contributions to genetics were profound, from the technical advances that allowed her to provide the first physical evidence of chromosomal crossing over to her insight into the existence of mobile genetic elements (Figure 5.29). And yet she almost missed out on a career in science, due to family resistance to her gaining a college education. However, she was allowed to attend Cornell's College of Agriculture, from which she graduated in 1923. One of her lecturers recognized McClintock's promise and encouraged her to undertake postgraduate study in genetics.

The 1920s and '30s were a time of great excitement in genetics. Many of the foundations of the field were being built, particularly in the *Drosophila* labs, such as those of Thomas Hunt Morgan. With her background in plant breeding, McClintock developed maize as an alternative experimental system. She and her research group spent time in the maize fields, making experimental crosses and observing the progeny. At the time genetics consisted of inferring genes from the observable results of breeding experiments, rather than by direct observation of the genetic material. McClintock brought about a revolution in plant genetics by combining these breeding experiments with cytogenetics (observations of chromosomes). With her talent for fine-scale microscope work and her endless patience, she developed ways of staining and preparing maize chromosomes so that they could be recognized and followed across generations of maize. By identifying specific chromosomes and their banding patterns, and using this physical geography of the chromosomes to observe the results of crosses, McClintock played a key role in connecting genetics theory to physical observations of the behaviour of chromosomes, before the biochemical basis of heredity was known. Most significantly, she and her student Harriet Creighton were the first to witness recombination between chromosomes, which had been predicted on the basis of breeding experiments but never before directly observed. Furthermore, her observation of ring chromosomes led her to predict the presence of specialized ends of chromosomes (telomeres) that regulated chromosomal behaviour at cell division.

In the 1940s, McClintock's ground-breaking work in cytogenetics was recognized by a number of honours, including election to the National Academy of Sciences of the USA and the presidency of the Genetics Society. She moved to Cold Spring Harbour laboratories where she began the work she is now most famous for: the recognition and description of unstable patterns of inheritance in maize. Through a combination of recording colour patterns on corn cobs over several generations and observation of minute changes in chromosomal structure, she described the action of genetic loci that appeared to move around the genome, generating mosaic patterns in phenotype as they switched genes on in some cells, and off in others. Although McClintock published her evidence for 'jumping genes' and spoke about it at many meetings, the work was not well received and eventually she gave up trying to convince people of her heretical hypothesis. However, in the 1960s and 70s, McClintock's observations of genetic regulation (switching genes on and off) and mobile elements (genetic loci transposing across the genome) were vindicated by independent research by other scientists. She was retrospectively recognized as the discoverer of transposition, for which she received a Nobel Prize in 1983.

McClintock's work was characterized by an intimacy with her study system, such that she frequently 'knew' what the genetic mechanism was before she had the observational evidence to prove it. This intuitive approach to her research inspired some students and colleagues, but irritated others. However, her intuitions were often right, and the combination of her 'feeling for the organism' and her innovative techniques revealed many important aspects of chromosomal evolution, once controversial but now taken for granted in genetics.

> "I found that the more I worked with [chromosomes]… the bigger and bigger they got, and when I was working with them I wasn't outside, I was down there. I was part of the system. … It surprised me because I actually felt as if I was right down there and these were my friends"
>
> B. McClintock, quoted in *A feeling for the organism* (1986) E. F. Keller.

TECHBOX 5.1

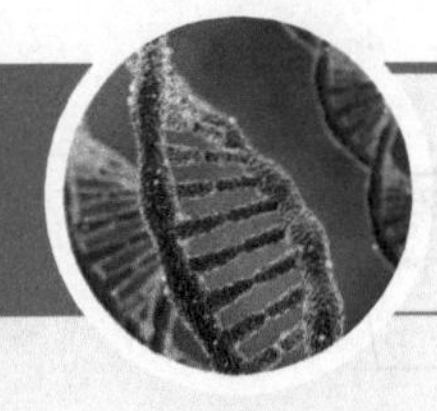

# Sanger sequencing

**RELATED TECHBOXES**

**TB 4.2:** DNA amplification

**TB 5.2:** High-throughput sequencing

**RELATED CASE STUDIES**

**CS 1.1:** Identification

**CS 5.2:** Genome

**Summary**

Classic DNA sequencing uses a clever combination of modified nucleotides to produce DNA fragments that each stop at a different base in the sequence.

**Keywords**

Sanger sequencing, chain termination, dideoxy, dNTP, primer, gel

The Sanger sequencing method, also known as dideoxy sequencing, is also known more generally as chain termination sequencing. Until recently, this was the main method used to sequence DNA, so was just known as 'sequencing', but now it is being overtaken by a range of high-throughput methods (**TechBox 5.2**). This rapid technological turnover has had the happy side effect that Fred Sanger's name is once again being used to describe the sequencing method he invented (**Hero 1**). The Sanger sequencing method is known as the chain termination method because it uses modified nucleotides to halt DNA synthesis at different points to produce an array of DNA fragments, each of which stops at a different nucleotide in the sequence. This method is also known as dideoxy sequencing because the modified nucleotides are missing their

OH group. Since nucleotides are added to the OH group of the last nucleotide in the chain (see **TechBox 4.1**), once a dideoxynucleotide is added, no more nucleotides can be added to that chain, so synthesis stops. To understand DNA sequencing, it is helpful to have a grasp of DNA structure (**TechBox 2.1**), DNA replication (**TechBox 4.1**) and DNA amplification (**TechBox 4.2**).

Sanger sequencing relies on having a vast number of copies of the DNA you wish to sequence, so the first step is usually to amplify the target sequence (**TechBox 4.2**). Then the amplified DNA is heated to separate it into single-stranded templates and mixed with polymerase (DNA copying enzyme), nucleotides (DNA 'letters'), and primers (short DNA sequences that serve as starting blocks for DNA synthesis: **TechBox 6.1**). The neat trick with Sanger sequencing is that some of the nucleotides are labelled with something that makes them detectable, such as fluorescent dyes or radioactive labels, and some of the nucleotides are modified so that whenever they are added to the new strand they stop the addition of any more nucleotides (hence 'chain termination').

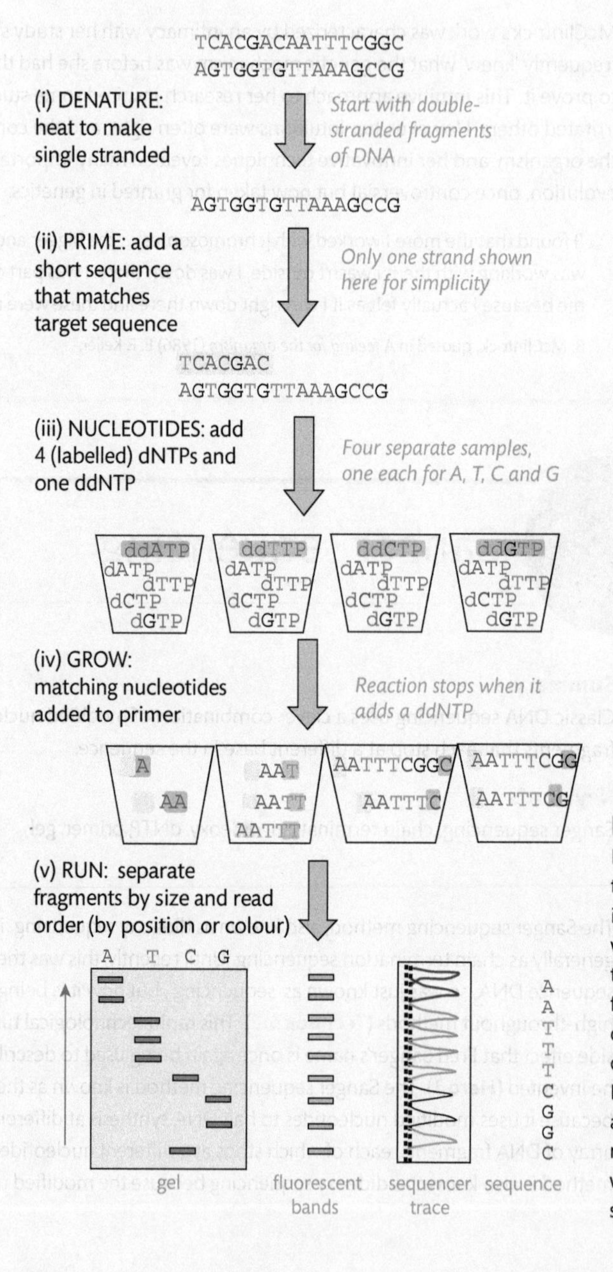


**Figure 5.30** A rough guide to DNA sequencing using the chain termination (Sanger) method, including several different ways of reading the order of the nucleotides. Traditionally, four separate reactions are conducted, one for each type of chain-terminating base. However, automated sequences avoid the need for separate reactions by labelling each base with a different fluorescent dye, so they can all be mixed together.

So, when all this lot is mixed together, the primers stick to the single-stranded DNA and the polymerase starts merrily making copies of the DNA in the sample, sticking the nucleotides together to form polynucleotides complementary to the template DNA (**Figure 5.30**). As the polymerase goes about its business, it will occasionally incorporate a labelled nucleotide in the growing nucleotide chain. And sometimes it incorporates a reaction-stopping dideoxy nucleotide, at which point the growth of that particular DNA chain stops. So throughout the reaction mixture, copies of the template DNA are being made by polymerase, and those copies are stopping at different points, whenever the polymerase happens to incorporate a reaction-stopping nucleotide. The result is a solution of DNA sequences of different lengths, each ending at a different 'letter' in the DNA sequence. If these sequences are 'read' in order of length, then the sequence of bases at the ends of the fragments provides the sequence for the template they were all copied from.

The specific details differ between methods, but the basic approach of Sanger sequencing is:

**(i) Denature:** The amplified DNA is heated, to separate the double-stranded DNA into single strands.

**(ii) Prime:** Short sequences that have a complementary sequence to the target sequence are added to the single-stranded DNA, which is cooled so that the primers can bind to the template (**TechBox 6.2**). The primer provides a starting block for synthesizing a new DNA strand, because it provides a free OH group for newly added bases to bind to (**TechBox 4.2**).

**(iii) Nucleotides:** In classic Sanger sequencing, the amplified DNA with attached primers is split into four samples (**Figure 5.30**). To each of these samples is added the four DNA bases (in the form of deoxynucleotides: dATP + dTTP + dCTP + dGTP) and DNA polymerase (the enzyme that makes a new DNA strand to match the template). In addition, each sample has a different chain-terminating dideoxynucleotide (ddATP or ddTTP or ddCTP or ddGTP).

**(iv) Grow:** As the polymerase moves along the template, it picks up and adds nucleotides to make a matching DNA strand. If it picks up a deoxynucleotide (dNTP), it continues to add nucleotides to the growing strand, but if it incorporates a dideoxynucleotide (ddNTP), chain elongation stops at that point. Using the example sequence given in **Figure 5.30**, in the ddCTP sample, the polymerase will add nucleotides to the growing strand (**Figure 5.31**).

**Figure 5.31**

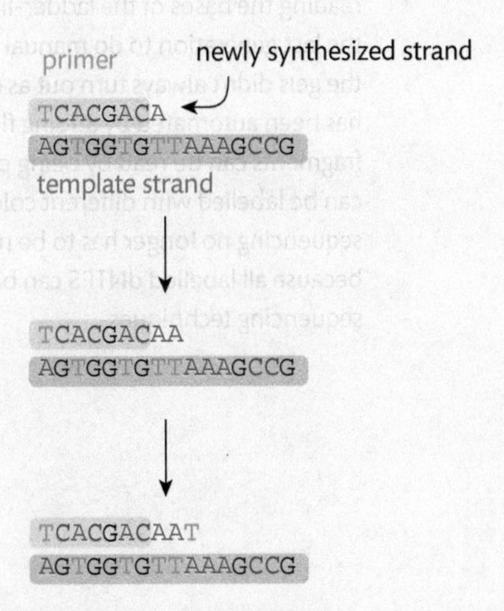


When it gets to a C it might incorporate a normal dCTP and keep going, or it might incorporate a chain terminating ddCTP and stop (**Figure 5.32**).

Figure 5.32

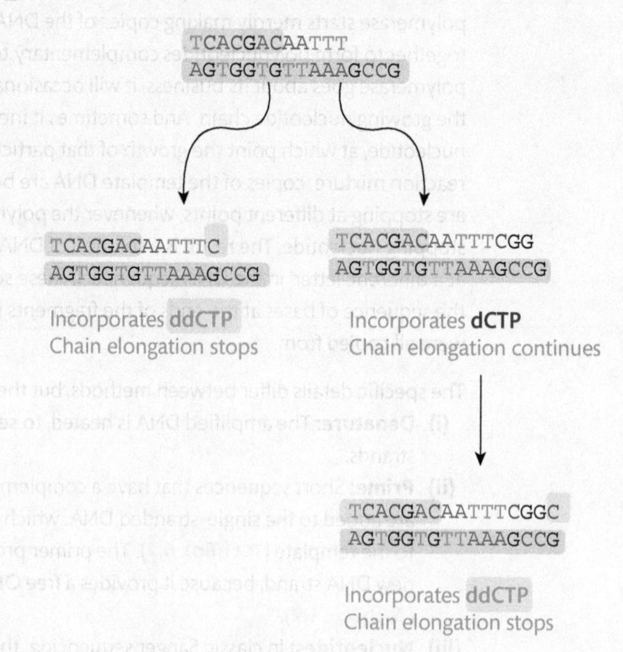


The upshot of this is that, in the ddCTP sample, there will be fragments of many different lengths, but they will all stop at a C. Similarly, the ddTTP sample will have fragments of different lengths that all end at a place in a sequence where there is a T.

**(v) Run:** To read the sequence, it is necessary to order the fragments with respect to size and report which nucleotide is at the end of each fragment. In traditional Sanger sequencing, the fragments are labelled radioactively (usually by labelling either the dNTPs or ddNTPs). The fragments are run through a gel, with each of the four samples run in a different lane (**Figure 5.33**). Since the longer sequences are heavier and travel more slowly through the gel, the bands will appear on the gel in order of length, so the nucleotides at the end of each fragment can be read in the order of the bands along the gel. This used to involve reading the bases of the ladder-like sequencing gels (**Figure 1.10**: although, being one of the last generation to do manual Sanger sequencing, I can tell you from experience that the gels didn't always turn out as neatly as the ones you see in books). But the process has been automated by adding fluorescent dyes to the primer sequences, so that the fragments can be read by being passed through an optical reader. Even better, the ddNTPs can be labelled with different coloured fluorescent dyes, which means that Sanger sequencing no longer has to be run in four different reactions (one for each nucleotide), because all labelled dNTPS can be added together. This is the basis of many automated sequencing techniques.

TECHBOX
5.2

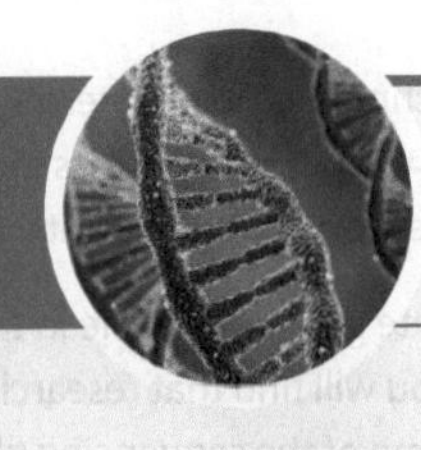

# High-throughput sequencing

**RELATED TECHBOXES**

**TB 5.1:** Sanger sequencing

**TB 6.2:** Primers and probes

**RELATED CASE STUDIES**

**CS 1.1:** Identification

**CS 4.2:** Hierarchies

**Summary**
New sequencing methods continue to be developed to allow DNA sequencing on a previously unimagined scale.

**Keywords**
next-gen sequencing (NGS), reads, pipeline, reduced representation, RNA-Seq, probe, primer, alignment, reads

Not that long ago, all DNA sequencing was conducted using the chain termination method (**TechBox 5.1**). These days you are unlikely to come across a lab still doing classic chain termination (Sanger) sequencing, with four dNTP reactions run in separate columns down a gel, but Sanger sequencing does form the basis of many automated sequencing techniques. However, DNA sequencing is increasingly relying on newer technologies, often referred to, rather futuristically, as 'next generation sequencing' (NGS). The technology evolves rapidly, so by the time this goes to print these 'next-gen' techniques may well be superseded, and indeed some people are already referring to 'next-next-gen sequencing'. So here I have opted to refer to these approaches by the more prosaic title of 'high-throughput sequencing' (HTS).

The breakthrough with high-throughput sequencing that made it faster and comparatively cheaper than Sanger sequencing was parallelization. This is analogous to speeding up computation by allowing many different calculations to be carried out at once rather than one after another. Sanger sequencing requires that strands of different lengths are separated, using electrophoresis through a gel or column, so that they are read sequentially to give the order of bases (**TechBox 5.1**). HTS does not rely on sequential reading of bases on size-separated fragments, instead the bases are read directly, using a variety of different methods. Many HTS methods are based on different ways of 'watching' the construction of a strand of DNA complementary to the target DNA, so most HTS methods require an amplification step that makes many copies of the DNA template (but some 'single molecule' techniques do not).

One important difference between Sanger sequencing and high-throughput sequencing is the role of bioinformatics. Sanger sequences are typically well-defined sequence fragments starting at specific primers. But HTS typically generates a massive number of short overlapping sequences (reads) that are not, in themselves, terribly useful, until they have been stitched together into longer sequences using computer programs. So in HTS, the bioinformatics 'pipeline' is as important as the machines and chemicals (**Case Study 4.1**). The pipeline typically involves moving the data from one program to another, which can involve quite a lot of fiddling around getting everything to work properly. Take heart that analysis of old-fashioned Sanger sequences used to be just as annoying until the development of user-friendly software that took away much of the pain and frustration. As the field advances, no doubt analysis of high-throughput sequences will become just as comfortable.

Here we will focus on the kind of approaches you might use to derive sequences for ecological or evolutionary analysis, rather than methods for constructing and analysing whole genome sequences. The speed of technological change makes it almost impossible for me to make any sort of useful description of methods here, so all I can practically do is wave vaguely in the direction of some basic approaches. As the technology gallops ahead, terminology hasn't settled down yet, so there will be multiple different terms used to describe the same functions or approaches. This makes a challenging field even more bewildering. So I will attempt here to describe general approaches rather than specific techniques, but the terms I use might not match exactly to descriptions in other sources. And, as with so many topics in this book, I can't possibly hope to cover all current methods, because new ones are constantly arising.

To be honest, you are much better off finding up-to-date guides available in the scientific literature and online than relying on a printed book. You will find that researchers will often refer to their sequencing approach using the brand name of the sequencing platform, because each of these uses a particular method and has a characteristic output. But the companies and equipment change just as fast as the actual methods, so we are going to stick to general descriptions here, and you can match them to whichever platforms happen to be available at the time you are reading this. Different methods may be useful for different applications, for example techniques that produce longer reads might be particularly desirable in some situations, but less important in others. Also remember that, as with so many things in molecular evolution and phylogenetics, choice of methods is subject to fashions and personal preferences, as well as practical considerations.

### 1 DNA extraction and fragmentation

The first step in many HTS methods is to break the DNA in the sample up into little bits. This is why HTS is so handy for poor quality samples, like those in ancient DNA and forensics, which tend to be quite fragmented. Some sequencing methods work best on particular fragment lengths, in which case some kind of filtering step is needed, to isolate fragments of the desired size.

### 2 Getting the data you need

Many of the applications of sequence data in evolution and ecology rely on comparing sequences between organisms, for example to gauge sequence variability in a population, or to trace evolutionary history using phylogenies. This requires comparison of homologous sequences—sequences that are similar by descent, not by chance. With Sanger sequencing, this is achieved by careful primer design that anchors the sequencing effort to a particular place in the genome (followed by alignment and analysis to spot any rogue sequences). While Sanger sequencing gives you less data, you generally know exactly what the data is, whereas HTS gives you heaps of data, but you have to identify the sequences that can be usefully compared between the particular individuals or lineages you want to study. One way of dealing with this crisis of excess is to reduce the sequenced fraction down to a more manageable set of sequences. Here are some ways to do this:

**(i)** ***Target amplification*** allows you to selectively amplify the bits of the genome you actually want to look at before sequencing the sample. This could be by PCR using single or multiple primers. Alternatively, instead of amplifying the bits you want, you can get rid of the bits you don't want. For example, molecular inversion probes have two end sequences that match the target locus, joined together with a linker sequence. This means when they stick to the target, they form a circle. Exonucleases, which digest linear DNA from the ends, will then chew up any unattached DNA in the sample, leaving just the circularized target DNA, which can be amplified by rolling circle replication. Another approach is to stick the probe sequences on little magnetic beads which can then be used to draw the target sequences out of the sample. For these methods to work, you need to start with some knowledge of the loci you wish to sequence, so that you can design probes or primers (**TechBox 6.2**).

**(ii)** ***Reduced representation*** is a way of selecting only part of the genome to sequence by chopping the genome up using a restriction enzyme that cuts at a particular sequence, then selecting only the fragments of a particular size class. Since restriction sequences will be distributed more-or-less randomly throughout the genome, but will tend to be inherited just like any other sequence, this approach will give you a repeatable set of sequences to work with. Remember that enzymes cut at particular recognition sequences, so if any of the recognition sequences undergo mutation they will not be recognized. This means you may not get exactly the same fragments in all individuals you sequence, and the more distantly related the genomes, the less likely they are to have matching sets of fragments. The handy thing about this approach is that, because you are not designing probes or primers, you don't need to have any prior knowledge about the sequences.

**(iii)** ***Transcriptome sequencing*** (RNA-Seq) targets mRNAs, so only sequences that are actively expressed will be sequenced. The advantage of this approach is that not only does it target only a fraction of the whole genome (since no cell expresses all its genomic DNA), it also tends to provide exons that are easier to align and analyse than many intergenic sequences. However, since expression is likely to vary over time, between individuals and between cells in the body, your results will to some extent depend on the sample. The sample has to be treated to preserve the RNA component, which is less stable than DNA therefore more likely to degrade over time without active preservation. Since you are getting the cell to do some of the amplification for you, in the form of RNA transcripts, you should expect very uneven proportions of different loci, so you could end up sequencing lots and lots of copies of the same highly expressed genes. This will be great for some applications (good coverage of conserved genes), but might not suit other jobs (where you might want a range of different loci).

**3 Creating a library**

Most sequencing platforms require you to fix adapter sequences to your DNA fragments to create a 'library' of fragments that are ready to be sequenced. Adaptors are little bits of DNA with known sequences that are stuck to the fragments of unknown sequences in your sample. This gives you a way of attaching a primer or probe to each fragment, and it can also contain features that cause the fragments to adopt a particular conformation, such as rolling into a circle or forming hairpin loops. DNA from different sources can be given different characteristic sequence tags (also referred to as barcodes or indexes), so that even when the samples are mixed together, DNA from different samples can be separated out during the analysis. This allows 'multiplexing' of samples, where DNA from different sources can be pooled and analysed in the same reaction. Be aware that any mixing of samples can result in uneven representation, either of particular sequences or different samples. Adaptor or barcode sequences need to be cut out of reads at the bioinformatic assembly stage (note that the term 'barcode' is used in a different sense in the field of DNA taxonomy: **TechBox 9.2**).

**4 Reading the nucleotide sequence of DNA fragments.**

In classic sequencing reactions, the reporter that allows the nucleotide sequence to be read is some kind of label attached to the products of DNA synthesis (**TechBox 5.1**). A similar approach is used in some HTS techniques, for example by washing labelled nucleotides (dNTPs) over templates as the complementary strand is extended. But in many high-throughput techniques, it is effectively the chain elongation reaction itself that is the reporter. For example, in pyrosequencing, luciferase (an enzyme that produces bioluminescence, like the enzyme that makes fireflies light up) is included in the reaction mixture, along with DNA, primers and polymerase. Each dNTP is washed over the reaction mixture in turn. The incorporation of each new nucleotide releases a flash of light, which identifies the position of that particular base in the DNA sequence. The sequence can be read off by monitoring whether light is released as each type of nucleotide is added. More subtly, ion semiconductor sequencing detects the release of a hydrogen ion as a dNTP is added to a new strand. Not all methods read the addition of single nucleotides. For example, sequencing by ligation starts by adding lots of short (<10bp) labelled probes of known sequences. Starting from an anchor sequence, ligase (an enzyme that joins the phosphate backbones of two adjacent DNA strands) will only join a probe to the growing strand if it matches the template strand.

**5 Assembly**

HTS produces a very large number of short sequences, referred to as reads. Depending on the method used, these reads may be only tens or hundreds of base pairs long. Raw reads will need to be trimmed to remove adaptor sequences, and because the ends of reads can be a bit shabby and error-prone. Trimming is an important part of the quality control of high throughput sequencing. Even so, given that the sequencing is never error-free, some of these trimmed reads will inevitably contain sequencing errors.

Now you have many thousands of short reads, many of which run over the same bits of the genome, and some of which contain mistakes. The challenge is to stick these reads together to form longer sequences, and to infer the consensus sequence from the many different reads. For species that have already have whole genome sequences available, reads can be aligned against the reference genome, because you expect individuals of a particular species to have identical sequences at most parts of the genome. If you don't have a reference genome then you are going to have to build your short reads into longer sequences by looking for areas of overlap, connecting them together like joining pieces of a puzzle to make a bigger picture. This is where the big computers come in. Actually many assembly procedures can be carried out on a decent desktop computer, but parallel processing will speed up the process, so access to high-end computing will be helpful. In addition, you are going to have so much data that file storage may be an issue.

Since HTS is error-prone, accuracy in sequencing relies on coverage, meaning that you need many copies of the same sequence before you can be sure that you have the correct nucleotide sequence and you are not being led astray by faulty reads. Note that the problem of contamination has not gone away with new technologies. Your sample may well contain DNA from many different sources, but the machines treat all DNA as being the same, and will try to align and assemble short reads even if they are not from your target organism. This is particularly a challenge with *de novo* assemblies where you don't have a reference genome. But even for well-studied organisms, contamination can be a problem: it has been suggested that an embarrassingly large amount of sequences from human genome sequencing projects are actually from contaminating microorganisms (see **TechBox 1.1**).

### *Which sequencing method should you use?*

You may have to weigh up a lot of different factors when you are thinking about which sequencing method is right for your project, including cost, read length, coverage accuracy, and number of samples. For example, do you want to maximize the number of individuals you sequence, or the number of loci you capture? Do you want more loci at lower coverage, or would you prefer few loci with higher coverage? What platforms are available to you at an acceptable cost? The right answer will change as the technology evolves, so keep your eye on the literature for new ideas and brand-spanking new techniques. It's an exciting time to be sequencing!

Finally, a word of caution: in all the excitement surrounding new sequencing technologies, you need to remember that large amounts of data will not magically solve all problems. All of the complexity of molecular evolution, the tangled histories of different parts of the genome, the complicated evolutionary stories of lineages are still there, and may indeed be magnified by having lots and lots of data to more fully capture the whole confusing mess. As always, remember that this is real biological data and must be interpreted with an eye to errors and misleading signals.

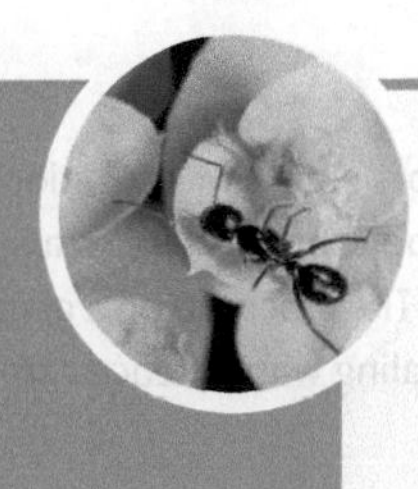

CASE STUDY 5.1

# Duplication: gene copies and microsatellites help aphids tolerate nicotine

**RELATED TECHBOXES**

**TB 3.1:** Single-nucleotide polymorphisms (SNPs)

**TB 6.2:** Primers and probes

**RELATED CASE STUDIES**

**CS 6.1:** Gene families

**CS 7.1:** Selection

Bass, C., Zimmer, C. T., Riveron, J. M., Wilding, C. S., Wondji, C. S., Kaussmann, M., Field, L. M., Williamson, M. S., Nauen, R. (2013) Gene amplification and microsatellite polymorphism underlie a recent insect host shift. *Proceedings of the National Academy of Sciences USA*, Volume 110, page 19460.

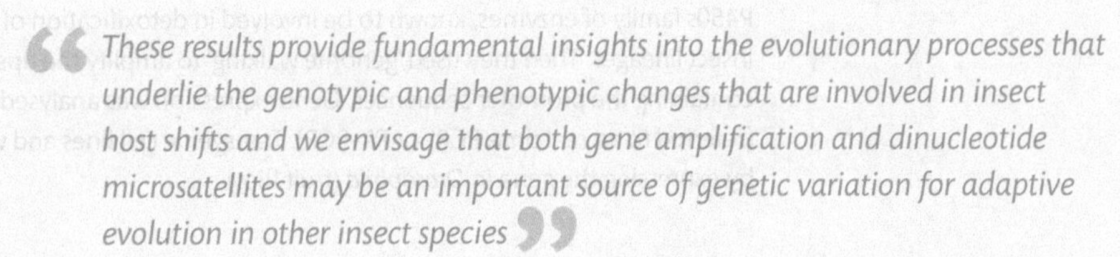

*These results provide fundamental insights into the evolutionary processes that underlie the genotypic and phenotypic changes that are involved in insect host shifts and we envisage that both gene amplification and dinucleotide microsatellites may be an important source of genetic variation for adaptive evolution in other insect species*

**Keywords**

microsatellite, gene duplication, gene expression, insecticide resistance, transgenenic, host shift, RT-PCR

## Background

Nicotine was not invented to make smokers feel good but to protect tobacco plants against insect attack. The peach-potato aphid, *Myzus persicae*, can feed on lots of plants (not just peaches and potatoes) but one particular lineage, *Myzus persicae nicotianae*, has adapted to live on tobacco (**Figure 5.33**). What genetic changes allowed them to exploit a new plant host?

**Figure 5.33** (a) Aphid v. (b) tobacco: who will win?

### Aim

These researchers combined DNA sequencing with biochemical investigations and transgenic experiments to identify the genetic changes and metabolic processes that allow *Myzus persicae nicotianae* to tolerate nicotine, and so live on tobacco. They focused on a gene from the cytochrome P450 family, known to be important in dealing with defence compounds produced by plants.

### Methods

They collected *Myzus persicae nicotianae* aphids from tobacco plants in Greece and Zimbabwe, and from eggplants in Japan (which actually contain high levels of nicotine), and *Myzus persicae s.s.* ('*sensu stricto*', or main species) from the United Kingdom and Germany. They used gene-specific primers to amplify and sequence the CYP6CY3 gene, which is part of the cytochrome P450s family of enzymes, known to be involved in detoxification of plant chemicals in other insect lineages. Then they used 'genome walking' to amplify the upstream regions of the gene containing the promoter sequences. Gene expression was analysed using quantitative PCR (referred to as 'real time PCR', or RT-PCR). Transgenic cell lines and whole organisms were made by expressing the gene in *Drosophila* (fruit flies).

### Results

The team confirmed that *Myzus persicae nicotianae* aphids could survive on a nicotine-enriched diet that was lethal to the *Myzus persicae s.s.* aphids. CYP6CY3 was significantly overexpressed in the *nicotianae* aphids compared to the *Myzus persicae s.s.* aphids. Whereas *Myzus persicae s.s.* aphids have only 2 copies of CYP6CY3, the tobacco-adapted lines have between 14 and 100 gene copies.

The tobacco-tolerant lines also had a single nucleotide change (SNP) and a microsatellite expansion in the upstream region of the CYP6CY3 gene, with 48 AC repeats compared to only 19 in *Myzus persicae s.s.* (though the Japanese-eggplant-derived aphid line did not have the microsatellite expansion). Reporter gene assays with increased numbers of upstream AC repeats increased gene expression. The transgenic cell lines and fruit flies that contained the aphid CYP6CY3 gene both had enhanced tolerance of nicotine.

### Conclusions

This project used a combination of approaches to build a convincing picture that a key cytochrome gene is associated with the ability to tolerate nicotine. Increase in gene expression may be connected with duplication of both the gene itself and in the upstream regulatory regions. While microsatellites are typically considered to be neutral markers, this study offers an example where microsatellite expansion could influence phenotype in a strongly selected trait. Interestingly, *nicotianae* aphids were also resistant to synthetic neonicotinoid insecticides, which are chemically similar to nicotine, even though these lines were isolated before neonicotinoids were introduced in the 1990s. This suggests that the resistance did not evolve in response to pesticide use, but was already present in the aphids (see also **Case Study 7.1**).

### Limitations

The nicotine-resistant strains were isolated from geographically separate populations from the nicotine-sensitive lines, so we would probably expect there to be many genetic differences between these different lineages. The Japanese nicotine-resistant line did not have the microsatellite amplification, so this change is not a necessary part of nicotine resistance in these aphids. Furthermore there was no simple correlation between number of repeats and expression of CYP6CY3. So while this study shows a convincing association between CYP6CY3 and increased nicotine resistance, it is not possible to prove that changes to this gene are the direct causal agent that permitted the host shift.

### Further work

The peach-potato aphid is an important crop pest worldwide and enhanced expression of cytochrome P450 has been implicated in the development of resistance to many insecticides[1]. Changes of expression of these genes may give some insects a handy toolkit not only for dealing with synthetic insecticides but also for increasing their host range[2]. It would be interesting to see if there is a more general evolutionary association between expression of these genes and host shifts, potentially leading to a higher diversification rate[3]. More practically, could these findings be used to help control aphids on tobacco or other crops?

### Check your understanding

1. What kind of genetic changes are associated with increased nicotine resistance in peach-potato aphids?
2. Why did they create *Drosophila* with aphid genes?
3. How could insecticide resistance occur in populations that have not been exposed to synthetic insecticides?

### What do you think?

Does this paper describe a case of speciation in action? What information would you need to consider whether these genetic changes have prompted the formation of a new species?

### Delve deeper

Host-plant shifts have been proposed as a means to sympatric speciation, where lineages become genetically distinct even though they continue to co-exist in the same area. How can distinct lineages form where populations co-exist? Won't interbreeding homogenize any genetic differences between populations? Why has sympatry been so controversial in the past, yet is gaining more acceptance now?

### References

1. Silva, A. X., Bacigalupe, L. D., Luna-Rudloff, M., Figueroa, C. C. (2012) Insecticide resistance mechanisms in the Green Peach Aphid *Myzus persicae* (Hemiptera: Aphididae) II: Costs and Benefits. *PLoS ONE*, Volume 7, page e36810.
2. Dermauw, W., Wybouw, N., Rombauts, S., Menten, B.R., Vontas, J., Grbisá, M., Clark, R. M., Feyereisen, R., Van Leeuwen, T. (2013) A link between host plant adaptation and pesticide resistance in the polyphagous spider mite *Tetranychus urticae*. *Proceedings of the National Academy of Sciences USA*, Volume 110, page E113.
3. Berenbaum, M. R., Favret, C., Schuler, M. A. (1996) On defining 'key innovations' in an adaptive radiation: cytochrome p450s and Papilionidae. *The American Naturalist*, Volume 148, page S139.

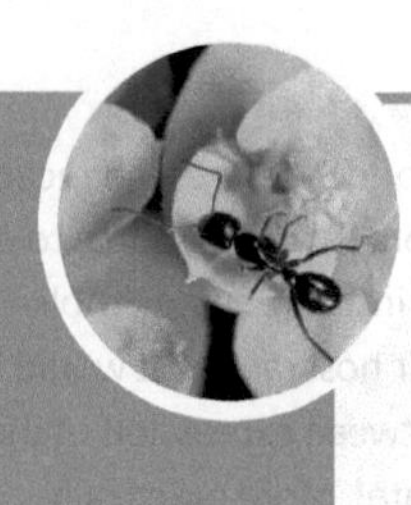

CASE STUDY 5.2

# Genome: a leukaemia virus becoming part of the koala genome

RELATED TECHBOXES

TB 4.2: DNA amplification

TB 5.1: Sanger sequencing

RELATED CASE STUDIES

CS 10.2: Horizontal gene transfer

CS 12.1: Epidemiology

Tarlinton, R. E., Meers, J., Young, P. R. (2006) Retroviral invasion of the koala genome. *Nature*, Volume 442, page 79.

*The cross-species spread of retroviruses, generating novel ERVs, can be considered a natural evolutionary force. It remains to be seen whether KoRV will belong to the category of benign viruses or whether its presence will compromise the ability of koalas to survive.*[1]

**Keywords**

endogenous retrovirus (ERV), Koala Retrovirus (KoRV), reverse transcriptase, epidemiology, pedigree cross-species transmission

## Background

All viruses replicate by infecting a host cell and hijacking its cellular machinery to make more copies of the virus. Retroviruses do this by making a DNA copy of their RNA viral genome (using the enzyme reverse transcriptase), then inserting this viral DNA into the host's own genome. Using strong promoters and enhancers, the virus convinces the host's transcription machinery to make many RNA copies of the viral genome. These RNA copies of the virus genome can be packaged into new virus particles or may be reverse transcribed into DNA and inserted into more places in the host genome. Any inserted viral genome that is not excised will be copied and inherited with the rest of the genome. Most animal and plant genomes contain sequences derived from endogenous retroviruses (ERV: **Figure 5.26**).

## Aim

Koala Retrovirus (KoRV) is a gammaretrovirus associated with a leukaemia-like illness[2], but is also found in asymptomatic koalas[3] (**Figure 5.34**). But is it an endogenous virus (inherited in the genome) or exogenous (gained by infection from other individuals) or both? These researchers looked for evidence that captive and wild koalas inherited KoRV in their genomes by comparing the retrovirus insertion sites between individuals.

## Methods

They digested koala DNA with a restriction enzyme that would cut a site in the middle of the KoRV genome (**Figure 5.35**), then hybridized the DNA fragments to probes that matched the *env* and *pol* genes of the KoRV genome. If KoRV viruses were only inherited vertically, then they should be found, more or less, in the same places in every koala genome, but if they were acquired from exogenous infections the KoRV sequences could be in different places in each infected individual's genome (and vary between tissues from a single individual). KoRV-specific PCR was used on blood samples from koalas from across their geographic range.

## Results

Most KoRV insertions were full-length viral genomes, consistent with recent infection from exogenous viruses. But they also showed that KoRV was present in germline cells, by both PCR and fluorescent *in situ* hybridization (FISH) of sperm cells. KoRV insertion sites varied between individuals. Pedigree analysis of captive animals showed a pattern of fragment sizes consistent with the inheritance of insertion sites as alleles across three generations, and

Figure 5.34 Though regarded as icons of cuddliness, koalas are plagued by a number of communicable diseases. Many koala populations have been significantly reduced by the sexually-transmitted disease chlamydia, and koalas are prone to leukaemia and lymphoma. Animals suffering from these diseases have been shown to have elevated levels of expression of the KoRV retrovirus.

Photographer: Erin Silversmith.

no novel bands in offspring, suggesting that KoRV was inherited in the germline. KoRV was found in nearly all koalas in in northern populations, but decreased in prevalence in the more southern populations. No evidence was found of KoRV in koalas from a population that had been isolated on an offshore island since the early 1900s. This observation was interpreted as evidence for a relatively recent spread of the virus throughout the mainland population, after the island population became isolated.

## Conclusions

The inheritance of insertion sites suggests that these animals are acquiring KoRV with their parents' genomes, not solely as an infection. However, the variability in the number and location of KoRV insertions, the intact KoRV genomes, the expression of KoRV in koala tissues and the association with disease all suggest an active virus capable of transposition and infection.

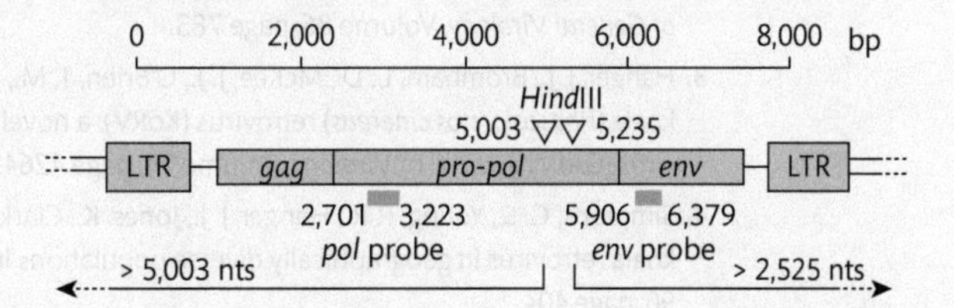


Figure 5.35 Tiny but effective: The KoRV (koala retrovirus) genome is packed into just over 8kb of DNA, containing only three distinct protein-coding genes: *gag* which makes core viral proteins, *pro-pol* which makes proteins for reverse transcriptase and associated functions, and *env* which makes the envelope proteins that form the outer layers of the virus. Luckily, there also happen to be some base sequences that are cut by the HindIII restriction enzyme.

### Limitations

The short timescale (a century or less) and the short sequences (KoRV genomes are only about 8,000bp long: **Figure 5.25**) make it difficult to reconstruct the relationships between viral strains using phylogenetic analysis, which is a common way to investigate the epidemic history of a virus (**Case Study 12.1**). The KoRV-free offshore population was taken as a sign of recent spread, but KoRV has since been detected in the island population[4], a reminder that interpreting absence of evidence as evidence of absence is sometimes risky. Analysis of DNA from koala skins from museums suggests that KoRV has been in koalas in northern Australia for at least 100 years, though possibly not in southern koala populations[5].

### Further work

Mysteriously, the closest relative of KoRV is the oddly named Gibbon Ape Leukemia Virus (GALV) which is only known in exogenous form, causing outbreaks of leukaemia in captive gibbons[3,4]. Both KoRV and GALV are able to infect other species, and their genome sequences are so similar that this seems likely to represent a recent host-jump. Since KoRV has been found in captive koalas it seems possible that this was a cross-species transfer between captive animals[6]. An alternative hypothesis is that wild animals that are sufficiently widespread act as vectors to transfer the virus between the non-overlapping populations (and across the not insubstantial barrier of the Arafura sea). Wider sampling and phylogenetic sequence analysis would shed some light on GALV origins.

### Check your understanding

1. What does the observation that virtually all KoRV sequences in the genome were full-length virus genomes tell you about the history and activity of KoRV?
2. Why would you expect bands of similar sizes in all individuals if KoRV is only transmitted vertically (inherited from parents) not horizontally (by infection)?
3. Is KoRV inherited, acquired by infection, or both?

### What do you think?

There are many well-studied exogenous retroviruses, and lots of bioinformatics studies of endogenous retroviruses, but this case study has been considered a rare example of endogenization in action. If there are so many exogenous retroviruses, and if genomes are full of endogenous retroviruses, then why aren't there more examples of one turning into the other?

### Delve deeper

For some retroviruses, the phylogeny of the viruses is very similar to the phylogeny of their hosts, but for others, including the Type-C family which includes GALV and KoRV, the virus phylogeny does not match that of their hosts. What can comparing host and virus phylogenies tell you about their evolution and behaviour?

### References


1. Stoye, J. (2996) Koala retrovirus: a genome invasion in real time. *Genome Biology*, Volume 7, page 241.
2. Tarlinton, R., Meers, J., Hanger, J., Young, P. (2005) Real-time reverse transcriptase PCR for the endogenous koala retrovirus reveals an association between plasma viral load and neoplastic disease in koalas. *Journal of General Virology*, Volume 86, page 783.
3. Hanger, J. J., Bromham, L. D., McKee, J. J., O'Brien, T. M., Robinson, W. F. (2000) The nucleotide sequence of koala (*Phascolarctos cinereus*) retrovirus (KoRV): a novel type-C retrovirus related to gibbon ape leukemia virus (GALV). *Journal of Virology*, Volume 74, page 4264.
4. Simmons, G. S., Young, P. R., Hanger, J. J., Jones, K., Clarke, D., McKee, J. J., Meers, J. (2012) Prevalence of koala retrovirus in geographically diverse populations in Australia. *Australian Veterinary Journal*, Volume 90, page 404.
5. Ávila-Arcos, MaC., Ho, S. Y. W., Ishida, Y., Nikolaidis, N., Tsangaras, K., Karin, H., Medina, R., Rasmussen, M., Fordyce, S. L., Calvignac-Spencer, S.B., Willerslev, E., Gilbert, M. T. P., Helgen, K. M., Roca, A. L., Greenwood, A. D. (2013) One hundred twenty years of koala retrovirus evolution determined from museum skins. *Molecular Biology and Evolution*, Volume 30, page 299.
6. Bromham, L. (2002) The human zoo: endogenous retroviruses in the human genome. *Trends in Ecology and Evolution*, Volume 17, page 91.

# Gene

## *Making an organism*

*"It is no hyperbolical figure that I use when I speak of Mendelian discovery leading us into a new world, the very existence of which was unsuspected before"*

**Bateson, W. (1908)** ***The methods and scope of genetics.*** **Cambridge University Press**

## What this chapter is about

Genes, on their own, are inert. It is only when they are transcribed (copied into RNA) that they can influence an organism's development, morphology, and behaviour. The expression of genes requires the co-ordinated interaction of a great many sequences, not only the gene itself but the regulatory sequences that control its expression and the genes that make the biochemical equipment needed for transcription and translation. Typically, for molecular phylogenetics, we are interested in the genotype (the information in the genome) rather than the phenotype (the form and function of the individual that develops from that genome). However, we need to understand the role of a sequence in the formation of phenotype in order to appreciate the different patterns of evolution we see in different parts of the genome.

### Key concepts

- Genes are expressed through transcription, where an RNA copy is made of the DNA sequence
- The function of a DNA sequence shapes the way it evolves

# Genes

*It is a well-established fact that language is not only our servant, when we wish to express, or even to conceal our thoughts, but that it may also be our master, overpowering us by means of the notions attached to the current words. This fact is the reason why it is desirable to create a new terminology in all cases where new or revised conceptions are being developed. Old terms are mostly compromised by their application in antiquated or erroneous theories and systems, from which they carry splinters of inadequate ideas not always harmless to the developing insight. Therefore I have proposed the terms 'gene' and 'genotype'... The 'gene' is nothing but a very applicable little word, easily combined with others... A 'genotype' is the sum total of all the 'genes' in a gamete or in a zygote. As to the nature of the 'genes' it is as yet of no value to propose any hypothesis; but that the notion 'gene' covers a reality is evident from Mendelism.*

Johannsen, W. (1911) The Genotype Conception of Heredity. *The American Naturalist*, Volume 45, page 129.

## What is a gene?

'Gene' is a surprisingly slippery concept. As we saw in Chapter 2, the modern concept of the gene is generally traced back to Mendel, who, in the 1870s, studied the inheritance of discrete 'factors' that influenced the characteristics of pea plants (**Figure 2.4**). But the word 'gene' was coined by Wilhelm Johannsen during the reflowering of Mendelian genetics in the early 1900s. Johannsen built upon ideas of Weismann (separation of soma and germline: **Hero 2**) and Mendel (discrete inheritance) to clarify key concepts in the study of heredity and evolution. One of the most important was the distinction between the inherited information (which he called the 'genotype') and the realization of that information in any particular individual (the 'phenotype').

Nowadays, DNA sequencing allows us to describe the genome in detail, so the word 'genotype' is often used to describe the particular set of alleles found in a given individual (see Chapter 3). But Johannsen could not observe the genotype directly, only indirectly through its effect on the phenotype. However, he knew from his work on pure-bred plant lineages that two individuals with the same genotype could grow up to have different phenotypes (**Figure 6.1**). To use a very unsophisticated analogy, if you give a recipe to your neighbour and ask them to cook it then pass the recipe on to the next neighbour, everyone in the street might end up with slightly different cakes due to variation in raw materials or baking conditions, but the recipe they hand on remains the same (if you live in a friendly neighbourhood and would like to try this experiment yourself, please do send me photos of the outcome).

Johannsen felt that new words were needed to make it clear that it was not the personal qualities of a particular individual (phenotype) that was transmitted to their offspring, but the information that they had inherited from their parents, passed on through the gametes (genotype). He chose the term 'gene' for the units of hereditary information that carried the genotype essentially unchanged from one generation to the next, to contrast with 'pangene', the hypothetical particle that transmitted information from the parent's body (phenotype) to its offspring (see Chapter 2).

The thing is, Johannsen didn't know what the gene actually was, neither did any of his contemporaries. Nor did any of the next generation of geneticists who developed key experimental systems for studying heredity, in fruit flies, bread moulds and bacteria. They could observe the effects of genes, describe their patterns of inheritance, localize them to chromosomes, mutate them, record their passage down generations. But the underlying physical structure of a gene was a mystery. The genetics pioneer Thomas Hunt Morgan declared, in his Nobel Prize acceptance speech in 1933: 'At the level at which the genetic experiments lie it does not make the slightest difference whether the gene is a hypothetical unit, or whether the gene is a material particle'. It is worth remembering, when one is tempted to feel sceptical about any science that relies on the assumption of unobservable or hypothetical phenomena, that such was the status of genetics in the first half of the twentieth century.

Because the gene was studied indirectly through its effect on phenotype, the perceived nature of the gene was somewhat dependent on the phenomenon being studied, or the means of observation. Classical Mendelian genetics considered genes as the smallest unit of heredity: their nature could be changed by mutation, they could be mixed by recombination, but the gene itself was the smallest possible building block of the genotype. But using fine-scale genetic mapping, Seymour Benzer (one of the bright young physicists attracted to the frontier world of molecular genetics in the 1950s) showed that recombination could occur within genes, and that mutation coincided with single nucleotides of the DNA sequence (**Figure 6.2**). Following the example of the division of the atom into the subatomic particles neutrons, electrons and protons, Benzer partitioned genes into mutons (the unit of mutation, which he resolved to the nucleotide level), recons (the smallest unit of recombination, also on the scale of nucleotides), and cistrons (the unit of biochemical function, typically the

**Figure 6.1** Mendel had peas, Johannsen had beans. (a) Wilhelm Johannsen introduced many of the words that define evolutionary genetics today, such as 'allele' for a heritable variant of a given gene. He produced 'pure-bred' lines of self-fertilizing beans (*Phaseolus*), inbreeding them until all allelic variation was eliminated. But although every plant in the pure-bred line received essentially the same genetic information (genotype), they did not grow into identical plants. (b) The phenotypes, such as bean size, varied around some mean value, an observation familiar to the biometricians such as Francis Galton. Johannsen demonstrated this variation by putting the beans from each of five pure-bred lines in glass tubes, sorted according to the bean's length. A neat way to produce a graph without computer software.

(a) Courtesy of Forest and Kim Starr, licensed under the Creative Commons Attribution 3.0 Unported license. (b) Reprinted from Johannsen, W. (1911) *The American Naturalist*, Volume 45(531), pages 129–59, courtesy of University of Chicago Press

protein-coding sequence). This naming scheme started a trend that continued for decades, and in this chapter you will meet more –ons, such as codons, exons and introns. In this way, Benzer helped to transform the concept of the gene as an indivisible Mendelian unit of heredity into the molecular genetic view of gene as a continuous nucleotide sequence.

It was not until it was shown how the DNA molecule could carry and copy information that the biochemical basis of the gene became known, as a series of nucleotide bases in a DNA strand that can be copied and translated into functional molecules (see Chapter 2). Yet the story does not stop there. As is so often the case in biology, the more we learn about gene structure and expression, the more complex the picture becomes, and the clearer it is that no one definition will capture all possible instances of the things we instinctively think of as 'genes'. The picture gets even more complicated as we consider the many ways that the word 'gene' is used in common discourse. There is no single definition of gene that can satisfy all customers, so in this chapter we are going to take the cowardly approach of avoiding any formal definition of 'gene'. The property of genes that we will be most interested in here is that they are sequences in the genome which act as a template for the production of a functional, biochemical product.

This book is aimed at people who want to be able to analyse DNA sequences to uncover evolutionary and ecological patterns and processes. Our focus is on the genotype as a carrier of useful information about lineage history, rather than on the role of the genome in the development of phenotype. So this is not the right place to learn about the biochemistry of gene action, nor to consider the many ways that biologists study the complicated association between

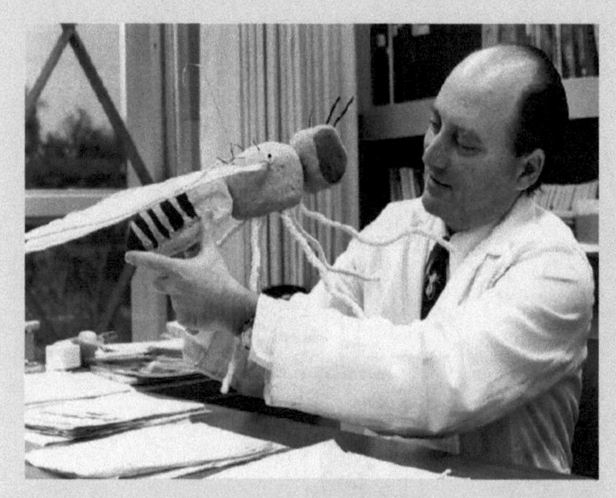

**Figure 6.2** Not a mutation experiment gone wrong. Trained as a physicist, Seymour Benzer was attracted to the American Phage Group, a cross-disciplinary network of scientists working on the nature of genes using experiments on bacteria and their viruses (bacteriophage: see also **Figure 2.5**). After producing one of the first fine-scale genetic maps, Benzer went on to challenge the idea that behaviour was too complex to be reduced to genes, by showing that single mutations could affect flies' ability to move, mate, see, learn, and remember (and in the process being at least partly responsible for naming genes such as *sluggish*, *drop-dead* and *freaked-out*). In a 1971 summary of his work he said 'Splitting the gene and running its map into the ground was exciting while it lasted, but molecular genetics, pursued to ever lower levels of organization, inevitably does away with itself: The gap between genetics and biochemistry disappears. Recently, a number of molecular biologists have turned their sights in the opposite direction, i.e. up to higher integrative levels, to explore the relatively distant horizons of development, the nervous system, and behavior ... Experience thus far with the fly as a model system for unraveling the path from the gene to behavior is encouraging. In any case, it is fun.'

Image courtesy of Harris, W. A. (2008) *PLoS Biology*, Volume 6(2):e41, 

the information in the genotype and its expression in the phenotype. Instead, our main goal here is to think about the way that different parts of the genome evolve and how that affects the kind of evolutionary signal we read from those sequences. Our consideration of genes will be necessarily brief, basic, and biased towards evolutionary analysis. Interested readers are urged to read a proper genetics textbook to appreciate the beauty and diversity of gene structure, function and expression patterns.

## Gene structure and expression

The genome contains an awful lot of DNA, three billion letters in the human case. But, as we saw in the previous chapter, relatively little of that DNA is made up of genes, which we will broadly consider to be sequences that are used as a template to make working bits of the cell. But the genes themselves are not doing any of the hard work of building a cell. As Johannsen surmised, the genotype carries the hereditary information, but it does not directly specify the phenotype. Instead, the genes in the genome can only influence form, development, and behaviour when an RNA copy is made of the gene (transcription).

Have another look at **Figure 2.6** to remind yourself how the DNA in the genome is transcribed into RNA then translated into protein. Here is a simplified version of the transfer of information from the genome to RNA to protein (**Figure 6.3**):

**Figure 6.3**

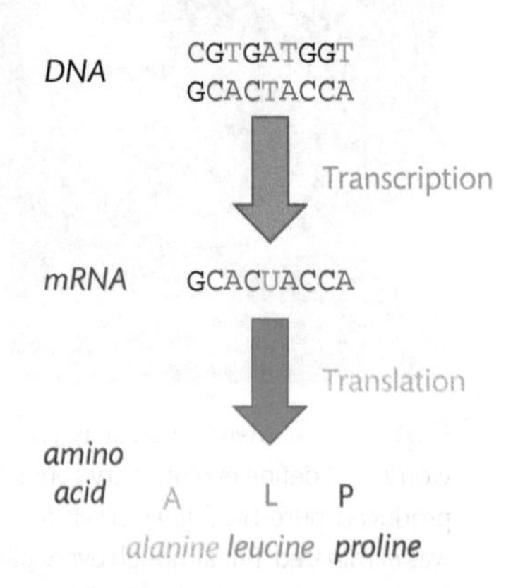


For some genes, the RNA molecule acts as a template for the production of an amino-acid sequence which then forms part of a protein (**Figure 6.3**). For other genes, it is the RNA copy itself that forms the functional product (**Figure 6.4**).

**Figure 6.4**

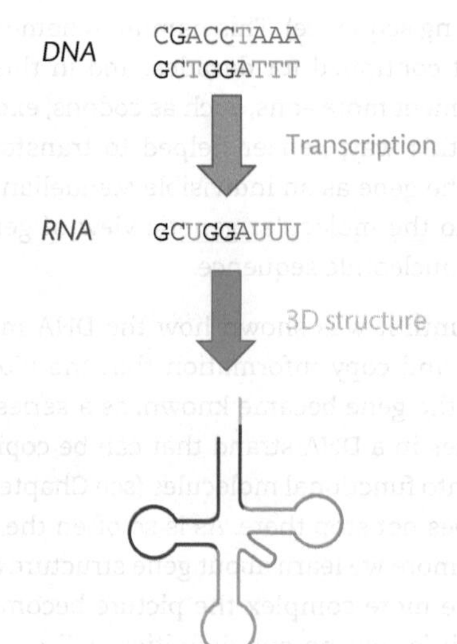

By and large, every cell in your body carries a whole copy of the genome, but it's unlikely to need to use every single gene. Your skin cells do not need to use their genes for making eye colour pigments, and your brain neurons will (all going well) not express genes for making hair. So for a cell to function it needs an efficient way of 'turning on' all those genes it needs at any given point, but not those it doesn't need. The monumental feat of organization needed for turning thousands of genes on and off at the right times is commonly referred to as the regulation of gene expression.

## Transcription and translation

For a gene to be expressed, it must first be used as a template for an RNA copy. RNA polymerase is the enzyme that makes an RNA copy from a DNA template (see Chapter 2). So to express a gene, an RNA polymerase needs to attach to the beginning of the gene, move along the gene making a complementary RNA copy, then stop when it gets to the end of the gene, ending the transcript.

Let's consider the beta-galactosidase gene in an *E. coli* bacterium. This gene, otherwise known as *lacZ*, produces an enzyme needed to metabolize certain sugars, cleaving molecules like lactose into simpler monosaccharides, which can then be used as an energy source. The heart of this gene is the protein-coding sequence. In the *lacZ* gene, this is a stretch of 3072 nucleotides. Here, for your edification, is the first 30 nucleotides of the coding sequence:

ATGACCATGATTACGGATTCACTGGCCGTC

The protein-coding part of the gene is written in the triplet-code (codons: **Techbox 6.1**). The 3072 nucleotides of the protein-coding part of the *lacZ* gene corresponds to a 1024 amino-acid sequence. When translated codon by codon, this first part of the protein-coding sequence reads:

| Codons: | ATG | ACC | ATG | ATT | ACG | GAT | TCA | CTG | GCC | GTC |
|---|---|---|---|---|---|---|---|---|---|---|
| Amino acids: | M | T | M | I | T | D | S | L | A | V |

Or, written in long form, the start of this protein is made of the amino acids methionine-threonine-methionine-isoleucine-threonine-aspartic acid-serine-leucine-alanine-valine.

Characteristically, the first codon of the *lacZ* gene is ATG. When this codon appears within a protein-coding sequence, it codes for the amino acid methionine, but it also functions as the 'start codon', signalling the beginning of the protein-coding part of the gene. Most genes start with ATG (though some bacteria like *E. coli* can use alternative start codons). In addition to beginning with a 'start codon', most genes end with a 'stop codon'. These codons do not have a corresponding amino acid, so when the sequence is translated, a stop codon creates a gap, preventing the addition of any more amino acids. *E.coli* has three alternative stop codons (called, for historical reasons, amber, ochre, and opal: **Figure 6.5**). The *lacZ* gene ends with the ochre stop codon TAA. Here are the last 30 nucleotides of the gene:

| CAT | TAC | CAG | TTG | GTC | TGG | TGT | CAA | AAA | TAA |
|---|---|---|---|---|---|---|---|---|---|
| H | Y | Q | L | V | W | C | Q | K | *stop* |

The sequence in the genome cannot be directly used to make a protein. It can only be expressed by being transcribed into RNA, then translated into an amino acid sequence. For the *lacZ* gene to be expressed, it relies on upstream sequences (before the 5' end of the gene) that initiate transcription, and downstream regions (after the 3' end of the gene) that terminate transcription. As it happens, *lacZ* is part of an operon, a set of genes that occur side-by-side and share a promoter so are transcribed into a single mRNA molecule, but are then translated into several distinct peptides, each defined by a start and stop codon. So even the simplest example of a gene, quite typically, turns out to be more complicated than it first appears (and, thanks to the craze started by Seymour Benzer, operon is also an example of another –on word in molecular genetics).

### *Gene names*

Genes often have a full-length formal name and a more commonly used abbreviation (both of which are usually written in italics). If the gene has a protein product it is given a related name (no italics, sometimes written in capitals). For example, the *Huntington's Disease* gene is commonly referred to as *HD*, or by the name of the protein product, Huntingtin. Gene names may reflect relationships to other genes, either within the same genome or in different species. For example, the gene *human period 2* (abbreviation *hPer2*: see Chapter 3) is one of three genes in humans that produce protein products related to the PERIOD family of proteins that control circadian rhythms in many animals (e.g. the mouse homolog is named *mPer2*). Sometimes genes are named for their similarity to other genes. The *hERG* gene codes for a potassium ion channel, but its full name—*Human Ether-à-go-go-Related Gene*—relates to its similarity to a gene in fruit flies. Flies with a mutant version of this gene have shaky legs when treated with ether, which

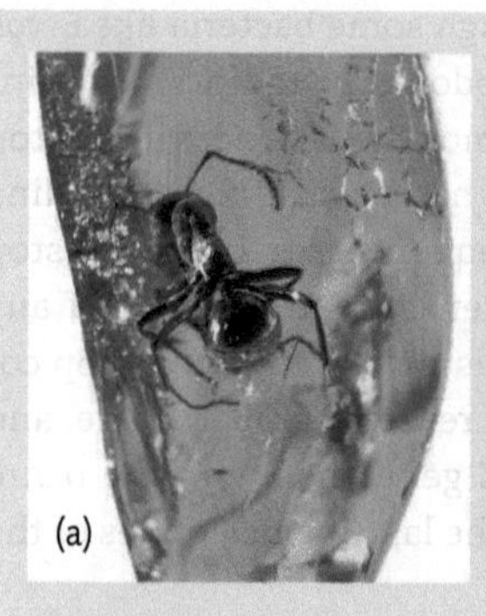

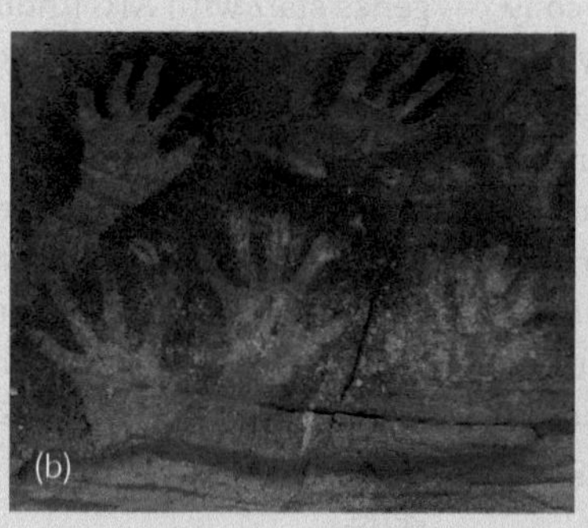

**Figure 6.5** It's not much of an excuse for a picture, but as it happens the three stop codons TAG, TAA and TGA were historically referred to as the (a) 'amber', (b) 'ochre', and (c) 'opal' codons. Why? Because members of the American Phage Group convinced their friend Harris Bernstein, a graduate student, to help them with the tedious work of sorting through 2,000 bacterial plaques to find potential mutants in return for naming a mutant after him—and one of the mutants they identified was the first recognized nonsense mutant (and Bernstein means 'amber' in German). Remember that next time they are calling for volunteers in the biology department. Then other workers followed the naming trend with ochre and opal. It's difficult to imagine, in these days of high-throughput sequencing and whole genome databases, how individual codons could be given such whimsical names. But this dates from a time when the genetic code was the wild frontier of biology. While genetics is every bit as exciting now, and there remains much mystery to entrance and allure the keen biologist, I can't help but feel there is perhaps a little bit less poetry than in the early days.



reminded the researchers who first described the gene in the 1960s of the dancing at the Whisky A Go-Go nightclub.

There is, as yet, no equivalent of Linnean nomenclature for genes. Whereas species taxonomy has a clear hierarchical structure, which aims to reflect evolutionary relationships, gene names are more haphazard. Related genes in different species may be given entirely different names. As an example, we can look at a very charismatic developmental gene which, in humans, is known as *paired box gene 6* (*Pax6*). This gene plays a key role in the induction of eye formation in a wide range of animals (it also has a number of other roles, particularly associated with the development of the central nervous system). One of the active sites of the Pax6 protein, the 60 amino-acid homeodomain, is so well conserved between animals that the sequence from a mouse can trigger eye formation in fruit flies—and not just in the usual place, but anywhere the gene is expressed (resulting in rather disturbing flies with eye tissue on their wings, legs, and abdomen). But although they are clearly versions of the same gene, inherited from a common ancestor, the *Pax6*-like gene is known by different names in different organisms. In *Drosophila* (fruit flies), there are two copies of this gene, called *eyeless* (*ey*) and *twin of eyeless* (*toy*). In the flatworm *C. elegans* it is known as *Variable ABnormal morphology* (*vab-3*). In the mosquito, *Anopheles gambiae*, the *Pax6*-related gene has the rather less catchy name of *AGAP000067-PA*. Furthermore, the *Pax6* gene may be sometimes referred to by its disease phenotypes, such as *Aniridia* (in humans) and *Smalleye* (in mouse).

Much like Benzer's labels for genes (muton, recon, cistron), which reflected the research strategies employed in gene discovery, fashions in gene names change with technical progress in genetics. Genes were originally identified through breeding experiments: mutants with noticeable phenotypic effects were identified and bred to isolate the genetic factor responsible for the change. Many of these mutants were knockouts—the mutation destroyed the normal functioning of the gene—so the genes were commonly named not for what they do when they function correctly, but what happens when they don't. For example, ordinary *Drosophila* have red eyes, but individuals in which the *white* gene (*w*) has been knocked out develop white eyes. So a working copy of the *white* gene is responsible for making eyes red. Confusing.

In the 1980s and 1990s, gene discovery rates began to increase massively. There was a palpable air of excitement surrounding gene discovery, and bright young things were attracted to work in genetics labs. They kept themselves amused dreaming up gene names with references to popular culture, one of the most famous being the *sonic hedgehog* gene (*SHH*) named after the indomitable hero of a computer game (*SHH* is a variant of the *hedgehog* gene (*hh*), the knockout of which causes spines to form on the embryo). Post-genomic gene discovery has lost a little of the enthusiasm of earlier days, and high-throughput sequencing has automated gene identification, so new genes tend to get assigned a systematic identifier (such as *AGAP000067-PA*) until someone finds the time, resources and inclination to work out what the gene actually does.

But a quick flick through genetics journals reveals the occasional quirky gene name. My personal favourite is the gene *makes caterpillars floppy* (*mcf*). Curiously, this is not a caterpillar gene at all, but a gene from a bacterium, *Photorabdus luminescens,* which lives as an essential symbiont in the guts of nematodes (such as *Heterorhabditis bacteriophora*). So how does it make caterpillars floppy? The nematode makes its living as an endoparasite, burrowing into insect larvae and digesting their tissues. The bacteria help the nematode by releasing the Mcf toxin which causes the larva to lose body turgor (i.e. go floppy) and die (**Figure 6.6**). *Photorabdus* also helps the nematode by releasing antibacterial and antifungal agents to prevent the larva being invaded by other competing pathogens. So *Photorabdus luminescens* is simultaneously a symbiont (of nematodes) and a parasite (of insects). This is important because *Heterorhabditis* nematodes are used as a biological control agent of insect pests, but it is the bacterial gene *mcf* that provides the lethal weapon. So the Mcf toxin might be developed as an insecticide, and the *mcf* gene could potentially be inserted into plants to confer defence against caterpillars (like the *Bt* gene from *Bacillus thuringiensis*: **Figure 4.24**).

**Figure 6.6** You make me go floppy. The caterpillar on the right is plump and healthy, the caterpillar on the left has gone floppy and is about to die, thanks to a bacterial toxin, Mcf, produced by a gene called *makes caterpillars floppy*. The bacteria do not directly infect caterpillars, but are carried in the guts of nematodes that do.



## Regulation of gene expression

If all genes were expressed all the time, the cell would be a noisy, inefficient mess. The cell needs to be able to turn genes on when and where their products are needed, and turn off transcription when their products are not needed. An *E.coli* cell only needs the beta-galactosidase enzyme when there is some lactose around to digest. The *lacZ* gene has an elegant mechanism for controlling expression, to make sure that the gene product is only produced when needed. Upstream of the coding sequence, at the 5' end of the gene, are sequences that are needed for the control of expression. These sequences are not written in triplet code and they won't be translated into protein. Instead they act as recognition sequences where control elements will dock, turning the gene on or off.

*See* **TechBox 2.1** *for an explanation of* 5' *and* 3'

In the case of *lacZ*, a different gene elsewhere in the genome makes a repressor protein that binds to the promoter region of the *lacZ* gene, preventing RNA polymerase from engaging with the start of the *lacZ* gene (**Figure 6.7**). As long as this repressor protein is blocking the promoter site, no RNA transcripts of the *lacZ* gene can be made, so no beta-galactosidase will be made.

**Figure 6.7**

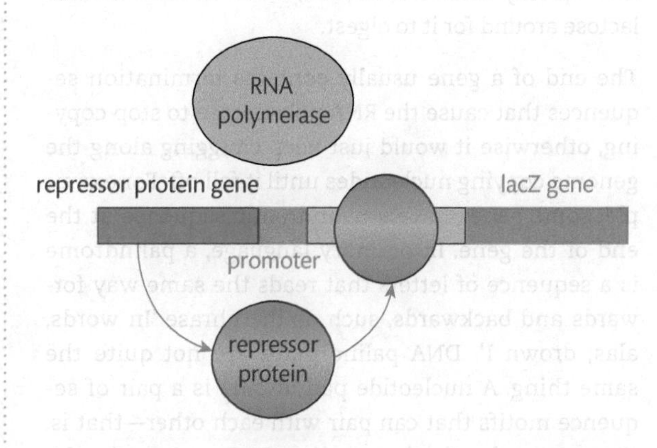


But—here is the really clever bit—the repressor protein responds to the presence of lactose by changing conformation and detaching from the DNA sequence. So when there is lactose about, the repressor protein drops off, the promoter sequence is exposed and made available to the polymerase enzyme (**Figure 6.8**).

**Figure 6.8**

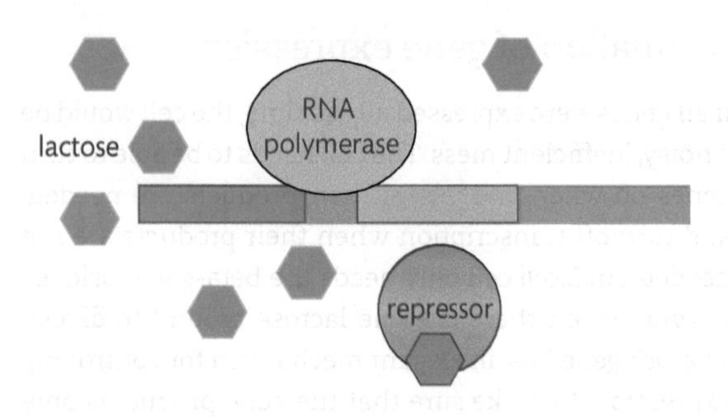


Now RNA polymerase can bind with the promoter sequence and begin moving along the gene, making a messenger RNA copy, which is then translated by the ribosome into an amino acid sequence, which folds into a beta-galactosidase molecule, which cleaves the lactose molecules into simple sugars (**Figure 6.9**).

**Figure 6.9**

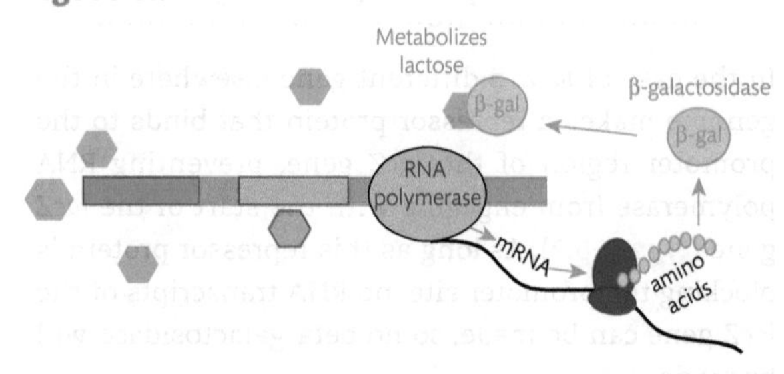


So this brilliant system turns on beta-galactosidase production only when it is needed, which is when there is lactose around for it to digest.

The end of a gene usually contains termination sequences that cause the RNA polymerase to stop copying, otherwise it would just keep chugging along the genome copying nucleotides until it fell off. For example, some genes have a palindromic sequence at the end of the gene. In ordinary language, a palindrome is a sequence of letters that reads the same way forwards and backwards, such as the phrase 'In words, alas, drown I'. DNA palindromes are not quite the same thing. A nucleotide palindrome is a pair of sequence motifs that can pair with each other—that is, they are each other's compliment. For example, the sequence CCGGACGCGTCCGG can pair back on itself like this (**Figure 6.10**):

**Figure 6.10**

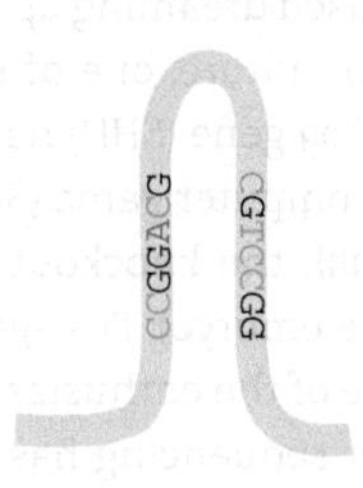


This self-pairing palindrome creates a hairpin loop in the RNA transcript, which can cause the transcription machinery to let go of the DNA, bringing transcription to an end. In some organisms, RNA transcripts are then finished off with a run of adenosine (A) residues, forming the poly(A) tail. So the end of the gene is often marked with a polyadenylation signal, like AATAAA, which acts as a binding site for the enzyme that cuts off the end of the transcript in preparation for the addition of the poly(A) tail.

## *Transcriptome: listening to the genome talk*

In Chapter 2, we saw how one of the really handy things about the genome, from the practical point of view of someone interested in using DNA to trace evolutionary history, is that every cell contains a copy of the entire genome. While the DNA from a cell contains the complete set of genes, the messenger RNA transcripts in the cell will reflect only those genes being actively transcribed. Therefore, sequencing and identifying the mRNA in a cell is one way of looking for the differences in gene expression that can transform phenotype. This book is not the place to learn about transcriptomics, which uses analysis of the RNA component of cells to compare expression levels of genes over time, between cell types, or among species. But I will briefly mention that the transcriptome (the set of all mRNAs in a cell at a given point in time) is being used by some evolutionary biologists as a way of generating useful sequences for phylogenetic studies, which, in a bold bid to push the boundaries of the number of syllables that can be reasonably jammed into one word, has been referred to as phylotranscriptomics.

One advantage of using messenger RNA as a source of phylogenetic information is that it is a way of getting the cell to do some of the work of amplification for you. A cell makes many RNA copies of highly expressed genes, including slow-changing house-keeping genes that are found in a wide range of organisms, which might be just what you want for your phylogenetic analysis. The cell will have even removed all the introns to make it easier to align and analyse the protein-coding parts of the genes.

This means you can get relatively high sequencing coverage of potentially useful gene sequences, without having to sequence large amount of intergenic or intronic DNA. But there are disadvantages too: the abundance of transcripts will be heavily biased toward certain highly expressed sequences, which might skew the signal in the sequences, and the expression levels of some sequences can change rapidly as the conditions the cell encounters change. Not only that, RNA degrades faster than DNA, so unless the sample was instantly preserved in an RNA-appropriate preservative, the RNA in your sample represents what is left after a decay process, however brief. Even a snap-frozen sample might lose RNA when it is thawed for sequence extraction.

# Sequence evolution

If you have picked up this book because you want to be able to make a molecular phylogeny for your favourite creatures, then you might be wondering why we have wandered off into basic genetics and cellular biochemistry. After all, if you are interested in tracing the evolutionary history of eusociality in stingless bees, or untangling the relationships between *Nothofagus* trees as a way of understanding the influence of plate tectonics on the flora of the southern hemisphere, then you might simply not care how *E.coli* manages its lactose metabolism. But the reason we have covered some basic concepts in gene structure and regulation is that these aspects of biochemistry constrain the way that DNA sequences evolve. And anything that constrains the way that sequences change over time will influence the recording of evolutionary history and processes in the genome. If you want to read the story from DNA, you need to know what forces have shaped the way that history is recorded in the genome. Now that we have considered some of the defining features of genes and their regulatory sequences, we can get to the bit that really matters for people wanting to construct phylogenies: how sequences evolve.

We have seen how a typical gene has a beginning (promoter sequences that engage the polymerase enzyme), a middle (the protein-coding section, written in triplet code, beginning with a start codon and ending with a stop codon, or the sequence of nucleotides that will form the functional RNA molecule), and an end (the downstream sequences that cause the polymerase enzyme to detach and trigger the addition of a poly(A) tail on the RNA transcript). The beginning, middle, and end of a gene correspond to three different stages of transcription: initiation, when the transcription machinery binds with the gene; elongation, where the machinery adds nucleotides to the growing transcript; and termination, where the transcription machinery stops copying and finishes off the end of the transcript. Mutations in any of these important sequences could affect a cell's ability to do something important, like metabolize lactose. A mutation that changed the promoter sequence, or any of the regulatory sequences scattered throughout the genome, could prevent the gene from being expressed, or cause it to be overexpressed when it is not needed. A mutation in the coding sequence could change the structure and function of the gene product. A mutation that changed the signals that end the protein might create an RNA molecule that is not transported correctly to the ribosome, or is translated into a protein that is too short or too long.

All of these different kinds of sequences involved in gene structure and expression are expected to have different patterns of sequence evolution. This makes DNA analysis complicated: we can't just expect to treat DNA sequence data like an homogeneous string of random letters. But these complicated patterns of sequence evolution are also what makes the genome such a wonderfully rich source of information on evolutionary past and processes. We can use the different patterns to tell us different stories.

## Divided genes: exons and introns

Not all genes are as neat as the *E. coli lacZ* gene. For a start, most protein-coding genes in eukaryotes are split into two basic kinds of sequences. Exons are sections of the gene that will be translated into amino acids, to make the protein. Introns are the intervening sequences that will not be translated into an amino acid sequence. For example, the human beta globin gene covers 1606 nucleotides of DNA, but only a third of it codes for the amino-acid sequence of the beta-globin protein (**Figure 6.11**).

**Figure 6.11**

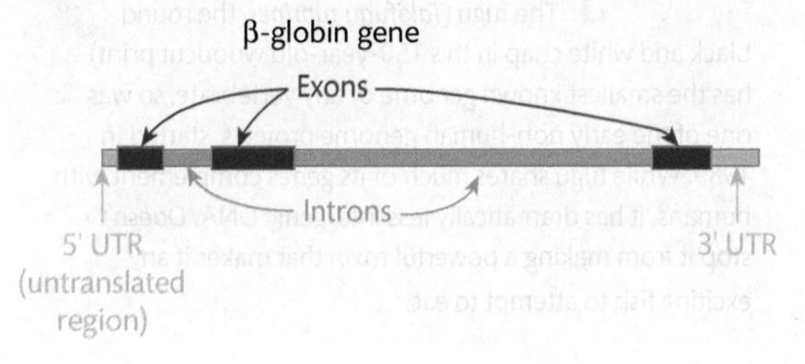

The introns have to be edited out of the RNA transcript before it can be translated into protein (Figure 6.12). In most cases, this involves an enzyme complex called the spliceosome which cuts out the introns at particular short nucleotide sequences and joins the exons together to make a continuous, protein-coding transcript (though some self-splicing introns can excise themselves).

**Figure 6.12**

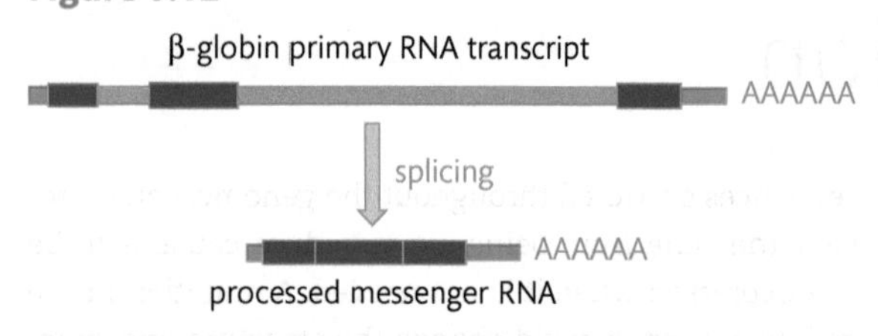


For many genes in large genomes, the intronic sequences are far greater in total length than the exonic sequences. For example, the coding part of human *Huntington's Disease (HD)* gene is split over 67 exons and makes up less than a tenth of the 170,000 nucleotides in the gene. Interestingly, the *Huntington's Disease* gene in fugu, a kind of pufferfish famous for having a compact genome, has the same 67 exons but the total gene length is only 23,000 nucleotides because the introns are much shorter (Figure 6.13).

## Exons

This division of genes into both protein-coding and non-protein-coding elements is important for the way we interpret patterns of sequence change. Exons have

**Figure 6.13** The fugu (*Takifugu rubripes*, the round black and white chap in this 150-year-old woodcut print) has the smallest known genome of any vertebrate, so was one of the early non-human genome projects, started in 1989. While fugu shares much of its genes complement with humans, it has dramatically less intergenic DNA. Doesn't stop it from making a powerful toxin that makes it an exciting fish to attempt to eat.

a distinct pattern of evolution. The DNA sequence of the exon is translated to an amino acid by a triplet-code: three nucleotides read to specify one amino acid (TechBox 6.1). For an exon to make sense, the triplet structure of the sequence needs to be maintained. If a base is deleted or inserted, the triplets get out of whack, and the sequence stops making sense. This is called a frameshift mutation. I can produce the same kind of effect by shifting the spaces in this sentence sot hatt hew ordsb ecomeq uited ifficultt oread. Because frameshift mutations tend to ruin the translation of the sequence, they usually wreck the gene, so are typically removed by natural selection. The upshot is that frameshift mutations in working protein-coding genes are rare, so the triplet structure is usually preserved.

*Chapter 7 discusses the way selection works to conserve genes against change*

Maintenance of the codon structure makes protein-coding exonic sequences much easier to align. You know that any gaps that do not preserve the codon structure, such as single or double base insertions or deletions, are unlikely to occur in a functional sequence. Furthermore, you can check that your alignment makes sense by translating the sequence into amino acids (Case Study 6.1). If you notice that stop codons are occurring throughout the sequence then you know you either have a non-functional protein or you have aligned it incorrectly (or your translation is out of frame, in which case it will make sense if you move the reading frame by one or two bases). A mutation that causes a stop codon in a protein-coding sequence is sometimes referred to as a nonsense mutation, because it destroys the sense of the message. Be warned that many automatic alignment programs ignore the fact that some sequences are constrained by the need to maintain coding sense. So these programs might insert single base pair gaps, wrecking the sense of the gene. As always, check all alignments carefully before you go ahead and analyse them or you could be wasting a lot of time making incorrect inferences from a misleading alignment.

*We discuss alignment in Chapter 10*

The exons of protein-coding genes also provide evolutionary biologists with sites that vary in their rate of molecular evolution in predictable ways. Not all changes to the exon sequence will change the amino acid sequence it codes for (Figure 4.21). In particular, the third codon position can often be changed without changing the amino acid specified. It is not surprising, then, that the third position of the codon can often be changed without affecting the protein-coding sequence (TechBox 6.1), so it tends to change much faster than the first and second codon positions. First and second codon positions of the exons might change too slowly to record recent events but carry historical signal of more ancient divergences. And we will see in Chapter 7 how the codon structure of exons produces patterns of changes that can reveal the action of natural selection.

*In Chapter 13, we consider how different positions in a protein-coding sequence change at different rates*

## Introns

Why on earth are most genes in eukaryotes broken up into little protein-coding pieces embedded in a lot of non-protein-coding DNA? Would it not be more sensible just to have the protein-coding sequences in blocks, like the *lacZ* gene, so the transcription machinery can start at the start, continue to the end, then stop, without bothering with all that post-transcriptional splicing and dicing? Many genes work perfectly well without introns. About three per cent of human genes have no introns, probably because these genes originated from a DNA copy of processed mRNA transcript. And the fugu seems to be able to go about its business with dramatically less intronic sequence than humans.

The debate about introns is similar to that concerning genome size as a whole (Chapter 5): why carry lots of extra DNA? As with genome size, there are two kinds of explanation typically discussed: one is that seemingly functionless DNA may in fact be useful and therefore maintained by selection, and the other that even useless sequences can persist in the genome if they are not sufficiently costly to be removed.

Some introns contain important signals that influence gene expression. These functional sequences can sometimes be detected as conserved sequences that change at a much slower rate than the rest of the intron. Differential splicing of introns can also allow different proteins to be generated from the same gene. Exon skipping allows different exons to be included or excluded in the processed mRNA transcript, creating 'mix and match' proteins. For example, the gene *doublesex* (*dsx:* Figure 6.14), crucial in determining gender in *Drosophila*, is spliced differently in males and females. The gene contains six exons, but in males, splicing cuts the fourth exon out of the transcript, while in females, a polyadenylation signal after exon 4 causes the transcript to be finished without exons 5 and 6. Expressing the male splicing pattern in a female fly causes her to develop male characteristics. There are even split genes

**Figure 6.14** This male butterfly knows he wants to be with this female even though she looks like another species. Female *Papilio polytes* come in many distinct forms, some of which mimic distasteful species. How can a species be polymorphic for distinct mimics that vary in many different characteristics, such as wing shape and colouring, without producing a range of intermediate forms that don't look much like their target species? Because the mimetic forms are inherited as single Mendelian locus, with no intermediate forms, it has been suggested that mimicry phenotypes are controlled by 'supergenes', sets of independent genes that are located close enough together that they can develop co-ordinated allele combinations that are nearly always inherited together, not re-assorted every generation. However, a recent study used backcrossing and genetic mapping to locate the mimicry locus to a section of the genome containing five genes. One of these, *doublesex*, is known to influences gender morphology and behaviour in fruit flies, being alternatively spliced to produce male and female specific transcripts. They suggest that *dsx* acts as a switch that triggers formation of distinct wing patterns, without necessarily requiring that genes be physically linked together.



where the coding sequence occurs at distant parts of the genome, then the RNA transcripts are joined together so that they can be translated as a single protein molecule (a phenomenon referred to as trans-splicing). But most alternative splicing occurs at a much finer scale. If introns are not always cut out at exactly the same site, then when the exons are spliced together there may be more or fewer nucleotides included in the message. Sequencing of transcriptomes demonstrates the presence of alternatively spliced variants of many genes.

Alternative splicing has been put forward as an explanation for the embarrassing observation that it takes roughly the same number of genes to make a tiny flatworm as it does to make magnificent you (Chapter 5). If many genes produce alternative transcripts, then the same number of genes could lead to a much more diverse proteome (the set of all proteins produced by a cell). However, as always, we must be cautious in assuming that all features of the genome must be useful and adaptive. Variation in splicing might often be the result of an imprecise splicing process generating somewhat sloppy outcomes. Not all alternative transcripts are functional, and many will be degraded in the cell before they are transcribed.

While some parts of introns clearly have a role to play in gene structure and expression, a lot of intronic sequence seems free to change and therefore evolves quite rapidly. This suggests that the actual nucleotide sequence in the introns is largely unimportant in the construction of phenotype. This observation lends support to the idea that introns may, by and large, be the excess baggage carried by large genomes. The human genome has twenty times as much sequence in introns as it does in exons. Given that the puffer fish manages quite well with a lot less, it does seem plausible that humans could trim a lot of intronic surplus and still operate just fine. As with the argument that increases in genome size reflect failure of selection to remove excess DNA (see Chapter 5), the interpretation of introns as accumulating without benefit has been supported by the observation that the amount of intronic sequence is related to population size.

*The influence of population size on the accumulation of slightly deleterious changes is explained in Chapter 8*

## Identifying genes

How might we find out if a particular stretch of genomic sequence contains any protein-coding genes? We could start by translating the nucleotide sequence into amino acids and see if it makes sense. Remember that not all organisms use the same genetic code, so you are going to need to use the right code for the genome (**TechBox 6.1**). Also remember that DNA is double-stranded, so the actual coding sequence might be on either strand, and might run forwards or backwards. So you will need to translate in both directions, and in complement (that is, the complementary base sequence, which is what you would find on the other strand). And, if you don't already know where the gene starts, then you will need to translate in all three reading frames.

Once you have translated the sequence in every frame, forward and reversed and complemented, you might find that in one of these orientations you get a continuous run of amino acids with no stop codons. Any decent stretch of amino acids with no stop codons is referred to as an open reading frame (ORF). So if you find a nice looking ORF, have you found a gene? Before you rush to name your new gene after your favourite cartoon character, stop and think about the likelihood that you could have got this ORF just by chance. To demonstrate this, I generated this sequence by just bashing away on the keyboard to get a meaningless string of ACTGs.

TGTCGATCGATCGTTTCTAGTGTGCATCGACTCTAGCTCTAGTCATGTCATCGA

If I translate this nucleotide gibberish into amino acids, it comes out like this:

C R S I V S S V H R L **Stop** L **Stop** S C H R

This doesn't look at all like a real protein-coding sequence, it's got multiple stop codons in it, so it could never be transcribed. But if I shift the frame one nucleotide I get this:

V D R S F L V C I D S S S S H V I

Here is an ORF of 17 amino acids. Now I might be fooled into thinking I had a bit of real protein-coding sequence. If I shift the frame again, I get this:

L S I D R F **Stop** C A S T L A L V **Met** S S

In this case, it doesn't look like a continuous protein-coding sequence because it has a stop codon in the middle and a start codon after that, although it might signal the end of one protein-coding sequence and the start of the next.

The point is that even meaningless nucleotide sequences could end up with short sections that look like coding sequence. So how can you tell when you have got a real gene sequence? In a randomly generated sequence of nucleotides, the chance of accidentally coming up with three bases in a row that happen to be the same as stop codon increases with the length of the sequence (averaging out to one every twenty-one codons for the standard genetic code). The more random codons you read, the more likely it is that eventually one of them will code for Stop. So if you translate your sequence into amino acids and it's chopped up with stop codons, it's probably not a functional protein-coding gene, but if you get a very long stretch with no stop codons, that looks much more promising.

Another way to check if a given sequence is likely to contain a gene is to compare it to known gene sequences. Since all genes evolve from other genes, your nucleotide sequence should resemble something else in the databases (**TechBox 1.2**). If it's a protein-coding gene, then when you align your putative protein-coding gene against other related genes, you shouldn't see any pesky frameshifting insertions or deletions—all gaps should be multiples of three nucleotides to avoid destroying the sense of the sequence. If it's an RNA-coding gene, then you would expect it to match more closely to conserved elements in other RNA-coding genes, and to conform to structural models of RNA folding.

But what these tests don't tell you is whether your gene is functional and important to phenotype. There are several ways to make this call. The classic way to find out whether a gene is important or not is to destroy it and see if the individual can still function normally. Even this doesn't always clarify function. In fact, geneticists working on model organisms such as yeast, mice and *Arabidopsis* have been perplexed to find that a substantial proportion of knockout mutants (individuals that lack a working copy of an apparently functional gene) have no obvious phenotypic change. We could compare our gene sequence to a transcriptome library, but that would only tell us if the gene has been transcribed in a particular tissue, it would not guarantee that those transcripts were being translated into proteins, nor that the proteins performed important functions in the cell. A simpler approach is to use the pattern of sequence evolution to give clues about the function of the sequence. If a sequence is evolving at a fast rate, accumulating changes evenly throughout the sequence and not just in a few key places, this suggests that it is likely to be essentially functionless. This is because important sequences are constrained in the changes they can carry and still retain their function. So we expect that a functional sequence will show some level of conservation (**Case Study 6.2**).

## *Not all genes code for proteins: RNA genes*

In this book, we are broadly considering 'gene' to refer to parts of the genome that are transcribed into RNA to make a functional product. Mostly, we have been discussing protein-coding genes, such as *lacZ*, where the RNA is used as a template to build a protein by matching nucleotide triplets to their corresponding amino acid (**TechBox 6.1**). But for some genes it is the RNA molecule itself that is the functional end-product (**Figure 6.15**). There are a whole host of RNA molecules produced by the cell. We have already met messenger RNAs (mRNA), transfer RNAs (tRNA), and ribosomal RNAs that contribute to the translation of protein-coding genes. But there are also many other roles for RNA in the cell. There are many small RNA molecules that act in gene expression, for example by blocking or degrading mRNA transcripts to prevent their translation. RNA molecules also contribute to the spliceosome complex.

The number and function of RNA-coding sequences in the genome is still a matter of debate, and analysis of small RNA molecules is something of a wild frontier in bioinformatics. But here we are going to stick to well-trodden ground. We will take as our example one of the most well-studied classes of RNA genes, because they are among the most commonly used genes in molecular phylogenetics. Ribosomal RNA (rRNA) genes are named after the size of the molecule they produce using the size measurement S (Svedburg, which reflects the rate of movement through a centrifuge gradient in seconds, named after the guy who invented the microcentrifuge). Ribosomes have two RNA units, the small subunit (made up of the 18S rRNA in eukaryotes, or the 16S rRNA in bacteria) and the large subunit (formed from three RNA molecules, such as 5S, 5.8S and 28S). The ribosomal genes that code for these subunits typically occur in clusters, separated by short spacer sequences. These spacers, like introns, are trimmed out of the initial RNA transcript, before it folds back on itself, creating a complex three-dimensional structure through complementary base pairing between different parts of the same sequence. In any one genome, there may be tens or hundreds of copies of each of the different rRNA genes, organized in clusters on different chromosomes.

Stop and think for a moment about how these few observations about rRNA genes influence the way that the sequence evolves. We might refer to sequencing '*the* 18S gene' of, say, a species of tardigrade (**Figure 6.16**, **TechBox 6.2**), but actually in most eukaryote genomes there are hundreds of copies of the 18S gene, and those copies are not necessarily identical. We saw in Chapter 5, tandem arrays of similar genes are prone to recombination between gene copies. This can lead to gene conversion, where the sequence of one copy replaces the sequence in another copy. Gene conversion erases any of the unique historical signal in one copy of the gene and replaces it with that of another copy. If the nucleotide sequence is occasionally 'reset' in this way, then this will disrupt the regular accumulation of DNA changes that we would expect to see in a single-copy gene. So while some copies of rRNA genes are kept the same by gene conversion, others accumulate unique changes. So, for example, if you are using rRNA sequences to gauge the diversity of species in a environmental sample, you need to remember that some of the different sequences might actually be derived from a single individual.

## *Stems and loops*

The three-dimensional structure of functional RNA molecules shapes the patterns of changes along RNA gene sequences. The structure of an RNA molecule is

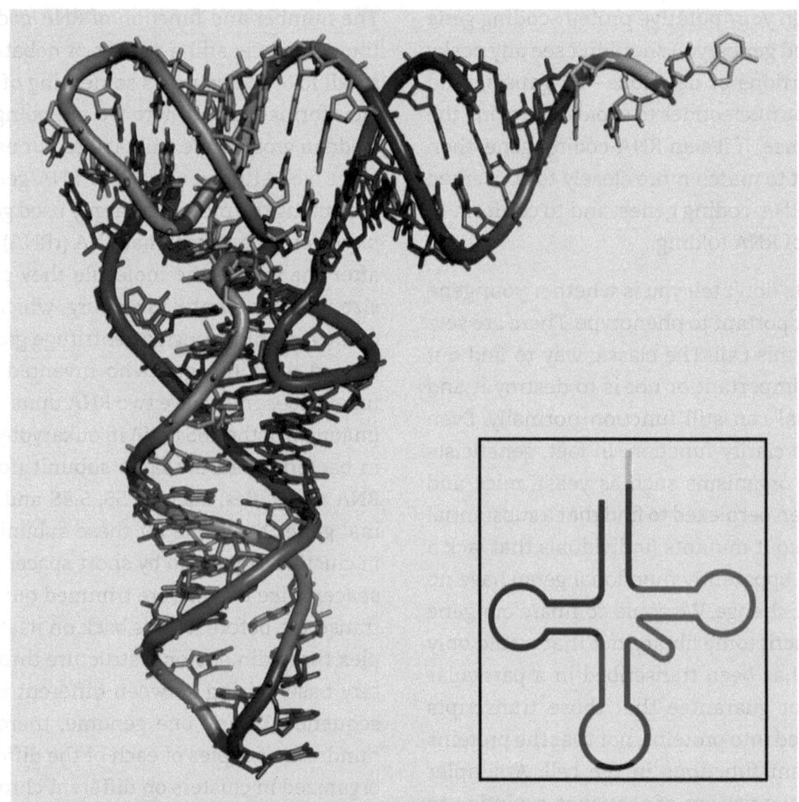

**Figure 6.15** Transfer RNA genes (tRNA) are very short, because they usually produce an RNA molecule around 70 nucleotides long that folds into a series of loops and paired stems. It is symbolically depicted by the classic 'clover leaf' shape (see inset). This is a good reminder that the way that we draw molecules is a kind of short-hand and often not a clear indication of their three-dimensional structure. Francis Crick (**Hero 6**) predicted the existence of transfer RNA, reasoning that there could be small 'adaptor' molecules that bound amino acids that recognized the DNA code. There are 64 different codons (**Figure 6.18**), but typically only around 30 different tRNA genes. Why? Because there are no tRNAs for the stop codons, and some tRNAs can recognize more than one codon due to the flexible 'wobble' pairing of the first base of the anticodon (which pairs with the often-redundant third position of the codon). In the human genome there are around 500 tRNA genes (plus 22 mitochondrial tRNA genes), and probably more than 300 tRNA pseudogenes.



maintained by complementary base pairing in the 'stem' regions, which holds two strands together by hydrogen bonds between paired bases. Any change to the nucleotide sequence of one strand could disrupt this pairing and wreck the molecule. But if a mutation in one side of the stem (say changing a C to an A) is matched by a change in the other strand that maintains the pair bond (a G is changed to a T), then the structure of the molecule may be preserved. Remember that mutations are undirected, so the presence of a mutation on one strand won't prompt the occurrence of just the right mutation on the other strand. But only sequences that maintain the structure of these important RNA molecules are likely to persist and be passed on. An rRNA molecule with two paired mutations is more likely to find its way to future generations than one with one mutation that destroys pairing in the stem. The need to maintain pairing in some parts of the rRNA molecule leads to interesting and important patterns of sequence change in rRNA genes. A change in one part of the gene sequence might tend to be associated with a change in another part of the gene, because they come together in the final product. This challenges the usual assumption in phylogenetic analysis that all the positions in a gene are changing independently of each other (see **Case Study 6.2** for another example of compensatory changes).

The unpaired loops in the RNA molecule may be more liberal with respect to the number of bases they contain,

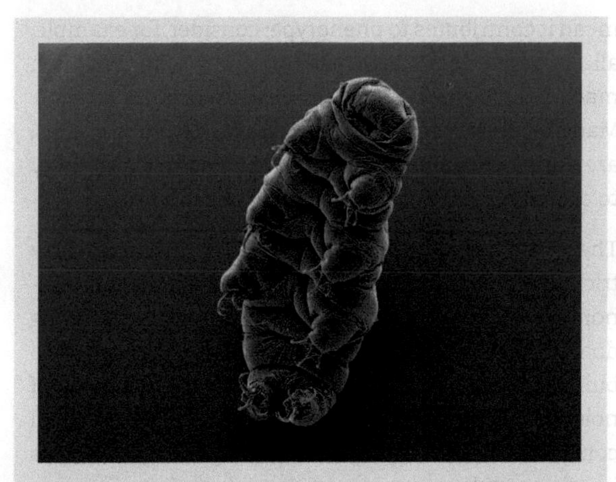

**Figure 6.16** Go on, tickle my tummy. Tardigrades are at the same time cute as kittens and tough as nails. While going by the rather adorable names of 'water bears' or 'moss piglets', they are virtually indestructible and can survive conditions that would demolish most organisms. They have been known to survive heating to 150°C and freezing to −200°C, dehydration, radiation and both high and low pressures (including the vacuum of space). An excellent companion for long journeys, they do not take up much luggage space, being less than a millimetre long.

Credit: Vicky Madden and Bob Goldstein, UNC Chapel Hill.

so they may vary in length, but any active sites within those loops are likely to be maintained by selection. The functional structure of the RNA molecule is reflected in the pattern of changes in the gene, and you will see this when you try to align RNA gene sequences: there will be variable regions with lots of change in both nucleotide identity and sequence length, and there will be conserved regions where the need to maintain structure keeps sequences from changing very much. So even when analysing DNA like a bioinformatician, you are going to need to think like an evolutionary biologist—how have my sequences evolved? What constraints have shaped the way that these sequences accumulate differences?

## Regulatory sequences

When I was an undergraduate, and the human genome project was getting under way, one of my genetics lecturers proclaimed with disgust what a waste of public money it was. It was, after all, an impossible dream to sequence the whole genome, and all that money would be better spent working on actual genes rather than trying to sequence all the useless and uninteresting parts of the genome. But, as it turns out, one of the advantages of whole genome sequencing is that you get to see all the bits of the genome that aren't genes.

We have seen how even the simplest gene needs many different parts of the genome to work together in order to produce a useful part of the phenotype. The bacterial *lacZ* is one of the simplest gene regulation examples we can describe, and yet it relies upon many different sequences working together. It requires the products of other genes, such as the repressor protein gene and all the genes needed for transcription and translation, such as RNA polymerase and transfer RNA genes. Successful expression of *lacZ* also relies upon many non-gene sequences in order to function, not only the promoter sequences and termination signals attached to the gene itself but all the regulatory sequences that turn on all of the other genes that make the transcription machinery. All these sequences involved in the regulation of gene expression are clearly of great interest to anyone who wants to know how genotype relates to phenotype. But regulatory sequences provide a much trickier target for investigation than genes.

The exonic parts of protein-coding genes are reasonably easy to recognize. The need to maintain a meaningful message written in triplet code constrains the kinds of changes that can occur. This leads to characteristic patterns of molecular evolution, such as the maintenance of frame by limiting the insertion or deletion of nucleotides to multiples of three, and third codon positions having a faster rate of change. Novel genes can sometimes be recognized by the predicted structure of their products, whether it's an amino acid sequence that forms recognized elements of proteins, such as beta-pleated sheets, or particular conformations of stems and loops for RNA.

Functional sequences that are not transcribed and don't produce proteins or RNAs are not as easy to recognize. This is because they are smaller, more dispersed, and more variable in structure than gene sequences. Consider the beta globin gene. Some of the regulatory sequences that control expression of the beta globin gene are right next to the coding part of the gene, for example the TATA box, a short promoter sequence usually found just 25bp upstream of the start of the coding region. But some other control elements that trigger expression of the beta globin gene are located over 6,000 bases upstream of the gene. Some genes can even be controlled by sequences on different chromosomes.

### *Identifying regulatory elements*

One way to identify regulatory sequences is to search the genome for sequences that match known elements by matching them to a curated database. This is trickier than it seems because regulatory sequences are usually short and often variable in nucleotide sequence. Take the common promoter sequence known as the TATA box, which is similar, but not identical, in many organisms. How similar does a short sequence have to be to the canonical TATA sequence (TATAAA) for us to recognize it as a match? In one gene the TATA box sequence might read TATGTT, in another it might be TAACTC or TATAAT. Some of these differences may confer functional differences in promoter strength, others might simply be random variation. Such short sequences have a high probability of turning up just by chance in any string of nucleotides. The TATA box might be recognized by position—usually just upstream from the start of the gene—but other enhancer sequences that influence the timing and level of gene expression can be located a long way from the gene they influence.

One way to identify non-gene DNA that has an important role in cell function is to find sequences that are unusually well conserved between species. Mutation is inevitable so any conservation of sequence suggests that mutations are being removed by natural selection. If natural selection is removing changes, that suggests that the organism needs that sequence to be just the way it is because it is doing something important.

The much-hyped ENCODE project (ENCyclopedia Of DNA Elements), built on the collaborative efforts of many different international research groups, sought to catalogue all the functional elements in the genome by combining a number of different tests to infer biochemical activity, looking for DNA that is transcribed or methylated, or for sequences that match known binding sites or show evidence of interaction with other sequences. By considering any sequence that passed at least one of these tests to be functional, they came to the conclusion that while the vast majority of the human genome doesn't code for genes, most of this non-gene DNA is nonetheless functional in some sense. This declaration was greeted with delight by some researchers, and derision by others. At this stage of the game, it is still a matter of debate exactly how the tests used by ENCODE relate to the intricate web of connections between phenotype and genotype. For example, just because something is transcribed doesn't mean it contributes to phenotype: consider, for example, all the genomic parasites that co-opt the transcription machinery for their own nefarious purposes (Chapter 5). Many regulatory sequences, such as DNA binding sites, are quite short nucleotide sequences, so they are likely to occur by chance throughout a large genome.

The controversy surrounding ENCODE illustrates how tricky it is to decipher the non-gene sequences in the genome. Identifying bits of DNA that seem to do something doesn't necessarily tell you what is actually needed to build and run an organism. There is bound to be a lot of noise and cross-talk in the genome. Hearing all the real conversations between genes and regulatory elements is going to take some more fine-tuning of our listening skills. But given that there are some non-gene sequences in the genome that have important, if mysterious, roles, we can't assume that the nucleotide sequences in the vast wildness of intergenic space evolve without constraint.

Note that there is an unhelpful tendency to refer to any sequence that is not destined to be translated into protein as 'non-coding DNA'. While this may be true in the strict sense of the genetic code which specifies the relationship between nucleotide and amino acid sequences, the phrase 'non-coding' is prone to misinterpretation. Some people would describe any situation where the information in one sequence is transferred to another as a coding relationship. In this sense, a ribosomal RNA gene (written in DNA) can be said to code for a functional ribosomal component (constructed out of RNA). You could even say that a promoter sequence in the upstream region of the gene 'codes for' a signal that says 'start transcription here'. Under this definition, 'non-coding DNA' should only refer to sequences that don't carry any useful information at all. And then that leads us to the whole big noisy argument about how much of the genome is non-functional. The point is not to get into a philosophical debate about the meaning of 'code' with respect to genetics (and believe me there have been many long debates on this very topic), but simply to follow Johannsen's advice in making sure that the words we use match neatly to the concepts we want to convey.

## Conclusion

In this chapter, we have emphasized the distinction between the information in the genome (genotype) and the expression of that information to form a unique individual (phenotype). Only genotype can be inherited stably across many generations, but genotype can only influence phenotype through transcription of DNA into RNA. Distinction between genotype and phenotype is important for anyone who wants to analyse DNA sequences. We need to consider not only the influence of genotype on the formation of phenotype through gene expression and development, but also of the influence of phenotype on the way that the genotype evolves, shaping the kind of changes that we see in the genome.

There is a good chance that, if you are primarily interested in analysing DNA to uncover evolutionary history, population processes or the relationships between species, your focus will be entirely on the genotype. Indeed, you may deliberately seek parts of the genome that appear to have no effect on phenotype, such as microsatellite loci, intergenic regions or non-functional SNPs. But even those who see the genome as a story book rather than a recipe book need to understand the way that the function of different sequences influences the way that those sequences evolve. Any evolutionary process that shapes the way the nucleotide sequence changes will impact on the recording of evolutionary stories. In fact, it creates additional layers of richness to those stories which we can learn to read. So in the next two chapters we explore the two main engines that drive sequence evolution: selection (Chapter 7) and drift (Chapter 8).

# Points to remember

### Genes

- Genes are sequences of DNA that act as a template for the formation of a RNA transcript that either folds to become a functional RNA molecule, or acts as a template for a chain of amino acids that will form a protein.
- Most cells carry a copy of every gene in the genome but will express only a subset of those genes, so the regulation of gene expression is an important part of developing phenotype.
- A typical gene has a beginning (promoter that attracts and binds the transcription machinery), a middle (the section that becomes the functional RNA molecule or the messenger RNA that specifies protein-formation), and an end (sequences that cause the transcription machinery to stop copying and trigger the processes that prepare the message for transport and translation).
- Even the simplest examples of gene expression require the co-ordinated action of a large number of sequences, throughout the genome.
- While the genome of every cell contains every gene, the transcriptome (the set of all RNA molecules found in the cell) contains only those sequences that are being actively transcribed.

### Sequence evolution

- In many genes in eukaryotes, protein-coding sequences (exons) are broken up by relatively larger non-protein-coding sequences (introns) which are spliced out of the RNA transcript before it is translated.
- The triplet genetic code of exons creates distinct patterns of nucleotide change, such as a faster rate of change in third codon positions.
- Protein-coding sequences can be recognized by their similarity to known proteins, and the presence of long sequences that make coding 'sense' without stop codons or insertions and deletions that destroy the triplet-structure of the sequence.
- RNA genes can exist in multiple copies within the genome, and their patterns of evolution are shaped by the three-dimensional structure and function of the RNA molecule.
- Gene names, usually written in italics, typically have a long form and a short abbreviation, while the corresponding protein name or abbreviation is often written in capitals.
- Because many different sequences are required to manage the expression of any gene, many sequences in intergenic regions may have important functions in the regulation of expression.

# Ideas for discussion

**6.1** Could a genetic system function without a genotype/phenotype distinction? For example, could we find an organism where functional RNA or protein molecules were copied directly to pass on to offspring? What implications would this have for evolution?

**6.2** Transfer RNA molecules are short RNA sequences (on average around 78 nucleotides long) that form a stem-and-loop structure, where one of the loops carries the anticodon that binds with the appropriate codon in the messenger RNA. At least 30 different tRNA genes are required by each cell to specify translation, but each of these tRNA genes can occur in multiple copies in a single genome, often co-located in clusters. How might these features of tRNA structure and function influence the way that tRNA genes evolve?

**6.3** How could you tell which non-coding sequences in the genome do something important, in the sense that they have a tangible effect on phenotype?

# Sequences used in this chapter

Table 6.1: LacZ gene used in Chapter 6

| Description from GenBank entry | Accession |
|---|---|
| *lacZ* from *Escherichia coli* str. K-12 substr. MG1655, complete genome (362455–365529, complement) | NC_000913 |

# Examples used in this chapter

Baxendale, S., Abdulla, S., Elgar, G., Buck, D., Berks, M., Micklem, G., Durbin, R., Bates, G., Brenner, S., Beck, S. (1995) Comparative sequence analysis of the human and pufferfish Huntington's disease genes. *Nature Genetics*, Volume 10, page 67.

Daborn, P. J., Waterfield, N., Silva, C. P., Au, C. P. Y., Sharma, S., ffrench-Constant, R. H. (2002) A single *Photorhabdus* gene, *makes caterpillars floppy* (*mcf*), allows *Escherichia coli* to persist within and kill insects. *Proceedings of the National Academy of Sciences USA*, Volume 99, page 10742.

Kunte, K., Zhang, W., Tenger-Trolander, A., Palmer, D., Martin, A., Reed, R., Mullen, S., Kronforst, M. (2014) *doublesex* is a mimicry supergene. *Nature*, Volume 507, page 229.

HEROES OF THE GENETIC REVOLUTION

6

# Francis Crick

*"I will always remember Francis for his extraordinarily focused intelligence and for the many ways he showed me kindness and developed my self-confidence.... Being with him for two years in a small room in Cambridge was truly a privilege"*
James Watson (2004) *Nature* (online) 29 July 2004.

**EXAMPLE PUBLICATIONS**

Watson, J. D., Crick, F. H. C. (1953) Molecular structure of nucleic acids. *Nature*, volume 171, page 737.

Crick, F. H. C., Barnett, L., Brenner, S., Watts-Tobin, R. J. (1961) General nature of the genetic code for proteins. *Nature*, Volume 192, page 1227.

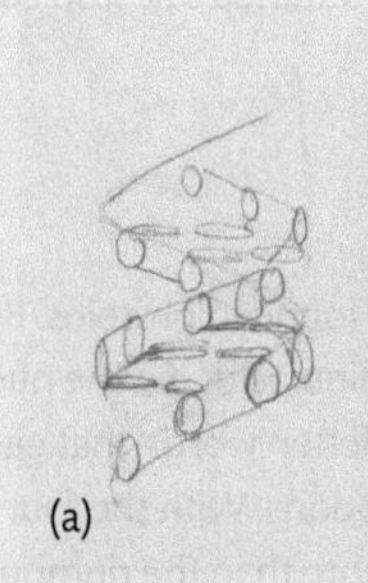
(a)

(b)

**Figure 6.17** (a) An early sketch of the structure of DNA. (b) Francis Harry Compton Crick (1916–2004).



Although Francis Crick (**Figure 6.17**) is best known as one of the co-discoverers of the double-helix structure of DNA, for which he shared a Nobel Prize in 1962 with James Watson and Maurice Wilkins, his contributions to the development of molecular genetics were legion. He brought his training in physics to understanding biological systems, using X-ray crystallography to explore the structure of proteins. Crick's commitment to modelling biochemical structures, drawing together information from other scientists' experiments, provided the key to unlocking the secret of DNA. In their famous *Nature* paper of 1953, Francis Crick and James Watson demonstrated that DNA consisted of a double helix with interconnecting pairs of complementary bases. Rarely has a single scientific paper changed a scientific field so rapidly. Watson and Crick's structure suggested a way that DNA could carry genetic information through complementary base pairing. In a second *Nature* paper, published the following month, they showed how the double-helix structure provided a means of self-replication by template copying.

Crick took this discovery forward, proposing an 'adaptor hypothesis' to explain how the DNA code could be translated into proteins (later proved correct by the discovery of transfer RNA molecules). Working with other scientists at the Cavendish Laboratory in Cambridge, Crick conducted genetic experiments on viruses to demonstrate that the code was formed of triplets of nucleotides, and that the code was degenerate (many different triplets could code for the same amino acid). His pioneering work in molecular genetics led Crick to deduce the 'central dogma' of molecular biology—one-way transfer of information from nucleic acid to protein—which remains one of the key principles of modern evolutionary biology ('dogma' was intended to indicate central importance, rather than unchallengability).

Crick wrote on a wide variety of topics, from selfish DNA to directed panspermia (the hypothesis that life on earth has an extraterrestrial origin). He devoted much of his research energy into understanding consciousness, exploring topics such as the function of dream sleep and the neural activity underlying visual awareness.

> "I think what led me into biological research was really because I felt there was a mystery which I thought to be explained scientifically. And one of these areas was the borderline between the living and the nonliving and the other one was the problem of how the brain works —the problems of consciousness. Of course nowadays we call those areas molecular biology and neurobiology, but those terms weren't known at that time."

TECHBOX
6.1

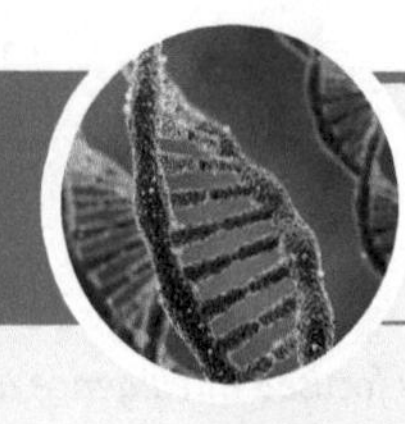

# Genetic code

**RELATED TECHBOXES**

**TB 2.1:** DNA structure

**TB 7.2:** Detecting selection

**RELATED CASE STUDIES**

**CS 6.1:** Gene families

**CS 6.2:** Genetic code

**Summary**

DNA sequences are translated into amino acid sequences three bases at a time.

**Keywords**

amino acid, translation, protein, gene, codon, redundancy, stop codon, nuclear, mitochondrial, start, stop, four-fold degenerate

## Degeneracy of the code

There are 20 different amino acids commonly found in proteins. But DNA has only four different nucleotides with which to specify all of the amino acids. So the nucleotide sequence in a protein-coding gene is read in triplets—three-letter 'words' that each specify a particular amino acid. These triplets of nucleotides are called codons. With the four-letter alphabet of DNA, there are 64 possible three-base codons, which is more than the number of amino acids, so many amino acids are represented by more than one codon. Having multiple codons specifying the same amino acid is known as the redundancy of the genetic code. This redundancy leads to some interesting patterns of sequence evolution. In particular, if you look closely at the genetic code (**Figure 6.18**) you will see that often it doesn't matter which base is in the third codon position. For example, if you have C's in the first two positions of the codon, then that codon will code for Proline (P), no matter whether it has a T or C or A or G in the third position. So the third codon position is often much freer to change nucleotide without changing the amino acid, and therefore without doing damage to the protein. Codons where all four combinations of third codon base specify the same amino acid are referred to as 'four-fold degenerate'.

Third codon positions are often used to estimate the neutral rate of change—the rate of change expected at a nucleotide position which is not subject to selection. But look carefully at the

| | 2nd | | | | |
|---|---|---|---|---|---|
| 1st | T | C | A | G | 3rd |
| T | TTT F Phe Phenylanaline | TCT S Ser Serine | TAT Y Tyr Tyrosine | TGT C Cys Cysteine | T |
| | TTC F Phe | TCC S Ser | TAC Y Tyr | TGC C Cys | C |
| | TTA L Leu Leucine | TCA S Ser | TAA *Stop* | TGA – *Stop* | A |
| | TTG L Leu | TCG S Ser | TAG – *Stop* | TGG W Trp Tryptophan | G |
| C | CTT L Leu | CCT P Pro Proline | CAT H His Histidine | CGT R Arg Arginine | T |
| | CTC L Leu | CCC P Pro | CAC H His | CGC R Arg | C |
| | CTA L Leu | CCA P Pro | CAA Q Gln Glutamine | CGA R Arg | A |
| | CTG L Leu | CCG P Pro | CAG Q Gln | CGG R Arg | G |
| A | ATT I Ile Isoleucine | ACT T Thr Threonine | AAT N Asn Asparagine | AGT S Ser Serine | T |
| | ATC I Ile | ACC T Thr | AAC N Asn | AGC S Ser | C |
| | ATA I Ile | ACA T Thr | AAA K Lys Lysine | AGA R Arg Arginine | A |
| | ATG M Met Methionine | ACG T Thr | AAG K Lys | AGG R Arg | G |
| G | GTT V Val Valine | GCT A Ala Alanine | GAT D Asp Aspartic acid | GGT G Gly Glycine | T |
| | GTC V Val | GCC A Ala | GAC D Asp | GGC G Gly | C |
| | GTA V Val | GCA A Ala | GAA E Glu Glutamic acid | GGA G Gly | A |
| | GTG V Val | GCG A Ala | GAG E Glu | GGG G Gly | G |

**Figure 6.18** The standard genetic code, sometimes also referred to as the universal genetic code. Note that these codons are written in DNA nucleotides as they would appear in the gene. Some code tables will write codons in RNA nucleotides using U (uracil) instead of T (thymine). The colours of the one-letter codes represent the chemical properties of the amino acids: Small nonpolar (G, A, S, T: Orange), Hydrophobic (C, V, I, L, P, F, Y, M, W; Green), Polar (N, Q, H; Magenta), Negatively charged (D, E; Red), Positively charged (K, R; Blue).

code: while some third codon positions are four-fold degenerate, others are not. In the standard code, the codon TGT codes for cysteine, but if I change the third position to a G, then I swap cysteine for tryptophan. Worse still, if I change the third codon position to A then I get a stop codon and almost certainly ruin my protein. So most estimates of the neutral rate of change from protein-coding sequences focus only on four-fold degenerate sites. Be aware, though, that just because the amino acid remains the same, doesn't mean all four-fold degenerate codons are selectively equal. There is some evidence of weak selection that results in some codons being more frequently used than others, a phenomenon referred to as codon usage bias.

### Start and stop codons

Most cells use the same 'Start' codon, ATG, which also codes for the amino acid methionine, though there are occasional alternative start codons. However, the number and identity of the 'Stop' codons vary between genetic codes (**Figures 6.18** and **6.19**). There are no tRNAs that match stop codons, so when the ribosome reaches a stop codon, no more amino acids are joined to the growing strand. Instead, a release factor binds to the ribosome and allows the completed amino acid chain to be released. A mutation that causes a stop codon to occur in the middle of a gene will give rise to a truncated protein product which will probably be non-functional (**Figure 6.5**).

### Amino acids

Amino acids can be represented by their name (e.g. isoleucine), a three-letter abbreviation (Ile) or a single letter (I). The three-letter abbreviations are written with a capital and two lowercase letters—e.g. Thr for threonine or Asn for asparagine. When an international convention for symbols for the genetic code was established in 1968, there was an attempt to make the single letter symbols of amino acids memorable using mnemonic associations. Some of the single letter codes are obvious, such as C for cysteine or M for methionine. Where more than one amino acid shared the same starting letter, it was given to the most commonly used amino acids—e.g. A for alanine but R for arginine. Other letter assignments are phonetic, like F for phenylalanine. And then the assignments start getting tenuous—for example, W for tryptophan because it's a big letter for a big double-ring molecule. U and O weren't assigned because they can be confused with V and zero, and J was left out because it's not used in some languages.

| | 2nd | | | | | | | | | | | | |
|---|---|---|---|---|---|---|---|---|---|---|---|---|---|
| 1st | T | | | C | | | A | | | G | | | 3rd |
| T | TTT | F | Phe | TCT | S | Ser | TAT | Y | Tyr | TGT | C | Cys | T |
| | TTC | F | Phe | TCC | S | Ser | TAC | Y | Tyr | TGC | C | Cys | C |
| | TTA | L | Leu | TCA | S | Ser | TAA | - | Stop | TGA | W | Trp | A |
| | TTG | L | Leu | TCG | S | Ser | TAG | - | Stop | TGG | W | Trp | G |
| C | CTT | L | Leu | CCT | P | Pro | CAT | H | His | CGT | R | Arg | T |
| | CTC | L | Leu | CCC | P | Pro | CAC | H | His | CGC | R | Arg | C |
| | CTA | L | Leu | CCA | P | Pro | CAA | Q | Gln | CGA | R | Arg | A |
| | CTG | L | Leu | CCG | P | Pro | CAG | Q | Gln | CGG | R | Arg | G |
| A | ATT | I | Ile | ACT | T | Thr | AAT | N | Asn | AGT | S | Ser | T |
| | ATC | I | Ile | ACC | T | Thr | AAC | N | Asn | AGC | S | Ser | C |
| | ATA | I | Met | ACA | T | Thr | AAA | K | Lys | AGA | - | Stop | A |
| | ATG | M | Met | ACG | T | Thr | AAG | K | Lys | AGG | - | Stop | G |
| G | GTT | V | Val | GCT | A | Ala | GAT | D | Asp | GGT | G | Gly | T |
| | GTC | V | Val | GCC | A | Ala | GAC | D | Asp | GGC | G | Gly | C |
| | GTA | V | Val | GCA | A | Ala | GAA | E | Glu | GGA | G | Gly | A |
| | GTG | V | Val | GCG | A | Ala | GAG | E | Glu | GGG | G | Gly | G |

**Figure 6.19** Spot the difference: the vertebrate mitochondrial code differs from the standard code at four codons.

### Alternative genetic codes

**Figure 6.18** gives all 64 possible codons with the corresponding amino acid for the code used in the majority of genomes. This particular set of codon assignments is, somewhat unhelpfully, usually referred to as the 'universal genetic code'. Actually there are many known variants to the code, and more are being discovered. One of the alternative codes you are likely to encounter is the vertebrate mitochondrial code, which differs from the 'standard' code at three codons (**Figure 6.19**). But there are many other variants, for example echinoderm and flatworm mitochondrial, mycoplasma nuclear, and plant plastid codes. Each of these codes varies from the standard code in only one or a few codon assignments, so while the standard code is not 'universal', it is nearly identical in all species. In addition to these consistent code differences, there may be minor occasional differences, such as the use of alternative start codons in some genes.

Appreciation of the variation in genetic coding schemes is important for two reasons. The practical reason is that if you want to make sense of a protein-coding gene, you will need to know what code is right for that genome. If you translate the sequence using the wrong code you are likely to get stop codons popping up in odd places, and the amino acid sequence may seem more divergent than otherwise expected. But there is another reason this is important, which is a bit more theoretical: variations in the genetic code illustrate that even the most fundamental aspects of organisms are subject to evolutionary change. Stop and think about the consequences of reassigning a codon: such a change affects not just one gene but, potentially, every protein-coding gene in the whole genome. In many cases, codon reassignment seems to have accompanied extreme genome reduction when a single-celled organism becomes an endosymbiont. It's not surprising that code changes are very rare, but the fact that they exist at all provides a fascinating insight into molecular evolution.

There are many slight variants on the genetic code (**Figure 6.19**). Listed below are some of the alternative codon translations used in various genomes. Note that this is not an exhaustive list and that the alternative codes may not be used by all organisms in that taxon. Some organisms do not use all codons in their code, and it is reasonably common for some organisms to use alternative start codons.

#### *Alternative nuclear codes*

- Blepharisma (a common ciliate protist): Nuclear Code: TAG = Q
- Ciliate (protists such as *Paramecium*): TAA = Q; TAG = Q
- Dasycladacea (green algae): TAA = Q; TAG = Q
- Hexamita (a flagellated protozoan that may cause 'Hole in the head' disease in fish): TAA = Q; TAG = Q
- Euplotid (ciliate protists): TGA = C
- Mycoplasma (bacteria): TGA = W

#### *Alternative mitochondrial codes*

- Arthropods, nematodes, molluscs: AGA = S; AGG = S; ATA = M; TGA = W
- Ascidian; AGA = G; AGG = G; ATA = M; TGA = W
- Chlorophycea (green algae including *Volvox*): TAG = L
- Echinoderm: AAA = N: AGA = S; AGG = S; TGA = W
- Fungi: TGA = W
- Protozoa: TGA = W
- *Scenedesmus obliquus* (a unicellular green alga): TCA = stop; TAG = L
- Trematode: TGA = W: ATA = M; AGA = S; AGG = S
- Vertebrate: AGA = stop; AGG = stop; A TA = M; TGA = W
- Yeast: ATA = M; CTT = T; CTC = T; CTA = T; CTG = T; TGA = W

# TECHBOX 6.2

# Primers and probes

**RELATED TECHBOXES**

**TB 5.1:** Sanger sequencing

**TB 5.2:** High-throughput sequencing

**RELATED CASE STUDIES**

**CS 6.1:** Gene families

**CS 9.2:** Barcoding


**Summary**

Short nucleotide sequences allow you to sequence a targeted part of the genome.



**Keywords**

PCR, high-throughput sequencing, *Tm*, annealing temperature, melting temperature GC content, degenerate primers, exploratory probes, targeted enrichment, oligonucleotide, sequence capture


How do you sequence a particular gene? The classic approach is to use PCR to amplify exactly the sequence you want (**TechBox 4.2**), then use Sanger sequencing to determine the base sequence in the sequence you have amplified (**TechBox 5.1**). The beauty of this approach is that you (hopefully) get only the gene you wanted. The limitations of this approach are that you need to know what the sequence you want is so you can design a primer that will stick to only that particular place in the whole genome, and it can take a lot of time to build up a database of useful size.

The alternative approach is to use high-throughput sequencing (HTS) on genomic DNA (**TechBox 5.2**). You can either select the sequences you want to analyse at the bioinformatics analysis stage or use 'targeted enrichment' to amplify only those parts of the genome that you want to analyse. The advantage of the HTS approach is that you can rapidly get heaps of potentially useful loci to analyse. The disadvantage is that more data does not necessarily mean better data: the error rate in HTS is higher than in Sanger sequencing, so unless you have a high degree of coverage of your target loci, you will have to take sequencing error into account in your analysis. Furthermore, including a very large number of sequences in your analysis may obscure the differences in signal between loci, potentially supporting a false conclusion.

These two approaches are ends of the spectrum and there is an endless number of possibilities in between. For example, many studies of bacterial diversity use high-throughput sequencing to sequence a single locus for all the taxa in the sample, particularly the 16S rRNA gene which has traditionally been used in bacterial systematics. Whichever approach you use, be aware of the errors inherent in the system, such as contamination, amplification of paralogs (gene copies that may have slightly different evolutionary histories), horizontal gene transfer (for example nuclear copies of mitochondrial genes), and remember that there is room for uncertainty at both the wet-lab stage (particularly for the targeted Sanger sequencing) as well as the bioinformatics stage of analysis (particularly for the genomic high-throughput approach).

Here we will briefly consider how primers and probes are used to select targeted parts of the genome for molecular phylogenetic analysis, whether a single gene or a set of useful loci for phylogenetic analysis. We won't be considering the way that primers and probes are designed for amplifying or sequencing whole genomes.

**Primers** are short nucleotide sequences that you add to your DNA sample in order to amplify a particular target sequence (Chapter 4). They work by binding to a matching sequence in the sample by complementary base pairing, thus providing a free 3' end for polymerase to work from to make a complementary copy of the adjacent sequence. Careful primer design is the primary way that PCR can be optimized to a particular task (that is, to amplify a specific sequence in the context of a particular sample). Typically, primers are designed in pairs—a forward primer defines the beginning of the sequence to be amplified, and a reverse primer defines the end.

**Probes** are similar to primers in that they are designed to stick to a particular bit of DNA in a sample, but the function of a probe is to 'capture' the sequence rather than to provide a starting block for synthesis. Probes are typically labelled with something that allows you to detect the target sequence or to separate it from the rest of the sample, for example, by making it stick to a surface while the non-target DNA is washed away. There is a huge diversity of approaches to designing probes for different high-throughput sequencing methods and we are not going to attempt to cover them all here. Instead, our goal is to broadly consider some of the features that are relevant to designing an oligonucleotide (short nucleotide sequence) that targets a particular sequence, so we are going to focus on a few general features that mostly apply to both primers and probes.

If you want to come up with a short nucleotide sequence that will stick only to your target sequence and nowhere else, then you first need to think about the chance of any sequence occurring at random in the genome. If all bases are equally frequent, then we would expect a particular sequence of four bases, say ACTG, to occur, on average, every $4^4$ bases (or once every 256 bases). So we would expect to find hundreds of thousands of instances of this four-letter sequence in the human genome. But any given 17 base sequence has a probability of occurring only once in every $4^{17}$ bases (over 17 billion), so it is less likely to occur twice in the genome simply by chance. You can experiment by coming up with random short nucleotide sequences and blasting them against GenBank and seeing how many perfect or near-perfect matches you get (**TechBox 1.2**). Incidently, length of oligonucleotides is sometimes reported in the somewhat perplexing units of 'mer'—so just as one can have a monomer or a polymer, you might see a sequence referred to as a '12-mer' (a sequence of 12 bases).

## Primer design

It is possible to design primers and probes manually, but most researchers use computer programs to come up with an optimum primer design and to check it against all known sequences to make sure it matches only what you want it to (**TechBox 1.2**). As with most techniques in this book, you should consult papers and protocols published within the last year or two to find the most up-to-date approaches and to get hints on the best primer design strategies for your particular organisms or question. But we can consider some basic principles to keep in mind when designing primers:

- **Sequence:** If you wish to target a particular locus in the genome, then your oligonucleotide should match a sequence that is unique to the target DNA (that is, a sequence that is not repeated elsewhere in the genome), so that it will bind by base pairing to the target DNA. It is also worth avoiding repeat sequences, runs of Cs and Gs, or any self-complementary sequences that could bind with another part of the primer sequence (e.g. **Figure 6.10**). If your primer sequence is prone to knotting itself up into a little self-complementary bundle then it is not going to be very efficient at sticking to your target DNA.
- **Length:** The shorter the oligonucleotide, the more chance that it will match a random sequence in the genome. But the longer the oligonucleotide, the more likely it is to have sequence features that cause it to adopt secondary structures that interfere with binding to target DNA. For PCR, longer primers have higher annealing temperatures, which means that the primer may begin annealing at the wrong point in the PCR temperature cycle (e.g. the primers may bind indiscriminantly during the elongation phase, resulting in non-target sequences being amplified).
- **Melting temperature:** For techniques that use temperature cycling, the temperature at which the oligonucleotide will stick to the target DNA is determined primarily by its sequence and length. The simplest formula for estimating the melting temperature (*Tm*) of a primer is the Wallace formula: add together 4° Celsius for every G or C and 2° Celsius for every A or T. This is because GC pairs are held together by three hydrogen bonds, so they require more energy to melt than AT pairs which are held together by only two bonds. However, primer design programs will usually give a more accurate indication of *Tm*, by including the influence of other factors such as salt concentration. Accurate estimation of *Tm* is important: if the *Tm* is lower than the annealing temperature used in a PCR reaction, the primer will stick to many non-target sequences. If it is too high, it may fail to anneal to the right sequence. The two members of the primer pair should have similar melting

temperatures. Because base composition affects the melting temperature of double helix DNA, the base composition of the primer sequence should be optimized to the procedure. For example, for PCR, a GC content between 30 and 60 per cent of the primer sequence will help to ensure that the primer anneals and melts at an appropriate temperature.

**Degenerate primers and exploratory probes**

Sometimes it is useful to add many different versions of a primer or probe, each one differing at one or two positions, so that they can stick to a range of similar sequences. This may be necessary when designing primers from a protein sequence: because of the degeneracy of the genetic code (see **Techbox 6.1**), a particular amino acid sequence could be coded for by many different nucleotide sequences. So you may need to include primers corresponding to each of the possible nucleotide sequences that could code the amino acid sequence of the gene you wish to amplify. If you don't know the exact sequence of the gene you are looking for, but you do know what the corresponding sequence is in related species, then you might design degenerate primers that cover alternative versions of the DNA sequence. Similarly, exploratory probes allow the construction of arrays than can detect related sequences in a range of different taxa, potentially including undescribed species, so may be very useful in assessing biodiversity in microbiomes.

CASE STUDY 6.1

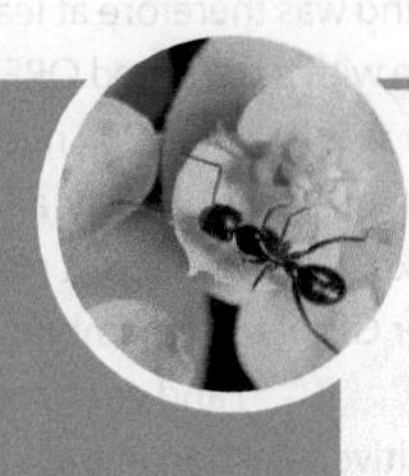

# Gene families: duplication and loss of taste receptor genes

**RELATED TECHBOXES**

TB 6.1: Genetic code

TB 7.2: Detecting selection

**RELATED CASE STUDIES**

CS 5.1: Duplication

CS 7.1: Selection

Shi, P., Zhang, J. (2006) Contrasting modes of evolution between vertebrate sweet/umami receptor genes and bitter receptor genes. *Molecular Biology and Evolution*, Volume 23, page 292.

“*What you eat clearly shapes the evolution of your genes*[1]”

**Keywords**

BLAST, positive selection, deletion, pseudogene, paralog, intron, exon, open reading frame (ORF), cDNA, whole genome

**Background**

Taste is an important sense for most vertebrates, allowing them to detect useful nutrients and avoid dangerous toxins. There are five basic tastes: sweet, sour, salty, bitter, and umami (a savoury taste, due to glutamates found in aged or fermented foods such as Parmesan cheese and soy sauce). Detection of these tastes is discrete: particular cells can detect particular tastes[2]. Taste receptor cells use specific G-protein-coupled receptors (GPCR) to detect various chemicals (a vast array of GPCRs are involved in other stimulus-response pathways, such as vision, smell, behaviour, and immune response). In particular, the T1R family of receptor proteins allow detection of sweet and umami tastes, and the T2R receptors allow detection of bitter tastes.

**Aim**

The authors aimed to reconstruct the evolutionary history of these taste receptor genes in vertebrates by identifying all the type 1 and type 2 taste receptor genes (*Tas1R* and *Tas2R*, respectively) in the ten complete (or near complete) genomes that were available at the time: five mammals (mouse, rat, human, dog, opossum), a bird (chicken), an amphibian (Western clawed frog), and three fish (zebrafish, pufferfish and fugu). They wished to contrast the pattern of evolution seen in the *Tas1R* genes which are found in roughly the same numbers in all vertebrates, to *Tas2R* genes which occur in much greater diversity.

**Methods**

The researchers used a five-stage strategy for locating taste-receptor genes in the ten vertebrate genomes. Firstly, the researchers found the genomic location of putative *Tas1R* and *Tas2R* genes by blasting known gene sequences against the whole genomes. *Tas2R* genes have a relatively simple structure, consisting of a single short coding sequence, which makes them ideal query sequences for BLAST searching (**TechBox 1.2**). But *Tas1R* genes have a more complex structure, with protein-coding sequences (exons) interspersed with non-protein-coding sequences (introns). Although the protein-coding exons tend to be conserved by negative selection to maintain protein function, the non-coding introns evolve more rapidly so may differ more between lineages. Secondly, they aligned the putative *Tas1R* genes they found against cDNAs, which are sequenced from messenger RNA transcripts. cDNA transcripts contain only exons, so this alignment allowed them to identify the protein-coding parts of the sequence. Thirdly, they translated each sequence into amino acids and checked for the presence of seven transmembrane domains known to be functionally important for T1R and T2R proteins. Fourthly, they checked that each putative gene sequence had a continuous sequence of translatable codons (an open reading frame, or ORF) and was therefore at least theoretically capable of producing a working protein. Any sequence with a disrupted ORF, or that was missing any of the seven trans-membrane domains, or was less than 200 nucleotides long, was considered to be a pseudogene (a non-functional copy of a gene). Finally, all putative *Tas1R* and *Tas2R* sequences were blasted against GenBank to make sure their closest matches were genes from the T1R and T2R families, not sequences for other GPCR transmembrane proteins. The researchers then aligned amino acid sequences (**TechBox 10.1**) and constructed phylogenetic trees (see Chapter 11). They tested for evidence of positive selection in the T1R genes by statistically comparing the fit of two alternative models of sequence evolution to their data. In one model, all codons had a ratio of nonsynonymous to synonymous substitutions (dN/dS, otherwise known as ω) of one or less than one. In the alternative model, some codons had a dN/dS greater than one, which is considered to be evidence of positive selection (see **TechBox 7.2**).

**Results**

Comparison of the number of genes in each species shows that the two families of genes show different evolutionary histories. The genes for the sweet and umami receptor proteins arose by two gene duplication events in the ancestor of all vertebrates. However, all three *Tas1R* genes appear to have been deleted from the Western clawed frog genome, and the sweet-receptor gene *Tas1R2* has apparently been lost from the chicken genome. Although many birds, such as hummingbirds, have the ability to detect sweetness, electrochemical studies of chicken tongues suggest they cannot taste sugar or saccharine, and chickens show no preference for sweet foods. *Tas1R2* is in the process of becoming a pseudogene in the cat lineage: a deletion of part of exon 3 and stop codons in exons 4 and 6 show that *Tas1R2* from cats, tigers and cheetahs cannot produce a functioning T1R2 receptor protein[3]. It seems likely that loss of sweet taste-reception does not decrease the ability of a meat-eater to feed itself successfully, so is selectively neutral in the cat lineage[3]. The *Tas1R* genes have apparently undergone selection to diversify function, with an excess of nonsynonymous changes affecting the N-terminal region of the protein, which is probably associated with the ability to bind sweet substances.

**Conclusions**

Given that taste is made up of discrete receptors, the duplication and loss of taste receptor genes can add or remove recognition of particular tastes, such as sweetness, without influencing the ability to taste other chemicals. The number of *Tas1R* genes have remained constant across vertebrate evolution, except for losses in a number of species that appear to be 'sweet-blind' (frogs, chickens, cats: **Figure 6.20**). The *Tas2R* genes have diversified greatly, and vary

**Figure 6.20** What do these animals have in common? They can't taste sugar. (a) Clawed frogs (*Xenopus*) seem to be missing the *Tas1R* genes that provide sweet taste receptors, in (b) chickens the critical gene seems to have been deleted, and in (c) cats the gene is in the process of decaying into a pseudogene.



dramatically in number between species, possibly to allow detection of a variety of bitter tastes. Members of this gene family show evidence of positive selection on both T1R and T2R proteins, suggesting functional diversification[4].

### Limitations

The authors note that their phylogenies suggest some odd relationships between species, such as humans appearing to be more closely related to dogs than to mice. While the focus of this study was the evolutionary history of genes, not species, these results suggest that the evolutionary story told from the data may need some refinement.

### Further work

Surveying genes may provide a fast way of cataloguing taste abilities in species. More detailed survey of the *Tas1r2* gene in carnivores shows that sweet receptors have been independently pseudogenized in several different lineages of meat-eaters due to mutations that interrupt the coding sequence of the gene. For example, the *Tas1r2* genes in dolphins all contain mutations that destroy protein sense, such as unexpected stop codons, frameshifts and lack of start codons[5]. But in some carnivorous species that do respond to sugar, such as the spectacled bear, the gene is intact[5]. Experimental modification of taste receptor genes may provide a useful case study for studies of stimulus-response pathways. For example, researchers created a mouse with a modified T1R2 taste receptor molecule that bound to a tasteless synthetic substance instead of sugar. These mice responded to water containing the tasteless molecule as if it was sugar water[6], showing that stimulating the taste pathway initiated specific feeding behaviour.

### Check your understanding

1. What could the researchers find out by aligning putative genes against cDNA?
2. What is a pseudogene, and how were they identified in this study?
3. How can the pattern of sequence changes be used to detect whether a gene has undergone positive selection?

### What do you think?

Does the rate at which pseudogenes accumulate changes simply reflect the mutation rate? Or could selection promote changes that knock out unneeded gene products?

### Delve deeper

Researchers can predict which species can respond to sugar by examining whether species have intact *Tas1r2* genes. What other behavioural or phenotypic capacities might be predicted by looking for working copies of genes in the genome of different species? How would you verify the predictions made from genomic analyses?

**References**

1. Caspermeyer, J. (2014) Avoiding poisons: a matter of bitter taste? *Molecular Biology and Evolution*, Volume 31, page 498.
2. Scott, K. (2004) The sweet and the bitter of mammalian taste. *Current Opinion in Neurobiology*, Volume 14, page 423.
3. Li, X., Li, W., Wang, H., Cao, J., Maehashi, K., Huang, L., Bachmanov, A. A., Reed, D. R., Legrand-Defretin, V., Beauchamp, G. K., Brand, J. G. (2005) Pseudogenization of a sweet-receptor gene accounts for cats' indifference toward sugar. *PLoS Genetics*, Volume 1, page 27.
4. Shi, P., Zhang, J., Yang, H., Zhang, Y-P. (2003) Adaptive diversification of bitter tase receptor genes in mammalian evolution. *Molecular Biology and Evolution*, Volume 20, page 805.
5. Jiang, P., Josue, J., Li, X., Glaser, D., Li, W., Brand, J. G., Margolskee, R. F., Reed, D. R., Beauchamp, G. K. (2012) Major taste loss in carnivorous mammals. *Proceedings of the National Academy of Sciences USA*, Volume 109, page 4956.
6. Zhao, G. Q., Zhang, Y., Hoon, M. A., Chandrashekar, J., Erlenbach, I., Ryba, N. J., Zuker, C. S. (2003) The receptors for mammalian sweet and umami taste. *Cell*, Volume 115, page 255.

CASE STUDY **6.2**

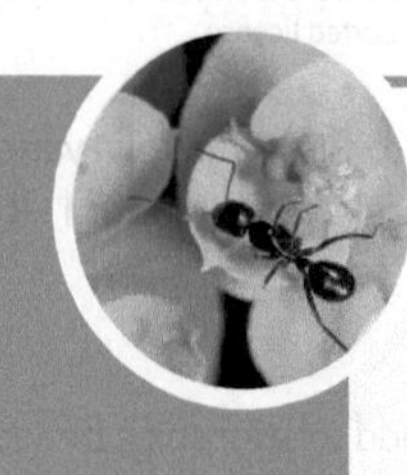

# Genetic code: multiple proteins from a single human gene

**RELATED TECHBOXES**

TB 6.1: Genetic code

TB 7.2: Detecting selection

**RELATED CASE STUDIES**

CS 5.1: Duplication

CS 7.1: Selection

Nekrutenko, A., Wadhawan, S., Goetting-Minesky, P., Makova, K. D. (2005) Oscillating evolution of a mammalian locus with overlapping reading frames: an XLαs/ALEX relay. *PLoS genetics*, Volume 1 (2):e18.

**Keywords**

overlapping genes, reading frame, stop codon, nonsynonymous, GC content, imprinting, alternative splicing, CpG, start codon, exon

**Background**

If *E.coli lacZ* illustrates genes at their simplest, *GNAS* illustrates the more complex end of the spectrum of gene structure and expression. This gene has four different promoters and is methylated such that the paternally and maternally derived copies of the gene have different expression patterns[1]. Alternative splicing generates different proteins by joining different first exons to shared terminal exons to produce functionally different proteins. Two of these proteins, XLαs and ALEX, are structurally unrelated but they bind together. But although these two proteins are entirely dissimilar in amino acid sequence, they are produced from the same DNA sequence, by translating the same mRNA sequence for exon 1 in two different reading frames from alternative start codons. Exon 1 consists of a 13-amino acid motif repeated 13 times, although some individuals carry an allele with 14 repeats. Heterozygotes with one 13-repeat allele and one 14-repeat allele are normal as long as the 14-repeat allele was inherited from the mother. But heterozygotes with a paternally derived 14-repeat allele suffer a range of severe ill effects including mental retardation, hypertrichosis (excessive hairiness) and growth deficiencies. That's because the paternally derived 14-repeat ALEX protein doesn't bind as well to XLαs, so unbound XLαs builds up in the cell. Mutations that impair the function of the XLαs protein are also highly deleterious.

### Aim

What are the evolutionary consequences of having two interacting proteins encoded by the same gene? Typical protein-coding sequences show a distinctive pattern of sequence evolution, where every third position changes faster than the first and second codon positions, because fewer third-codon position changes result in a change to the amino acid sequence. But what would happen to these patterns in a gene with overlapping reading frames?

### Methods

These researchers sequenced exon 1 of the GNAS locus from eight primate species, and analysed the non-repeat sequences. Standard analytical methods for inferring patterns of change in protein-coding sequences wouldn't work for this locus, because a third codon position mutation that might be synonymous (silent) in one frame might induce a stop codon when it is in the first or second position read in the alternative frame. So they adapted the methods to take into account changes that might be silent in one frame but not in the other. This allowed them to compare the ratio of nonsynonymous rates in the XLαs and ALEX proteins in each of the species they sequenced.

### Results

Species differed in the number of repeats in the XL exon, with humans having the fewest repeats. Both reading frames contain much a higher representation of GC-rich codons than most human genes. High GC content might reduce the chance of mutations to A and T, which have a greater chance of creating a stop codon in either frame (since stop codons are TAA, TGA and TAG: **Figure 6.18**). But high GC comes at a cost of generating more CpG sites, which are prone to mutation (see Chapter 3). Changes to functionally important genes are usually discouraged by selection so that the sequences evolve relatively slowly. Yet the ALEX protein has evolved so rapidly that it is almost unalignable between mouse and human, being similar at only half of the amino acid positions. The authors used analysis of the DNA sequence to infer amino acid changes in both of the proteins, and found that the ratio of the nonsynonymous changes in both proteins was not significantly different from one. From this they concluded that the amino acid changes in one protein were matched by a compensatory change in the other protein.

### Conclusions

By sharing a single mRNA transcript, the expression of the two interacting proteins, ALEX and XLαs, will be tightly coupled, so they are produced in the same amounts in the same place at the same time. To maintain the binding affinity between the protein, a change in one protein may need to be matched by a compensatory change in the other protein. The authors describe high GC content as 'a blessing and a curse': while it helps to maintain coding sense of both proteins, it might also lead to higher mutation rates.

### Limitations

Although the researchers hypothesize 'oscillating evolution', where each change in one protein must be matched by a compensating change in its sister protein, this is inferred from high rates of change rather than by direct observation of compensatory changes. The sequences compared have evolved over tens of millions of years and the compensatory substitutions would have happened close together, so direct signal of paired changes would be lost.

### Further work

There are two other confirmed cases of mammalian genes that produce multiple proteins from alternative reading frames, but bioinformatic analysis has been used to identify 40 additional mammalian genes that could potentially encode multiple proteins in overlapping reading frames[2]. 'Overprinting' has been proposed as a way of generating novel genes through alternative reading frames of existing genes[3] (**Figure 6.21**).

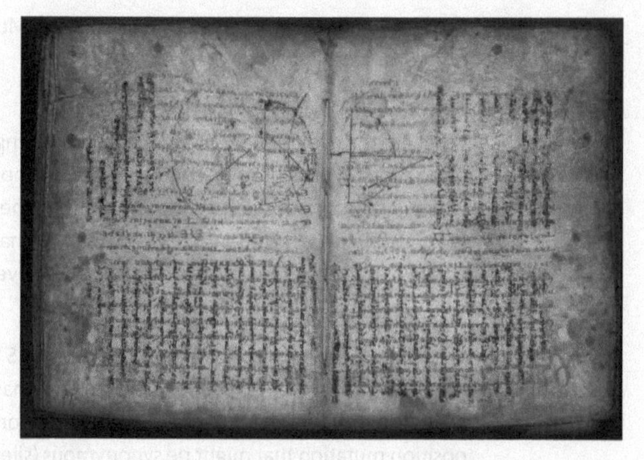

**Figure 6.21** A palimpsest is a manuscript where the original text has been scraped away so that the writing surface can be reused for a different text, sometimes leaving the ghost of past writing visible behind. In this case, a previously unknown work of Archimedes of Syracuse was detected under a thirteenth-century text. Palimpsest has also been used repeatedly as a metaphor in evolutionary biology to express the way that biological changes are often 'overwritten' on existing structures and functions, such that previous states can sometimes be discerned from features with very different current functions. It is usually assumed that new genes must always evolve by tweaking or duplicating existing genes. The surprising finding of human genes with overlapping reading frames had suggested the possibility that a new gene be formed by reading an existing sequence in a new way, to produce an entirely different product.

### Check your understanding

1. How can the researchers tell that XLαs and ALEX have a rapid rate of change, and why is this observation surprising?
2. How does the codon structure of protein genes influence patterns of protein-coding-sequence evolution?
3. Why does high GC content help reduce the chance of nonsense (stop codon) mutations, but increase the rate of mutation across the gene?

### What do you think?

Why do both the ALEX and XLαs sequences evolve rapidly? Would an alternative solution to the problem of sequence-sharing and interacting sister proteins be for them both not to change much at all?

### Delve deeper

Overlapping reading frames are found in a range of organisms, most notably viruses, but are also being discovered in a growing number of eukaryote genomes. How might these be detected? How could predicted alternative reading frames be verified? What impact would the existence of overlapping reading frames have on bioinformatic analysis of genome sequences?

### References

1. Wadhawan, S., Dickins, B., Nekrutenko, A. (2008) Wheels within wheels: clues to the evolution of the *Gnas* and *Gnal* loci. *Molecular Biology and Evolution*, Volume 25, page 2745.
2. Chung, W-Y., Wadhawan, S., Szklarczyk, R., Pond, S. K., Nekrutenko, A. (2007) A first look at ARFome: dual-coding genes in mammalian genomes. *PLoS Computational Biology*, Volume 3, page e91.
3. Neme, R., Tautz, D. (2013) Phylogenetic patterns of emergence of new genes support a model of frequent *de novo* evolution. *BMC Genomics*, Volume 14, page 117.

# Selection

## *Descent with modification*

*" . . . can we doubt (remembering that many more individuals are born than can possibly survive) that individuals having any advantage, however slight, over others, would have the best chance of surviving and procreating their kind? On the other hand, we may feel sure that any variation in the least degree injurious would be rigidly destroyed. This preservation of favourable individual differences and variations, and the destruction of those which are injurious, I have called Natural Selection . . . "*

Darwin, C. (1872) *On the Origin of Species by Means of Natural Selection, or the Preservation of Favoured Races in the Struggle for Life*, 6th edn. John Murray.

## What this chapter is about

Mutation creates differences among individuals in a population. Some of these mutations will be lost when their carriers fail to reproduce. But some mutations will increase in frequency until they replace all alternative alleles in the population. The process of molecular evolution is, at its most basic, the substitution of one base in the DNA sequence for another, so that all members of a population carry a copy of the same mutation. Some mutations influence their own chance of being copied to the next generation and these mutations can increase by positive selection until they replace all alternative alleles or decrease by negative selection until they are lost from the population. The process of natural selection is evident when we compare patterns of changes in DNA sequences.

### Key concepts

- Heritable variation is ubiquitous in natural populations
- Mutations that increase the chance of successful reproduction, relative to other alleles in the population, will tend to increase in frequency
- Fitness of an allele is a measure of its relative increase in frequency in a given environment

# Variation

You will notice that quotes from Charles Darwin keep popping up in this book (Figure 7.1). This is partly due to the unshakeable hero worship that I, along with many evolutionary biologists, have for Darwin. It is hard to think of anyone who has had a greater impact than Darwin on the way that humans understand themselves and the world around them. But more importantly, the frequency of quotes from Charles Darwin is a demonstration of the quality of Darwin's writing on evolutionary biology, and the extent of his remarkable insight into the workings of the natural world. In fact, why don't you put down this book now and go and find a copy of *The Origin of Species* and read that instead. Much has changed in evolutionary biology in the 150 years since Darwin wrote *The Origin of Species*, most importantly the discovery of the molecular basis of inheritance. Yet one of the best books on evolution you can read is the book that started the whole field, a book usually referred to, rather appropriately, as *The Origin*.

The *Origin of Species* is not Darwin's only great work, though it is his most famous and arguably his most important. Before he wrote *The Origin*, he wrote a popular account of his five-year voyage on the *Beagle*, a survey ship in which he circumnavigated the world. This journey gave Darwin many valuable opportunities to make observations on geology and natural history. After *The Origin*, he tackled the thorny subject of human evolution in *The Descent of Man* (1871), which is notable for its exploration of the influence of sexual selection (the influence of mate choice

Figure 7.1 (a) Charles Darwin at Down House, his home in the village of Downe, Kent, England. Darwin conducted countless experiments in the house, gardens and greenhouses, and used observations of the surrounding countryside to develop his ideas. (b) In an adjoining wood, he had a path constructed, known as the sand walk, where he would walk every day and cogitate. You can visit Down House and walk the Sand Walk to help you develop your own ideas in evolutionary biology. Darwin often employed his children's help in making natural history observations around their home in Kent. For example, Darwin studied the path-following behaviour of bees for many years, describing the way bees stopped at particular sites (which he termed buzzing places): "I was able to prove this by stationing five or six of my children each close to a buzzing place, and telling the one farthest away to shout out 'here is a bee' as soon as one was buzzing around. The others followed this up, so that the same cry of 'here is a bee' was passed on from child to child without interruption until the bees reached the buzzing place where I myself was standing". A nice example of inclusive fitness.

on the evolution of morphological and behavioural traits). As long as you can cope with typically lengthy Victorian prose, Darwin's books are often entertaining as well as enlightening. This is particularly true of *The Expression of the Emotions in Man and Animals* (1872) which is a delight to read (and, incidentally, one of the first scientific publications to make extensive use of photographs).

Darwin's curiosity about the natural world was boundless, as was his capacity for careful and thorough observation: any one of his books is a primer on how to be an effective scientist. For example, in *The Formation of the Vegetable Mould, Through the Action of Worms, With Observations On Their Habit* (1881), Darwin reports his systematic investigations into the contribution of earthworms to the formation of soil through an exhaustive series of behavioural experiments and observations in the field (he concludes that 'worms have played a more important part in the history of the world than most persons would at first suppose' and that 'worms, although standing low in the scale of organization, possess some degree of intelligence'). But, unlike *The Origin* or *The Descent of Man*, the majority of Darwin's books—on domestication, formation of coral reefs and islands, orchids, climbing plants, carnivorous plants, and pollination biology—are rarely read nowadays.

One such work is Darwin's four-volume taxonomic treatise on barnacles, published between 1851 and 1854 (**Figures 1.7** and **3.8**). This work represents eight years' worth of careful dissections of barnacles and detailed examinations of fossils, collected by Darwin himself on the *Beagle* voyage or sent to him by his global network of correspondents. Darwin's barnacle work became such an all-consuming project that one of his children is said to have asked a friend 'where does your daddy do his barnacles?', on the assumption that every father spent their days absorbed in the dissection of tiny crustaceans. Darwin, whose close observations of the natural world continued to instil in him 'awe before the mystery of life', described the form of his 'beloved barnacles' in such rapturous terms that his children likened his words to an advertisement. However, by the end of the eight years he grew understandably weary of the little creatures and declared 'I hate a Barnacle as no man ever did before, not even a Sailor in a slow-sailing ship' (**Figure 7.2**). Which leads us to an important and perplexing question: why would one of history's greatest thinkers devote such a large amount of time to dissecting barnacles and describing the minute differences between them? Shouldn't he have been working on something more important and interesting?

**Figure 7.2** In his essay 'Why steal beetles?' J. B. S. Haldane (**Hero 7**) wrote on the value of taxonomic collections which includes the following comment on the importance of determining whether there are appropriate pollinators for crops of sunflowers: 'It may seem ridiculous that the measurement of bees' tongues under a microscope could decide whether sunflowers can or cannot be grown profitably in a particular area. I hope it did not seem ridiculous to the people responsible for these schemes. If it did, the results may be pretty serious. A vast amount of public money will have been wasted and we shall have less margarine in 1950. But this simple example shows that a collection of insects has use value as well as a rarity value like postage stamps. Of course the majority of insects have no economic importance and are not likely to have one. But you do not know this beforehand. It would be short-sighted and impracticable to try to ignore animals not known to be of economic importance. Darwin's barnacles are a good example. If we knew how to stop barnacles growing on ships' bottoms we could save a vast amount of coal and oil. Among the things which Darwin discovered in barnacles was the apparatus by which they cement themselves on to rocks or ships.'



Thomas Henry Huxley, a friend of Darwin's and staunch promoter of evolutionary biology, wrote to Darwin's son Francis that 'in my opinion your sagacious father never did a wiser thing than when he devoted himself to the years of patient toil that the Cirripede [barnacle] book cost him'. Huxley considered that, while Darwin had gained field experience in geology and natural history from the *Beagle* voyage, he needed a solid basis in anatomy and development on which to build his theory of natural selection. Darwin's work on barnacles was critically important to the formation of evolutionary theory because it was a key demonstration of the ubiquity of variation in natural populations. Variation is the raw material of evolution. It is because individuals within a population are not all the same that evolution occurs.

## Substitution

In Chapters 3, 4, and 5, we saw how mutation is responsible for generating heritable variation between members of a population, through base changes, insertions, deletions, and rearrangements. Although DNA is copied with extraordinary accuracy and nearly all damage to DNA is repaired, the occasional mutation slips through the net and becomes a permanent change to an individual genome. Mutation occurs sufficiently frequently that, for many types of organisms, each individual has a unique genome. Furthermore, these mutations can be passed to the offspring, so that related individuals will share many of the same mutations. Mutation creates a wealth of heritable variation.

Let's think about the life cycle of a mutation. Any mutation must begin its existence in a single molecule of DNA in a single cell in a single individual (though the same mutation may also occur, by chance, in other individuals too). For the sake of argument, let's follow the fate of mutation that happens to occur in the genome of an individual named Sina. If this mutation happens to be in a cell that will form a new individual, then the mutation can be passed from Sina to her offspring, and so make it to the next generation. Our mutation now finds itself in a new individual, Sina's offspring. What happens now? That offspring may grow up to have offspring of their own, in which case that mutation has a chance to make it into Sina's grandchildren (remembering that if Sina is a sexually reproducing organism, her offspring get only half of the alleles carried by each parent, so her new mutation may or may not make it into her kids and grandkids: **Figure 5.6**). If the mutation continues to be copied into new generations, then we can say that the population is polymorphic, because many individuals, all descendants of Sina, carry a copy of her original mutation.

Now there are three possible fates of our mutation. One is that the population may continue to be polymorphic. The second fate is that Sina's mutation may decline in frequency until eventually a generation is born where no individuals carry the mutation. Now the mutation has been irrevocably lost because there are no more parents carrying the Sina mutation to pass on to their offspring. The third fate is that the Sina mutation may increase in frequency, with every generation having more and more individuals carrying a copy of the mutation that originally occurred in Sina herself, until eventually a generation arises where everyone carries only copies of this mutation and there are no other variants. Now the mutation has become fixed in the population so that all members of the population now carry a copy of that mutation. We have gone from a mutation in a single individual to a polymorphic population to fixation. In molecular evolutionary terms, this process is referred to as substitution—one variant has been substituted for another. In this chapter and the next, we look at how a mutation (a permanent change in a particular genome) becomes a substitution (when that mutation is now carried by all members of a population).

We should start by clarifying some of the terms we will be using in this chapter. I think the most useful definition of evolution is Darwin's phrase: descent with modification. 'Descent' emphasizes continuity of heritable information: that change over many generations is what we are concerned with, not the changes that happen to an individual in its lifetime. 'Modification' reminds us that it is change itself that is the stuff of evolution, which need not always be adaptation. We also need to describe units of genetic variation, and the most flexible term we can adopt is 'allele' (see Chapter 2). An individual chromosome carries one possible allele (genetic variant) at any given locus (point in the genome). Mutation generates a new allele which can be inherited by its carrier's descendants. Each new allele starts as a mutation in a single individual, but through the process of descent can come to be carried by an increasing (or decreasing) proportion of individuals in the population with each passing generation. The basic process of molecular evolution is the substitution of one allele for another, which occurs when a single allele increases in frequency to the point where it replaces all other alleles at that locus. At this point, all members of a population carry a copy of the same allele, and we say the allele has reached fixation (or become fixed). Substitution is commonly used to describe both the process whereby one allele replaced another, and the allele itself (e.g. estimating the number of substitutions that have occurred in a sequence).

To understand the process of substitution, we need to think about the effect a particular mutation has on the organism that carries it and the processes that can cause the frequency of a particular mutation to increase or decrease in the population. Specifically, we need to think about the roles of a natural selection and chance events in determining allele frequencies. This chapter focuses on selection, and Chapter 8 focuses on chance effects (more commonly known as genetic drift). But it is important to realize that this does not mean that these two mechanisms are mutually exclusive. Actually, random sampling is always acting on the passing of alleles from one generation to the next, so chance effects are ever-present. You need to keep both selection and chance in mind when considering any evolutionary patterns.

## Natural selection

Darwin was not the first to propose that the living world was the product of change. The debate about the history of the natural world had raged in both academic circles and in the general public for decades. But Darwin brought about a revolution in two ways. Firstly, he amassed an impressive catalogue of evidence in favour of transformation of species, from his observations of fancy pigeon breeds to his description of finding fossils of giant ground sloths near their modern-day relatives in South America. Secondly, Darwin did what no one had done before: he provided a plausible mechanism for the transformation of species, by linking mutations that make offspring different to their parents to the variation ever present in populations to the diversification of biological lineages.

Darwin's argument can be summarized as follows. Individuals in a population vary, in a myriad of ways, from the most trivial differences in minor features to critical alterations to important traits. These variations are often heritable, so that offspring tend to resemble their parents. Any heritable variation that increases the chance of an individual producing successful descendants will tend to be perpetuated down the generations, because successful reproducers will tend to produce offspring with the same characteristics that promoted their own reproductive success. Because they result in more successful offspring, these advantageous heritable variants will increase in frequency in the population at the expense of other less successful variants, eventually replacing all other variants. And that's natural selection. Simple, isn't it?

Natural selection is the name we give to the process that occurs when heritable variation influences the number of copies of different variants. It can apply to any population of things that have heredity (they copy) and variation (not all copies are identical). In common with the rest of life on earth, you are a template-copier because you use your own genome to make a copy of your genetic information to give to your offspring. Mutations that change the template will be passed on to all copies made from that template. Any change to the template that increases the number of copies made from it will be present in more copies, naturally. So the world becomes full of copiers that are good at copying. This simple logic that explains so much of our world is just as mind-blowing today as it was one hundred and fifty years ago when *The Origin* was published.

Consider a heritable change that makes baby spiders feed on their mother's body. Matriphagy (mother-eating) might give those spiderlings a nutritional head start in life and thereby increase the chances that they will survive to maturity and become mothers themselves (and get eaten by their own offspring who have also inherited the mother-eating tendency). This reproductive success could result in an increase in the number of mother-eating spiders next generation, each of which will also stand a good chance of having baby spiders that survive to eat their own mothers. By and by, within that particular spider population, spiders that eat their mothers may outnumber those that don't by virtue of their nutritional advantage in early life. Evidently, this has not happened to all spider species. But it has happened in several species in which all offspring now consume their mothers (**Figure 7.3**). Are mother-eating spiders 'better' than their non-cannibalistic relatives? That can only be a matter of personal opinion. But we can definitely say that matriphagy is, in this particular case, an adaptation, because it is a heritable variant that increases its own chance of transmission. This is a case

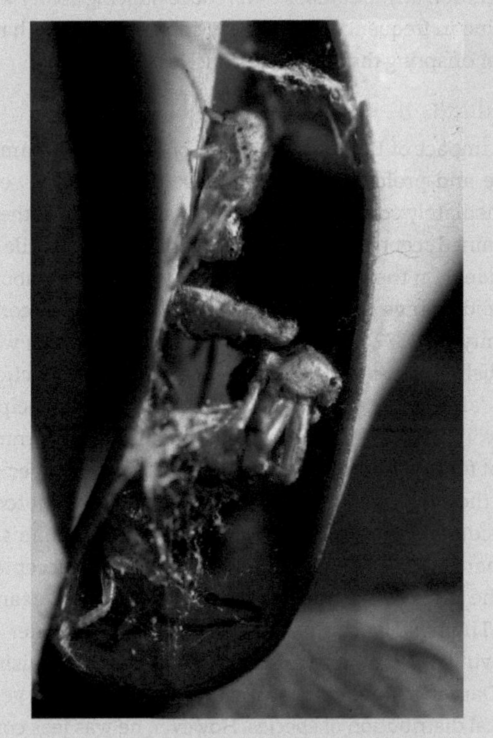

**Figure 7.3** All spiders provide some degree of maternal care, such as constructing a safe environment for eggs to hatch or bringing prey items for spiderlings to eat. In the extreme case, the mother allows herself to be cannibalized by her own offspring. This *Diaea* female has been consumed by her young, who sucked her body fluid out from her elbows. Obviously, species with suicidal maternal care produce only one brood of eggs per female, but the increased mass and dispersal ability of the offspring increases reproductive success of the cannibalized female. Photograph courtesy of Theo Evans.

of positive selection: a mutation that has a positive effect on its own chances of being copied is a mutation that will tend to increase in frequency each generation.

The flipside of positive selection is negative selection: a mutation that reduces the chances of its carrier reproducing will tend to decrease in the population with each passing generation. For example, when a female *Drosophila subobscura* (fruit fly) encounters a male, she tests his mettle in a dancing competition. The female moves rapidly side to side, and the male must follow her movements in order to keep facing her. If the female out-dances the male, she rejects him. If he keeps up with her fancy footwork, she might mate with him. So a male fruit fly may be possessed of all the faculties he needs to survive and flourish, he may be as fertile as any other fly, and devilishly handsome to boot. But if he can't dance, his genes are going nowhere. Any mutation that reduces dancing ability is less likely to find its way into a baby *Drosophila*. So with every passing generation any mutations that reduce dancing ability will decline in frequency, as the lousy dancers will tend have fewer offspring than the good dancers.

## Gradualism

The impact of Darwin's *The Origin of Species* was immediate and profound. Many scientists who read it were immediately convinced of its explanatory power. Others remained sceptical, or in some cases downright hostile to the idea. On the whole, though, *The Origin* brought about a rapid change in the worldview of the biological community. By 1872 when the last edition of *The Origin* was published, Darwin could state with some satisfaction that 'almost every naturalist admits the great principle of evolution'. But although *The Origin* marked a turning point for the acceptance of the transformation of species and the evolutionary history of the natural world, it took considerably longer for Darwin's proposed mechanism of change—natural selection—to be universally accepted as the primary agent of adaptive evolution. For example, Thomas Henry Huxley, that tenacious defender of Darwinism, was 'prepared to go to the stake if requisite' for Darwin's arguments about the geographic and geological distribution of species. However, he was less convinced by Darwin's insistence on diversification by the accumulation of very many small heritable variations.

Darwin based his theory of evolution by natural selection on the principle of gradualism: that descent with modification occurred by continuous accumulation of many small changes. Darwin followed the uniformitarian principles of the founding father of modern geology, Sir Charles Lyell, in explaining deep history in terms of processes that could be witnessed today. Lyell's central argument can be summarized as 'the present is the key to the past'. He explained major geological events in the past, such as mountain building and change in the course of rivers, in terms of the gradual and continuous action of forces we can observe in action today, such as sedimentation, erosion, and uplift. Similarly, Darwin explained the massive changes in the biological world—the transformation of species over time—in terms of the accumulation of many small heritable variations leading to the formation of distinct varieties and races, processes that any naturalist could witness working on a small scale today. It is important to note that hypothesis of gradualism does not concern the speed of change. Darwin expected that different lineages would change at different rates, but that the process of modification was one of step-wise, gradual change. Just as Lyell had done with geology, Darwin insisted that we can explain the large biological changes of the past by considering the cumulative effect of many tiny differences, continuously accumulated over thousands of generations.

*We explore the pace of evolutionary change in Chapter 13*

But this gradualist explanation did not satisfy some scientists, including Huxley. He emphasized that there was no experimental evidence for the origin of new species, and thus that it was impossible to rule out a role for heritable changes of large effect in the process of adaptation and the formation of new kinds. It was not until after Darwin's lifetime that the power of natural selection to bring about evolutionary change was demonstrated by a combination of laboratory experiments, field studies and mathematical models. Laboratory studies, such as those on the fruit fly *Drosophila*, showed that new mutations arise constantly and that these mutations can rise in frequency under particular conditions or due to selective breeding. Field studies, such as those on predation of stripy-shelled snails in English hedgerows, showed that heritable variations can influence the chances of survival and reproduction of individuals in the wild (**Figure 7.4**). From the point of view of understanding patterns of molecular evolution, the most important advances came from setting population genetics within a mathematical framework.

## The power of selection

Natural selection is a numbers game, a consequence of a population of reproducing entities having heritable variations that result in different average numbers of copies. Yet while *The Origin* included a vast catalogue of observations and inferences, it contained no maths, not one single equation (and only one diagram, but

**Figure 7.4** Nice and stripy: the variation in banding patterns on the snail *Cepaea nemoralis* were initially interpreted as neutral variation, but classic studies in the 1950s showed that some banding patterns were more likely to end up as thrush food than others, suggesting a role for banding and colour in predator protection.



more of that in Chapter 11). In the 1920s and '30s, a triumvirate of scientists was instrumental in putting Darwin's theory of evolution by natural selection into a firm mathematical framework: R. A. Fisher, founder of many of the statistical techniques biologists use today, also a keen promoter of eugenics; Sewall Wright (**Figure 7.5**), whose analysis of agricultural breeding programmes led to new analytical methods and to the enduring 'fitness landscape' model of adaptive evolution; and J. B. S. Haldane, a brilliant and outrageous polymath who illuminated the implications of selection and whose socialist beliefs led him to be a committed popularizer of biology (**Hero 7**).

The mathematical theory of natural selection is based in probabilities. That is, it describes the likely set of outcomes of a repeated process, rather than predicting with certainty the outcome of any particular case. The important thing about thinking in terms of probabilities is that you cannot predict the fate of a particular mutation but you can make statements about the

**Figure 7.5** Sewall Wright and John Maynard Smith. According to Richard Lewontin, Wright (left) claimed that the theoretical work for which he is justly famous was really a diversion from his first love, which was guinea pigs.



overall expected outcome of particular kinds of mutations within a given population.

Let's think about the fate of a single mutation. A mutation arises in a single individual's genome. If this individual breeds, then that mutation has a chance of being copied to the next generation. If the individual does not reproduce, or all their offspring die, then that mutation will disappear. The loss of a new mutation might occur because the mutation itself reduces the chance of that individual successfully reproducing (negative selection). But many new mutations will be lost by chance events, when their carrier is hit by a bus, falls into a swamp, or fails to find sufficient water in a drought. Because every mutation must start at very low frequency, very many will be lost when their carrier fails to reproduce, regardless of the effect of the mutation on phenotype.

However, if a new mutation is not lost, and is copied to more individuals in the subsequent generations, then its fate is subject to both random and deterministic forces. The carriers of any given allele are subject to the same blind chance that affects us all, which can increase or decrease the frequency of an allele. But in addition, natural selection results in the increase in frequency of particular alleles at the expense of others due to the effect of those alleles on the probability of survival and reproduction. We are not going to look at any of the details of the mathematical theory of natural selection here, suffice to say that it was demonstrated that, in a large population, even a mutation with only a tiny relative advantage could be driven to fixation by natural selection.

In summary, mutations occur continuously, but most will be lost when their carriers fail to pass them on. Of those that are not lost, most will be either deleterious or harmless, but some will, by a lucky chance, confer slight benefits. We can expect some of these beneficial mutations to go to fixation by natural selection. The substitution of a mutation that makes an almost immeasurable difference to phenotype may seem a hopelessly inefficient means to evolutionary change. But the effect of natural selection plays out in populations consisting of hundreds, thousands or even millions of individual genomes, each of which has a small chance of producing a beneficial mutation, and each beneficial mutation has a small chance of going to fixation. And the ages of species are measured in hundreds of thousands of generations, so even a small rate of fixation would allow a species to accumulate a large net amount of change, even if each selected mutation had only a tiny effect.

*Why are most mutations harmful? See Chapter 3*

## *Novel mutations and standing variation*

We can actually witness the process of substitution in organisms with rapid generation turnover and high selection pressures. Human immunodeficiency virus (HIV) has an average mutation rate of 0.2 mutations per genome copy. This means that, on average, one in every five new HIV genomes is likely to carry a novel mutation. These mutations will normally be harmful to their carrier (reduce the virus' chance of replication) or occasionally harmless (have no effect on the chance of the virus replicating). But every now and then, one of these new mutations might increase the chance of the mutant genome being copied more than the alternative genomes in the population. For example, a mutation in the HIV genome that makes the viral particle unrecognizable to the immune system will allow its carrier to replicate freely, untrammelled by its host's defences, and therefore produce more descendants than virus genomes carrying alternative alleles that allow them to be recognized and suppressed by the immune system.

Large populations have more chance of beneficial mutations arising, simply because they contain more genomes, each of which has a chance of producing a new, potentially useful, mutation. Compare an infected human with one hundred HIV genomes in their body to a human with one million HIV genomes. The large population of HIV will produce ten thousand times the number of new mutations per day than the small population will. So even if the chance of any given mutation being beneficial remains the same, there is more chance that at least one of those million viruses will produce a mutant offspring that can outwit the immune system or resist the effect of drugs and go on to produce a whole new generation of HIV viruses.

But selection does not always need to wait for a new mutation to occur. The response to selection will often be to promote the frequency of variants already in the population: that is, selection can act on standing variation. A change in the environment may give existing variants a new advantage (Case Study 7.1). For example, in the famous case of Darwin's finches on the Galapagos islands, in a drought year the average beak size increased because big-beaked birds could crack harder seeds and could therefore eat seeds that the smaller-beaked birds couldn't (Figure 7.6). The drought did not cause those larger-beaked variants to occur, they were already

**Figure 7.6** Does my beak look big in this drought? This unprepossessing little black bird is a hero of evolutionary biology. This is partly because the medium ground finch (*Geospiza fortis*) is one of the famous Darwin's finches, a classic example of an adaptive radiation. Darwin's finches are all descended from a single species that colonized the Galapagos Islands and gave rise to over a dozen species that diversified into a range of ways of life including seed-eaters, insectivores, cactus-feeders and, this is barely believable but it is true, vampire finches that drink the blood of living seabirds. But Darwin's finches are also important in evolutionary biology because careful observation of *G. fortis* over many years has provided a classic demonstration of the power of natural selection. In a drought year most of the population died, but those that survived had larger than average beaks, so they could crack larger, tougher seeds. This differential survival of larger-beaked individuals resulted in a heritable increase in beak size in only a few generations. Then, a few decades later when the island was invaded by a larger species of Darwin's finch (*G. magnirostris*), beak sizes shrank again as competition with the larger birds reduced the availability of larger seeds.

there, even in the good years. But in a drought year, they stood a better chance of reproducing than their small-beaked comrades, so there were relatively more copies of alleles contributing to large-beakedness in the generation following the drought. Strong directional selection can reduce the supply of standing variation in the population because, as selection pushes a single variant to fixation, all alternative versions of that trait are lost. Many artificial selection experiments report a decline in the response to selection over time, which may reflect a depletion of the available genetic variation in the population.

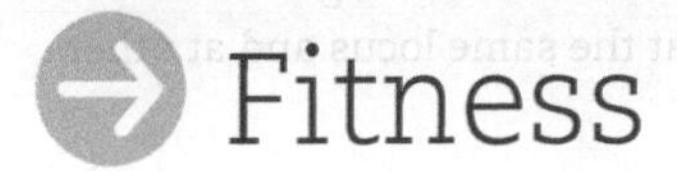

# Fitness

John Maynard Smith (**Figure 7.5**) was one of the leading evolutionary thinkers of the last century. His poor eyesight prevented him from joining the army to fight in the Second World War, so instead he contributed to the war effort by designing aircraft, a field that would lead him to evolutionary biology through consideration of flight in animals. He commented that, in his generation, bad eyesight was a selective advantage in that it prevented him from getting shot. The point of this story is that whether a trait is 'good' or 'bad' depends on the circumstances. Fitness is relative: the selective advantage of an allele must be considered relative to other alternative alleles in the population and the particular environment in which it arises (**TechBox 7.1**). The 'environment' of an allele has many aspects. We immediately think of the external conditions of the population—climate, predators, food sources—but environment also includes other members of the population and even the other mutations in the genome. And if the environment changes, the fitness of the mutation may change with it (**Case Study 7.2**).

We can see the dramatic effect of changing environment on the fitness of an allele by considering the human metabolic disorder, phenylketonuria (PKU). PKU is caused by mutation in the phenylalanine hydroxylase gene (*PAH*). There are over 400 different mutations in the *PAH* gene that cause PKU, but the phenotype is the same for all of them. A child born without a working copy of *PAH* cannot make the phenylalanine hydroxylase enzyme and therefore cannot metabolize the amino acid phenylalanine, which builds up in their brain causing the gradual onset of mental retardation. Since a person affected with PKU is unlikely to reach reproductive age, the mutation will be subject to negative selection. But in the 1960s, the Guthrie (heel-prick) test for PKU was introduced (**Figure 3.15**). Now, in many parts of the world, every newborn child is tested for PKU. Those that test positive are put on a special diet with no phenylalanine, thus avoiding brain damage. In places where the Guthrie test is used to detect PKU and the environment modified accordingly, there is no reason someone carrying the PKU mutation cannot reproduce, so there is no longer strong negative selection against it.

One aspect of an allele's environment that can vary considerably from one generation to the next is the other individuals in the population. In some situations, the selective advantage of a mutation can change with the frequency of other alleles, for example a particular variant may be advantageous when it is rare, but disadvantageous when common. Frequency-dependent selection can maintain polymorphism: if one allele is only advantageous when it is relatively rare, it will increase by selection only until it becomes too common, then its selective advantage will cease.

**Figure 7.7** Males of the side-blotched lizard *Uta stansburiana* come in three kinds, each characterized by a mating strategy and a throat colour. The selective advantage of each morph depends on the relative frequency of the others: in a population with lots of orange-throats it's better to be a yellow than a blue, but when there are a lot of blues around, it is better to be an orange than a yellow. Frequencies of the three morphs fluctuate on a four to five year cycle. When Barry Sinervo and Curtis Lively discovered this, they 'looked at each other and . . . said "Dude! It's the rock-scissors-paper game!"'
Photograph courtesy of Barry Sinervo.

Maynard Smith once hypothesized that a population could maintain a polymorphism if the variants were like the 'rock, scissors, paper' game (rock wins against scissors, scissors win against paper, paper wins against rock). Several natural examples of this have since been described (**Figure 7.7**). For example, some strains of the bacterium *Escherichia coli* produce colicin, a chemical which kills competing *E. coli* strains by interfering with their cell membranes. Colicin-producing strains (C) will have a higher reproductive output in a population of colicin-sensitive strains (S), because fewer Ss will survive to reproduce. Colicin-resistant strains (R) can survive in the presence of colicin due to changes in their cell membrane receptors. In a population of Cs, Rs have a reproductive advantage, because they avoid the cost of producing colicin. But the modified receptor proteins that confer resistance to colicin are less efficient at other metabolic tasks, so resistant strains (R) have a slower growth rate than sensitive strains (S). Therefore, in a population of Ss, Rs will be at a selective disadvantage, because they can't grow as fast and have no advantage in the absence of colicin. So now we see that C beats S, S beats R, and R beats C. This non-transitive loop means that if any one strain increases in frequency it will decrease the frequency of the next strain in the chain, which will cause an increase in frequency of the next in the chain, which will cause a decrease in the first strain (**Figure 7.8**). This probably explains why all three variants can coexist in natural populations without any variant going to fixation.

## Genetic background

The environment of a mutation must also include the genetic context within the individual because the influence of a mutation on an individual's chance of reproduction can depend on which other mutations are present in the same genome. The genetic environment within an individual (referred to as the genetic background) includes alleles at the same locus and at other places in the genome.

We can illustrate the influence of other mutations in the genome on response to selection using examples from the human haemoglobins (proteins that carry oxygen in the blood). There are multiple haemoglobin genes in the human genome and the effect of a mutation in one gene can be influenced by mutations carried in other globin genes. Each haemoglobin molecule is made of four amino acid chains, two alpha and two beta (**Figure 7.9**). There is a family of related genes that produce these alpha and beta chains. Mutations in these genes can cause a range of illnesses, collectively termed haemoglobinopathies, due to inadequate transport of oxygen around the body. When considered on a global scale, haemoglobinopathies are the most common single-gene diseases of humans. It is estimated that more than five per cent of the world population carries a mutation for a form of haemoglobinopathy. But the clinical outcome of mutations in haemoglobin genes depends both on the type of mutation and the genetic background that the mutation occurs in.

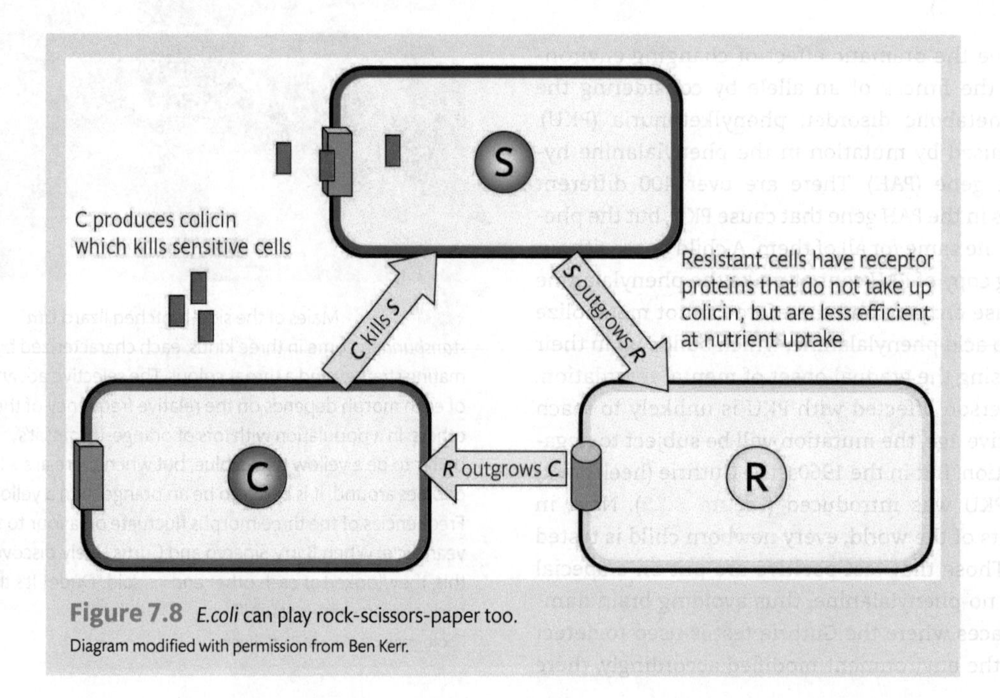


**Figure 7.8** *E.coli* can play rock-scissors-paper too.
Diagram modified with permission from Ben Kerr.

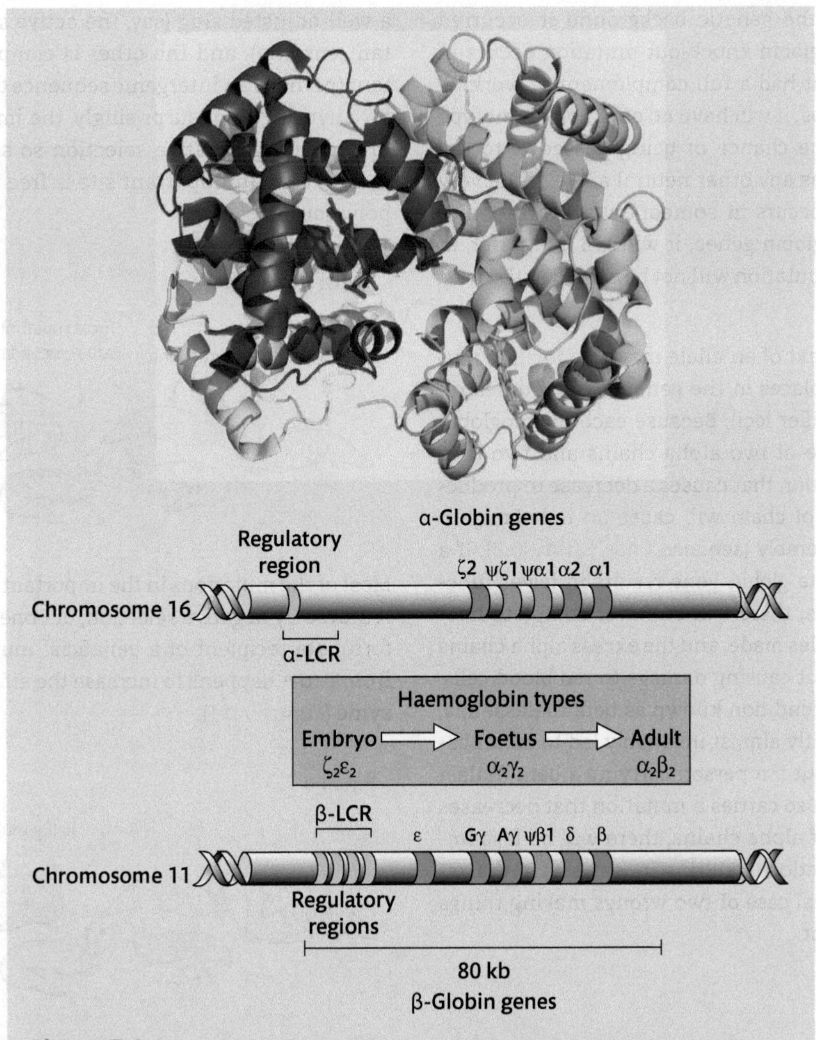


**Figure 7.9** Haemoglobin, the protein that carries oxygen in your blood, consists of four chains of two types, but the composition of haemoglobin changes throughout development. The genes in the two globin clusters in humans are activated in sequence. In the alpha globin cluster, the zeta gene is expressed in early embryogenesis, then alpha 1 and alpha 2 take over. In the beta globin gene cluster, epsilon is expressed in embryogenesis, gamma during foetal development, then beta from infancy onward (small amounts of delta are produced in both children and adults).

Humans normally carry four copies of the alpha globin gene, two on each copy of chromosome 16 (**Figure 7.9**). But because alpha globin genes are tandem (side-by-side) copies, they are prone to deletion by unequal crossing over (**Figure 5.10**). Individuals with four or three copies of the gene produce enough alpha chains to have sufficient haemoglobin function. But individuals with only two working copies of the alpha globin gene do not produce enough alpha chains, so cannot manufacture sufficient haemoglobin. Without enough haemoglobin to transport oxygen in the blood, individuals with only two functioning alpha globin genes can suffer from anaemia.

Someone with only one functioning alpha globin gene will have severe anaemia, and individuals with no working copies of alpha globin will usually die before birth. So the effect of a mutation that ruins or removes one copy of the alpha globin gene will depend on how many other working copies of the alpha globin gene are in that genome.

Now consider the effect of a new mutation that knocks out an alpha globin gene in a particular individual. Can you tell me what the effect of this mutation will be on the individual—harmless, severe, lethal? In this particular case, you can't tell me the effect of the mutation until you know the genetic background it occurred in. If this alpha globin knock-out mutation occurs in an individual that had a full complement of working alpha globin genes, it will have no effect on phenotype and has the same chance of being passed onto the next generation as any other neutral allele. If the very same mutation occurs in someone with three other defective alpha globin genes, it will kill its carrier, in which case that mutation will not be copied to the next generation.

The genetic context of an allele may also include mutations at other places in the genome (sometimes referred to as modifier loci). Because each haemoglobin molecule is made of two alpha chains and two beta chains, any mutation that causes a decrease in production of one type of chain will cause an imbalance in haemoglobin assembly (see also **Case Study 6.2**). If a mutation in a beta globin gene results in fewer functional beta chains, there will be fewer complete haemoglobin molecules made, and the excess alpha chains will precipitate out causing damage to red blood cells. This produces a condition known as beta thalassemia, which until recently almost invariably led to death before adulthood. But if a person carrying a beta thalassemia mutation also carries a mutation that decreases the production of alpha chains, there will be less imbalance in production, resulting in less severe thalassemia—an unusual case of two wrongs making things a bit closer to right.

## Linkage

The environment of a mutation also includes its physical location in the genome. Recombination shuffles alleles among chromosomes (**Figure 5.6**), but mutations that sit next to each other in the genome will tend to be inherited together. We take advantage of the fact that closely occurring alleles are inherited together when we use genetic markers to identify genes associated with particular traits or diseases (**TechBox 3.1**). We see in Chapter 3 that some SNPs are informative markers for disease not because they have any causal role in the disease, but because they occur close enough to the relevant disease allele that they are usually inherited along with it (see Chapter 3). The corollary of this is that the fate of a particular SNP can depend on neighbouring alleles in the genome.

Say we have a population of slugs and we want to track allele frequencies at two sites in the slug genome, which are close enough together that alleles at these two loci are almost always inherited together (**Figure 7.10**). One of these sites is important for being a well-adjusted slug (say, the active site of an important enzyme), and the other is completely unimportant (perhaps an intergenic sequence that doesn't code for anything). Not surprisingly, the important locus is under strong negative selection so substitutions are rare, but the unimportant site is free to change so it's polymorphic.

**Figure 7.10**

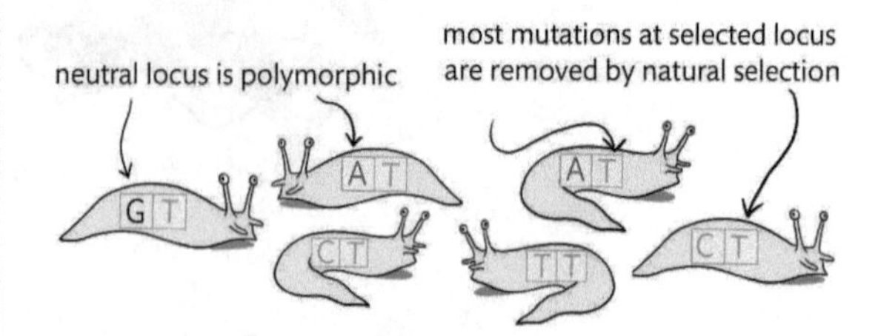


Most of the mutations in the important sequence will be removed by negative selection, but one lucky slug is the fortunate recipient of a beneficial mutation: a change from T to A happens to increase the efficiency of the enzyme (**Figure 7.11**).

**Figure 7.11**

This happy mutation could have occurred in any slug but, by chance, it just happened to occur in a slug that had an A at the neutral locus. The neutral A still doesn't make any difference to the slug, but it has the good fortune to be attached to the selected A that gives the slug an advantage over its comrades. So next generation, this slug leaves more descendants that those with T at the selected locus (**Figure 7.12**).

**Figure 7.12**

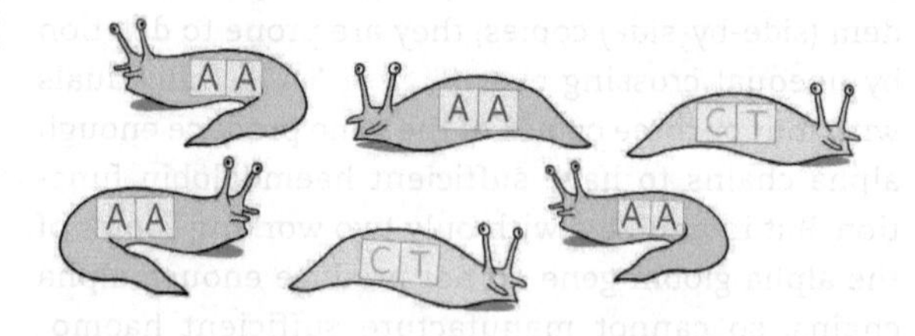


Eventually, the beneficial mutation goes to fixation, and because it takes the linked neutral allele along with it, the polymorphism at both sites is lost (**Figure 7.13**).

Figure 7.13

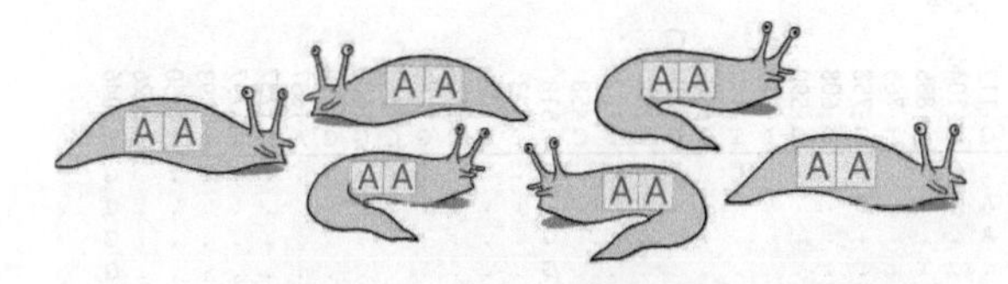


Of course, in future, both sites may be subject to further modification and, in particular, the neutral site will begin to accumulate substitutions. But the selective sweep has reduced the amount of allelic diversity in the linked sites—not only do all the slugs carry the selected A but also the neutral A. So, compared to unlinked neutral sites elsewhere in the genome, the neutral sites linked to a locus under selection will have reduced nucleotide diversity. This hypothetical slug example illustrates how a mutation occurring next to a beneficial allele can be swept to fixation along with it, hitchhiking on the positive selection of its neighbour. Conversely, any mutation occurring next to a negatively selected allele may hitchhike its way to oblivion, being removed from the population when carriers of the negatively selected allele fail to reproduce.

We can see the effects of hitchhiking in patterns of variation across the human genome. For example, there are a number of alleles of the human beta globin gene that confer some resistance to malaria. One of these, known as the C allele (or HbC), contains a mutation from G to A, which changes one amino acid in the beta globin chain from glutamic acid to lysine. HbC appears to be a fairly recent mutation, possibly only 3,000 years old, but it reaches high representation in some parts of West Africa where malaria is prevalent, where up to half of the population can carry the HbC variant of the beta haemoglobin gene. Someone who carries two copies of HbC (i.e. homozygous for HbC at the beta globin locus) is much less likely to get sick with malaria than someone with alternative versions of the beta globin gene. Heterozygotes, with only one copy of the HbC allele, also have some protection against malaria. The HbC allele differs from other variants of the beta globin gene at a number of different sites (**Figure 7.14**). The malaria-resistant mutation seems to have originally occurred on a version of the beta globin gene that, by chance, happened to have a C instead of a T twenty bases upstream of the start of the gene. As the HbC allele is increasing in frequency due to positive selection of the A at site 16, it is taking the C at site –2906 with it. It has been predicted that the HbC allele will go to fixation due to its high selective advantage, in which case the C at site –2906 is likely to become fixed too, even if it has no effect on fitness (**TechBox 7.2**).

## Are humans still evolving?

> ". . . *Endless forms most beautiful and most wonderful have been, and are being, evolved.*"
>
> Darwin, C. (1872) *On the origin of species by means of natural selection, or the preservation of favoured races in the struggle for life* 6th edn. John Murray.

I once attempted to learn Latin, but, hampered by a striking ineptitude for languages, I seem to have retained only two Latin phrases in my memory. My favourite is *auribus teneo lupum*—I hold the wolf by the ears. This phrase describes being in a tricky situation: while you are holding the wolf by the ears, it can't bite you, but in order to run away you need to let go of the ears . . . and then the wolf can bite you. Admittedly, that's not much to show for a year of Latin classes (the only other phrase I remember—*Sum bos triturans* (I am the threshing ox)—is of no use whatsoever). But *auribus teneo lupum* nicely describes the trepidation I feel when faced with discussions of human evolution, because they are prone to misinterpretation, liable to offend, and hampered by deeply held beliefs. However, I think that the examples we have used in this chapter, which have centred on human diseases, lead us naturally to the broader topic of molecular evolution in our own species. Humans make a useful case for reviewing what we have learned about molecular evolution, because we will have to consciously step back from the phenotype and ask ourselves what is happening at the level of the genotype. The point of the following discussion is to summarize and reinforce the conclusions of this chapter: molecular evolution is critically dependent on the environment (including external conditions, population size and composition, and genetic background), so the fitness of a particular variant can only be meaningfully judged with respect to the particular situation in which it is found.

***We look at the influence of population size on patterns of substitution in Chapter 8***

There is a common notion that human beings stopped evolving in the Pleistocene, because when they developed tool use, symbolic communication, and abstract thought, they rose above natural selection (sadly, this argument is all too often used to claim an innate propensity for uncivilized behaviour). But if we use Darwin's definition of evolution—descent with modification—then we could only say that humans had stopped evolving if we knew that allele frequencies were no longer going to change. As far as I can see, the only way to achieve such evolutionary stasis at a molecular genetic level is to have such a severe and sustained degree of inbreeding that all variation is removed from

Sabe

| | Sample | Country/Ethnicity | | -3092 | -3072 | -3062 | -2917 | -2903 | -2883 | -2801 | -2589 | -2400 | -2181 | -1944 | -1934 | -1927 | -1873 | -1847 | -1588 | -1491 | -1124 | -1044 | -835 | -765 | -758 | -608 | -590 | -543 | -392 | 6 | 18(HbC) | 85 | 187 | 458 | 518 | 523 | 702 | 1108 | 1256 | 1301 | 1603 | 1654 | 1657 | 1737 | 1797 | 1893 | 1950 | 2026 | 2046 |
|---|---|---|---|---|---|---|---|---|---|---|---|---|---|---|---|---|---|---|---|---|---|---|---|---|---|---|---|---|---|---|---|---|---|---|---|---|---|---|---|---|---|---|---|---|---|---|---|---|---|
| | | | Consensus | A | C | C | C | T | T | T | G | C | C | C | A | T | T | G | C | T | G | C | A | T | T | T | C | A | T | C | G | T | T | C | T | C | A | T | G | C | G | G | A | C | T | A | C | T | T |
| | Ive05 | Ivory Coast | | . | . | . | . | . | . | . | . | . | . | . | . | G | . | . | . | . | . | G | . | . | G | . | T | C | C | . | . | . | . | . | G | . | . | . | . | T | . | . | . | . | . | . | . | . | G |
| | Cmr13 | CAR | Baka | . | G | . | . | . | . | . | . | . | . | . | . | . | . | . | . | . | A | . | . | . | G | . | . | C | . | . | . | . | . | . | G | . | . | . | . | . | . | . | . | . | . | . | . | . | G |
| | Sen42 | Senegal | Mandinka | . | G | . | G | . | C | . | . | . | G | A | . | . | . | . | T | C | A | . | . | . | . | . | . | . | . | . | . | . | C | . | G | . | . | . | . | . | . | . | . | . | . | . | . | . | G |
| | Dgn37 | Mali | Dogon | . | G | . | G | . | C | . | . | . | G | . | . | . | . | . | . | C | . | . | T | . | . | . | . | . | . | . | . | . | . | . | G | . | . | . | . | . | . | . | . | . | . | . | . | . | G |
| | Ghn023 | Ghara | Ga | . | G | . | G | . | C | . | . | . | G | . | . | . | . | . | . | C | . | . | T | . | . | . | . | . | . | . | . | . | . | . | G | . | . | . | . | . | . | . | . | . | . | . | . | . | G |
| | Ghn40 | Ghara | Ga | . | G | . | G | . | C | . | . | . | G | . | . | . | . | . | . | C | . | . | T | . | . | . | . | . | . | . | . | . | . | . | G | . | . | . | . | . | . | . | . | . | . | . | . | . | G |
| | Ghn195 | Ghara | Ewa | . | G | . | G | . | C | . | . | . | G | A | . | . | . | . | T | . | A | . | . | . | C | C | . | . | . | . | . | . | . | . | G | . | . | . | . | . | . | . | . | . | . | . | . | . | G |
| | Ghn133 | Ghara | Famie | . | . | . | . | . | . | . | . | . | . | . | . | . | . | . | . | . | . | . | T | . | C | . | . | . | . | . | . | . | . | . | G | . | . | . | . | . | . | . | . | . | . | . | . | . | G |
| | Sen31 | Senegal | Mandinka | . | G | . | G | . | C | . | . | . | G | A | . | . | . | . | T | C | . | . | . | . | . | . | . | . | . | . | . | . | . | . | G | . | . | . | . | . | . | . | . | . | . | . | . | . | G |
| | Dgn83 | Mali | Dogon | . | . | . | . | . | . | . | . | . | . | . | . | G | . | . | . | . | . | . | . | . | . | C | . | . | . | . | . | . | . | . | G | . | . | . | . | . | A | . | . | . | . | . | . | . | G |
| | Cmr15 | CAR | Baka | . | . | . | . | . | . | . | . | . | . | . | . | . | . | . | . | . | . | . | . | . | . | C | . | . | . | . | . | . | . | . | G | . | . | . | . | . | A | . | . | . | . | . | . | . | G |
| | Gna27 | Ghana | | . | . | . | . | . | . | . | . | . | . | . | . | . | . | . | . | . | . | . | . | . | . | C | . | . | . | . | . | . | . | . | G | . | . | . | . | . | A | . | . | . | . | . | . | . | G |
| | S782 | Nigaria | | . | . | . | . | . | . | . | . | . | . | . | . | . | . | . | . | . | . | . | . | . | . | C | . | . | . | . | . | . | . | . | G | . | . | . | . | . | A | . | . | . | . | . | . | . | G |
| | Sen50 | Senegal | Mandinka | . | . | . | . | . | . | . | . | . | . | . | T | . | . | . | . | C | . | G | . | . | . | . | T | C | . | . | . | . | . | . | . | . | . | . | . | . | . | . | . | . | . | . | . | . | G |
| | Dgn06 | Mali | Dogon | . | . | . | . | . | . | . | . | . | . | . | . | . | . | . | . | . | . | . | . | . | . | C | . | . | . | . | . | . | . | . | . | . | G | . | . | . | . | . | . | . | G | . | . | . | . |
| | Dgn67 | Mali | Dogon | G | . | . | . | . | . | . | . | . | . | . | . | C | . | . | . | . | . | . | . | . | C | . | . | . | . | . | . | . | . | . | . | . | . | . | . | . | . | . | . | . | . | . | . | . | . |
| | Cmn097 | Cameroon | Ngoumba | . | G | . | . | . | C | C | . | . | G | A | . | . | . | . | . | C | . | G | . | . | C | C | . | . | C | T | . | . | . | . | . | . | . | C | . | . | . | . | . | . | . | . | . | . | . |
| HbA | JK1033 | DRC | Mbuti | . | G | . | . | . | C | . | . | . | G | A | . | . | . | . | T | . | . | G | . | . | C | C | . | . | C | T | . | . | . | . | . | . | . | C | . | . | . | . | G | . | . | . | . | . | . |
| | Nov24 | Egypt | | . | . | . | . | . | . | . | . | . | . | . | . | . | . | . | . | . | . | . | . | . | . | . | . | . | C | T | . | C | . | G | . | T | . | C | . | . | . | C | . | A | . | . | . | . | . |
| | Ghn017 | Ghana | Ga | . | G | . | G | . | C | . | . | . | G | A | . | . | . | T | . | . | . | G | . | G | . | . | . | . | C | T | . | . | . | G | . | T | . | C | . | . | . | C | . | A | . | . | . | . | . |
| | Ghn009 | Ghana | Ga | . | G | . | . | . | C | . | . | . | G | A | . | . | . | . | T | . | A | G | . | . | G | C | . | . | . | . | . | . | . | . | . | . | . | . | . | . | . | . | . | . | . | T | . | . | . |
| | Ivc11 | IvaryCoast | | . | G | . | . | . | . | . | . | . | G | A | . | . | . | . | T | C | . | G | . | . | G | C | . | . | . | . | . | . | . | . | . | . | . | . | . | . | . | . | . | . | . | T | . | . | . |
| | G25 | Gambia | | . | . | . | . | . | . | . | . | . | . | . | . | C | . | . | . | . | . | . | . | . | C | C | . | . | . | . | . | . | . | . | . | . | . | . | . | . | . | . | . | . | . | T | . | . | . |
| | Sen10 | Senegal | Weler | . | G | G | G | . | C | . | . | . | G | A | . | . | . | . | T | . | . | . | . | . | C | C | . | . | . | . | . | . | . | . | . | . | . | . | . | . | . | . | . | . | . | T | . | . | . |
| | Dgn52 | Mali | Dogon | . | G | . | G | . | C | . | . | . | G | A | . | . | . | . | T | C | . | . | . | G | C | . | . | . | . | . | . | . | . | . | . | . | . | . | . | . | . | . | . | . | . | T | . | . | . |
| | G08 | Gambia | Jela | . | G | . | G | . | C | . | . | . | G | A | . | . | . | . | T | C | . | G | . | . | . | . | . | . | . | . | . | . | . | . | . | . | . | . | . | . | . | . | . | . | . | T | T | . | . |
| | Ivc16 | Ivary Coast | | . | G | . | . | . | . | . | . | . | . | . | . | . | . | . | . | . | . | . | . | . | . | . | . | . | . | . | . | . | . | . | . | . | . | . | . | . | . | . | . | . | . | T | . | . | . |
| | Ivc04 | Ivary Coast | | . | G | G | G | . | C | . | . | . | G | A | . | . | . | . | T | . | . | . | . | . | C | . | . | C | . | . | . | . | . | . | . | . | . | . | . | . | . | . | . | . | . | T | . | . | . |
| | Ghn117 | Ghana | Ewe | . | . | . | . | . | . | . | . | . | . | . | . | C | . | . | . | C | . | . | . | . | C | . | . | C | . | . | . | . | . | . | . | . | . | . | . | . | . | . | . | . | . | . | . | . | . |
| | Cmn087 | Camroon | Ngoumbe | . | . | . | . | . | . | . | . | . | . | . | . | . | . | . | . | . | . | . | . | . | . | . | . | C | . | . | . | . | . | . | . | . | . | . | . | . | . | . | . | . | . | . | . | . | . |
| | Dgn66 | Mali | Dogon | . | . | . | . | . | . | . | . | . | . | . | T | . | . | . | . | C | . | G | . | . | C | . | T | C | . | . | . | . | . | . | . | . | . | . | . | . | . | . | . | . | . | . | . | . | . |
| | G37 | Gambia | | . | G | . | G | . | C | . | . | . | G | A | . | . | . | . | T | C | . | . | . | . | C | . | T | C | . | . | . | . | . | . | . | . | . | . | . | . | . | . | . | . | . | . | . | . | . |
| | Dgn58 | Mali | Dogon | . | . | . | . | . | . | . | . | T | . | . | . | . | . | . | . | . | . | . | . | . | C | . | T | C | . | . | . | . | . | . | . | . | . | . | . | . | . | . | . | . | . | . | . | . | . |
| | Ivc18 | Ivory Coast | | . | . | . | . | . | . | . | . | . | . | . | . | G | . | . | . | . | . | . | . | . | . | . | T | C | . | . | . | . | . | . | . | . | . | . | . | . | . | . | . | . | . | . | . | . | . |
| | Dgn99 | Mali | Dogon | . | . | . | . | . | . | . | . | . | . | . | . | . | . | . | . | . | . | . | . | . | . | . | T | C | . | . | . | . | . | . | . | . | . | . | T | . | . | . | . | . | . | . | . | C | . |
| | Dgn06 | Mali | Dogon | . | . | . | . | C | . | . | A | . | . | . | . | . | . | . | . | . | . | . | T | . | . | . | T | C | . | . | A | . | . | . | . | . | . | . | . | . | . | . | . | . | . | . | . | . | . |
| | Dgn31a | Mali | Dogon | . | . | . | . | C | . | . | A | . | . | . | . | . | . | . | . | . | . | . | T | . | . | . | T | C | . | . | A | . | . | . | . | . | . | . | . | . | . | . | . | . | . | . | . | . | . |
| | Dgn31b | Mali | Dogon | . | . | . | . | C | . | . | A | . | . | . | . | . | . | . | . | . | . | . | T | . | . | . | T | C | . | . | A | . | . | . | . | . | . | . | . | . | . | . | . | . | . | . | . | . | . |
| | Dgn37 | Mali | Dogon | . | . | . | . | C | . | . | A | . | . | . | . | . | . | . | . | . | . | . | T | . | . | . | T | C | . | . | A | . | . | . | . | . | . | . | . | . | . | . | . | . | . | . | . | . | . |
| | Dgn52 | Mali | Dogon | . | . | . | . | C | . | . | A | . | . | . | . | . | . | . | . | . | . | . | T | . | . | . | T | C | . | . | A | . | . | . | . | . | . | . | . | . | . | . | . | . | . | . | . | . | . |
| | Dgn58 | Mali | Dogon | . | . | . | . | C | . | . | A | . | . | . | . | . | . | . | . | . | . | . | T | . | . | . | T | C | . | . | A | . | . | . | . | . | . | . | . | . | . | . | . | . | . | . | . | . | . |
| | Ghn009 | Ghana | Ga | . | . | . | . | C | . | . | A | . | . | . | . | . | . | . | . | . | . | . | T | . | . | . | T | C | . | . | A | . | . | . | . | . | . | . | . | . | . | . | . | . | . | . | . | . | . |
| | Ghn017 | Ghana | Ga | . | . | . | . | C | . | . | A | . | . | . | . | . | . | . | . | . | . | . | T | . | . | . | T | C | . | . | A | . | . | . | . | . | . | . | . | . | . | . | . | . | . | . | . | . | . |
| | Ghn023 | Ghana | Ga | . | . | . | . | C | . | . | A | . | . | . | . | . | . | . | . | . | . | . | T | . | . | . | T | C | . | . | A | . | . | . | . | . | . | . | . | . | . | . | . | . | . | . | . | . | . |
| | Ghn154a | Ghana | Famle | . | . | . | . | C | . | . | A | . | . | . | . | . | . | . | . | . | . | . | T | . | . | . | T | C | . | . | A | . | . | . | . | . | . | . | . | . | . | . | . | . | . | . | . | . | . |
| HbC | Ghn154b | Ghana | Famle | . | . | . | . | C | . | . | A | . | . | . | . | . | . | . | . | . | . | . | T | . | . | . | T | C | . | . | A | . | . | . | . | . | . | . | . | . | . | . | . | . | . | . | . | . | . |
| | Ivc16 | Ivory Coast | | . | . | . | . | C | . | . | A | . | . | . | . | . | . | . | . | . | . | . | T | . | . | . | T | C | . | . | A | . | . | . | . | . | . | . | . | . | . | . | . | . | . | . | . | . | . |
| | Nov24 | Egypt | | . | . | . | . | C | . | . | A | . | . | . | . | . | . | . | . | . | . | . | T | . | . | . | T | C | . | . | A | . | . | . | . | . | . | . | . | . | . | . | . | . | . | . | . | . | . |
| | Ghn195 | Ghana | Ewe | . | . | . | . | G | . | . | . | . | . | . | . | . | . | . | . | . | . | . | T | . | . | . | T | C | . | . | A | . | . | . | . | . | . | . | . | . | . | . | . | . | . | . | . | . | . |
| | Ghn117 | Ghana | Ewe | . | . | . | . | . | . | . | . | . | . | . | . | . | . | . | . | . | . | . | T | . | . | . | T | C | . | . | A | . | . | . | . | . | . | . | . | . | . | . | . | . | . | . | . | . | . |
| | Ghn133 | Ghana | Famle | . | . | . | . | . | . | . | . | . | . | . | . | . | . | . | . | . | . | . | T | . | . | . | T | C | . | . | A | . | . | . | . | . | . | . | . | . | . | . | . | . | . | . | . | . | . |
| | Ghn40 | Ghana | Ga | . | . | . | . | . | . | . | . | . | . | . | . | . | . | . | . | . | . | . | T | . | . | . | T | C | . | . | A | . | . | . | . | . | . | . | . | . | . | . | . | . | . | . | . | . | . |
| | S762 | Nigeria | | . | . | . | . | . | . | . | . | . | . | . | . | . | . | . | . | . | . | . | T | . | . | . | T | C | . | . | A | . | . | . | . | . | . | . | . | . | . | . | . | . | . | . | . | . | . |
| | Dgn66 | Mali | Dogon | . | G | . | G | . | C | . | . | . | G | A | . | . | A | . | . | . | . | . | T | . | . | . | T | C | . | . | A | . | . | . | . | . | . | . | . | . | . | . | . | . | . | . | . | . | . |
| | Dgn99 | Mali | Dogon | . | . | . | . | . | . | . | . | . | . | . | . | . | . | . | . | . | . | . | . | . | G | . | . | C | . | . | A | . | . | . | . | . | . | . | . | . | . | . | . | . | . | . | . | . | . |
| | Gdn83 | Mali | Dogon | . | . | . | . | . | . | . | . | . | . | . | . | C | . | . | . | . | . | . | . | . | . | . | . | . | . | . | A | . | . | . | . | . | . | . | . | . | . | . | . | . | . | . | . | . | . |
| | Chimp | | | . | . | . | . | . | . | . | . | . | G | A | . | . | C | . | . | . | A | . | T | . | . | . | T | G | . | . | . | . | . | . | . | . | . | . | . | . | . | . | . | . | . | . | . | . | . |

**Figure 7.14** This alignment shows a short region of the β-haemoglobin gene, sequenced from over 50 people from different parts of Africa. The HbC allele provides some protection from malaria in the heterozygous state and relatively mild haemoglobinopathy in the homozygous state (unlike the HbS allele which gives protection when heterozygous but sickle-cell anaemia in the homozygous state). You can see that there is a lot of genetic variation in the β-haemoglobin gene, even in this very short sequence of an important gene. HbA is the wild-type (most common, or 'normal' variant) of β-haemoglobin. HbC differs from HbA by a mutation that causes an amino acid change in the sixth position of the protein, but it also has a number of other nucleotide differences. This suggests that the advantageous β6Glu→Lys mutation occurred on a chromosome that had a number of other mutations that are now sweeping to fixation by hitchhiking on the selective advantage of the HbC mutation.



the population, producing a true-breeding line of genetically identical clones, then somehow prevent the occurrence or establishment of any new mutations. In addition to producing a very dull species, I suspect the human race would expire from the effects of homozygous recessive diseases long before such homogeneity was achieved.

We began this chapter by considering the ubiquity of variation in natural populations. Genetic variation is fundamental to evolution. It is the continuous generation of variation by mutation, and the substitution of some fraction of these mutations by selection and drift, that drives molecular evolution. SNP databases show us there is plenty of genetic variation in human populations (**TechBoxes 3.1** and **3.2**). In fact, we can confidently predict that, with very few exceptions, each of us has a unique genome. What happens to all this variation? Just as in any other species, the fate of alleles in human populations is determined by chance and selection. As with any other population of template copiers, any heritable difference that reduces the chance of reproduction (such as a deleterious mutation in the alpha globin gene) will tend to decrease in frequency, and any heritable difference increasing the chance of making copies of itself (such as a beta globin allele that provides protection against malaria) will tend to increase in frequency.

*Chance effects on allele frequencies drift are covered in Chapter 8*

If you doubt that selection is still acting to shape human evolution, think about the global impact of malaria (**Figure 7.15**). Half of the world's population are at risk of malaria. Hundreds of millions are infected every year. Each year around half a million people die of malaria, over half of them children. This is equivalent to one child dying of malaria every minute. In some countries malaria is the leading cause of death. It has been suggested that the scale of impact of malaria is such that it has been one of the major selection pressures on many human populations for thousands of years. We have seen how malaria exerts a tangible effect on human genetic variation: consider the relatively recent mutation that gave the HbC allele its benefits against malaria, which is well on its way to fixation in some parts of world.

But even rapid changes in allele frequency, such as the rise of the HbC allele, occur over many generations. This means that current allele frequencies in a population may carry the imprint of past selection. For example, in the 1940s it was noted that Italians in New York had a surprisingly high frequency of beta thalassemia. Some researchers interpreted this pattern as evidence that some populations suffered a higher mutation rate than others, giving rise to a greater frequency of heritable disease. But J. B. S. Haldane (**Hero 7**) offered an alternative explanation. He proposed 'the malaria hypothesis': that the high frequency of certain blood disorders was a result of heterozygote advantage (also known as heterosis). Alleles that caused illness in the homozygous state could be maintained in the population due to the relative advantage that heterozygote carriers had in resisting malaria. Given that malaria was endemic in Mediterranean Europe until the 1940s, the higher fitness of heterozygotes for beta thalassemia alleles could result in carriers having relatively more children. Some of these children would be healthy heterozygotes who inherited their parents' malaria resistance, but unfortunately some children would end up with two copies of the allele, which would give them thalassemia. Sickle cell anaemia is the classic textbook example of heterozygote advantage for the same reason: heterozygotes for the sickle cell allele of beta haemoglobin (HbS) are more resistant to malaria, driving up the frequency of this allele in populations exposed to malaria in Africa. When people moved from Africa to America they took their HbS alleles with them, even though they no longer had any advantage in the malaria-free environment.

**Figure 7.15** You can help save lives: give money to a reputable charity that helps to provide mosquito nets (a), malaria prevention information, and effective treatment to people living in malaria-affected areas. (b) These people in Senegal are receiving insecticide treated bed nets. However, any control programme requires careful strategy and monitoring, because the use of insecticides in the fight against malaria, even in the form of these treated nets, can contribute to resistance in mosquitoes.

Image (b), credit: Maggie Hallahan.

In fact, all human populations vary in their allelic composition, so allele frequencies can be a reflection of population history, carried with people as they move around the world. PKU alleles are found at higher frequency in people of Irish descent, wherever they are found in the world. The Australian island state of Tasmania has a disproportionately high incidence of Huntington's disease, and the majority of cases can be traced to a single individual, Mary Cundick, who migrated with her 13 children from Cornwall to Tasmania in 1848 (see also Figure 5.16). In some populations with a high frequency of a particular disease-causing mutation, genetic testing can be used to help people to make decisions about their own reproduction. For example, a genetic screening service operating in Jewish communities has been credited with reducing the incidence of serious genetic disorders, particularly Tay-Sachs, a neurodegenerative disease. Members of the community can be screened for a range of alleles and the results stored on a database. The database can advise if two people share the same recessive Mendelian disease alleles, in which case they have a 25 per cent chance of having an affected child together. Importantly, the service does not reveal directly who carries which alleles, for fear of stigmatizing carriers. This contemporary case of active intervention in human breeding in order to reduce the incidence of heritable disease brings us to a topic usually avoided in polite company, yet one which may help us bring important concepts in molecular evolution into sharp focus. Now that we have the wolf of human evolution by the ears, why don't we gather our courage and leap straight into its jaws. Let us consider the far more dangerous topic of eugenics.

**Figure 7.16** R. A. Fisher, a giant in statistics as well as one of the founders of modern evolutionary biology, was also one of eugenics' most enthusiastic promoters. He wore thick glasses, yet still considered himself fit enough stock to have nine children. Still, like John Maynard Smith (Figure 7.5), in Fisher's case poor eyesight may have been a selective advantage as it preventing him from going to war and getting shot.

Reproduced courtesy of Fisher Memorial Trust.

## *Can we manipulate human evolution?*

R. A. Fisher, one of the founders of population genetics, was deeply concerned about human health and prosperity, and this concern motivated much of his work (Figure 7.16). He was a strong proponent of eugenics, which is the active intervention into human breeding to improve heritable traits. Having worked in agricultural genetics, Fisher stated that 'if the methods of the stockyard were applicable to mankind the human race could be improved in any desired direction, within a short historical period, to an extent exceeding existing differences between widely different races'. Fisher gave a clear voice to the decades-old concern that, in post-industrial society, natural selection was being turned on its head. Since Victorian times, many people felt that the people with the least desirable traits, such as 'feebleness of mind', were often those that had the most offspring, so with every passing generation those of poor stock would out-reproduce their betters. Could feeble-mindedness thus go to fixation at the expense of higher intelligence, leaving the human race full of blithering idiots? (Some might suggest this has already happened. Several times.)

More recently, some evolutionary geneticists have expressed fears that modern medicine will weaken the human race by allowing a build-up of deleterious mutations, through providing the means for individuals who would once have died young to survive and reproduce. For example, the great evolutionary biologist Bill Hamilton worried that the availability of caesarean sections meant that individuals who would otherwise have died in childbirth now survive, and potentially pass on a propensity for obstructed labour to their own offspring. If so, the frequency of traits associated with caesareans (say, small pelvises or large-headed babies) could increase in the population with each passing generation. Should we worry that the human race could become unable to bear children without surgery? H. J. Muller, a pioneer in the study of mutations, bleakly predicted the accumulation of so many deleterious mutations that the human race would slide into decay: 'instead of people's time and energy being mainly spent in the struggle with external enemies of a primitive kind such as famine, climatic difficulties and wild beasts, they would be devoted chiefly to the effort to live carefully, to spare and to prop up their own feeblenesses, to soothe their inner disharmonies and, in general, to doctor themselves as effectively as possible. For everyone would be an invalid, with his own special familial twists.'

Many, many books have been written on the perceived rights or wrongs of eugenic thinking, and it is not my intention here to examine these issues in detail, nor to promote any particular conclusion. I just thought it was an interesting case in point to think about some of the concepts that we have covered in taking a molecular genetic view of natural selection. Firstly, when it comes to humans we have an instinctive feeling for what we mean by 'fitness'—a fit individual is someone gorgeous, clever, talented, healthy and, above all, desirable. But this is not the molecular genetic definition of fitness. At the beginning of this chapter we saw that natural selection is a numbers game, simply the consequence of differential reproduction of heritable variants. How do we identify the fittest variant? It is the one that increases in frequency with respect to other variants in the population. So, in evolutionary terms, we would identify any trait associated with a higher chance of reproduction as having a selective advantage. If feeble-mindedness was heritable, and if the feeble-minded out-reproduced their brainy betters, then from a genetic point of view we would have to view 'feeble-mindedness' as a positively selected trait, because it resulted in its carriers having more offspring of their own kind.

We also saw that fitness can only be measured with respect to a particular environment. Mutations that cause phenylketonuria (PKU) are strongly deleterious in humans that ingest phenylalanine, but selectively neutral when phenylalanine is absent from the diet. The only measure of fitness is whether a carrier of a particular mutation in a particular environment tends to have more or fewer offspring than other members of the population. In an environment where surgical intervention is unavailable, carriers of a mutation that causes larger-than-average baby heads may reproduce less due to deaths in childbirth, but the mutation may be effectively neutral in an environment where surgery is available. It may be that the selective value of a trait will change over time—if civilization collapses and hospitals disappear, then carriers of big-baby-head alleles might reproduce less. But evolution has no foresight, and mutation frequencies cannot prepare for future change. The only meaningful molecular genetic measure of fitness is the influence of a given mutation on its own frequency in subsequent generations. Selection operates relative to current conditions.

In some cases the influence of the environment may outweigh underlying genetic variation in human phenotypes. We can see the effect of environment by considering one of the favourite concerns of eugenicists, variation in human intelligence. Intelligence has some degree of heritability, in that children are more likely to resemble their parents than they are to resemble random members of the population. And there are genetic factors that can influence intelligence (for example, genetic diseases like PKU can cause mental retardation). However, these observations do not necessarily imply that variation in intelligence is due to genetic factors. It may be the case that smart parents have a tendency to raise smart children, even if those children are genetically unrelated to their parents. We know that environmental factors can reduce intelligence (for example, alcohol-related brain damage) or increase intelligence (for example, education). Nonetheless, there have been various attempts to set up sperm banks exclusively based on donations from those perceived to have high intelligence, such as Nobel prize winners, in order to create a genetic resource for selectively breeding from the smartest individuals. Could we increase the average intelligence of the population by such selective breeding? Even if we could, we would be wasting our time. Whether or not there is genetic variation in levels of intelligence, the observation of massive changes in measures of intelligence in some populations over the course of a single generation demonstrates the strong influence of environment on intelligence scores, for it cannot be due to change in allele frequencies. This leads to the very practical conclusion that, if your goal is to improve human intelligence, you would guarantee higher returns on your investment of time and money by improving the educational environment of as many people as possible than you would from breeding from Nobel prize winners. Let's give the last word to Alfred Russel Wallace, who, along with Darwin, discovered and developed the concept of natural selection:

> *"Why, never by word or deed have I given the slightest countenance to eugenics. Segregation of the unfit, indeed! It is a mere excuse for establishing a medical tyranny. . . . Give the people good conditions, improve their environment, and all will tend towards the highest type."*
>
> A. R. Wallace in an interview published in *The Millgate Monthly*, August 1912

## Conclusion

The study of heritable variation has provided the building blocks of evolutionary theory. Darwin showed that variation was ubiquitous, and he proposed that cumulative selection of small variants could lead to the formation of phenotypically distinct populations. He emphasized the continuous accumulation of tiny modifications, such that populations were in a constant but insensible state of flux. If some of the heritable variation ever present in populations confers any increased chance of successful reproduction, then, over many generations, we expect those variations to increase in representation in the population. But while the basic process of selection is wonderfully simple—the world becomes full of good copiers—the outcomes of selection plays are complicated by many influences, including the environment, population composition, genetic background, and linkage between alleles. These complicating factors can make fitness of single alleles far from simple to assess.

But not all changes in allele frequencies in populations are driven by selection. We have seen that, in human populations, alleles can be present in relatively high frequency even if they are not advantageous, simply because those alleles were overrepresented in founder populations. In fact, chance effects are pervasive in molecular evolution. So now that we have considered selection, in the next chapter we will consider the other great force acting on allele frequencies: genetic drift.

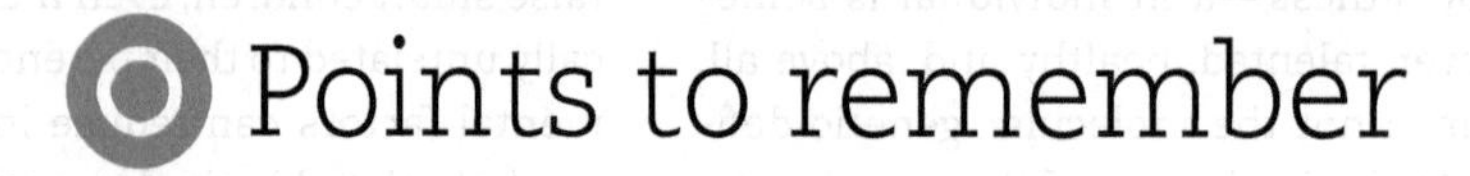

## Points to remember

### Variation

- Genetic variation is ubiquitous in natural populations, because individuals all carry different mutations in their genomes.
- An allele is a heritable variant at a particular locus in the genome, such that members of a population can carry alternative versions of that nucleotide sequence.
- Selection can act on standing variation, alleles already present in the population that become advantageous due to a change in environment.
- Substitution is the process whereby alleles increase in frequency in the population over generations, until all members of the population carry a copy of that mutation, and all other alleles at that locus have disappeared.
- Natural selection can operate in any system where entities make copies of themselves (replication) that can inherit individual variations (heredity) which can influence the chance of successful copying (selection).
- Positive selection occurs when a heritable variant increases the chance of successful reproduction, relative to other alleles in the population, so it will tend to increase in frequency.
- Negative selection occurs when a heritable variant decreases the chances of successful reproduction, so ends up in fewer descendants than alternative alleles.
- Gradualism is the assumption that most evolutionary change occurs through the incremental accumulation of many variations, each of relatively small effect.

### Fitness

- Fitness is a reflection of an alelle's probability of successful replication, so reflects increase or decrease in representation under current conditions, rather than any absolute notions of good or bad.
- The selective environment of an allele depends on other alleles present, including those at the same locus (polymorphism) and those at other loci (genetic background), both in the individual and in the population.
- A mutation can 'hitchhike' to fixation if it is closely enough linked to an allele under selection that they will be inherited together.

- Human populations carry the imprint of selection, with allele frequencies influenced by their effect on survival, either in the current environment or as a result of past selection pressures.
- Heritability is not always the sign of genetic determination, and genetic predispositions may be overwhelmed by environmental influence.

# Ideas for discussion

**7.1** What would happen if an allele under positive selection was linked to a deleterious mutation?

**7.2** Should we be worried about low genetic variation in small populations of endangered species?

**7.3** High-throughput (next-generation) sequencing techniques are making 'community genetics' programmes more widely applicable, because cheap tests can be used to screen population members for asymptomatic heterozygote carriers of many different disease alleles. What are the potential benefits and costs of this community genetics approach?

# Sequences used in this chapter

Table 7.1: Human haemoglobin sequences used in Figure 7.14

| Description from GenBank entry | Accession |
|---|---|
| *Homo sapiens* isolate beta globin (HBB) gene | DQ126270–DQ126325 |

# Examples used in this chapter

Evans, T. A., Wallis, E. J., Elgar, M. A. (1995) Making a meal of mother. *Nature*, Volume 376, page 299.

Grant, P. R., Grant, B. R. (2006) Evolution of character displacement in Darwin's Finches. *Science*, Volume 313, page 224.

Hedrick, P. (2011) Population genetics of malaria resistance in humans. *Heredity*, Volume 107, page 283.

Kerr, B., Riley, M. A., Feldman, M. , Bohannan, B. J. M. (2002) Local dispersal promotes biodiversity in a real-life game of rock-paper-scissors. *Nature*, Volume 418, page 171.

Kohn, M. (2004) *A reason for everything: natural selection and the English imagination.* Faber and Faber, London.

Nisbett, R. E., Aronson, J., Blair, C., Dickens, W., Flynn, J., Halpern, D. F., Turkheimer, E. (2012) Intelligence: new findings and theoretical developments. *American Psychologist*, Volume 67, page 130.

Wood, E. T., Stover, D. A., Slatkin, M., Nachman, M. W., Hammer, M. F. (2005) The β-globin recombinational hotspot reduces the effects of strong selection around HbC, a recently arisen mutation providing resistance to malaria. *The American Journal of Human Genetics*, Volume 77, page 637.

HEROES OF THE GENETIC REVOLUTION

7

# J. B. S. Haldane

*"J. B. S. Haldane, in some ways the cleverest and in others the silliest man I have ever known"*

Medawar, P. (1986) *Memoir of a thinking radish: an autobiography*. Oxford University Press.

**EXAMPLE PUBLICATIONS**

Haldane, J. B. S. (1932) *The causes of evolution* (reprinted 1990, Princeton University Press).

Haldane, J. B. S. (1957) The cost of natural selection. *Journal of Genetics*, Volume 55, pages 511–524.

**Figure 7.17** John Burdon Sanderson Haldane (1892–1964). This is a copy of the photo that John Maynard Smith (**Figure 7.5**) kept in a frame in his office, as a constant reminder of his supervisor and mentor. Maynard Smith never tired of telling stories of Haldane—outrageous and wonderful, many possibly apocryphal or at least greatly enriched by time—just as those who were fortunate enough to know Maynard Smith will do, and so on down the generations, each adding their own little mutations to increase the fitness of the stories, promoting their longevity and copy frequency. Such stories as these improve with the telling.

J. B. S. Haldane was born to be a scientist (**Figure 7.17**). He worked with his father on physiology experiments from an early age, and it was from his father that he inherited the rather dramatic habit of using himself as an experimental subject, a practice both enlightening and dangerous. His first scientific publication was a collaboration with his sister Naomi, reporting pioneering findings on genetic linkage from experiments they had conducted at their home in Oxford by breeding hundreds of guinea pigs and mice. He studied mathematics, zoology and 'greats' (classics) at Oxford. Haldane claimed to have quite enjoyed fighting in the First World War, despite receiving several serious injuries while serving with the Black Watch (he later pondered whether he had not actually died in the war and imagined the rest of his 'rather outrageous' life). His frontline service, like many of his physiology experiments, was characterized by taking well-calibrated risks, for example when he rode a bicycle across the front line, accurately predicting that the German soldiers would be too stunned to shoot him. Haldane ran a bomb-making workshop, staffed by men of a similar bomb-proof constitution, where smoking was compulsory. They would occasionally toss lighted bombs to each other before lobbing them into the enemy trenches. This, Haldane insisted, was not reckless behaviour: 'provided you are a good judge of time, it is no more dangerous than crossing the road among motor traffic, but it is more impressive to onlookers'. After the war he returned to Oxford, then worked at Cambridge, before settling at University College London until emigrating to India in 1957 and working at the Indian Statistical Institute in Calcutta (he gave many reasons for moving to India, one of which was that 60 years of wearing socks was enough).

In addition to his pioneering work on genetic linkage, Haldane developed ways of measuring the intensity of selection and the speed of evolutionary change. He was one of the first to consider the costs of natural selection: this classic paper shows his modus operandi, beginning with a common-sense observation then putting it into algebra ('In this paper I shall try to make quantitative the fairly obvious statement that natural selection cannot occur with great intensity for a number of characters at once unless they happen to be controlled by the same genes'). He developed the idea of heterozygote advantage (originally referred to as the 'malaria hypothesis'). He played a leading role in shaping the mathematical treatment of population genetics that transformed evolutionary biology into a quantitative science. This work helped to lay the foundation for the 'modern synthesis' of evolution, and paved the way for the gene-centred view of evolution. In addition to evolutionary genetics, much of his work concerned physiology and biochemistry: for example, his hypothesis about the origins of life is the basis of the commonly used phrase 'primordial soup'.

Haldane's impact was not confined to scientific research. He was committed to the dissemination of scientific knowledge to wider society. His prodigious output of essays and articles, published in periodicals including *The Daily Worker* and the *Rationalist Annual*, covered a broad range of biological topics, from 'Why I admire frogs' to 'What "hot" means' (**Figure 7.2**). One of Haldane's most famous essays 'On Being the Right Size', a perspicacious view of the physiological consequences of body size in animals, should be required reading for all biologists. But Haldane also wrote on a much wider range of topics, from other fields of science (e.g. astronomy), biographical essays (e.g. Archimedes) and social issues (e.g. factory ventilation). He even dabbled in future prediction, in *Daedalus, or Science and the Future.* He made authoritative contributions to the discussion of eugenics, presenting clear arguments from population genetics why human eugenic schemes were unlikely to achieve their goals.

It is simply impossible, and actually quite frustrating, to try to provide anything like an adequate summary of even Haldane's major achievements in only one small page (did I mention he wrote a popular children's book?). All I can do is implore you to use this as a starting point for reading more about him, because he is not only one of the most important biologists of the genetic revolution, but also one of the most entertaining. If you read only one biography of a scientist, for entertainment's sake choose one about Haldane.

# TECHBOX 7.1

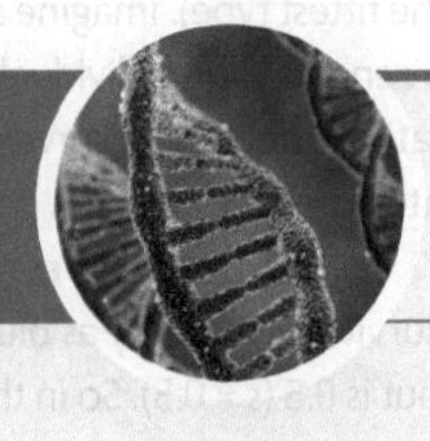

# Fitness

**RELATED TECHBOXES**

TB 7.2: Detecting selection

TB 8.2: Population size

**RELATED CASE STUDIES**

CS 7.1: Selection

CS 7.2: Variation

**Summary**

The fitness of an allele is a reflection of its change in frequency relative to other alleles in the population.

**Keywords**

selection, selective coefficient (*s*), population genetics, resistance, selective sweep, substitution

There are many ways of defining fitness. For example, we may use the word fitness to describe the way that organisms are suited to particular aspects of their environment. Alternatively, we may use the term fitness to report the net reproductive output of individuals, or the change in frequency of particular traits in a population. Confusion between the different definitions has led to misunderstandings in evolutionary biology. In this chapter, we are concerned with substitution: the process whereby a particular mutation increases in frequency from generation to generation until it replaces all other alternative alleles and becomes fixed in that population. In order to make statements about this process, it is useful to consider the term fitness as representing the effect that the properties of a mutation have on its chance of becoming a substitution in a given population and a particular environment. But remember that you will come across many alternative definitions of fitness, which may be appropriate to different topics.

Probably the best place to start in taking a molecular genetic perspective on fitness is to abandon the phrase 'survival of the fittest', because stating that the fittest are those that survive is not very helpful to our understanding. Instead, we need to think of fitness as a statement about probabilities. If we have a population of copiers which have heritable variation, we say that a particular heritable variant has a higher fitness if it has properties that make it more likely to make more successful copies of itself than alternative variants. The reason we find it useful to have a term with this definition is that we expect fitter variants to increase in frequency in the population with each passing generation. In population genetics, fitness is most commonly expressed algebraically, but in this book the aim is to provide explanations of molecular evolution in words rather than equations (this is not to say that you should make a habit of ignoring equations, which are there for a reason—they can be the clearest statement that can be made about a principle, and you would do well to make sure you understand them).

The important thing to remember when taking a molecular genetic perspective of fitness (that is, one aimed at understanding the process of substitution) is that fitness of a mutation is relative to the other mutations in the population. Remember that the expected relative rate at which a given variant will be copied into the next generation must depend on two things: the effect of the mutation, and the environment in which it finds itself. 'Environment' must be taken here in the broadest sense, to include the external conditions, the composition of the population in which the mutation finds itself (whether there are any other alleles that might do relatively better), and the genetic background (the influence of other mutations in the same genome). The relative selective advantage or disadvantage of each mutation can be considered by comparing the rate at which it is copied to the next generation, with respect to alternative alleles at the same locus.

The relative advantage or disadvantage of a particular mutation is sometimes represented by the selective co-efficient, denoted by the letter *s*. Fitness is (somewhat mystifyingly) often denoted by *W*. If we say that the fittest mutation in the population has a fitness of 1 ($W = 1$), then all other mutations at the locus must either be of equal fitness ($W = 1$) or must have a lower fitness ($W = 1-s$, where *s* is the difference between this type and the fittest type). Imagine a polymorphic population of beetles, with equal numbers of yellow, green and blue individuals (**Figure 7.18**). We follow this population over some number of generations, noting the survival and reproductive rates of the different morphs. At the end of the observation period we find a population with no yellows, and twice as many blues as greens (**Figure 7.18**). If blues and greens have the same average number of offspring, but half as many greens survive to maturity as blues, then we would say that the difference in the relative reproductive output is 0.5 ($s = 0.5$). So in this case, blue has a *W* of 1 and green has $W = 1-0.5 = 0.5$. If yellows are just as likely to survive as blues but are infertile, then they have no chance of reproducing so they have an *s* of 1, and *W* of 0.

Note that just as fitness has various definitions and usages, so does the parameter representing selection. *s* may be called the selection differential, and it is sometimes reported as the proportional advantage over other alleles in the population. Furthermore, *s* is sometimes reported for genotypes, which, for diploid organisms, are a combination of two alleles: the principle is the same but the equations will be slightly different, and we might have to add in another parameter to describe the heterozygotes for traits that are not completely dominant (such as the HbC and HbS alleles of β-haemoglobin: **Figure 7.14**). As with fitness, since there are several different definitions of *s*, you need to get the intended meaning from the context it is written in.

The key point is that we need to stop ourselves from thinking of fitness as 'good' or 'bad'. All that matters is relative copy numbers in a particular context. A mutation that confers antibiotic resistance may seem like a huge advantage to a bacterium in a hospital. But if it slows the growth of bacterium by reducing the rate of uptake of nutrients, then a resistant strain may still be at a selective disadvantage. If the antibiotic-sensitive strain suffers 80 per cent more mortality than the resistant strain, but each survivor produces ten times as many descendants as the resistant strain, then it will still reproduce at twice the rate. We could say that the antibiotic-resistant strain had *s* of 0.5: if there are 100 sensitives for every 10 resistants, all 10 resistants

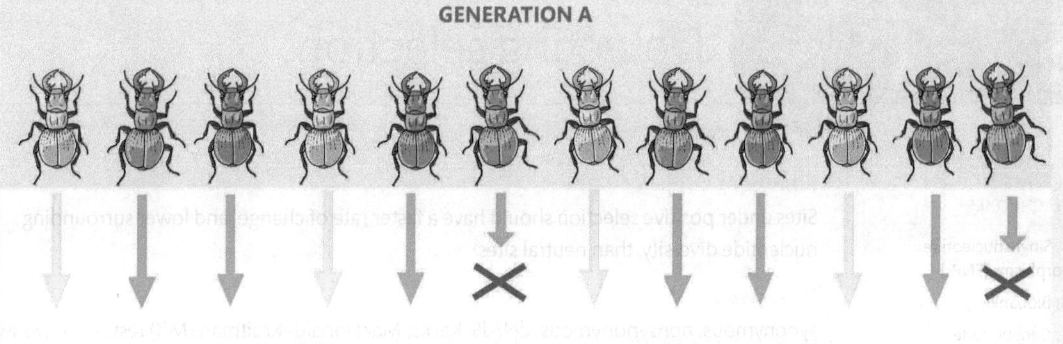




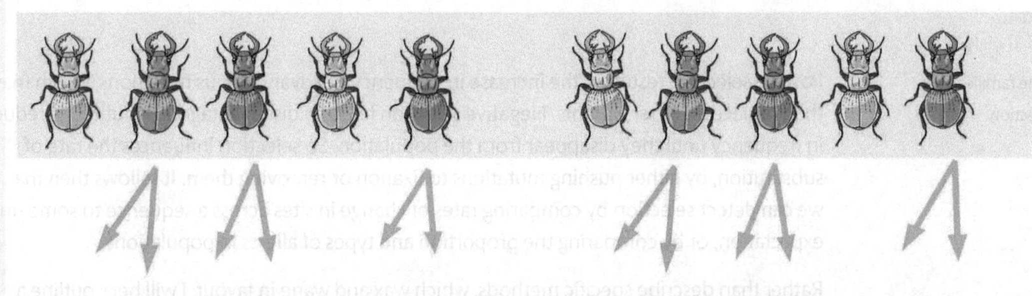

**Figure 7.18** There are many ways to think about fitness, but here we are interested in the relative proportion of alleles in a population over multiple generations.

survive but only 20 of the sensitives, then there will still be twice as many sensitives as resistants. So even though we instinctively think of antibiotic resistance as advantageous in the presence of antibiotics, in this particular case it has a lower selective co-efficient.

Fitness can be estimated for particular alleles in real populations. For example, an international team including researchers from Laos, Thailand, Vietnam and Myanmar found that an allele associated with resistance to the anti-malarial drug pyrimethamine had increased rapidly in *Plasmodium* (malaria) parasites in South East Asia[1]. They estimated a single origin of the resistant allele, which then spread throughout the region since the introduction of pyrimethamine treatment in the 1970s. They estimated the selection co-efficient *s* in two ways—from the decline in efficacy of pyrimethamine treatment, and from the reduction of genetic diversity around the allele (indicating a selective sweep pushing this allele to fixation at the expense of other alleles in the population: see **Figures 7.10–7.14**). Both approaches suggested that the resistant allele had around 10 per cent advantage over alternative alleles.

## References

1. Nair, S. , Williams, J. T. , Brockman, A. , Paiphun, L. , Mayxay, M. , Newton, P. N. , Guthmann, J-P. , Smithuis, F. M. , Hien, T. T. , White, N. J. (2003) A selective sweep driven by pyrimethamine treatment in Southeast Asian malaria parasites. *Molecular Biology and Evolution*, Volume 20 (9), Pages 1526–36.

TECHBOX
7.2

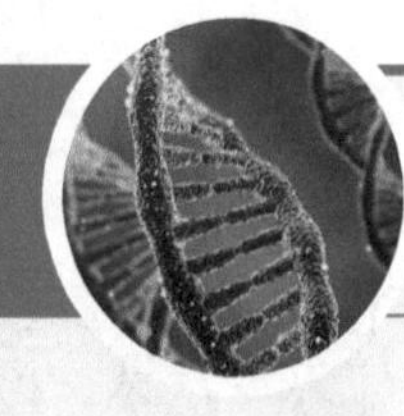

# Detecting selection

RELATED TECHBOXES

TB 3.1: Single nucleotide polymorphisms (SNPs)

TB 3.2: Biobanking

TB 6.1: Genetic code

RELATED CASE STUDIES

CS 6.1: Gene families

CS 7.1: Selection

**Summary**

Sites under positive selection should have a faster rate of change, and lower surrounding nucleotide diversity, than neutral sites.

**Keywords**

synonymous, nonsynonymous, dN/dS, Ka/Ks, MacDonald–Kreitman (MK) test, selective sweep, haplotype, polymorphism, codon usage bias, intergenic

Positive selection results in the increase in frequency of advantageous mutations, which may then replace all other variants. Negative selection leads to disadvantageous mutations reducing in frequency until they disappear from the population. So selection influences the rate of substitution, by either pushing mutations to fixation or removing them. It follows then that we can detect selection by comparing rates of change in sites across a sequence to some null expectation, or by comparing the proportion and types of alleles in populations.

Rather than describe specific methods, which wax and wane in favour, I will here outline a couple of alternative approaches to detecting selection. There will be other methods out there that I haven't covered, and new methods are constantly being devised. The method you choose will depend on the kind of data you have and the question you want to ask (and, to an unfortunate degree, your tolerance of not always entirely friendly computer programs). We are going to concentrate on the kind of methods used by people who analyse DNA sequence data to answer questions in evolutionary biology, rather than focusing on medical or population genetics.

Many methods start by assuming that one class of mutations are always neutral (TechBox 8.1). This neutral class might be all the mutations occurring in particular sequences, such as pseudogenes, introns, or intergenic regions, or it may be particular mutations within a sequence, such as synonymous mutations within a protein-coding gene. The rate of substitution in these presumed neutral sites gives a baseline against which to look for the signature of selection (Techbox 8.1). If we observe patterns of substitution that are very different from the expected patterns under neutral evolution, then we might conclude that selection is responsible for at least some of the substitutions we observe. For such tests, it is important to remember that not all introns or intergenic sequences are neutral (there are conserved functional sequences between genes and within introns) and there is also evidence of weak selection acting on synonymous mutations (codon usage bias favours some synonymous codons over others).

1. **Comparing synonymous and nonsynonymous substitutions.** This approach starts with an alignment of protein-coding genes and then estimates the relative frequency of two different kinds of substitutions. Synonymous mutations are base changes that don't change the amino acid specified by the codon (TechBox 6.1), so they change the DNA sequence but not the amino acid sequence that it codes for. Since synonymous mutations are phenotypically silent, they are assumed to be under no (or at least very little) selection, therefore their substitution rate will be determined by random fluctuations in frequency (drift: TechBox 8.1). Nonsynonymous changes are base changes that do change the amino acid sequence, so are more likely to have an effect on survival and reproduction. Since most random changes to a protein will be deleterious, we expect that most nonsynonymous mutations are under negative selection so will decrease in frequency until they disappear from the population. So under negative selection, the rate of nonsynonymous substitution per nonsynonymous site (often

termed *dN* or Ka) will be less than the rate of synonymous substitution per synonymous site (*dS* or *Ks*). In this case the ratio of the two rates (*dN/dS* or Ka/*Ks*) will be less than one. If all changes in a sequence are neutral, for example in a pseudogene that has no effect on phenotype, then both synonymous and nonsynonymous mutations will go to fixation at the same rate, and *dN/dS* will be around one. If positive selection is driving some mutations to fixation faster than they would under drift, then *dN/dS* may be greater than one. *dN/dS* is sometimes referred to as ω (omega). This kind of test can identify when a particular sequence is under selection in a given lineage (**Case Study 5.1**). Or it can be used to identify particular sites that are under selection, by testing whether the ratio of nonsynonymous to synonymous substitution rates varies across individual codons. These type of tests typically have relatively low power, so a failure to detect selection does not mean that evolution at that locus has been neutral. But a *dN/dS* significantly greater than one is often considered convincing evidence of positive selection.

2. **Comparing polymorphisms in a population to differences between species.** The life cycle of a substitution is that it is born as a mutation in one individual (or a few if it is a common mutation), at which point it is a low frequency polymorphism (because other individuals in the population have a different base at that point in the sequence). If it is deleterious, it will tend to decrease in frequency and disappear, but if it is advantageous it will tend to rise in frequency, persisting as a polymorphism until it displaces all alternative alleles in the population. So a mutation must initially exist as a polymorphism, then become a substitution when all alternative versions have disappeared. At this point, all members of the population carry a copy of the same mutation, and we say it has become fixed in that population.

Deleterious mutations, those that decrease in representation, spend relatively little time as transient polymorphisms on the way down to zero. Advantageous mutations increase in frequency until they reach fixation, replacing all other alleles, spending little time as transient polymorphisms on the way up. But neutral mutations can either increase or decrease in frequency, so are left to fluctuate as polymorphisms, occasionally going to fixation by chance. Because polymorphisms of alleles under selection are expected to be short-lived (on their way up or down), our chance of observing them is small. So, on the whole, we expect the majority of observed polymorphisms to be effectively neutral. If positive selection is a significant driver of evolution at a particular locus, then it will be responsible for some of the substitutions, but few of the polymorphisms.

MK (McDonald–Kreitman) tests take advantage of this imbalance, by comparing the ratio of nonsynonymous to synonymous polymorphisms, derived from within-population samples, to the ratio of nonsynonymous to synonymous substitutions, derived from comparing sequences to a related species. An excess of nonsynonymous substitutions at the species level is taken as evidence of positive selection, because it is assumed they got there by rapid fixation, spending relatively little time as polymorphisms. MK-style tests require data on allele diversity within a species and sequence differences between species. Bear in mind that the rate of fixation of mutations of small selective coefficients varies with population size (**TechBox 8.2**), and this can skew the ratios of nonsynonymous to synonymous polymorphisms and substitutions, so MK tests are not reliable for populations that have fluctuated in size (e.g. undergone a recent bottleneck or expansion).

3. **Identifying selective sweeps.** Both of the above approaches are generally only applied to protein-coding data, because they rely on defining synonymous (silent) and nonsynonymous (potentially selected) changes. Genomic data offers an alternative approach that is not limited to protein-coding sequences. When an allele increases under positive selection, it takes any linked mutations with it to fixation (**Figures 7.10–7.14**). So many members of the populations will all have not only the allele under selection, but also all the linked alleles, whether or not they are also under selection. The closer a linked allele is to the site under selection, the less likely that recombination will break up the association between the alleles. This means that one of the signatures of positive selection is a reduction in nucleotide diversity, not only in the locus under selection but in sites closely linked to it. Selective sweeps can be detected by identifying regions

of the genome with unexpectedly low nucleotide diversity. Over time, mutations appear near the selected site, but they will be in relatively low frequency compared to polymorphisms in the rest of the genome. So when you compare individuals within a population, you might expect to see fewer differences between them in haplotype blocks under selection. Haplotype blocks under selection will also contain a relatively greater number of differences caused by rare alleles, present in low numbers in the population. As time goes on, recombination events will start to separate the selected site from the accompanying mutations in its haplotype block. So the more recently an allele has been fixed, the larger the block of low-diversity will be. Long-range haplotypes, that preserve a particular combination of mutations in an allele, can be a sign of a relatively recent and rapid rise of a positively selected mutation.

CASE STUDY 7.1

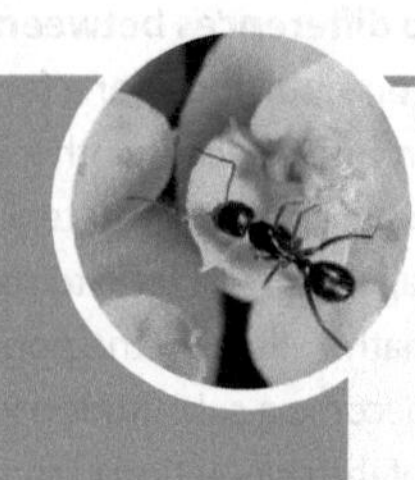

# Selection: tracking insecticide resistance over time using museum specimens

RELATED TECHBOXES

TB 4.2: DNA amplification

TB 7.2: Detecting selection

RELATED CASE STUDIES

CS 2.2: Ancient DNA

CS 5.1: Duplication

Hartley, C. J., Newcomb, R. D., Russell, R. J., Yong, C. G., Stevens, J. R., Yeates, D. K., LaSalle, J., Oakeshott, J. G. (2006) Amplification of DNA from preserved specimens shows blowflies were preadapted for the rapid evolution of insecticide resistance. *Proceedings of the National Academy of Sciences USA*, Volume 103, page 8757.

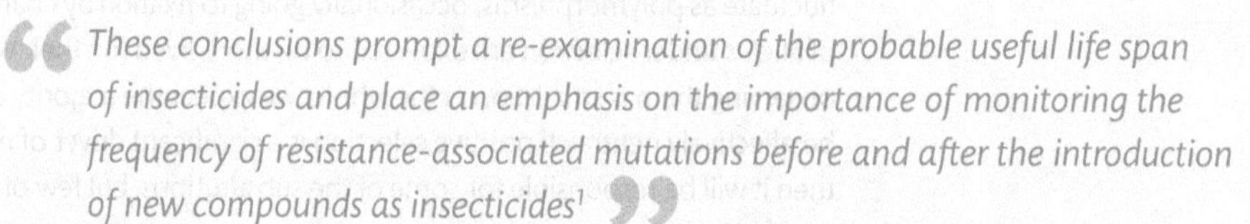

*“These conclusions prompt a re-examination of the probable useful life span of insecticides and place an emphasis on the importance of monitoring the frequency of resistance-associated mutations before and after the introduction of new compounds as insecticides[1]”*

**Keywords**

selection, ancient DNA, mutation, variation, polymorphism, standing variation, selective sweep

## Background

The evolution of insecticide resistance has been frequently used to study the dynamics of selection at the molecular level, due to the impact of resistance on human economies and health, and because the rapid timescales and large populations make it a tractable experimental system. One important question, with both theoretical and practical significance, is whether the response to a new selective regime tends to come from standing variation (promoting alleles already in the population), or from selection of novel mutations that arise after selection is applied. In the case of insecticide resistance, it is often assumed that alleles conferring resistance are costly to individuals, so they will not persist in untreated populations, therefore must arise after treatment starts.

## Aim

This study focused on two species of sheep blowfly, *Lucilia cuprina* and *Lucilia sericata*, that are agricultural pests around the world. Organophosphates (OPs) have been used to control these pests since the 1950s, but resistance to OPs arose within a decade of their introduction. Now virtually all *L. cuprina* in Australasia are OP resistant and resistance is common in *L. sericata*. Resistance to different forms of OPs (marketed under different names) occurs through changes to the *αE7* gene that codes for the Esterase 3 enzyme (E3). A single amino acid change, from glycine to aspartate

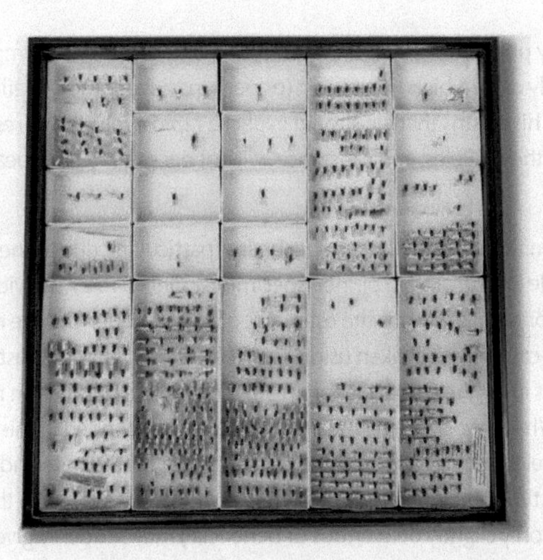

**Figure 7.19** Pole dancing for science: pinned specimens of the sheep blowfly *Lucilia cuprina* from the Australian National Insect Collection, maintained by CSIRO. Started in 1928, this collection contains more than 12 million specimens. Resources such as insect collections might have an old-fashioned feel to them, but the value of these whole-organism non-electronic biodiversity databases is becoming even more obvious in the post-genomic bioinformatic era as we begin to be able to access the vast amounts of biological information stored in dead bugs. (**Figure 7.2**)



at residue 137 of the E3 protein, gives a high degree of resistance to the OP insecticide Diazinon. This mutation is referred to as Asp-137. A different single amino acid change, from tryptophan to leucine at residue 251 of E3 (Leu-251) provides a similar degree of resistance to the OP Malathion. The aim of this study was to determine the distribution of Asp-137 and Leu-251 mutations in populations of *Lucilia cuprinia* and *Lucilia sericata* with different histories of OP exposure.

### Methods

Blowfly specimens were taken from collections made in Australasia, Africa, Europe, Asia, North America and Hawaii, including stored specimens of *Lucilia cuprina* that were sampled in Australasia before the introduction of OP insecticides in 1950 (**Figure 7.19**). For the fresh samples, DNA was extracted from these specimens, and PCR was used to amplify a 1.2kb region encoding part of the *αE7* esterase gene. For the old samples, sterile 'ancient DNA' techniques were used to extract DNA (e.g. in a lab that was not used for other E3 amplifications to prevent cross-contamination) and two short regions, each around 150bp spanning the site of a key resistance mutation (Asp-137 and Leu-251), were amplified.

### Results

There was no evidence of the allele that confers Diazinon resistance (Asp-137) in the pre-1950 specimens, which suggests either that this allele arose by mutation after the introduction of OP insecticides, or that it was at much lower frequency in the pre-OP populations and therefore was undetected in these samples. The Asp-137 mutation allows the E3 enzyme to break down OPs, but they reduce the enzyme's effectiveness at other tasks, so while the Diazinon resistance mutation confers high benefits in the presence of OP insecticides, it carries a high cost. It is therefore less likely to be maintained in a non-treated population, but could rapidly go to fixation if it arises in a population that is exposed to OPs. The authors observed lower nucleotide diversity both at the E3 locus and in surrounding loci in their samples of post-OP *L. cuprina* from Australasia, which they interpret as the signature of a selective sweep due to the strong selection for the Diazinon-resistance mutation. The Malathion-resistance (Leu-251) mutation tells a different story. Two out of twenty-one pre-OP flies sampled carried this resistant allele, indicating that it was present in the population long before the flies were treated with OP insecticides (another resistant allele—Thr-251—was also detected in the pre-OP samples). The Leu-251 mutation retains more of the E3 enzyme's other activities, so the Malathion-resistance mutation might be less costly, which may explain why it can persist in a population that is not treated with insecticides.

### Conclusions

This study seems to provide examples of both selection on standing variation (Malathion resistance) and selection on novel mutations (Diazinon resistance). Furthermore, insecticide treatment resulted in selection for the same mutation (Asp-137) independently in two different species—*L. cuprina* and *L. sericata*. The existence of the Malathion-resistant allele at detectable frequencies in the pre-treatment fly populations suggests that, contra to a common assumption, resistance alleles do not necessarily carry a high fitness cost (or perhaps have some benefit even in the absence of insecticide)[1]. This means that we should expect a rapid evolution of resistance in populations already carrying these alleles, and resistance will not necessarily disappear when treatment ceases.

### Limitations

This study was limited to relatively few specimens (particularly due to the low success rate of extracting useable DNA from old pinned specimens). Only 5 to 27 individual flies were sampled from each category, which means the estimates of allele frequencies are not very precise. On a semantic note, care must be taken using a term like 'preadaptation', just in case it could be misinterpreted as implying that alleles are maintained in the population for the sake of their future benefit. While that is not the meaning the authors intended in this paper, it is important to remember that selection has no foresight, it can only act on the here and now. Selection coefficients can change over time and when they do, a mutation already existing in the population (previously either neutral or only slightly deleterious or beneficial) may suddenly give its carriers a selective advantage: its selective co-efficient has taken a sudden leap due to a change in the environment.

### Further work

These findings are consistent with results from other studies of insecticide resistance that demonstrate that resistance is often based on predictable mutations in particular genes[1]. For example, the same mutation to the E3 esterase—Leu-251—has been shown to confer OP resistance in the house fly *Musca domestica*[2]. New sequencing methods that allow a wide range of loci to be sequenced from ancient DNA specimens increase the scope of using museum specimens to look for genes under selection[3], and are permitting genome-wide searches for resistance genes.

### Check your understanding

1. Why did these researchers use insects from museum collections rather than catching fresh specimens in the field?
2. Why did they conclude that Diazinon resistance is due to a relatively recent mutation?
3. Why did the researchers suggest that only some insecticide resistance mutations carry a physiological cost?

### What do you think?

Would it be possible to use genomic surveys for resistance alleles to judge the likelihood of a population developing resistance to particular insecticides? Could you extend this approach to other important cases of resistance, for example antibiotic resistance in bacteria? What genetic or population features of resistance might help to make such an approach successful?

### Delve deeper

Applying any agent with high mortality will give a huge selective advantage to any resistant individuals, promoting rapid evolution of resistance. Is it possible to evolution-proof insecticides? One possible approach is to specifically target later life stages of the vector so that they kill vectors responsible for disease transmission with relatively little selective cost[4]. How effective do you think this strategy would be? What alternative strategies might reduce the likelihood of resistance?

### References

1. ffrench-Constant, R. H. (2006) Which came first: insecticides or resistance? *Trends in Genetics*, Volume 23, page 1.
2. Claudianos, C., Russell, R. J., Oakeshott, J. G. (1999) The same amino acid substitution in orthologous esterases confers organophosphate resistance on the house fly and a blowfly. *Insect Biochemistry and Molecular Biology*, Volume 29, page 675.
3. Wandeler, P., Hoeck, P. E. A., Keller, L. K. (2007) Back to the future: museum specimens in population genetics. *Trends in Ecology & Evolution*, Volume 22, page 634.
4. Read, A. F., Lynch, P. A., Thomas, M. B. (2009) How to make evolution-proof insecticides for malaria control. *PLoS biology*, Volume 7, page e 1000058.

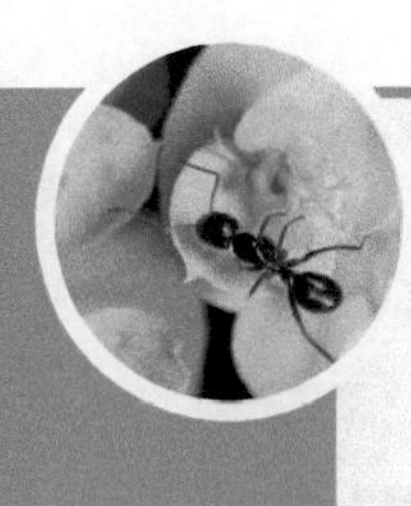

CASE STUDY 7.2

# Variation: experiments on evolutionary rescue under environmental change

**RELATED TECHBOXES**

TB 7.1: Fitness

TB 8.2: Population size

**RELATED CASE STUDIES**

CS 3.2: Mutation rate

CS 7.1: Selection

Lindsey, H. A., Gallie, J., Taylor, S., Kerr, B. (2013) Evolutionary rescue from extinction is contingent on a lower rate of environmental change. *Nature*, Volume 494, pages 463–466.

*... Rapid environmental change closes off paths that are accessible under gradual change*

## Keywords

mutation, selection, *E. coli*, *rpoB*, fitness, antibiotic resistance, epistasis, extinction, adaptation

## Background

In **Case Study 7.1**, we considered a system where a single amino acid change in an enzyme can make flies resistant to insecticide, giving that mutation a huge selective advantage. But some adaptive changes require multiple mutations, each building on the effect of previously acquired mutations (a phenomenon referred to as epistasis). 'Evolutionary rescue' can occur through stepwise adaptation that allows a population to persist under conditions that would have been lethal to its recent ancestors[1]. But if a population is put under sudden and severe stress, then the probability that the right series of mutations will arise may be low, especially if the population is declining[2].

## Aim

These researchers used an experimental evolution approach, growing 1,255 replicate populations of *Escherichia coli* bacteria through serial transfers (populations perpetuated through a controlled series of sampling and re-growing in fresh medium: **Figure 7.20**). They subjected separate populations to different regimes of antibiotic treatment. All treatment lines ended up at the same high concentration of the antibiotic rifampicin, but in the Sudden treatment the high dose was applied immediately, while in the Gradual treatment the dose was slowly increased until it reached the high dose, and the Moderate had a more rapid rise in concentration (**Figure 7.21**). Would bacteria in these treatments follow similar or different evolutionary paths to resistance?

## Methods

The population replicates were all started from a single *E. coli* culture, so were genetically identical at the beginning of the experiment. They were grown initially without rifampicin, to allow build-up of mutations in the different populations, then each population was subjected to one of the three treatments. Samples from each population were periodically frozen to provide snapshots of the evolutionary process. Populations showing no evidence of growth were declared extinct and no further transfers made, so the number of replicate lines declined throughout the study. At the end of the experiment, the *rpoB* gene was sequenced from 30 Gradual populations, 30 Moderate populations, and all 13 surviving Sudden populations. Frozen samples were used to identify mutations that appeared during the experiment, which were then engineered into a common genetic background in various combinations. By gauging the growth rate and competitive ability of these engineered strains across a range of drug concentrations, the fitness effect of mutations in different genetic backgrounds and different drug levels could be evaluated.

## Results

Treatments differed in the population extinction rate: less than two per cent of the populations in the Sudden treatment persisted to the end of the experiment, whereas approximately 45 per

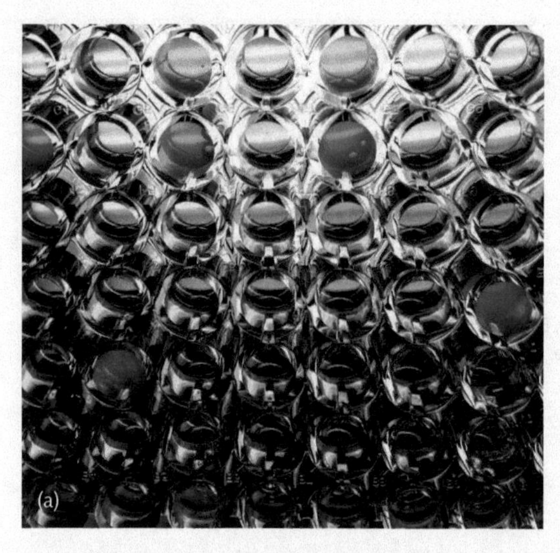

Figure 7.20 (a) Replicate bacterial populations are grown in microtitre plates: each of the little round wells in the plates contains just enough glucose to keep the *E. coli* population growing, and every 48 hours a sample is taken from each and put in a new, fresh well, so it can keep on growing without exhausting its resources. (b) For some experiments involving lots of bacterial replicates, a specialized robot can be programmed to make all the right transfers, but in this case the transfers were done by real people with pipettes.

Photographs by Sarah Hammarlund.

cent of the populations in the Moderate treatment, and 90 per cent of the Gradual treatment populations, made it to the end (Figure 7.21). Populations that survived the maximum dose of antibiotic differed in their growth rates at the end of the experiment, suggesting that they carried different mutations. Sequences revealed many nonsynonymous changes, but Gradual and Moderate populations had more mutations and a greater diversity of changes than the Sudden populations, which tended to each carry only a single mutation (Figure 7.22). For the set of Gradual and Moderate populations possessing multiple mutations, sequencing frozen samples revealed that only one population had the same first mutation as any of the Sudden populations. Strains engineered to have only the early mutations were not able to grow under the high antibiotic dose, which suggests that the early mutations under gradual drug increase conferred low-level resistance, but did not in themselves confer resistance to the high dose. However, secondary mutations in the presence of these early mutations (but not in their absence) could promote resistance to higher concentrations of the drug.

### Conclusions

Populations exposed to sudden, large change could only become resistant if they carried a single mutation that conferred high-level resistance, so most populations went extinct for want of an appropriate mutation. But some of the populations under gradual increase in dose acquired a series of mutations that each acted as stepping stones toward developing higher resistance as the dose increased. In this case, evolutionary rescue relied on epistasis (where the selective effect of one mutation depends on the presence of another mutation), so it was the particular series of mutations, rather than any one mutation itself, that allowed the evolution of resistance to high doses. Because the stepping-stone mutations were selected under low levels of the drug, slower rates of drug increase may raise the likelihood that mutational combinations occur that can rescue the bacteria from higher concentrations of antibiotics.

### Limitations

Only a single gene was analysed in this study, so the effect of mutations in the rest of the genome could not be gauged. Due to the low survival of the Sudden treatment populations, there were fewer than half as many Sudden sequences as Gradual or Moderate (Figure 7.21), so

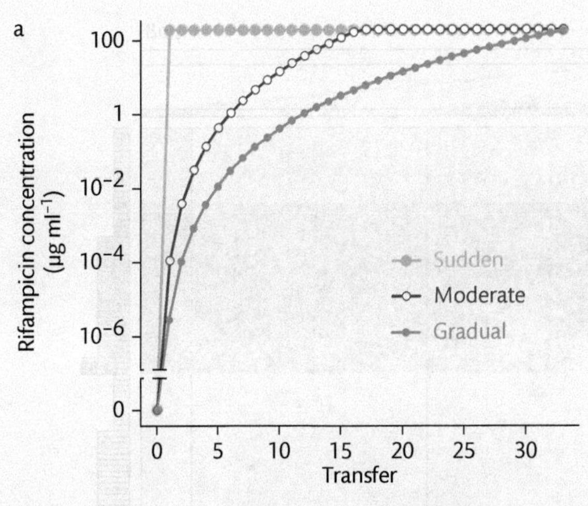


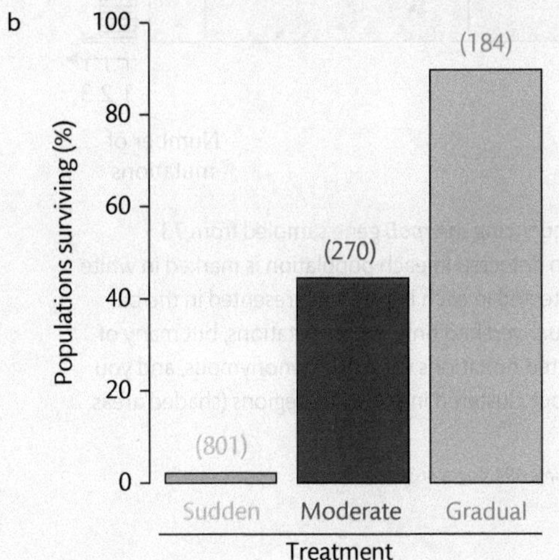


**Figure 7.21** (a) Replicate populations of bacteria were exposed to one of three different treatments of the antibiotic rifampicin, differing in how quickly they reached the maximum dose. Even though all populations were ultimately exposed to the same dose, population extinction rate differed between treatments (b) with the majority of populations surviving in the gradual increase in dose, but most populations going extinct when the maximum dose was applied suddenly without a gradual buildup.



this might result in a less diverse sample of possible rescuing mutations. However, sample size was explicitly taken into account in the statistical analysis.

### Further work

Low levels of antibiotics are present in many environments due to contamination from human waste or agricultural systems. This study suggests that these low levels could prime bacterial populations to more readily develop resistance when exposed to higher doses. Given that the same mutations seen in this study are also detected in populations of *Myobacterium tuberculosis*, these results suggest that the design of the tuberculosis treatment regime could play a role in the likelihood of developing resistant strains.

### Check your understanding

1. What did the researchers do with the frozen samples?
2. Why were all of the detected mutations nonsynonymous changes? Would you expect to see the same pattern in other genes in the *E. coli* genome?
3. Given that the experimental populations all ran for the same amount of time, why did the sequences from the Sudden treatment carry fewer mutations than those in the Gradual treatment?

### What do you think?

What relevance do these findings have for the development of antibiotic resistance? Could they be used to inform treatment strategies?

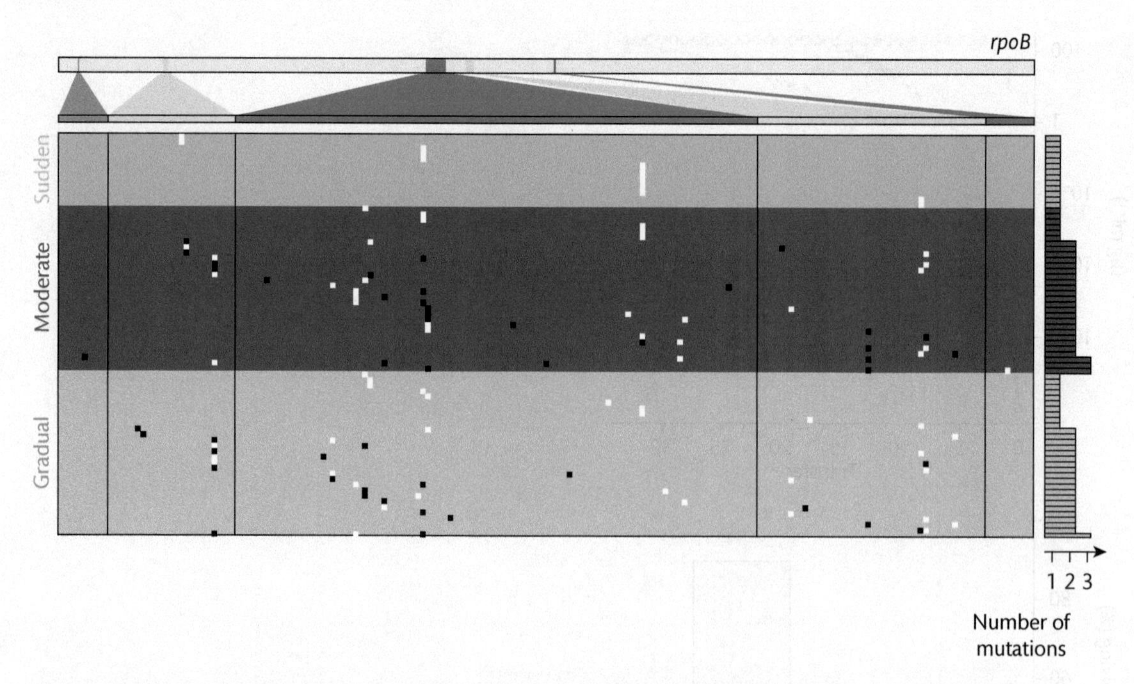


**Figure 7.22** This figure shows the location of mutations identified by sequencing the *rpoB* gene sampled from 73 populations of bacteria from the three different treatments. The first mutation detected in each population is marked in white and subsequent mutations are marked in black. The number of mutations detected in each isolate is represented in the bar chart on the right-hand side—you can see that the Sudden populations that survived had only single mutations, but many of the Moderate and Gradual populations had two or three mutations. All detected mutations were non-synonymous, and you can also see that the mutations were not evenly spread over the whole gene but clustered in particular regions (shaded areas on the gene map at the top of the diagram).



### Delve deeper

Much research effort is being directed at developing tools that help to predict the likely fate of populations and species under changing environmental conditions. Should such models take genetic change into account? Do you think that evolutionary rescue will have a tangible effect on species' responses to climate change? How will population size of endangered species influence the chances of evolutionary rescue?

### References


1. Bell, G. (2013) Evolutionary rescue and the limits of adaptation. *Philosophical Transactions of the Royal Society B: Biological Sciences*, Volume 368, page 20120080.
2. Gonzalez, A., Ronce, O. I., Ferriere, R., Hochberg, M. E. (2013) Evolutionary rescue: an emerging focus at the intersection between ecology and evolution. *Philosophical Transactions of the Royal Society B: Biological Sciences*, Volume 368, page 20120404.

# Drift

## Chance and necessity

*"Variations neither useful nor injurious would not be affected by natural selection, and would be left either a fluctuating element, as perhaps we see in certain polymorphic species, or would ultimately become fixed, owing to the nature of the organism and the nature of the conditions"*

Darwin, C. (1872) *On the origin of species by means of natural selection, or the preservation of favoured races in the struggle for life*, 6th edn. John Murray.

## What this chapter is about

Not all changes to the genome are driven by selection. Many alleles in a population make no difference to an individual's chances of reproduction. Instead of being removed by negative selection or fixed by positive selection, selectively neutral alleles will fluctuate in frequency over generations, until they are eventually either lost from the population or replace all other alleles at that locus. Neutral theory also allows us to predict the patterns of substitution we expect to see in the absence of selection. We can use this expectation to identify sites that are conserved by negative selection or mutations that have been promoted by positive selection. Neutral theory also provides insight into the way that even complex patterns and traits do not necessarily demand an adaptive explanation.

### Key concepts

- Substitution by drift occurs when random fluctuations in allele frequency result in the loss of all alleles at a locus but one
- The fate of a mutation is determined not only by its effect on the organism, but also by chance events, and the size and structure of the population in which the mutation occurs
- Comparing rates and patterns of substitution at different sites in the genome allows us to detect which sites are positively or negatively selected or neutral

# Neutral theory

## *Creatures of fortune*

Once, at a party on the roof of a biology department, up among the crocodiles wallowing in wading pools and birds conducting mate choice experiments in aviaries, I went around asking each of the biologists a question: if you were given a magic hotline that allowed you to get the true answer to one of the questions you spend so much time studying, what would you ask? Some scientists found it hard to answer this, possibly because for a lot of scientists it is the thrill of the chase that makes science so exhilarating, so an instant answer might take some of the fun out of it. But one researcher answered, with some degree of resignation, that he wanted to know for sure that the patterns in the evolution of bird colouring he had put so much effort into studying were not just the result of random, meaningless variation.

This reaction encapsulates how many scientists feel about neutral evolution: that finding that an intriguing pattern is actually the result of stochastic processes alone ruins a good evolutionary story, because it is considered less beautiful an answer than finding out that it's all explained by adaptation. This attitude may explain why there was such an outcry about the development of the neutral theory of molecular evolution (**TechBox 8.1**). Neutral theory suggests that most of the changes in the genome do not represent adaptations, mutations that became fixed because they conferred some natural advantage on their carriers. Instead, neutral theory proposes that molecular evolution is predominantly driven by random mutation producing variation and chance effects sending some of these variants to fixation. When it was first proposed, this non-adaptive explanation of genome evolution affronted many people. It was even referred to as non-Darwinian evolution, which is odd because Darwin was one of the first to describe the process.

***Chapter 7 introduced the concepts of substitution and fixation***

## Genetic load

Just as observations of phenotypic variation in natural populations were critical to the formation of Darwin's theory of natural selection, so observations of variation at the molecular level were instrumental in the formation of the neutral theory. The theory of natural selection was developed with phenotypic variation in mind—the physical or behavioural traits that could be observed by a naturalist, such as Darwin patiently dissecting his barnacles (Chapter 7). The mathematical theory of natural selection took a population-level view, imagining the fate of mutations swimming in the gene pool. But, at the time, there was very little raw data against which the theory could be tested. Genetic variation in natural populations could only be guessed at, or inferred indirectly by observing the fluctuating proportions of phenotypic variants, such as shell banding patterns in snails (**Figure 7.4**). Richard Lewontin, one of the pioneers of the study of genetic variation, described early population genetic theory as 'like a complex and exquisite machine, designed to process a raw material that no-one had succeeded in mining'.

The molecular genetic revolution changed that. DNA sequencing techniques and protein electrophoresis allowed scientists their first glimpse at the genetic variation in populations. And what they saw was surprising and, at first, inexplicable. A large proportion of the genetic loci examined showed detectable genetic variation. In some populations, at least a third of loci were polymorphic. This means that individuals in those populations would all differ slightly in the forms of at least some of their proteins. This hidden world of genetic variation was completely unexpected. If selection was the prime driver of substitution, then we would only expect to see a polymorphism if we just happened to observe the population while one mutation was going (rapidly) to fixation, so on the whole we would expect most sites to be uniform in all members of the population (**TechBox 7.2**). If all of that variation was subject to selection, then the chance of any individual being lucky enough to receive the fittest version of every single polymorphic locus would be very low, so all members of the population would suffer a reduction in fitness at some loci. This is often referred to as the genetic load of a population.

You can appreciate the power of genetic load if you talk to an animal breeder (as Darwin himself often did). For example, I once bought some rather handsome Wyandotte chickens (**Figure 8.1**) from an enthusiastic backyard breeder, who was trying to create a very specific combination of plumage colours and a particular pattern of 'lacing' on the feathers. Every year he breeds hundreds of chickens, but he selects only half a dozen of the best, those closest to his ideal Wyandotte, to be the breeding stock for the next generation. The problem is that if you try to select for several different traits at once (such as colours and lacing) then you have to generate a very large number of individuals that have non-ideal

**Figure 8.1** (a) Ida, (b) Grace, and (c) Julia.
Photographs by Lindell Bromham.

combinations of alleles in order to produce any individuals with just the right combination. Strong selection on multiple traits is wasteful in that it produces a lot of suboptimal individuals (which is OK if you can find someone like me who is happy to take the less-than-show-quality chickens). As J. B. S. Haldane (**Hero 7**) said in his 1957 paper on the costs of natural selection: 'It is well known that breeders find difficulty in selecting simultaneously for all the qualities desired in a stock of animals or plants. This is partly due to the fact that it may be impossible to secure the desired phenotype with the genes available. But, in addition, especially in slowly breeding animals such as cattle, one cannot cull even half the females, even though only one in a hundred of them combines the various qualities desired.'

So if all the variation uncovered by genetic methods was under selection, the cost of selection would be enormous: hardly any individuals would have the most fit combination of alleles, so most would be of lower fitness. In a sexual population, where alleles are sampled and remixed every generation, even those lucky few individuals with the best alleles would produce few offspring that were just as fit as themselves. A number of researchers, most notably Motoo Kimura (**Figure 8.2**), explained this excess of variation by postulating that most of the observed genetic variation was selectively neutral. If these different alleles were all functionally equivalent, they would (as Darwin foresaw) be left to fluctuate in frequency until one mutation became fixed by chance.

The rise of neutral theory led to vigorous debates concerning how much of the genome was evolving under selection and how much by drift. This debate is ongoing. We caught a glimpse of this debate in Chapter 5, when we discussed the evolution of genome size, and in Chapter 6 when we discussed introns. Just as the

**Figure 8.2** Motoo Kimura (木村資生: 1924–1994) began his studies at Kyoto Imperial University in 1944, studying botany as a way of avoiding military service, but focusing on studies of chromosomes (cytogenetics). There he was introduced to mathematical population genetics by reading the papers of J. B. S. Haldane (**Hero 7**) and Sewall Wright (**Figure 7.5**). With no way of duplicating the single copy of Wright's papers, Kimura wrote out his own copies by hand, adding his own notes and derivations as he did so. Kimura went to the United States on a Fullbright fellowship. On the ship on the way from Japan to the USA, he wrote a paper on fluctuating selection coefficients, which replaced the application of complex differential equations with the more tractable form of equations for heat conduction (published in the journal *Genetics* in 1954). He later brought a similar transformation to the description of the process of substitution by applying diffusion equations to the problem. He studied at the University of Wisconsin from 1954 to 1956, under the direction of Jim Crow (**Hero 3**). It is remarkable to note that Kimura published his mathematical solution for neutral genetic drift, then extended his models to allow multiple alleles, mutation, migration and selection, and wrote papers on Fisher's fundamental theorem and selection on linked loci, all during the two years it took him to complete his PhD.
Courtesy of the American Philosophical Society Library, USA.

biologist at the beginning of the chapter did not want to find out that magnificent plumage was the result of neutral processes, so many biologists have an instinctive dislike of the suggestion that much of the genome consists of 'junk DNA' that serves no particular purpose. How could something as wonderful and intricate as a working genome be described as being full of junk? But there is no need to be disappointed if the patterns you study turn out to be explicable by neutral processes. Instead, you should marvel at just how much variety and complexity can be generated by a simple, stochastic process being left to run over evolutionary time. In this chapter we will consider just what extraordinary ends can be achieved through the simple action of chance.

## Genetic drift

> *Fitness is a property, not of an individual but of a class of individual . . . Thus the phrase 'expected number of offspring' means the average number, not the number produced by some one individual. If the first human infant with a gene for levitation were struck by lightning in its pram, this would not prove the new genotype to have low fitness, but only that the particular child was unlucky*
>
> J. Maynard Smith (1998) *Evolutionary genetics*. Oxford University Press.

We have seen how the relative frequency of a particular allele can increase or decrease by selection. But we also have emphasized that the action of selection is probabilistic. Carriers of an advantageous allele may tend to have more descendants, but this is a statement about average outcomes, not about the fate of each individual with that allele. An individual that carries a mutation that makes its metabolism more efficient at high altitudes may die without issue when it is unluckily buried in an avalanche. Similarly, an individual with a mutation that reduces its oxygen transport efficiency may just be lucky enough to find a cache of food that allows it to keep its offspring alive over winter, despite its metabolic inefficiency. Life is full of surprises, so allele frequencies are always subject to chance events. We use the term genetic drift to describe the way that allele frequencies drift up and down due to random variation in survival and reproductive success.

Genetic drift comes about as the result of the sampling of alleles that occurs every generation when offspring are produced from the parent generation. In general, not all individuals in one generation will produce offspring. So the alleles in one generation are likely to be an incomplete sample of the alleles in the previous generation. Chance events may influence which individuals contribute alleles to the next generation and which don't. Understanding the role of chance is so critical to understanding patterns of molecular evolution that it is worth starting from the basics. So let's play with some elementary exercises in probability.

Take a coin out of your pocket. If it is a fair coin and is thrown without bias, then it has an even chance of landing on heads or tails. And, since each coin toss is independent of the last, when you toss it again, you again have a 50/50 chance of getting heads or tails. So, to state the obvious, you expect half of all tosses to result in heads (H), and half to result in tails (T). But any particular sample may deviate from this expectation, simply by chance. Try it now, toss your coin ten times. I did, and I got 60 per cent tails, 40 per cent heads:

THTTHHHTTT

Now let's turn our Hs and Ts into reproductive entities so that we can start to imagine how populations of template copiers behave. We have a polymorphic population of Ts and Hs, and we allow each one to have two offspring of its own kind, giving 20 offspring:

TTHHTTTTHHHHHHTTTTTT

Not all of them will survive to reproduce, so let's randomly kill 10 offspring (you can use a coin toss to decide who dies). In this instance, I just happen to have killed seven Ts and three Hs:

~~TT~~H~~H~~T~~TTT~~HHHH~~H~~HT~~T~~T~~T~~T

So the next generation we are left with 50 per cent H and 50 per cent tails:

HTHHHHTTTT

So the frequencies have changed in a single generation due to random events (**Figure 8.3**). I used this sample to produce a third generation by the same process and got 60 per cent heads and 40 per cent tails. At no point

**Figure 8.3**

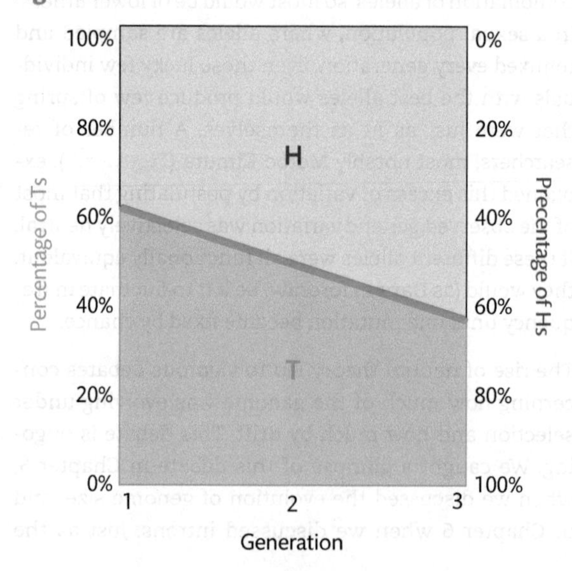

did Hs have a better survival probability than Ts, they just got lucky.

Allele frequencies fluctuate, sometimes up and sometimes down. But how is it possible for random fluctuation to result in substitution? Won't the frequencies just randomly drift up and down forever? Imagine we keep breeding from our Hs and Ts: each generation we randomly choose the survivors who will each have two offspring of their own kind. The frequencies of Hs and Ts in one generation are never very different from the frequencies in the previous generation. But it is possible to occasionally have a run of chance events that all go in the same direction, as we did in the first three generations when the percentage of Hs happened to increase. If it happens that there are more generations in which Hs get overrepresented, and Ts underrepresented, then the proportion of Ts will undergo a decline, and may even go as far as 0 per cent (**Figure 8.4**).

Substitution occurs because loss of a variant is irreversible (barring a new mutation). Unlike a coin toss, which is endlessly repeatable, in a population of reproducing entities if the frequency of one kind goes to zero, it's Game Over for that variant. Our population is no longer polymorphic. There are no more Ts to produce a new generation of Ts. There are only Hs that produce more Hs. H has gone to fixation at the expense of T. Substitution is the loss of all other variants at a particular genetic locus, leaving only a single type left in the population. (As an aside, extinction of a species represents the irrevocable loss of a large number of unique alleles all at once. Like the loss of single allele, once the genetic inheritance of a species is lost, it's gone forever: Chapter 9.)

No individual is immune to chance, and all alleles are subject to drift. The effect of drift will be more dramatic when an allele is at low frequency because it is at great danger of being lost from the population if its few carriers fail to reproduce. So when we consider the fate of mutations we always need to consider the effects of both selection and drift. If selection is very strong on a particular allele, then it may overwhelm the effects of drift. But, critically for understanding molecular evolution, when selection is weak or absent, drift can be a major determinant of allele frequencies. This is important when we consider the fate of neutral alleles.

## Neutrality

Mutations that confer on their carriers a relatively higher rate of successful reproduction will tend to increase at the expense of other variants until all members of the population carry copies of the same mutation at that locus. Mutations that decrease the chance of successful reproduction are likely to dwindle with each generation and eventually disappear. But, as Darwin noted (in the quote at the beginning of this chapter), some heritable changes will make no difference whatsoever to chances of surviving and reproducing. Mutations that have no effect on their own chances of successful reproduction are commonly referred to as being neutral with respect to fitness.

To illustrate the concept of neutrality, let's consider a hypothetical example. Imagine there is a population of wild geraniums that are polymorphic for flower colour. Some individuals produce red flowers, but some carry a mutation that means they produce a slightly different pigment, so they have dark pink flowers. Both types of flowers are equally attractive to their pollinators, so the two types of flowers are equally likely to be pollinated. Both pigments require the same amount of resources to make, so there is no metabolic advantage to producing one pigment rather than the other. In this case, it simply doesn't matter which colour a flower has: there isn't the slightest advantage to having one or the other. The mutation for the pink pigment has no influence on its chances of getting into the next generation. So the evolutionary fate of such a trait cannot be determined by selection.

**Figure 8.4**

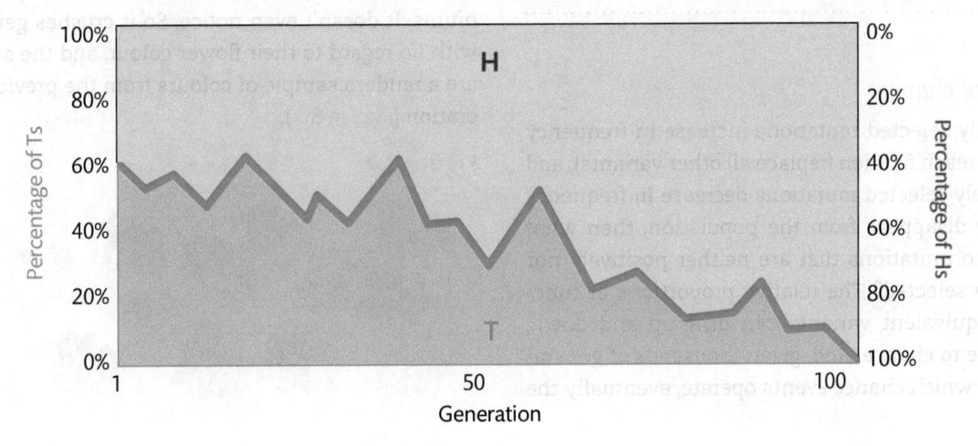

It is important to note selectively neutral alleles do not imply functionally unimportant sequences. Selective neutrality refers to functional equivalence, such that one variant could be exchanged for another with no effect on fitness. Flower colour is of critical importance to these geraniums. A mutation that ruined the flower pigment gene would result in colourless flowers that would not attract pollinators, so could set no seed. Such a mutation would be unlikely to find its way into the next generation: it would be subject to strong negative selection. But pink and red floral pigments are, in this example, functionally equivalent, and will make no difference to the chance of setting seed.

Neutrality is an important concept in molecular evolution because many heritable changes to the genome have no apparent effect on their carrier's chances of successful reproduction (**TechBox 8.1**). We have already seen some examples of neutral genetic changes. Microsatellites—runs of short tandem repeats—often occur in parts of the genome that don't contain any genes or regulatory elements. Different numbers of repeats in these sequences may be used to distinguish individuals, but they typically do not make any noticeable difference to morphology or behaviour. Consequently, an individual with six CA repeats at a particular microsatellite locus may not have any advantage or disadvantage compared to an individual with ten CA repeats. Even changes to functional sequences can be neutral. Some amino acids in proteins can be exchanged for similar amino acids with no apparent effect. These substitutions generate isozymes—functionally equivalent proteins—that can be detected biochemically, but have no effect on the individual's phenotype. But, as the geranium example above illustrates, a neutral trait does not need to be one that has no effect on morphology or behaviour. It can have a very noticeable effect, such as flower colour, but if it has no influence on relative reproductive success it will be selectively neutral.

*See Chapter 5 for an explanation of microsatellites*

## *The role of chance*

If positively selected mutations increase in frequency until they reach fixation (replace all other variants), and if negatively selected mutations decrease in frequency until they disappear from the population, then what happens to mutations that are neither positively nor negatively selected? The relative proportions of functionally equivalent variants can drift up and down, simply due to chance. And, given thousands of generations over which chance events operate, eventually the relative proportions may wander all the way down to zero (so that a variant disappears from the population) or all the way up to 100 per cent (a variant becomes fixed in the population so all members carry the same variant).

To illustrate the change in frequencies of functionally equivalent variants, let's go back to that hypothetical population of geraniums: some were pink, some were red, and it made no difference to their chances of survival and reproduction. Imagine a quiet hillside on which a patch of geraniums is growing (**Figure 8.5**). There are equal numbers of red and pink variants.

**Figure 8.5**

Lo and behold, along comes a moose and lies down on the patch of geraniums, creating random carnage (**Figure 8.6**). 60 per cent of the geraniums are crushed.

**Figure 8.6**

The moose doesn't care if it lies on red or pink geraniums. It doesn't even notice. So it crushes geraniums with no regard to their flower colour, and the survivors are a random sample of colours from the previous generation (**Figure 8.7**).

**Figure 8.7**

Once the moose has gone, the uncrushed survivors are pollinated and set seed, repopulating according to their kind (**Figure 8.8**). Red geraniums have red offspring, pink geraniums have pink offspring, and, though some individuals have more offspring and some have fewer, neither colour has consistently more offspring than the other.

**Figure 8.8**

The following year there is a drought, and the ground is so dry that few geraniums survive (**Figure 8.9**). The drought selects for plants with thick cuticles, long roots and good moisture retention. None of these traits have any connection to flower colour, and neither red nor pink is better at surviving drought than the other.

**Figure 8.9**

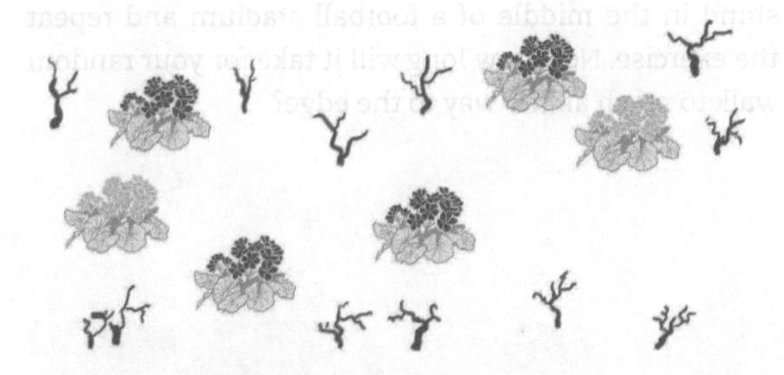

When the drought finally breaks, the rain collects in hollows in the ground. The geraniums set seed. The seeds drop on the ground, near the parent plant. Any seed that has the good fortune to land on a patch of damp ground has a good chance of germinating. Seeds on dry ground do not grow (**Figure 8.10**).

**Figure 8.10**

The next year, the seeds that germinated on the damp group grow up and set seed, according to their kind (Figure **8.11**).

**Figure 8.11**

Through a series of chance events—being squashed by a moose, subjected to a drought, seeds falling on dry ground—the pink variant has disappeared. In the first generation, the frequency of pinks was 50 per cent. After the moose attack, it was 40 per cent. After the drought it was 33 per cent. Of the seeds that set after drought, the percentage of pinks was 0 per cent. With respect to flower colour, we have gone from a polymorphism (pink + red) to a homogenous population (all red). There is no more possibility of fluctuation in percentages now. No matter how you sample this population, whether they get eaten by grasshoppers or buried under snow, all the offspring will have red flowers. The pink variant has been lost, so the red variant has gone to fixation.

Let's review the roles of chance and selection in this hypothetical scenario. The indifferent moose was a random disaster. It caused a lot of destruction but may not have had a selective effect on survival: it is hard to imagine selecting for moose-proof geraniums (though perhaps the ability to re-sprout when crushed would help). The drought was also a chance event but surviving the drought was not entirely due to luck. Those plants better able to withstand desiccation were more likely to survive, so the post-drought population had, on average, stronger cuticles and longer roots than the pre-drought population: a clear case of selection (see also **Figure 7.6**). Chance also played a role in the recovery after drought because some seeds were lucky enough to fall on fertile ground, while others were not (though perhaps certain adaptations might increase the chances of distributing over a wider area and having more chance of falling on damp ground). But, throughout all these events, there was no selection on flower colour. At no point did the reds have a better survival chance than the pinks. So why did the reds go to fixation, if they weren't any better? This came about simply because the population was finite, not all individuals reproduced successfully, so not all variants in the population were passed on to the next generation.

This kind of process is going on all the time, throughout the genome. Think, if we may get so personal, about the polymorphic sites in your own genome. Let's say you have a SNP somewhere on chromosome 4: one copy of chromosome 4 carries an A at this position, and the other carries a T, and both are equally viable. Now (getting very personal) when you make gametes, each gamete has an even chance of getting a chromosome 4 with an A or a chromosome 4 with a T. Very few of your gametes will go on to form an embryo, but those that do are a random sample, some have A and some have T. Imagine you have five children, 2 A-carriers and 3 T-carriers. You have just skewed the gene frequencies in favour of T, even though T was no better than A. Next generation the frequencies might even out again (say, if one of the T-carriers joins a monastery and doesn't have offspring) or it might get more skewed (if one of the T-carriers has 14 children). This random sampling happens every generation for every polymorphism. If the SNP has no effect on reproduction, then it will fluctuate in frequency. Eventually, by pure chance, the fluctuation will hit 0 per cent of A and 100 per cent of T, or vice versa. At that point, the polymorphism is gone, so there is no more fluctuation in frequencies of A and T.

# Population size

The effect of drift on allele frequencies will be greatest when the number of parents contributing alleles to the next generation is small. We can see the effect of sample size in our simple coin tossing experiments. Pick up your coin again, if you didn't lose it under the furniture in the last exercise, and go and stand in the middle of the room you are currently in (if you are lucky enough to be reading this book on a beach, then just draw yourself a room-sized square in the sand and stand in the middle of that). Toss your coin and take a step to the left if you get heads, or a step to the right if you get tails. Toss it again, and take another step, left for heads, right for tails. Since you have an even chance of getting heads or tails, chances are you will oscillate around the midpoint of the room. But you just might get a run of heads and move off to the left, or you might get a run of tails and move off to the right. If you keep doing this for long enough there is a chance that this random walk will take you all the way to the edge of the room, at which point you stop. How many coin tosses did it take you before you hit the left or right wall? (Figure 8.12). Now stand in the middle of a football stadium and repeat the exercise. Now how long will it take for your random walk to reach all the way to the edge?

**Figure 8.12** Genetic drift: Most molecular evolution textbooks contain a standard diagram showing the consequences of a random walk in a small population. However, I decided it was more fun to employ a team of small children to generate a similar distribution with coloured chalk, each child tossing a coin and moving to the left when they got a head, to the right when they got a tail (they were bribed by being allowed to keep the dollar coin they used to generate the random walk). You can see that although most lines wander around the 50 per cent mid-line, two lines hit the right-hand side (representing fixation of T and loss of H) and one reaches the left-hand side (representing fixation of H and loss of T). If you are the serious type you may wish to consult a population genetics textbook for a proper diagram. Otherwise, you can try this at home yourself, and at least you will get more fresh air than writing a computer simulation (though, in reality, the three-year-old couldn't draw straight lines and kept shouting 'I won!' at random junctures, the four-year-old took the coin and ran away, and only the six-year-old had the patience to persist to fixation).

Translating this somewhat tedious exercise into molecular genetics, mutations that have no effect on fitness are not subject to selection, so they will be randomly sampled every generation, and their frequencies will fluctuate up and down. In a small population, a run of random changes is more likely to result in substitution because it takes fewer steps in the same direction to get to 0 per cent or 100 per cent. In a large population it would take a really unlikely run of changes in the same direction to take you all the way to 0 per cent or 100 per cent. Given enough generations, even unlikely events will happen, so we expect substitution by drift to occur occasionally in large populations, but less often than in small populations. So now we get to a very important conclusion. We have seen how the fate of a mutation depends on its influence on the chance of reproduction, and on the environment in which it finds itself (Chapter 7). Now we can see that the fate of a mutation is also dependent on the size of the population in which the mutation occurs.

Because selection is a matter of probabilities, not certainties, even a selectively advantageous trait can be lost by chance events. If a mutation has a slight advantage, but it is in a small population where allele frequencies can vary substantially each generation due to chance, then the positive influence of selection on the frequency of that allele could be easily derailed by random sampling. Imagine running the geranium example again but this time allowing pink geraniums to have a 1 per cent selective advantage. If 100,000 geraniums are growing on our hillside, then we would expect the 1 per cent advantage to result in hundreds of extra pink offspring compared to reds. In this case, losing a few pinks to a wayward moose won't make any appreciable difference to the relative proportions of pinks and reds in the populations, and pinks will steadily rise in frequency each generation until they replaced reds. But in a small population, this slight selective advantage translates into a small number of extra offspring. In a population of 100 individuals, there might be one extra pink. An unlucky accident could easily erase this selective advantage—the moose's foot might just fall on that extra pink geranium. So the fate of a mutation depends not only on its effect on the individual's chance of reproduction, but also on the size of the population in which it finds itself (**TechBox 8.2**).

Genetic drift occurs because not all variants present in the parent population will make it into the next generation. So what we really mean by 'population size' is the number of parents that successfully get their alleles into the next generation. This will depend not only on the number of individuals in the population, but on the relative mating success of those individuals. There are many situations in which only a small fraction of the individuals in a population have the chance to contribute alleles to the next generation (**Figure 8.13**). We

**Figure 8.13** Who's a pretty boy then? This gorgeous chap belongs to a species known as the Lesser Bird of Paradise (which is no less fancy than the Greater Bird of Paradise, just a bit smaller). Male Lesser Birds of Paradise form leks, each occupying a specially-prepared branch in a mating arena. When a female approaches, the males shimmy and shake, showing their magnificent plumage to best advantage. Despite all their efforts to look good, few males get to mate. In one lek, observed over 18 days in which the males spent several hours each day displaying, there were 99 visits by females resulting in 26 matings. All but one of those matings were with a single male. Most of the males had nothing to show for all that expenditure on appearances.

saw an environmental example of this in our hypothetical geranium example, where the seeds of only a few individuals fell upon the damp ground and germinated. The mating system of a species will also influence the number of successful parents (Case Study 8.1). For example, there may be thousands of individuals in a honeybee hive, but usually only one female and a handful of males will give rise to all of the next generation. The relatively small number of parents may explain why eusocial bees and wasps (those with high reproductive division of labour) have a relatively fast rate of substitution. We can see the genetic consequences of reduction in the number of parents when we consider how inbreeding influences the process of substitution.

## Inbreeding

When I was a teenager my older sister gave me some advice. She said 'If you ever find yourself in the situation where you are the last woman on earth, and there is a single remaining man, whatever you do, do not touch him! If you start the human race again from just two people, your descendants will be so horribly inbred that it is better for the human race to die out at that point than to carry on.' This is, of course, just the kind of advice a growing girl needs.

A common cause of small effective population size is the foundation of a new population from only a few individuals. If a population starts from a small number of parents, then each subsequent generation must mate with close relatives if the population is to persist. But why should inbreeding (mating with relatives) result in unhappy offspring? After all, Darwin married his cousin and Darwin's children were not noticeably less well adapted than other children in the population. The effect of inbreeding is, like most patterns in molecular evolution, a matter of probabilities, not certainties. Mating between relatives increases the chances of offspring having two copies of the same mutation. If your grandfather carried a particular allele, his children and grandchildren may have inherited the allele from him. If you marry your cousin, your children could get one copy of that allele from their mother (who inherited it from her grandfather) and one copy of the same allele from their father (who also inherited it from the very same grandfather). In this way, inbred lines become progressively more homozygous, as with each generation there is more chance of each offspring inheriting two copies of the same allele, which was ultimately copied from a single ancestor.

Breeders take advantage of this progressive loss of variation in inbreeding by crossing related individuals to produce true-bred lines, pushing desirable traits to fixation (Figure 6.1). But as we saw in Chapter 3, most mutations are deleterious. The increased risk of having offspring which are homozygous for deleterious traits contributes to 'inbreeding depression' (loss of fitness due to shared ancestry of parents). Both of these processes are observable in pure-bred dog breeds, each of which is characterized by a suite of desirable traits and also a set of characteristic illnesses. For example, Newfoundland dogs were kept by fishing communities in Canada and were often used to pull in fishing nets. Newfoundland dogs have a number of heritable traits that adapt them to this lifestyle, such as good swimming ability, webbed paws and a thick waterproof coat (Figure 8.14). But the breed is also characterized by a number of less desirable heritable traits, such as being prone to bad hips and bladder stones.

We can extrapolate the effects of inbreeding to all small populations. Each mutation begins at low frequency, so in a large, randomly mating population, there is relatively little chance that two individuals with the same mutation will mate and produce offspring homozygous for that mutation. But the chance of getting two copies of a particular mutation will be greatly increased if there is a tendency to breed with related individuals. The likelihood of mating between relatives increases the smaller the population gets, so the chance of offspring being homozygous for any particular alleles rises as population size drops (Case Study 8.2). Eventually, for any given locus, all individuals are homozygous for a particular variant, and all other variants are lost. At this point, all individuals in the population carry a copy of

**Figure 8.14** Newfoundland dogs were bred to be at home in the water and many will happily jump into any water source they come across. This dog has been trained to rescue people from rivers.

the same mutation that originally occurred in a single ancestral individual. And in a small population there are relatively few genomes that might undergo mutation, so fewer new alleles are generated (**TechBox 8.1**). So, over many generations, small populations will undergo rapid substitution, and therefore loss of genetic variation.

Random sampling also plays an important role in allele frequencies in human populations. The role of chance in allele frequencies is particularly evident for small, intermarrying populations. For example, mutations that cause the recessive genetic disease Ellis–van Creveld syndrome (EvC, previously known as six-fingered dwarfism) are extremely rare in most populations, so it is unlikely that two people with EvC mutations will come together and have an affected child. But EvC is very common in a particular Amish population, who tend to only marry members of their own community. All 50 people with EvC in this population are descendants of a single couple who joined the community over 250 years ago, and twelve per cent of this population carry the EvC mutation as unaffected heterozygotes. This is an example of a founder effect: when a small group of individuals start a new population, they are unlikely to carry all of the alleles present in the parent population, and the alleles they do carry are effectively now at much greater frequency in the new population. The frequency of EvC mutations in the general population has been estimated at around 1 in 60,000 but because one person in the small founding population carried that mutation, it immediately jumped up in frequency by several orders of magnitude.

Some conservation biologists worry about the genetic effects of inbreeding in small populations. For example, there is some evidence that inbreeding can reduce fertility in populations reduced to a very small number of individuals, presumably through the increase in homozygosity of deleterious mutations (**Case Study 8.2**). Concerns have also been raised about the long-term effects of loss of heterozygosity in small populations, on the grounds that a low pool of standing variation may limit the ability of an endangered population to respond to future environmental change. It is difficult to evaluate the hypothesis that reduced genetic variation increases extinction risk, because small population size itself is a primary risk factor for extinction. Small populations are more likely to go extinct for the same reason that neutral alleles go to fixation in small populations. Random fluctuations in population size are more likely to reach zero for a small population, at which point it's Game Over for the population, as there are no individuals left to produce a new generation.

# Patterns of substitution

Neutral theory allows us to predict the patterns of molecular change we should see if there is no selection acting on mutations (**TechBox 8.1**). This is important. In previous chapters, we have seen how we expect the action of selection to influence the rate of change at different sites in the genome. Negative selection slows down the rate of change, because most mutations are removed. Positive selection increases the rate of change, because it drives advantageous mutations to substitution. So we can identify sites that might be under selection if we can find sites that are evolving faster or slower than expected if there was no selection (**TechBox 7.2**). Let's revisit a number of examples from earlier chapters to illustrate how different patterns of substitutions can be produced by the action of selection and drift. Given what we have learned, we can now make some predictions about the rate of substitution at different sites in the genome.

## *Conservation of sequences is due to negative selection*

We saw in Chapter 2 that nearly all organisms have the same amino acid sequence (RFGEME) in the active site of the RNA polymerase beta subunit, an essential enzyme involved in gene expression (**Figure 8.15**).

The DNA sequence that codes for the RFGEME part of the enzyme is just as likely to undergo mutation as any other, so why hasn't it changed? RNA polymerase is so critical to cellular function that any inhibition of its function would be disastrous, so we can guess that the chance acquisition of a mutation that alters the active site of this enzyme is likely to put its carrier at a reproductive disadvantage (to put it mildly). When we see a DNA or protein sequence that has a relatively slow rate of change, we can usually invoke negative selection: the sequence is functionally important to the organism's survival and reproduction, and so any mutation that disturbs that function is likely to disappear from the population. Although mutations continue to occur in this sequence, we tend not to see them, because their carriers did not have descendants. So negative selection is evident in the conservation of sequences.

**Figure 8.15**

| | |
|---|---|
| Homo sapiens | RDGGLRFGEMERDCQIA |
| Rattus norvegicus | RDGGLRFGEMERDCQIA |
| Drosophila melanogaster | RDGGLRFGEMERDCQIS |
| Neurospora crassa | RDGGLRFGEMERDCMIA |
| Oryza sativa | YGGGIRFGEMERDALLA |
| Escherichia coli | QFGGQRFGEMEVWALEA |

Even within functionally important sequences, some changes are more likely to be harmful than others. You can see in the alignment shown in Figure 8.15 that the human (*Homo sapiens*) and rat (*Rattus norvegicus*) sequences have exactly the same amino acid sequence (GRSRDGGLRFGEME). The fly (*Drosophila melanogaster*) and the mould (*Neurospora crassa*) differ at only one amino acid from human and rat (GRARDGGLRFGEME). But when we look at the DNA sequences we see far more changes. Here is the same section of the RNA polymerase gene, this time showing the underlying nucleotide sequence (Figure 8.16).

Now we can see that although the amino acid sequences of human and rat are identical, their DNA sequences vary at four positions (Figure 4.21). Human and fly differ by only one amino acid, but their DNA sequences differ at eight positions. And rat and mould have only one amino acid difference but 14 nucleotide differences. If the DNA sequence specifies the amino acid sequence, how is it possible to have so much change in the DNA sequence, but so little change in the protein?

The answer lies in the redundancy of the genetic code. In this sense, 'redundancy' does not mean useless or superseded, but refers to the way that several different codons can code for the same amino acid (TechBox 6.2). There are 64 different codons but only 20 common amino acids, so most of the amino acids are specified by more than one codon. Different codons that specify the same amino acid are referred to as synonymous codons. Just as synonyms are different words with the same meaning (e.g. 'injurious' and 'deleterious' are synonyms that both mean 'bad for the organism'), synonymous codons are different triplets that code for the same amino acid (e.g. ACC and ACT are synonymous codons that both code for threonine).

Changing the DNA sequence from one synonymous codon to another will not change the protein it produces, so, on the whole, we expect synonymous changes to be functionally equivalent and not subject to selection (there may be biochemical reasons for preferring some synonymous codons, leading to bias in codon usage, but we will gloss over that here for the sake of simplicity). Look at the first triplet in the alignment shown in Figure 8.16. Both GGT and GGC code for glycine (G). In the second triplet, both AGA and CGT code for arginine (R). So the nucleotide sequence differs but the amino acid sequence stays the same. If synonymous changes make no difference to phenotype, then it is considered to be selectively neutral. Since many third codon positions are free to vary without changing the amino acid sequence, they usually have a relatively fast rate of change. Conversely, most mutations in the first and second codon positions change the amino acid assignment of that codon. Since changes to the amino acid sequence of the protein are often deleterious, these will be removed by negative selection, giving a slower rate of change at first and second codon positions.

### *Neutral substitution rate is determined by the mutation rate*

If a mutation is neutral, then it will be neither promoted by positive selection, nor removed by negative selection. So the chance of any neutral mutation going to fixation is the same as any other neutral mutation. The rate of substitution of neutral mutations should be primarily determined by the mutation rate—the more mutations occur, the more substitutions there will be (TechBox 8.1). The substitution rate will be faster in neutral sites than it is in sites under negative selection, because none of the mutations in neutral sites are removed by selection. This prediction is borne out by inspection of non-functional copies of genes, known as pseudogenes. We saw in Chapter 5 that gene duplication is an important factor in genome evolution, because spare copies of genes may be free to evolve new functions (Figure 5.10).

**Figure 8.16**

| | |
|---|---|
| Homo sapiens | GGTAGATCTCGTGATGGTGGCCTGCGTTTTGGAGAAATGGAA |
| Rattus norvegicus | GGCAGATCGCGTGATGGTGGCCTGCGCTTTGGAGAAATGGAG |
| Drosophila melanogaster | GGTCGTGCTCGTGATGGTGGCTTGCGTTTCGGTGAGATGGAG |
| Neurospora crassa | GGTCGTGCCAGAGACGGTGGTCTGCGTTTCGGTGAAATGGAA |

**Figure 8.17**

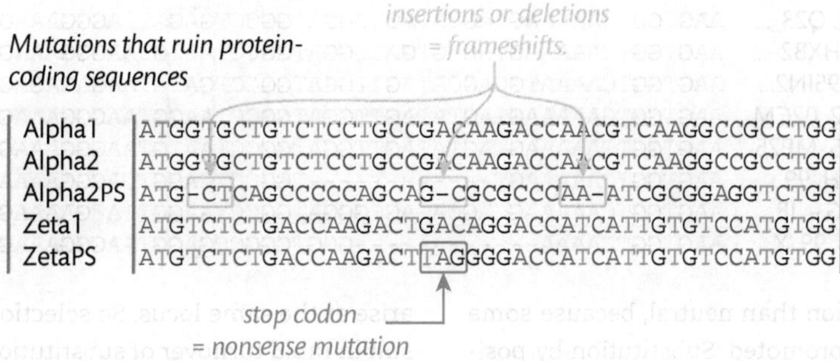

*Mutations that ruin protein-coding sequences*

*insertions or deletions = frameshifts.*

```
Alpha1    ATGGTGCTGTCTCCTGCCGACAAGACCAACGTCAAGGCCGCCTGG
Alpha2    ATGGTGCTGTCTCCTGCCGACAAGACCAACGTCAAGGCCGCCTGG
Alpha2PS  ATG-CTCAGCCCCCAGCAG-CGCGCCCAA-ATCGCGGAGGTCTGG
Zeta1     ATGTCTCTGACCAAGACTGACAGGACCATCATTGTGTCCATGTGG
ZetaPS    ATGTCTCTGACCAAGACTTAGGGACCATCATTGTGTCCATGTGG
```

*stop codon = nonsense mutation*

However, if a duplicated gene is excess to requirements, so that it makes no difference to its carrier's chance of reproduction whether it functions or not, then any mutation that destroys the function of the sequence will not be subject to negative selection. Since all mutations in the excess copy are selectively neutral, these non-functional gene copies have a very rapid rate of substitution (**Case Study 6.1**).

The globin gene families are a good example of gene duplication (**Figure 7.9**). All of the different globin genes are similar enough that we can tell they are homologous, originating by duplication from a single ancestral gene. In the human genome, there is a cluster of alpha globin genes on chromosome 16. There are two copies of the alpha globin gene, alpha1 and alpha2, both producing the alpha chains that combine with beta chains to make haemoglobin. A related gene, zeta globin, makes an alpha-like chain expressed in early embryonic development. There are also several pseudogenes, two of which are non-functional copies of the alpha2 and zeta genes. Here is an alignment of the first 45 nucleotides of these genes (**Figure 8.17**).

The genes in this alignment are clearly homologous, because they are similar at far more positions than we would expect by chance. The pattern of substitutions in the pseudogenes (Alpha2PS and ZetaPS) is different from the other genes in a number of respects. Critically, we can see that the pseudogenes have collected substitutions that destroy the ability of the sequence to produce a working protein. This becomes obvious when we translate the sequence into amino acids (**Figure 8.18**).

**Figure 8.18**

```
Alpha1    MVLSPADKTNVKAAWGKVGAHAGEYGAEALE
Alpha2    MVLSPADKTNVKAAWGKVGAHAGEYGAEALE
Alpha2PS  M?QRPG?AP?IAQVWDLIAGHEAQFGAELLL
Zeta1     MSLTKTERTIIVSMWAKISTQADTIGTETLE
ZetaPS    MSLTKT*GTIIVSMWAKISTQADTIGTETLE
```

In the zeta pseudogene, a GAG codon (glutamic acid, E) has acquired a substitution that has turned it into a TAG stop codon (marked with * in the alignment: **Figure 8.18**). Construction of an alpha chain from this gene would be terminated at this point, making a useless protein with only the first six amino acids. In the alpha2 pseudogene, insertions and deletions (indels) have disrupted the triplet coding structure of the sequence (marked with '?' in the alignment). These indels would result in a frameshift mutation that would destroy the amino acid sequence downstream of the mutation.

More generally, we can see that the pseudogenes have a higher rate of substitution than the functional genes. Furthermore, the substitutions in the pseudogenes are not biased towards silent changes (those that don't change the amino acid sequence of the protein). In a working gene, any mutations that cause a change in the amino acid sequence are likely to change the resulting protein for the worse, so are unlikely to become substitutions. But in a pseudogene, it doesn't matter what kind of mutations occur, because neither silent (synonymous) nor replacement (nonsynonymous) mutations make a difference to their carrier's chances of reproduction. Neutral sites are often used to estimate the baseline substitution rate in the absence of selection (**TechBox 7.2**). This is a useful thing to measure because when you see sites that have slower rate of substitution than the neutral rate, you can assume they are under negative selection. When you see sites with a much faster rate of substitution than the neutral rate, they may well be evolving under positive selection (**Case Study 10.2**).

## *Positive selection results in rapid substitution*

Negative selection results in slower substitution rates (most mutations are removed by selection). Neutral sites have a higher rate of substitution (mutations are not removed by selection, so by chance some will go to fixation). Sites under positive selection can have an even

**Figure 8.19**

```
A1. KE. 94. Q23_...    AAGTGGTCAAAAAGTAGCATAGTGGGATGGCCTGAGATTAGGGAAAGAATG
B. FR. 83. HXB2-...    AAGTGGTCAAAAAGTAGTGTGATTGGATGGCCTACTGTAAGGGAAAGAATG
C. IN. 95. 95IN2...    AAGTGGTCAAAATGCAGCATAGTTGGATGGCCTGATATAAGAGAGAGAATG
F2. CM. 02. 02CM..     AAGTGGTCAAAAAGTAGTATAGTTGGATGGCCTAAGGTAAGGGAAAGAATG
F2. CM. 95. MP25...    AAGTGGTCAAAAAGTAGTATAGTTGGATGGCCTAATGTAAGGGAAAGAATG
15_01B. TH. 99...      AAGTGGTCAAAAAGT------------TGGCCTCAGGTCAGGGAAAAAATA
02_AG. NG.-. IB...     AAGTGGTCAAAAAGCAGCATAGTGGGATGGCCTAAGGTTATGAAAAGAATG
14_BG. ES. 99. X..     AAGTGGTCAAAA------------GGGTGGGCCGAGGTAAGGGAAAGAATG
```

higher rate of substitution than neutral, because some mutations are actively promoted. Substitution by positive selection will tend to be faster than substitution by drift, because, unlike neutral mutations whose frequencies wander up and down, the frequency of an advantageous mutation is expected to steadily increase until it reaches fixation, replacing all other alleles at that locus.

Have a look at this alignment of sequences from the *nef* gene of the HIV genome (**Figure 8.19**). Thanks to a stupendously high mutation rate, large population sizes and a short generation time, HIV genomes change so fast that the virus evolves within a single infected individual over time. Not surprisingly, then, HIV sequences differ between patients. Each sequence in this alignment represents a virus from a different infected person.

There is one codon that stands out as being highly variable across the sequences. Does this tell us that this particular site must be neutral, so accumulating changes faster than the other sites in this important sequence? Look closely and you will notice a curious pattern. In most protein-coding sequences, we expect most of the changes to be in the third codon positions and relatively few changes in the first and second codon positions. But here we see there are more changes in the first codon position than in the third. Why?

It turns out that this site is within the CTL epitope (**Figure 8.20**). Cytotoxic T lymphocytes (CTLs), also known as killer T cells, are part of the human immune system. CTL response plays an important role in suppressing HIV replication in the human body. CTLs recognize specific viral protein sequences, called epitopes. A mutation in the HIV genome that occurs in a sequence coding for a CTL epitope may make that virus unrecognizable to the immune system. Any mutation that allows HIV to replicate unrecognized by the immune system will be rapidly amplified because carriers of that mutation will massively out-reproduce other members of the population. But the selective advantage of a particular mutation can change over time: once the immune system recognizes it, the allele will no longer have a reproductive advantage, and it will be replaced by any new immune-escape mutation that happens to arise at the same locus. So selection for novelty can result in rapid turnover of substitutions.

**Figure 8.20**

```
A1. KE. 94. Q23_...    KWSKSSIVGWPEIRERM
B. FR. 83. HXB2-...    KWSKSSVIGWPTVRERM
C. IN. 95. 95IN2...    KWSKCSIVGWPDIRERM
F2. CM. 02. 02CM..     KWSKSSIVGWPKVRERM
F2. CM. 95. MP25...    KWSKSSIVGWPNVRERM
15_01B. TH. 99...      KWSKS----WPQVREKI
02_AG. NG.-. IB...     KWSKSSIVGWPKVMKRM
14_BG. ES. 99. X..     KWSK----GWAEVRERM
                               CTL epitope
```

Unlike most protein-coding sequences, where most changes to the amino acid sequence are unlikely to help the organism, changing the amino acid in the key position in the CTL epitope may effectively hide the virus from the immune system, giving it a massive selective advantage. So selection for novelty creates a fast rate of amino-acid changing substitutions at this site. There have been many studies that use these kinds of patterns in order to identify sites in the HIV genome that might be evolving under positive selection, in the hope that these sites will provide useful targets for antiretroviral therapy or an effective vaccine.

## Chance and complexity

> “ *... familiarity with the 'patterns' that random processes create is an essential piece of a scientist's mental furniture* ”
>
> Sean Nee (2006) Birth-Death Models in Macroevolution. *Annual Reviews in Ecology, Evolution and Systematics*, Volume 37, pages 1–17

If we want to identify sites under selection, we will usually need to do so by comparing our observations to what we would expect to see if there was no selection operating. Is that site changing rapidly due to natural selection? If it is changing much faster than it could if it was neutral, then maybe it is. Why are those sites conserved between species? If they are much more similar than we would expect from neutral sites, then we can consider the possibility that they are being maintained by negative selection.

In this sense, the neutral theory provides us with a null model. A null model is a way of generating predictions about patterns what we might see when the process we are interested in detecting is not in operation. This is important, because before we can get excited about spotting a pattern in the genome that we think is the result of natural selection, we first need to consider whether we could get exactly the same pattern through random processes. And you would be surprised what you can create with nothing more than random processes (**Figure 8.21**).

We saw in Chapter 5 that the debate about genome size has challenged notions of the evolution of complexity, because sometimes the biggest genomes belong to apparently less complex species, and simpler organisms can have many more genes than their more impressive relatives (**Figure 5.3**). It has been argued that one of the driving forces in the evolution of genome size is not selection for bigger and better, but random processes generating excess DNA which then goes to fixation by drift. There are several key assumptions underlying this hypothesis. One is that increases in DNA happen more often than decreases. The second assumption is that many increases in amount of DNA are slightly, but not devastatingly, harmful.

There are two kinds of costs to carrying more DNA than is strictly necessary. One is that copying and maintaining DNA takes time and energy, so a bigger genome costs more to make and run. The other potential cost is that the more DNA you have, the more chance you have for something to go wrong. A mutation in any part of the genome could end up doing you a lot of harm. This is obvious when we think about having more essential genes, a mutation in any one of which might spell the end for an individual's reproductive potential. But it also applies to non-coding DNA. Mutations in apparently useless DNA might create something that actually impacts on phenotype, and not necessarily in a good way. For example, gain-of-function mutations might cause previously non-translated sequences to be transcribed and translated into useless peptides that can have toxic consequences for the cell. Or mutations in intergenic regions may cause overexpression of genes that should be downregulated or turned off.

So excess DNA might be a bit of a burden on the carrier's economic balance and it might impose a slightly increased risk of mutation, but perhaps many organisms can get by almost as well as their neighbours when carrying a little bit more DNA. Mutations like this, which aren't great but aren't really terrible (that is, they have small selective coefficients: **TechBox 7.2**), are referred to as 'nearly neutral' (**TechBox 8.2, Hero 8**). And here is the critical point: the efficiency of natural selection at removing these slightly deleterious mutations depends on the effective population size. In a big population, even a small fitness cost will tend to result in a steady reduction in frequency of a nearly neutral mutation until it disappears from the population. But in a small

**Figure 8.21** Many theories have been devised to explain the high diversity of tropical rainforests, and the factors that determine which species are found where. What if the mix of species found at a given site is largely determined by chance? This proposal was so confronting that, at a conference on tropical biodiversity in 1998, I saw biologists literally on the edge of their seats listening to Stephen Hubbell explain his 'unified neutral theory of biodiversity' which models the patterns we would expect to see if species composition at a site is determined by stochastic processes rather than selection. This is not a claim that selection does not act to adapt species to particular niches, but a theoretical framework for generating predicted distributions against which to compare observed patterns of diversity. Although there has been much debate about whether species distributions fit the predictions of this neutral model, and how to interpret distributions that do, neutral models in ecology are not a new phenomenon. For example, the island biogeography theory of Macarthur and Wilson (1967) predicted the expected number of species in an island given the balance between immigration rate (new species colonizing the habitat) and the extinction rate (survival probability of colonizing taxa).

Image courtesy of Neil Palmer/CIA, licensed under the Creative Commons Attribution Share Alike 2.0 Generic license.

population, where random sampling can derail the action of selection, these slightly costly mutations can increase in frequency, simply by chance, until they are present in all members of the population. So in a small population, little increases in DNA amount can become fixed, and so genome size can creep up, and up, and up.

Non-adaptive increases in complexity need not be confined to puffing out the genome with non-coding DNA, repeat sequences and transposons. For example, we need to consider the possibility that increases in gene number can also occur by neutral processes. We saw in Chapter 5 that whole gene sequences can be duplicated, and in Chapter 6 we learned that gene copies can also be created when DNA copies of messenger RNA transcripts are reinserted into the genome. Some of these extra gene copies decay into pseudogenes, some add to the output of the original gene, and some change and take on new, or more specialized, functions. But, if gene copies are constantly being generated, and as long as extra copies are not very costly, then they may also go to fixation by drift. So just because a gene family has multiple copies, we can't necessarily infer that the greater complexity makes the organism work better that it would have done with only one.

### Not as neutral as you think

Just as we should not assume that apparently functional traits evolve by natural selection when they might have arisen by chance, so we should be careful not to assume that apparently useless variation is entirely neutral when it might have some selective value that we haven't identified. In Chapter 5, we briefly discussed the current debate over how much of the intergenic DNA in the genome is functionless, in the sense that it makes no useful contribution to host phenotype. There are short stretches of nucleotide sequences within these intergenic regions that are remarkably conserved between distantly related species. For example, some short sequences, around 200 nucleotides long, are nearly identical between humans and mice, despite many tens of millions of years of independent evolution (these sequences are sometimes referred to as ultra conserved elements, or UCEs). We would expect any non-functional sequences to have accumulated quite a few substitutions in that time, so their resistance to change suggests that selection is removing any mutations from these sequences, which in turn implies that they have an important function. Indeed, some of these conserved elements seem to be associated with human diseases, suggesting that they have a key regulatory role in the genome. Similar conserved blocks can be found in some introns. This reminds us to think carefully about assuming that any sequences represent the neutral rate of evolution. For example, the synonymous substitution rate is often interpreted to reflect the mutation rate on the assumption that a 'silent' change to a protein-coding gene can't be under selection. But evidence is growing that some synonymous codons are used more than others, implying a degree of selection on synonymous sites. This selection is probably quite weak, so it might not have a practical impact on estimates of rates. Nonetheless, codon bias is a further reminder not to jump to conclusions about the neutrality of apparently selection-free mutations.

The trick is to think of ways of testing between alternative scenarios in order to weigh up whether particular genome features are card-carrying adaptations or generated primarily by random processes. For example, humans use multiple beta globin genes to make haemoglobin, switching between versions at different developmental stages (**Figure 7.9**). Is this so that selection can fine-tune the different globin genes to work more exactly at particular stages of development, or is it the sign of a crazy, overcomplicated construction, using multiple genes where one beta globin would do the job just as well? If you think of a way of testing this, then please go and find the data you need to do it and let me know when you have an answer. Then I can put it in the next edition of this book.

So it may be that, at least in some cases, complexity increases due to a failure to stay simple. After all, who says that being bigger and more complex makes you more successful? Hats off to unicellular organisms that dominate the biosphere with much tinier genomes than yours. If you think your genome is better than theirs, try to maintain that smug feeling next time you suffer a bacterial infection.

## Conclusion

Not all changes at the molecular level are driven by selection. Random sampling of alleles can drive some mutations to fixation by chance, particularly in small populations. Because of this, small populations tend to undergo rapid substitution and loss of genetic variation. Genetic drift is the prime determinant of allele frequencies for mutations that have little or no impact on chances of survival and reproduction. We can use our expectation of the rate of substitution of neutral mutations to interpret patterns of substitution in the genome. Sites under negative selection, for which most mutations are harmful, will tend to evolve more slowly than neutral sites, but sites under positive selection, where some mutations are actively promoted, will often evolve faster than neutral sites. However, we should be cautious when making assumptions about which kinds of heritable changes are neutral. In some cases apparently trivial changes may be under selection, in other cases complex features that have been interpreted as adaptations may actually have been generated by purely stochastic processes.

Selection and drift are the engines of evolution, driving mutations from their origins in an individual genome to increase in frequency over many generations until copies of that original mutation are found in all members of a population. One of the tenets of the modern evolutionary synthesis, which is the refinement of Darwin's theory in light of population genetics, is that there is no qualitative difference between the evolutionary changes that happen with populations (microevolution) and the differences that distinguish different species (macroevolution): differences between species are simply the outcome of a long period of population divergence. The accumulation of genetic differences in populations through the process of substitution occurs continuously and the longer two populations are separated the more different substitutions they acquire, so the more they diverge from each other. In the next chapter, we will learn how to compare DNA sequences from different populations in order to identify consistent genetic differences between them.

## Taking stock: what have we learned so far and where are we going next?

In the first six chapters, we primarily focussed on individual genomes, considering the way the genetic information is expressed, altered and passed on to offspring. Then in Chapters 7 and 8 we took the leap from individuals to populations, to consider how a mutation that occurs in a single individual can become a standard part of the genome of all members of a population. We covered the two processes that govern which mutations go to fixation in the population: selection and drift.

Now we are going to build on our understanding of the way mutations in individuals can rise in frequency until they become substitutions carried by the whole population, in order to begin to take the long view. We will consider how the process of substitution generates differences between populations. In Chapter 9, we move from the population to the species, as we learn how to compare the genetic differences between species and higher taxa. Chapter 10 leads us from the species to the lineage, by considering how the substitution creates hierarchies of similarities that can be used to reconstruct the evolutionary history of biological lineages. Then in Chapters 11 and 12 we look more closely at the way that we use the information in DNA to discover the relationships between lineages and test ideas about evolutionary patterns and processes.

# Points to remember

### Neutral theory

- There is more genetic variation between individuals in populations than would be expected if all alleles were under selection.
- Selective differences between all alleles would result in high genetic load through the production of many individuals with sub-optimal allele combinations.
- Some neutral alleles don't affect phenotype, others create differences that are functionally equivalent, so that they have no fitness advantage or disadvantage.

- Neutral alleles can go to fixation by drift if chance fluctuations in frequency result in the loss of all alternative alleles at a locus.
- Neutral theory proposes that most substitutions are the result of random fluctuations in allele frequency, rather than being driven by selection.

### Population size

- Allele frequencies are always subject to chance events.
- In a small population, it takes fewer chance increases in frequency for an allele to go to fixation, so drift can overwhelm selection for mutations of small selective effect.
- The effective population size, which reflects the chance of alleles being passed to the next generation, is reduced by unequal mating opportunities or inbreeding.

### Patterns of substitution

- Neutral theory provides a null model against which to compare the observed patterns of molecular evolution to those expected if selection is not operating.
- Every neutral mutation has the same chance of going to fixation by drift, so the rate of substitution is determined by the rate at which they are generated by mutation.
- Functional sequences typically have lower substitution rates than the neutral rate, because many mutations are removed by negative selection.
- Sites under positive selection can have a high rate of substitution, as some mutations will rise rapidly in frequency due to fitness advantage.
- Genomes could increase in size and complexity if increases in DNA amount or genome interactions are common and only slightly costly, so that they go to fixation by drift, particularly in small populations.

## Ideas for discussion

**8.1** Some genes are present in multiple copies in the genome, others are represented by only single copies. Why don't organisms maintain multiple 'back-up' copies of all important genes, so that if a deleterious mutation ruins one gene, it can continue to function using the spare copy?

**8.2** The different versions of the alpha globin gene in the human genome were created by a process of gene duplication. Looking at **Figures 8.17** and **8.18**, is it possible to tell which of the pseudogenes was created first, and which is more recent?

**8.3** Is the influence of population size on the rates and patterns of molecular evolution likely to have a significant impact on effort to save critically endangered species from extinction? How would you test the impact of small population size on long-term success of conservation programmes? What approaches could be taken to mitigate any predicted or observed effects?

# Sequences used in this chapter

**Table 8.1: Human alpha and zeta globin gene sequences used in Figure 8.17**

| Description from GenBank entry | Accession |
|---|---|
| *Homo sapiens* hemoglobin, alpha 1 (HBA1), mRNA. | NM_000558 |
| *Homo sapiens* hemoglobin alpha 2 (HBA2) gene, complete cds | DQ499017 |
| *Human alpha*-2 pseudogene of alpha-globin gene cluster | X03583 |
| *Homo sapiens* hemoglobin, zeta (HBZ), mRNA | NM_005332 |
| *Homo sapiens* HBZP pseudogene, exons 1, 3, 2 and complete cds. | HUMHBA3 |

**Table 8.2: HIV *Nef* gene sequences used in Figure 8.19**

| Description from GenBank entry | Accession |
|---|---|
| HIV-1 isolate Q23-17 from Kenya, complete genome | AF004885 |
| Human immunodeficiency virus type 1 (HXB2), complete genome; HIV1/HTLV-III/LAV reference genome | K03455 |
| Human immunodeficiency virus type 1 (HXB2), complete genome; HIV1/HTLV-III/LAV reference genome | AF067155 |
| HIV-1 isolate 02CM.0016BBY from Cameroon gag protein (gag) and pol protein (pol) genes, partial cds; vif protein (vif), vpr protein (vpr), tat protein (tat), rev protein (rev), vpu protein (vpu), and envelope glycoprotein (env) genes, complete cds; and nef protein (nef) gene, partial cds | AY371158 |
| Human immunodeficiency virus type 1 proviral mRNA for partial GAG protein, partial POL protein, partial ENV protein and partial NEF protein isolate 95CM-MP255 | AJ249236 |
| HIV-1 isolate CRF01_AE/B from Thailand gag protein (gag) gene, complete cds; pol protein (pol) gene, partial cds; and vif protein (vif), vpr protein (vpr), tat protein (tat), rev protein (rev), vpu protein (vpu), envelope glycoprotein (env), and nef protein (nef) genes, complete cds | AF516184 |
| Human immunodeficiency virus type 1 gag polyprotein (gag), pol polyprotein (pol), vpr protein (vpr), vif protein (vif), rev protein (rev), vpu protein (vpu), env polyprotein (env), and nef protein (nef) genes, complete cds | L39106 |
| HIV-1 isolate X397 from Spain, complete genome | AF423756 |

# Examples used in this chapter


Beehler, B. (1983) Lek behavior of the lesser bird of paradise. *The Auk*, Volume 100, page 992.

Bromham, L., Leys, R. (2005) Sociality and rate of molecular evolution. *Molecular Biology and Evolution*, Volume 22, page 1393.

Hubbell, S. (2001) *The unified neutral theory of biodiversity and biogeography*. Princeton University Press.

Lynch, M. (2007) *The origins of genome architecture.* Sinauer Associates.

MacArthur, R. H., Wilson, E. O. (1963) An equilibrium theory of insular zoogeography. *Evolution*, Volume 17, page 373.

McKusick, V. A. (2000) Ellis-van Creveld syndrome and the Amish. *Nature Genetics*, Volume 24, page 203.

HEROES OF THE GENETIC REVOLUTION

8

# Tomoko Ohta (太田 朋子)

*"Kimura's theory was simple and elegant, yet I was not quite satisfied with it, because I thought that natural selection was not as simple as the mutant classification the neutral theory indicated, and that there would be border-line mutations with very small effects between the classes. I thus went ahead and proposed the nearly neutral theory of molecular evolution in 1973. The theory was not simple, and much more complicated, but to me, more realistic, and I have been working on this problem ever since"*

Tomoko Ohta (2012) *Current Biology*, Volume 22, Issue 16, pR618–R619

**EXAMPLE PUBLICATIONS**

Ohta, T. (1973) Slightly deleterious mutant substitutions in evolution. *Nature*, Volume 246, page 96.

Ohta, T. (1980) *Evolution and variation of multigene families.* Springer Verlag.

**Figure 8.22** Tomoko Ohta (born 1933) has enriched our understanding of the complexities of molecular evolution by exploring the interplay between selection and drift on mutations of small selective effect.



Tomoko Ohta (**Figure 8.22**) went to school during the post-war period, a time of hardship in Japan when many children were encouraged to work rather than study. But it was also a time when educational opportunities were opening up, so Ohta was among the first generation of Japanese women allowed to study at the good universities, and she was one of the first women to graduate as a scientist in Japan. Although she had enjoyed mathematics, she didn't feel it offered good employment prospects, so she studied horticulture. After graduating, she worked first in publishing and then in genetics of crop plants. A Fullbright scholarship gave her the opportunity to study in the United States, where she took a PhD in population genetics.

When she returned to Japan in 1967 there was only one research group working on population genetics, so she persuaded Motoo Kimura (**Figure 8.2**) to take her on as a postdoctoral researcher, though it took two years for her to be officially hired. She worked with Kimura to marshal evidence for his revolutionary neutral theory of molecular evolution (**TechBox 8.1**), and, together with Kimura, published a suite of papers on theoretical population genetics, exploring the role of mutation and drift in shaping molecular evolution.

Kimura's elegant neutral theory classified mutations as either neutral, good or bad, but Ohta felt that natural selection was unlikely to be that simple. She proposed that, between these three simple classes, there would be many mutations of small selective effect. Ohta explained that, just as the revision of the neo-Darwinian view of evolution in the light of evidence of the high levels of genetic variation had led to the development of the neutral theory, so the neutral theory itself needed refinement when considering the range of possible selective effects of mutations. She felt that although this added complexity to theoretical population genetics, it was a more realistic view of molecular evolution.

Ohta worked to find empirical support for the neutral and nearly-neutral theories, for example providing one of the early tests of the generation time effect on rate of molecular evolution, which can be interpreted as evidence that the supply of mutation is a major driver of the rate of sequence change. Much of her subsequent work explored the interaction between drift and selection in the evolution of gene families, particularly the role of weakly selected (nearly neutral) mutations in driving the diversification of duplicate genes, and the role of gene conversion in evolution. When Ohta began her career there was very little data against which to test theories of population genetics. She is now turning her attention to developing ways of analysing the great waves of data emerging from genomics. Although she officially retired from the National Institute of Genetics in 1996, she continues to publish scientific articles, for example, exploring the relevance of nearly neutral theory to the evolution of gene regulatory networks. Ohta's work has been recognized by many awards including the Japan Academy Award, Saruhashi Prize, and Foreign Membership of the National Academy of Science of the United States. She was recognized as a 'Person of Cultural Merit' (文化功労者) by the Japanese emperor in 2002. Ohta received the 2015 Crafoord Prize for Biosciences, the equivalent of a Nobel Prize, for 'pioneering analyses and fundamental contributions to the understanding of genetic polymorphism'.

> "You can't have an intelligent discussion about genome evolution, adaptationism, molecular evolution, or junk DNA without a firm grasp of Nearly Neutral Theory. It's a shame there's no Nobel Prize for evolution."
>
> Laurence A. Moran (2012), Sandwalk.blogspot.com

TECHBOX
8.1

# Neutral theory

**RELATED TECHBOXES**

**TB 7.2:** Detecting selection
**TB 8.2:** Population size

**RELATED CASE STUDIES**

**CS 8.1:** Substitution
**CS13.2:** Diversification

> *The neutral theory asserts that the great majority of evolutionary changes at the molecular level as revealed by comparative studies of protein and DNA sequences, are caused not by Darwinian selection but by random drift of selectively neutral or nearly neutral mutants. The theory does not deny the role of natural selection in determining the course of adaptive evolution, but it assumes that only a minute fraction of DNA changes in evolution are adaptive in nature, while the great majority of phenotypically silent molecular substitutions exert no significant influence on survival and reproduction and drift randomly through the species*
>
> Motoo Kimura (1983) *The neutral theory of molecular evolution*. Cambridge University Press

**Summary**

Neutral theory was devised to explain the high and constant rate of molecular evolution by suggesting that most changes to the genome are driven by drift not selection.

**Keywords**

selection, substitution, mutation, nearly neutral theory, Kimura, Ohta

The idea that some mutations might have no influence on fitness, and therefore not be subject to selection, is not a new one. Not only had Darwin discussed how neutral variation could remain as a polymorphism until fixed or lost, early writers on molecular evolution in the early

1960s had also considered the possibility that many features of molecular evolution, such as differences in base composition between species, might be the result of random processes rather than selection. But the person that did the most to bring about the transformation in the view of the role of chance in molecular evolution was Motoo Kimura (**Figure 8.2**).

Kimura developed the neutral theory as an explanation of the patterns emerging from the growing collections of homologous protein sequences (at that time limited to the amino acid sequences for a dozen or so proteins from a small number of species, a far cry from the billions of nucleotides of DNA now available in GenBank: see **Hero 10**). Firstly, there was much, much more variation between protein-coding genes than had been expected from a selection-based view of molecular evolution. Secondly, evidence was emerging that the rate of amino acid substitution was too high to be reasonably accounted for by natural selection alone, given the amount of genetic load resulting from each substitution. Thirdly, not only was the amino acid substitution rate higher than expected, it also appeared to be remarkably constant in different species (an observation sometimes referred to as the molecular clock: see Chapter 14). These observations did not match the patterns expected if molecular change was predominantly driven by selection, in which case it might be expected to be characterized by occasional bursts of substitution, connected with phenotypic change, and lineages with a more rapid pace of morphological change should have faster molecular evolution than species that changed little over millions of years. Instead, it seemed that most proteins steadily acquired changes over evolutionary time.

Just as Darwin had explained many puzzling facts of phenotypic variation—such as the geographic and temporal association of similar organisms, adaptation to the environment, and hierarchical structuring of phenotypic similarities across species—with a single elegant theory (evolution by natural selection), so Kimura explained many puzzling facts about molecular variation—such as a steady and surprisingly rapid rate of molecular change and high frequency of polymorphic loci—with a single elegant theory (that most change at the molecular level is driven by drift not selection). Many of the analyses described in this book are ultimately based on the neutral theory (or its offspring, the nearly neutral theory: **Hero 8**).

Since neutral alleles have no advantage or disadvantage over each other, each one has the same chance of going to fixation. So to estimate the rate of substitution of neutral mutations we need to know how many neutral mutations are produced each generation and what percentage of them is expected to go to fixation. Both of these quantities depend on the population size. The smaller the population, the more likely it is that random sampling of alleles will result in the loss of all alleles but one, so the percentage of neutral mutations going to fixation will go up as population size goes down. But, the smaller the population is, the fewer mutations are produced each generation, simply because there are fewer genomes to undergo mutation. A larger population produces more mutations, but the probability of any one mutation being fixed is lower because random sampling error is less likely to result in fixation. Increasing population size increases the number of mutations but decreases the proportion of mutations that will become substitutions. This leads to the beautiful conclusion that, for purely neutral mutations, the effect of population size rather neatly cancels out to leave the substitution rate determined solely by the mutation rate[1].

This simple model was used to explain not only the high level of polymorphism in natural populations (if most mutations are neutral they will not be removed by selection), but also the apparently constant rate of molecular evolution (if substitution rate is driven by the mutation rate, then as long as the mutation rate remains constant, substitutions will accumulate at a fixed rate over time). The neutral theory also led to predictions about the patterns of molecular evolution across the genome. If most substitutions were driven by positive selection, then we would expect parts of the genome most critical to survival and reproduction to accumulate the most substitutions. Instead, sites that have no apparent impact on phenotype, such as microsatellites in non-coding DNA, non-functional pseudogenes, and silent sites in protein-coding sequences, evolve faster than functionally important sequences such as genes and regulatory sequences. These observations support a prominent role for drift in molecular evolution.

The neutral theory was further refined by Tomoko Ohta, a colleague of Kimura's (**Hero 8**). Where the neutral theory was based on the assumption that the majority of mutations were either strongly deleterious or strictly neutral, Ohta considered the effect of having many mutations with only slight effects on fitness[2]. The majority of these 'nearly neutral' mutations would be slightly deleterious, so in a very large population, they would be removed by selection. But in small populations, random sampling effects have a greater influence on allele frequencies, and so drift can overwhelm selection for mutations of small selective coefficients (**TechBox 7.2**). So the proportion of nearly neutral mutations going to fixation should rise as population size decreases. We can measure this effect in real populations. If we assume that, for a protein-coding gene, synonymous changes (that do not change the amino acid sequence) are likely to be neutral, but some nonsynonymous changes (that do change the amino acid sequence) will fall into the 'nearly neutral' category, then we would expect

**Figure 8.23** What's the story? Even apparently obvious cases of selection might be less clear cut than they first appear. Prince of Wales heath flowers (*Erica perspicua* (a) and (b)) occur in polymorphic populations, with pink and white flowers (here shown in photos from the Fernkloof reserve in South Africa). Aviary experiments show that sunbirds (*Anthobaphes violacea*, (c)) prefer the pink ones. Since sunbirds like to visit pink flowers more than white ones, we would expect pink-flower alleles to have a selective advantage due to higher pollination rates. But, as it happens, pink flowers seem to set seed at the same rate as whites. Maybe, even if sunbirds prefer pink, they end up going to the nearest flower, pink or white, erasing any advantage of pink attractiveness. Morning glories (*Ipomoea purpurea*, (d)) also occur in colour polymorphic populations, and have been an important case study for investigating the genetic basis of flower colour variation and its effect on pollinators. Yet an analysis of genes involved in flower colour found no evidence of selection for wild populations of morning glory: all allele frequencies matched the expectation from neutral theory.

*Erica perspicua* photographs by Christine Wakfer. Sunbird image courtesy of Michael Clarke, licensed under the Creative Commons Attribution Share Alike 2.0 Generic license.

smaller populations to have a higher ratio of nonsynonymous to synonymous substitutions. This effect has been observed in fruit flies with smaller population sizes[3], species confined to islands[4], and even tiny populations of bacteria and fungi sequestered in the guts of insects[5] (**TechBox 8.2**).

In addition to having a revolutionary impact on the way we view the process of evolution at the genomic level, the neutral and nearly neutral theories form an important basis for many of the analyses described in this book. It is important to note that neutral and nearly neutral theories do not deny the role of natural selection in generating adaptations or conserving functionally important sequences. The debate concerns the relative proportion of changes in the genome that are due to drift or selection. For many years there was a hearty debate between 'selectionists', who considered that the patterns of molecular evolution were best explained by the action of natural selection, and 'neutralists', who considered the neutral theory to provide a more plausible explanation. But the field is less black and white these days, and the terms 'selectionist' and 'neutralist' are now rarely used. Although an early paper described neutral evolution as 'non-Darwinian', there is no contradiction between neutral models of molecular evolution and Darwinian evolution: the neutral theory recognizes that all adaptations are produced by natural selection, but considers that these account for only a small percentage of observed substitutions. However, there is still a vigorous debate about the relative contributions of selection and drift to molecular evolution, and this is the focus of much ongoing research[6,7] (**Figure 8.23**).

### References


1. Kimura, M. (1983) *The neutral theory of molecular evolution*. Cambridge University Press.
2. Ohta, T. (1995) Synonymous and nonsynonymous substitutions in mammalian genes and the nearly neutral theory. *Journal of Molecular Evolution*, Volume 40, pages 56–63.
3. Petit, N. Barbadilla, A. (2009) Selection efficiency and effective population size in *Drosophila* species. *Journal of Evolutionary Biology*, Volume 22(3), pages 515–26.
4. Woolfit, M. Bromham, L. (2005) Population size and molecular evolution on islands. *Proceedings of the Royal Society of London*, Volume B 272, pages 2277–82.
5. Woolfit, M. Bromham, L. (2003) Increased rates of sequence evolution in endosymbiotic bacteria and fungi with small effective population sizes. *Molecular Biology and Evolution*, Volume 20, pages 1545–55.
6. Gonzales, A. M., Fang, Z., Durbin, M. L., Meyer, K. K., Clegg, M. T., Morrell, P. L. (2012) Nucleotide sequence diversity of floral pigment genes in Mexican populations of *Ipomoea purpurea* (morning glory) accord with a neutral model of evolution. *Journal of Heredity*, Volume 103, page 863.
7. Heystek, A., Geerts, S., Barnard, P., Pauw, A. (2014) Pink flower preference in sunbirds does not translate into plant fitness differences in a polymorphic *Erica* species. *Evolutionary Ecology*, Volume 28, page 457.


# TECHBOX 8.2

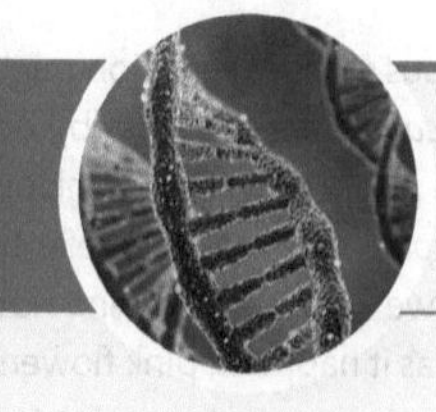

# Population size

**RELATED TECHBOXES**


**TB 7.2:** Detecting selection

**TB 8.1:** Neutral theory

**RELATED CASE STUDIES**


**CS 8.1:** Substitution

**CS 8.2:** Population size



### Summary

The fate of a mutation depends not only on the effect of the mutation on its carrier but also on the size and structure of the population it is in.



### Keywords

effective population size ($N_e$), substitution, selection, drift, mutation rate, polymorphism, hitch-hiking, inbreeding, linkage, homozygosity, heterozygosity

Population size is a constant concern of conservation biologists and there are various techniques for estimating the total number of individuals in an interbreeding set of organisms from different kinds of survey data. Molecular geneticists are also concerned with population size, but have a rather different perspective. From the standpoint of molecular evolution, we are primarily interested in the movement of alleles from one generation to another. Since not all individuals in a population will survive and breed, only a subset of the population contributes alleles to the next generation. So rather than estimating the total population count (N) which represents all individuals in the population, geneticists are generally interested in the effective population size ($N_e$), which represents the contribution of alleles to the next generation. Effective population size is an important component of models of molecular evolution, because it is a measure of the degree to which allele frequencies will be influenced by random sampling, or in other words, how effective selection is likely to be at a particular locus. Effective population size may be influenced by many different factors, such as sex ratio bias or variation in offspring numbers between individuals. The term effective population size was introduced by Sewall Wright (**Figure 7.5**), who said that it represents 'the number of breeding individuals in an idealised population that would show the same amount of dispersion of allele frequencies under random genetic drift or the same amount of inbreeding as the population under consideration'. So a population with many individuals (large N), few of whom will contribute alleles to the next generation (small $N_e$), behaves, from a population genetic point of view, as if it were a population of many fewer individuals.

Imagine two populations of birds, both of which contain the same number of individuals, and equal numbers of males and females. In one species, males and females form long-lasting pair bonds, mating almost exclusively with each other throughout their lives. In the other species, males display together in leks (**Figure 8.24**). Females come to the lek, choose their favourite male and mate with him, then go off and raise the chicks on their own. Fashion being what it is, the females all tend to prefer the same males, so in this lekking species a small number of fashionable males get all the matings, and most males miss out (**Figure 8.13**). So even if there are lots of males displaying on the lekking ground, the effective number of males is only those lucky few who mate and pass on their alleles to the next generation. If these two populations of birds have the same absolute number of individuals (N), the lekking species will have a much lower effective population size ($N_e$) because many of the males have not participated in handing alleles from one generation to the next.

Now let's think about the fate of a slightly advantageous mutation occurring in a male in each of these species—say, a modification of the globin gene that makes him slightly less likely to die from avian malaria. In the monogamous species, most males who survive to maturity get to pair up with a female, so a novel advantageous mutation in a male has a good chance of making it to the next generation. But in the lekking species, very few males get the opportunity to mate. As it happens, our male with the anti-malarial mutation is, alas, not one of lucky few fashionable males, and he dies without reproducing. So this particular mutation will disappear from the population when its unfortunate male carrier dies. Alternatively, if a mutation occurs in one of the fashionable males who gets all the matings, then that mutation will end up in a

**Figure 8.24** Pick me! Pick me! Black Grouse males (*Tetrao tetrix*) doing their best to impress in a lek.

large proportion of the next generation even if it is slightly deleterious (say, it slightly reduces the efficiency of a metabolic enzyme). So unequal mating opportunities increase the influence of random sampling error on allele frequencies, and thus decrease the effectiveness of selection at promoting advantageous alleles or removing deleterious alleles.

In this example, the slightly advantageous mutation for malaria resistance was lost because it was unlucky enough to occur in a male who lacked the fashionable accoutrements to triumph in the mating arena, and the slightly deleterious mutation for a less efficient enzyme increased in frequency because it was lucky enough to occur in a male who had lots of offspring. We can see that having the good fortune to be inherited along with a positively selected allele, or the bad luck to be bundled with a negatively selected allele, can make an important difference to the eventual fate of a new mutation.

We can take this same principle down to the genomic level to examine the influence of genetic linkage on the fate of mutations. If a slightly advantageous new mutation happens to be linked to an allele that reduces its carrier's chance of reproduction, then negative selection on the linked allele may prevent the positive allele from going to fixation (a phenomenon known as background selection). Alternatively, if a slightly deleterious mutation occurs next to a favourable allele, then it may be swept to fixation by positive selection acting on the linked locus (known as hitchhiking: **Figure 7.12**). Just as skewed mating frequencies reduce effective population size at the level of individuals, genetic linkage reduces the effective population size at the level of genomic loci. So we can consider that different parts of the genome have different values for $N_e$.

## Estimating population size from genetic data

Because population size influences both the rate of substitution and the types of substitutions that occur, we can use observed patterns of genetic variation to estimate effective population size. Firstly, we expect population size to influence whether the polymorphic loci we observe tend to have synonymous or nonsynonymous alleles. Imagine a gene that codes for a protein. We might assume that a nonsynonymous mutation (a nucleotide change that alters the amino acid sequence of the protein) is more likely to be subject to selection than a synonymous mutation (one that doesn't change the amino acid sequence: **TechBox 6.1**). In a large population, selection will efficiently remove the majority of nonsynonymous mutations (most of which will be deleterious), so we would expect that most mutations that remain in the population as polymorphisms will be synonymous. In a small population, selection is less efficient at removing those nonsynonymous mutations that have relatively small selective coefficients, so there will be more of these floating around in the gene pool. So one way to estimate effective population size is to use SNP data to compare the ratio of nonsynonymous polymorphisms to synonymous polymorphisms, on the assumption that there will be a higher proportion of non-synonymous mutations persisting in small populations (see also **TechBox 7.2**).

Secondly, we expect small populations to have lower levels of genetic variability than large populations. If a population is reduced to a small number of parents during a population bottleneck, then during that period it will experience rapid substitution due to drift, resulting in loss of many alleles. Small populations increase the chance of individuals mating with relatives that share a similar complement of alleles, and this inbreeding increases the homozygosity of the population (number of individuals carrying two copies of the same mutation), and reduces the heterozygosity (number of individuals carrying two different alleles at the same locus). Low genetic variability is therefore often taken as an indication of a past or current reduction in $N_e$.

Thirdly, because small populations lose alleles by random sampling, we expect genetically isolated populations to become fixed for different substitutions. This is useful for diagnosing population sub-division, when an apparently large or widespread population is actually divided into smaller genetically isolated subpopulations. If the whole population were behaving like an idealized, randomly mating population, then each subpopulation should contain a random sample of the available alleles. But if the population is divided into small isolated groups, then each of these groups may undergo substitution for different alleles, and loss of alternative alleles, producing non-random representation of alleles in the different subpopulations. The degree of population subdivision can be gauged by comparing the genetic variation within subpopulations to the variation in the population as a whole (using measures commonly referred to as 'F-statistics').

It is important to remember that estimation of population size from genetic data is usually based on a number of key assumptions. Most importantly, these models typically assume that most mutations are either neutral or deleterious. If a significant proportion of mutations are advantageous, then an estimate of population size based on measures of genetic variability may be incorrect. This is because, unlike slightly deleterious mutations, the rate of fixation of slightly advantageous mutations increases with $N_e$, because natural selection can override drift in larger populations and drive advantageous mutations to fixation. If we knew what proportion of mutations were positive or negative, we could estimate the overall effect on rates, but we rarely (if ever) have a way of measuring the distribution of fitness effects of spontaneous mutations. Some estimates of population size from genetic data are based on the assumption that mutation rate will be the same in all individuals, all lineages and at all loci. For reasons we explore later (Chapter 11), even closely related species can have different mutation rates, so borrowing the mutation rate of one species to estimate the population size of another can be problematic. Furthermore, the mutation rate probably varies across the genome, so you cannot safely extrapolate results from one locus to another. So when using genetic data to estimate population size, it's a good idea to be aware of the assumptions you are making and keep reminding yourself of them as you consider the answers you get.

CASE STUDY 8.1

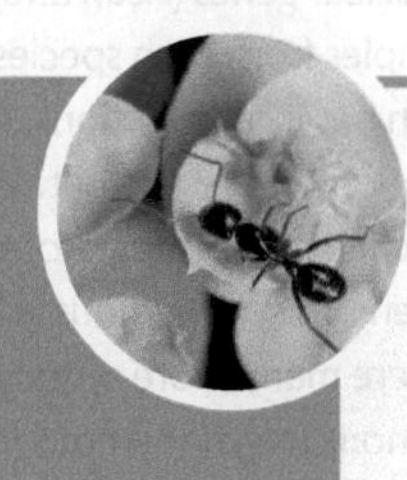

# Substitution: mutation accumulation in asexual walking stick insects

**RELATED TECHBOXES**

**TB 7.2:** Detecting selection

**TB 12.1:** Maximum likelihood

**RELATED CASE STUDIES**

**CS 3.2:** Mutation rate

**CS 4.1:** Replication

Henry, L., Schwander, T., Crespi, B. J. (2012) Deleterious mutation accumulation in asexual *Timema* stick insects. *Molecular Biology and Evolution*, Volume, 29, page 401.

*It is widely believed that mutation accumulation plays an important role in limiting the long-term persistence of all-female lineages ... deleterious mutations may increase the extinction rate of asexual lineages relative to their sexual congeners, thereby contributing to a presumed 'twiggy' distribution of asexuality in phylogenetic trees*

**Keywords**
dN/dS, maximum likelihood, nonsynonymous, synonymous, Muller's ratchet, deleterious mutations, sexual reproduction

**Background**

Why have sex? Why would any well-adjusted individual want to gamble on its reproductive legacy by shuffling its alleles and mixing them with someone else's to produce offspring different to their parents? Would it not be more sensible to save the bother of mucking around with mating and just reproduce asexually, making exact copies of your proven-to-be-successful genome? The repeated evolution of asexual lineages shows it is possible for animals to reproduce without sex, and yet the majority of animal species are sexual, which suggests that asexual lineages must be at some disadvantage. One possible advantage of sexual reproduction is that genetic recombination and re-assortment every generation breaks up allele combinations, allowing selection to work more efficiently at removing deleterious mutations. Deleterious mutations that occur in asexual individuals, if they are not harmful enough to prevent reproduction, will be passed to all offspring along with the rest of the genome. So asexual lineages are expected to accumulate deleterious mutations, without having a means of 'purging' them from the genome. This effect is referred to as Muller's Ratchet (a ratchet effect is one where increases cannot be reversed by decreases, and Hermann Muller was a pioneering geneticist).

### Aim

*Timema* (walking stick insects: **Figure 8.25**) are phasmids, part of the insect group that includes the more charismatically camouflaged stick insects. They are a useful test case for studying the consequences of asexuality because there are about half a dozen asexual species, each of which seems to have evolved independently from a sexual ancestor[1]. Previous work by this team showed that the asexual lineages have persisted without recombination for a long time, based on the high level of divergence between individuals within the asexual populations[2]: without sexual reproduction to shuffle and blend alleles from different individuals, descendant lineages within a population get more and more different from each other with every generation. In this study they compare gene sequences from asexual *Timema* species to their sexual sister species to look for evidence that the asexual lineages accumulate deleterious mutations at a significantly faster rate than sexual lineages.

### Methods

These researchers analysed DNA sequences for two nuclear genes (*Actin* and *Hsp70*) and one mitochondrial gene (*COII*), including multiple samples from each species. These are 'housekeeping' genes that produce proteins involved in basic cell functions, so they are unlikely to be involved in selection for ecological or morphological differences between species (see Chapter 10). First, they looked for evidence that most changes to these proteins were deleterious by comparing the ratio of nonsynonymous (replacement) to synonymous (silent) substitutions in the sexual *Timema* species. They found that there were many more synonymous substitutions than nonsynonymous: in fact, the ratio of nonsynonymous to synonymous (d*N*/d*S*) was close to

**Figure 8.25** Peekaboo. A walking stick insect (*Timema*).

zero, suggesting that most amino acid changing mutations are removed from the population by selection (**TechBox 7.2**). They created mini-phylogenies consisting of the sequences from one asexual species, its closest sexual sister species, and an outgroup (a species more distantly related to both of them). They then used a maximum likelihood method to estimate the d*N*/d*S* ratio for each branch, which allowed them to compare the relative number of nonsynonymous changes in the asexual species to its nearest sexual relative. They tested for a significant difference across all sexual/asexual comparisons using an analysis of variance (ANOVA). They also compared the fit of different models of molecular evolution across the phylogeny of all species, testing whether allowing the asexual lineages to have a higher d*N*/d*S* than the sexuals was a better description of the data than one where all lineages had the same ratio.

### Results

In all three genes, in each of the six independently derived asexual lineages, the rate of nonsynonymous substitutions was on average seven times faster than in their sexual sister species. The model that allowed sexual lineages to have a different d*N*/d*S* was a better fit than the single rate model, suggesting that asexual lineages have different patterns of substitution. They also found that, while nonsynonymous changes in the sexual species typically involved exchanging functionally similar amino acids, which might limit the disruptive effects on protein function, many nonsynonymous substitutions in the asexual species change to an amino acid with different hydrophobicity, which is likely to result in a greater impact on protein function (see **Figure 6.18**).

### Conclusions

These results provide evidence that the asexual lineages of *Timema* accumulate more nonsynonymous substitutions than their sexual sister species. On the assumption that most amino acid changes in these housekeeping genes will be deleterious in both the sexual and asexual species, this finding supports the hypothesis that selection is less efficient at removing non-lethal deleterious mutations in asexual lineages, potentially providing a long-term cost to uniparental reproduction.

### Limitations

This study relies on the assumption that higher d*N*/d*S* is a reflection of the accumulation of deleterious mutations, but a similar result could be obtained if asexual lineages experience more relaxed selection. Given that asexuality changes the rates and patterns of substitutions, the phylogenetic estimates of sister relationships and times of divergence may be influenced by rate differences (and the mitochondrial and nuclear genes support different relationships between the species, which may reflect different patterns of inheritance or phylogenetic uncertainty). One of the sexual species was the closest relative of three different asexuals, which means that measurements from that species influenced multiple datapoints in the analysis, introducing some statistical non-independence between datapoints (see Chapter 13).

### Further work

There are many other groups with multiple origins of asexual lineages, so it would be interesting to see if these results are replicated in other taxa. One intriguing finding of this study is that the asexual lineages also had slightly lower synonymous substitution rates. Could this indicate that the asexual lineages have lower mutation rates, perhaps due to selection pressure to decrease the generation of deleterious mutations which are difficult to purge (**Case Study 3.2**)?

### Check your understanding

1. Why do sexual lineages have so few nonsynonymous changes in these genes? Is it because fewer protein-changing mutations occur in sexual species?
2. Why do the authors assume that the nonsynonymous substitutions they observe are deleterious? Could the higher d*N*/d*S* in asexual lineages be a sign of stronger positive selection in these species?
3. Why is the accumulation of mutations in asexual lineages referred to as a ratchet? What releases the ratchet in sexual species?

**What do you think**

Do you think it is likely that asexual lineages could evolve lower mutation rates in order to slow the accumulation of deleterious mutations? How could you test this hypothesis?

**Delve deeper**

Asexual lineages are often described as occurring on the twigs of the tree of life, because most asexual lineages have a reasonably recent origin[3,4] (although there are a small number of 'scandalous' ancient asexual lineages that have persisted without sexual reproduction over tens of millions of years[5,6]). What could this 'twiggy' distribution of asexuality tell us about the evolution of sexual reproduction?

**References**


1. Schwander, T., Crespi, B. J. (2009) Multiple direct transitions from sexual reproduction to apomictic parthenogenesis in *Timema* stick insects. *Evolution*, Volume 63, page 84.
2. Schwander, T., Henry, L., Crespi, B. J. (2011) Molecular evidence for ancient asexuality in *Timema* stick insects. *Current Biology*, Volume 21, page 1129.
3. Williams, G. C. (1975) *Sex and evolution*. Princeton University Press.
4. Maynard Smith, J. (1978) *The evolution of sex*. Cambridge University Press.
5. Maynard Smith, J. (1986) Evolution: contemplating life without sex. *Nature*, Volume 324, page 300.
6. Judson, O. P., Normark, B. B. (1996) Ancient asexual scandals. *Trends in Ecology & Evolution*, Volume 11, page 41.


CASE STUDY 8.2

# Population size: inbreeding depression in an endangered species

**RELATED TECHBOXES**

TB 7.1: Fitness

TB 8.2: Population size

**RELATED CASE STUDIES**

CS 7.2: Variation

CS 8.1: Substitution

Brekke, P., Bennett, P. M., Wang, J., Pettorelli, N., Ewen, J. G. (2010) Sensitive males: inbreeding depression in an endangered bird. *Proceedings of the Royal Society B: Biological Sciences*, Volume 277, page 3677.

> *This study adds support to the caution expressed in the conservation genetic literature of the insidious effects inbreeding can have on population fitness ..., especially when using reintroduction management*

**Keywords**

microsatellite, inbreeding coefficient, relatedness, fitness, homozygosity, conservation

**Background**

Inbreeding is a concern for conservation biologists because, almost by definition, endangered species tend to have reduced population sizes. Mating between close relatives can reduce fitness of offspring through the expression of deleterious recessive alleles. The potential loss of fitness through inbreeding is important to consider in management of endangered species, especially when establishing small isolated populations through reintroductions. New Zealand has a diverse and distinct endemic bird fauna, but it has suffered a high rate of extinction due to habitat modification and introduced predators. Many endemic bird species now persist only on offshore islands where they have suffered less impact of land clearance, disease, rats and

**Figure 8.26** Hihi (stitchbird, *Notiomystis cincta*) are socially monogamous but genetically promiscuous. This means that they form pair bonds where a female (a) and male (b) jointly raise offspring, however extra-pair matings are common so the social father of the clutch may not be the genetic father of all the offspring he raises.



other introduced species. Many endangered species recovery programmes establish additional protected populations on other islands or predator-protected sanctuaries. Could this strategy increase the effects of inbreeding due to small founding populations?

### Aim

The hihi or stitchbird (*Notiomystis cincta*: **Figure 8.26**) was once widespread in New Zealand but declined after Europeans arrived until, by the late 1880s, they were reduced to a single population on Little Barrier (Hauturu) Island. One hundred years later, 51 hihi from this remaining population were reintroduced to another island, Tiritiri Matangi (**Figure 8.27**). This reintroduced population has grown nearly exponentially and is now used as a source of individuals for other reintroductions: removing birds to other populations keeps the Tiritiri Matangi population at around 150 adults. Researchers genotyped individuals in the populations and monitored reproductive success over a single breeding season in order to quantify the degree and effect of inbreeding in this population.

### Methods

All individuals were genotyped at 19 microsatellite loci from across the nuclear genome, assumed to be selectively neutral. Each sample was amplified at least twice to make sure that allele calls were consistent across replicates. The birds all used nesting boxes, so the researchers could

**Figure 8.27** Tiritiri Matangi island, near the New Zealand capital Auckland, is managed as an 'open sanctuary'. Native forest is being re-established and endemic bird populations established through reintroduction.

monitor clutches, weighing and banding all surviving nestlings at 21 days old, and collecting any unhatched eggs or dead nestlings. This meant that, unlike many population studies, they could genotype both surviving and unsuccessful offspring, including unhatched embryos. They used software developed by one of the authors to estimate the relatedness and inbreeding coefficients from multilocus data. These measures compare the observed combinations of alleles shared to the expected patterns of shared allele combinations that would be generated in a large randomly mating population. When individuals share a more recent common ancestor than would be expected from a randomly mating population, they will have more shared alleles. For example, first cousins (shared grandparent) are expected to have a coefficient of relatedness of 12.5 per cent (shared alleles at an eighth of all loci), but fourth cousins (shared great-great-great-grandparent) have a coefficient of relatedness of only 0.2 per cent. They also checked whether their results were driven by only a few loci by dividing the microsatellite markers into two sets and calculating the inbreeding coefficients of individuals on each set, repeating the process on different random samples of markers to test for consistency in estimates. Genetic load was evaluated using a method for estimating the number of lethal equivalents, where the increased mortality due to inbred matings is used to estimate the number of deleterious recessive alleles in the general population: this reflects the additional mortality in the population due to the combined effects of exposure of deleterious alleles in the homozygous state.

### Results

The average level of inbreeding was high, with nearly a fifth of individuals having an inbreeding coefficient of 0.125 or higher (i.e. the level expected for offspring whose parents are cousins). Hatching failure was also comparatively high in this population, with around one third of eggs failing to hatch. Statistical analysis showed that the strongest correlates of survival probability were an individual's level of inbreeding and sex. Male offspring had a greater level of inbreeding, and a lower survival probability if inbred. However, the inbreeding level of the mother was not correlated with nestling survival. They also estimated a moderate level of lethal equivalents, suggesting that each offspring carries an average of 6.9 recessive lethals per haploid genome, which indicates the potential for fitness loss through breeding with close relatives.

### Conclusions

This study shows that inbreeding does influence survival in hihi, and that the effects of inbreeding are male-biased. Why are males more inbred than females? This could be because inbred females die at an early developmental stage and are thus removed from the breeding population. However, it may also be because closely related pairs have more male offspring. Why do male hihi suffer more from inbreeding? Hihi males are larger than females so must grow faster than female offspring, so potentially suffer a greater disadvantage from reduced fitness due to inbreeding.

### Limitations

Although these researchers were able to genotype failed embryos and nestlings, this will not pick up cases of infertility or early stage lethal mutations that prevent embryo formation[1]. As with any study looking at species characteristics, it is difficult to pull apart causal relationships from statistical correlations. Traits other than sex and inbreeding also influenced survival probability, most notably clutch size and maternal age. Repeating the study over multiple breeding seasons may help to tease apart the influence of individual traits and environmental factors on fitness.

### Further work

'Genetic rescue' strategies are employed in some reintroduced populations, for example by moving individuals between populations to artificially create gene flow between populations. Management teams are considering establishing gene flow between reintroduced populations by ongoing periodic translocations of individuals between populations. This strategy is already being employed in managed populations of the kakapo, a quirky flightless parrot from New Zealand. In some endangered populations, active intervention in mate choice has been used to break up pairs of closely related individuals to encourage breeding between more genetically dissimilar individuals: this strategy has been employed in managed populations of the New

Zealand takahē, a flightless rail previously thought to be extinct, now occurring in one natural population and five reintroduced populations (including on Tiritiri Matangi).

### Check your understanding

1. Why does inbreeding increase the level of homozygosity?
2. Why would increased homozygosity result in lower survival?
3. Why did they choose to sequence microsatellite loci, and why did they need many loci from different parts of the genome?

### What do you think?

Will continued inbreeding in the reintroduced population result in population extinction, or will it be sustainable despite higher mortality, or will inbreeding eventually result in deleterious recessive mutations eventually being purged from the population reducing the fitness cost due to genetic load?

### Delve deeper

Establishing insurance populations is an important part of some conservation programmes, for example setting up isolated disease-free populations of Tasmanian Devils to prevent the entire species from being wiped out by the infectious facial tumour that is decimating wild populations. But genetic homogeneity is often considered to be a risk factor in disease vulnerability, as there may be insufficient genetic variation in the population to allow selection for resistance. How could you weigh up the benefits and risks of establishing insurance populations for endangered species? What strategies could reduce the vulnerability of small populations?

### Reference

1. Hemmings, N., West, M., Birkhead, T. R. (2012) Causes of hatching failure in endangered birds. *Biology Letters*, Volume 8, page 964.

# 9 Species

## *Origin of species*

*"In considering the Origin of Species, it is quite conceivable that a naturalist, reflecting on the mutual affinities of organic beings, on their embryological relations, their geographical distribution, geological succession, and other such facts, might come to the conclusion that species had not been independently created, but had descended, like varieties, from other species"*

**Darwin, C. (1872)** *On the origin of species by means of natural selection, or the preservation of favoured races in the struggle for life,* **6th edn.** John Murray.

## What this chapter is about

Genetically isolated populations that cannot exchange alleles with other populations accumulate a distinct set of substitutions, due to the random nature of mutation, the chance effects of genetic drift, and different alleles being fixed by selection. When genomes from different daughter populations have accumulated sufficient distinct substitutions that they cannot be combined together to produce healthy hybrids, due to genomic incompatibility or adaptation to different niches, we can say they have become distinct species. Taxonomy aims to identify and name these distinct species, while also recognizing that all populations form a hierarchy of relatedness. This provides a natural basis for classification and allows prediction of similarities, because relatives are likely to be more similar in many different traits, due to their shared ancestry. DNA sequence data is increasingly being used to identify species, and to place these species within a broader systematic framework that reflects patterns of descent with modification.

## Key concepts

- Species are formed when genetically isolated populations accumulate substitutions that make them distinct from all other such populations
- Defining and identifying species is a key part of describing biodiversity and understanding how it is generated
- DNA can be used to identify samples to species on the basis of sequences differences unique to that species

# Taxonomy

### *Naming names*

Biologists are very particular about names. The procedure for naming a new species is arcane and has changed relatively little in the three centuries since the system of binomial nomenclature was established. Each species that has been formally described in the scientific literature is given a unique double-barrelled name, in the form *Genus species* (in italics or underlined, with a capital for the genus name and a lower case for the species name). A genus is a group of related species, so binomial nomenclature is similar to the practice of referring to people by Firstname Familyname. I share the family name Bromham with my parents and sisters, but I am the only member of my family with the first name Lindell, so the name Lindell Bromham is a unique double-barrelled identifier for me. Similarly, there are many species in the genus *Canis* (dogs and wolves), but *Canis simensis* refers only to the Ethiopian Wolf, and *Canis rufus* refers specifically to the Red Wolf (**Figure 9.1**).

**Figure 9.1** Two members of the genus *Canis* (dogs and wolves), (a) the Red Wolf *(Canis rufus)* and (b) the Ethiopian Wolf (*Canis simensis*). Both of these species are endangered.

Scientific names are designed to standardize the labels given to species. Technically, a species can have only one scientific name, although it may be known by any number of 'common names' (the names used in everyday discussions). So, for example, the scientific name of the thylacine is *Thylacinus cynocephalus* (which translates to something like pouched dog-headed animal). But the thylacine has been known by many different common names including Tasmanian wolf and Tasmanian tiger, reflecting its recent distribution on the Australian island of Tasmania, the tiger-like stripes on its hindquarters, and its wolf-like appearance and hunting behaviour (**Figure 9.2**). But thylacines are not wolves or tigers. They are marsupials, belonging in the same taxonomic group as kangaroos and koalas. If the dog-headed thylacine looks like a wolf and acts like a wolf, then why is it grouped with the very un-wolf-like marsupials? In addition to giving us a unique name with which to identify a particular species, modern taxonomy aims to reveal patterns of relatedness. The thylacine is more closely related to other Australian marsupials, and only distantly related to wolves. As far as taxonomy is concerned, family relationships trump superficial resemblance. A marsupial in wolf's clothing is still a marsupial.

Basing taxonomy on descent gives us a natural basis for classification. But it also creates some practical difficulties. Descent with modification is a continuous process, so species typically are formed by the gradual accumulation of many differences. This means that there may not be a clear line between a descendant species and its ancestor, or between two closely related species. In fact, early evolutionary biologists such as Lamarck and Darwin used the continuum of differences between populations, races, subspecies, and species as evidence that species are the product of descent, not each created separately. In this chapter, we will consider how species are classified in a taxonomic hierarchy. We then take a genome-level view of species formation and ask how this evolutionary perspective informs how we define species, before considering some of the practical aspects of defining and detecting different species.

**Figure 9.2** The thylacine was driven to extinction by hunting by humans, environmental change, impact of introduced dogs, and possibly even disease. There was a bounty on thylacines from 1830 to 1914. When it became clear they had been nearly exterminated, they were given official protection in 1936, two months before the last known living thylacine died in a zoo.

## Classification

With at least 1.5 million named species (and maybe 8 million or more still to be described), a sensible and accessible system of classification is an essential foundation for biology. The taxonomic system used today is the direct descendant of the brilliantly simple strategy set out by the Swedish botanist Carolus Linnaeus in his *Systema Naturae*, first published in 1735. Linneaus is most strongly associated with the binomial naming system (though he was not its originator), whereby each species is known by its genus and species name (e.g. *Homo sapiens*). More importantly, Linneaus established the principle of hierarchical classification. Species are grouped together into genera on the basis of shared key characteristics, and related genera are grouped together into families, which are grouped into orders, then classes, then phyla and finally into the five (or so) kingdoms into which the living world is divided. So as you travel down the taxonomic hierarchy, from kingdom to phylum, to class, order, family and genus, you find collections of ever more-closely related species.

As an example of the modern application of Linneaus' hierarchical taxonomy, here is the classification of the Australian mountain ash, the tallest flowering plant species in the world (**Figure 9.3**), according to the International Code of Botanical Nomenclature (**Figure 9.4**). This species illustrates the value of formal scientific names for species. *Eucalyptus regnans* is sometimes known by different common names, and many entirely unrelated species are also known by the common name 'mountain ash', including a type of ash tree from Texas (*Fraxinus texensis*) and the European Rowan tree (*Sorbus aucuparia*). But the scientific name *Eucalyptus regnans* is an unambiguous label for only one particular species.

**Figure 9.3** Mountain ash: (a) A stand of *Eucalyptus regnans*, just under a century old, growing above a treefern understorey. *E. regnans* has a remarkably growth rate, capable of increasing in height by more than a metre a year. It is considered by many to be the tallest species in the world, because trees felled by loggers in the late 1800s were found to be (or at least had been) taller than any other tree known, including the Californian redwoods. However, thanks to the dauntless spirit of those who felled the mighty mountain ashes with hand-saws (in some cases, the mountain ash were cleared to grow strawberries), the tallest currently surviving mountain ashes are just under a hundred metres. (b) This tree, unimaginatively named 'The Big Tree', is one of the tallest surviving individuals. It is shown here with a human for scale (though admittedly not a very big human).

Photographs by Lindell Bromham.

**Figure 9.4**

| Taxonomic level | Formal name | Common name |
|---|---|---|
| Kingdom | Viridiplantae | green plants |
| Phylum (Division) | Embryophyta | land plants |
| Subphylum | Tracheophytina | vascular plants |
| Class | Angiospermopsida | flowering plants |
| Order | Myrtales | myrtles & allies |
| Family | Myrtaceae | myrtles |
| Genus | *Eucalyptus* | gum trees |
| Species | *regnans* | mountain ash |

However, although the basic principle of hierarchical classification is a foundation of modern biology, there is no one taxonomy accepted by all biologists. Authorities may differ in the taxonomic divisions used, or the names given to levels in the taxonomic hierarchy, or how many different hierarchical divisions there are. There is often disagreement about what name a species should be known by, or which higher group a given species should be placed in. For example, the classification of *Eucalyptus regnans* looks quite different on GenBank (TechBox 1.1), which provides more levels of classification and provides alternative names for some of the divisions. Differences between taxonomic schemes arise because taxonomy is in a constant state of revision and debate. In particular, the increasing use of DNA sequence data to define taxonomic units and assign species to different taxa has resulted in a great deal of rearrangement of taxonomic hierarchies.

## Relatedness

> *As descent has universally been used in classing together the individuals of the same species, though the males and females and larvæ are sometimes extremely different; and as it has been used in classing varieties which have undergone a certain, and sometimes a considerable amount of modification, may not this same element of descent have been unconsciously used in grouping species under genera, and genera under higher groups, though in these cases the modification has been greater in degree, and has taken a longer time to complete? I believe it has thus been unconsciously used; and only thus can I understand the several rules and guides which have been followed by our best systematists*
>
> Darwin, C. (1859) *On the origin of species by means of natural selection, or the preservation of favoured races in the struggle for life*. John Murray

There are two important benefits of a hierarchical taxonomy based on relatedness. Firstly, it is a logical and manageable way of organizing information about the ever-increasing number of known biological species. Secondly, and more importantly, taxonomy based on the natural hierarchy of the evolved world can be used to predict the degree of similarity of any two species. For example, without any special knowledge of the organisms, I can make an educated guess that *Secale* (rye) and *Triticum* (wheat), which are both members of the tribe Triticeae, are likely to be more similar to each other than either is to *Poa* (bluegrass), which is a member of the tribe Poeae (Figure 9.5). But *Triticum*, *Secale* and *Poa* are all members of the family Poaceae, so they are likely to be more similar to each other than they are to papyrus (*Cyperus papyrus*), which is in the family Cyperaceae.

These species are all contained within the order Poales. All monocots, including these species and, for example, the banana (*Musa*), are more similar to each other than they are to all dicots, such as a gum tree like *Eucalyptus regnans*.

If our hierarchical classification accurately represents patterns of descent then, in general, we expect most heritable characteristics to broadly follow the same hierarchy of similarity (although, like most things in biology, there will always be exceptions). This applies to all levels of biological organization, including the genome. So, for example, the taxonomic relationships between these five plant species are reflected in the relative percentage difference between sequences of the *rBCL* gene (Figure 9.5). To explore why genetic data is so useful in identifying species, let's take a genome-level view of the formation of species. To do this, we need to start at the very beginning of genetic divergence, considering the fate of a new mutation in a DNA molecule in a single individual. This is where evolution starts.

**Figure 9.5** Classification forms a nested hierarchy. The tribe Triticeae is contained within the family Poacea which is contained within the order Poales, which is in the monocots, which are given the class designation Liliopsida in some taxonomic schemes. The blue numbers record the average percentage difference between the *rBCL* sequences of these species. The *rBCL* gene codes for the protein ribulose 1,5-bisphosphate carboxylase/oxygenase, a photosynthesis enzyme better known as RuBisCO. RuBisCO is one of the most abundant proteins on Earth. As an important enzyme in the carbon-fixing metabolic cycle in green plants and algae, RuBisCO plays a key role in global carbon cycles.

# Divergence

*Life cycle of a mutation*

Let's review what we have learned about molecular evolution so far. In Chapters 3 and 4, we saw how mutation is inevitable. Damage to DNA occurs, and replication errors happen. Although DNA repair is breathtakingly efficient, given the vast numbers of nucleotides that must be maintained and copied in every genome, some changes to the DNA sequence slip through the net and get passed on to the next generation. Most new mutations are lost when their carrier fails to leave descendants. But some persist, copied from one generation to the next, increasing in frequency by chance (drift) or because there is something about the mutation that increases its chances of being copied and passed on to successful offspring (selection). So now the population is polymorphic, because many members of the population carry a copy of this mutation, all copied from that original mutation that occurred in a single genome. The frequency of this mutation, relative to other variants at that site in the genome, may go up and down, as its carriers have more or less offspring. If, on average, it increases in frequency, then eventually it may replace all other alternative alleles, so that all genomes in the population include a copy of that mutation. At this point our mutation has become a substitution, characteristic of the whole population.

This process of mutation → polymorphism → substitution is occurring continuously. From a molecular evolution perspective, we can view the process of speciation as one of genetic divergence. When a population splits to produce two or more daughter populations, each inherits a sample of the genetic variation of the parent populations, so they must start off with similar genomes. But genetic change is continuous, and the longer these populations are separated from each other, the more substitutions they will accumulate. Since they will tend to acquire different substitutions, related populations start the same, but get progressively more different over time. At some point, these populations become so different that we call them different species.

## Speciation

The mechanisms and consequences of speciation are among the most hotly debated topics in evolutionary biology, and there is a rich and voluminous literature on the topic. Here we are taking a very simple, genome-level view and we are ignoring all of the detail concerning the mechanisms whereby populations become separated, whether physically or genetically or behaviourally. This is a fascinating field and, if you are interested in evolution (or life in general), you should read about it. But, for the purposes of this book, the main point that we need to consider is that once two populations have become separate to the point where they cannot interbreed, a mutation that occurs in one population cannot be passed into the next population, and so they will continue to independently accumulate different sets of substitutions.

Both chance and adaptation contribute to the genetic divergence between populations. Mutations arise constantly, so that many individuals carry unique changes to their genomes. Mutations are chance events, so a mutation that occurs in one population might not arise in a neighbouring population. We expect each population will have a different set of mutations arising each generation. Chance also plays a role in which of these mutations persist and which are lost. Genetic drift, due to random variation in allele frequencies each generation, can cause some mutations to be lost from the population and drive other mutations to fixation. This is particularly true for mutations with little or no effect on fitness, each of which has only a small chance of going to fixation. Even if two populations started with exactly the same set of neutral mutations, only a small, random sample of these will go to fixation by chance, so it is unlikely that exactly the same mutations will be fixed in each of the populations.

But it is not only the undirected influence of mutation and drift that drives genetic divergence between populations. Selection can drive different substitutions in different populations. This is most obvious when two neighbouring populations experience different selective pressures. For example, populations of deer mice living at high altitude have genetic variants in their haemoglobin genes that make their blood more efficient at carrying oxygen. But even if two populations are under exactly the same selection pressure, they may 'solve' the same problem with different substitutions (see **Hero 9**). Which brings us to a very important point: even when the external environment of neighbouring populations is identical, one of the most important aspects of an allele's 'environment' is likely to differ, which is the presence and frequency of other alleles in the population.

Some alleles that differ in frequency between populations could have a direct influence on mate choice and interbreeding. Imagine a population of frogs in which males vary slightly in the length of mating calls and females vary in their mate-call preferences. Now place a barrier to mating down the middle of your imaginary frog population. It might just happen that in one isolated half of the population there was a slightly higher proportion of females carrying the 'I like long calls' allele than the other. In this half of the population, males with long calls have a slightly higher chance of mating than short-call males. So we should expect, next generation, there will be an increased frequency of long-call males and long-call-preferring females in this isolated subpopulation. Which means that next generation there will be even more opportunities for long-call males to mate, and so the allele frequencies will continue to rise, potentially leading to the fixation of both the long-call and long-call-preference alleles in this subpopulation.

Two populations might also diverge if their different alleles cause them to occupy different parts of the habitat, for example to choose different host plants. An individual carrying the allele for preferring to feed on plant A is less likely to meet an individual carrying the allele for preferring plant B, because, naturally, the plant A types are all hanging out on plant A, and the plant B types are all over there on plant B. Similarly, any allele that shifts flowering time, or results in attracting different pollinators, might limit the extent to which gametes are shared between the two types, providing an opportunity for two genetically isolated types to diverge. But many alleles that differ between populations are not directly linked to traits concerned with reproduction, such as mate preference or flowering time. Substitutions in any part of the genome, even in the 'housekeeping' sequences shared by all species, can contribute to the isolation of two populations through the development of genomic incompatibility.

## Hybrid incompatibility

The upshot of these three influences—the randomness of mutation, substitution by drift, and different responses to selection—is that if you took a population, divided it in half, and let each half evolve independently of the other, then after some time you would see that they had both acquired different substitutions. Now what will happen if we bring the two halves of the population back together again?

One possibility is that they will mate with each other and produce heterozygote offspring carrying alleles from both subpopulations. If these heterozygotes are just as good at reproducing as their parents, then eventually the two populations may completely mix together, once again sharing a common pool of alleles. But if these hybrid offspring are in some way disadvantaged, then there will be fewer successful offspring that carry alleles from both subpopulations. For example, if the two sister populations have become adapted to different environments, then hybrid offspring might represent an intermediate state between the two types, not particularly well adapted to either.

Reduced hybrid fitness is not always a result of offspring being maladapted to either of the parent's environments. Substitution occurs throughout the genome, not just in the genes involved in adaptation to the environment, but also in the genes responsible for the ordinary workings of the organism, like metabolic enzymes, growth hormones, or immune function. Any new mutation, wherever it occurs in the genome, must pass two tests. The first is that any new mutation has to work in the individual it arises in, or it's not going to have much of an evolutionary future. But, in a sexually reproducing population, being compatible with the other alleles in your natal genome is not enough, because next generation, that mutation will find itself in a different genome with a different set of alleles. So the next test comes when the alleles in the population are mixed and resampled to create different combinations in new individuals, so each allele will end up with different genomic neighbours than in the previous generation. A mutation that worked just fine in the genome it arose in is not going to last long if it results in lower fitness when combined with other common alleles in the population. Any allele that does not play well with others will, by definition, end up in fewer successful offspring, so tend to decrease in frequency. In this way, by weeding out the poor team-players, each population builds a set of co-adapted alleles.

Now think about what happens when a stranger from a neighbouring population comes and attempts to produce offspring with one of the locals. This stranger carries a set of alleles that are all co-adapted to each other, but they haven't been tested against alleles from other populations. So, for example, the neighbouring population might have acquired substitutions that change both the active site of a regulatory protein and the sequence of the corresponding binding site. The hybrid offspring might end up with the regulatory protein sequence from the neighbours but the binding site sequence from the local population, and the two might not match. Alas, without matching regulatory protein and binding site, the hybrid can't switch on an important gene at the right time, so won't have a chance to produce offspring of its own. In this example, the populations are not physically isolated, because individuals from each population can meet, and they are not behaviourally isolated, because they can mate. But they are genetically isolated because they can't produce viable, fertile hybrid offspring, so there is no way for alleles in one population to be included in the next generation of offspring in the other population.

If individuals that mate with members of another population are less likely to have well-adjusted offspring, then any mutation that gives its carrier a tendency to mate only with members of its own population will have a selective advantage, by preventing its carrier from wasting its reproductive effort producing hopeless hybrids. Thus selection may drive the substitution of mutations that contribute to reproductive isolation mechanisms, favouring traits that increase the chance that individuals will mate with their own kind. This process is known as reinforcement, because it favours the evolution of traits that reinforce the reproductive separation of the populations (**Case Study 9.1**). Examples of reproductive isolation mechanisms include different mating calls, separation of flowering times or attracting different pollinators.

This Darwinian view of molecular evolution, where the longer two populations are separated the more different they will become, provides us with both useful tools and conceptual challenges. On the one hand, because change is continuous and ubiquitous, we can use the pattern of substitutions that accrue throughout the genome to identify members of a particular population. Furthermore, because each population carries the genetic inheritance of its ancestors, we ought to be able to use patterns of shared substitutions to reconstruct the evolutionary history of these populations (**TechBox 9.1**). But this Darwinian view also tells us that it won't always be easy to define exactly what is, or isn't, a species.

# Species

> *No one definition has as yet satisfied all naturalists; yet every naturalist knows vaguely what he means when he speaks of a species*
>
> Darwin, C. (1859) *On the origin of species by means of natural selection, or the preservation of favoured races in the struggle for life*. John Murray

The debate over what, if anything, is a species has been rumbling along for centuries. As the number of possible 'species concepts' continues to increase, it could be argued that the main outcome of the debate has been to conclusively demonstrate that there is unlikely to ever be a single universally acceptable definition that encompasses all of the entities that we instinctively refer to as species. The most intuitive species concept is to group together recognizably similar individuals that differ consistently from members of all other such groups. The difference between some species may be obvious: no-one is going to mistake a lion for a leopard, even though they are closely related sister species. Since they last shared a common ancestor several million years ago, leopards have become distinct from lions in their size and coat patterning and colour (**Figure 9.6**).

**Figure 9.6** Where do we draw the lion? While there is a clear difference in body size and coat markings between (a) lions (*Panthera leo*) and (b) leopards (*Panthera pardus*), like many big cat species, lions and leopards have been known to produce fertile hybrid offspring in captivity.

But physical dissimilarity may not always delineate natural groups. On the one hand, members of the same species may look rather different. For example, females of the African Mocker Swallowtail (*Papilio dardanus*) come in 14 distinct forms, many of which mimic unrelated species that taste nasty (**Figure 9.7**). On the other hand, different species may seem entirely indistinguishable. *Chiloglottis* orchids attract male wasp pollinators by mimicking a female wasp (**Figure 9.8**). The orchids use specific chemicals to attract different species of wasp pollinators. So co-existing individuals can belong to two reproductively isolated populations, because the pollen from the type that attracts one wasp species will not be carried to the other type, which attracts a different wasp species. Genetic studies have revealed that, in some areas, there may be several different orchid species present, indistinguishable to the human eye but smelling very different to their insect pollinators.

Instead of relying on morphological similarity, we could consider that a species is a group with a common genetic heritage, sharing a distinct set of alleles through interbreeding. This is the basis of the 'biological species concept', one of the most well-known and widely accepted species definitions. It makes evolutionary sense: if two populations are prevented from interbreeding with each other, then they will have different evolutionary fates. A new mutation can become fixed in one interbreeding population, but it cannot be shared with the other population, so reproductively isolated populations

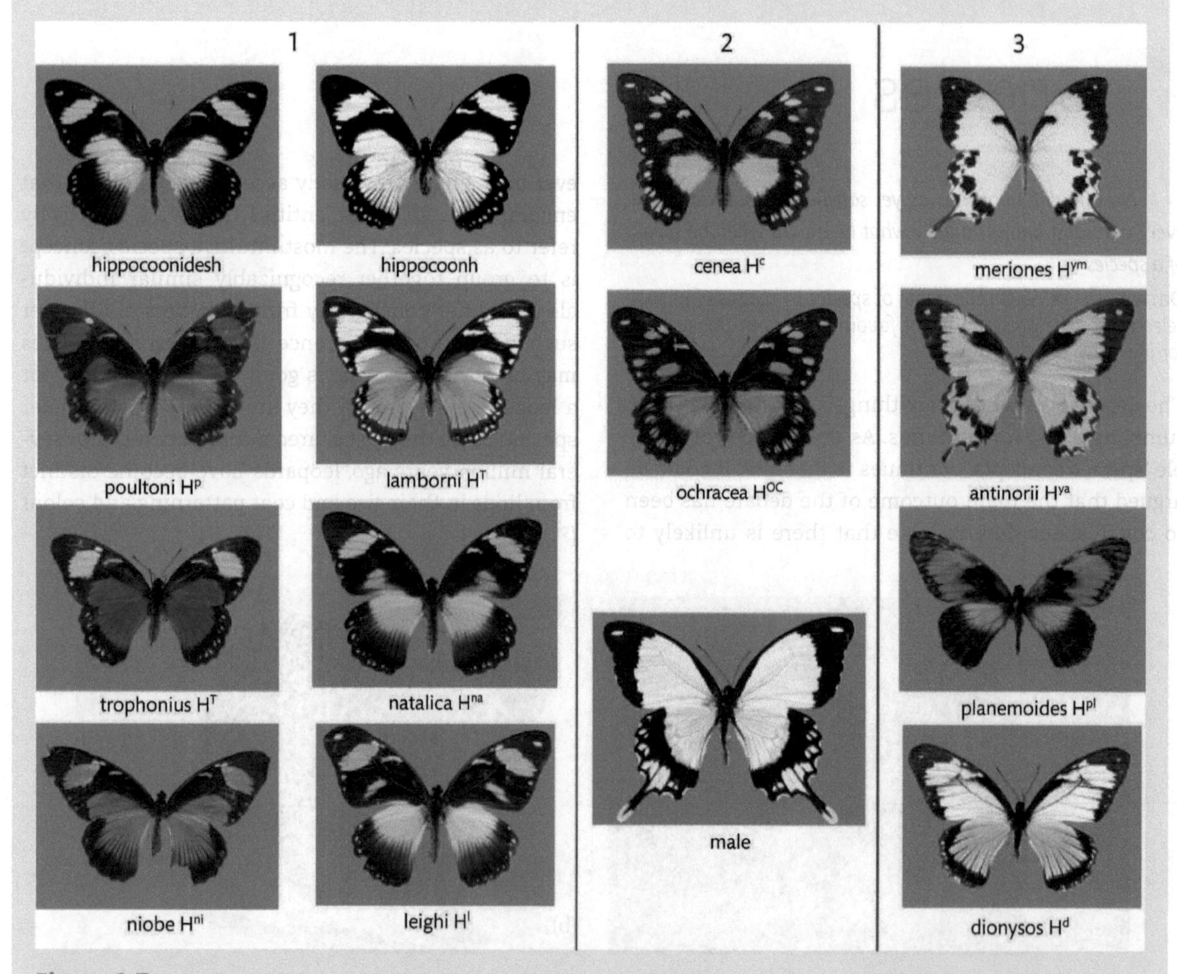


**Figure 9.7** The mocker swallowtail (*Papilio dardanus*) has been described as the most interesting butterfly in the world. Populations of this species, found throughout sub-Saharan Africa, can contain up to six morphologically distinct types of females. There are 14 different female morphs in total, some of which are mimics of different distasteful butterfly species, while others mimic males. Males, on the other hand, are the same, and are not mimics.

**Figure 9.8** Sexually deceptive orchids, such as this *Chiloglottis* flower, attract wasp pollinators (*Neozeleboria cryptoides*) through a combination of visual and chemical signals that make the flowers sexually irresistible. A male wasp, blinded by love, attempts to copulate with the flower, and in the process gets a bundle of pollen slapped on his back. Then the disappointed wasp goes off to try his luck elsewhere, taking the pollen to other sexually deceptive orchids.
Reproduced courtesy of Rod Peakall.

will gradually diverge from each other. Eventually, this accumulation of genetic changes will lead to the physical or ecological distinctness that we will recognize as species.

But, as with morphological dissimilarity, interbreeding does not always draw a clear line between natural groups. The Crimson Rosella and Yellow Rosella (*Platycercus flaveolus*) are clearly morphologically distinct, have different calls and occupy different habitats (**Figure 9.9**). However, some biologists feel that the Yellow Rosella cannot be considered a separate species because where its distribution overlaps with the Crimson Rosella, they can interbreed to form viable hybrids. In fact, these rosellas form part of a ring species, a series of morphologically distinct types with a widespread distribution, in which each population can interbreed with its near neighbours. In fact, the biological species concept cannot be easily applied to much of the earth's biodiversity, because it is not clear how it applies to asexual organisms, or populations that exchange genetic material through lateral gene transfer, or populations that maintain distinct characteristics despite genetic exchange with other populations.

The practical difficulties of applying the criterion of interbreeding to delineate species have led to the increasing use of phylogenetic species concepts, which consider a species to be a natural group whose members share a unique evolutionary history. Phylogenetic species can be defined on the basis of shared traits that indicate a common descent. Increasingly, analysis of

**Figure 9.9** (a) The Crimson Rosella and (b) Yellow Rosella were considered to be two separate species but evidence of hybridization in nature has resulted in the Yellow Rosella being downgraded in some taxonomies to a subspecies of the Crimson Rosella (*Platycercus elegans flaveolus*).
© John Milbank 2007.

molecular data is used to demonstrate that a given group forms a distinct evolutionary lineage (Case Study 9.2). Conversely, phylogenetic analysis has been used to reject species status for populations that do not form a single unique lineage. For example, the Cape Verde Kite (*Milvus milvus fasciicauda*) has been considered to be one of the rarest birds of prey, yet analysis of DNA sequences from the contemporary population and from historical museum specimens were used to suggest that individuals from this population do not form a distinct lineage, and instead should be considered to be members of the more widely distributed Red Kite (*Milvus milvus*). It has also been suggested that the individuals recognized as *M. milvus fasciicauda* might be hybrids between the Red Kite (*Milvus milvus*) and Black Kite (*Milvus migrans*). If the Cape Verde Kite is not recognized as a distinct taxon, then it can't have a specific conservation status. Sadly, whether the Cape Verde Kite represents a species, subspecies, population, hybrids or none of the above, it now appears to be extinct in the wild.

### *Where do we draw the line?*

Whether based on similarity, interbreeding or shared descent, species concepts face some common problems due to the nature of descent. Even if reproductive isolation occurs suddenly—for example, when continuous forest is subdivided by logging—we expect the genetic changes associated with speciation to accumulate gradually. If we accept that species change over time, then where do we draw the line between closely related but independently evolving populations? And exactly where do we divide an ancestral species from its modified descendants? Is *Homo sapiens* a different species to *Homo floresiensis*, the diminutive people whose remains have been found on the Indonesian island of Flores? Can we be sure that *Homo erectus* could not have interbred with a (fairly open-minded) member of our own species?

At this point the reader may, as many thousands have done before, give up in despair and state that there is no universal species concept, so we should stop fussing about it. That would be an attractive proposition, were it not for the great practical importance of defining species, both in terms of scientific research and management of natural resources. Biologists who study patterns of biodiversity rely on species lists being accurate, for example to identify areas with high numbers of endemic species that should receive special environmental protection, or to study the driving forces behind high species diversity in biodiversity hotspots. Species definitions also have important legal implications. For example, the US Endangered Species Act is influenced by the biological species concept, so a species can be removed from the protected list if it is found to interbreed with any other population. Genetic analysis has demonstrated that there is ongoing hybridization between the critically endangered North American Red Wolf (*Canis rufus*: Figure 9.1) and coyotes (*Canis latrans*). This had led to calls to remove the Red Wolf from the endangered species list and to cease spending such large amounts of money on the Red Wolf recovery programme. In fact, hybridization with coyotes is cited by the International Union for Conservation of Nature as one of the primary threats to the persistence of red wolves in the wild.

The most pragmatic approach to the mire of species definitions is to consider that, just as there are many processes that lead to species formation, so there are many different kinds of species. A species definition that works for one group may not create clear divisions in another group. We should use utility as our guide—what do we want to define species for? If we are concerned about conservation, we should, in an ideal world, make sure we give protection to any unique groups that we would be sorry to lose, whatever we decide to call them.

## Defining species

“*There are almost as many concepts of species as there are biologists prepared to discuss them.*”

Isaac, N. J. B., Mallet, J., Mace, G. M. (2004) Taxonomic inflation: its influence on macroecology and conservation. *Trends in Ecology and Evolution*, Volume 19, page 464.

Taxonomy is the bedrock of biology. Many of the case studies in this book rest critically on the ability to assign an individual organism to a particular species. For example, recognition and treatment of infections relies on the ability to correctly identify the infective agent. Species identification also plays an important role in many legal proceedings, such as the trade in endangered species (e.g. Case Study 1.2). Many conservation programmes rest on the recognition of species as unique biological forms that should be protected and preserved. More broadly, if we want to study or communicate ideas about biological diversity, we need units of measurement with which we can quantify species richness and monitor changes in biodiversity.

If our primary goal in naming species is to provide a reliable and useful way of delineating real biological units, then it would be helpful to have a clear and practical definition of what a species actually is. Although, at the broadest level, many biologists feel that a species is a natural unit of biodiversity which can be separated from all other such units on the basis of reproductive isolation, and/or unique adaptations, the key aspects of population biology and ecology of a species are not always easy to measure in practice. A species may contain individuals that vary markedly in appearance or behaviour, yet are still members of a single interbreeding gene pool (**Figure 9.7**, **Case Study 9.1**). Conversely, members of genetically isolated populations could be identical in morphology, behaviour and ecology (**Figure 9.8**, **Case study 9.2**). It could take a substantial amount of genetic analysis to prove that a recognized species never interbreeds with any other, particularly as genetic isolation is not an all-or-nothing phenomenon. In fact, many recognized species undergo low levels of gene flow with other species. For example, there are over 700 species of gum trees (*Eucalyptus*), adapted to a wide range of niches from alpine areas to the desert. *Eucalyptus* species have recognizably distinct morphological and ecological traits, but many eucalypts can form occasional hybrids with other species (**Figure 9.10**). If we were being strict on interbreeding as a rule, we might have trouble drawing many hard boundaries between many eucalypt species, despite their obvious differences in appearances and ecology.

Actually, taxonomic descriptions of new species typically don't evaluate gene flow at all. Formal taxonomic description of a species is typically an account of the unique characteristics that allow the members of a species to be distinguished from all other species. Rather than using direct observation of breeding and ecology to delimit species, most taxonomy is based on more easily measured characteristics, such as leaf shape, bone structure, or biochemical components of cell walls. The defining characteristics of a species are often not features we would easily notice, such as the iridescent blue colour of a beetle or acorn-collecting habits of a squirrel. Instead, the formal classification may be based on less exciting characteristics such as number of bristles on the forelimbs, or the shape and size of holes in the skull.

It may seem odd that species could be classified using such apparently unimportant characteristics, but it follows from the principle that, if all species have descended from shared ancestors, then related species will have a great many shared characteristics. So we should expect to be able to look at nearly any characteristic and find it is more similar between close relatives. However, this principle will be least robust for characteristics that are adaptations to particular niches or locations, such as particular colour patterns or foraging behaviours, because these are just the sort of characteristics that we expect to be shaped by selection acting on particular species (which is a shame, because these are usually the most interesting and obvious characteristics of a species). The use of DNA sequences to identify species is essentially an extension of this principle.

**Figure 9.10** Hybrid advantage: The tree in the centre of this photo, growing on the Murray floodplain in south eastern Australia, is a hybrid, the result of a naturally occurring cross between a Black Box (*Eucalyptus largiflorens*) and a Red Mallee (*Eucalyptus gracilis*), two very distinct tree species that tend to grow in different habitats. The hybrids don't have a scientific name but are referred to as Green Box. Green Box trees are noticeably different in appearance from either parent species. They also differ in physiology and growth patterns, and have enhanced tolerance to drought and salinity. Due to a combination of drought and diversion of water from the Murray River (particularly for use in irrigation projects), the Murray floodplain has become increasingly dry and saline. Under these conditions, the Black Box trees are suffering from dieback, as you can see in this photo, from the rather scruffy looking Black Box trees either side of the Green Box. Their enhanced tolerance to changed environmental conditions led to a programme to propagate the Green Box using tissue culture, to plant out onto the floodplain.

Photograph: Freya Thomas.

Members of a species will tend to share particular DNA sequences, even in genes that are not in any way associated with the morphology or behaviour unique to that species.

## DNA taxonomy

Molecular data is, in many ways, an ideal form of information for identifying species. Just as mutation makes individual genomes unique, so substitution makes populations unique. The longer populations remain genetically isolated, the more unique substitutions they will acquire. Therefore we ought to be able to use consistent genetic differences to identify distinct species.

There are many advantages of using DNA sequences to identify species. Most importantly, DNA is universal to all living things, so molecular taxonomy can be applied to any kind of organism. Molecular data can be used to distinguish species that have relatively few distinctive morphological characteristics, such as many microbes. If you consider each nucleotide in the genome to be a potential diagnostic characteristic, then even the smallest viral genome has thousands of directly comparable characters, far more than you could derive from physical features of the organism. And, since the genome is contained in almost every cell, DNA taxonomy can be applied to partial specimens, from living or dead organisms (Case Study 1.1). Even tiny samples of biological material can yield enough DNA to make an unambiguous identification. This is invaluable for species that cannot be directly observed, so has been useful for monitoring populations of elusive organisms (e.g. Case Studies 1.2 and 2.1), and solving many case studies in cryptozoology (the study of mysterious or mythical creatures: see Case Study 1.1). DNA taxonomy is being increasingly used to describe biodiversity in some bacterial communities where taxa cannot be easily cultured or described using traditional microbiological methods. DNA can also help identify species whose different life cycle stages are morphologically dissimilar, for example allowing plant species to be identified from seeds, or animal species from larvae.

One of the practical advantages of molecular taxonomy, though it might ultimately be regarded as a disadvantage for biology as a whole, is that it requires a great deal less training to be able to identify species using DNA data (Case Study 9.1). It can be incredibly frustrating to peer down a microscope and try to determine whether a pickled beetle has a diagnostic characteristic such as a 'pronotum with transverse furrow and with or without lateral depressions', or to follow a botanical key only to find that you are missing one of the key diagnostic characteristics, because you only have a leaf and cannot count petal number. But someone who has been trained to perform DNA extraction, amplification and sequencing would technically be able to apply those skills to any biological sample (though in practice, lab protocols differ somewhat between sample types). Furthermore, DNA sequencing is becoming increasingly efficient and cost-effective.

All of these advantages of DNA-based identification have led to attempts to develop universal 'DNA barcodes', loci in the genome that can be sequenced to reliably identify any biological sample to species level (TechBox 9.2). DNA barcoding is not a new technique, as it uses the same procedures described throughout this book: sequencing a specific locus then looking for matches in a DNA database. The term DNA barcoding really describes an attitude: that it is both possible and desirable to identify and catalogue species using DNA sequence databases. One of the goals of DNA barcoding is to provide unambiguous diagnostic characteristics that identify species, yet can be read automatically, without requiring years of specialist training. Although the holy grail of barcoding is to find a short gene segment that can be used to identify any species on earth, it may be impossible to find a universal barcode that will work for all taxonomic groups.

### *Defining species with DNA*

It is important to make the distinction between using DNA to identify species and using DNA to define species. If we know that all members of a certain species share particular bases in a specific DNA sequence, then we could use that sequence to prove an organism belonged to that particular species. In Figure 4.24, we saw that members of the species *Bacillus anthracis* all appear to have the same sequence for a particular region of the RNA polymerase II beta subunit gene, which is different from their close relatives, so this sequence is likely to provide a useful diagnostic test for the presence of anthrax.

But, more controversially, DNA analysis is increasingly being used to define species. Taxonomy often relies on using easily measured characteristics as a proxy for reproductive isolation or ecological distinctness. Not only are DNA sequences a very convenient way to measure similarity, they can also reveal distinct populations that would not have been detected using morphological data alone. Genetic analysis has led to the recognition of many 'cryptic species', which are clearly reproductively separate populations despite a lack of morphological divergence (Figure 9.8). Many zooplankton (minute aquatic organisms) have been considered to belong to cosmopolitan species, found all over the globe, but DNA analysis has revealed many ancient cryptic species, whose lack of morphological differentiation belies substantial genetic diversification. For example, DNA sequences taken from the rotifer *Brachionus plicatilis* collected all over the world show that this species actually consists of at least seven distinct lineages, sometimes co-existing in the same habitats, that may have been separated for tens of millions of years (**Figure 9.11**).

***The relationship between genetic and morphological divergence is discussed in Chapter 13***

The relative ease of DNA analysis, compared to traditional taxonomy, has spurred a huge increase in the use of DNA sequences to define species limits. However, molecular definition of species is problematic in practice because there is no clear cut-off for what is a 'real' species and what is a genetically diverse interbreeding population. If we want species to represent distinct populations characterized by unique traits, then identifying differences in gene sequences may be of little value if we do not know how those differences relate to other important features such as ecology and behaviour. So DNA is not an instant solution to taxonomy. And, as we will see in the next chapter, molecular data is not free of the problems of determining which similarities are the result of shared ancestry, and which are superficial resemblance due to convergent evolution. But molecular data does provide a very useful and accessible tool for estimating relatedness, which is becoming more and more valuable in recognizing species and placing them within an evolutionary framework.

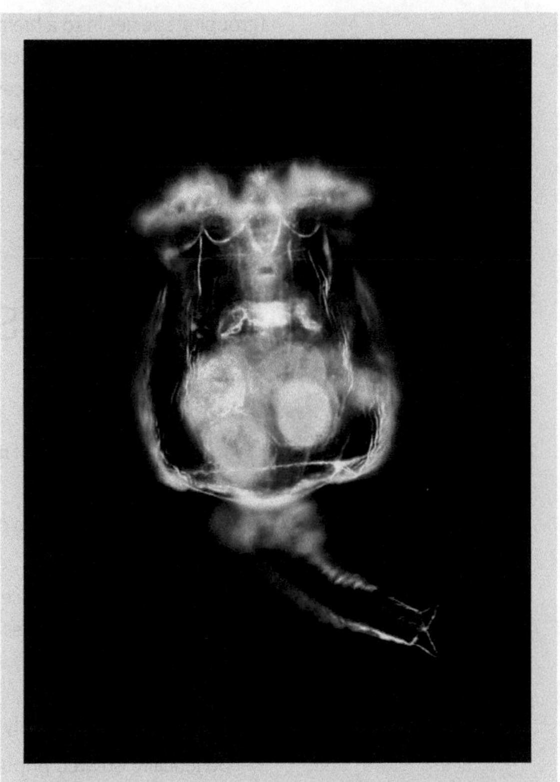

**Figure 9.11** Rotifers are microscopic aquatic animals. They were formerly known as wheel animalcules on account of the circular band of cilia that powers their locomotion, and draws food into their mouths. Rotifers are famous for their powers of abstinence: in their dormant state, they can go without food or water for extended periods, and some lineages of rotifers have apparently gone without sex for tens of millions of years.

# Conclusion

It may seem odd that biologists can't agree on how to define species. How can a scientific discipline function without agreement on the identity and definition of one of its most fundamental units? But when we consider species formation, typically by the gradual acquisition of many substitutions, it is hardly surprising that it is often difficult to find a clear dividing line that separates biological diversity into neat units. No species definition provides an unambiguous way to delineate all natural groups, as taxa may vary in their degree of morphological dissimilarity, amount of gene flow with related taxa, or adaptation to different niches.

If we want to use DNA sequences to recognize or define species, then we need to know that we are reliably identifying sites that differ consistently between species. To do this, we need to be able to compare DNA sequences that were identical in a common ancestor, but have since acquired distinct substitutions. Correct sequence alignment is essential for identification, classification (grouping species in to a hierarchy of similarity that reflects evolutionary history), and phylogeny (reconstructing the evolutionary history of populations). In the next chapter, we will see how alignments of DNA sequences allow us to reconstruct the history of life. This is probably the most important topic in the whole of bioinformatics, because virtually all analyses described in this book rest upon the sequence alignment being a reliable indicator of evolutionary descent.

# Points to remember

### Taxonomy

- Taxonomy, the system of naming species and other natural groups, aims to reflect relatedness, not just similarity.
- Species names have two parts: *Genus* (group of related species) and *species* (unique identifier for that species).
- If taxonomy reflects descent, then many different characteristics will show the same hierarchical patterns of similarities between relatives, including DNA sequences.

### Divergence

- Populations continuously acquire substitutions by drift and selection.
- Genetically isolated populations cannot exchange alleles with other populations so they accumulate a distinct set of substitutions, due to the random nature of mutation, the chance effects of genetic drift, and different alleles being fixed by selection.
- Populations build sets of co-adapted alleles, because mutations that don't work well with other common alleles will have a reduced chance of being passed on to the subsequent generations.
- Hybrid incompatibility can occur when the substitutions in one population do not work well in combination with the substitutions from another population.
- If hybrid taxa have lower fitness then gene flow between populations will be reduced, and selection may favour traits which decrease mating between individuals from different populations, reinforcing genetic isolation.
- When two sister populations have acquired alleles that make them distinct or prevent them from interbreeding, we recognize them as separate species.

### Species

- No species definition provides an unambiguous way to delineate all natural groups, as taxa may vary in their level of morphological dissimilarity, degree of gene flow with related taxa, or adaptation to different niches.
- Since genetic differences accumulate continuously, there may be no clear line that separate a species from its relatives or from its ancestor.
- Molecular taxonomy can aid species identification if a particular nucleotide sequence at a given locus is known to be characteristic of a given species.
- DNA sequences can also be used to define species based on shared history, by defining all the descendants of a particular ancestor that form a clade that is distinct from all other such lineages.

# Ideas for discussion

**9.1** If species are defined by lack of exchange of alleles, so that they accumulate unique substitutions, then should we define species by opportunity for gene flow? Should grey squirrels in the US and UK be considered separate species because they have no opportunity to interbreed?

**9.2** The HeLa cell line was established from a sample of human cancer cells in 1951. HeLa cells have so successfully adapted to cell culture that they frequently invade and replace other cell lines in laboratories. Since HeLa cells are now genetically distinct, having a different set of genes and chromosomes from normal human cells, should they be considered a separate species? If so how should they be classified, as a new species of *Homo*? Or, given their morphological and ecological dissimilarity to apes, as an entirely new group?

**9.3** The idea of sympatric speciation—the divergence of a population of co-existing individuals into two distinct non-interbreeding species—has been controversial. Why might some biologists consider sympatric speciation to be unlikely? What factors might promote the formation of a new species that overlaps in space and time with its sister species?

# Examples used in this chapter

Gomez, A., Serra, M., Carvalho, G. R., Lunt, D. H. (2002) Speciation in ancient cryptic species complexes: evidence from the molecular phylogeny of *Brachionus plicatilis* (Rotifera). *Evolution*, Volume 56, page 1431.

Johnson, J. A., Watson, R. T., Mindell, D. P. (2005) Prioritizing species conservation: does the Cape Verde kite exist? *Proceedings of the Royal Society B: Biological Sciences*, Volume 272, page 1365.

Koerber, G. R., Anderson, P. A., Seekamp, J. V. (2013) Morphology, physiology and AFLP markers validate that green box is a hybrid of *Eucalyptus largiflorens* and *E. gracilis* (Myrtaceae). *Australian Systematic Botany*, Volume 26, page 156.

Mant, J., Bower, C. C., Weston, P. H., Peakall, R. (2005) Phylogeography of pollinator-specific sexually deceptive *Chiloglottis* taxa (Orchidaceae): evidence for sympatric divergence? *Molecular Ecology*, Volume 14, page 3067.

Nijhout, H. F. (2003) Polymorphic mimicry in *Papilio dardanus*: mosaic dominance, big effects, and origins. *Evolution and Development*, Volume 5, page 579.

Storz, J. F., Sabatino, S. J., Hoffmann, F. G., Gering, E. J., Moriyama, H., Ferrand, N., Monteiro, B., Nachman, M. W. (2007) The molecular basis of high-altitude adaptation in deer mice. *PLoS Genetics*, Volume 3, page e 45.

HEROES OF THE GENETIC REVOLUTION

9

# Rosemary Gillespie

*"What I really like is exploring. Fieldwork often requires a lot of hard hiking and crossing swollen rivers –it's just terrific fun . . . "*

Rosemary Gillespie (2008, Tennessee Alumnus Magazine)

**EXAMPLE PUBLICATIONS**

Gillespie, R. G., Oxford, G. S. (1998) Selection on the color polymorphism in Hawaiian happy-face spiders: Evidence from genetic structure and temporal fluctuations. *Evolution*, Volume 52(3), pages 775–783.

Gillespie, R. G. (2004) Community assembly through adaptive radiation in Hawaiian spiders. *Science*, Volume 303 (5656), pages 356–359.

**Figure 9.12** Rosemary Gillespie in her natural habitat, collecting spiders in the field.

Image courtesy of Rosemary Gillespie.

I was once lucky enough to go nocturnal spider hunting with Rosemary Gillespie on the Big Island of Hawaii. She and her husband George Roderick each carried one of their children on their backs, picking their way across blackened lava flows in the dark to reach an isolated island of remnant forest. There, we looked for the glint of torchlight on silk and the shine of tiny eyes, collecting the spiders in small vials as they spun down on silken threads from the foliage (and, happily, I still have the scars on my leg from falling through the rough larval rocks on the way back as a permanent reminder of this adventure).

Rosemary Gillespie was born and raised in Scotland, and completed her undergraduate degree in zoology at the University of Edinburgh in 1980 (Figure 9.12). She worked on a group of long-jawed spiders for her PhD at the University of Tennessee, and, when she moved to the University of Hawaii as a postdoc, she found the same group of spiders in the Hawaiian Islands. She used molecular phylogenies to reconstruct the evolutionary history of these *Tetragnatha* spiders, demonstrating that this diverse radiation arose from multiple colonization events, which then diversified into a greater variety of forms and behaviours than the mainland lineages. These adaptive radiations played out independently on each of the Hawaiian islands, giving rise to convergent sets of 'ecomorphs' on each island. For example, one lineage has abandoned web-building to become an active hunter: this lineage is found in four distinct ecomorphs on each island (Green, Big brown, Little brown, Maroon), and yet each of the ecomorphs evolved independently on each island. It's as if each island is a repeated experiment in evolution, and all have similar outcomes. Similarly, Gillespie has also shown that the striking colour polymorphisms in the 'happy face spiders' (Theridiidae) have evolved repeatedly on different islands, though the underlying genetics of the colour patterns may differ even when the morphological appearance is very similar.

Gillespie has used the radiation of spiders in Hawaii and other islands to examine the general principles governing community assembly: how regional biodiversity is generated through a combination of colonization, speciation and extinction. In this way, Gillespie is an example of a common pattern in evolutionary biology: someone who uses their love and understanding of a particular taxonomic group or ecosystem to build a more general understanding of evolutionary and ecological processes. She has also worked to encourage citizen science, through public engagement in digitizing entomological collections, and is involved in programmes to increase appreciation of invertebrate biodiversity in school children. Rosemary Gillespie is currently the Schlinger Chair of Systematics at the University of California Berkeley Campus, director of the Essig Museum of Entomology, and president of the International Biogeography Society.

TECHBOX 9.1

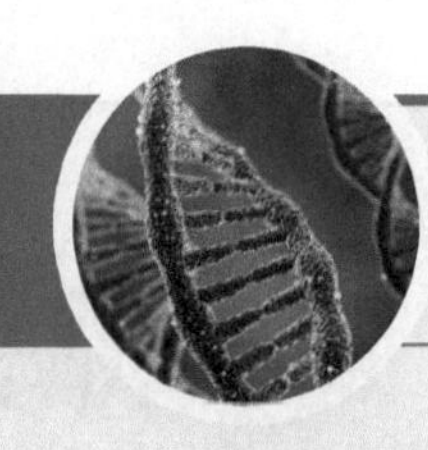

# Phylogeography

**RELATED TECHBOXES**

**TB 8.1:** Neutral theory

**TB 8.2:** Population size

**RELATED CASE STUDIES**

**CS 2.1:** Environmental DNA

**CS 9.2:** Barcoding

> *The foundation of the Darwinian theory is the variability of species, and it is quite useless to attempt even to understand that theory, much less to appreciate the completeness of the proof of it, unless we first obtain a clear conception of the nature and extent of this variability*
>
> Wallace, A. R. (1889) *Darwinism: an exposition of the theory of natural selection with some of its applications.* Macmillan and Co.


### Summary

Considering spatial patterns in genetic variation can reveal the degree of connection between populations and uncover evolutionary histories and processes.



### Keywords

allele, mitochondrial, high-throughput sequencing, gene tree, species tree, coalescent, landscape genetics, evolutionary significant units (ESU), environmental niche modelling (ENM), mutation rate, population size


Phylogeography uses patterns of genetic variation sampled across the distribution of closely related lineages to distinguish populations and to generate hypotheses about their recent evolutionary history. Phylogeography is often applied at the intraspecific level, comparing different populations within a recognized species. Therefore data chosen for phylogeographic studies will often be fast-evolving loci that are expected to be polymorphic within populations, such as mitochondrial sequences, microsatellite loci or introns. Phylogeographic studies typically focus on spatial patterns of variation, comparing populations across an area to see which ones are most closely related, determining whether populations are connected by gene flow or genetically isolated, and identifying patterns that might reveal the way the landscape form and history has shaped patterns of biodiversity.

The basic premise of historical phylogeography is much the same as that for phylogenetics as a whole: descent leaves a hierarchy of similarities, so analysing the differences between lineages can reveal past patterns of descent. Populations that were recently connected by gene flow are expected to share more alleles than those that have long been separated. These patterns of similarities are sometimes displayed in cheerily coloured lollipop diagrams, where the allelic complement of each population is represented by coloured circles, and the degree of difference between populations represented by lines connecting the circles[1] (**Figure 9.13**). Remember that these diagrams are not in themselves documents of population history: they are descriptions of patterns in the data from which you might generate a hypothesis about the history that shaped those patterns of genetic differences.

Phylogeographic analysis is sometimes used to identify 'evolutionary significant units' (ESU): populations that may not differ in their observable physical characteristics but represent distinct genetically isolated lineages[2]. The focus of a study may be on developing conservation strategies to conserve all of the genetic diversity in a species, or it may be on assessing the degree of connection between isolated populations in order to judge the viability of the metapopulation and the potential for future gene flow between them. Assessment of population-level genetic variation and gene flow has also been used to predict the future resilience of populations, for example their ability to adapt to changing environmental

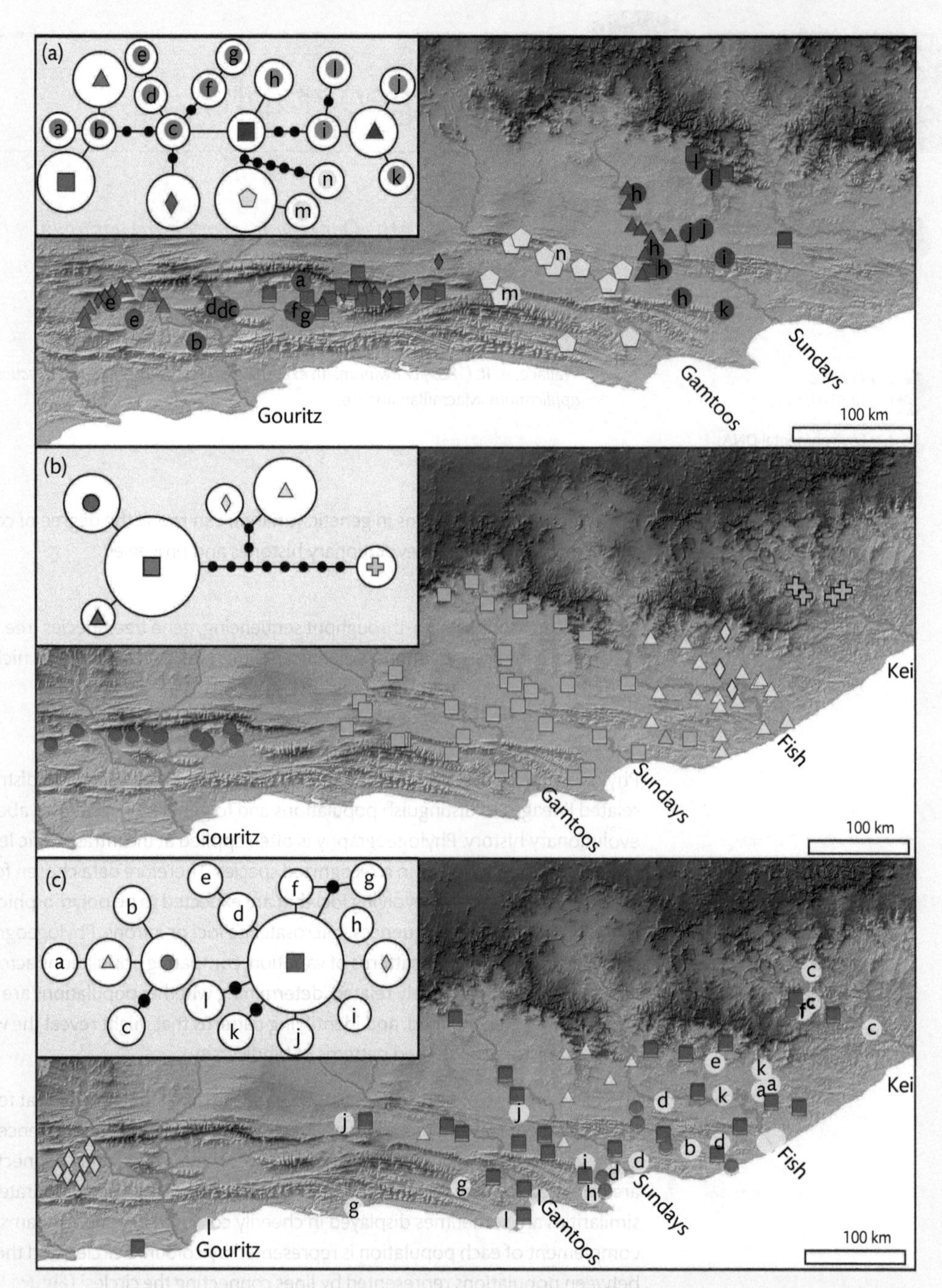


**Figure 9.13** Trees can travel by pachyderm: Phylogeographic patterns in genetic variation can reveal differences in dispersal ability. These tree species in South Africa show different patterns of genetic variability across the deeply incised landscape. The tree species whose seeds are dispersed by wind (*Nymania capensis*, a) or birds (*Pappea capensis*, b) show clear evidence of populations being confined to river basins, but the species whose seeds are dispersed by large herbivores such as elephants (*Schotia afra*, c) show more gene flow between neighbouring basins. Further work by the same research team suggests that the these distribution patterns are also driven by Pleistocene climate cycles which cause the frost-sensitive subtropical thicket communities to retract into river basins.

conditions, on the assumption that lack of genetic variation will inhibit potential for future adaption. For example, responses to climate change are usually considered on a species-wide basis, using current distribution as a guide to the range of environmental tolerance, but if a species actually consists of genetically isolated populations that are subject to different environmental conditions, then populations may differ in their environmental tolerance limits[3]. Phylogeographic analysis of variation is increasingly being combined with environmental niche modelling to enrich predictions concerning both past and future species distributions under changing environmental conditions[4]. Bear in mind, though, that phylogeographic studies typically survey neutral markers, on the grounds that they will be useful indicators of population history and demographic parameters. Such data do not provide a direct evaluation of alleles that might be subject to selection, current or future, but they are used as an indicator of general levels of genetic variability.

### Example: the Wet Tropics World Heritage Area

To illustrate the application of phylogeographic analysis, let's briefly consider research aimed at understanding the factors shaping the high biodiversity in the Wet Tropics region of Northern Australia (**Figure 9.14**). Researchers used mitochondrial sequences to identify regions with putative ESUs—that is, areas containing populations that differed from all other related populations by virtue of unique sequence differences[5]. They inferred that the genetic uniqueness of the populations in these areas was due to lack of genetic exchange with other areas. Then, to gain a measure of the degree of isolation, they estimated the number of substitutions unique to each ESU. Deep divergences between mitochondrial sequences led

**Figure 9.14** Where the rainforest meets the reef: much of the Wet Tropics of northeastern Australia is protected as a World Heritage Area, but ongoing management decisions are critical to conserving the unique flora and fauna. Craig Moritz and colleagues have used a comparative phylogeographic approach—comparing patterns of genetic variation across the landscape for a large number of species including frogs, lizards, birds, snails and beetles—to link genetic patterns to the history of habitat expansion and contraction inferred from the fossil record and climatic models. Moritz has used the conclusions of these studies to promote a dynamic approach to conservation, not simply protecting existing species but aiming to preserve the potential for future evolutionary diversification.

Reproduced courtesy of Marcel Cardillo.

to the proposal that most vertebrate species in the Wet Tropics are considerably older than the Pleistocene, the last great period of climate change (approx. 2 Myr ago). The genetic data suggests that rainforest vertebrates were restricted to isolated refugia when the rainforest contracted during late Pleistocene, then expanded again as the climate warmed in the Holocene (the last 10,000 years). They concluded that the Wet Tropics are 'museums' that preserve vertebrate species rather than 'cradles' that create them. Thus the present genetic diversity reflects past habitat distribution. For example, many species showed a major genetic disjunction across the Black Mountain corridor, which separated rainforest refugia in the past but is today connected by continuous habitat. These patterns led these researchers to conclude that any conservation strategy should aim to 'keep all the pieces': even where ESUs show little physical differences they should be preserved as unique gene pools that cannot be recovered once lost.

### Choosing loci

Phylogeographic studies have traditionally relied on mitochondrial or chloroplast loci, being sufficiently variable among populations and relatively convenient to sequence. But these markers may not always accurately represent the recent demographic history of the populations, especially if there have been fluctuations in population size[2]. High-throughput sequencing (HTS) techniques are now making it possible to detect and analyse population-level variation in a wide range of nuclear markers[6]. These techniques can be used to target fast-changing sequences, such as introns or intergenic regions, or to sample a large number of slower-changing loci that each have fewer polymorphisms[7]. Needless to say, you need to be sure that you are comparing homologous sequences between individuals and populations (that is, sequences representing the same locus in the genome all originally copied from a single common ancestor) so HTS techniques that target specific loci are likely to be the most practical[6]. The selection of appropriate loci may occur at the 'wet' stage (targeted amplification of DNA from the sample) or at the 'dry' stage (bioinformatic selection of homologous loci from sequence data).

One of the advantages of high-throughput sequencing is that it potentially provides a large number of independent genetic markers of population history. This is important because analysing any single sequence gives you information about the shared common ancestor of that particular locus, but each separate locus may tell a slightly different story. This is because the shared common ancestor of any given allele may have existed before the populations divided, or (in the case of gene flow) after the populations became physically separate. So if you sequence a single locus, you can't be sure if the history of that sequence matches the history of lineages exactly. In other words, any given 'gene tree' (phylogeny based on a single locus) might not accurately estimate the 'species tree' (history of the lineages under consideration: **Figure 9.15**[8]).

But the histories of different alleles can be explicitly incorporated into analyses using a coalescent framework, which takes a population genetic approach to reading evolutionary history from genetic data. Starting with the sample of alleles from the contemporary population, coalescent methods work backwards, using a neutral model of population genetics (**TechBox 8.1**) to predict how many generations ago these alleles were copied from a common parent (that is, the age of the last common ancestor). These methods incorporate parameters to describe key population genetic processes, such as the mutation rate and effective population size. Remember that as populations change over time, these parameters may in themselves change. For example, any barriers to gene flow in the ancestral population will shorten the coalescent times compared to an undivided population of the same size: 'It is important to remember that each ancestral node in a phylogeny represents a population, so we must keep in our minds' eye the population processes that impinge on that ancestral species'[8].

A note on terminology: you may come across papers that treat the pursuits of phylogeography, landscape genetics and population genetics as entirely different fields, and yet others will view them as essentially parts of the same discipline. To some extent, the label that you choose to describe your work in this area will be influenced by cultural factors (what scientific sub-tribe

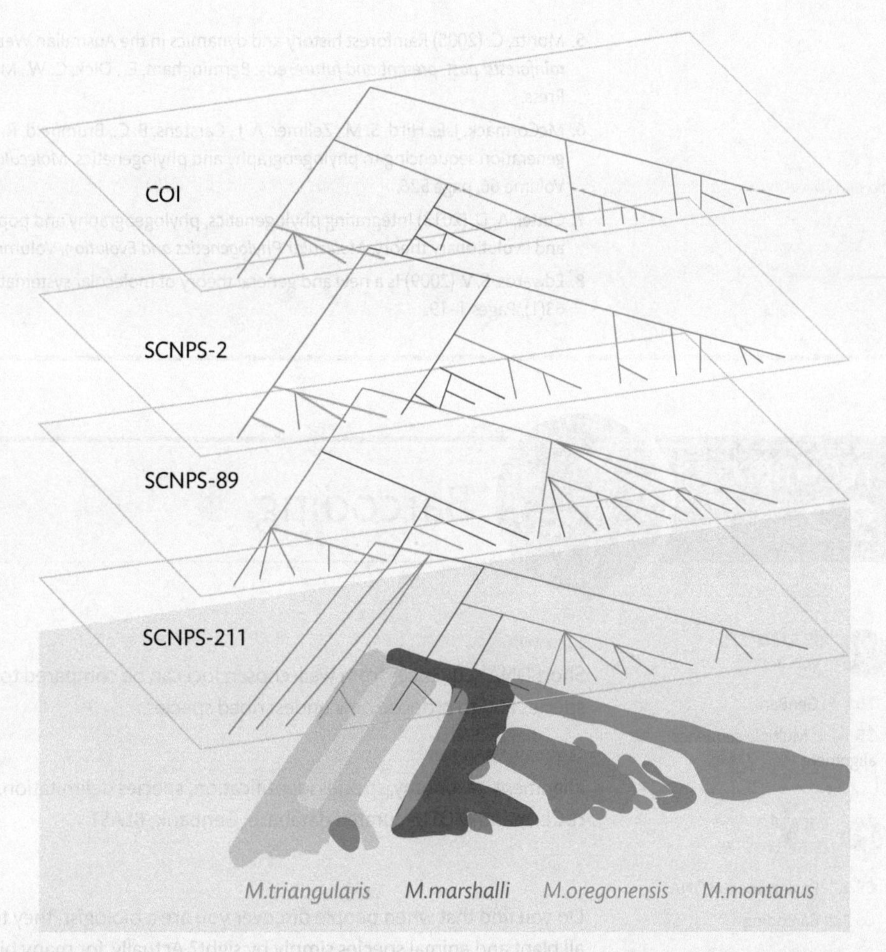


**Figure 9.15** Alternative histories: gene trees based on different loci (COI, SCHPS-2, -89 and -211) may not exactly match the species tree, which represents the population history of a set of distinct lineages (in the lower panel). But since all of the gene trees must be contained with the species history, it is possible to use the variation or concordance between different gene trees to estimate population history.



you identify with), the data analysed (e.g. whether you target neutral loci or protein-coding sequences) and the analytical approach (whether you are interested in gene flow, history, adaptation to environmental conditions, and so on). At the risk of appearing blasé, I would venture that it doesn't particularly matter what you want to call your research as long as you are clear on the question you are asking, you take care to choose appropriate data and scrutinize them carefully, and you are mindful of the strengths and weaknesses of each analytical approach.

## References

1. Potts, A. J., Hedderson, T. A., Cowling, R. M. (2013) Testing large-scale conservation corridors designed for patterns and processes: comparative phylogeography of three tree species. *Diversity and Distributions*, Volume 19, page 1418.

2. Moritz, C. (1994) Defining 'Evolutionarily Significant Units' for conservation. *Trends in Ecology and Evolution*, Volume 9, page 373.

3. D'Amen, M., Zimmermann, N. E., Pearman, P. B. (2013) Conservation of phylogeographic lineages under climate change. *Global Ecology and Biogeography*, Volume 22, page 93.

4. Alvarado-Serrano, D. F., Knowles, L. L. (2014) Ecological niche models in phylogeographic studies: applications, advances and precautions. *Molecular Ecology Resources*, Volume 14, page 233.

5. Moritz, C. (2005) Rainforest history and dynamics in the Australian Wet Tropics. pp. 313–21. In *Tropical rainforests: past, present and future*. eds: Bermingham, E. , Dick, C. W., Moritz, C., University of Chicago Press.
6. McCormack, J. E., Hird, S. M., Zellmer, A. J., Carstens, B. C., Brumfield, R. T. (2013) Applications of next-generation sequencing to phylogeography and phylogenetics. *Molecular Phylogenetics and Evolution*, Volume 66, page 526.
7. Cutter, A. D. (2013) Integrating phylogenetics, phylogeography and population genetics through genomes and evolutionary theory. *Molecular Phylogenetics and Evolution*, Volume 69(3), Pages 1172–85.
8. Edwards, S. V. (2009) Is a new and general theory of molecular systematics emerging? *Evolution*, Volume 63(1), Pages 1–19.

TECHBOX
9.2

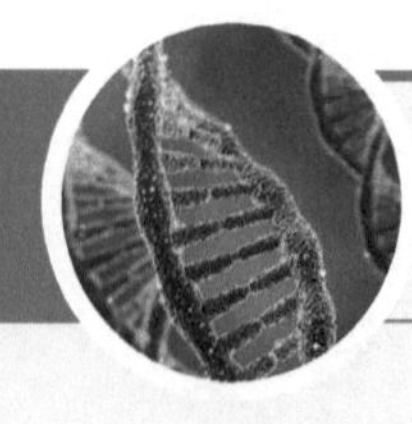

# Barcoding

**RELATED TECHBOXES**

**TB 1.1:** GenBank

**TB 10.2:** Multiple sequence alignment

**RELATED CASE STUDIES**

**CS 2.1:** Environmental DNA

**CS 9.2:** Barcoding

**Summary**

Short DNA sequences from well-chosen loci can be compared to curated databases to identify species and to propose new undescribed species.

**Keywords**

alignment, taxonomy, species identification, species delimitation, barcoding gap, loci, COI, ITS, rBCL, matK, MOTU, curated database, GenBank, BLAST

Do you find that when people discover you are a biologist, they think you will be able to identify all plant and animal species simply by sight? Actually, for many biological groups, species identification is very difficult and can only be undertaken by someone with deep knowledge and experience of that particular group. Wouldn't it be nice if there was an easier way, if you had some technology that could scan an organism and tell you exactly what species it belongs to and where it fits in the whole taxonomic hierarchy. Such is the hope of DNA barcoding.

The ideal DNA barcode is a short DNA sequence (less than 1,000 bases long) that can be compared between a wide range of organisms, with key sequence differences that will unambiguously identify any sample to species level. Barcoding projects are set up as curated databases of sequences against which a sample from the focus group can be matched. DNA barcoding is not a different method of sequence analysis: it uses the same basic techniques and analyses that you will find described throughout this book. But the primary goal of barcoding is taxonomic identification, rather than identifying relationships between taxa, tracing evolutionary history, or describing processes of molecular evolution. And, unlike other sequence repositories, barcoding databases are specifically constructed for taxonomic identification, with the aim of reducing the chances of misidentification that could arise from simply blasting a sequence to GenBank and seeing what the best match is.

There are two broad ways that DNA barcoding is used: firstly, as a means of identification through comparing a DNA sequence to those of known species; and secondly as a means of species discovery, where the barcode sequences from many different specimens are compared to identify how many putative species they represent.

## Species identification

### 1. Locus

The first step of any barcoding project is to select one or more loci (specific parts of the genome) that have several key properties. A barcoding locus must be sufficiently similar among members of the target group that primers can be designed to amplify that sequence from a wide range of different organisms with a consistently high degree of success[1]. The resulting sequences must be able to be confidently aligned so that differences can be reliably identified, yet the sequences must also be sufficiently variable that different species will have different nucleotide sequences at that locus. While there are some genes that can be compared between any species, from bacteria to blue whales (see **Figure 2.2** for an example), these highly conserved loci may be insufficiently variable between closely related species to be useful barcodes. So most barcoding projects target a particular locus that is useful for the group in question: many animal barcoding projects use part of the mitochondrial *cytochrome oxidase* gene (*COI*); many plant projects use the chloroplast sequences *rBCL* and *matK*; nuclear internal transcribed space sequences *ITS* have been used for fungi; and small subunit ribosomal RNA (*SSU*) for microbial taxa. None of these sequences is without problems. For example, there can be copies of organelle genes in the nucleus, and multiple variable copies of nuclear sequences that may either diverge from each other, or be homogenized by gene conversion.

### 2. Database

DNA barcode databases resemble traditional taxonomic collections in that unique species names should be attached to 'vouchered' specimens: like the traditional use of a type specimen, each reference sample should be identifiable to the original sample from which it was taken, with information about its collection (place, date, collector, etc.). Databases may contain additional information, such as morphological data, images or videos[2]. Many barcoding projects are represented by international consortia, consisting of independent research groups that share the task of sequencing and cataloguing diversity to build a reference database. For example, the fungal barcoding project website contains information on primers for the key loci, some of which are optimized to different fungal groups, an alignment and search tool, database of different fungal strains, and a sequence deposition portal. Alternatively, sequences intended for use in taxonomic identification can be deposited in existing databases. Submissions to GenBank can be marked as a barcode, in which case they should also include a valid species name, an unambiguous sequence at lease 500 nucleotides long from the 5' end of the COI gene, and reference to a particular voucher specimen and collector. In any contributed database, quality control will be an issue. For example, it may happen that two independent researchers both submit sequences from the same taxon but assign it to different species names[3]. Mistakes in identification could potentially be propagated if people use matches to that sequence to identify their own sequences which then will be deposited in the database under the same incorrect taxonomic assignment.

The use of DNA barcoding for species identification is only as good as the reference database, because you can't identify a species that has not yet been entered in the database. The ideal database would have unique barcodes for every species on earth, but, naturally, representation remains incomplete, with many groups having a relatively small fraction of known species included in barcoding databases[4]. Therefore failure to find a match does not necessarily mean that you have discovered a new species, it may be that the species you have sequenced has not yet been added to the database. Barcoding for species identification will be most reliable for species whose genetic diversity has been well surveyed.

### 3. Identification

The simplest way to identify a sequence by comparison to a database is to use a tool like BLAST (**TechBox 1.2**). This is fine if your sequence has an exact match to one and only one sequence in the reference database, but if there is no exact match, or several different close matches, then the species diagnosis may be a matter of opinion on what constitutes an acceptable match (remembering that the right species may not even be in the database). Alternatively, a diagnostic set of nucleotides can be defined, such that having those particular combinations of nucleotides acts as characters for species identification, similar to the approach in morphological identification[5]. More often, some kind of measure of genetic distance is used to find the nearest match to the query sequence (**TechBox 11.1**). In this case, a decision needs to be made about how close the match needs to be to be deemed a reliable identification, either on the basis of similarity or relatedness on a phylogeny. As with most techniques in this book, preferred methods change quite rapidly, and you will need to consult the recent literature for discussions about the best way to match barcode sequences to databases.

### Species discovery

It is important to note the difference between using a curated DNA database as an aid to identifying species, and using barcode sequences to delineate undescribed species. DNA sequences, whether short 'barcodes' or longer sequences, are increasingly being used as a tool in surveying biodiversity. The sequences from many unidentified specimens can be compared with the aim of detecting how many different species they belong to (**Case Study 9.2**). This may be based on a defined level of difference, such that two lineages that differ by a particular percentage of sites in the sequence are deemed to be from different species. This approach creates arbitrary divisions which may divide a continuous array of divergent lineages. The rate of acquisition of substitutions depends on many different factors, such as demography and mutation rate, so the percentage difference that seems to correspond to species-level differences in one set of lineages may not work effectively at dividing natural groups in another set of lineages. In some cases, the proposed level of difference cut-offs will be too conservative, such that many previously recognized species will be lumped together[6]. In other cases the distance between apparently identical taxa will be much greater than that expected. An alternative approach is to try to define 'molecular operational taxonomic units' (MOTUs) by finding clusters of sequences that are clearly different from other such clusters. A suite of methods aims to identify the 'barcoding gaps' between these clusters, which are assumed to represent natural divisions between species.

The use of DNA barcoding for species discovery has been controversial, with a degree of scepticism that any single locus can provide a 'quick fix' for species discovery. In addition to the challenges of deciding how to divide samples into discrete species, the divergence at one particular locus may not accurately represent the divergence across the genome. For example, organelle genes can have different evolutionary histories from genes in the nucleus, particularly in the face of gene flow and hybridization. A study of pairs of unicellular taxa suggested that the 18S rRNA sequence commonly used in barcoding projects was often unrepresentative of the degree of divergence in protein-coding parts of the genome, such that pairs of taxa with identical 18S sequences might show differences measured from the proteome that were equivalent to the differences seen between vertebrate classes[7]. However, DNA barcoding can be looked upon as a first step in species discovery, a way of 'generating hypotheses of new taxa in need of formal taxonomic treatment'[5].

### The future of DNA barcoding

There is no clear line between DNA barcoding and other forms of DNA sequence analysis. While the term 'DNA barcoding' was originally applied to species identification through matching a short universal sequence (COI) to a database, you will also find the term barcoding applied to examples using a wide range of sequences and techniques at every level of biological differentiation, from individuals to kingdoms. Some researchers feel that DNA barcoding will become irrelevant with the uptake of high-throughput sequencing, because there will be no advantage to targeting a single short locus for sequencing. However, others have embraced the potential for high-throughput sequencing to extend the reach of barcoding[8], allowing the use of longer sequences or multiple loci to aid identification[9]. Longer sequences provide more variable sites, and may allow different taxonomic levels to be distinguished. But the more DNA is sequenced, the more samples will vary, such that no two individuals will have exactly the same sequence: in this case, decisions must be made on identifying criteria that correspond to species-level differences[10] (**Figure 9.16**). Naturally, sequencing errors rates must be taken into consideration when identifying taxa from high-throughput sequencing programs, which can artificially increase estimates of diversity of samples[11].

**Figure 9.16** Checking the chocolate: DNA barcoding can be used to distinguish lineages within species. Researchers used whole chloroplast genomes to characterize different varieties of cacao (*Theobroma cacao*) with the aim of being able to distinguish different varietal origins of chocolate. Since every individual sampled had a unique sequence, diagnostic criteria are needed to assign any individual to a particular strain or variety.

## References


1. Schoch, C. L., Seifert, K. A., Huhndorf, S., Robert, V., Spouge, J. L., Levesque, C. A., et al. (2012) Nuclear ribosomal internal transcribed spacer (ITS) region as a universal DNA barcode marker for Fungi. *Proceedings of the National Academy of Sciences USA*, Volume 109, page 6241.
2. De Ley, P., De Ley, I. T., Morris, K., Abebe, E., Mundo-Ocampo, M., Yoder, M., Heras, J., Waumann, D., Rocha-Olivares, A., Jay Burr, A. H., Baldwin, J. G., Thomas, W. K. (2005) An integrated approach to fast and informative morphological vouchering of nematodes for applications in molecular barcoding. *Philosophical Transactions of the Royal Society London: Biological Sciences*, Volume 360, page 1945.
3. Collins, R., Cruickshank, R. (2013) The seven deadly sins of DNA barcoding. *Molecular Ecology Resources*, Volume 13, page 969.
4. Kvist, S. (2013) Barcoding in the dark?: a critical view of the sufficiency of zoological DNA barcoding databases and a plea for broader integration of taxonomic knowledge. *Molecular Phylogenetics and Evolution*, Volume 69, page 39.
5. Goldstein, P. Z., DeSalle, R. (2011) Integrating DNA barcode data and taxonomic practice: determination, discovery, and description. *Bioessays*, Volume 33, page 135.
6. Moritz, C., Cicero, C. (2004) DNA barcoding: promise and pitfalls. *PLoS biology*, Volume 2(10):e354.
7. Piganeau, G., Eyre-Walker, A., Grimsley, N., Moreau, H. (2011) How and why DNA barcodes underestimate the diversity of microbial eukaryotes. *PLoS ONE*, Volume 6:e16342.
8. Shokralla, S., Gibson, J. F., Nikbakht, H., Janzen, D. H., Hallwachs, W., Hajibabaei, M. (2014) Next-generation DNA barcoding: using next-generation sequencing to enhance and accelerate DNA barcode capture from single specimens. *Molecular Ecology Resources*, Volume 14, page 892.
9. Li, X., Yang, Y., Henry, R. J., Rossetto, M., Wang, Y., Chen, S. (2014) Plant DNA barcoding: from gene to genome. *Biological Reviews*, Volume 90, page 157.
10. Kane, N., Sveinsson, S., Dempewolf, H., Yang, J. Y., Zhang, D., Engels, J. M. M., Cronk, Q. (2012) Ultra-barcoding in cacao (*Theobroma* spp.; Malvaceae) using whole chloroplast genomes and nuclear ribosomal DNA. *American Journal of Botany*, Volume 99, page 320.
11. Bik, H. M., Porazinska, D. L., Creer, S., Caporaso, J. G., Knight, R., Thomas, W. K. (2012) Sequencing our way towards understanding global eukaryotic biodiversity. *Trends in Ecology & Evolution*, Volume 27, page 233.

CASE STUDY 9.1

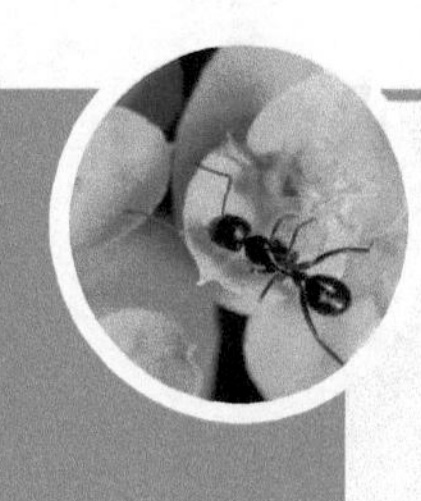

# Speciation: evolution of novel flower colour through reinforcement

RELATED TECHBOXES

TB 7.2: Detecting selection

TB 8.1: Neutral theory

RELATED CASE STUDIES

CS 7.1: Selection

CS 8.2: Population size

Hopkins, R., Levin, D. A., Rausher, M. D. (2012) Molecular signatures of selection on reproductive character displacement of flower color in *Phlox drummondii*. *Evolution*, Volume 66(2), pages 469–85.

> *Although reinforcement has been studied primarily in animals, our work indicates that it may also be an important contributor to speciation in plants. If so, this phenomenon may provide a partial explanation for the tremendous diversity of floral color, floral morphology, and inflorescence structure that characterize flowering plants*[1]

**Keywords**

microsatellite, regulatory, allele, population, selection, hybrid, selective sweep, neutral diversity, pollinator, sympatry

## Background

As populations diverge, they accumulate substitutions that may make the genomes from one population incompatible with the genomes in related populations. In such cases, cross-species mating can produce hybrids that are less fit than either of the parents, so an individual that has traits that reduce the chance of mating with members of another population, thus avoiding wasting reproductive effort on less-fit hybrid offspring, should have greater reproductive success. This process, referred to as reinforcement, can lead to sister populations that differ in distinct mating characteristics that act as barriers to mating with members of neighbouring populations. Theoretical studies have suggested that reinforcement is most likely to occur when it can act on one or few genes of large effect[2], but it has been remarkably difficult to demonstrate the genetic basis of reinforcement in naturally occurring species.

## Aim

In most places in which it is found, *Phlox drummondii* has pale blue flowers. But in one small part of its range, where it overlaps with a closely related species (*P. cuspidata*) that also has blue flowers, *P. drummondii* has red flowers (**Figure 9.17**). Hybrids between the two species have reduced fertility[3]. Red-flowered *P. drummondii* hybridize with *P. cuspidata* less often than the blue flowered form does, so they should have a selective advantage where the two species co-exist. But was the evolution of the red-flowered form driven by selection for reinforcement, or could the red flowers simply be a product of genetic drift driving rare alleles to fixation in some populations?

## Methods

Two genetic changes contribute to the difference in flower colour between the red and blue phlox flowers. One changes the hue by downregulating the expression of an enzyme that produces the blue pigment, leaving only the red pigment expressed. The other changes colour intensity by upregulating the production of pigment. Both these changes occur in the regulatory regions of the genes. But since the location and nature of the relevant regulatory sequences is unknown, this study focused on the protein-coding sequences of both the hue and intensity genes. Researchers took 605 samples of *P. drummondii* from 39 different populations across its range, taking equal numbers of each flower colour in mixed populations. Microsatellites were

**Figure 9.17** Wild populations of light-blue (A) and dark-red (B) *Phlox drummondii*. To evaluate pollinator behaviour, researchers used experimental arrays of flowers of light-blue and dark red *P. drummondii* as well as light-blue flowers of *P. cupsidata* (smaller flowers at bottom right in C). The main pollinator is the swallowtail butterfly pollinator *Battus philenor*, seen visiting a light-blue inflorescence of *P. drummondii* in (D).



used as a measure of neutral genetic variation, in order to represent the background rate of divergence between populations. They also took samples along a 38 km transect that spanned the transition from blue- to red-flowered populations. Variation in the DNA sequences from the two target genes were evaluated along this transect.

### Results

Microsatellite loci showed no detectable differences with flower colour across blue, mixed or red populations, and there was no clear pattern of genetic difference with distance, supporting the existence of gene flow between neighbouring populations despite differences in flower colour. There was much less genetic variation in the hue gene in the red population than in the dark blue population: this is consistent with a selective sweep, where selection on one allele drives fixation of linked sites, reducing genetic variation around the selected site. However, tests for the signature of selection on the coding sequence of the hue gene show no significant departure from neutrality. The intensity gene showed no significant evidence of selection, as there was no consistent difference in genetic diversity between the blue and red populations.

### Conclusions

The researchers concluded that the difference in flower colour cannot be explained by genetic drift leading to the chance fixation of the red colour morph, because the pattern of variation in colour pattern does not match the pattern of variation of the microsatellite alleles. They also ruled out a recent founder effect in the establishment of the red populations, because the red populations had the same levels of microsatellite diversity as the blue and mixed populations.

The lack of genetic differences in neutral alleles across the species range suggests that the red colour morph must be maintained in the sympatric populations in the face of substantial gene flow. They concluded that the low genetic diversity at the hue locus is evidence that the red colour has been under strong selection in populations of *P. drummondii* where they overlap with their blue-flowered relatives.

### Limitations

The molecular results supporting selection on the hue gene but not the intensity gene are somewhat at odds with field trials that suggest that darker pigmentation, whether blue or red, reduced the production of hybrid progeny, giving the high intensity allele a strong selective advantage, with no obvious selection on hue[3]. However, the absence of a signal of selection in the coding sequence does not imply that no selection is taking place. The genetic changes affecting hue and intensity are in the regulatory sequences of the genes, rather than being determined by changes to the coding sequence of these enzymes.

### Further work

This study provides supporting evidence for the hypothesis that selection has resulted in the fixation of the red flowered allele in the sympatric *Phlox* populations, but it doesn't provide a clear reason why. Although subsequent experiments suggest that pollinators are less likely to move between darkly pigmented flowers of *P drummondii* and *P. cuspidata*, thus limiting the potential for hybridization between the two, pollinators are not more attracted to red than blue flowers[1]. The authors offer the tantalizing suggestion that the red flowers may be adapted to a past pollinator, no longer present.

### Check your understanding

1. What role does reinforcement play in speciation? Do all speciation events involve reinforcement?
2. Why did the researchers include microsatellite loci that are not involved in the determination of flower colour?
3. Why is reduced genetic variation in the hue gene in the red population taken as a sign of selection on flower colour?

### What do you think?

Why are some hybrids between species of equal or greater fitness than their parents (e.g. **Figure 9.10**) while others produce hybrids with lower (or zero) fitness?

### Delve deeper

How can populations diverge and become separate species when they co-exist and there is gene flow between them? Won't any differences that arise be diluted by mating between individuals of the population, and any combinations of alleles broken up, long before enough differences can accumulate to bring about reproductive isolation? 'Magic traits' offer a possible way around this[4]: if there is a single trait that is both under diversifying selection and drives non-random mating, then two non-interbreeding populations can form sympatrically. What kind of traits might cause speciation in this way? Is this likely to be a common driver of speciation?

### References

1. Hopkins, R., Rausher, M. D. (2012) Pollinator-mediated selection on flower color allele drives reinforcement. *Science*, Volume 335, page 1090.
2. Pannell, J. R. (2012) Speciation Genetics: reinforcement by shades and hues. *Current Biology*, Volume 22, page R299.
3. Hopkins, R., Rausher, M. D. (2011) Identification of two genes causing reinforcement in the Texas wildflower *Phlox drummondii*. *Nature*, Volume 469, page 411.
4. Servedio, M.R., Doorn, G., Kopp, M., Frame, A. M., Nosil, P. (2011) Magic traits in speciation: 'magic' but not rare? *Trends in Ecology & Evolution*, Volume 26, page 389.

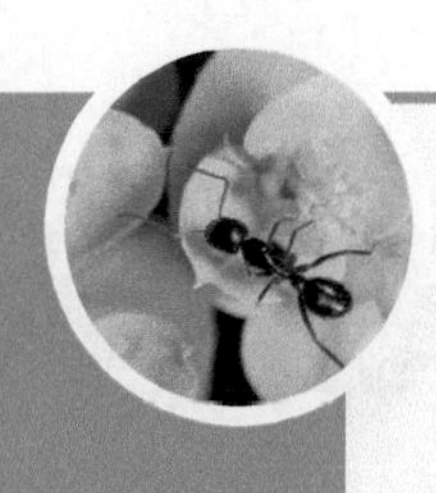

CASE STUDY 9.2

# Barcoding: cataloguing weevil diversity in New Guinea

RELATED TECHBOXES

TB 9.2: Barcoding

TB 11.1: Distance methods

RELATED CASE STUDIES

CS 1.1: Identification

CS 2.1: Environmental DNA

Tänzler, R., Sagata, K., Surbakti, S., Balke, M., Riedel, A. (2012) DNA barcoding for community ecology—how to tackle a hyperdiverse, mostly undescribed Melanesian fauna. *PLoS One*, Volume 7:e28832.

> *If it is our purpose to protect the biodiversity of Melanesia, Trigonopterus weevils are surely a valuable part of that biodiversity, both in the numbers of species and in terms of quality as an indicator. The tools are ready; it is now up to conservationists to use them*


### Keywords

COI, beetle, species, biodiversity, conservation, indicator, uncorrected p-distance, general mixed Yule clusters (GMYC), taxonomy, parataxonomist, morphospecies, non-destructive DNA sampling


### Background

Highly diverse invertebrate groups are sometimes those with the least formal taxonomic treatment, partly due to the mammoth task of sampling and describing a daunting number of undescribed species. Biodiversity surveys are increasingly relying on DNA sequencing to estimate how many species occur in an area, and to identify distinct lineages that may represent undescribed species.

### Aim

*Trigonopterus* weevils are extremely diverse in New Guinea, with marked turnover of species along altitudinal gradients. They have good diagnostic morphological characteristics that allow expert taxonomists to group them into 'morphospecies' based on physical characteristics. However, previous DNA sequencing showed high levels of divergence between recognized species[1]. This project aimed to evaluate the utility of DNA barcoding as a tool for surveying diversity of *Trigonopterus* weevils, compared to traditional classification measures used by taxonomists.

### Methods

About 6,500 weevils were collected, across seven different sites across New Guinea, over four years. These were sorted by an expert taxonomist into morphospecies (**Figure 9.18**), with several specimens per morphospecies selected for DNA extractions (including some unique specimens not assigned to morphospecies). DNA was non-destructively sampled from whole individuals: this is important because it preserves morphological characteristics of the voucher specimen for later investigation[2]. COI sequences from 1,002 individuals were aligned and used to estimate a maximum likelihood phylogeny. Two approaches were taken to group these samples into putative species. The first method used a predefined level of sequence divergence to define clusters of sequences deemed to correspond to within-species differences. For this approach, they used uncorrected p-distances: counts of number of sites that differ between a pair of sequences (otherwise known as a Hamming distance). The second approach was to use a model of the branching process in phylogenetic trees to detect lineages that show the signature of switching from within-population patterns of lineage coalescence to between-species patterns of lineage divergence[3] (known as general mixed Yule clusters, or GMYC).

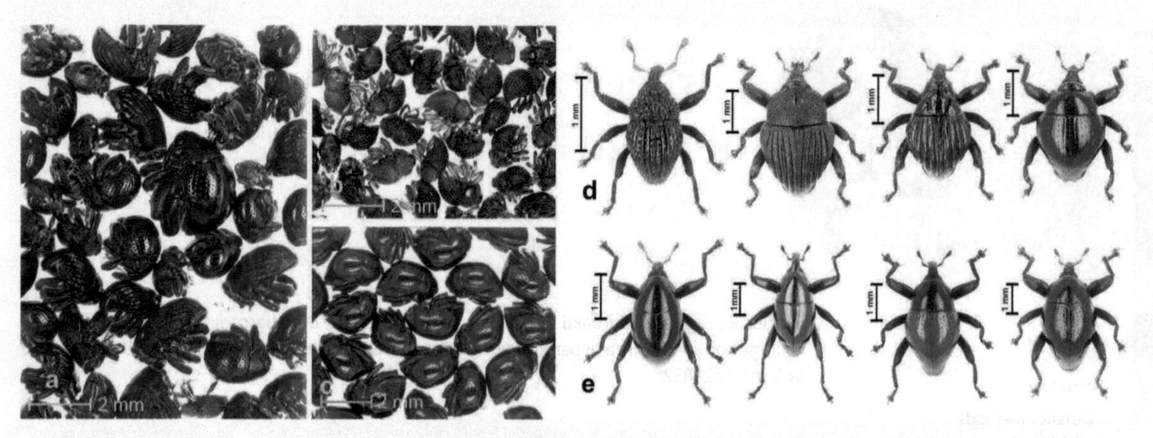


**Figure 9.18** A multitude of weevils: *Trigonopterus* weevils were collected and stored in alcohol (a), then sorted as wet specimens into morphospecies (examples in (b) and (c)). The morphospecies designations were refined through closer examination of the anatomy and the specimens then dry-mounted ((d) and (e)) and samples of DNA taken for barcoding. Incidentally, *Trigonopterus* have the first identified biological screw mechanism in their leg joint[5]. This was identified from close examination of museum specimens, which serves as an example of the surprises that can come out of biodiversity surveys.



### Results

Using a cut-off value of three per cent sequence difference resulted in more clusters (324) than morphospecies (274), but a sequence difference of eight per cent gave 247 clusters, 91 per cent of which matched to one of the morphospecies. The tree-based GYMC clustering identified 328 clusters, 227 of which matched the morphospecies, and 101 generated splits within 41 of the morphospecies (**Figure 9.19**).

### Conclusions

Alternative clustering methods identified different putative species: clustering at three per cent difference and the GYMC approach had similar results, with an 86 per cent match to recognized morphospecies, but also identifying 16 per cent more species than morphological sorting. Taking this result at face value suggests that the samples contain over 250 undescribed species, some of which are cryptic species that are not readily distinguishable on morphological characters. Clustering at eight per cent difference produced a higher match to the morphospecies, finding the same number of putative species, though not always drawing the boundaries between species in the same places. *Trigonopterus* weevils show high alpha diversity at all sites: that is, there appear to be many co-existing taxa. They also show high beta diversity: different sites have different sets of weevil taxa.

### Limitations

Since these methods rely on the interpretation of sequence differences or branch lengths on phylogenies, differences in rates of molecular evolution could influence the results (see Chapter 13). The study concluded that a 'parataxonomist' (a trained but non-expert taxonomist) had a higher error rate of species detection because they often failed to recognize the distinction between species recognized by an expert taxonomist or by the DNA-based methods. But they only tested one parataxonomist, who was actually a professional beetle taxonomist with expertise in a different group of beetles. It would be good to repeat the test on a large sample of the more typical 'parataxonomist' who are often field workers employed from local communities in high-diversity regions.

### Further work

In this study, 1,002 sequences were sampled from 279 morphospecies, so each putative species is represented by an average of less than four sequences. Sampling few specimens per species from discrete locations may inflate the gaps between putative species by failing to sample

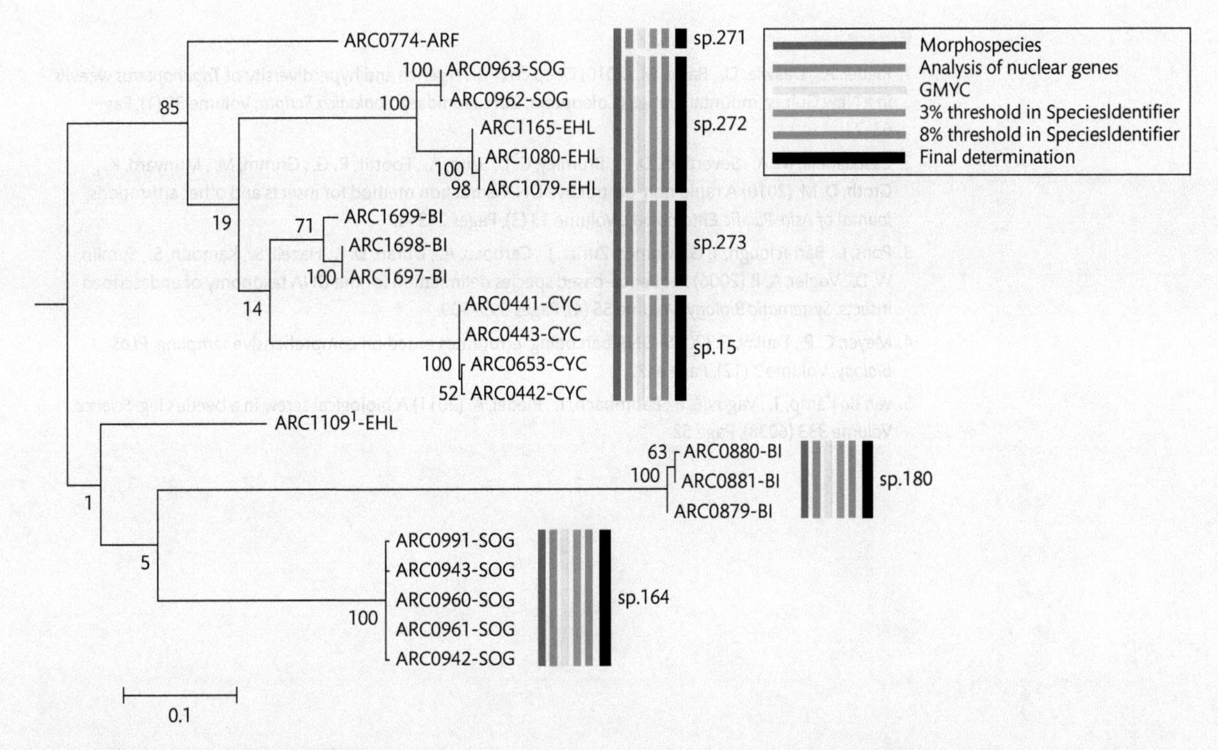


**Figure 9.19** Illustration of the relationship between morphospecies and species proposed from barcoding analysis. Here you can see that different specimens within the same morphospecies cluster together into separate clades. The three-letter codes correspond to different sample regions in New Guinea (ARF = Arfak, SOG = Sogeri, EHL = Eastern highlands, BI = Biak, CYC = Cyclops), showing that these clusters are typically defined by regions. The bootstrap values on the nodes are high for groupings within these clusters, but very low for relationships between clusters, suggesting a lack of resolution from these barcoding sequences for the relationships between lineages in a morphospecies or relationships between morphospecies.



the full range of genetic diversity in the intervening regions. Given that low levels of sampling may inflate the perception of a 'barcoding gap' (discrete difference) between putative species[4], further sampling would be needed to test whether the species identified are robust to sampling strategy.

### Check your understanding

1. What is a morphospecies? How does it differ from a formally described species?
2. Why does changing the threshold value in clustering by distance result in different numbers of species identified?
3. Why do some of the barcoding analyses propose more species than were identified by the expert taxonomist? Does this mean the barcoding assignments are in error?

### What do you think?

Biodiversity studies on faunas containing many undescribed species often use the approach of counting the diversity and abundance of morphospecies (often using trained but non-professional parataxonomists to assist with sorting and identification). Do you think this is more or less valuable than using DNA barcodes to assess species richness?

### Delve deeper

It has been suggested that *Trigonopterus* weevils would make a good indicator taxon[1]: that is, that surveying the number of species of *Trigonopterus* in an area may provide a proxy for the species richness of other groups in the area, and may serve to delineate different ecosystems. Are indicator species a useful and reliable way to assess biodiversity? Can indicators be used to assess the 'ecological health' of regions in the face of ecosystem change?

## References

1. Riedel, A. , Daawia, D. , Balke, M. (2010) Deep cox1 divergence and hyperdiversity of *Trigonopterus* weevils in a New Guinea mountain range (Coleoptera, Curculionidae). *Zoologica Scripta*, Volume 39 (1), Pages 63–74.
2. Castalanelli, M. A. , Severtson, D. L. , Brumley, C. J. , Szito, A. , Foottit, R. G. , Grimm, M. , Munyard, K. , Groth, D. M. (2010) A rapid non-destructive DNA extraction method for insects and other arthropods. *Journal of Asia-Pacific Entomology*, Volume 13 (3), Pages 243-8.
3. Pons, J. , Barraclough, T. G. , Gomez-Zurita, J. , Cardoso, A. , Duran, D. P. , Hazell, S. , Kamoun, S. , Sumlin, W. D. , Vogler, A. P. (2006) Sequence-based species delimitation for the DNA taxonomy of undescribed insects. *Systematic Biology*, Volume 55 (4), Pages 595–609.
4. Meyer, C. P. , Paulay, G. (2005) DNA barcoding: error rates based on comprehensive sampling. *PLoS Biology*, Volume 3 (12), Page e422.
5. van de Kamp, T. , Vagovič, P. , Baumbach, T. , Riedel, A. (2011) A biological screw in a beetle's leg. *Science*, Volume 333 (6038), Page 52.

# Alignment

10

## *Same but different*

*"We have no written pedigrees; we have to make out community of descent by resemblances of any kind. Therefore we choose those characters which, as far as we can judge, are the least likely to have been modified in relation to the conditions of life to which each species has been recently exposed . . . . We care not how trifling a character may be—let it be the mere inflection of the angle of the jaw, the manner in which an insect's wing is folded, whether the skin be covered by hair or feathers—if it prevail throughout many and different species, especially those having very different habits of life, it assumes high value; for we can account for its presence in so many forms with such different habits, only by its inheritance from a common parent."*

Darwin, C. (1859) *On the origin of species by means of natural selection, or the preservation of favoured races in the struggle for life.* John Murray.

## What this chapter is about

Descent with modification creates hierarchies of similarities, such that close relatives will be more similar in many respects, including their genomes. So a systematic hierarchy based on descent allows prediction of similarity across many different characteristics. Therefore taxonomic groups should be based on similarity by descent (homologies) not incidental similarities that evolved independently in different lineages (analogies). To use DNA sequences to trace evolutionary past and processes, you must first establish homology through sequence alignment, where sites all originally copied from the same ancestral sequences are compared across lineages to detect evolutionary changes.

## Key concepts

- To uncover evolutionary relationships, we must distinguish characteristics that are similar by descent (homologies) from those that have evolved independently in different lineages (analogies)
- Sequence alignment is the process whereby homologies are distinguished from analogies, so it is the most important part of any evolutionary analysis of DNA sequences

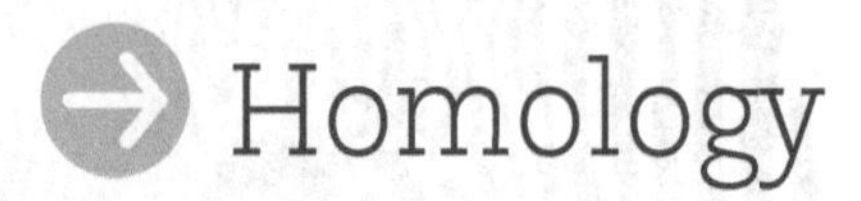

# Homology

*De-extinction*

Regret can be a powerful driver of behaviour, as people can make heroic attempts to right past wrongs. But, almost by definition, regret tends to kick in after the wrong has been horribly, and often irreversibly, done. Sadly, such is often the case with species extinction. Serious attempts at conservation often begin when the last viable population has already been obliterated. The thylacine (*Thylacinus cynocephalus*: **Figure 9.2**) was given protected status 59 days before the last known individual died in a zoo. In little more than a century, the passenger pigeon (*Ectopistes migratorius*) went from one of the most abundant birds on earth to extinction by hunting. Bills limiting hunting of passenger pigeons were introduced in some state legislatures only a few years before the last known wild individual was shot by a boy with a pellet gun. Similarly, the quagga (*Equus quagga quagga*) was once the most common type of zebra in southern Africa, so much so that it was considered a nuisance, on the grounds it might compete with domestic animals for grazing. Quaggas were vigorously hunted for their meat and their skins, which were exported in great numbers and used locally for the production of grain sacks. Quagga hunting was banned in 1886, three years after the last quagga died in an Amsterdam zoo (**Figure 10.1**).

**Figure 10.1** If only this last surviving quagga had known that in three years' time it would be declared a protected species.

Given that quaggas were hunted to extinction partly for their skins, it is ironic that only 23 skins remain. These remaining pelts allowed the extraction of proteins and DNA, so quaggas were the first extinct species to be included in molecular phylogenies (**TechBox 11.1**). But while DNA can be derived from dried skins, whole cell nuclei needed for cloning cannot (**TechBox 10.1**), so sadly there is little chance of reviving quaggas through cloning technology. But an organization in South Africa is determined to bring quaggas back by selective breeding.

The quagga is now generally regarded as a subspecies of the highly variable Plains Zebra, so by targeting variants of coat colour found in living Plains Zebras that most closely resemble those of the quagga, the breeders can generate the Quagga phenotype. After several generations they have succeeded in producing animals that look a lot like quaggas, with the distinctive tan colour and a reduced number of stripes (**Figure 10.2**).

Given continued targeted breeding, the Quagga phenotype is likely to be convincingly recreated. But the Quagga genotype has not been recreated. If you were to analyse sequences the genomes from these half-stripey brown zebras, you would find that the sequences clustered with the Plains Zebras from which they were derived. Even if key alleles for coat colour are similar to the original quaggas, the rest of the genome won't be any more similar to the original quaggas than to the genome of the zebras from which the new quaggas were derived.

**Figure 10.2** Khumbu is the product of several generations of selective breeding by The Quagga Project, which aims to revive the extinct quagga by breeding from Plains zebras that most closely resemble quagga coat patterns.
Photograph courtesy of The Quagga Project.

These rebred quaggas are analogous to the original Quaggas—although they are similar in appearance, the similarity is not due to a unique ancestry shared with members of the subspecies *E. quagga quagga*. Whatever name we give them, their path of descent connects them more closely to Plains Zebras than to Quaggas.

## Systematic hierarchy

In the previous chapter, we saw how the aim of modern taxonomy is not simply to recognize and name all biological species, but also to organize those species into a hierarchy that reflects their relatedness. A hierarchy is a useful way of organizing the very large, and ever increasing, number of recognized species. This hierarchy of shared characteristics can be used to design a taxonomic key: a nested set of questions that will progressively narrow down the number of possible species to which a given organism might belong (**Figure 11.2**). But the taxonomic hierarchy is more than just a fancy filing system. The systematic hierarchy should reflect the actual pattern of evolutionary descent that produced the species. This provides a logical and natural system of classification.

More importantly, an evolutionary classification has great predictive power, because we expect patterns of relatedness to generate patterns of similarity at all levels of biological organization (**Figure 9.5**). Since we expect the hierarchy of differences in the genome to be broadly indicative of how closely related two individuals are, whether these are individuals of the same species or different species, DNA sequence comparison is a very useful tool for estimating relatedness. Measures of relatedness, in turn, allow us to make predictions about similarities and differences, because we expect closely related individuals or species to have more in common than those which are more distantly related.

*The hierarchy of similarities arising from the copying process at the heart of evolution is discussed in Chapter 4*

Despite initial appearances, thylacines are actually a very good example of the hierarchical nature of biological characteristics (**Figure 9.2**). Although aspects of their external appearance may be similar to wolves, almost everything else about thylacines is more similar to other marsupials. Marsupials are one of the three major divisions of mammals. These three groups can be distinguished by a number of defining characteristics, the most obvious of which is reproductive mode. Monotremes (platypuses and echidnas) lay eggs (**Figure 10.3**). The embryos of placental mammals (such as horses and bats) are carried internally until an advanced stage of development. The embryos of marsupials (such as possums and kangaroos) are born at a very early stage and undergo most of their development in a pouch on the outside of the mother's body.

Although they have the external appearance of placental wolves, thylacines have all of the classic defining characteristics of marsupials, including pouch-based development. In fact, the name *Thylacinus* is derived from a Greek word for pouch, though in this case it may equally refer to the pouch around the testes on male thylacines, an unusual aspect of their biology. But thylacines also have other, less obvious, key marsupial characteristics. Female thylacines, like all other marsupials, have two vaginas (and most male marsupials have a forked penis to match). As with other marsupials, the penis on male thylacines is behind the scrotum, not in front of it as in placental mammals. There are also skeletal characteristics that separate marsupials such as thylacines from placentals such as wolves, including the epipubic bones (attached to the pelvis), and small curved triangles that stick out of the jawbones (formally referred to as medially inflected angular processes of the lower jaw).

Just as most physical characteristics of thylacines are more similar to other marsupials than they are to wolves, so too is their DNA. Here is part of an alignment from the 12S ribosomal RNA gene, which codes for an RNA molecule that makes an essential part of the genomic system (**Figure 10.4**).

**Figure 10.3** Monotremes are the most charming mammals you could ever hope to meet. There is an odd notion that monotremes are somehow primitive or less evolved than other mammals. Naturally, the monotreme lineage had just the same amount of evolutionary history as their sister orders, the marsupials and placentals, since they diverged from their last common mammalian ancestor at least 130 million years ago. In that time monotremes have evolved some pretty neat tricks like electrodetection of prey. (a) Platypuses are also one of the only venomous mammals, as males have a spur on their hind flippers that can inject poison: while non-lethal to humans, a platypus sting is apparently excruciatingly painful. (b) The echidna is also much tougher than you might suppose when you see it ambling along with its amiable rolling gait. In fact, it is the only native mammal found all over Australia, from the deserts to the rainforests, from the mountains to the coasts.



**Figure 10.4**

| | |
|---|---|
| Thylacine | GGTCCTGGCCTTACTGTTAATTCTTATTAGACCTAC |
| Quoll | GGTCCTGGCCTTACTGTTAATTTTTATTAGACCTAC |
| Dunnart | GGTCCTGGCCTTACTGTTAATTTTTATTAGACCTAC |
| Marsupial mole | GGTCCTAGCCTTATTATTAATTATTGCTAGTCCTAC |
| Dog | GGTCCTAGCCTTCCTATTAGTTTTTACTAGACTTAC |
| Cat | GGTCCTGGCCTTTCCATTAGTTATTAATAAGATTAC |
| European mole | GGTCCCAGCCTTTCTATTAGCTGTCAGTAAAATTAC |
| Shrew | GGTCCTAGCCTTCCTATTAGTTGTTAGTAAACTTAC |

Four of the species included in this alignment—thylacine, quoll, dunnart and marsupial mole—are marsupials. The other four are placentals—dog, cat, shrew and European mole. The sequences from the marsupials are all more similar to each other than any of them is to any of the placental sequences (**Figure 10.5**). The same cannot be said for the external morphology of these species, which varies greatly between the four marsupials. Not only that, but the external morphological features of each of the marsupials are far more similar to one of placental species than it are to the other marsupials. Why are the patterns of similarity in the DNA sequences not reflected in observable features of the organisms?

The eight species in **Figure 10.5** illustrate the concept of convergent adaptation, where separate biological lineages independently evolve similar adaptations to similar niches (ways of living). For example, both the marsupial mole and the European mole are adapted to a life of burrowing through soil hunting for worms and grubs: both have the typical mole-like characteristics of elongated body, flat paddle-like feet for digging, no external ears and only rudimentary eyes. But marsupial and placental moles don't share their mole-like characteristics because they both inherited these characteristics from the same mole-like ancestor. Both have independently evolved very similar adaptations to the same mole lifestyle. The same can be said for marsupial dunnart and placental shrew, which are independently adapted to eating insects and other small animals, and for the nocturnal predators, the marsupial quoll and the placental cat.

It would be possible for convergent adaptations such as these to confuse taxonomy, causing species to be grouped on the basis of superficial similarity rather than true relatedness. However, if we look at enough different characteristics, particularly concentrating on those not associated with adaptation to a particular niche, then we ought to be able to see the pattern of hierarchical differences resulting from the copying process at the heart of evolution. This hierarchy of similarities should apply to all inherited traits, including DNA sequences. For example, we can see that the DNA sequence from the thylacine is more similar to the sequences from the other marsupial species than it is to the morphologically similar dog (**Figures 10.4** and **10.5**). This illustrates

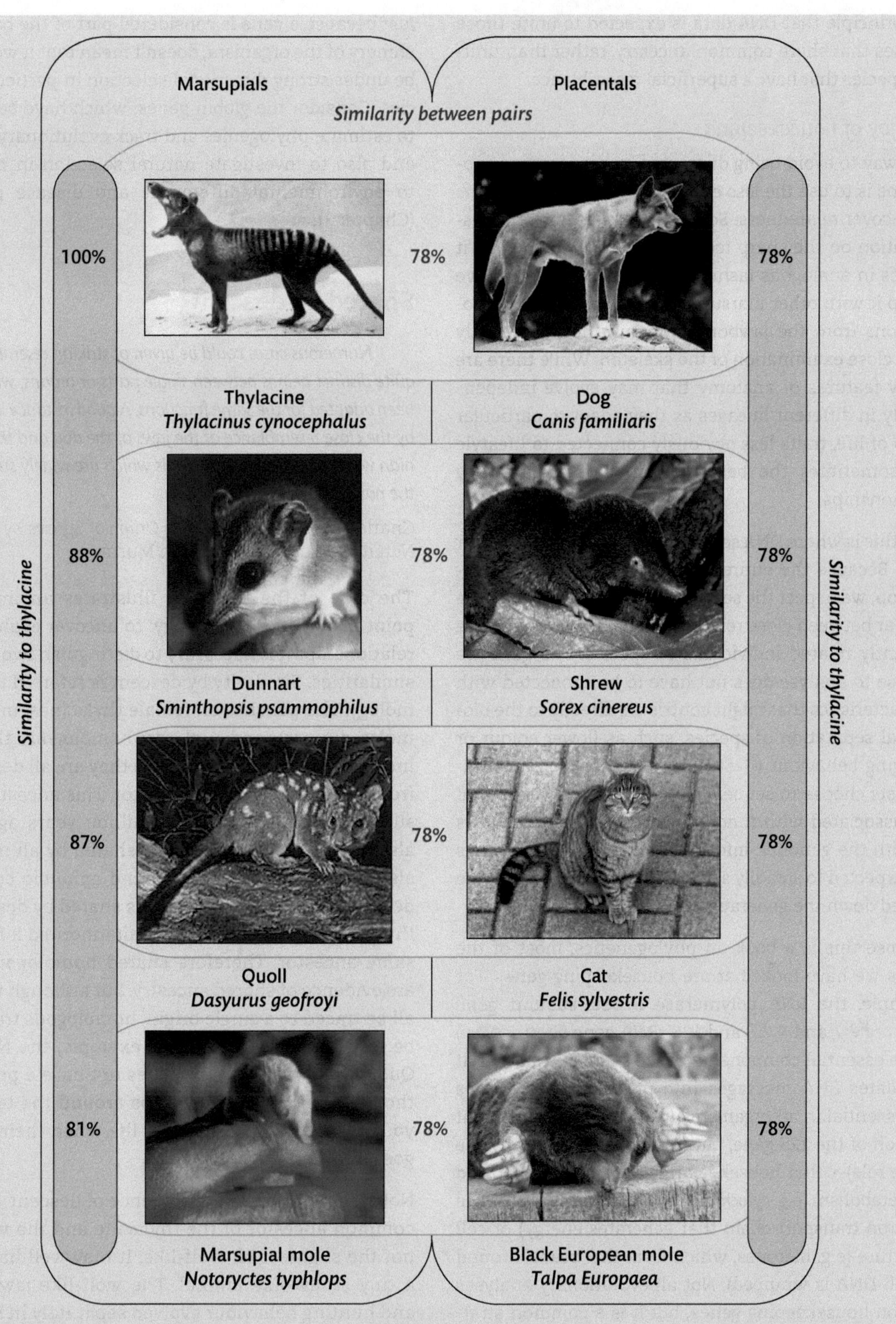


**Figure 10.5** Similarity of 12S sequences between 8 mammal species. The grey columns compare each species to the thylacine, the blue figures are the similarities between pairs of species. Marsupials and placentals last shared a common ancestor over 125 million years ago, while the last common ancestor of the Australian marsupials is around 65 million years. So why is the difference between the placentals and marsupials not twice as much as the difference between marsupial species? Keep reading, because you will find out why in Chapter 11 . . .

Shrew photograph: Patrick Coin; cat photograph: Tilo Hauke, Germany; dingo photograph reproduced with kind permission from www.giveusahome.co.uk; black European mole photograph courtesy of Hannes Grobe, licensed under the Creative Commons Attribution Share Alike 2.5 Generic license.

the principle that DNA data is expected to unite those species that share common ancestry, rather than uniting species that have a superficial resemblance.

### The joy of housekeeping

One way to avoid being distracted by convergent adaptations is to use the less exciting features of organisms to uncover relatedness. So rather than basing our classification on the sharp teeth of the thylacine, which it shares in analogous fashion with other carnivores, we group it with other marsupials that share the tiny protrusions from the jawbone, a feature noticeable only after close examination of the skeleton. While there are many features of anatomy that may evolve independently in different lineages as they adapt to particular ways of life, traits less obviously connected to lifestyle are sometimes the best indicators of evolutionary relationships.

And this is where DNA sequences really come in to their own. Because the entire genome is copied every generation, we expect the sequence of any gene to be more similar between close relatives than it is between more distantly related individuals or species. The gene we choose to analyse does not have to be connected with characteristics that might contribute directly to the biological separation of species, such as flower colour or foraging behaviour (**Case Study 9.1**). In fact, most biologists choose to sequence 'housekeeping' genes that are associated with fundamental metabolic pathways or with the genomic information system. These genes are expected to steadily accumulate changes as they are copied down the generations.

Because this is a book on phylogenetics, most of the genes we have looked at are housekeeping genes. For example, the RNA polymerase II beta-subunit gene (**Figures 2.2** and **4.22**) and 12S rRNA gene (**Figure 10.4**) make essential components of the cellular system that translates RNA messages into proteins. These genes are essential in all organisms (bacteria have a different version of the 12S gene, but it performs essentially the same role). Other housekeeping genes may be involved in metabolism (e.g. *cytochrome b*, which forms part of the electron transport chain that generates energy) or cell structure (e.g. histones, which form the 'beads' around which DNA is wrapped). Not all evolutionary analyses rely on housekeeping genes, but it is a common strategy when the aim is to uncover evolutionary relationships, because it is a handy way of finding homologous sequences common to all species, and avoiding the problems of analogous changes due to adaptation to a specific niche. But be aware that there is no clear dividing line between 'housekeeping' genes and other genes. Just because a gene is considered part of the basic machinery of the organism, doesn't mean that it won't ever be under strong directional selection in particular species. Consider the globin genes, which have been used to estimate phylogenies and track evolutionary history, and also to investigate natural selection in response to environmental differences and disease pressure (Chapter 7).

## Shared by descent

> *Numerous cases could be given of striking resemblances in quite distinct beings between single parts or organs, which have been adapted for the same functions. A good instance is afforded by the close resemblance of the jaws of the dog and the Tasmanian wolf or Thylacinus,—animals which are widely sundered in the natural system*
>
> Charles Darwin (1872) *On the Origin of Species by Means of Natural Selection*. 6th edn. John Murray

The case of the thylacine illustrates an important point in evolutionary biology: to uncover evolutionary relationships it is necessary to distinguish two types of similarities. Similarity by descent is referred to as homology. The pouches on female thylacines, marsupial moles, dunnarts and quolls are homologous: these animals all have pouches because they are all descended from the same pouched ancestor. This ancestral marsupial, which lived over 80 million years ago, must also have had the other traits shared by all marsupials, such as a double vagina and epipubic bones. By definition, a homologous trait is shared by descent, so that all the species with that trait inherited it from the same ancestor. Therefore shared homologous traits are evidence of shared ancestry. But although they can all be traced to a single origin, homologous traits may be lost during evolution. For example, the Northern Quoll (*Dasyurus hallucatus*) does not have a pouch: although a fold of skin develops around the teats, the young must hang on for dear life when their mother goes hunting.

Not all similar traits are evidence of descent. The last common ancestor of the thylacine and the wolf was not the slightest bit wolf-like. It may well have been a tiny shrew-like animal. The wolf-like jaws, teeth and hunting behaviour evolved separately in both the canid and thylacine lineages (and, incidentally, also in the long-extinct borhyaenid lineage of American marsupials: **Figure 10.6**). Similar traits that have evolved independently in separate lineages are often described as analogies, to distinguish them from homologies. Analogous traits have evolved more than

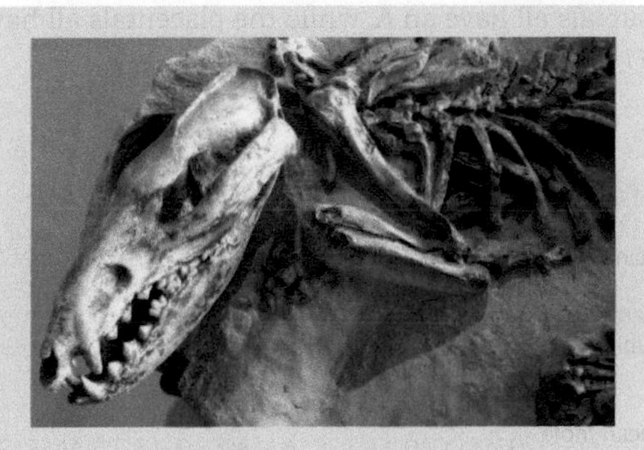

**Figure 10.6** Another non-wolf: Marsupials didn't only invent one wolf-like form, they did it twice. The Australian lineage produced the thylacine (*Thylacinus cynocephalus:* **Figure 10.16**), and the South American marsupial lineage gave rise to the wolf-like borhyaenids.

once. While analogies are very interesting they are, by definition, traits that are counter to the hierarchy of similarities. So while analogous traits are critical for understanding the processes of adaptation to a particular niche, they are no help at all in uncovering descent.

## DNA can reveal homology

Loss of homologous traits, or gain of analogous traits, can confuse classification. One of the advantages of molecular taxonomy is that it can reveal when species are erroneously grouped together on the basis of analogous similarities. For example, honeybees and stingless bees were traditionally grouped together on the basis of shared morphological characteristics, and because their way of life was so similar: both make hives where large numbers of non-reproductive workers store pollen and honey and raise the queen's offspring (**Figure 10.7**). But DNA sequence analysis has been used to suggest that honeybees are more closely related to the solitary (non-social) orchid bees than they are to the eusocial stingless bees, suggesting that eusociality has evolved multiple times in bees. Similarly, molecular data also implies that eusociality has arisen multiple times in shrimps, in rodents, and in wasps (in fact, sociality has arisen independently in more than 25 different biological lineages, including beetles, termites, and ants). Furthermore, molecular data shows that eusociality has been lost in multiple lineages in ants and bees. So molecular data reveals a complex pattern of the evolution of eusociality, with multiple gains and losses. These multiple origins would have been more difficult to detect from morphology alone.

DNA analysis can also reveal surprising affinities between apparently unrelated species. This is particularly helpful for organisms that have lost the usual defining homologous traits of a group. For example, parasites often evolve a simplified morphology, jettisoning characteristics needed for independent life. Some intracellular parasites take this simplification to extremes, even losing their mitochondria (energy-generating

**Figure 10.7** Molecular data suggest that the very similar hive-based societies of these two types of bees—(a) honeybees (*Apis mellifera*) and (b) stingless bees (*Melipona*)—are analogous (evolved independently) not homologous (similar by descent). While most of the global honey industry is based on honeybees, honey from stingless bees has been traditionally used in Australia, Central America and South America. Concerns have been raised in many parts of the world about negative effects on native stingless bees of competition with introduced honey bees.

organelles). Microsporidia are intracellular parasites that infect a wide range of species: the resulting diseases have a large economic impact on industries from apiculture to aquaculture. Because of their apparently simple morphology, microsporidia were briefly, in the 1980s, grouped with other simple single-celled parasitic taxa into a new kingdom called the Archezoa. The Archezoa was thought to represent an early offshoot of the eukaryote lineages, having split before the acquisition of the mitochondrion so preserving a pre-mitochondrial state.

But DNA sequence analysis clearly demonstrates that microsporidia are actually dramatically reduced fungi, so they belong in the fungal kingdom with their mushroom relatives. It seems that the similarity between the 'archezoan' parasites is due to superficial resemblance, not shared ancestry. Most of the archezoans have evidence of mitochondrial sequences in their genomes, so they must have evolved from ancestors that had mitochondria, then lost them along the way. DNA analysis has now scattered the archezoans around the eukayote phylogeny. The kingdom Archezoa was based on analogy not homology, so this taxonomic grouping has now been abandoned.

## Alignment establishes homology

The distinction between homologous (shared by descent) and analogous (independently evolved) traits is just as critical when using molecular data to uncover relationships. In fact, it is even easier to be fooled by analogy with DNA sequences, because there are only four possible character states. An adenine will look just the same whether it is a new mutation or inherited from an ancestor: an A is an A.

Look at the alignment of the 12S rRNA gene in Figure 10.4. We have already seen that, for this gene, the marsupial species have more nucleotides in common than any of them shares with a placental species. Now we can look more closely at the patterns of similarities. For example, at one position in the sequence, we can see that marsupials all have an A, while the placentals all have a G (Figure 10.8).

**Figure 10.8**

| | |
|---|---|
| Thylacine | GGTCCTGGCCTTACTGTTAATTCTTATTAGACCTAC |
| Quoll | GGTCCTGGCCTTACTGTTAATTTTTATTAGACCTAC |
| Dunnart | GGTCCTGGCCTTACTGTTAATTTTTATTAGACCTAC |
| Marsupial mole | GGTCCTAGCCTTATTATTAATTATTGCTAGTCCTAC |
| Dog | GGTCCTAGCCTTCCTATTAGTTTTTACTAGACTTAC |
| Cat | GGTCCTGGCCTTTCCATTAGTTATTAATAAGATTAC |
| European mole | GGTCCCAGCCTTTCTATTAGCTGTCAGTAAAATTAC |
| Shrew | GGTCCTAGCCTTCCTATTAGTTGTTAGTAAACTTAC |

There are several possible explanations for the presence of A at this position in all four marsupial species. The marsupial mole, quoll, thylacine and dunnart may all have inherited it from a common ancestor that had an A at position 32, in the same way that they all inherited the pouch and the epipubic bones. In this case, we could say the A was homologous: shared by descent. Alternatively, the common ancestor of the marsupials could have had, for example, a G at that position, but subsequently there was a change from a G to an A in each of the four lineages. In this case, the As would be similar by analogy: they had evolved independently many different times. It would look exactly the same. You may think 'who cares if it's homologous or analogous? An A is still an A!'. But the reason we wish to know whether shared nucleotides are similar due to homology or analogy is that only homologies reveal relationships.

The only way we can tell whether a shared A is a homology or an analogy is to consider it in the context of the rest of the sequence. This is similar to recognizing that the wolf-like characteristics in thylacines are analogous to those in placental wolves because they are found in an otherwise entirely marsupial context. For DNA sequences, this context is revealed by alignment, which is the process of arranging DNA sequences so that all of the homologous nucleotides are lined up in columns, so the differences and similarities can be meaningfully compared to reveal evolutionary history and processes.

# Alignment

"*One of the strengths of scientific inquiry is that it can progress with any mixture of empiricism, intuition, and formal theory that suits the convenience of the investigator*"

Williams, G. C. (1966) *Adaptation and Natural Selection: A critique of some current evolutionary thought*. Princeton University Press

Alignment is the most important stage in most evolutionary analysis of DNA sequences. This is because alignment is the step that establishes homology, which is the basis for recognizing and understanding changes in DNA sequences over time. Incorrect alignment will lead to false conclusions. But alignment can also be very difficult, due to the limited number of possible character states in DNA (or protein) sequences (**TechBox 10.2**).

For physical characteristics, it will often be possible to distinguish homology from analogy by careful inspection. At first sight, the thylacine's teeth seem remarkably similar to a wolf's, with their prominent canines and sharp, slicing molars. But closer inspection reveals important differences: the molar teeth of the thylacine, for example, differ in size, number and structure from those of the wolf. But this approach won't work for nucleotides—no matter how closely we look at an A, it will look the same as all other As. Instead, we rely on making inference of homology from probability: it is more likely that all four marsupials inherited the same A at the same position than it is that each one of them just happened to evolve that A independently. Weighing up the probabilities of different possible evolutionary explanations for the nucleotide sequences we see is the basis of sequence alignment.

In **Figure 10.8**, we can be quite confident that we are looking at homologous sequences, because they are clearly far too similar to be due to chance. Over half the bases in the sequence are the same for all of the sequences, far greater than we would expect if the sequences were not related. But not all parts of the *12S* gene can be so easily aligned. The parts of *12S* that are trickier to align have higher rates of change, so that more nucleotide positions vary, and variable numbers of nucleotides, so that alignment requires the insertion of gaps to make homologous positions line up.

## Alignment gaps

So far, we have concentrated on nucleotide substitutions where one base in the nucleotide sequence is exchanged for another. For example, in the alignment below (**Figure 10.9**), you can see that at position 836, a G has been substituted for an A in the quoll lineage.

**Figure 10.9**

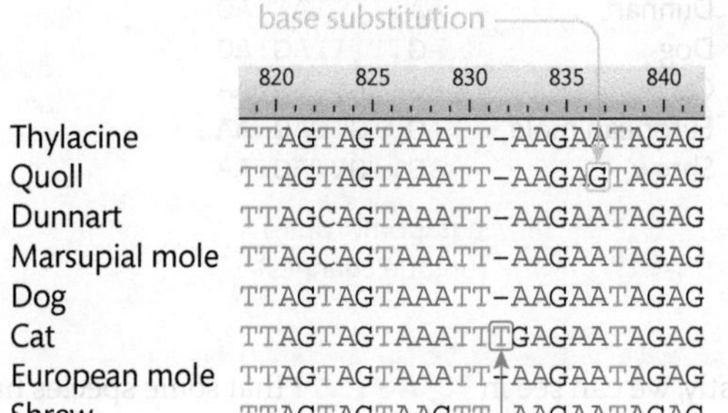

```
                base substitution
               820  825  830  835  840
Thylacine      TTAGTAGTAAATT-AAGAATAGAG
Quoll          TTAGTAGTAAATT-AAGAGTAGAG
Dunnart        TTAGCAGTAAATT-AAGAATAGAG
Marsupial mole TTAGCAGTAAATT-AAGAATAGAG
Dog            TTAGTAGTAAATT-AAGAATAGAG
Cat            TTAGTAGTAAATTTGAGAATAGAG
European mole  TTAGTAGTAAATT-AAGAATAGAG
Shrew          TTAGTAGTAAGTT-AAGAATAGAG
                             indel
```

But not all changes to sequences involve changing one base to another. Sometimes extra bases are inserted into the sequence, and sometimes bases in the sequence are deleted. Insertions and deletions, collectively known as indels, may be any size, ranging from addition or subtraction of a single nucleotide to the loss or gain of large sections of DNA (**Case Study 10.1**). In this alignment, we can see that the cat lineage appears to have gained an extra T at position 831 (**Figure 10.9**). So to align sequences that are similar, but not identical, we have to consider both nucleotide substitutions (one base is exchanged for another) and indels (bases are inserted or deleted).

*Insertions, deletions and duplications are discussed in Chapter 5*

Gaps are not read from the sequence itself: there are no spaces in real DNA molecules, as each nucleotide must be joined to its neighbour (see Chapter 2). Instead, these gaps are inferred as part of the alignment procedure. They represent a place where a sequence has one or more bases missing compared to other sequences in the alignment, or where a sequence has got extra bases that others don't have. So to make all the sequences line up into columns of homologous nucleotides, it may be necessary to add gaps. Often the same symbol (usually a dash) is used to signify missing data, where the nucleotides are not known for that part of the sequence.

But how many gaps should we add? After all, by inserting more gaps, we could make more bases match. To appreciate the influence of gap placement on alignment, let's look at a small section of the *12S* gene, and to simplify the example we will include just three species: two marsupials (thylacine and quoll) and one placental (dog: **Figure 10.10**).

**Figure 10.10**

```
Quoll      ACC-TAATTAGAATACGCTAAAAA----GAG
Thylacine  ACC-TAATACGAATACG-TAAAAAA---GAG
Dog        ACCATA-TTAACTTAA-CTAAAACACAAGAG
```

Here we can see that the sequences are still roughly similar, but there are fewer nucleotide matches than in the sections of the *12S* gene we looked at in **Figures 10.8** and **10.9**. Not only are there more nucleotide differences, we can see that the alignment contains gaps (shown as a grey dash), where the sequence lacks a matching base. There are many parts of the 12S ribosomal RNA molecule that can vary in length, as insertions or deletions of bases can change the shape of the molecule without necessarily destroying its function. The combination of fewer matching nucleotides and variable length of sequences due to indels makes this part of the sequence trickier to align. We have to make decisions about which alignments are more likely to represent homologous positions: for example, should we align an A with a C, representing a substitution, or should it be aligned with a gap, representing an insertion?

To illustrate how the decisions about alignments influence the story we read from our DNA sequences, consider what would happen if we aligned these sequences differently. When we compare the nucleotides that are aligned against each other in **Figure 10.10**, we see that in all but two positions, the thylacine sequence has the same base as the quoll. When we compare aligned nucleotides between the thylacine and dog, we note that there are seven bases that don't match. So we could conclude from this alignment that the thylacine is more closely related to the quoll that it is to the dog.

However, this is not the only way of aligning those sequences. Instead of aligning sequences as we did in **Figure 10.10**, we could have chosen to align the same sequences as shown in **Figure 10.11**:

**Figure 10.11**

```
Quoll      ACC-TAATTAGAATACGCTAAAAA---------GAG
Thylacine  ACC-TAAT--AC-GAA--TA---CG-TAAAAAAGAG
Dog        ACCATA-TTAACTTAA-CTAAAACACAA-----GAG
```

And now a curious thing has happened. For exactly the same sequences, but with a different alignment, the thylacine sequence is now more similar to the dog (only three nucleotide differences) than it is to the quoll (five differences). The two species that share analogous physical traits are now united by analogous nucleotides. In this case, most biologists would probably consider the first alignment with fewer gaps (**Figure 10.10**) to be a more believable assignment of the homology of those sequences than the second alignment with more gaps (**Figure 10.11**). But this example illustrates that different alignments can change the conclusions reached from examining DNA sequence data. Since gaps are inferred from the alignment rather than read directly from the sequences, we need a way of choosing between alternative alignments.

## Homologous sites

We need to do two things in order to distinguish a meaningful (homologous) match from a coincidental (analogous) match. Firstly, we need to establish homologous sites. A site is a position in a sequence. Each site could hold any one of the four nucleotides, or a gap. When we align DNA sequences we are attempting to find sites that were all originally copied from the same site in the common ancestor of all the sequences in the alignment.

The 12S genes of the marsupial mole, thylacine, dog and cat were all ultimately copied from a single ancestral 12S gene in their shared mammalian ancestor. So the 12S gene in each of these species, taken as a whole, can be considered homologous (similar by descent). The 12S gene has just under 1,000 sites. We can see that the sites in **Figure 10.4** are homologous, because they all have very similar sequence of nucleotides—all of the species share at least 50 per cent of the nucleotides in this alignment. Just as we argued the RFGEME motif in RNA polymerase II beta would be unlikely to have arisen by chance in all species (**Figure 2.2**), so too we consider that the match between mammal species for this stretch of the 12S gene is far too similar to be simply due to chance. Instead, we assume that these sequences are similar because they were all copied from the same ancestral copy of the gene. Once we have established homologous sites, we can identify changes that have occurred along particular lineages.

### *Homologous nucleotides*

If we are sure that we are looking at homologous sites, we can begin to consider which actual nucleotides are homologous—that is, which shared nucleotides reflect shared ancestry. Since that last common ancestor of marsupials and placentals, the 12S gene has been copied over a hundred million times. Not surprisingly, it has accumulated various changes in different lineages. So, although the sites are clearly homologous, the sequences are no longer identical. Let's focus on a shorter segment of the 12S alignment to illustrate the types of differences we can observe between sequences.

**Figure 10.12**

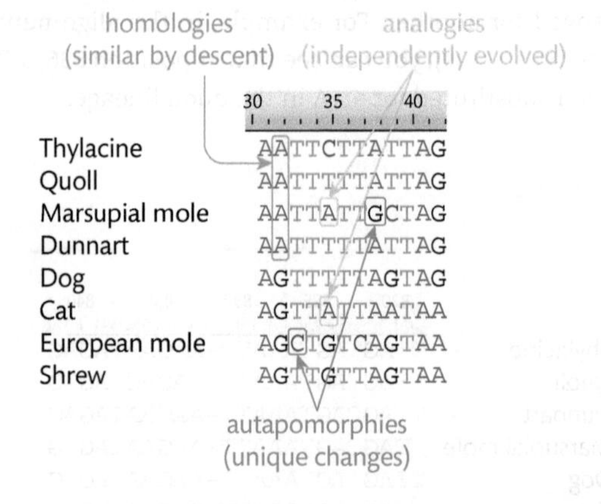


Firstly, we can see in **Figure 10.12** that some species have unique changes which are not shared with any of the other sequences in the alignment. Only the European mole has a C at site 32, all other species have a T, so it is fair to conclude that, at some time in the history of the European mole lineage, since it split with its common ancestor with the other mammals in this alignment, the T at site 32 was changed to a C. Similarly, only the marsupial mole has a G at site 37. A nucleotide substitution in only one species in the alignment is sometimes referred to as an autapomorphy. Autapomorphies are useful for two things. Firstly, just as unique physical characteristics are a key part of species taxonomy, unique sequence changes are exactly what we need if we want a diagnostic 'DNA barcode' to identify a species (**TechBox 9.2**). Secondly, autapomorphies can tell us about relative rates of change in particular biological lineages. For example, the marsupial mole has a relatively high number of autapomorphies, indicating either that it has a long, independent evolutionary history, or that it has a particularly rapid rate of change (or both).

*We look at rates of sequence change in Chapter 13*

But autapomorphies are no help in establishing relationships because, by definition, they are not shared by any of the other species in the alignment, so we can't use them to group related species together. To classify species, we need to group them on the basis of shared characters (formally referred to as synapomorphies). More specifically, we need to distinguish homologous and analogous shared characteristics. All marsupials in this alignment share an A at site 31, which can reasonably be interpreted as a homology, inherited from the marsupial ancestor. But not all shared nucleotides are evidence of shared ancestry. This short section of the alignment contains at least one example of an analogy: two species having independently evolved the same nucleotide (**Figure 10.12**). The cat and the marsupial mole share an A at position 34, but we can guess that their shared common ancestor did not have an A at this position, because none of the other descendants of that ancestor have an A. It seems most likely that at some point in the marsupial mole's history this site changed to an A, and, entirely independently, this site also changed to an A in the cat lineage. We can recognize this A as an analogy from the context of the rest of the alignment: at most sites, cats are similar to dogs not to marsupial moles, so this shared A between the cat and marsupial mole is anomalous.

## Manual alignment

How do we decide upon the correct alignment for a set of sequences? There are two approaches. One is to align sequences manually, usually with the help of a computer program that makes it easy to slide the nucleotides back and forth (just as a word processing package makes it easy to rearrange words on a page). A biologist can use a manual sequence editor to produce different possible sequence alignments, then use their understanding of molecular evolution to judge which alignment is most likely to represent homologous characters. The alternative alignments in this chapter were all created manually in a sequence editor, and most biologists would judge the alignment in **Figure 10.10** to be more reasonable than the alignment in **Figure 10.11**.

Manual sequence alignment has a number of advantages. Humans are actually rather good at alignment, because they can quickly learn to spot meaningful patterns and they can interpret alternative alignments in terms of what they know about sequence evolution. For example, a knowledgeable human will approach the alignment of a ribosomal RNA gene (with its stem and loop structure) very differently from a protein-coding gene (with its codon structure) or an internal spacer or intron (which have lots of highly variable positions with the occasional island of conserved sequence). They might spot particular patterns that help them make sense of the alignment, such as similarities between close relatives, or that some sequences seem to have a faster rate of change than others.

Manual sequence alignment allows biologists to actively engage with the assignment of homology in their sequences. Before you even begin your analysis you will have a good appreciation of the signal in your data, the level of variability between sequences, and how the conservation of sequence changes along the locus and between taxa. By making decisions about which nucleotide sequences to align, you will be forced to actively consider whether homology can be reliably established for the variable regions of your alignment or whether you have to accept that the historical signal is lost from the highly changeable regions. Furthermore, manual alignment represents a very important error checking stage. If you notice that one of your sequences is strangely different from the others, you might want to go back and look at that sequence carefully, and wonder about whether it should be resequenced.

One advantage of manual sequence alignment that is often overlooked is that biologists of all kinds benefit from close observation of nature, to develop 'a feeling for the organism' (a phrase coined by the pioneering geneticist Barbara McClintock: **Hero 5**). Close observation of sequence data is essential for developing a feeling for DNA and an appreciation of molecular evolution. The more you look at alignments, the more you become familiar with the kinds of changes that are most likely to occur. This feeling for DNA sequence evolution is the key to recognizing a reasonable alignment.

But manual sequence alignment can be time-consuming. High-throughput sequencing, and the growth of available sequences in GenBank, has resulted in a massive increase in the amount of sequence data included in many phylogenetic analyses. The increase in the size of datasets, and the improvement in the speed and performance of computational alignment programs, has resulted in many researchers relying entirely on automated alignment to establish homology in their datasets. Some people do not even check their alignments before analysing them, piping them straight from alignment program to analytical software without ever actually seeing what their data looks like. This carries with it the potential for disaster. Big data is worth nothing if the alignments aren't any good. When you are seeking the answers to biological questions, it is better to have a smaller amount of reliable information than a very large amount of nonsense. Whether using big or small amounts of sequence data, be careful with your alignments to make sure that you are getting your story right. Check the veracity of your alignments before your analysis or you risk telling false tales from your data.

## Automated alignment

Alignment programs use an algorithm (an ordered series of instructions) to create and evaluate alternative alignments according to some objective criteria (**TechBox 10.2**). Automated alignment programs are fast, they can spot shared sequences that a human might miss, and they can handle very large amounts of data. Best of all, they can be asked to complete what is often a tedious task without (on the whole) complaining. The disadvantage is that, unlike a human who can blend intuition, experience and knowledge of evolutionary processes to make a judgement about which sequence alignments reliably represent homologous positions, a computer can only work to rule. The alignment they produce will be a product of particular criteria they have been given to decide which alternative alignment is best. In some cases, the rules they follow will produce a faultless alignment, but in other cases, it will produce a mess. And you need to know how well it has worked for your sequences before you go any further.

### *Scoring alignments*

DNA sequences are just the kind of information that computers are good at manipulating: strings of repeating characters. It is a relatively simple matter to write a computer program that can read in two DNA sequences and tell you at how many sites they have the same nucleotide. This similarity score could then be used to choose between alternative alignments. We could score an alignment, change it slightly, then score it again, repeating this process until we have found the highest scoring version of the alignment.

How do we make sure our score will be highest for the most biologically realistic alignment? We could use a very simple score that simply gives the alignment one point for every matching nucleotide and subtracts one point for every mismatch. Or we could have a more sophisticated scoring system where different kinds of mismatch were awarded different scores. For example, programs that align protein sequences can use a matrix which assigns higher scores to mismatches between chemically similar amino acids, on the grounds that changing to a functionally equivalent amino acid will be a less serious structural change to a protein (**Hero 10**).

Making matches at all costs is not reasonable, otherwise we could just slide letters across until we found a matching one, creating two perfectly matching sequences with an absurd number of gaps. For example, compare the following sequences from the thylacine and dog 12S sequences (**Figure 10.13**):

**Figure 10.13**

```
Thylacine  CAAGTTTCCGCGC
Dog        CAAGCCTCCACGC
```

There are three mismatches between these sequences, suggesting at least three substitution events. But if we could introduce as many gaps as we liked, then we could make all the bases match (**Figure 10.14**).

**Figure 10.14**

```
Thylacine  CAAGTT--TCCG-CGC
Dog        CAAG--CCTCC-ACGC
```

The second alignment has more bases matching, but we might consider it a less likely explanation of the evolution of these sequences because we would have to infer at least four insertion or deletion events, but no substitutions. This is why most alignment programs penalize gaps, by reducing the score for every gap added, to discourage indiscriminate inclusion of indels.

We should also expect the probability of a gap occurring in a sequence to be related to the biochemical function of the sequence. For example, the product of the 12S gene is an RNA molecule that assumes a tertiary structure by twisting around itself: in some places two parts of the strand pair with each other, in other places the single strand loops out (**Figure 10.15**). Indels that disrupt the matching of bases along a paired region are likely to be more damaging than indels that slightly extend or reduce the length

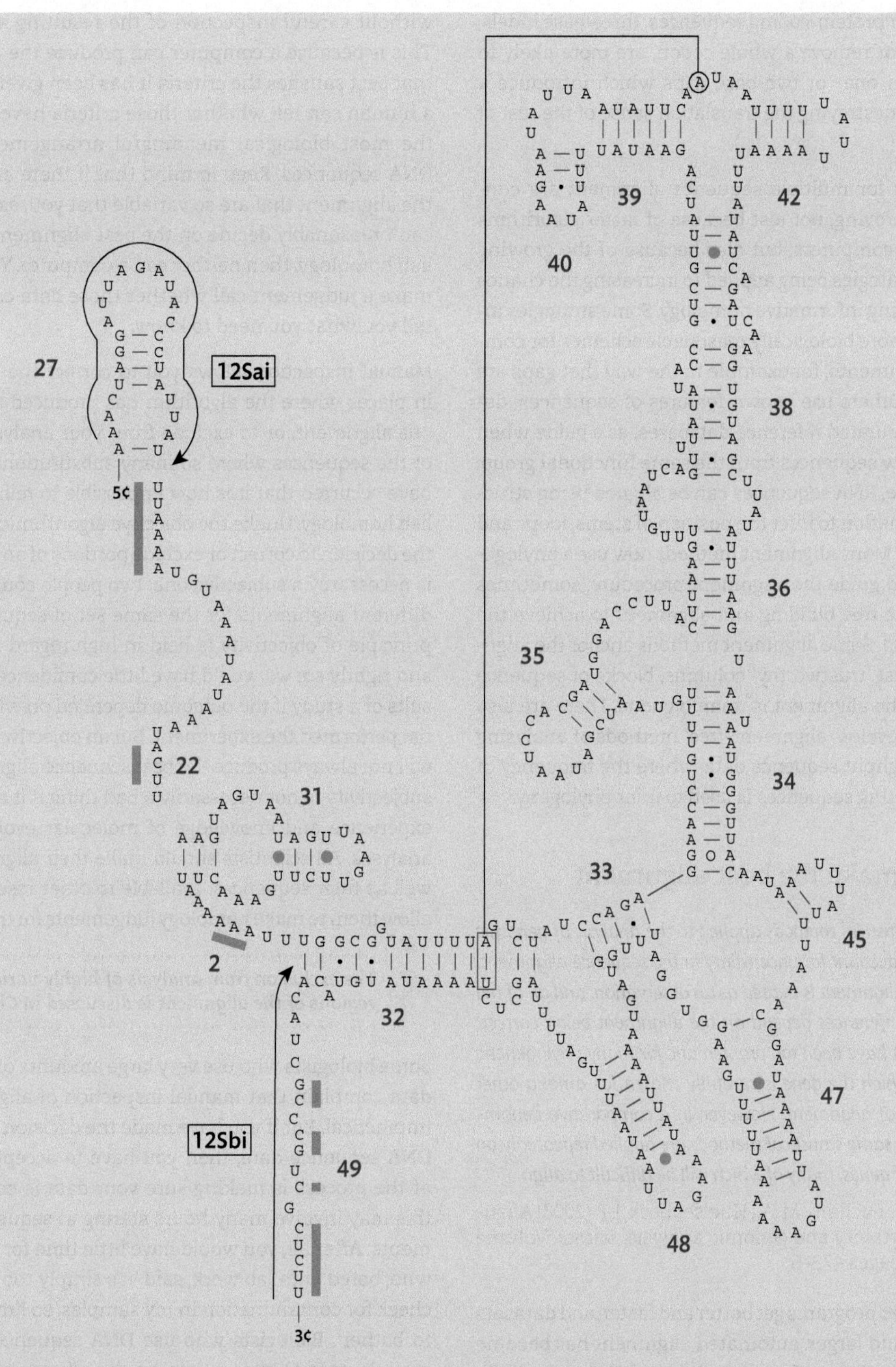


**Figure 10.15** RNA molecules, such as the ribosomal RNA produced by the *12S* gene, can form functional secondary structures by base-pairing between different parts of the RNA polynucleotide strand. This diagram of part of the 12S RNA molecule from a fruit fly (*Drosophila*) shows double-stranded stems made of paired bases and loops made of single strands of bases.

From Page, R. D. M. (2000) *Nucleic Acids Research*, Volume 28(20), pages 3839–3845, by permission of Oxford University Press.

of a loop. For protein-coding sequences, three-base indels, which add or remove a whole codon, are more likely to persist than one- or two-base gaps which introduce a frameshift, destroying the translation sense of the rest of the protein (Case Study 10.1).

Techniques for multiple sequence alignment are constantly improving, not just because of faster algorithms and bigger computers, but also because of the growing range of strategies being applied to increasing the chance of establishing informative homology. Some strategies incorporate more biologically reasonable schemes for comparing alignments, for example in the way that gaps are weighted. Others use known features of sequences, derived from curated reference databases, as a guide when aligning new sequences from the same functional group. For example, RNA sequences can be aligned using structural information to infer the position of stems, loops and active sites. Many alignment methods now use a phylogenetic tree to guide the alignment procedure, sometimes iterating the tree building and alignment to achieve the best solution. Some alignment methods anchor the alignment against 'trustworthy' columns, blocks of sequence for which the alignment is unambiguous. There are also moves to develop 'alignment-free' methods of analysing high-throughput sequence data, where the frequency of short matching sequences is used to infer phylogeny.

## How to make the best alignment

*"The statistical methods applied to the analysis of genomic data do not account for uncertainty in the sequence alignment. Indeed, the alignment is treated as an observation, and all of the subsequent inferences depend on the alignment being correct. This may not have been too problematic for many phylogenetic studies, in which the gene is carefully chosen for, among other things, ease of alignment. However, in a comparative genomics study, the same statistical methods are applied repeatedly on thousands of genes, many of which will be difficult to align"*

Wong, K. M., Suchard, M. A., Huelsenbeck, J. P. (2008) Alignment uncertainty and genomic analysis. *Science*, Volume 319(5862), pages 473–6

As computer programs get better and faster, and datasets get larger and larger, automated alignment has become a fundamental part of most analyses of DNA sequences. It's simply not practical to manually align megabases of sequences from thousands of taxa. However, you may be surprised, disappointed or pleased to hear this, but computers do not make better biologists than humans do. Biologists are generally better able to judge what is a reasonable hypothesis for the evolution of DNA sequences than a computer is. Computer-generated alignments should never be used in an evolutionary analysis without careful inspection of the resulting alignment. This is because a computer can produce the alignment that best satisfies the criteria it has been given, but only a human can tell whether those criteria have produced the most biological meaningful arrangement of the DNA sequences. Keep in mind that if there are parts of the alignment that are so variable that you, as a human, can't reasonably decide on the best alignment to establish homology, then neither can a computer. You have to make a judgement call whether those data can reliably tell you what you need to know.

Manual inspection allows you to correct the alignment in places where the algorithm has produced an erroneous alignment, or to exclude from your analysis regions of the sequences where so many substitutions or indels have occurred that it is now impossible to reliably establish homology. Unlike the objective algorithmic approach, the decision to correct or exclude portions of an alignment is necessarily a subjective one: two people could produce different alignments for the same set of sequences. The principle of objectivity is held in high regard in science, and rightly so: we would have little confidence in the results of a study if the outcome depended on which scientist performed the experiment. But an objective algorithm will not always produce the best sequence alignment and subjectivity is not necessarily a bad thing if it arises from experience and knowledge of molecular evolution and analyses. All scientists should make their alignments, as well as their sequences, available to other researchers to allow them to make homology judgements for themselves.

***The exclusion from analysis of highly variable regions of the alignment is discussed in Chapter 13***

Some biologists who use very large amounts of sequence data complain that manual inspection of alignments is impractical. But if you have made the decision to analyse DNA sequence data, then you have to accept that part of the process is making sure your data is correct, and this may involve many hours staring at sequence alignments. After all, you would have little time for a biologist who, bored with lab work, said 'it's simply too tedious to check for contamination in my samples, so I'm not going to bother'. Biologists who use DNA sequence data but are reluctant to spend time on the alignment phase of the analysis need only remind themselves of the basic maxim: 'rubbish in, rubbish out'. For example, poor alignment can result in falsely inferring high rates of change at particular sites, which might be mistaken for the signal of positive selection (TechBox 7.2). Poor alignment can also lead to strong statistical support for an incorrect phylogeny: in other words, it can make you feel very confident you have the right answer, even when you are wrong.

With the massive outputs of high-throughput sequencing, it's hardly surprising that most people these days just trust their automated alignments, leaving computers to do the spade-work of DNA analysis. But stop and ask yourself this: how would you feel if you did all that work, analysed your sequences, interpreted the results, presented them at conferences and sent them to journals, then someone pointed out that actually the alignments you used were so terrible that you hadn't actually been analysing homologous sequences. The patterns you detected in your data weren't biological signal at all, they were noise amplified to resemble signal by poor alignment and therefore your results were meaningless. You were telling a story based on false information. You might think that a couple of days spent staring at alignments is a big waste of time, but the aim is to prevent you from wasting even more time working from rubbish alignments. If you care about the answers you are getting in your research (and if you don't, then please leave now), then take the time to check your alignments. Computers are an essential part of the analysis of DNA sequences and we wouldn't get very far without relying on increasingly sophisticated and marvellous analysis programs. But, in the end, you are way smarter than a computer.

### *The fate of the thylacine*

There are many reasons why all biologists should learn about thylacines, not just because they are an excellent case study in convergent evolution. More importantly, the thylacine was one of the most beautiful and fascinating mammal species on earth. It is heartbreaking to think that no-one reading this book will ever see a thylacine alive. For, although some optimistic souls still hope that they may be rediscovered in the wilds of Tasmania one day, the thylacine officially went extinct on 7 September 1936. That is the day that the last known thylacine died in a Tasmanian zoo.

Many different factors may have contributed to the thylacine's demise. That they went extinct due to the influence of European settlers seems an inescapable conclusion. From 1803 to the present day, these settlers cleared the forests of Tasmania for agriculture or forestry, reducing the thylacine's habitat. European settlers may also have inadvertently introduced a disease that decimated native animal populations. But, most tragic of all, thylacines were actively hunted: killed by dogs, strangled in snares, poisoned, and shot. The vendetta against the thylacine was justified as a stock protection measure, though there is little evidence that thylacines ever caused major losses of poultry, sheep, or cattle (in fact, it seems likely that most predatory attacks on sheep were due to dogs, another destructive influence brought by the settlers). Nonetheless, private grazing companies, landholder associations, and the Tasmanian government itself all offered bounties for thylacine kills. These bounties may have been the nail in the coffin for a species already under threat.

The thylacine was mourned as soon as it was lost. It was officially declared a protected species in July 1936, 59 days before the last lonely thylacine died in captivity. This stupid and irreversible loss of such an extraordinary species should remind us to think carefully about the destructive consequences of human activity. It will not, alas, be the last preventable extinction of an irreplaceable species.

Movies of the last thylacine show a beautiful animal, pacing its cage, opening its jaws to display its distinctive gape (probably a threat warning—indeed, this thylacine bit the photographer on the buttock shortly afterwards: **Figure 10.16**). These movies never fail to fill me with a great sense of sorrow and loss. I would dearly love to see a live thylacine, so it is tempting to hope that there is some chance of success for the thylacine cloning project launched in 1999, based on extracting DNA from a thylacine pup preserved in a jar of alcohol in 1866 (**TechBox 10.1**). However, the Australian Museum officially abandoned the project in 2005, stating that the DNA was too degraded. Although other research groups intend to continue attempts to clone the thylacine, they face apparently insurmountable technical obstacles. However, in the rapidly moving field of molecular genetics, it is unwise to ever make sweeping claims about what it will be possible to achieve in the future. So we can only hope that unexpected advances in ancient DNA sequencing and cloning technology will one day cause us to look back and laugh at our earlier cynicism, as we stroke our resurrected thylacine puppies.

**Figure 10.16** One of the ways in which the thylacine differed from wolves was in its impressive gape: it could open its jaws wider than any other mammal, perhaps greater than 80°. This is often interpreted as a threat display.

## Conclusion

In this chapter, we have considered the importance of homology in establishing evolutionary relationships. We can trace the evolutionary history of species by considering traits that are shared through common descent. In order to read the history of species, we must consider traits that are truly homologous (inherited from a common ancestor) not analogous (independently evolved). Distinguishing homology from analogy is just as critical for DNA sequences as it is for morphology. Alignment, which is the process of establishing homologous sites, is therefore the most important step in evolutionary analysis of DNA sequences. Only by comparing homologous sites can we begin to read the story of descent in DNA sequences. In the next chapter, we will begin to see how we can use the differences we see in homologous sequences to map the path of descent as a phylogenetic tree.

Alignment rests critically on the concept of homology, or similarity by descent. By definition a homologous trait has evolved only once. The corollary of this is that once genetic information is lost, it is gone forever. When a species goes extinct, it takes with it the unique genetic information that informed its development and recorded its history. The dream of resurrecting extinct species from the DNA they have left behind is a distant hope—best not to lose them in the first place.

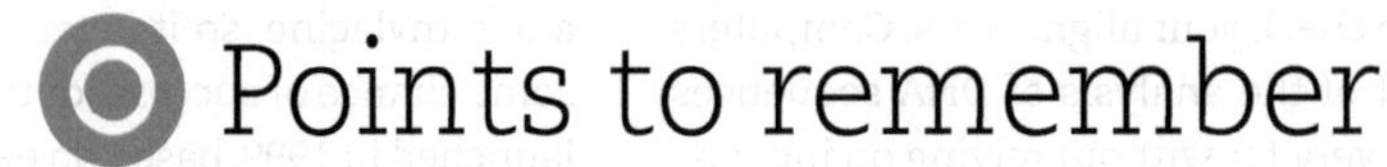

## Points to remember

### Homology

- Systematic hierarchies are expected to broadly predict similarity in many different traits, including DNA sequence similarity.
- Fundamental aspects of anatomy or 'housekeeping genes' that perform basic cellular functions are often used to uncover patterns of relatedness because they are directly comparable across many different species and are less likely to have evolved particular features to match specific ways of life.
- To uncover patterns of relatedness, it is important to distinguish homologies (similarity by descent) from analogies (independently evolved).
- When comparing DNA sequences, the only way to establish homology from analogy is alignment.
- Extinction represents the loss of all living carriers of the unique genetic information that defines a species.

### Alignment

- Homologous sequences were copied from the same ancestral sequence, so any differences in the sequences have evolved since their shared common ancestor.
- Gaps in alignments, representing insertions or deletions, are not read from DNA sequences, they are inferred to bring homologous sites into alignment.
- Any set of sequences could be aligned in many different ways, so alternative alignments need to be compared to choose the one that maximizes the chance of comparing homologous sites.
- Homologous sites are positions in an alignment that were all copied from the same ancestral sequence, even if they now have different character states (A, C, T, G or gap).
- Character states not shared with other sequences (autapomorphies) can distinguish individual sequences but cannot establish relationships.
- Manual alignment allows a researcher to judge which arrangement of bases and gaps provides the most reasonable assignment of homology.
- Automated multiple alignment uses a set of instructions for producing alternative alignments and a score to compare those alignments.

- Optimality scores for alignments may penalize inclusion of too many gaps, or favour some kinds of substitutions as being more likely than others.
- Never analyse alignments without first checking them carefully to satisfy yourself that they are a reliable representation of homology between sequences.

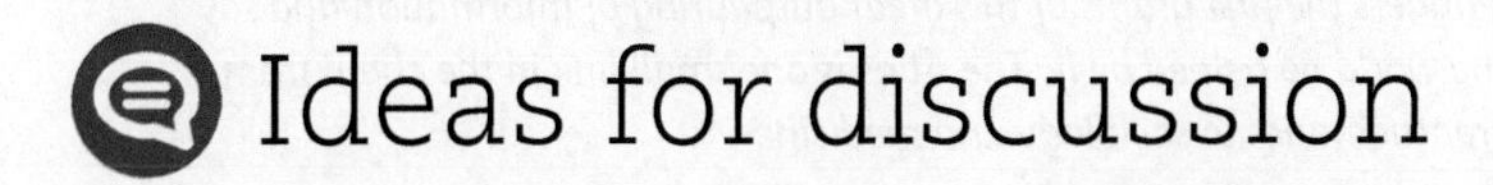

# Ideas for discussion

**10.1** The Quagga project aims to eventually reintroduce animals into the wild. On their website they say: '. . . since the coat-pattern characteristics are the only criteria by which the Quagga is identified, re-bred animals that demonstrate these coat-pattern characteristics could justifiably be called Quaggas.' How should these animals be formally recognized–as a colour variant of the Plains Zebra (*Equus quagga*) or a recreation of the extinct quagga subspecies (*E. quagga quagga*)? What conservation status should these released animals have? Does rebreeding offer hope for reviving other extinct species?

**10.2** What could cause a sequence to have more autapomorphies than any other sequence in the alignment?

**10.3** Should conservation funding be used to support de-extinction programmes?

# Sequences used in this chapter

Table 10.1: 12S rRNA gene sequences used in Figure 10.4

| Description from GenBank entry | Accession |
|---|---|
| *Thylacinus cynocephalus* 12S ribosomal RNA gene, complete sequence | TCU87405 |
| *Dasyurus geoffroii* 12S ribosomal RNA gene, mitochondrial gene for mitochondrial rRNA | AF009891 |
| *Sminthopsis psammophila* 12S ribosomal RNA gene, mitochondrial gene for mitochondrial RNA | AF088974 |
| *Notoryctes typhlops* 12S ribosomal RNA gene, complete sequence | NTU21179 |
| *Canis familiaris* mitochondrial 12S rRNA gene. | Y08507 |
| *Felis domesticus* mitochondrial 12S rRNA gene. | Y08503 |

# Examples used in this chapter

Cameron, S. A., Mardulyn, P. (2001) Multiple molecular data sets suggest independent origins of highly eusocial behavior in bees (Hymenoptera: Apinae). *Systematic Biology*, Volume 50, page 194.

Duffy, J. E., Morrison, C. L., Ríos, R. (2000) Multiple origins of eusociality among sponge-dwelling shrimps (*Synalpheus*). *Evolution*, Volume 54, page 503.

Jordan, G., Goldman, N. (2011) The effects of alignment error and alignment filtering on the sitewise detection of positive selection. *Molecular Biology and Evolution*, Volume 29, page 1125.

Keeling, P. J. (1998) A kingdom's progress: Archezoa and the origin of eukaryotes. *BioEssays*, Volume 20, page 87.

HEROES OF THE GENETIC REVOLUTION

10

# Margaret Oakley Dayhoff

*"We shift over our fingers the first grains of this great outpouring of information and say to ourselves that the world be helped by it. The Atlas is one small link in the chain from biochemistry and mathematics to sociology and medicine."*

Margaret O. Dayhoff to Susan Tideman, 18 October 1968, National Biomedical Research Foundation Archives

**EXAMPLE PUBLICATIONS**

Dayhoff, M. O. (1965) Computer aids to protein sequence determination. *Journal of Theoretical Biology*, Volume 8, pages 97–112.

Dayhoff, M. O. (1965) *Atlas of Protein Sequence and Structure*. National Biomedical Research Foundation.

**Figure 10.17**
Margaret Oakley Dayhoff (1925–1983) was one of the founders of bioinformatics. She created the first protein sequence database and pioneered the use of computers in analysing sequences and estimating phylogenies. She had to work against a common prejudice against 'theoretical' researchers (those who did not do lab work) to convince biochemists to allow their sequences to be included in the database, as a resource for other researchers.

I remember when GenBank used to come in the post on a CD. Although that may seem primitive now, the first ever comparative database of protein sequences was actually printed as a book. Margaret Dayhoff's *Atlas of Protein Sequence and Structure* was first produced in 1965. There were a total of 70 protein sequences, each given its own page with notes about its origin and characteristics, plus some alignments of homologous proteins from different species. It was no small feat to collate these sequences. Some sequences were hunted down in papers published in various scientific journals, and other sequences remained unpublished because some researchers did not want to release them for others to use (this possessive attitude persisted into the early days of GenBank, as some researchers resented their hard-won sequences being analysed by freeloaders who had not done the lab work). The *Atlas* went through a series of increasingly hefty editions over the following decade, then eventually moved to magnetic tape and went online in 1980 (accessible through telephone lines). Dayhoff's *Atlas* provided a template from which GenBank was eventually built (**TechBox 1.1**).

Margaret Dayhoff was one of the first bioinformaticians, building the early interconnections between biochemistry, computing, information science, and evolutionary biology (**Figure 10.17**). For her PhD in quantum chemistry, Dayhoff pioneered the use of computational analysis in understanding organic molecules. At this time, multiple research facilities would all share the same computer, which took up a whole room, and data were input on paper punched cards. Her early research career included work on astrobiology, using computer simulations to explore the formation of macromolecules in prebiological atmospheres. This led her to an interest in protein chemistry, building upon new ideas that molecular sequences could be read as documents of history. She used analysis of patterns of homology in amino acid sequences, combined with an understanding of biochemistry of proteins, to reconstruct the evolution and diversification of proteins from predicted ancestral sequences.

In 1966, Margaret Dayhoff and Richard Eck published one of the first attempts to use a computational analysis to infer a molecular phylogeny from amino acid sequences using an evolutionary model (considering the number of changes required to fit the data to a given phylogenetic tree). For these early analyses, all amino acid changes were weighted equally, but to improve estimates of evolutionary distance from protein sequences, Dayhoff estimated the frequency of different amino acid substitutions from protein sequence data, leading to the 'Percent Accepted Mutation' (PAM) matrix (now usually called the Dayhoff Matrix). She developed and improved computational methods for studying protein evolution, including tracing the origin and diversification of gene families, and constructing deep phylogenetic trees for the origins of the kingdoms. Dayhoff used phylogenies from several different proteins to support the endosymbiont theory for the origin of mitochondria and chloroplasts, demonstrating that the organelle proteins were derived from a bacterial lineage. This was an early example of showing that discordance between phylogenies based on different loci can show that genes from the same species can have different evolutionary histories.

> *"There is a tremendous amount of information regarding evolutionary history and biochemical function implicit in each sequence and the number of known sequences is growing explosively. We feel it is important to collect this significant information, correlate it into a unified whole and interpret it"*
>
> Margaret Dayhoff to Carl Berkley, 27 February 1967, National Biomedical Research Foundation Archives

TECHBOX **10.1**

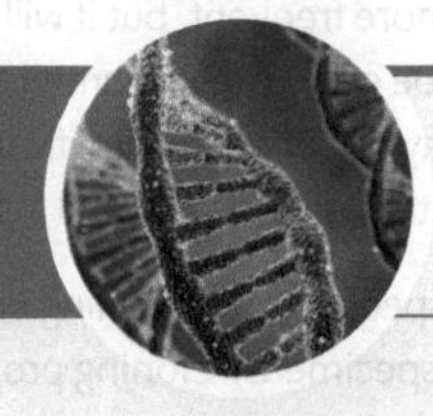

# De-extinction

**RELATED TECHBOXES**

TB 2.2: DNA extraction

TB 3.2: Biobanking

**RELATED CASE STUDIES**

CS 2.2: Ancient DNA

CS 12.1: Epidemiology


## Summary

New genetic technologies offer hope of bringing recently extinct species back to life.



## Keywords

cloning, gene bank, high-throughput sequencing, somatic nuclear cell transfer, seed bank, genome editing, endangered, reproductive technology, frozen zoo


## Cloning and conservation

In the broadest sense, cloning refers to reproduction where the genetic material of one individual is copied to make a new genetically identical individual (note that the word 'cloning' is also used to describe the process of making multiple copies of a gene). Many organisms produce natural clones: for example, tuberculosis bacteria divide by fission, new strawberry plants can grow from a runner (extended stem) from one parent, and identical twins are genetic clones that have developed from a single fertilized egg (**Figure 10.18**). The technology of reproductive cloning generates an embryo from the genome of one parent, rather than by fusing gametes from two different parents. Usually, this is achieved by nuclear transfer. For example, Noah, the first cloned gaur (*Bos gaurus*), was created in 2001. The gaur is a species of a wild woodland ox from south Asia, classified as vulnerable to extinction (you may recognize the gaur as the logo on Red Bull drinks). Noah was created by taking the nucleus from a gaur skin cell, then inserting it into an egg cell from a domestic cow. Sadly, Noah died when only two days old (the cause of death—dysentery—may have been unconnected to the cloning procedure). Similarly, frozen tissue from the last bucardo (the Pyrenean ibex *Capra pyrenaica pyrenaica*) was used to create a cloned embryo that was gestated in a goat, but the kid died shortly after birth[1].

Cloning is now being explored as a last-ditch form of captive breeding for endangered species. In populations with very few remaining individuals cloning could increase the number of individuals in the population (but, by definition, it will not increase the genetic diversity of the population). Conveniently, cells for cloning can be moved long distances, and even frozen for later use. There are now several genetic resource banks that store cells from endangered species for future reproductive cloning[2]. In this sense, cloning technology can be seen as an extension of current assisted breeding strategies, such as sperm banking and artificial insemination. However,

**Figure 10.18** Human clones are already with us. Identical twins arise when a single embryo splits to form two genetically identical embryos. Unlike clones produced by recent technology based on nuclear transfer, natural clones share both their nuclear and mitochondrial genomes.



at this point in time, the utility of cloning for rescuing endangered species is low. As cloning technology improves, successes are likely to become more frequent, but it will be difficult to fine-tune the procedures for rare animals that cannot be easily experimented on. For most endangered species recovery programmes, cloning may not offer any significant advantages over commonly applied methods of assisted reproduction, such as artificial insemination.

But assisted reproduction is not an option for extinct species. Could cloning be used to resurrect extinct species using genetic material from preserved specimens? Cloning programmes have been initiated for a number of extinct species, including thylacine (**Figure 10.16**) and woolly mammoth (**Figure 2.28**), though many scientists are highly sceptical of the chances of success of these programmes. Current reproductive cloning techniques require intact nuclei from live donor cells, which is unlikely to be an option for most extinct species. Very recently extinct species, such as the bucardo, may provide whole cells from which nuclei can be extracted. Attempts to extract whole gametes from frozen mammoths have so far met with no success.

### Cloning by sequencing

Given that intact cell nuclei will not be available for most extinct species, some programmes to clone extinct species are resting their hopes on extracting and sequencing DNA from long-dead remains, assembling the entire genome, then somehow persuading a living cell from another species to use the genome to develop an embryo. There are many challenges confronting these attempts at de-extinction:

**1 DNA:** High-throughput sequencing (**TechBox 1.2**) has made it possible to sequence large amounts of genomic DNA for recently extinct species, such as the mammoth genome[3]. But, for example, only a handful of genes have been sequenced from the thylacine[4] as the DNA is too degraded to allow useful amounts of the genome to be amplified.

**2 Genome:** even if all of the DNA from the an extinct species could be sequenced, assembling it into a working genome would be challenging without a reference genome. For example, the thylacine mitochondrial genome was assembled against the genome from the Tasmanian devil, even though these lineages are separated by tens of millions of years of evolution. Even if the genome of an extinct animal could be fully assembled, the DNA would somehow need to be arranged into chromosomes that would function within a living cell.

**3 Development:** A reconstituted genome would need to be placed in a cell that could interpret the genetic information and use it to build a functioning embryo. A cloning programme for an extinct species must necessarily use an egg from a related species. Researchers implanted nuclei from frozen tissue of the extinct gastric brooding frog (**Figure 10.19**) into embryos of frog from a different subfamily (the Great Barred Frog, *Mixophyes fasciolatus*), but the embryos did not survive for more than a few days. Even if development proceeds, somatic nuclear transfer effectively makes the cloned individuals a hybrid, with the nuclear DNA of one species, and the mitochondrial DNA of a different species.

**4 Gestation:** By definition, an extinct species has no living mothers to gestate or raise a cloned embryo. The Mammoth Project plans to use a female elephant, even though elephants are separated from mammoths by millions of years of evolution. The closest relative of the thylacine is the Tasmanian devil, half the size of the thylacine and strikingly dissimilar in morphology and ecology (although development of young in a pouch may make marsupials more amenable to cloning than placental mammals). This will, of course, be less of a problem for species with external fertilization and no parental care.

**5 Environment:** Concerns have been raised over the wisdom of recreating a species that may have gone extinct through loss of habitat, though this need not be a practical problem if the resurrected clones are destined for life in captivity (as demonstrated by the presence of polar bears in tropical zoos). If the aim of cloning an extinct species is its reintroduction to the wild, then some candidate species will have more suitable habitat remaining than others[5]. The recent extinction of the Yangtze river dolphin (or baiji, *Lipotes vexillifer*) was driven by intense river development including modern fishing techniques[6]. It has been argued that the baiji may not be a viable case for de-extinction while the Yangtze remains heavily polluted and industrialized, as there may be no suitable habitat remaining in which a resurrected population can live[5].

### Genome editing

Somatic nuclear cell transfer requires intact cell nuclei from the donor species and sufficiently closely related recipient cells. An alternative approach is to use DNA sequencing of material from extinct species as a source of genes which can be engineered into a closely related living species. In this case, the resulting individual would not be a clone with the same genome as the extinct species, since its genome would be a combination of donor and recipient species genes. Instead, it would be a form of genetically modified organism containing genes from more than one species. Which species would this individual be assigned to?

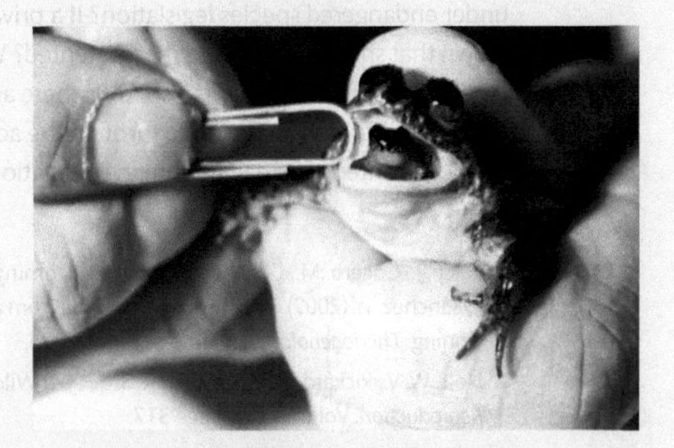

**Figure 10.19** The gastric brooding frog (*Rheobatrachus*) was discovered in the 1970s. Females had the highly unusual habit of swallowing their fertilized eggs and gestating their offspring in their stomach. During this gestation, the mother did not produce stomach acid and could not eat. Fully formed froglets would eventually emerge from the mother's mouth. But this wonderful and unique frog was lost almost as soon as it was found, being last seen alive in the 1980s. A research group located a frozen sample of tissue from a gastric brooding frog and has attempted cloning nuclei into the embryos of a distantly related frog species from the same family, but have been unsuccessful so far. Should the remaining tissue be preserved for a future age when cloning has been refined and success rates are higher? Or should the researchers continue to work on this species in the hope of a breakthrough? Or should scientific effort be directed instead to saving species on the brink of extinction rather than on expensive and time-consuming efforts to resurrect the dead? Incidentally, time is running out to save another quirky amphibian, Darwin's mouth brooding frog (*Rhinoderma darwinii*), from extinction.

### Preserving genetic resources

Given the concern about loss of unique genomes through extinction, and the hope of preserving endangered species or reviving lost taxa through cloning or other technologies, many organizations have established genome resource collections (often referred to as gene banks). For animal species, gametes (eggs and sperm) or body tissues may be cryopreserved: for example the 'Frozen zoo' at San Diego Zoo has material from over 800 species stored at −196 °C. Many seed banks have been set up to preserve plant genetic diversity, some focused on the wild relatives of crop plants as a genetic resource for agricultural development. For example, the Svalbard Global Seed Vault is constructed on a geologically stable island, 1,300 km from the North Pole. It has been built into the permafrost, above the worst-case scenario for sea level rise. It acts as a global back-up for other seed banks around the world, containing replicate samples of accessions held in other collections. It currently contains over three quarters of a million samples, held at −18 °C.

### Could we? Should we?

The rise of new genomic technologies has generated excitement about de-extinction, through using biological material from extinct species to create living animals. Given the current high cost and low success rates, some conservation biologists feel that resources would be better directed to endangered species management than revival of extinct species. Currently, most cloning programmes for extinct species rely on private funding, therefore there is a tendency to focus on relatively charismatic species such as woolly mammoths. This means that the pioneering programmes might not be focused on the most tractable cases, but those that can most easily attract support[7] It has been suggested that resources should instead be concentrated on those species most likely to be able to be returned to the wild to establish a viable population.

Some conservationists worry that the hope of reviving extinct species through cloning will compromise the prima facie case for conservation because it suggests that extinction is not an irreversible loss. The legal status of resurrected species has also been discussed[8]: would they fall under endangered species legislation? If a private consortium recreates an extinct species, does it own that species, and could it be patented? Would there be restrictions on their release, as for genetically modified organisms?[7] Is there any risk involved in releasing previously extinct species to the wild, for example through the activation of ancient pathogens, impacts on other species, or interference with human populations?[5]

### References

1. Folch, J., Cocero, M., Chesné, P., Alabart, J., Domínguez, V., Cognié, Y., Roche, A., Fernández-Arias, A., Martí, J., Sánchez, P. (2009) First birth of an animal from an extinct subspecies (*Capra pyrenaica pyrenaica*) by cloning. *Theriogenology*, Volume 71, page 1026.
2. Holt, W. V., Pickard, A. R., Prather, R. S. (2004) Wildlife conservation and reproductive cloning. *Reproduction*, Volume 127, page 317.
3. Miller, W., Drautz, D. I., Ratan, A., Pusey, B., Qi, J., Lesk, A. M., Tomsho, L. P., Packard, M. D., Zhao, F., Sher, A. (2008) Sequencing the nuclear genome of the extinct woolly mammoth. *Nature*, Volume 456, page 387.
4. Menzies, B. R., Renfree, M. B., Heider, T., Mayer, F., Hildebrandt, T. B., Pask, A. J. (2012) Limited genetic diversity preceded extinction of the Tasmanian Tiger. *PLoS ONE*, Volume 7: e 35433.
5. Seddon, P. J., Moehrenschlager, A., Ewen, J. (2014) Reintroducing resurrected species: selecting DeExtinction candidates. *Trends in Ecology & Evolution*, Volume 29, page 140.
6. Turvey, S. T., Pitman, R. L., Taylor, B. L., Barlow, J., Akamatsu, T., Barrett, L. A., Zhao, X., Reeves, R. R., Stewart, B. S., Wang, K., Wei, Z., Zhang, X., Pusser, L. T., Richlen, M., Brandon, J. R., Wang, D. (2007) First human-caused extinction of a cetacean species? *Biology Letters*, Volume 3, page 537.
7. Jones, K. E. (2014) From dinosaurs to dodos: who could and should we de-extinct? *Frontiers of Biogeography*, Volume 6, page 20.
8. Sherkow, J. S., Greely, H. T. (2013) What if extinction is not forever? *Science*, Volume 340, page 32.

TECHBOX
**10.2**

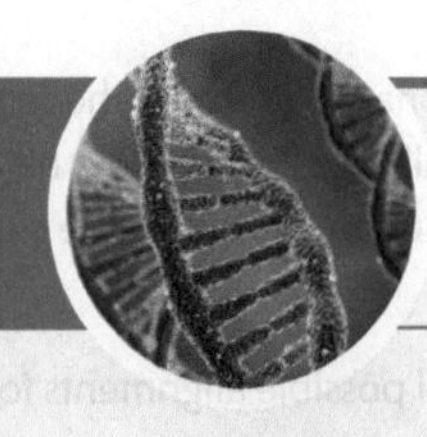

# Multiple sequence alignment

RELATED TECHBOXES

**TB 1.2:** BLAST
**TB 11.1:** Distance methods

RELATED CASE STUDIES

**CS 1.2:** Forensics
**CS 9.2:** Barcoding


## Summary

Alignment establishes shared ancestry of DNA sequences so that evolutionary changes can be identified.



## Keywords

automatic alignment, manual alignment, score, gap penalty, affine, penalty function, PAM, BLOSUM, indel, MSA, RNA


The aim of multiple sequence alignment is to find the arrangement of a set of DNA sequences that maximizes the chance of comparing homologous sites. There are many alignment programs freely available, either as downloadable programs or server-based applications, and new improved methods are published regularly. Rather than discussing any particular program currently available, we will take a very simple overview of what multiple alignment programs do.

**Automated alignment** has two basic components: the score (a numerical function designed to rate the biological reasonableness of an alignment) and a search strategy (a way of changing the alignment to create alternative arrangements whose scores can be compared).

**1 Score:** The score reflects the goodness of match between a set of aligned sequences, according to some defined model that expresses the probability of different types of sequence change. The score for the whole alignment is essentially a weighted sum of scores for each site in the alignment plus a penalty for each gap. For a multiple alignment, the overall score is calculated either as a sum of scores between all pairs of sequences in the alignment, or from each sequence to a consensus sequence[1]. The key features of the score (or penalty function) are:

- **Substitution score:** Some nucleotide alignment programs use a basic score of +1 for a match and −1 for a mismatch. But, particularly for protein-coding sequences, some mismatches may be considered more likely to occur than others. For example, exchanging one hydrophobic amino acid for another may be less disruptive to a protein than changing it to a hydrophilic amino acid. So the penalty function may give positive scores for exact matches, intermediate scores for conservative differences between similar amino acids, but negative scores for non-conservative differences or gaps. Amino acid alignment programs often use a matrix of substitution probabilities, such as PAM or BLOSUM, derived from the frequency of substitutions in known sequence alignments (**Hero 10**).

- **Gap penalty:** Gaps are sites where one or more sequences have no matching base, representing an insertion or deletion event (or missing data). These are typically represented by inserting a dash (-) into the nucleotide sequence. Gaps will generally accrue a negative score, to discourage matching nucleotides by introducing too many gaps. A linear gap penalty scores all gaps equally, so three one-base gaps would have the same penalty score as one three-base gap. An affine gap penalty includes a score for opening a gap plus a length-dependent score for extending the gap. In this way, affine gap penalties favour extension of existing gaps over the introduction of new gaps, for example preferring one three-base gap over three one-base gaps.

The penalty function should ideally reflect the evolutionary patterns of the sequences. When you use a multiple alignment program, it is usually possible to input different values for parameters that reflect the model of sequence evolution. Altering these parameters can produce different alignments.

**2 Search:** It is not possible to score and compare all possible alignments for a set of sequences. For example, for two sequences of 300 bases, there are over a googol of possible alignments[2] (that is, more than the number of particles in the universe). Therefore, any program needs an efficient heuristic method for searching alternative alignments. Most multiple alignment methods are progressive[3]: they begin by making pairwise alignments which are used to cluster similar sequences into a 'guide tree', which is then used to select the most closely related pair of sequences to be aligned against each other. These aligned pairs are then held constant in a profile alignment that is then aligned against the next most closely related sequence, and so on, until all sequences are aligned. Progressive alignment dramatically reduces the search space (and therefore the computation time), but does not guarantee to find the best alignment: an incorrect guide tree may lead to a poor end result. The reliability of progressive alignment can be improved by using more information to make profile alignments (such as structural information or libraries of short local alignments[4]) or iterative refinement (randomly sampling sequences to create and score alternative profile alignments[5]). Aligning new sequences against a reference alignment, for example from a curated database of structural alignments, can reduce the search time and increase the accuracy of the alignment[6]. Similarly, starting with a guide tree based on assumed relationships between species can make alignment more efficient, but the flipside of this is that your alignment may have a tendency to support the relationships that you have specified in the guide tree, even if these are not correct[7].

**Manual sequence alignment** allows you to employ your understanding of molecular evolution to choose the best alignment. For example, you would be reluctant to introduce a one- or two-base gap in a protein-coding sequence, because that would cause a frameshift, destroying the translation sense of the rest of the protein. This should be obvious in a manual alignment editor which allows you to toggle between nucleotide and amino acid sequences. But some automatic alignment methods do not penalize frameshifts, and this can lead to unrealistic alignments of protein-coding sequences. Manual alignment editors also make it easy to delete sections of the alignment where there has been too much sequence change to allow confident assignment of homology. The limitations of manual alignment are that it can be time consuming and it is best suited to relatively small datasets. Manual alignment editing can be applied to any dataset, to remove incorrectly aligned or uninformative sequences, or to adjust regions where the automatic alignment has produced a questionable arrangement. Removing highly variable regions that can't be reliably aligned can improve the outcome of phylogenetic analyses[8].

## High-throughput sequencing

Classical Sanger sequencing produces sequences with a defined beginning, middle, and end. If you have designed your primers well, then you begin your alignment knowing that all your sequences are homologous, so you just have to work out which alignment best represents the evolutionary history of those sequences. Which is not to say that alignment of Sanger sequences is easy, but at least you have a head start in determining homology.

High-throughput sequencing techniques often start with no such knowledge of homology. Many of the new sequencing techniques produce a vast array of relatively short, overlapping sequences, each one covering one small part of the target DNA sequence. For most analyses, the first step is to try to stitch these fragments together to form a continuous sequence (**TechBox 5.2**). This process, usually referred to as sequence assembly, is also a form of homology assignment, because the aim is to identify which bits of sequence were all read from the same target sequence, rather than being a coincidental match copied from somewhere else in the genome. Whether this is done by matching to a reference sequence or to other sequences in the sample, sequence assembly relies on the same kind of algorithms as conventional sequence alignment: match, score, modify, score again, until you get the alignment with the highest score. The more data you have, the longer this will take. If you are using a sequencing technique that samples sequences from across the genome, then you will need a way of ensuring that when you compare sequences from different samples, you are only comparing homologous sites, related by descent from a common ancestor, and not analogously similar sequences from other parts of the genome (**Case Study 10.2**).

**Always check your alignments.** No matter what alignment method you use, you must very carefully inspect every part of your alignment before you analyse it, to make sure that the alignment you have chosen is a plausible hypothesis for the homology of sites in your sequences. Remember that an alignment program is just following orders. It will find best possible alignment for any set of sequences, given a set of rules and a scoring function. What the program cannot do is make an assessment of the biological meaningfulness of the resulting alignment. If you put in randomly generated sequences or sequences from unrelated genes, the program will still give you the best alignment, even though it will in no way reflect homology between those sequences. Therefore, while you can ask a machine to do the hard work of alignment for you, only you can assess the results. This is why manual inspection of computational alignments is essential for any analysis of DNA sequences.

Whether manually or automatically generated, you must look very carefully over all parts of an alignment to make sure there are no areas where one or more sequences are out of alignment. More specifically, you must always check alignments for:

- **frameshifts and stop codons:** indels in protein-coding sequences can destroy the translation sense of the protein. If there are stop codons within protein-coding sequences then your alignment is unlikely to be correct, unless your sequences are from non-functional genes (also check that you are using the correct genetic code for your sequences: **TechBox 6.1**). Note that frameshifts and stop codons are not relevant to sequences that do not code for a protein, such as RNA genes, pseudogenes, introns or intergenic DNA.
- **excessive introduction of gaps:** automatic alignments can sometimes insert gaps in apparently nonsensical places, for example isolating single bases in the middle of long gaps. Try modifying the gap penalty, or editing the alignment by hand, or removing these regions altogether.
- **regions of poor alignment:** sequences that have had a high rate of nucleotide substitutions or indels may be effectively randomized, so that the historical signal in the sequence is lost. There is no golden rule for deciding which regions to exclude. You must ask yourself if you can be sure that each position in your alignment represents true homology, because if it doesn't, any inference from that position is spurious. You can repeat your analysis with and without the questionable regions to test whether it has any effect on your results.
- **aberrant sequences:** you might notice that one or more of your sequences seem to be surprisingly different from the others in the alignment. Are they from distantly related species? Does this sequence have a fast rate of change, for example because it is from a taxon with a known high rate of change, or because this gene is non-functional in this particular taxon? Perhaps the sequence has been placed in the alignment in the wrong orientation (such as reverse or complementary). Is the sequence poor quality? Could it be a non-homologous gene, for example a paralog (gene copy from a different locus in the genome)? Is it a nuclear copy of an organelle gene? If you don't notice aberrant sequences at the alignment stage, you might identify them when you make a phylogenetic tree, as these sequences tend to 'stick out the top' of the tree, because the high number of differences compared to other sequences in the alignment may result in a long branch

length (**TechBox 13.2**). In any case, when you notice an odd sequence, do what you can to work out why. It's far better to detect any problems now when you can still fix it than later when your analysis is finished.

**Alignment as hypothesis.** It should be obvious from the discussion above that the decisions you make, such as how to weight gaps or mismatches, or the way you conduct the search, will influence the outcome of your alignment procedure. Therefore any given alignment should be read as one possible hypothesis of homology for a set of sequences. For some sequences, particularly those produced by targeted sequencing of conserved protein-coding genes, the alignment may be unambiguous. But for many other alignments, particularly those involving large amounts of variable sequence data, there may be many alternative alignments that are of equal or near-equal plausibility. Just as many phylogenetic methods are now embracing uncertainty in tree inference (**TechBox 13.2**), there are emerging methods for dealing with uncertainty in alignment by comparing the outcome of analyses using many alternative alignments.

Whatever approach you decide is best for your data, always remember that your alignment is an inference, not a fact. Some sequences will have an unambiguous alignment, but for most there are alternative alignments that you could choose which might influence your analysis. You always make a decision about the most likely alignment, whether that decision is made by default, by accepting the output of a program, or more actively, by inspecting and adjusting your alignments. Regard your alignment as a hypothesis about the homology and history of your sequences. Try changing your alignment and seeing if the results of your analysis are affected. If they are, then you will need to think about the strategies you will use to test whether you have read the right story.

## References


1. Batzoglou, S. (2005) The many faces of sequence alignment. *Briefs in Bioinformatics*, Volume 6, page 6.
2. Eddy, S. R. (2004) What is dynamic programming? *Nature Biotechnology*, Volume 22, page 909.
3. Higgins, D. (2003) Multiple alignment. pp. 45–71 in: Salemi, M., Vandamme, A-M. eds. *The phylgoenetic handbook: a practical approach to DNA and protein phylogeny*. Cambridge University Press.
4. Notredame, C., Higgins, D. G., Heringa, J. (2000) T-Coffee: A novel method for fast and accurate multiple sequence alignment. *Journal of Molecular Biology*, Volume 302, page 205.
5. Katoh, K., Kuma, K., Toh, H., Miyata, T. (2005) MAFFT version 5: improvement in accuracy of multiple sequence alignment. *Nucleic Acids Research*, Volume 33, page 511.
6. Pruesse, E., Peplies, J., Glöckner, F. O. (2012) SINA: accurate high-throughput multiple sequence alignment of ribosomal RNA genes. *Bioinformatics*, Volume 28, page 1823.
7. Kumar, S., Filipski, A. (2007) Multiple sequence alignment: in pursuit of homologous DNA positions. *Genome Research*, Volume 17, page 127.
8. Talavera, G., Castresana, J. (2007) Improvement of phylogenies after removing divergent and ambiguously aligned blocks from protein sequence alignments. *Systematic Biology*, Volume 56, page 564.

CASE STUDY **10.1**

# Alignment: identifying insertions and deletions in endosymbiont genomes

**RELATED TECHBOXES**

**TB 6.1:** Genetic code

**TB 10.2:** Multiple sequence alignment

**RELATED CASE STUDIES**

**CS 3.2:** Mutation rate

**CS 5.1:** Duplication

Williams, L. E., Wernegreen, J. J. (2013) Sequence context of indel mutations and their effect on protein evolution in a bacterial endosymbiont. *Genome Biology and Evolution*, Volume 5, page 599

*"Indel mutations play key roles in genome and protein evolution, yet we lack a comprehensive understanding of how indels impact evolutionary processes. Genome-wide analyses enabled by next-generation sequencing can clarify the context and effect of indels . . ."*

**Keywords**

selection, synonymous, neutral, mutation, substitution, genome assembly, high-throughput sequencing, pseudogene, intergenic

## Background

Insertions and deletions (indels) are important drivers of genome evolution. But while large-scale gains and losses of DNA sequences have been well-studied, less is known about small-scale changes that alter only one or a few nucleotides. Whole genome sequencing of close relatives offers a chance to study the frequency and spectrum of small indels. Endosymbiotic bacteria provide an interesting case study for changes in DNA sequence length as they typically undergo reduction in genome size. The gamma proteobacterium *Blochmannia* is an obligate symbiont of carpenter ants (*Camponotus:* **Figure 10.20**). The symbionts live within the ant's cells and are transferred vertically from each ant mother to her offspring. Like many other endosymbionts, *Blochmannia's* genome is reduced compared to its free-living relatives, only around 800 kb[1] (so the whole genome is smaller than the largest human genes).

## Aim

By sequencing the genome of *Blochmannia chromaiodes*, an endosymbiont of the ant *Camponotus chromaiodes*, researchers could compare it to a previously sequenced genome from a close relative *Blochmannia pennsylvanicus* (an endosymbiont of *Camponotus pennsylvanicus*). Alignment of the two genomes allows the identification of putative insertion and deletion events, represented by gaps where one sequence is missing one or more bases found in the other sequences.

## Methods

Two samples of *Camponotus chromaiodes* ants were collected from the same colony, 12 months apart. Three different sequencing methods were used. Pyrosequencing (454) was used for *de novo* assembly of the draft sequence (**TechBox 5.2**), and any gaps closed with Sanger sequencing (**TechBox 5.1**). But 454 has a high error rate in areas of homopolymers (runs of the same nucleotide). Since indels are most likely to occur in these homopolymeric runs, it was important for this project to sequence these areas carefully. So the researchers resequenced the genome using a dye-based sequencing platform (Illumina) which has a lower error rate for homopolymers[2]. These reads were aligned against the 454 assembly, and used to correct the draft genome sequence. The genome was annotated to identify genes, using both automated and manual approaches. They used two different alignment programs to align the genome sequences of *B. chromaiodes* and *B. pennsylvanicus*, then inspected the alignments to identify areas where the two methods gave different solutions to the alignment. They selected the best

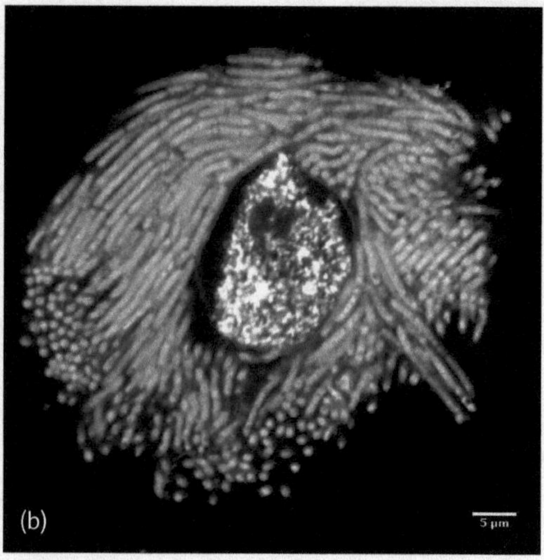


**Figure 10.20** (a) *Camponotus* is a diverse genus of ants, commonly known as carpenter ants as many live in decaying or hollow wood in forests or buildings. All *Camponotus* have the obligate bacterial symbionts, *Blochmannia*. These endosymbionts are able to recycle nitrogen and synthesize some amino acids and other nutrients, so it seems likely that they contribute to their host's nutrition: the ants are less healthy when given antibiotics that kill the *Blochmannia*, but the exact nature of the relationship between ant and bacteria is still not clear. The evolution of the bacteria closely tracks that of the ants, so that genes from *Blochmannia* predict the same phylogeny as genes from their *Camponotus* hosts. The bacteria are housed in special intracellular structures called bacteriocytes in particular cells in the midgut and ovaries and are passed vertically from mother to offspring: the image (b) shows the host cell nucleus at centre with the rod-shaped *Blochmannia* filling the cytoplasm. At what point do we stop calling something an 'endosymbiont' and start referring to it as an 'organelle'?

Photograph (a) by Adam B. Lazarus. Image (b) by Erika del Castillo.

solution by considering the reading frame of the gene, and choosing the alignment with the fewest inferred substitutions and indels. To estimate the neutral substitution rate (**TechBox 8.1**), they aligned protein-coding genes according to their amino acid sequences then estimated the rate of synonymous substitutions.

### Results

The alignment of the nucleotide sequences suggested that the two *Blochmannia* genomes were 98.0 per cent similar, with 13,389 substitutions and 1,051 indels. Substitution rates in pseudogenes (non-functional gene copies) and intergenic regions were higher than the rate of change in genes, but lower than the neutral substitution rate estimated from synonymous changes in proteins. The rate of occurrence of indels in pseudogenes and intergenic regions (between five and seven indels per 1,000 bases) was ten times higher than in genes. The majority of indels were single-base changes, and almost all were less than seven bases. Indels occurred almost exclusively in repeat regions, either of homopolymers (runs of the same nucleotide) or microsatellites (repeated sequence motifs). Two thirds of the indels in protein-coding genes occurred in multiples of three bases, maintaining the triplet coding structure. The other third occurred either at the 3' end of the genes or were balanced by compensatory indels (two indels that occur close to each other that restore the reading frame over the rest of the gene: **Figure 10.21**).

### Conclusions

Sequences where there are runs of the same base or repeats of short nucleotide motif are indel hotspots, consistent with the hypothesis that polymerase slippage during DNA replication is a common mechanism for changes in DNA sequence length. Most of the indels in protein-coding sequence maintained the triplet reading frame, either because they are three or six bases long or because they are balanced by nearby compensatory indels. Intergenic regions do *rimM*

**Compensatory indel hypothesis**

```
                       G  C  V  V  I  T  V  Q  G  V  L  L  G  D  I  I
Bpenn      227752 AGGATGTGTAGTAATCACCGTACAAGGG-GTCCTTTTAGGAGATATTATCA 227801
                  |||||||||||||||||||.|||||||| || |||||||||||.|||||||
Bchrom     227598 AGGATGTGTAGTAATCACTGTACAAGGGAGT-CTTTTAGGAGAAATTATCA 227647
                   G  C  V  V  I  T  V  Q  G  S  L  L  G  E  I  I
```

**Substitution hypothesis**

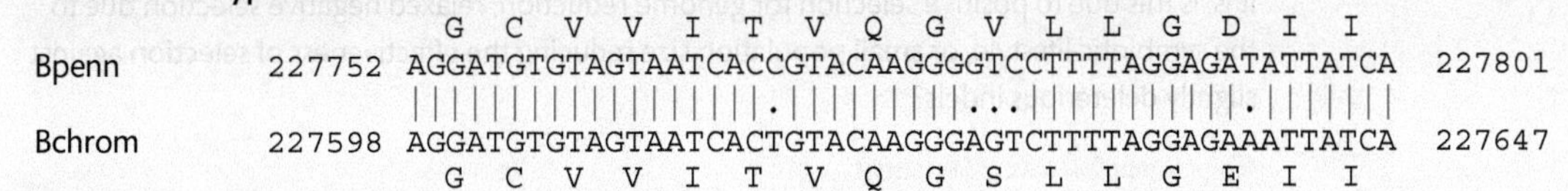

```
                   G  C  V  V  I  T  V  Q  G  V  L  L  G  D  I  I
Bpenn      227752 AGGATGTGTAGTAATCACCGTACAAGGGGTCCTTTTAGGAGATATTATCA  227801
                  |||||||||||||||||||.|||||||...||||||||||.|||||||
Bchrom     227598 AGGATGTGTAGTAATCACTGTACAAGGGAGTCTTTTAGGAGAAATTATCA  227647
                   G  C  V  V  I  T  V  Q  G  S  L  L  G  E  I  I
```

*yraP*

**Compensatory indel hypothesis**

```
                            N  T  S  C  H  I  S  Q  A  L  L  L  L  F  S  I
Bpenn revcomp     64298 TGAA-TACTTCATGTCATATATCACAGGCATTATTAATTTTATTTTCTATA 64249
                        |||| || ||||||||||||||||||||||||||||||||||||||||||
Bchrom revcomp    64216 TGAAATA-TTCATGTCATATATCACAGGCATTATTAATTTTATTTTCTATA 64167
                            K  Y  S  C  H  I  S  Q  A  L  L  I  L  F  S  I
```

**Substitution hypothesis**

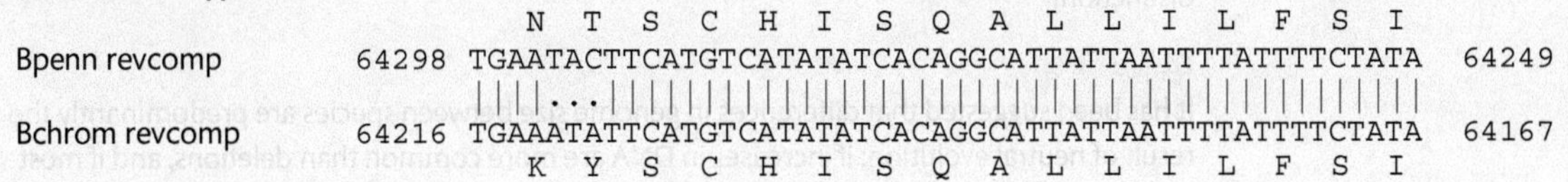

```
                            N  T  S  C  H  I  S  Q  A  L  L  I  L  F  S  I
Bpenn revcomp     64298 TGAATACTTCATGTCATATATCACAGGCATTATTAATTTTATTTTCTATA  64249
                        ||||...|||||||||||||||||||||||||||||||||||||||||||
Bchrom revcomp    64216 TGAAATATTCATGTCATATATCACAGGCATTATTAATTTTATTTTCTATA  64167
                            K  Y  S  C  H  I  S  Q  A  L  L  I  L  F  S  I
```

*BCHR0640_042/znuB*

**Compensatory indel hypothesis**

```
                    F  V  I  Y  K  K  *
Bpenn      51727 TTTGTTATATA-GAAAAAATAAAATTTTTTTCGTGTTTATTTAATTAAAT 51775
                 ||||| |||||  |||||||||||||||||.|||||||||||||||||||
Bchrom     51652 TTTGT-ATATAACAAAAAATAAAATTTTTCTCGTGTTTATTTAATTAAAT 51700
                    F  V     Y  N  K  K  *
```

**Substitution hypothesis**

```
                    F  V  I  Y  K  K  *
Bpenn      51727 TTTGTTATATACAAAAAATAAAATTTTTTTCGTGTTTATTTAATTAAAT  51775
                 |||||.....||||||||||||||||||.||||||||||||||||||||
Bchrom     51652 TTTGTATATAACAAAAAATAAAATTTTTCTCGTGTTTATTTAATTAAAT  51700
                    F  V  Y  N  K  K  *
```

**Figure 10.21** Proposed compensatory indels in three genes from two endosymbiotic bacteria, *Blochmannia pennsylvanicus* (Bpenn) and *Blochmannia chromaiodes (*Bchrom). Insertion of gaps in a sequence is a decision made to maximize the chance of comparing homologous positions across sequences, so for an alignment involving gaps, there are always many possible alignments. In this figure, the authors compare two alternative alignments for the same sequences from each gene region—one inferring compensatory indels, and one with no indels and therefore more inferred substitutions. Which of the alternative alignments do you think is more plausible?



not appear to be entirely neutral, but are instead under some selection to maintain nucleotide sequence and length, suggesting that these regions can contain functional elements.

### Limitations

High-throughput sequencing is error-prone, and any misreads or mistakes in assembly could be erroneously interpreted as substitutions or indels. Any alignment procedure involves choosing the arrangement of bases that is deemed most likely to represent presumed homologous sites, so different choices made during this analysis would have changed the number and type of sequence changes inferred. The rate of substitution in pseudogenes may be underestimated if they have lost functionality since the last common ancestor of the two species.

### Further work

Rates of nucleotide substitutions in endosymbionts have been shown to be higher than their free-living relatives[3], potentially due to relaxed selection pressure or small effective population sizes[4]. Is the rate of indels in endosymbionts greater than in their free-living relatives? And if it is, is this due to positive selection for genome reduction, relaxed negative selection due to the symbiotic lifestyle, or small population size reducing the effectiveness of selection against slightly deleterious indels?

### Check your understanding

1. How can alignment be used to infer insertion or deletion events?
2. Given that most indels detected in this study were only one base long, why are the majority of indels in protein-coding sequences three or six bases long?
3. Why do the non-triplet indels detected in protein-coding sequences occur mostly at the 3' end of the gene?

### What do you think

Is it possible to tell whether a gap in the alignment of these two genomes is due to an insertion in one sequence or a deletion in the other? What information would you need to make this distinction?

### Delve deeper

It has been suggested that differences in genome size between species are predominantly the result of neutral evolution: if increases in DNA are more common than deletions, and if most increases have only a very slight fitness cost, then amount of DNA in the genome may increase until the cumulative cost of all the extra DNA exerts significantly negative selection pressure to overcome stochastic accumulation of insertions[5]. How could you test the components of this hypothesis: that insertions are more common than deletions, that increases in DNA typically have low fitness costs, and that species might differ in the amount of extra DNA due to differences in the action of selection reducing genome accumulation?

### References


1. Wernegreen, J. J., Lazarus, A. B., Degnan, P. H. (2002) Small genome of *Candidatus Blochmannia*, the bacterial endosymbiont of *Camponotus*, implies irreversible specialization to an intracellular lifestyle. *Microbiology*, Volume 148, page 2551.
2. Luo, C., Tsementzi, D., Kyrpides, N., Read, T., Konstantinidis, K. T. (2012) Direct comparisons of illumina vs. roche 454 sequencing technologies on the same microbial community DNA sample. *PLoS ONE*, Volume 7, page e 30087.
3. Degnan, P. H., Lazarus, A. B., Brock, C. D., Wernegreen, J. J. (2004) Host-symbiont stability and fast evolutionary rates in an ant-bacterium association: Cospeciation of *Camponotus* species and their endosymbionts, *Candidatus Blochmannia*. *Systematic Biology*, Volume 53, page 95.
4. Woolfit, M., Bromham, L. (2003) Increased rates of sequence evolution in endosymbiotic bacteria and fungi with small effective population sizes. *Molecular Biology and Evolution*, Volume 20, page 1545.
5. Lynch, M. (2007) *The origins of genome architecture*. Sinauer Associates.

CASE STUDY 10.2

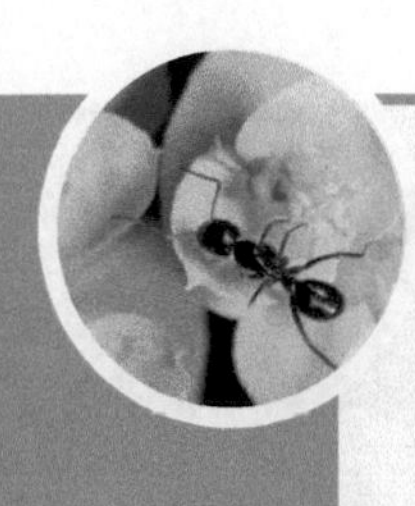

# Horizontal gene transfer: genes in parasitic plants derived from their hosts

**RELATED TECHBOXES**

TB 5.2: High-throughput sequencing

TB 12.2: Bootstrap

**RELATED CASE STUDIES**

CS 12.2: Prediction

CS 13.2: Diversification

Xi, Z., Wang, Y., Bradley, R. K., Sugumaran, M., Marx, C. J., Rest, J. S., Davis, C. C. (2013) Massive mitochondrial gene transfer in a parasitic flowering plant clade. *PLoS Genetics*, Volume 9: e1003265

> *"Now that it is clear that plant mitochondria exchange genes relatively frequently, caution is necessary when interpreting plant phylogenies from one or even a few mitochondrial genes as they may not reflect the underlying organismal phylogeny[1]"*

**Keywords**

HGT, phylogeny, maximum likelihood, bootstrap, Approximately Unbiased (AU) test, transgene, mitochondria, synteny

## Background

Genes are normally inherited vertically, from parent to offspring. But sometimes a gene can be gained horizontally, transferred between unrelated individuals. Horizontal gene transfer (HGT) was once thought to be a rare phenomenon, or confined to unicellular organisms. But evidence is growing that it is a surprisingly common feature of evolution, even for many multicellular lineages. HGT is often detected by finding unexpected relationships between genes that do not match the pattern of similarities predicted from the relationships between the species. For example, it was noticed that in molecular phylogenies based on the mitochondrial *nad1* gene, the parasitic plant *Rafflesia* clustered with its host plant *Tetrastigma* rather than with its own relatives[2]. The detection of other anomalous relationships lent support to the hypothesis that the intimate relationship between parasite and host tissues promoted horizontal gene transfer between species[3, 4].

## Aim

*Rafflesia* produces the largest known flowers, up to a metre in diameter and ten kilograms in weight (**Figure 10.22**). By comparison, the rest of the plant is insubstantial: no leaves, stems or roots[5]. This is because *Rafflesia* is entirely parasitic on a vine (*Tetrastigma*) so its body consists of a fungus-like haustorium that grows within its host's tissues. Previous studies have identified possible cases of horizontal gene transfer between *Rafflesia* and *Tetrastigma* for both mitochondrial and nuclear genes. This study used mitochondrial genome sequences from parasitic plants and their hosts to estimate the prevalence of HGT.

## Methods

Since parasites grow inside their host's tissues, care must be taken to avoid cross-contamination. DNA was extracted from different samples of three closely related parasitic plants (*Rafflesia cantleyi*, *Rafflesia tuanmudae*, and *Sapria himalayana*). High-throughput sequencing was used to sequence the mitochondrial genomes which were assembled *de novo* (that is, using the overlap between sequence fragments: **TechBox 5.2**). Assembling the entire mitochondrial genome was deemed to be too difficult due to repetitive sequences, but 38 genes were identified by comparison to assembled mitochondrial genomes from other plant families. The genes were present in high copy numbers, consistent with mitochondrial origins (since mitochondrial genomes are present in hundreds or thousands of copies per cell). They also sequenced the same 38 genes from three members of the grapevine family: *Tetrastigma rafflesiae* (the host of

Figure 10.22 *Rafflesia* flowers take up to 12 months to grow then flower for less than a week. Because they are unisexual, a flower will only be pollinated if a flower of the opposite sex flowers within fly-flight distance within that 5–7 day window. Given that *Rafflesia* are rare (and getting rarer), this is fairly unlikely. I was once staying in a village in Sumatra when word got round that there was a *Rafflesia* in flower. We hiked up the rim of the crater lake with a group of locals who were as keen to see it as we were. It was a magnificent beast of a flower and, despite a reputation for stinking like a rotting buffalo carcass in order to attract flies to act as pollinators, the *Rafflesia* we saw smelled just fine.

Photograph: Lindell Bromham.

*R. cantleyi* and *R. tuanmudae*), *Tetrastigma cruciatum* (host of *S. himalayana*) and the more distantly related *Leea guineensis*. They also got sequences for the same genes from 27 species from across the angiosperms and included all sequences in a phylogenetic analysis using maximum likelihood. They used measures of phylogenetic support to detect potential cases of HGT by identifying any gene placed outside its family of origin with a bootstrap value of greater than 70 per cent (see **TechBox 12.2**). This bootstrap probability (BP) test was supplemented by an Approximately Unbiased (AU) test[6], which is a multiscale bootstrap test that adjusts for the sample size of trees and sequences compared.

**Results**

Of the 38 genes tested, 21 could be placed in the phylogeny with a high degree of statistical support. But of these well-supported groupings, up to half of them (between 5 and 11 genes) placed the parasitic plant sequence outside the Rafflesiaceae. Strong support for the placement of a gene outside the lineage it was sampled from was interpreted as evidence that those sequences had been derived from an unrelated lineage by horizontal gene transfer. In most cases, the identified donor lineage was the host plant (*Tetrastigma*) or its relatives, but in several cases the inferred donor was from a different lineage. Note that three of the genes from autotrophs (non-parasitic species) were also placed outside of accepted relationships with >70 per cent support. For each of the putative cases of HGT, the transferred gene either replaced an existing gene in the parasite's mitochondrial genome, or the species had both host and parasite copies of the gene. There were no cases of novel genes acquired through HGT. The putative cases of HGT identified maintained gene synteny, as they were found in the same places on the chromosome as the local copy of the gene (that is, they were not randomly inserted somewhere else in the genome). Most of the putative transgenes had intact reading frames and were found in the transcriptome of *Rafflesia cantleyi*, suggesting that they are transcribed.

**Conclusions**

In this phylogenetic analysis, at least a quarter of the mitochondrial genes in these three parasitic plant species group with other angiosperm lineages, particularly with the host plant *Tetrastigma*. This was interpreted as a sign of substantial amounts of horizontal gene transfer between mitochondrial genomes of different plant species. They conclude that genes are typically transferred by close physical contact, allowing homologous recombination that either exchanges the gene copy in the genome for a transferred copy or inserts a new copy next to the native gene.

### Limitations

Without assembling the whole mitochondrial genome, they can't be sure that these genes are all mitochondrial in origin[7]. It is important to consider the bootstrap percentages in context: 70 per cent bootstrap support means that 30 per cent of phylogenies estimated from the resampled data did not support that grouping. Furthermore, bootstrap test is dependent on the models of molecular evolution used[8]. Parasitic plants tend to have faster rates of molecular evolution in the nuclear, chloroplast and mitochondrial genomes which can complicate the phylogenetic placement of parasitic taxa[9]. An unexpected phylogenetic grouping is not necessarily the sign of horizontal gene transfer. For example, discordance between phylogenetic trees estimated from different genes has been taken as a sign of recent hybridization in species of parasitic dodder (*Cuscuta*)[10], but could it be due to rate variation, horizontal gene transfers or gene rearrangements?

### Further work

A formal hypothesis testing approach, asking whether it is possible to reject vertical descent as an alternative explanation for the patterns of similarity in the sequence data, could be used to examine the effect of assumptions of the analysis on the conclusions reached. Such a test could be devised to take into account the higher rates of molecular evolution in parasitic taxa. Ideally, it would be good to use more comprehensive sequencing of both the mitochondrial and nuclear genomes to determine the number of different transfer events, and the direction of transfer between parasite and host. Repeating this analysis for other parasitic plants and their hosts would test whether horizontal gene transfer is a common feature of the parasitic lifestyle in plants.

### Check your understanding

1. Why do parasitic plants seem to undergo high rates of horizontal gene transfer?
2. How were bootstrap values used to identify cases of horizontal gene transfer?
3. What effect would cross-contamination of host and parasite tissue have on this analysis?

### What do you think?

When is it fair to conclude that strange groupings on phylogenies are evidence of horizontal gene transfer rather than sequence error or an artefact of analysis? How would you test whether a phylogenetic grouping that does not seem to reflect descent is a sign of HGT?

### Delve deeper

Many reported cases of horizontal gene transfer in plants involve mitochondrial genes. Why?

### References


1. Richardson, A. O., Palmer, J. D. (2007) Horizontal gene transfer in plants. *Journal of Experimental Botany*, Volume 58, page 1.
2. Nickrent, D., Blarer, A., Qiu, Y-L., Vidal-Russell, R., Anderson, F. (2004) Phylogenetic inference in Rafflesiales: the influence of rate heterogeneity and horizontal gene transfer. *BMC Evolutionary Biology*, Volume 4, page 1471.
3. Davis, C. C., Wurdack, K. J. (2004) Host-to-Parasite gene transfer in flowering plants: phylogenetic evidence from Malpighiales. *Science*, Volume 305, page 676.
4. Mower, J. P., Stefanović, S., Young, G. J., Palmer, J. D. (2004) Plant genetics: gene transfer from parasitic to host plants. *Nature*, Volume 432, page 165.
5. Nikolov, L. A., Tomlinson, P., Manickam, S., Endress, P. K., Kramer, E. M., Davis, C. C. (2014) Holoparasitic Rafflesiaceae possess the most reduced endophytes and yet give rise to the world's largest flowers. *Annals of Botany*, Volume 114, page 233.

6. Shimodaira, H. (2002) An approximately unbiased test of phylogenetic tree selection. *Systematic Biology*, Volume 51, page 492.
7. Molina, J., Hazzouri, K. M., Nickrent, D., Geisler, M., Meyer, R. S., Pentony, M. M., Flowers, J. M., Pelser, P., Barcelona, J., Inovejas, S. A., Uy, I., Yuan, W., Wilkins, O., Michel, C-I., LockLear, S., Concepcion, G. P., Purugganan, M. D. (2014) Possible loss of the chloroplast genome in the parasitic flowering plant *Rafflesia lagascae* (Rafflesiaceae). *Molecular Biology and Evolution*, Volume 31, page 793.
8. Buckley, T. R., Simon, C., Chambers, G. K. (2001) Exploring among-site rate variation models in a maximum likelihood framework using empirical data: effects of model assumptions on estimates of topology, branch lengths, and bootstrap support. *Systematic Biology*, Volume 50, page 67.
9. Bromham, L., Cowman, P. F., Lanfear, R. (2013) Parasitic plants have increased rates of molecular evolution across all three genomes. *BMC Evolutionary Biology*, Volume 13, page 126.
10. Stefanović, S., Costea, M. (2008) Reticulate evolution in the parasitic genus *Cuscuta* (Convolvulaceae): over and over again. *Botany*, Volume 86, page 791.

# Phylogeny

## Tree of life

*"The time will come I believe, though I shall not live to see it, when we shall have very fairly true genealogical trees of each great kingdom of nature"*

**Darwin, C. (1857) Letter to Thomas Henry Huxley, 26 September 1857.**

## What this chapter is about

As populations divide and diverge, their DNA sequences become increasingly different from each other. DNA sequences sampled at the end point of this process carry the historical signal of population divergence. We can use alignments of DNA sequences to draw evolutionary trees that display the similarities between related lineages and the paths of descent of species. Sometimes the patterns in the sequences reveal a hierarchical history of populations dividing again and again. Because of this, we can use patterns of differences between DNA sequences from contemporary populations to reconstruct their evolutionary history. But sometimes the patterns are more complicated, if the historical signal has been erased, or when previously distinct lineages join back together through hybridization or gene transfer. We can represent many different histories in a network diagram.

### Key concepts

- Phylogenies represent a history of population divergence that leaves a hierarchy of differences in DNA sequences
- Patterns of similarities between these sequences can reveal the population history

# History

*The Origin of Species* was a best seller. The first edition sold out on the day of publication and the second edition sold out six weeks later. *The Origin* went through six editions in Darwin's lifetime, and has remained in print ever since. The phenomenal success of *The Origin* is certainly not due to lavish illustration, because there is only a single black and white picture in the whole book. Not a drawing of a fancy pigeon, not a sketch of a fossilized giant ground sloth, not even a portrait of the author. The sole illustration in *The Origin* is a simple line drawing (**Figure 11.1**), showing the diversification of an imagined set of species.

In order to explain the diversity of life, Darwin had to demonstrate not just how a species could be modified by natural selection but how this process could give rise to a great diversity of organisms. To do this, he needed to show how natural selection could produce two separate species from a single stock. The one and only diagram in *The Origin* depicts the following evolutionary story to illustrate the process of diversification: eleven related species (A to L) each occupy a particular habitat (**Figure 11.1**). Each species is a group of individuals that vary slightly from each other. Through the process of natural selection, advantageous variations tend to be preserved and propagated. Darwin suggested that variations at the extremes of a distribution—those that are most different from the rest of the population—will often be those that flourish: 'the more diversified the descendants from any one species become in structure, constitution and habits, by so much will they be better enabled to seize on many and widely diversified places in the polity of nature, and so be enabled to increase in numbers'. Eventually sufficient variation accumulates in these divergent lineages to warrant them being called varieties or species. After a long interval of time has elapsed (at point XIV in the diagram), some lineages have gone extinct (such as D), some have persisted without change (such as F), and some have given rise to many descendant lineages (A gave rise to 8 new lineages). The net result is that there are more lineages at time XIV, and they occupy a greater range of variation. This process could have taken place over any time-scale—each horizontal line might represent one thousand generations, or one million generations, or successive geological strata. And the lineages could represent any level of biological organization, varieties, species, genera, families and so on. Darwin proposed that the process of diversification is essentially similar for all.

*Natural selection is discussed in Chapter 7*

Diversification is a fundamental process of evolution. It is usually represented by a branching diagram, referred to as an evolutionary tree or phylogeny. Although we

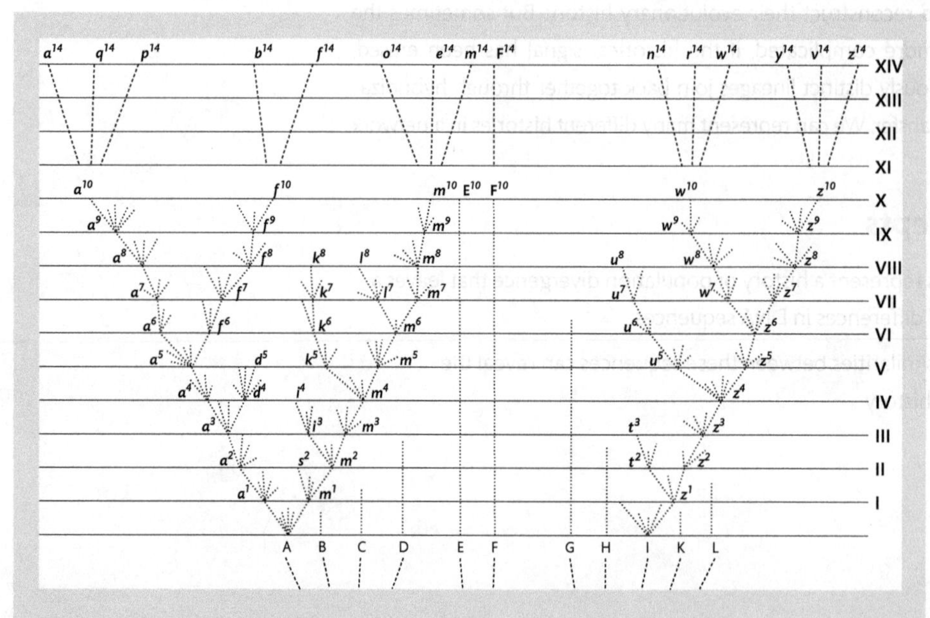


**Figure 11.1** Darwin's tree, the only illustration in *The Origin of Species*.

might be tempted to think of Darwin's one and only diagram as the first modern evolutionary tree, people had been drawing branching diagrams to represent biological diversity for centuries. This is because scientists had long recognized that biological diversity has a hierarchical nature, and they used this observation to create logical and useful classifications. For example, long before the discovery of evolution, the diversity of bird species had been variously divided into nested categories (forgive the pun). Characteristics used to divide bird species into natural groups included morphology (e.g. Pliny used the shape of the foot), behaviour (e.g. Aristotle divided birds into those that took dust baths or water baths or both or neither) and ecology (e.g. ground birds, river birds, etc.).

*The hierarchical system of taxonomy used today is outlined in Chapter 9*

The pioneering systematist John Ray produced an ornithological classification in 1676 that resembles modern taxonomy in that species are arranged in a hierarchical list. This list is structured by using a key characteristic to divide birds into groups, then another character chosen to divide each of those groups into smaller groups, and so on (**Figure 11.2**). Ray divided all birds into land and water birds; water birds were divided into those that swim and those that feed by water; those that swim were divided into those with cloven (separate-toed) feet or whole (webbed) feet; those with whole feet were divided into those with short or long legs; the short-legged ones were divided into those with three toes or four toes, and so on. So a pelican is grouped with other birds with four connected toes, which are then grouped with a larger set of four-toed birds, and these are grouped with all of the short-legged aquatic birds. All the short-legged birds are then grouped with all the whole-footed aquatic birds, which are then grouped with all aquatic birds, and so on upwards until we have a group of all birds (**Figure 11.2**).

Branching diagrams are a logical way to represent a hierarchical structure, and you will see them in many non-biological situations: for example, charts illustrating levels of governance in organizations such as universities. But the discovery of evolution provided an explanation of why biodiversity, in particular, has a hierarchical organization. And this is what makes Darwin's diagram from *The Origin* special. It does not just describe the hierarchical nature of biological diversity, it provides an explanation for it. Descent with modification gives rise to a nested hierarchy of individuals, populations, species and lineages, in which close relatives tend to be more similar to each other than more distant relatives. Darwin's tree represents the process of diversification which leaves a nested pattern of similarities and differences.

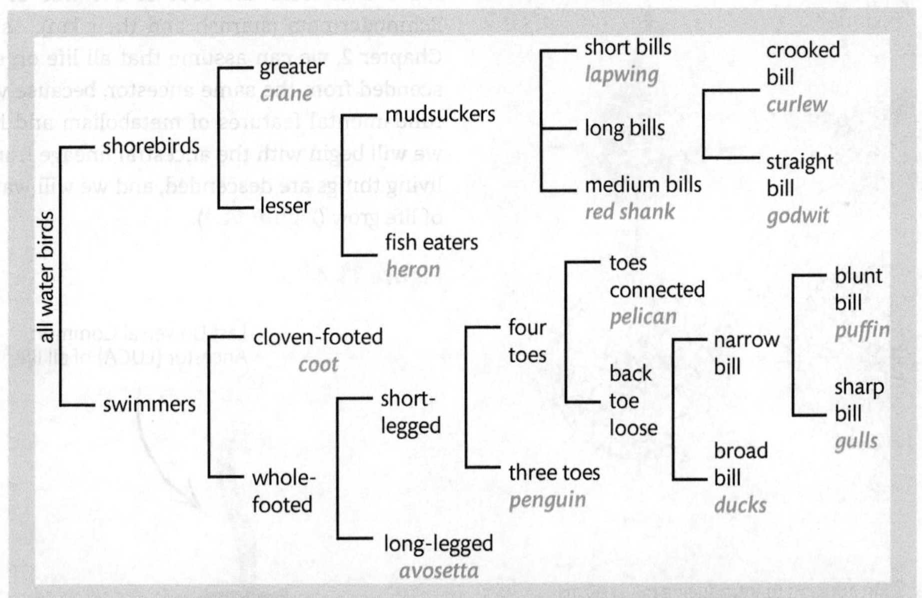


**Figure 11.2** John Ray used nested sets of shared characteristics to divide species into groups containing ever-more similar species. Part of his hierarchical classification of birds, published with his student Francis Willughby in 1676, is shown as a branching diagram here, with groups of birds labelled in black, and specific examples in blue. Actually, Ray wrote out his classifications in the same manner as identification keys today, as a table of nested categories, not in tree-like branching diagrams.

Based on information given in Bedall (1957).

## Life as a tree

Phylogenies are a way of organizing ideas about evolution. We can read the information in phylogenies in several different ways. We can use phylogenies to display biological relationships: lineages most closely related to each other will be more closely connected on a phylogeny, because they share a more recent common ancestor. Phylogenies also represent a record of the evolutionary past: we can use the connected paths of the phylogeny to reconstruct the evolutionary history of a lineage, as populations divided and diverged. More generally, phylogenies allow us to generate and test hypotheses about the patterns, processes and causes of diversification.

The word 'phylogeny', meaning evolutionary tree, was coined by Ernst Haeckel, a naturalist famous for his exquisite drawings of microscopic creatures (**Figure 11.3**).

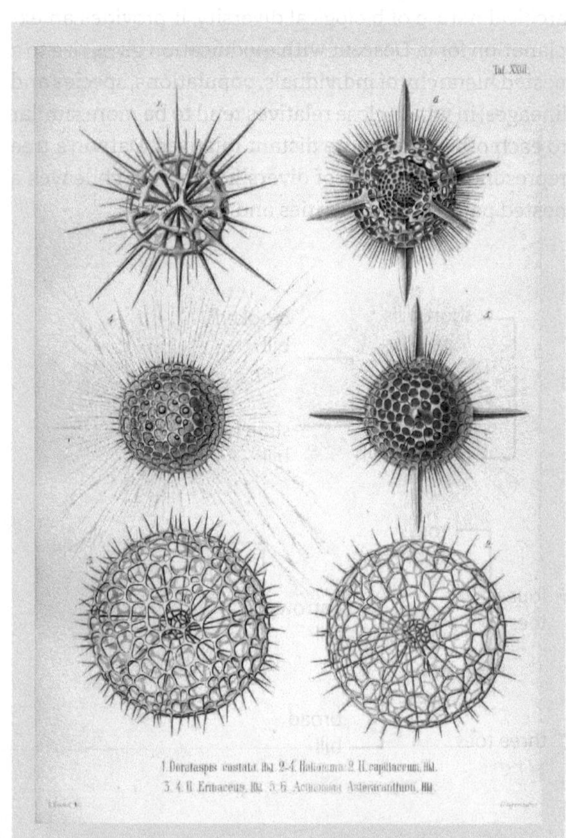

**Figure 11.3** In addition to introducing many terms we use today, such as phylum, phylogeny and ecology, Ernst Haeckel produced beautiful taxonomic illustrations. These drawings were not simply aesthetic, they recorded important information about species that, at that time, could not be photographed. This is an illustration from Haeckel's *Die Radiolarien* published in 1862.

He saw that the detailed drawings he had made reflected patterns of similarity that could be explained by Darwin's theory of diversification from a common origin. Perhaps more than any other scientist, Haeckel is responsible for embedding the iconography of the evolutionary tree into modern biology. For not only did Haeckel describe the relationships between living groups using branching diagrams, he drew these branching diagrams as actual trees (**Figure 11.5**).

If you could watch a tree grow from seed, you would see it start as a single tiny stem, which then branches into two thin twigs, and as the twigs grow thicker they branch to form more twigs and so on. When you look at a tree, the particular pattern of branches and twigs represents the historical process that gave rise to the tree. Instead of seeing the tree as a static object, we can consider that the tree we see now represents a time series of branching events, from the original stem (now represented by the trunk from which all other branches arise) to the most recently branched twigs. We could follow the series of branching events in two ways: from the bottom up (from the past to the present) or from the top down (from the present to the past).

### *Reading the tree of life*

As an illustration, let's follow the series of branching events from the root of the tree of life to the Echinodermata (starfish and their kin). As we saw in Chapter 2, we can assume that all life on earth is descended from the same ancestor, because we all share fundamental features of metabolism and heredity. So we will begin with the ancestral lineage from which all living things are descended, and we will watch the tree of life grow (**Figure 11.4**).

**Figure 11.4**

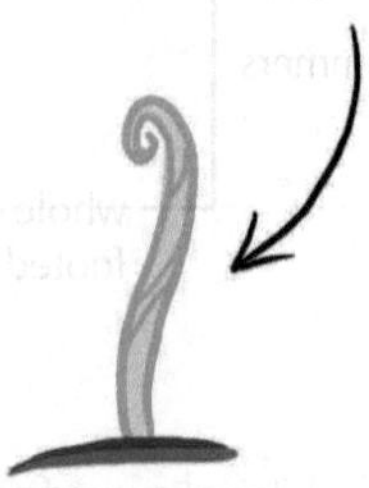


This ancestral lineage split into the lineages that would go on to form the basis of the major kingdoms of life. Of course, at the time they split from the common stock, they would have been very similar to each other, acquiring their different characteristics over time as they diverged. So

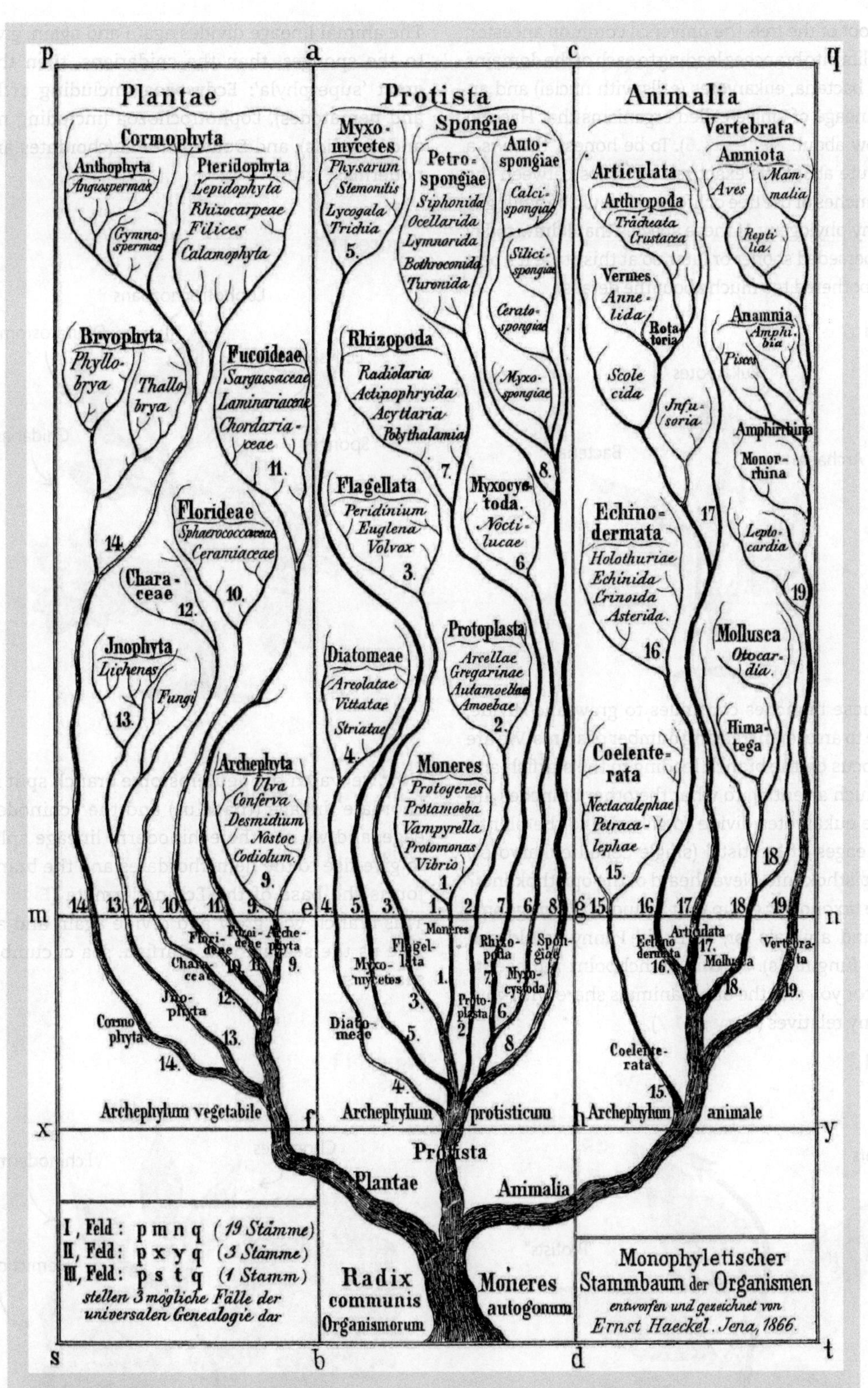


**Figure 11.5** One of Ernst Haeckel's evolutionary trees. Most of the relationships depicted here have since been revised, and continue to be revised. There is an ongoing and vigorous debate about the relationships at the deeper levels of the tree of life.

Ernst Haeckel, 1866.

from the root of the tree, the universal common ancestor, the tree splits into branches leading to each of the domains of life, the bacteria, eukaryotes (cells with nuclei) and archaea (a lineage of single-celled organisms that Haeckel didn't know about: Figure 11.6). To be honest, there is a lot of dispute about the exact relationships between the deeper branches of the tree of life, and I would be willing to bet that any phylogeny of these species that I draw today will be superseded sooner or later, so at this stage it's best not to get bothered too much about the details.

**Figure 11.6**

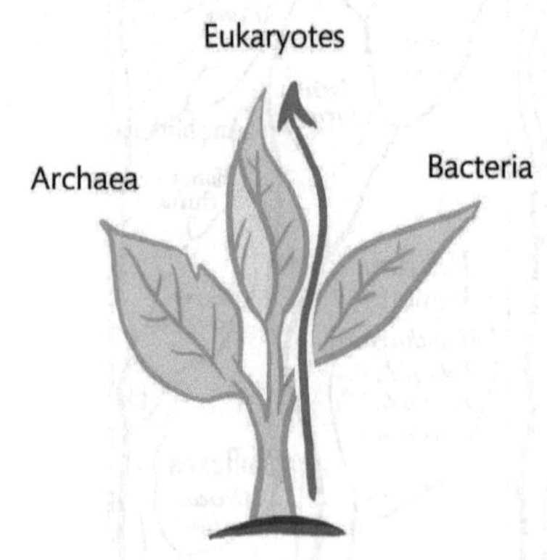


Each of these branches continues to grow and divide, giving rise to an ever increasing number of stems. We are going to focus on the branch leading to the starfish and not pay much attention to what the other branches are doing. The eukaryotes divide to give rise to the plants, lots of lineages of 'protists' (single-celled eukaryotes) and the Opisthokonta. Never heard of the opisthokonts? This is the taxonomic group that includes the kingdoms of fungi and animals (or, as David Penny would call them, the fungimals). So this branchpoint represents the ancestor you and the other animals share with your mushroomy relatives (Figure 11.7).

**Figure 11.7**

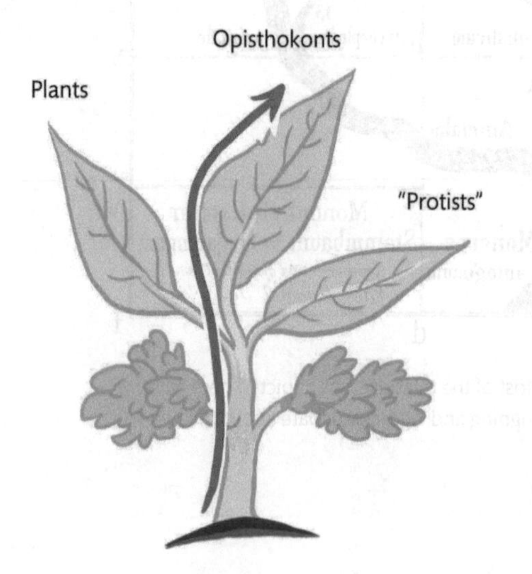


The animal lineage divides again and again, giving rise to the sponges, then the cnidarians, then the three great 'superphyla': Ecdysozoa (including arthropods and nematodes), Lophotrochozoa (including molluscs and annelids), and Deuterostoma (chordates and echinoderms: Figure 11.8).

**Figure 11.8**

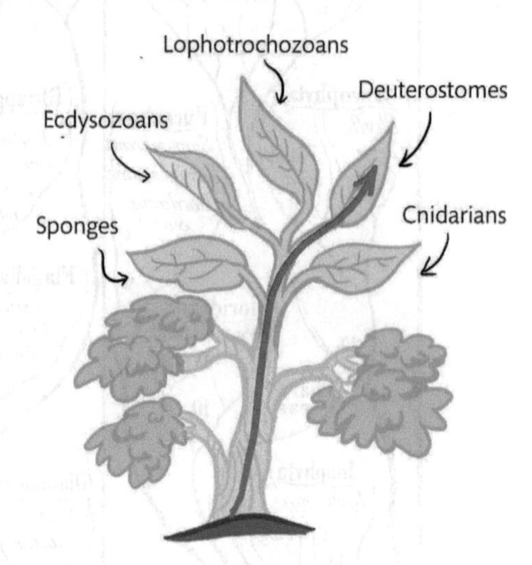


Now we watch the deuterostome branch split into the chordate lineage (that's us) and the echinoderm lineage, and we see the echinoderm lineage split again to give rise to the hemichordates and the branch that forms the base of the Echinodermata (Figure 11.9). This branch will grow and divide again and again to give us the sea urchins, starfish, sea cucumbers and sea stars.

**Figure 11.9**

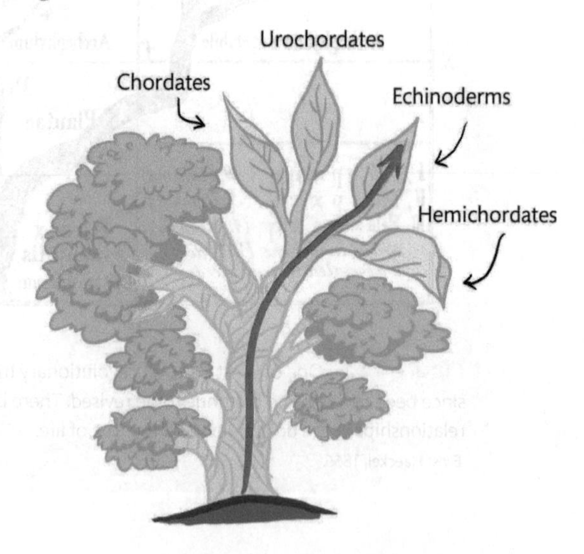

The point of all that is to remind ourselves that a tree that is a static object (a set of branches) can also be interpreted as a record of history (a series of branching events). In phylogenetic parlance, the point at which a branch divides to give rise to two or more new branches is called a node. We could trace the history of any twig at the tip of the tree by following it down along branches and traversing nodes to eventually reach the root of the tree where it all began. Just for fun, you can try tracing the same journey, from the root to the tip of the echinoderms, on Haeckel's tree in **Figure 11.4**. Do you pass through the same set of branching points? If not, which are different?

Rarely do we watch a tree grow from root to tip. Most of the time when we look at phylogenies, we are considering present-day species arrayed on the tips and tracing their origins back through past branches and nodes. So let's start at one of the tips of Haeckel's tree and trace its evolutionary path back to the root. In his book *The Ancestor's Tale*, Richard Dawkins used the analogy of following a path back in evolutionary time and meeting the ancestors at various cross roads (nodes in the tree), thus being joined by a growing band of evermore distant relatives who travel down their own lineages to meet you at the common ancestral nodes. Dawkins began his evolutionary journey with *Homo sapiens*, but let's be a bit more exotic. We will begin our journey from the tip of Haeckel's tree to the root with the single-celled *Vampyrella* (**Figure 11.10**).

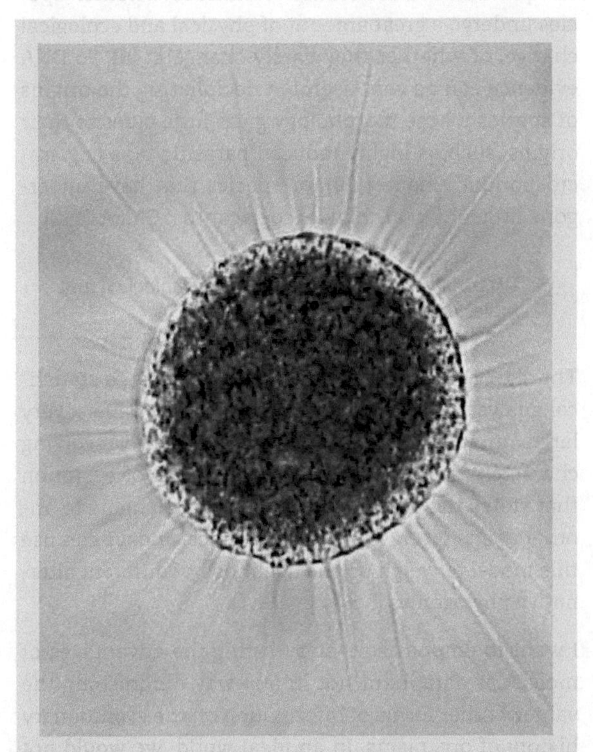

**Figure 11.10** *Vampyrella* is a protist (single-celled eukaryote) that lives in bog-pools. Its name is presumably derived from its habit of stabbing hapless algal victims with its spines then sucking out their chloroplasts.
Image courtesy of Giuseppe Vago, licensed under the Creative Commons Attribution 2.0 Generic license.

In Haeckel's tree, *Vampyrella* groups with a strange collection of single-celled beasties that he terms the Monera. This taxonomic group is now essentially defunct. Haeckel's Monera included organisms now thought to be only distantly related, such as amoebae, *Vibrio* (bacteria of such diverse habits as causing human cholera and making jelly-fish glow: **Case Study 11.1**) and *Protomonas* (slime nets, once thought to be fungi due to their saprophytic habits, some of which have evolved to use their mycelium-like nets for ectoplasmic gliding). These diverse organisms would not be grouped together today, but we are using Haeckel's tree as an illustration of how we derive information on branching points, rather than the latest word on systematics.

Travelling down the tree from *Vampyrella* (**Figure 11.11**), the first node we meet is the branching point between the moneres and the flagellates (single cells that swim with whip-like tails: node *b*), then the combined moneres + flagellates lineage is joined by the slime-mould lineage (myxomycetes: node *c*), then by all of the other protist lineages (node *d*, including the sponges, which today are considered to be in the animal kingdom), and then the diatoms (node *e*). As we travel down the tree, each node connects to another related lineage. We keep collecting progressively more distantly related lineages until we arrive at the root, the ancestor shared by all the lineages in the tree (node *g*).

## Molecular phylogenetics

Phylogenies can be based on any information that reveals similarity or descent. Haeckel's phylogeny is based on his observations of the morphology and development of living species. Some phylogenies use palaeontological information to reconstruct the relationships between extinct species (**Figure 11.12**). But, of course, this book is about the way we use DNA sequences to reconstruct evolutionary history, so naturally that is what we will focus on here.

Molecular data have a number of important advantages for estimating phylogeny. Firstly, the genome provides a unifying framework for estimating phylogeny, because DNA sequences can be compared between all species. By

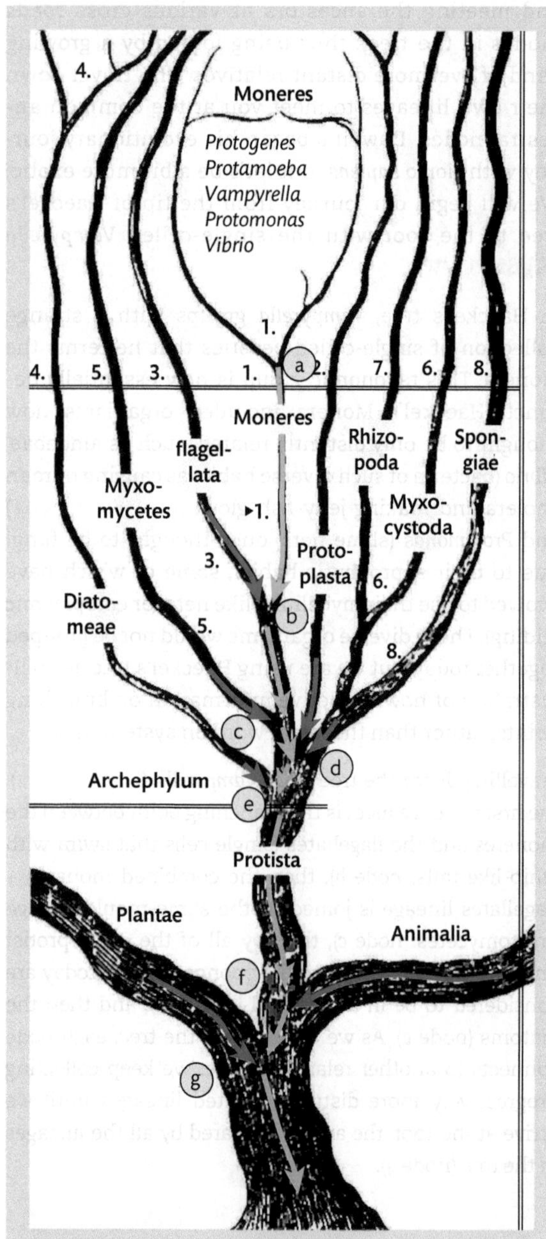


**Figure 11.11** Travelling down Haeckel's tree, from *Vampyrella* to LUCA, the direct line of descent (orange arrows) joins other related lineages (red arrows) at nodes in the tree (branching points where lineages split).
Based on a diagram by Ernst Haeckel, 1866.

choosing appropriate sequences, DNA can provide information at all depths of evolutionary divergence. As we saw in Chapter 2, all living species share the same genomic system. All genomes are written in the directly comparable language of DNA (or RNA) bases, so DNA sequences can be compared between all organisms, from individuals (Case Study 1.2) to populations (Case Study 9.1) to species (Case Study 9.2) to genera and families (Case Study 2.2), all the way up to kingdoms. For closely related lineages, we compare rapidly changing sites of the genome (e.g. using microsatellites to track paternity: Chapter 5). For distantly related organisms we will use genes for highly conserved enzymes (e.g. comparing RNA polymerase II beta across different kingdoms: Chapter 2). There is something to be compared between all living organisms when we look at the level of the genome. Morphological characteristics are less easily compared between groups: the phenotypic characteristics that are informative for one set of organisms may be meaningless for another. The phylogeny of dinosaurs in Figure 11.12 was generated by analysis of the similarities and differences in fossil bones: for example, skull shape, leg bone dimensions, and so on. We could not use the same set of characteristics to classify starfish (that have no bones) or lungfish (that have different bones).

Secondly, DNA provides a record of evolutionary history independent of many other sources of historical information, such as the fossil record, comparative morphology, or biogeography. DNA sequences are therefore very useful for uncovering the evolutionary history of organisms that have no fossil record, such as viruses (Case Studies 5.2 and 12.1). Furthermore, the process of substitution continues whether species undergo a great amount of physical and ecological change, or whether they barely change at all. So DNA evidence can be very useful in deciphering the origins of species whose morphology gives little clues to their origins, such as highly reduced parasitic taxa (e.g. microsporidia: Chapter 10), or species that have undergone little physical change (coelacanths: Chapter 13).

*We consider the decoupling of morphological and molecular evolution in Chapter 14*

Thirdly, DNA sequence data is ideally suited to statistical analysis. Even the smallest genome contains a very large number of essentially independently evolving characteristics. We can describe a model of evolution that states the probability of one base changing to another at any given site in the sequence, and we can use this model to weigh up the likelihood of different alternative phylogenies (TechBox 13.1).

I want to emphasize that by listing the advantages of molecular data, I am not in any way diminishing the value of other forms of information on the evolutionary history of organisms. In an ideal world, we would not rely on DNA evidence alone to build evolutionary trees, for a number of reasons. Although DNA (or RNA) is a universal feature of life on earth, sequences are not available for all species. In particular, DNA is not available for most extinct species. Phylogenies based on DNA are not always an accurate representation of the history of those species. This might be because not all genes in the

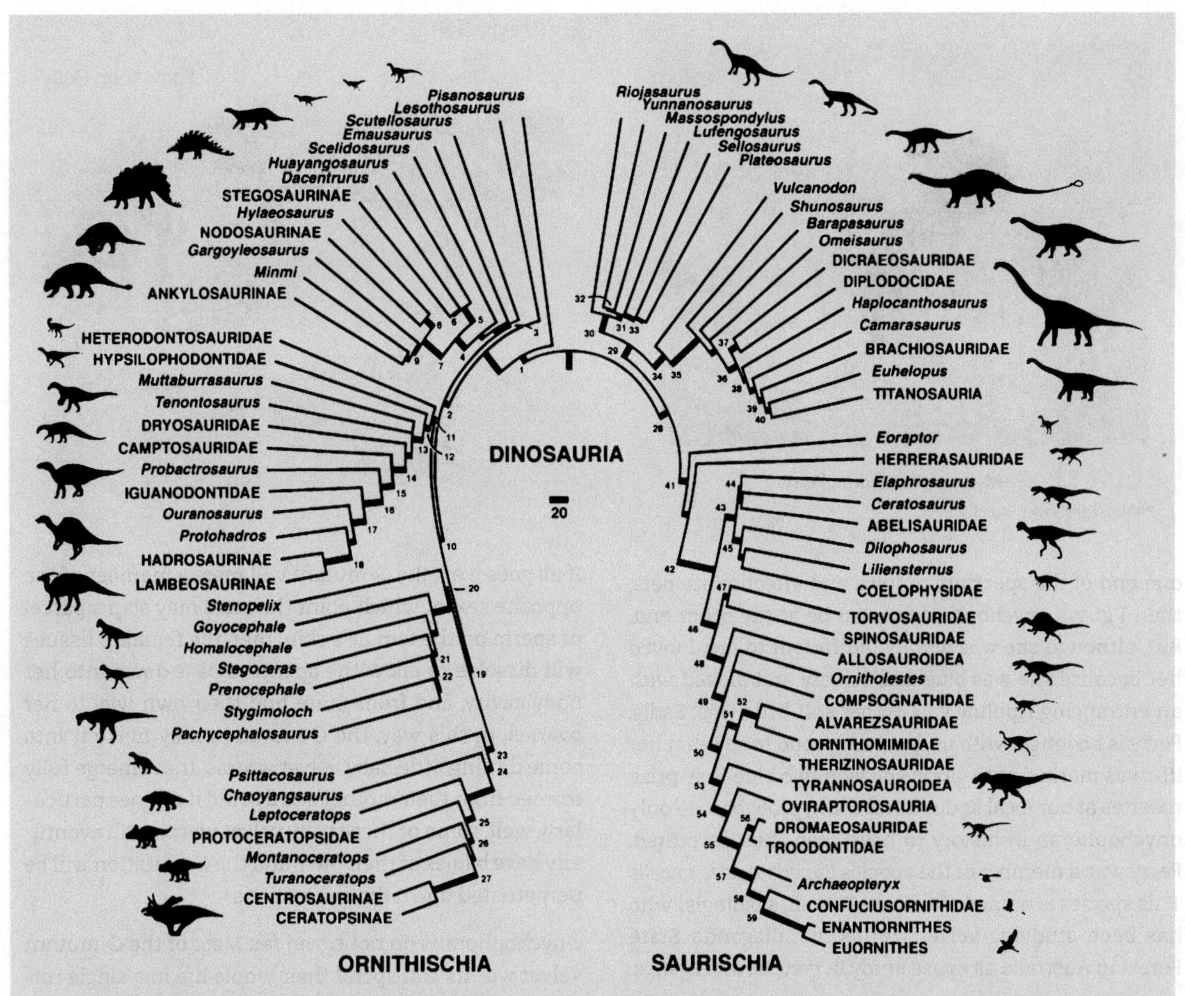


**Figure 11.12** This phylogeny of dinosaurs was constructed by palaeontologist Paul Sereno, using parsimony analysis of hundreds of skeletal characteristics, each coded using single digits to represent character states. For example, one of the characteristics used to analyse the Ornithischia (a group including the ceratopsids and stegosaurs) is 'Premaxillary tooth number' which in this dataset has five possible states (3, 4, 5, 6 or 7 teeth).

From Sereno PC (1999) The Evolution of Dinosaurs. *Science*, Volume 284(5423), pages 2137–2147. Reprinted with permission from AAAS.

genome share the same history (e.g. **Case Study 10.2**) or it may simply be due to uncertainty in phylogeny construction. It is quite common for two different research groups to publish contradictory molecular phylogenies for the same group of species. How can you tell which one is correct? Where possible, evolutionary hypotheses generated using molecular data should be compared to those derived from other sources of information, such as morphology, biogeography and fossil evidence (e.g. **Case Studies 4.2** and **14.1**). Finally, although we can derive a great deal of information from molecular phylogenies, they cannot tell us everything. All the DNA sequences in the world will not tell you that there were once pterodactyls with ten-metre wingspans. However, while previous decades were witness to a war of 'molecules vs morphology', it is increasingly the case that DNA provides the main source of phylogenetic information for an ever-growing number of taxa.

## *Substitutions track relationships*

I was once fortunate enough to have a pet onychophoran (**Figure 11.13**). Onychophora is an enigmatic group of caterpillar-like animals commonly known as velvet worms (or peripatus). They are beautiful little creatures, but, admittedly, they do not make very responsive pets. My velvet worm, Perry, lived in a lunchbox in the fridge and sulkily refused to eat the termites that I lovingly provided for her. The closest she ever came to interacting with me in any meaningful way was once, when I surprised her, she spat glue at me (velvet worms are predators and they hunt by shooting a jet of superglue at their prey to immobilize them). If labradors are at

**Figure 11.13** My pet velvet worm, Perry.
Photograph: Lindell Bromham.

one end of the spectrum of loyal and affectionate pets, then I guess onychophorans must be at the other end. But, although she was wholly indifferent to me, I loved her because she was blue and velvety and moved with an entrancing rippling gait on her soft little legs. Sadly, Perry is no longer with us, but I am proud to say that her life was marked with glory: she won two blue first-prize rosettes at our local agricultural show, probably the only onychophoran in history to have been thus decorated. Perry was a member of the species *Euperipatoides rowelli*. This species is named after Dave Rowell, a biologist who has been studying velvet worms in Tallaganda State Forest in Australia as a case study in genetic divergence.

We are going to imagine a population of velvet worms living in a mythical part of Tallaganda to allow us to follow the process of genetic divergence as populations divide and diversify over time. We will picture the velvet worms living in a relatively small area, on the forested ridge of a low hill with tree-lined gullies on either side, one gully dominated by Scribbly Gums (*Eucalyptus rossii*) and one by Snow Gums (*Eucalyptus pauciflora*). A human would walk through this whole area in minutes, but it represents a vast distance to a tiny, soft-footed velvet worm. We will track just one tiny part of that genome, a six base sequence (and for the sake of a simple story, we will also ignore the fact that onychophorans are diploid, and focus on only one haploid allele). Like most of the genome, this sequence is the same for most members of the population. But it just so happens that one of our onychophorans is a mutant, because its fifth base is a G instead of a C (**Figure 11.14**). By chance, this mutation is neutral. It has no effect on its carriers' chances of reproduction.

***Neutral alleles, that make no difference to fitness, are discussed in Chapter 8***

**Figure 11.14**

If all goes well, the G-mutant will meet a member of the opposite sex, at which point the male may slap a parcel of sperm on the female's skin, then the female's tissues will dissolve to allow the sperm to pass down into her body cavity, and from there find their own way to her ovaries. In this way, the G-mutation may make it into some darling little baby velvet worms, that emerge fully formed from their mother's body. And if all goes particularly well, some of these baby velvet worms will eventually have babies of their own, and the G-mutation will be perpetuated down the generations.

Onychophorans do not travel far. Most of the G-mutant velvet worms will spend their whole life in a single rotting log. Their soft bodies lose moisture quickly so they risk desiccation if they attempt to cross the open spaces between logs. But when it is wet, some brave individuals venture forth across the forest floor in search of new logs to colonize. If one of our mutants makes it to a new log, it will bring its G-allele with it, and if it finds a mate and successfully reproduces, then that G-allele can spread throughout a new rotting log. In this way, generation by generation, log by log, a new mutation can spread throughout the forest. Thus an interconnected population shares a common pool of alleles.

However, one summer a fire burns through the forest, torching the forests along the ridge but leaving the wet forest in the gullies untouched (**Figure 11.15**). Velvet worms survive in the unburnt gullies but are unable to cross the dry ridges to get to the neighbouring gully. Now, a velvet worm from Scribbly Gum will never get close enough to a velvet worm from Snow Gum to exchange gametes. This means that the G-allele that is found in Snow Gum Gully is never going to end up in the offspring of a Scribbly Gum Gully velvet worm. The populations have become genetically isolated from each other.

**Figure 11.15**

Over time it happens that, by chance, the G goes to fixation in the Snow Gum population (**Figure 11.16**). A new mutation, from A to T, arises in the Scribbly Gum population. This new mutation affects the production of a pheromone. Pheromones are chemicals that influence the behaviour of other individuals (examples include the chemical trails laid by ants, scent marking of territories by carnivorans, or mating pheromones released by moths). Some onychophorans release pheromones that attract other members of their species. The change from an A to a T in this pheromone gene changes the chemical signal, so that the A-allele velvet worms are not attracted to T-allele velvet worms. Therefore the T-allele is less likely to end up in the offspring of an A-allele velvet worm.

**Figure 11.16**

But the T-allele pheromone is more powerful than the A-allele version, so T-allele carriers are more effective at finding each other, so on the whole they mate more often and have more offspring. Eventually the T-allele goes to fixation in the Scribbly Gum population by positive selection due to this reproductive advantage. Now when the forest on the ridge grows back and velvet worms can once more move through the valley (**Figure 11.17**), the A-allele velvet worms from the Snow Gum Gully are not attracted to the T-allele velvet worms from the Scribbly Gum Gully. So although the individual velvet worms can now move from one gully to the next, the alleles from the Snow Gum population can no longer enter the Scribbly population.

**Figure 11.17**

A storm sweeps through the forest, knocking down old trees in the Snow Gum Gully. The damage is extensive. A tiny population of velvet worms survives in a lucky stand of Snow Gums, but they cannot cross the open ground to the ridge, nor reach the Scribbly Gum Gully (**Figure 11.18**). So now we have three separate populations, each genetically isolated from the other two. The two Scribbly Gum populations are separated by a biological mechanism (different pheromones) and the Snow Gum population is isolated by a physical factor (can't cross open ground). Now any mutation in one population cannot move to a different population. Each population will continue to accumulate different substitutions, by drift or selection. For example, the T-allele Scribbly velvet worm population undergoes further selection-driven molecular change affecting habitat choice (changing a T to a C). The remnant Snow Gum population accumulates another substitution by drift (from an A to a T): being such a small population, drift is an important driver of substitutions in this population (see Chapter 8).

**Figure 11.18**

Figure 11.19

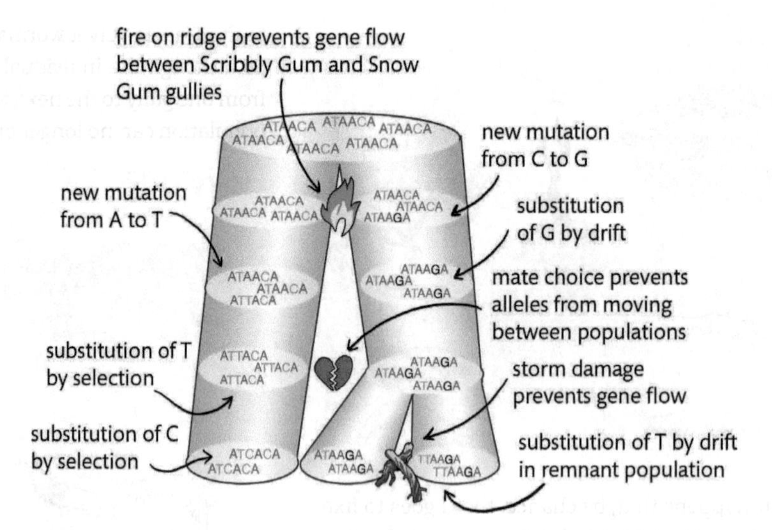


Now let's put all of these events together in one diagram. For simplicity, we will take away the velvet worms and just look at their haploid DNA sequences (Figure 11.19). Every split in the population is marked by a division in the lines of descent.

We can simplify this diagram even further, simply marking the substitutions on lines of descent (Figure 11.20).

Figure 11.20

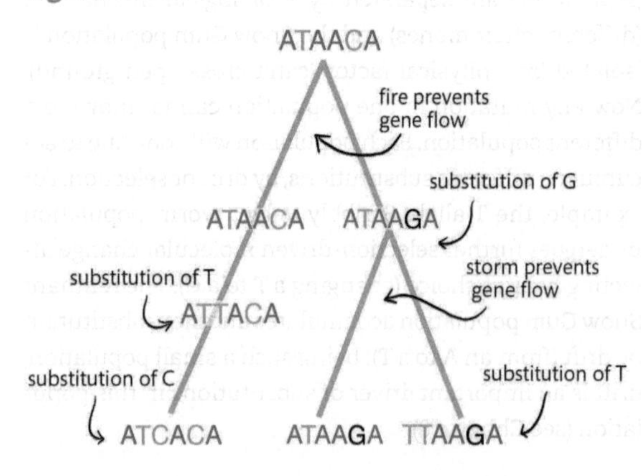


We have produced a branching diagram that represents the evolutionary history of these populations. Such a diagram is usually referred to as an evolutionary tree, or, more formally, a phylogeny.

If we let those three populations evolve even longer, so that they accumulate more substitutions and undergo further population subdivisions, we might see the tree grow like this (Figure 11.21).

Before we go on, it is helpful to have a standard set of terms for describing trees. We are going to call the end-states of the phylogenetic tree 'tips' (Figure 11.22). For DNA-based phylogenies, the tips are actually sequences from our

Figure 11.21

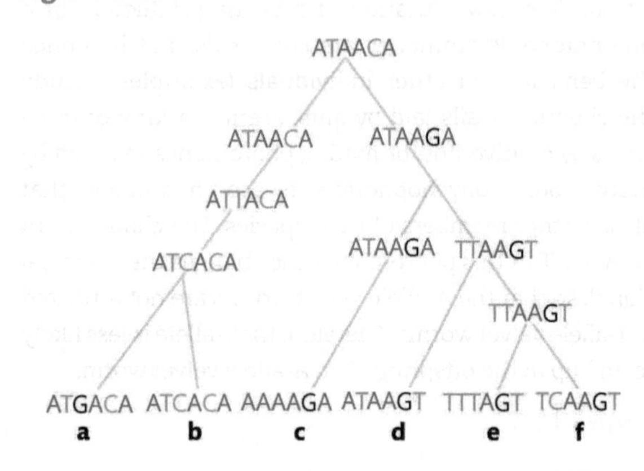


alignment, but we expect that these tips represent the taxa from which the sequences were sampled (taxon is an all-purpose term for 'biological group' that might be a population, a species, a phylum, etc.). Nodes are the branch-points, where a single line of descent splits to give rise to two or more lineages. We can interpret nodes as representing the last shared ancestor of the lineages that split at that point. Branches, also known as edges, are the lines that connect nodes together (or connect the tips to internal nodes). So a branch (or an edge) is any of the lines in a phylogeny. A clade is all of the descendants of a particular node. Because phylogenies represent a hierarchy of relationships, clades are nested within phylogenetic trees, just as taxonomy is nested (Chapter 9). In this case, the tip *f* is nested within the *e-f* clade, which is nested in the *d-e-f* clade, which is nested in the *c-d-e-f* clade.

## Phylogeny reconstruction

Let's quickly recap what we have learned so far. Mutations occur in individuals (Chapters 3 and 4). Some mutations become substitutions carried by all members

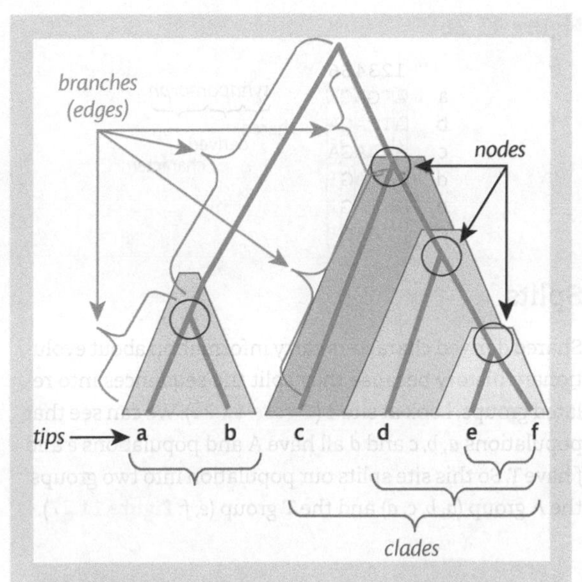


**Figure 11.22** Anatomy of a tree: some terms used in describing phylogenies. Tips are the taxa sampled at the end point of the process. In a molecular phylogeny these are usually DNA sequences sampled in the present day, though some phylogenies contain ancient DNA or stored samples of pathogens that represent earlier points in time (these are still drawn at the tips of the tree). Nodes are the branch points in the tree. Node is usually used to refer to internal nodes which represent the last common ancestor of two descendant lineages, but you might also hear tips referred to as terminal nodes. Lines of descent (blue) are usually referred to as branches, or more formally as edges. While branch is used more loosely, an edge connects two nodes, whether that's a node to a tip, or two nodes. Clades are all the lineages descending from a particular node, so represent overlapping sets of tips on the tree.

of a population (Chapters 7 and 8). The DNA sequences from closely related populations will be more similar to each other than they are to DNA from more distantly related populations, so when we align DNA sequences we can usually identify hierarchies of shared substitutions (Chapter 9 and 10). Now we are going to explore how we can use these hierarchical patterns of shared substitutions to reconstruct evolutionary history.

In the onychophoran example above, we were able to follow the substitutions through evolutionary time, as the populations divide and diverge (**Figures 11.14** to **11.18**). This is equivalent to watching a tree start from a single shoot, then grow more and more branches (**Figures 11.5** to **11.9**). But we can rarely watch evolution happen. In most situations, all we can directly observe is the end points of the process: the DNA sequences at the tips of the tree. So the aim of phylogenetics is usually to use information from the twigs to reconstruct the series of branching events leading back to the root.

If we collected some onychophorans from our hypothetical populations at the end point of this diversification process (**Figure 11.21**), then took them back to the lab and sequenced their DNA, could we accurately reconstruct the series of evolutionary events that produced the present-day genetic data? We would have six sequences, sampled from populations *a* to *f*. Our first task would be to align those sequences (Chapter 10). In this case we can align six homologous sites (**Figure 11.23**).

**Figure 11.23**

```
  123456
a ATGACA
b ATCACA
c AAAAGA
d ATAAGT
e TTTAGT
f TCAAGT
```

In Chapter 10, we saw that when we compare the patterns of nucleotides at homologous sites in an alignment, we can recognize different patterns of similarity and difference. We can categorize these sites by the kind of information they give us about evolutionary relationships.

*Chapter 10 explains the importance of correct alignment to recovering evolutionary history from DNA sequences*

Some sites contain the same base in all sequences. For example, all sequences in this alignment have an A in the fourth site (**Figure 11.24**). The most likely explanation for this pattern is that all of these lineages inherited this A from a common ancestor (see Chapter 9), so we refer to the similarity between them as ancestral. Sites that are the same in all taxa do not reveal relationships within the group. If you are fond of polysyllabic jargon, then you might like to know that shared ancestral character states are known as symplesiomorphies.

**Figure 11.24**

```
  123456
a ATGACA   symplesiomorph
b ATCACA   shared
c AAAAGA   ancestral
d ATAAGT   character
e TTTAGT
f TCAAGT
```

Some sites in the alignment have a particular nucleotide state that is found in one sequence but not in any of the others. For example, at position 2, most populations have T but in population *c* this was substituted with an A and in population *f* the T was substituted with a C (**Figure 11.25**). Because they have arisen anew in a particular population, we refer to these substitutions as being derived. Unique derived changes, found in one sequence but not shared with any others, are also known as autapomorphies (see also **Figure 10.12**). Unique derived characters tell us about evolution in specific lineages, but, again, they do not help us split the sequences into related groups.

**Figure 11.25**

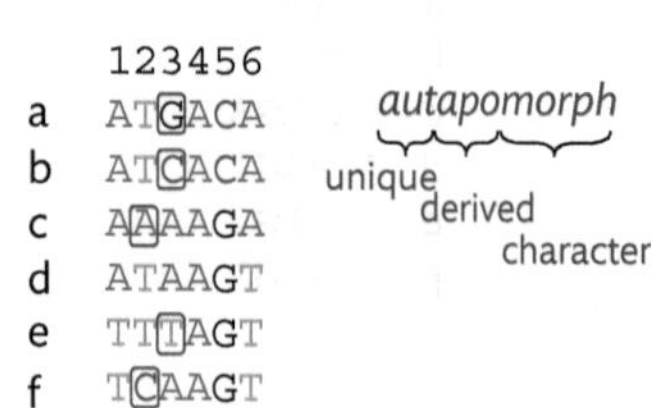


So shared ancestral characteristics do not help divide our sequences into related groups, and neither do unique derived characteristics. What we need is sites that give us information about which populations are more closely related to each other. In other words, we need characteristics that split the sequences into groups, just as Ray used physical and behavioural characteristics of birds to split them into related groups (**Figure 11.2**).

When a population divides to give rise to two or more new populations, each new daughter population inherits the genetic variation of the parent population. Any derived substitution in the parent population is now shared between two related populations. And if one of these populations splits again, then that shared substitution will be passed on again. We can see this process of inheritance of substitutions in the example above. The Snow Gum population underwent a substitution of a G at site 5 (**Figure 11.16**). When the storm divided the Snow Gum population into two populations at Stage 5, both populations inherited this G (**Figure 11.18**). As the lineages continued to diverge, that G was inherited by their descendants (lineages *d*, *e* and *f*: **Figure 11.21**). The process of descent leaves a trail of shared derived characters, each of which arose in a particular lineage (derived), then was inherited by its descendants (shared). So sites where a particular nucleotide state is shared by some but not all of the sequences in our alignment can provide information on descent. Shared derived characters are also known as synapomorphies (**Figure 11.26**).

**Figure 11.26**

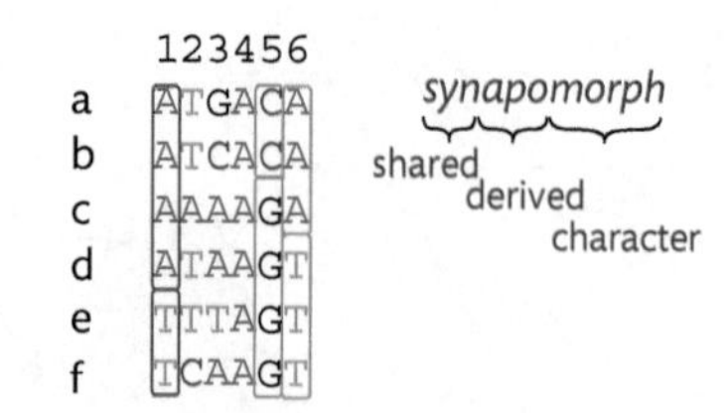


## Splits

Shared derived characters carry information about evolutionary history because they split the sequences into related groups. Look at site 1 (**Figure 11.26**). We can see that populations *a*, *b*, *c* and *d* all have A and populations *e* and *f* have T. So this site splits our population into two groups: the A group (*a*, *b*, *c*, *d*) and the T group (*e*, *f*: **Figure 11.27**).

**Figure 11.27**

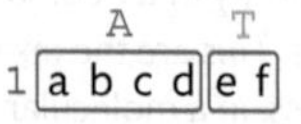


We could interpret this split as a branching event—the lineage that led to populations *e* and *f* diverged from the rest of the lineages and on the way an A was changed for a T. In other words, *e* and *f* share a more recent common ancestor than either does with *a*, *b*, *c* or *d*, and we mark this on our tree with a node connecting *e* and *f*, and an edge along which the substitution occurred (**Figure 11.28**).

**Figure 11.28**

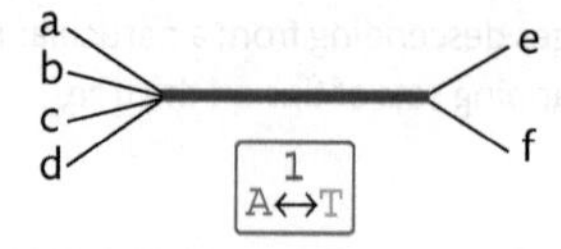


We can interpret site 5 in a similar way. Populations *a* and *b* are characterized by having a C at this site, and populations *c*, *d*, *e* and *f* are characterized by having a G, so site 5 splits populations *a* and *b* from all the others (**Figure 11.29**).

**Figure 11.29**

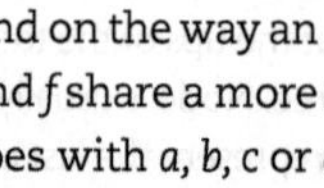

Just as before we can interpret this split as an edge in our tree (**Figure 11.30**).

**Figure 11.30**

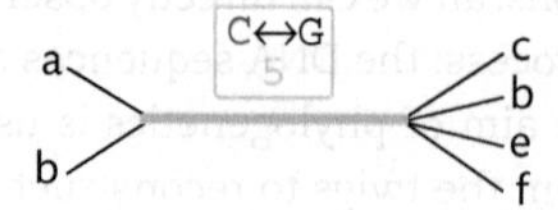

But now we can add this information to what we already know from site 1. Let's take the diagram we drew from site 1 (*e* and *f* split from the rest) and add to it the split inferred from site 5 (*a* and *b* split from the rest: **Figure 11.31**).

**Figure 11.31**

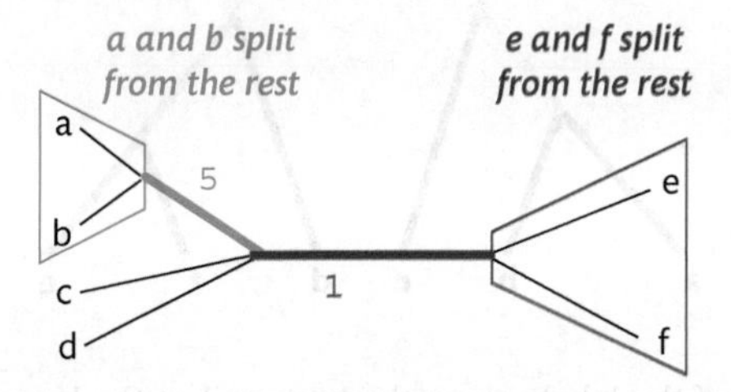


We have combined the splits from two different nucleotide sites. Now let's add site 6, which splits populations *a*, *b* and *c* from populations *d*, *e* and *f* (**Figure 11.32**).

**Figure 11.32**

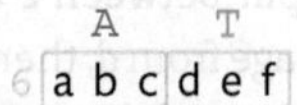

We already know that site 1 splits *e* and *f* from the rest, site 5 splits *a* and *b* from the rest, and now we can put an edge between the *a-b-c* group and the *d-e-f* group (**Figure 11.33**).

**Figure 11.33**

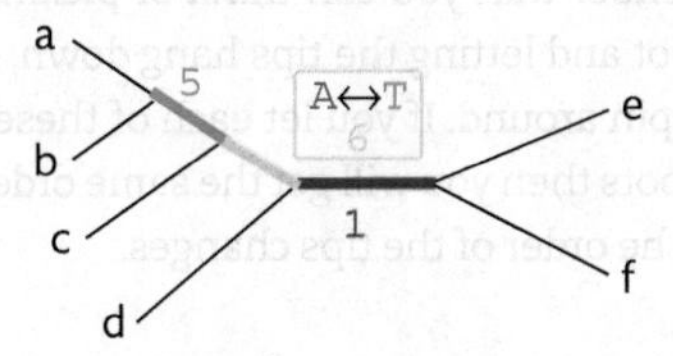

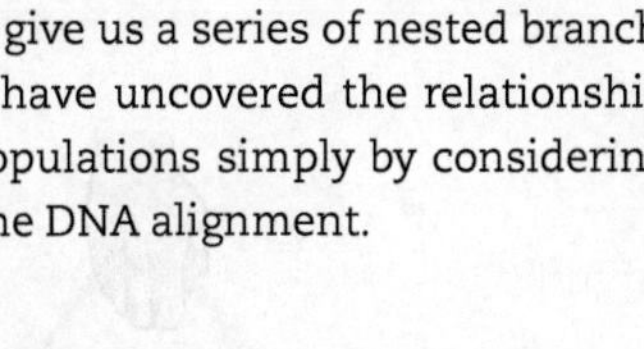

On their own, each of the splits (1, 5 and 6) simply divided the populations into two groups, but taken together they give us a series of nested branching events. So now we have uncovered the relationships between these six populations simply by considering the information in the DNA alignment.

### *Reading trees*

Why doesn't this tree (**Figure 11.33**) look like the one in **Figure 11.21**? Trees contain information about relationships, and that information can be displayed in a variety of ways. In this section we will spend some time rearranging our phylogeny so that we can see that trees that look very different from each other may contain the same information. Firstly, we need to distinguish between rooted and unrooted trees. In Darwin's and Haeckel's trees (**Figures 11.1** and **11.4**), we could begin at the root (ancestral lineage) and follow the branching patterns forward in time and reach the tips (descendant lineages). With an unrooted tree, we have all the branching events that divide the taxa into groups, but we don't know which edge represents the ancestral lineage from which all the others arose. In other words, unrooted trees don't have a starting point.

**Figure 11.33** is an unrooted tree: it shows the relationship between lineages, but it doesn't reveal which of these series of splits happened first. To make **Figure 11.33** look like the tree in **Figure 11.21**, we need to give it a starting point (root), by identifying the edge along which the first split happened. In this case, we happen to know that the very first split divided the lineage leading to *a* and *b* from the lineage leading to *c*, *d*, *e* and *f*. On the unrooted tree, this is the edge marked 5. Imagine that our unrooted tree diagram is actually a mobile made of string. Pick up the unrooted tree on the edge marked 5 (**Figure 11.34**).

**Figure 11.34**

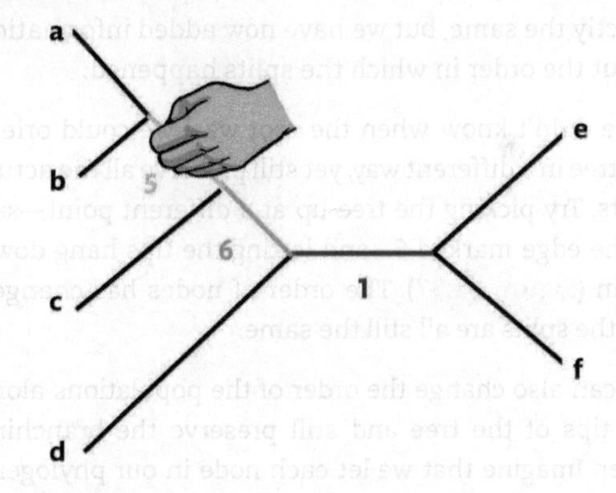


Let the rest of the branches hang down, as if the letters on the tips were heavy (**Figure 11.35**).

**Figure 11.35**

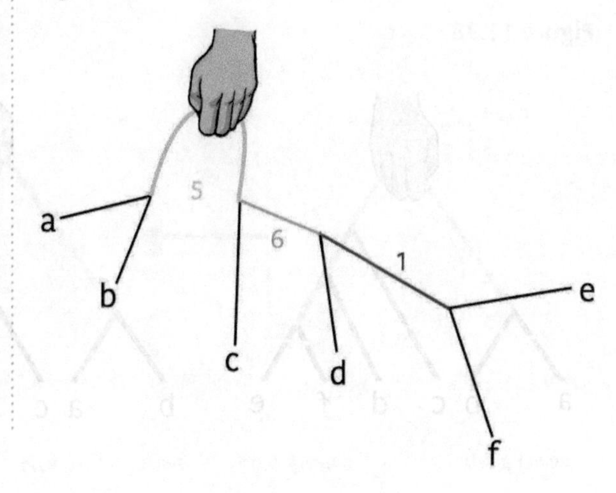

And now we have a rooted tree that looks like Figure 11.21. Our tree now begins at the root with a branch point that splits *a* and *b* from the lineage leading to *c*, *d*, *e* and *f* (Figure 11.36).

Figure 11.36

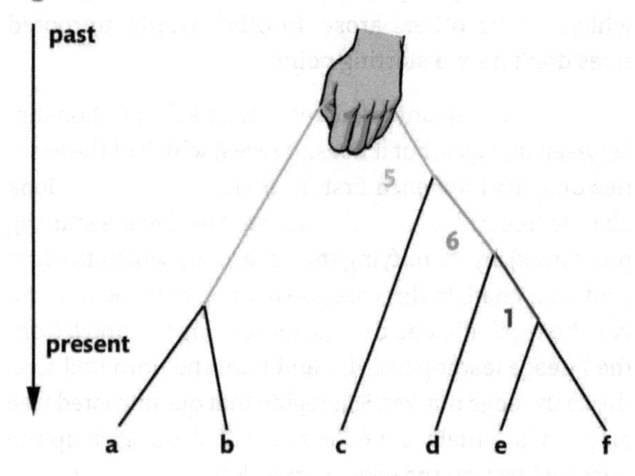


We have added the direction of time to our phylogeny. The information about which splits occurred is exactly the same, but we have now added information about the order in which the splits happened.

If we didn't know when the root was, we could orient the tree in a different way, yet still preserve all the actual splits. Try picking the tree up at a different point—say, at the edge marked 6—and letting the tips hang down again (Figure 11.37). The order of nodes has changed but the splits are all still the same.

We can also change the order of the populations along the tips of the tree and still preserve the branching order. Imagine that we let each node in our phylogeny mobile swing around (Figure 11.38). We will change the order of the labels along the bottom, but the branching order will be the same.

Figure 11.37

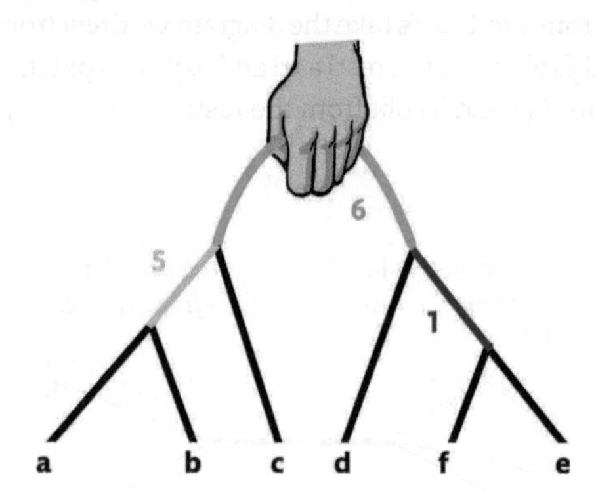


The information provided by the splits is the same in all these trees. You can check this by starting at a population at the tip of the tree and asking which nodes you travel through to get to the root. Try starting at *f*. The first node you come to is the split between *e* and *f*, then the next node splits the *e-f* lineage from *d*, then the next node splits *e*, *f* and *d* from *a*, *b*, and *c*, and then you reach the root.

Here are three more alternative ways of displaying the same phylogenetic relationships (Figure 11.39). Satisfy yourself that each one of these trees carries exactly the same information about branching order by starting at any tip and following the lineage through to the root, checking that you get the same series of splits in each of the three trees. Remember that you can think of picking the trees up by the root and letting the tips hang down, so that the nodes can spin around. If you let each of these trees spin from their roots then you will get the same order of nodes, even when the order of the tips changes.

Figure 11.38

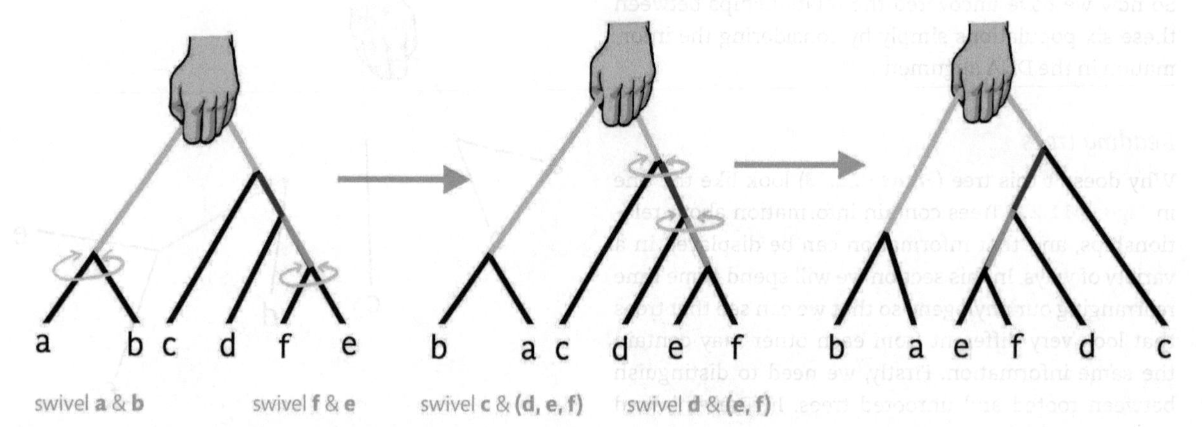

**Figure 11.39**

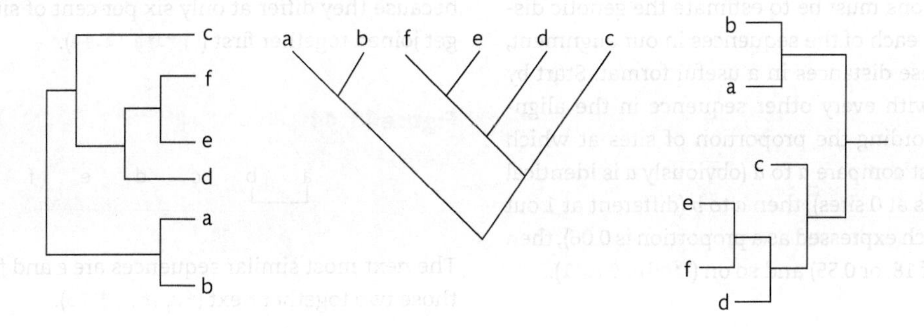


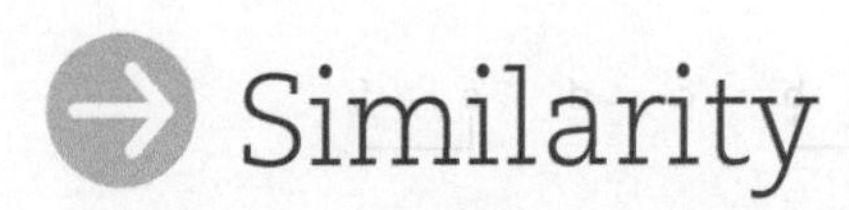

# Similarity

We have shown how the pattern of substitutions in an alignment of homologous sequences can be used to reconstruct the historical process that created the sequences. This was easy to do with such a small alignment: only six sequences with six aligned sites. But most molecular phylogenies are based on sequences with thousands of aligned sites. There are two key elements that make it possible to infer phylogenies from much larger datasets: automation and statistics.

The rise of molecular phylogenetics coincided with the rise of computers (see **Heroes 10 and 11**), and it is difficult to see how molecular phylogenetic inference could have developed without computing. Computers are very good at solving problems that involve tedious repetitive tasks: they have good memories and they don't get bored. But computers cannot think for themselves. For a computer to solve a problem, it needs to be given an unambiguous set of instructions that break the task down into a series of steps. If you can describe the process of constructing phylogenies as a series of computational steps (an algorithm), then you can get a computer to do it. An algorithm is a set of instructions that can be followed to solve a problem. We use a range of types of algorithms in our lives, for example when you follow the set of instructions in a recipe. Rather than explaining any particular phylogenetic method, we are going to consider a few fundamentally different approaches to solving phylogenies using algorithms. In this chapter, we will look at distance methods and networks, and in the next two chapters we will focus on maximum likelihood and Bayesian statistics.

## Distance methods

The simplest approach to estimating phylogenies from DNA sequence data is to start with the proposition that since the process of descent leaves a hierarchy of similarities, so similarities should tell us about descent. If we compare DNA sequences sampled from different populations, then we expect sequences sampled from more closely related populations to be more similar to each other than they are to sequences from more distantly related populations. We can use these measures of differences between sequences to draw a phylogenetic tree by clustering sequences together in the order of most similar (therefore probably the most closely related) to most different (therefore probably the most distantly related). Phylogenetic methods that use measures of difference between sequences to draw evolutionary trees are generally known as distance methods (**TechBox 11.1**).

Let's go back to our hypothetical onychophorans, but this time we will take a slightly longer alignment (**Figure 11.40**: you can see that the first 6 sites are the same ones as the alignment in **Figure 11.23**).

**Figure 11.40**

```
a  ATGACAATATGACAGACA
b  ATCACAATATGACAGACA
c  AAAAGAACAAAAGAATGA
d  ATAAGTACATAAGTAAGT
e  TTTAGTACATAAGTAAGT
f  TCAAGTACATAAGTAAGT
```

Count how many sites differ between *a* and *b*. There is just one difference between *a* and *b* out of 18 sites in the alignment: *a* has a G at the third position whereas *b* has a C. Now compare *a* to *c*: 10 differences out of 18. So on this information alone we would guess that *a* is more closely related to *b* than it is to *c*. The assumption we are making is that the reason that *a* is more similar to *b* than to *c* is that *a* and *b* share a more recent common ancestor.

To automate the process of using measures of sequence similarity to draw a tree, we need an algorithm that describes a simple ordered set of instructions. Our first set of instructions must be to estimate the genetic distance between each of the sequences in our alignment, and record these distances in a useful format. Start by comparing *a* with every other sequence in the alignment, and recording the proportion of sites at which they differ. First compare *a* to *a* (obviously *a* is identical to *a* so it differs at 0 sites), then *a* to *b* (different at 1 out of 18 sites, which expressed as a proportion is 0.06), then *a* to *c* (10 out of 18, or 0.55) and so on (**Figure 11.41**).

**Figure 11.41**

| | a | b | c | d | e | f |
|---|---|---|---|---|---|---|
| a | 0 | 0.06 | 0.55 | 0.56 | 0.61 | 0.66 |

Repeat the process for sequence *b*. We have already compared *a* and *b* (distance = 0.06), so we don't need to do it again. So now we record the proportional difference between sequence *b* and the other sequences (**Figure 11.42**).

**Figure 11.42**

| | a | b | c | d | e | f |
|---|---|---|---|---|---|---|
| b | – | 0 | 0.55 | 0.61 | 0.61 | 0.67 |

We can do the same for every sequence in the alignment, building up a matrix where each line represents the difference between one sequence and all the other sequences (**Figure 11.43**). This matrix is triangular because we don't need to record any comparisons twice. For example, by the time we get to the last row, we have already compared *f* to all other sequences and don't need to do it again.

**Figure 11.43**

| | a | b | c | d | e | f |
|---|---|---|---|---|---|---|
| a | 0 | 0.06 | 0.55 | 0.56 | 0.61 | 0.66 |
| b | | 0 | 0.55 | 0.61 | 0.61 | 0.67 |
| c | | | 0 | 0.33 | 0.39 | 0.39 |
| d | | | | 0 | 0.28 | 0.22 |
| e | | | | | 0 | 0.17 |
| f | | | | | | 0 |

Now that we have measured and recorded genetic distances between our sequences, we need a set of instructions for turning this information into a tree. Most distance methods build a tree step by step by finding the most similar sequences and clustering them together, then repeating the process until all the sequences are joined to each other. Look at the matrix (**Figure 11.43**).

We can see that the most similar sequences are *a* and *b* because they differ at only six per cent of sites. So they get joined together first (**Figure 11.44**).

**Figure 11.44**

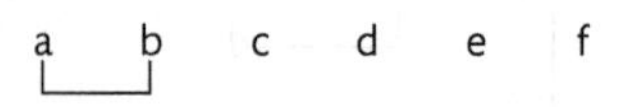


The next most similar sequences are *e* and *f*, so we join those two together next (**Figure 11.45**).

**Figure 11.45**

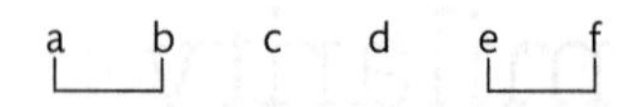


The next most similar entries in the matrix are the difference between *d* and *e* (0.28) and *d* and *f* (0.22). So we can infer from that that the lineage that connects *e* and *f* is most closely related to lineage *d* (**Figure 11.46**).

**Figure 11.46**

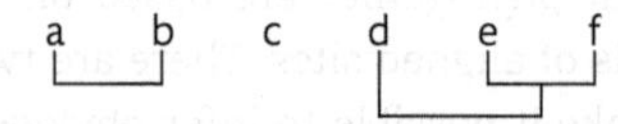


We continue to look for the next most similar lineage and group them together. *c* differs from *d*, *e* and *f* by around 30 per cent, so we assume it is the sister lineage to the *d-e-f* clade (**Figure 11.47**).

**Figure 11.47**

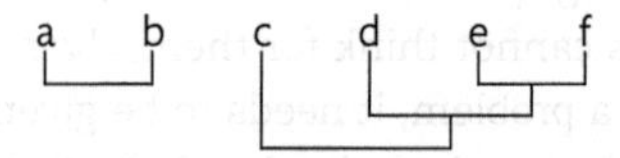


Then *c*, *d*, *e* and *f* are all roughly the same distance from *a* and *b*, so we are placing the split between these two groups at the root of the tree (**Figure 11.48**).

**Figure 11.48**

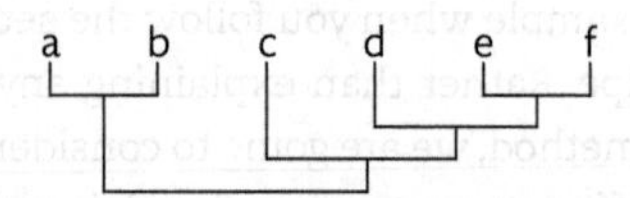


And now we have a tree! Compare this tree to **Figure 11.21** and you will see that we have correctly uncovered the evolutionary history of these populations using nothing more than the sequences we sampled at the end point of the process.

I'm sure you will appreciate that we have used a very simple example. Not only did we use a small number of very short sequences, but we also used the simplest possible measure of genetic distance, and took a very basic approach to estimating phylogeny. In practice, most distance methods use more sophisticated means of estimating genetic distances, and more complicated ways of using those distances to build a tree (see **TechBox 11.1**). But the basic approach we have taken illustrates the essential elements of most phylogenetics programs based on genetic distance: begin with a set of aligned sequences, generate a distance matrix that represents the genetic difference between the sequences, then use these distances to progressively cluster the sequences into a phylogeny.

Distance methods are usually relatively fast, and they will in many cases correctly infer the evolutionary history of a set of sequences. Distance methods are used commonly in some applications, such as DNA barcoding (**TechBox 9.2**). These days, relatively few published phylogenies are distance trees, perhaps due to two main problems. The first problem is a practical one: they tend to return an incorrect phylogeny under several common scenarios (for example when rates of molecular evolution vary between lineages: see **TechBox 11.1**). The second problem is more philosophical: distance methods are essentially non-evolutionary. A distance tree is just a way of displaying information about similarities and differences. It may reflect evolutionary relationships, because descent with modification tends to leave a hierarchical pattern of differences. But just because we can draw a tree from a distance matrix does not mean we have uncovered evolutionary history. In fact, you can take any set of distances and display them as a tree. You could stop half a dozen shoppers as they left the supermarket and compare the contents of their shopping trolleys, use this information to create a measure of difference (say, one minus the ratio of the number of shared items to the total number of items compared), then draw a tree that shows which shoppers bought the most similar groceries. This would in no way imply that the reason two shoppers had similar trolley contents was because they were somehow more closely related to each other. As with most phylogenetic methods, you need to be keenly aware of exactly what any given method is telling you about your sequences, and use your understanding of molecular evolution to judge what meaning it has for your samples.

## Conflict

In the example we followed above, the sequence data had clear hierarchical structure. We could use the splits in the data to specify an unambiguous branching order on our tree. If only it was always that neat, then this book could be much shorter. But reality is often much messier. Very often we find that not all the splits in the data will support the same conclusions. To illustrate this conundrum, imagine adding one additional site to our alignment from **Figure 11.23**.

**Figure 11.49**

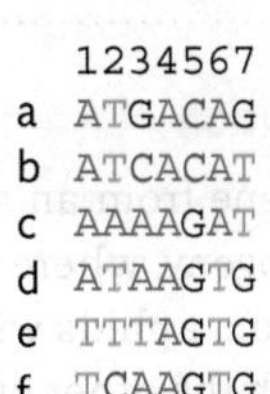


Site 7 apparently gives us another shared derived character, with two lineages sharing a T and four sharing a G (**Figure 11.49**). But the split defined by this site—*b* and *c* split from the rest—doesn't fit neatly into the tree we have already drawn. We can't have *b* and *c* forming a pair split from the rest, when we already have *a* and *b* split from the rest. These two splits contradict each other, ruining our nice hierarchical structure (**Figure 11.50**).

**Figure 11.50**

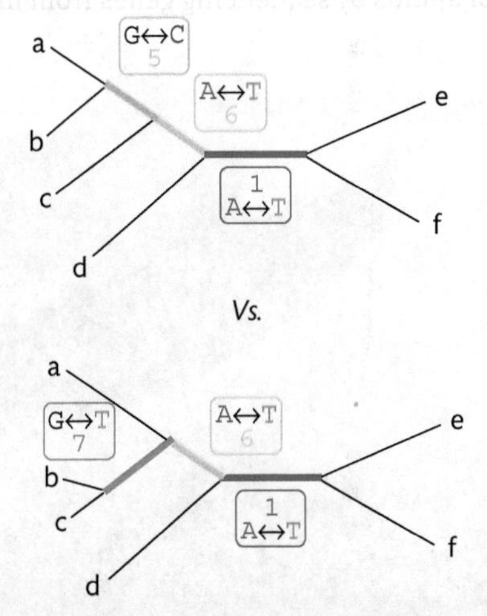


What are we to make of this contradictory information? There are a number of different ways we could explain this conflict in our data. It may be that the sites are telling different stories because the real history of tree-like splitting of lineages has been somehow erased, either by a sequencing error or by multiple hits overwriting the signal of past changes. But we should also consider the possibility that the reason our data doesn't fit a neat tree is that it wasn't all produced by a single process of a lineage splitting into two again and again. Maybe our data doesn't fit a single tree because the history of the data doesn't follow a single tree (**TechBox 11.2**). For example, if there was a recombination event between site 6 and 7 (see Chapter 4), we might have combined together two alleles with different histories. Horizontal gene transfer also results in genomes with mosaic histories (Case Study 10.2).

*Multiple hits are explained in Chapter 12*

## Non-tree-like evolution

If you sequence a gene from an aphid and include it in a molecular phylogeny, where would you expect it to fit? Most genes from aphids will place them in the animal kingdom with the other bugs (Hemiptera). But if you use the aphid's carotenoid biosynthesis gene in your phylogeny, the aphid sequences will come out as members of the fungal kingdom. Actually, most animals can't synthesize carotenoids and have to get them from their diet, but aphids managed to borrow genes for carotenoid biosynthesis from a fungus by horizontal gene transfer (Figure 11.51). So these genes trace their own particular history back through the fungi, whereas other aphid genes trace their histories back through the animal kingdom. As it happens, you could also uncover the evolutionary history of aphids by sequencing genes from members of an entirely different kingdom. *Buchnera* are bacteria that live their entire lives inside aphids, as endosymbionts. When aphid lineages diverge, they take their *Buchnera* genes with them, so genes from the bacteria and the aphids both track the same history (Case Study 10.1).

While this chapter has focused on lineages splitting and diverging, it is also possible for separate lineages to come together and merge. Hybridization has been an important force in the evolution of biodiversity. Many plant species have a hybrid origin, so that their genome contains DNA sequences from two or more different lineages. For example, conflict between phylogenies based on different gene copies in tetraploid orchids from the genus *Polystacha* revealed patterns of hybridization, where new lineages were formed from the fusion of two different diploid species (Figure 11.52). This means that if we were to construct a phylogeny from two different sequences we might get conflicting phylogenies, because the sequences themselves have travelled different evolutionary paths to end up in that genome. So while some species have a history that can neatly be described by a bifurcating tree (one where each lineage splits into two daughter lineages), for other species

**Figure 11.51** Where do you get your carotenoids? Some animals use carotenoids for pigmentation, to produce orange and yellow colours and for other biochemical roles such as antioxidants. But, unlike bacteria and fungi, most animals can't manufacture their own carotenoids so they need to get them by eating carotenoid-containing foods. But these two creatures have acquired the ability to synthesis carotenoids by borrowing the genetic apparatus of other organisms. The pea aphid (*Acyrthosiphon pisum*, (a)) appears to have acquired carotenoid biosynthesis genes from a fungus by horizontal gene transfer, and the white fly (*Bemisia tabaci*, (b)) has endosymbiotic bacteria that have retained their carotenoid biosynthesis genes (despite having greatly reduced genomes as is typical of endosymbionts: see Case Study 10.2).

**Figure 11.52** *Polystachia bella* is an epiphytic orchid from high altitude rainforest in Kenya and Uganda. This species is tetraploid, which means that it has four copies of every chromosome rather than the more usual two (diploid). Tetraploids can form when two diploid lineages hybridize or through doubling of a single species genome.



there may be no one single tree that describes the history of all sequences in the genome. Instead, the history of some species might be considered reticulate, described by a net-like intertwining of lineages.

## Networks

*It seems axiomatic in biology that the more data one has that bears on a problem, the greater the chance of resolving that problem. It does not follow, however, that mere compilation of data will result in a more accurate analysis, especially when the phylogenetic signal is small relative to other variation that is present.*

Whitfield, J. B., Lockhart, P. J. (2007) Deciphering ancient rapid radiations. *Trends in Ecology & Evolution*, Volume 22, page 258.

In reality, nearly all datasets will contain some conflicting signal. Not all splits tell the same story. In some cases this may be telling us something really interesting about the history of those lineages. But in many cases, it will be telling us something we already know, which is that biological data are often messy. Either way, we need a strategy for coping with conflicting signal. One option is to accept that although no one tree provides a perfect fit to the data, we can choose the best tree by selecting the one that has the greatest degree of support (see Chapter 12). We can then express the degree of conflict in the data with some measure of phylogenetic support, such as bootstrap values (**TechBox 12.2**). Or we may describe a set of alternative trees that all have similar degrees of support in the data (**TechBox 13.1**). Alternatively, we may choose to reflect the conflict in the data by displaying all the different splits in a single diagram. This diagram won't be entirely tree-like, because not all of the patterns in the data will fit the same branching patterns. Instead, it will be a network of interconnecting lineages.

Let's draw our little alignment (**Figure 11.49**) as a network. We are going to include both the split defined by site 5 (*a* goes with *b*) and the split defined by site 7 (*b* goes with *c*: **Figure 11.53**).

**Figure 11.53**

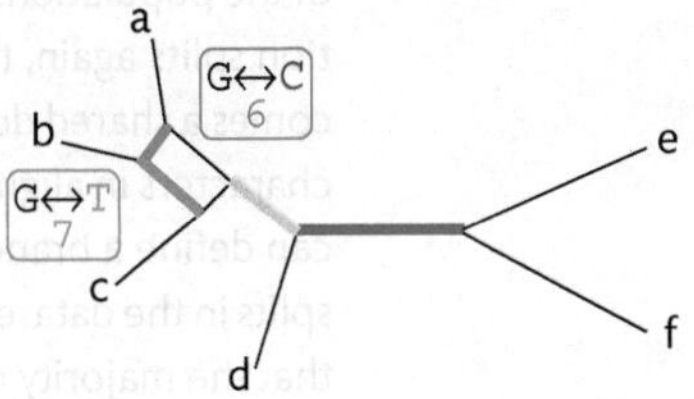

Before, when we followed paths through a bifurcating phylogenetic tree, there was only one set of edges and nodes we could follow, whether from the tip of the tree to the root, or between two tips. But in a network, we will sometimes be faced with alternative paths connecting tips. Here, if we want to find the path connecting *b* and *c*, we will get a different path depending on which part of the network we traverse. We could start at *b* then follow the orange path, defined by site 5, in which case we first meet the node joining *a* and *b* together then travel to *c*. Or, if we follow the blue path, defined by site 7, we travel from *b* through the node connecting *b* and *c* together to *c*. Our network is telling two alternative stories at once, one in which *a* and *b* are each other's closest relatives, and one in which *b* and *c* are each other's closest relatives.

If we had a large alignment, we could draw all of the different splits described by our data, and we could represent the proportion of splits supporting each grouping by relative length of the connecting lines (**Case Studies 11.1 and 11.2**). These networks look a bit like phylogenetic trees, but instead of each node being represented by a single lineage splitting into two (a bifurcation), many of the splits are represented as boxes, indicating conflicting signal in the data (Case Study 11.2). The conflicting splits might represent hybridization events, reconnecting previously separated lineages. Or they might represent incomplete lineage sorting, where the polymorphic alleles from the ancestral population are still present in some of the descendant lineages, not yet lost or fixed as substitutions. Or perhaps the conflicting, non-tree-like signal in the data is due to errors in sequencing or poor alignment, both of which can generate false stories in the data. Even if the evolutionary process that produced these sequences was perfectly tree-like, with a nested series of bifurcating splits, subsequent changes in the sequences could erase phylogenetic signal, effectively rewriting part of the story in the data. If there is a lot of conflicting signal in the data, or a high level of noise, then phylogenetic networks can end up looking more like string bags than branching trees. What kind of story is your data telling you?

# Conclusion

Distance trees provide a way of visualizing the patterns of similarity and difference in DNA sequences. We expect that these patterns will reveal the evolutionary history of the sequences, because the process of evolution creates hierarchies of similarity in species. When a population becomes isolated, any mutations that occur in that population will not be shared with neighbouring populations. Any mutation that becomes a substitution, carried by all members of the populations, may be a unique marker of that interbreeding population. If that population splits again, then its daughter populations will inherit that substitution, which now becomes a shared derived character that unites its descendants. We can identify shared derived characters in alignments as informative splits that can be used to group lineages. Sets of splits can define a branching tree that reveals the history of the sequences. Sometimes not all the splits in the dataset tell the same story. We can deal with conflicting splits by choosing the tree that the majority of splits support, or we can display all the information from the conflicting splits in a network which capture all of the signal in the data.

In some ways, distance methods are similar to the early classification systems, such as those of Ray or Linnaeus. These pre-evolutionary schemes described the hierarchical pattern of similarities between species without giving an explanation of why biological diversity was organized that way. These classification systems work well in cases where similarity is a good indicator of evolutionary history. But, just as similarity between morphological characteristics is not always a sign of shared ancestry (such as the thylacine and the wolf in Chapter 10), similarity between DNA sequences is not always a reliable indicator of evolutionary history. In the next chapter, we will discuss alternative methods that aim not just to describe but also to explain the patterns in the data. Just as Darwin's tree depicted the process of diversification, these methods use what we know about the process of molecular evolution to reconstruct the evolutionary history of the sequences.

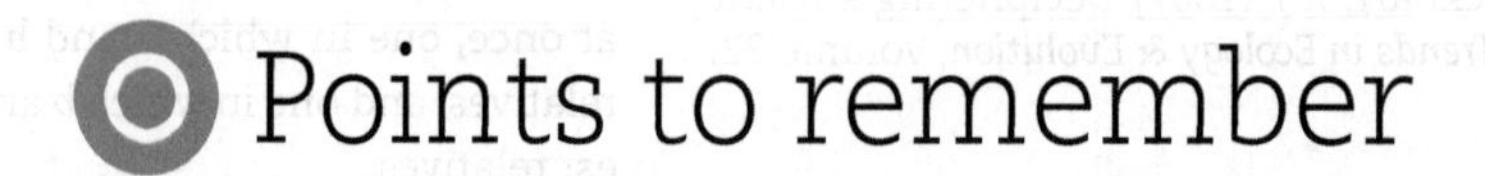

# Points to remember

## History

- Biological diversity is often displayed as a nested hierarchy, grouping together species into groups that share similarities.
- Phylogenetic trees can be read as evolutionary histories, tracing the paths of descent and divergence back to a common ancestor.

- DNA sequences are useful for estimating phylogenies because they are universal (comparable across all living things); they provide an alternative historical record to morphological characteristics or the fossil record; and DNA sequences are ideal for statistical analysis.
- When substitutions occur in isolated populations, they distinguish those lineages from their relatives, and are inherited by any descendant populations.
- Homologous sites in an alignment of DNA sequences show patterns of similarity and difference between taxa that we can interpret as signatures of descent.
- Shared derived characters which are the same in some, but not all, sequences in the alignment split the data into groups.
- By combining splits, we can produce nested hierarchies.
- Branches on a phylogeny, also known as edges, represent lines of descent from common ancestors to descendants.
- Sampled sequences define the terminal branches, or tips, of a phylogeny.
- Branching points in phylogenies, where two lineages split, are referred to as nodes.
- A clade is the group of lineages all descended from a particular node.
- Unrooted trees are branching diagrams that show the relationships between sequences.
- Rooted trees identify the initial split in the lineages, so indicate direction of descent.
- We can draw trees in many ways, reordering the tips, without changing the information on relationships defined by the order of nodes.

### Similarity

- Distance methods use a matrix of measures of similarity between pairs of sequences to progressively cluster together the most similar sequences.
- Conflict occurs when sites in the alignment do not all define a compatible set of splits that correspond to a single bifurcating tree.
- Contradictory splits in the data could arise from sequencing or alignment errors, multiple hits erasing phylogenetic signal, or reticulate evolution such as hybridization or horizontal gene transfer.
- Networks provide a way of displaying the conflicting signal in an alignment, representing the support for different splits as box-like connections between lineages.

# Ideas for discussion

**11.1** How would you decide whether to interpret the conflicting splits in your alignment as due to false signal (for example, the original patterns have been overridden by subsequent changes) or as true signal of a complicated evolutionary history (such as hybridization or horizontal gene transfer)?

**11.2** Is it possible for two populations to remain similar to each other, without diverging?

**11.3** Why don't all phylogenetic studies use networks? Should they?

# Examples used in this chapter

Beddall, B. G. (1957) Historical notes on avian classification. *Systematic Zoology*, Volume 6, pages 129–36.

Dawkins, R. (2004) *The Ancestor's Tale: A Pilgrimage to the Dawn of Life*, Weidenfeld & Nicholson.

Moran, N. A., Jarvik, T. (2010) Lateral transfer of genes from fungi underlies carotenoid production in aphids. *Science*, Volume 328, page 624.

Nováková, E., Hypša, V., Klein, J., Foottit, R. G., von Dohlen, C. D., Moran, N. A. (2013) Reconstructing the phylogeny of aphids (Hemiptera: Aphididae) using DNA of the obligate symbiont *Buchnera aphidicola. Molecular Phylogenetics and Evolution*, Volume 68, page 42.

Russell, A., Samuel, R., Klejna, V., Barfuss, M. H., Rupp, B., Chase, M. W. (2010) Reticulate evolution in diploid and tetraploid species of *Polystachya* (Orchidaceae) as shown by plastid DNA sequences and low-copy nuclear genes. *Annals of Botany*, Volume 106, page 37.

Sereno, P. C. (1991) Basal archosaurs: phylogenetic relationships and functional implications. *Journal of Vertebrate Paleontology*, Volume 11, page 1.

Sloan, D. B., Moran, N. A. (2012) Endosymbiotic bacteria as a source of carotenoids in whiteflies. *Biology Letters*, Volume 8, page 986.

Tsui, C. K. M., Marshall, W., Yokoyama, R., Honda, D., Lippmeier, J. C., Craven, K. D., Peterson, P. D., Berbee, M. L. (2009) Labyrinthulomycetes phylogeny and its implications for the evolutionary loss of chloroplasts and gain of ectoplasmic gliding. *Molecular Phylogenetics and Evolution*, Volume 50, page 129.

HEROES OF THE GENETIC REVOLUTION 11

# Joseph Felsenstein

*"I always felt that the important thing wasn't that you were you using a computer, it was that you were thinking evolutionarily, or you were thinking statistically."*

Joe Felsenstein (2012) UW Genome Sciences—Distinguished Faculty Interview Series

**EXAMPLE PUBLICATIONS**

Felsenstein, J. (1981) Evolutionary trees from DNA sequences: a maximum likelihood approach *Journal of Molecular Evolution*, Volume 17, page 368.

Felsenstein, J. (2004) *Inferring phylogenies*. Sinauer Associates, Sunderland, Massachusetts.

**Figure 11.54** Joe Felsenstein is a keen contributor to various blogs and discussions of evolutionary biology, so you may have the chance to chat with him.

Reproduced courtesy of Joe Felsenstein.

Joe Felsenstein has been a key player in setting problem-solving in molecular phylogenetics in a statistical framework (**Figure 11.54**). He began developing these tools when the sequence databases were beginning to get larger and computers were starting to get smaller. The processing speed of the mainframe computer he learned to program on in the 1960s is equivalent to the processing speed in a singing birthday card today. But in 1977, he had one of the first microcomputers in his genetics department (which, as it happens, was designed by his younger brother Lee Felsenstein). This computer apparently had a wooden case, ran off 8-inch floppy disks and was upgraded to have 64 kb of memory. It was on this computer that he first developed his pioneering phylogenetics software. As the number of available DNA sequences exploded, Felsenstein's statistical tools provided a robust statistical framework for producing and interpreting phylogenies.

Felsenstein was an undergraduate at University of Wisconsin, where he learned population genetics from the great Jim Crow (**Hero 3**), then he completed his PhD in 1968 with leading evolutionary biologist Richard Lewontin at the University of Chicago. After a short postdoc in Edinburgh, he joined the University of Washington in Seattle, where he has been ever since. Although he is best known for his work in the statistical inference of phylogeny, Felsenstein has published on a very wide range of topics. His early papers were in the field of theoretical population genetics, covering topics as diverse as mutation accumulation, speciation and the rate of loss of genetic variation from populations, but later he also examined models of ecological diversity and macroevolutionary change.

Felsenstein's contributions to molecular phylogenetics have defined the field. He introduced statistical techniques that have now become standard, such as bootstrapping sequence data (**TechBox 12.2**). In particular, he promoted the use of maximum likelihood (ML) (**TechBox 12.1**), developing algorithms that made ML tractable for phylogeny estimation. Felsenstein has also played a key role in testing the performance and reliability of phylogenetic methods, demonstrating the conditions under which particular methods will give you the wrong answer. For example, he showed that parsimony can be consistently misleading if evolutionary rates vary widely between lineages, a set of conditions now known as the 'Felsenstein Zone' (which he has commented is like having a black hole named after him). Importantly, he did not just demonstrate the logic of using statistical tools like maximum likelihood to phylogenetics, he also developed the means of making them easily applicable to biological data. He created one of the first really useful phylogenetics packages, PHYLIP, which has been continuously updated since its release in 1980, and currently has over 30,000 registered users. Felsenstein has championed accessibility of both the theory and the tools of molecular phylogenetics. PHYLIP is freely available as long as it is not used for commercial gain.

As you will no doubt discover, much of the molecular phylogenetics literature is, unfortunately, rather a dull read. And one of the things that makes it dull is the lack of a personal voice: most papers are written in an anonymous style that completely obscures the person behind the paper. One of the things I like about Felsenstein's writing is that it breaks this mould: in each paper, or book, or website, it feels very much like Joe is speaking directly to the reader. His common-sense approach has been summarized in 'Felsenstein's Law', which states that all true results are obvious—after the fact. This pragmatic approach is also evident in his contributions to long-standing disputes on the correct basis for systematics (Felsenstein claims to have 'founded the fourth great school of classification, the It-Doesn't-Matter-Very-Much school'). His magnum opus, *Inferring Phylogenies*, has been described as 'truly majestic', 'an outstanding achievement' and 'a classic by the time it was published'. Felsenstein was pleased to see a reviewer draw a parallel to the classic text on theoretical population genetics by Crow and Kimura: 'I modelled my level of presentation and writing style in great measure on theirs, and hoped to have an effect similar to theirs. It is gratifying to see that connection made.'

# TECHBOX 11.1

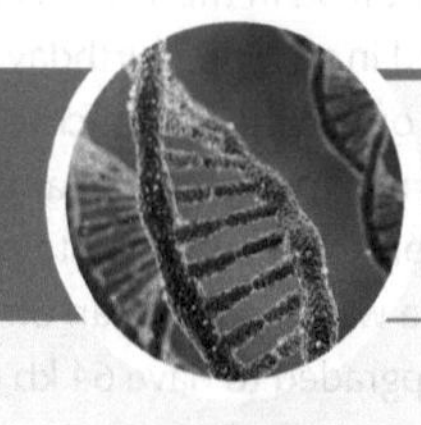

# Distance methods

**RELATED TECHBOXES**

TB 1.2: BLAST

TB 9.2: Barcoding

**RELATED CASE STUDIES**

CS 9.2: Barcoding

CS 11.2: Distance

**Summary**

Clustering sequences by similarity provides a fast way of estimating a phylogeny.

**Keywords**

immunological distance, DNA hybridization, neighbour joining (NJ), UPGMA, cluster, minimum evolution, additive, ultrametric

Distance methods are based on grouping sequences by some measure of their overall similarity. Clustering by distance is used in a lot of sequence analysis, such as BLAST (**TechBox 1.2**), some DNA barcoding methods (**TechBox 9.2**), and also many applications in gene expression and genomics (**Case Study 11.2**). But here we are specifically concerned with the use of clustering by distance to produce phylogenetic trees. These days, distance trees are less fashionable as a phylogeny estimation method than other methods, but they do provide a quick and easy way to assess the historical signal in the data. Most phylogenetics packages will include distance-based phylogeny estimation methods, such as neighbour-joining (NJ).

Distance methods begin with a matrix of genetic distances between a set of sequences, then find the tree that best describes the genetic distance data[1]. It's important to remember that these two steps are essentially separate from each other. There are many ways to generate a matrix of genetic distances, and once you have that matrix, you can apply many different ways of deriving a tree from that distance data.

### Distance data

Some types of molecular data begin as a matrix of distances. For example, the phlylogeny of horses in **Figure 11.55** is based on immunological distances[2]. The authors created antibodies that would bind specifically to proteins from nine different horse species (bizarrely as it may seem in this postgenomic age, these antibodies were created by injecting horse proteins into rabbits). Antibodies are specific to particular antigens, so they will bind most strongly to a protein that is identical to the proteins they were created against. Strength of binding is therefore a measure of protein similarity. Here, they generated a distance matrix by testing

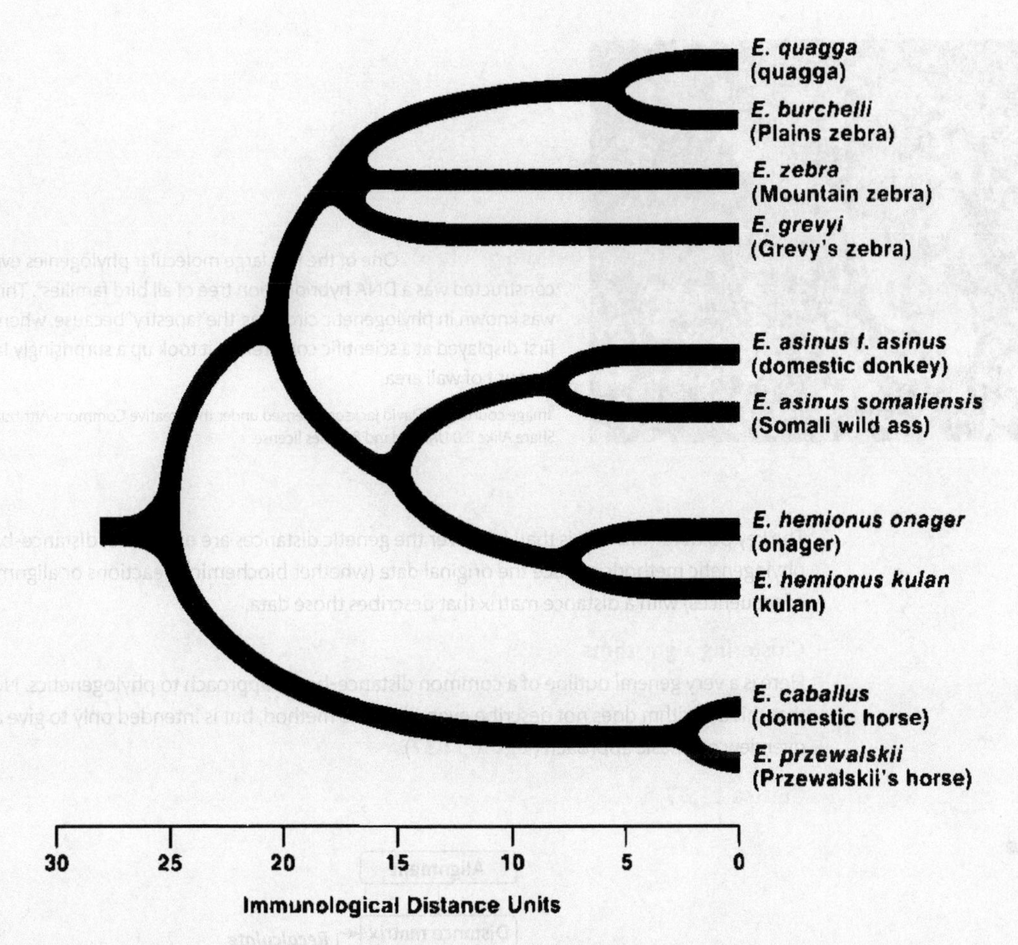


**Figure 11.55** Ancient ancient DNA: this is the first molecular phylogeny to include an extinct taxon. The phylogeny is based on immunological distance, which is measured by reacting antibodies against proteins.



each of the horse proteins against each of the anti-horse rabbit antibodies. They also placed the extinct quagga in this phylogeny by testing ground-up quagga skin against the nine different anti-horse antibodies (**Figure 10.1**). This may seem to be a very involved way to produce a molecular phylogeny (and not much fun for the rabbits), but it is worth noting that the results are similar to those produced by more recent phylogenies using mitochondrial DNA sequences from several quagga specimens[3]. DNA hybridization has also been used to produce phylogenies by measuring similarity of DNA sequences by the amount of energy it took to disassociate them: the raw data for these phylogenies was a matrix of melting temperatures that represented a measure of difference between each of the DNA samples[4] (**Figure 11.56**). But immunological distance and DNA hybridization are rarely used to estimate phylogenies today.

Nowadays, the most common approach is to estimate genetic distance from a sequence alignment. The simplest way to estimate genetic distance between two sequences is to count the number of positions in the alignment where they differ. But when we build phylogenetic trees that represent evolutionary history, we are interested in distances as a representation of the amount of evolutionary change that has occurred on these lineages since they split from each other. For sequence data, we may miss some of those changes when we count observed differences due to the problem of 'multiple hits'. If a single site changes multiple times, we can only observe a single difference. There are a number of corrections that aim to estimate the number of substitutions that have occurred from the observed differences (**TechBox 13.1**).

**Figure 11.56** One of the first large molecular phylogenies ever constructed was a DNA hybridization tree of all bird families[4]. This tree was known in phylogenetic circles as 'the tapestry' because, when it was first displayed at a scientific conference, it took up a surprisingly large amount of wall area.



The key point for this box is that, however the genetic distances are estimated, distance-based phylogenetic methods replace the original data (whether biochemical reactions or alignments of sequences) with a distance matrix that describes those data.

### Clustering algorithms

Here is a very general outline of a common distance-based approach to phylogenetics. Note that this algorithm does not describe every distance method, but is intended only to give an overview of a basic approach (**Figure 11.57**).

**Figure 11.57**

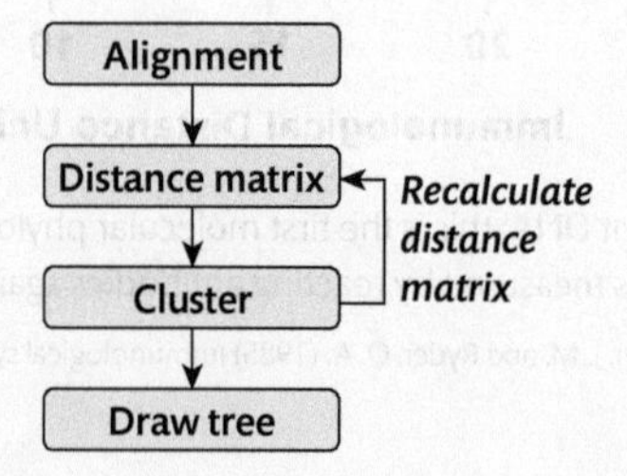


1. **Alignment:** construct an alignment of your sequences so that you know you are comparing homologous sites (**TechBox 10.2**).

2. **Distance matrix:** count the number of differences between each sequence and every other sequence, and convert this count data to a distance measure—this may be as simple as calculating the proportion of sites at which two sequences differ, or it may involve a more sophisticated correction for multiple hits (**TechBox 13.1**).

3. **Cluster:** use distances to progressively join together the sequences with the smallest distance between them. Here are two common clustering methods:

- UPGMA: the most similar sequences are clustered, then this cluster is treated as a single lineage in each subsequent clustering step. The distance matrix is recalculated after each step to give the average distance between each sequence and this new conjoined pair of sequences, then the smallest distance in this new matrix is used to cluster again, and so on. This algorithm assumes that the rate of change is equal in all lineages (under this assumption, distances are said to be additive, and the tree is ultrametric).
- Neighbour joining (NJ): like UPGMA, similar sequences are clustered then replaced by a single node in the distance matrix. However, NJ does not assume absolute rate constancy. The distance between each sequence and all other sequences is used to come up with a 'corrected' (average) distance.

4. **Draw tree:** the branching diagram in **Figure 11.48** indicates which lineages are more similar to which others. We could also have drawn the branches so that they were proportional to the amount of difference between the sequences. For example, we could draw an ultrametric tree displaying average distances, so that each path from tip (sequence) to root (basal split in the data) was the same (in this ultrametric tree, each tip is 0.6 from the root: **Figure 11.58**).

**Figure 11.58**

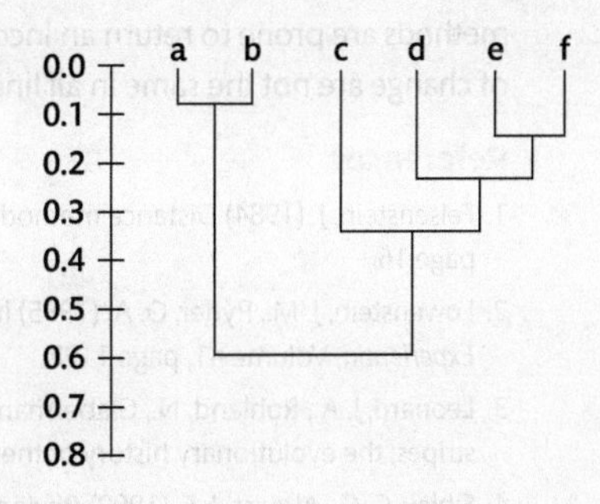


Or we could have shown the relative amounts of change on each lineage. For example you can see from the distance matrix in **Figure 11.43** that the distance between *d* and *e* is greater than between *d* and *f*, implying that there has been a faster rate of change in the lineage leading to *e*. So some sequences are separated from the root by more changes than others. We draw this as uneven heights of the tips of the tree (**Figure 11.59**). This should not be confused with a statement about time: in this case, all tips are sampled in the present, so *f* is not older than *e* just because it is lower on the tree.

**Figure 11.59**

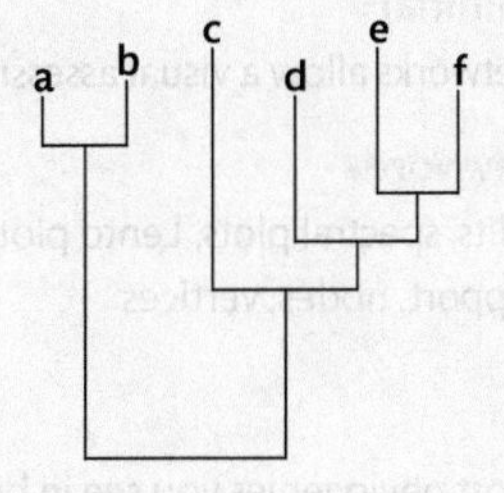


### Minimum evolution

There is an alternative way of deriving phylogenetic trees from distance data that does not rely on stepwise clustering and readjustment of the distance matrix. Like clustering methods, minimum evolution (ME) starts with a distance matrix. However, instead of grouping tips two at a time, ME essentially sums the distances along the branches of each possible phylogeny, then selects the phylogeny with the lowest total distance (otherwise known as the shortest tree). Because it involves calculating total branch lengths for every possible tree, ME is generally slower than clustering methods.

### Advantages and disadvantages

Distance methods are often described as being 'quick and dirty'. Most phylogenetics programs will return a distance tree almost immediately, whereas you may have to wait hours, days or weeks for other methods. It is common practice to use a distance-based analysis to give you a good starting point for your investigation. For example, you might estimate a NJ tree to see whether the patterns of similarity in the data are as you expected, or whether there are certain lineages for which the results seem counter-intuitive and therefore might require more investigation. You might spot something in your neighbour joining tree that requires checking before going ahead with a more time-intensive phylogeny estimation. For example, if one of your taxa 'sticks out the top' (has a much longer tip branch than other tips in the tree) then now would be a good time to go back to your alignment and check that there is nothing suspicious about that sequence and that it is all aligned satisfactorily. Many phylogenetic methods use a neighbour joining tree as a starting point for searching tree space. But, increasingly, distance trees are viewed as being poor cousins to the optimality-based phylogenetic methods, such as maximum likelihood (TechBox 12.1) and Bayesian methods (TechBox 13.1), because distance methods are prone to return an incorrect tree for some datasets, most importantly when rates of change are not the same in all lineages.

## References


1. Felsenstein, J. (1984) Distance methods for inferring phylogenies: a justification. *Evolution*, Volume 38, page 16.
2. Lowenstein, J. M., Ryder, O. A. (1985) Immunological systematics of the extinct quagga (Equidae). *Experientia*, Volume 41, page 1192.
3. Leonard, J. A., Rohland, N., Glaberman, S., Fleischer, R. C., Caccone, A., Hofreiter, M. (2005) A rapid loss of stripes: the evolutionary history of the extinct quagga. *Biology Letters*, Volume 1, page 291.
4. Sibley, C. G., Alquist, J. E. (1990) *Phylogeny and classification of birds: a study in molecular evolution*. Yale University Press.


TECHBOX
11.2

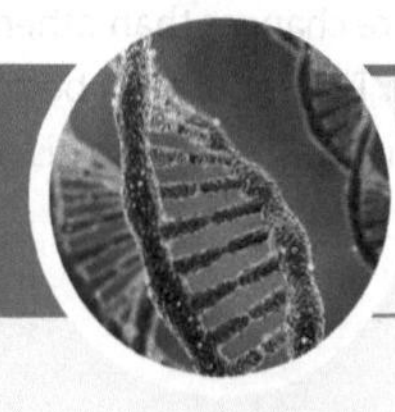

# Phylogenetic networks

RELATED TECHBOXES

TB 11.1: Distance methods
TB 13.2: Bayesian phylogenetics

RELATED CASE STUDIES

CS 11.1: Networks
CS 11.2: Distance


## Summary

Networks allow a visual assessment of both the signal and noise in phylogenetic data.



## Keywords

splits, spectral plots, Lento plots, graphs, horizontal gene transfer, hybridization, signal, noise, support, nodes, vertices


Most phylogenies you see in biology journals are bifurcating trees: each split in a lineage leads to two descendant lineages which diverge and never re-join. But not all evolutionary histories follow a strictly bifurcating path. Evolution is sometimes reticulate, meaning that lineages that have diverged may join up again. Reticulation may involve the whole genome: for example, hybridization can reconnect the branches across a phylogeny, blending genomes from different lineages to create a new taxon (Figure 11.52). Or it may involve just part of the genome: for example, horizontal gene transfer creates mosaic genomes, such that different genes can be described by different branching histories (Figure 11.51). One way of capturing these non-bifurcating modes of evolution is in a network, which is a depiction of the relationships between taxa as a series of interconnected lines.

Phylogenetic networks have been enthusiastically embraced by many biologically inclined mathematicians, and mathematically inclined biologists, who recognize them as graphs[1]. In this sense, a graph is a visual display of the connections between objects. The objects could be sequences or taxa, either at the tips of the networks (sampled taxa), or on the internal nodes (ancestral states). Nodes are also referred to as vertices. The lines connecting them, referred to as edges, indicate the relationships between them. A graph might be undirected, where the connections between the nodes do not imply a particular direction of relationship between the two: an unrooted tree or network is an undirected graph. Or they may be directed, so that you can follow a path from one node to another: a rooted tree, that defines ancestor-descendant relationships along some kind of temporal axis, is a directed graph. Just as we learned to follow a path through a bifurcating tree in this chapter (**Figure 11.11**), so you can follow a path connecting sequences through a network, from one tip to another. But in a network you may be able to follow alternative paths that suggest different connections between the taxa.

Networks are conceptually similar to distance trees in that they are a way of displaying information about the similarities and differences between your taxa. The major advantage of using networks rather than distance-based trees is that you can display incompatible splits—that is, you have a way of representing the parts of your data that are telling different stories. This conflict might be due to different patterns of inheritance (e.g. hybridization, horizontal gene transfer), or it may be due to incomplete lineage sorting such that different alleles have different histories[2], or it might be due to noise in the data overriding the signal of evolutionary history[3]. There are many different methods for creating these networks, and they can be applied to different kinds of data: for example, networks can be built from distance data[4], shared gene sets[5], sets of smaller trees (such as quartets of taxa[6]), sequence alignments[7], or they can represent the signal across a large sample of phylogenies for the same taxa (**Case Study 11.1**).

## Visualizing signal and conflict

Phylogeny estimation can be deceptively simple: you take your alignment, select a few options from a drop-down menu or type in some lines of instructions, go off and have a cup of coffee (or a meal, or a weekend away, or an overseas holiday, depending on the size of the tree and the nature of the data), then come back and hey presto there is a nice branching diagram of your sequences. But remember that you get what you ask for: if you ask for a bifurcating tree, that is what your loyal computer will give you, whether or not it accurately describes the true history of your sequences. So, if you are honourable, brave and dedicated to the pursuit of truth (and I am hoping you are), then you might want to ask just how well this tree describes your data. Have you unfairly whipped your data into a tree-shape when it is really a tangled web of connections? Have you optimistically focused your attention on the signal in the data and conveniently ignored the noise? There are a number of diagnostic tools you can use to test the support in your data for the tree the computer gave you (see for example **TechBox 12.2**). But one step you should consider, if your dataset is not horrendously large, is to visualize the phylogenetic structure in your data using a network or a plot of the splits in the alignment.

A plot of splits that speak for and against particular groupings of taxa (clades) is sometimes referred to as a spectral plot or a Lento-plot[8,9]. Admittedly, it is not used very often but it does give an intuitive way of visualizing the relative amount of signal and conflict in the data. Here are two examples. In the first example, sequences from a number of endogenous retroviruses were compared to try to determine the closest relative of the newly described Koala Leukemia Virus KoRV (**Figure 11.60**). Surprisingly, the koala virus was most closely related to a virus known only from gibbons (the tautologically named Gibbon Ape Leukemia Virus, GALV), despite the fact that their hosts' distributions don't overlap in nature (**Case Study 5.2**). To check how clearly this conclusion was supported by the data, we made a spectral plot, which revealed that there was strong support for the split that grouped KoRV with GALV, and little evidence of support for alternatively groupings[10].

The second example presents a noisier dataset[3]. Some of the splits in the best tree have relatively few supporting splits and many conflicting splits (**Figure 11.61**). So if there are many more sites speaking against a particular grouping than speak for it, why is it shown in the tree? The conflicting splits support many different alternative groupings, none of which on its own has more support than the one shown, which gets the most votes from the data. This plot suggests that were the data slightly different (more sequences, or a different sample of sites), we might well have ended up with a different phylogeny.

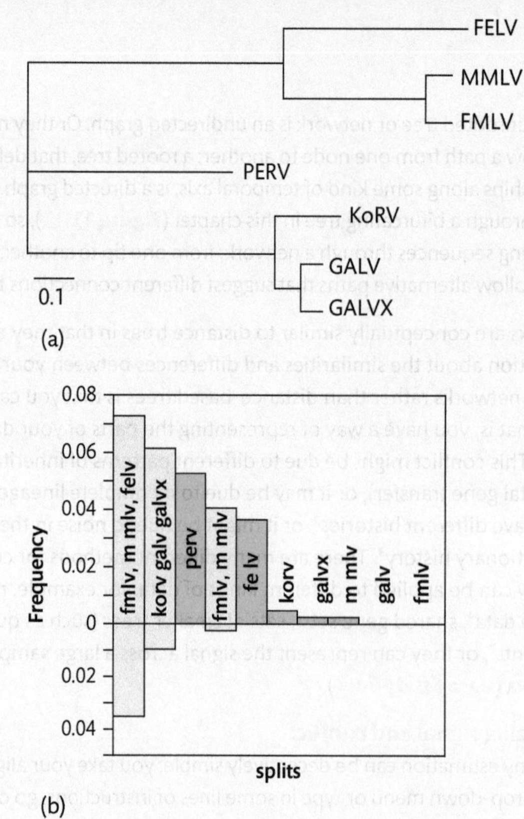


**Figure 11.60** (a) A tree of leukemia viruses from cats (FELV), mice (FMLV, MMLV), koalas (KoRV), pigs (PERV), gibbons (GALV) and cell cultures (GALVx). (b) The spectral plot (also known as a Lento-plot) displays the number of splits supporting a particular grouping above the *x*-axis, and splits that support a different grouping below. The columns in green represent the 'informative splits' because they group more than one taxon together. The columns in orange represent the 'trivial splits' because they define a tip with just one taxon—these represent the number of autapomorphs defining the edge leading to that taxon (therefore indicative of branch length).



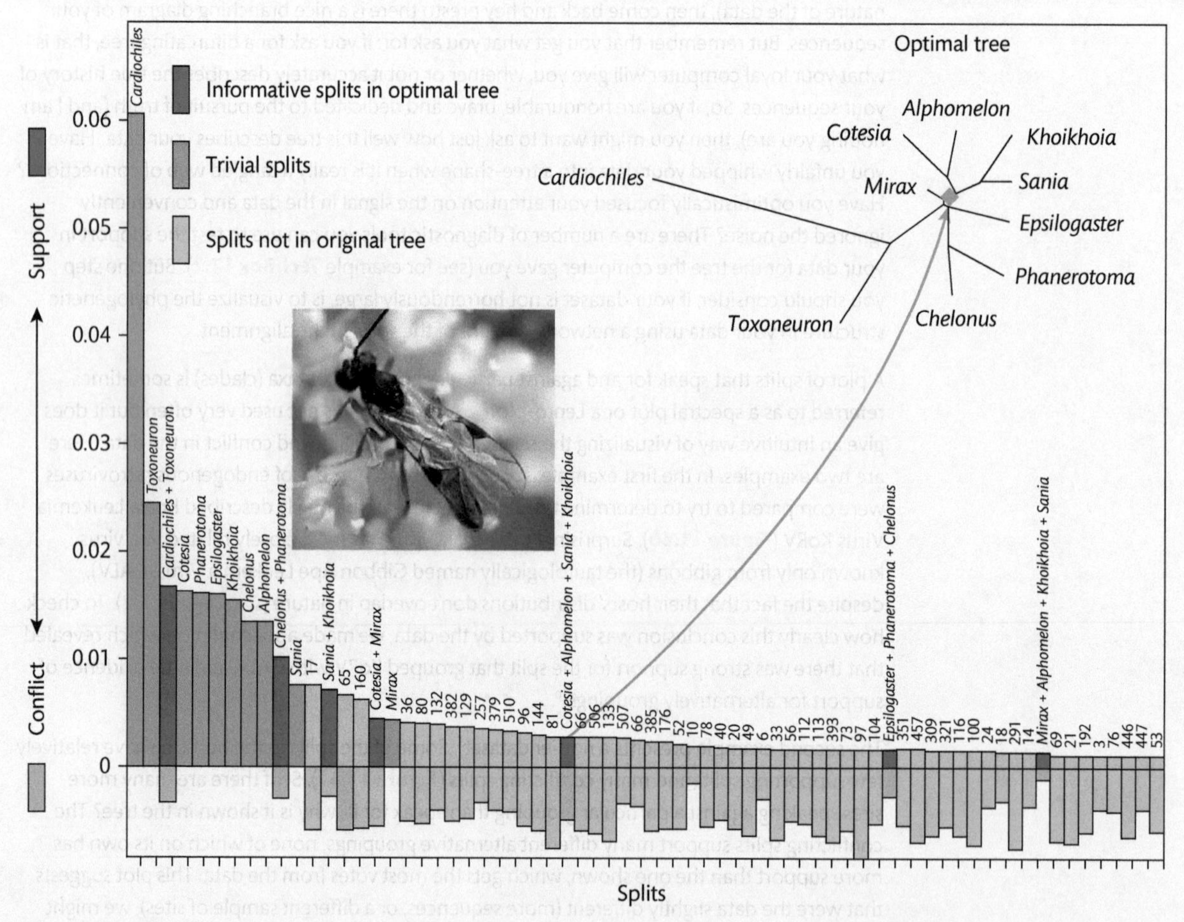


**Figure 11.61** An unrooted tree of ten genera of microgastrine wasp. One particular split in the tree is highlighted in blue: the grouping of four genera –*Alphomelon, Khoikhoia, Sania* and *Cotesia* (pictured, inset)—has relatively few supporting splits and quite a lot of conflicting splits.

The value of visualizing the signal and noise in your data is that it allows you to distinguish two different causes of low phylogenetic support. One is low signal, where there are relatively few informative sites that support a grouping, because the historical signal is weak (perhaps because the divergence is recent or the rate of change in that gene has been slow). For example, in **Figure 11.60**, the two mouse viruses (FMLV and MMLV) appear quite similar, so there are relatively few sites that support the grouping. If you were to perform a bootstrap resampling analysis you might miss the supporting sites and fail to recover the FMLV-MMLV group (**TechBox 12.2**). In the case of low signal, getting extra data might help you to test the support for a particular relationship by building up a greater number of informative sites. The second cause of low phylogenetic support is the presence of a lot of conflicting signal (high noise), so that even if there are many informative sites that support a particular grouping there are as many or more sites that support different groupings. We can see an example of this in the group indicated by the arrow in **Figure 11.61**. In this case resampling the data might tip the balance over to supporting a different phylogeny. If you have a high ratio of signal to noise, then there is no guarantee that getting more data is going to help, because it might just add lots more noise and relatively little extra signal.

Drawing a network offers an alternative way of viewing the supporting and conflicting signal in the data. If your data comes out looking like a fishing net–a whole lot of interconnecting lines with taxa scattered around the edge–then you might say your data are 'non-tree-like'. Your data don't contain a pattern of differences that define a neat hierarchical structure where close relatives are more similar across almost all sites. You could still go ahead and make a tree from that data, but you might just be kidding yourself.

### Limitations

Networks are currently most often applied at either the shallow end of phylogenetics, to describe the relationships between closely related taxa that frequently blend their genes, or at the deep end, to investigate the signal for ancient evolutionary events for which the signal to noise ratio may be less than ideal. Networks are less common in large-scale phylogenetic projects, partly because networks with lots of taxa can be hard to read and interpret, though they are increasingly being applied in phylogenomics[5]. At the moment, few of the statistical methods used to investigate patterns of evolution using phylogenies can handle networks, though that might change if they become more widely used.

Remember that networks are describing patterns in the data, which you might interpret in many different ways. If your data reveal a complex network of interconnections, maybe this is telling you that your taxa engage in rampant gene swapping, or that reticulate evolution is a better description of your data than bifurcation, or maybe it's telling you that noise outweighs signal in your data, or that you have accidently aligned paralogs rather than orthologs, or any number of things that could scramble the historical signal in your data or produce non-tree-like patterns. The network is the starting point for understanding your data: use it to generate some hypotheses then work out how you can investigate these ideas further to work out which story your data is trying to tell you.

### References

1. Huson, D. H., Rupp, R., Scornavacca, C. (2010) *Phylogenetic networks: concepts, algorithms and applications.* Cambridge University Press.
2. Posada, D., Crandall, K. A. (2001) Intraspecific gene genealogies: trees grafting into networks. *Trends in Ecology & Evolution*, Volume 16, page 37.
3. Whitfield, J. B., Lockhart, P. J. (2007) Deciphering ancient rapid radiations. *Trends in Ecology & Evolution*, Volume 22, page 258.
4. Bryant, D., Moulton, V. (2004) Neighbor-net: an agglomerative method for the construction of phylogenetic networks. *Molecular Biology and Evolution*, Volume 21, page 255.
5. Dagan, T. (2011) Phylogenomic networks. *Trends in Microbiology*, Volume 19, page 483.
6. Grünewald, S., Forslund, K., Dress, A., Moulton, V. (2007) QNet: an agglomerative method for the construction of phylogenetic networks from weighted quartets. *Molecular Biology and Evolution*, Volume 24, page 532.

7. Kloepper, T. H., Huson, D. H. (2008) Drawing explicit phylogenetic networks and their integration into SplitsTree. *BMC Evolutionary Biology*, Volume 8, page 22.
8. Lento, G. M., Hickson, R. E., Chambers, G. K., Penny, D. (1995) Use of spectral analysis to test hypotheses on the origin of pinnipeds. *Molecular Biology and Evolution*, Volume 12, page 28.
9. Huber, K. T., Langton, M., Penny, D., Moulton, V., Hendy, M. (2002) Spectronet: a package for computing spectra and median networks. *Applied Bioinformatics*, Volume 1, page 2041.
10. Hanger, J. J., Bromham, L. D., McKee, J. J., O'Brien, T. M., Robinson, W. F. (2000) The nucleotide sequence of koala (*Phascolarctos cinereus*) retrovirus (KoRV): a novel type-C retrovirus related to gibbon ape leukemia virus (GALV). *Journal of Virology*, Volume 74, page 4264.

CASE STUDY 11.1

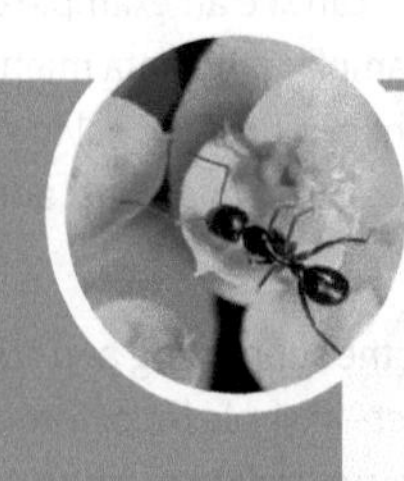

# Networks: bobtail squid and their symbiotic bioluminescent bacteria

RELATED TECHBOXES

TB 11.2: Phylogenetic networks
TB 13.2: Bayesian phylogenetics

RELATED CASE STUDIES

CS 10.1: Alignment
CS 11.2: Distance

Wollenberg, M. S., Ruby, E. G. (2011) Phylogeny and fitness of *Vibrio fischeri* from the light organs of *Euprymna scolopes* in two Oahu, Hawaii populations. *The ISME Journal*, Volume 6, page 352

*"this highly specific symbiosis ensues within hours of the host hatching from the egg, early developmental events of the partners occur within a couple of days, and, from onset throughout the host's life, the symbiosis is on a profound daily rhythm[1]"*

**Keywords**
phylogeny, network, splits, co-evolution, consensus, symbiont, selection, life-history tradeoffs

## Background

Hawaiian bobtail squid glow at night (**Figure 11.62**). That's because the squid (*Euprymna scolopes*) have special light-producing organs that contain bioluminescent bacteria (*Vibrio fischeri*). Unlike some symbioses where the host inherits their endosymbiotic bacteria from their parent, bobtail squid are born without their *Vibrio* partners. The bacteria colonize juvenile squid from the surrounding seawater, collecting on a special epithelial tissue, from where they move into internal crypts inside a special light-producing organ[1]. Each squid 'vents' its light-producing organ each day, flushing most of the bacteria back into the seawater. The crypt is then repopulated by the remaining bacteria.

## Aim

Phylogenies of vertically transmitted symbionts often track that of their hosts, but horizontally-acquired symbionts are resampled from the environment every generation. So do bobtail squid all have a random selection of *Vibrio* strains in the environment, or are there distinct lineages of symbionts associated with particular populations of hosts? A previous study suggested that one particular strain of *V. fischeri* identified from DNA fingerprinting, referred to as Group-A, was more abundant in light organs of bobtail squid collected in one particular place (Maunalua Bay)[2]. In this study, researchers set out to characterize the relationship of this strain to other *V. fischeri* strains, and to investigate whether its relatively high abundance in Maunalua Bay squid

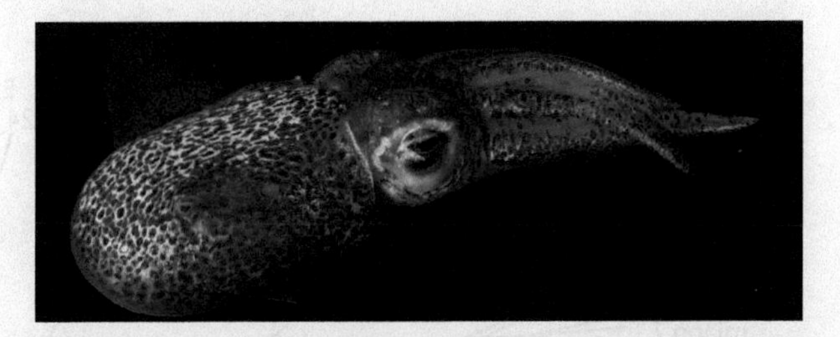

**Figure 11.62** Why do bobtail squid light up at night? The most widely accepted explanation is that it is for camouflage. This may seem counterintuitive that an animal would turn on its party lights to make itself harder to see, but the light may break up the silhouette of the animal when viewed from below, making it harder for predators to spot.



could be explained by its having particular advantages either in the open environment or within its host's light-producing organ.

## Methods

Forty-five different strains of *Vibrio fischeri* sampled from around Oahu (Hawaii) were included in the phylogenetic study, plus two outgroups from other species of *Vibrio* to root the tree. They sequenced four housekeeping genes and two genes involved in bioluminescence, and estimated phylogenetic trees using a number of different methods. They also used several methods that specifically permit non-tree-like patterns in the data, to allow them to detect conflict in the historical signal and to allow for the possibility of recombination between strains[3].

## Results

The maximum likelihood phylogenies estimated separately from each of the six genes were all different from each other, with few well-supported nodes in common between gene trees. The two genes associated with bioluminescence had twice as many informative sites as the housekeeping genes. Inspection of the data revealed that the phylogeny based on the *lux* gene (critical in bioluminescence) had the strongest support, because it had fewer distinct alleles in the population but a greater degree of sequence divergence. *lux* sequences had the highest number of changes to the base sequence but the lowest rate of recombination, so the historical signal was recorded the most reliably, and scrambled less often. This pattern of variation may be a result of the *lux* gene having undergone either strong selection or a bottleneck. A consensus network (which summarizes support from multiple runs of a phylogenetic analysis) indicates that the Group-A strains cluster together (**Figure 11.63**). Experiments showed that the Group-A strains were more efficient at colonizing the light organs of juvenile squid, but that they were less capable of thriving in unfiltered seawater over several days.

## Conclusions

The phylogenetic analyses confirm that Group-A *V. fischeri* cluster together, suggesting that they form a genetically distinct lineage. The lack of agreement between phylogenies based on single genes may indicate that recombination is common in *V. fischeri*. The authors give two alternative explanations for the identification of a distinct Maunalua Bay strain of *V. fischeri*, consistent with the results of their selection experiments. One explanation is that the Group-A strain has undergone co-evolution with the particular host population found in Maunalua Bay, so that it has become more efficient at colonizing those hosts than other strains in the environment. The other explanation is that the *V. fischeri* population in Maunalua Bay is distinguished by trade-offs, increasing its ability to colonize and grow in the nutrient-rich environments of squid light-organs at the expense of growth in nutrient-poor seawater.

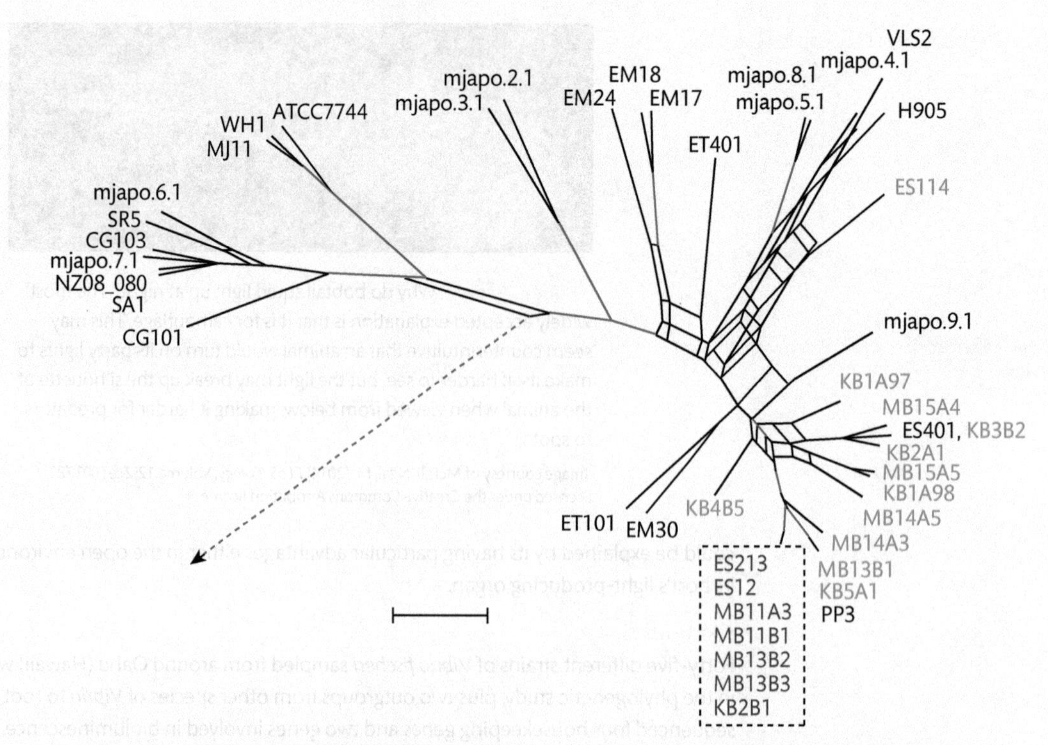


**Figure 11.63** Consensus network for strains of *Vibrio fischeri* samples from bobtail squid in Oahu. A consensus network summarizes phylogenetic information from multiple phylogenies. In this case, the sample of trees were all estimated from the posterior distribution of trees from an alignment of four housekeeping genes (see **TechBox 13.2**). The red branches are edges that were present in over 95 per cent of trees in the sample, while the black boxes represent alternative splits present in at least 20 per cent of trees. The coloured labels indicate strains identified using DNA fingerprinting. The strains in the dotted-line box are those identified as Group-A, which correspond to a particular strain. The dotted red line indicates the position of the root connecting to the other species of *Vibrio* included as outgroups.



### Limitations

This paper focuses on establishing whether the strain that was previously shown to be abundant in Maunalua Bay squid formed a distinct evolutionary lineage. It would be interesting to see if the Group-A *V. fischeri* are also more abundant in the Maunalua Bay environment, as well as being more abundant in the Maunalua Bay bobtail squid. The authors point out that they do not yet know the extent to which *Vibrio fischeri* strains are adapted to free-living existence, given that the population numbers within light-organs is assumed to be much greater than in the free-living population.

### Further work

It would be interesting to look for similar associations between other bobtail squid populations and specific *Vibrio fischeri* strains. Because both the squid and the bacteria can be grown efficiently in the lab, the authors suggest it may be possible to conduct multi-generation experiments, starting with a single genetic isolate, to track bacterial adaptation to hosts.

### Check your understanding

1. Why did the authors choose to display their data as a network rather than a bifurcating phylogeny?
2. Does the clustering of Group-A *Vibrio fischeri* strains together in the network indicate that they are better suited to environmental conditions in Maunalua Bay than other strains?
3. Does the presence of conflicting signal in the data prove that there is recombination between *V. fischeri* strains?

**What do you think?**

How could you distinguish the two possible hypotheses given by the authors for the overrepresentation of Group-A *V. fischeri* in the Maunalua Bay bobtail squid: co-evolution to local host versus life history trade-offs in the group-A lineage?

**Delve deeper**

What effect should the 'venting' of bioluminescent bacteria every day have on the genetic association between the hosts and symbiont? Would you expect a different genetic pattern if bioluminescent bacteria were passed from host parent to offspring?

**References**

1. McFall-Ngai, M. (2014) Divining the essence of symbiosis: insights from the Squid-Vibrio model. *PLoS Biology*, Volume 12: e1001783.
2. Wollenberg, M., Ruby, E. (2009) Population structure of *Vibrio fischeri* within the light organs of *Euprymna scolopes* squid from two Oahu (Hawaii) populations. *Applied and Environmental Microbiology*, Volume 75, page 193.
3. Didelot, X., Falush, D. (2007) Inference of bacterial microevolution using multilocus sequence data. *Genetics*, Volume 175, page 1251.

CASE STUDY 11.2

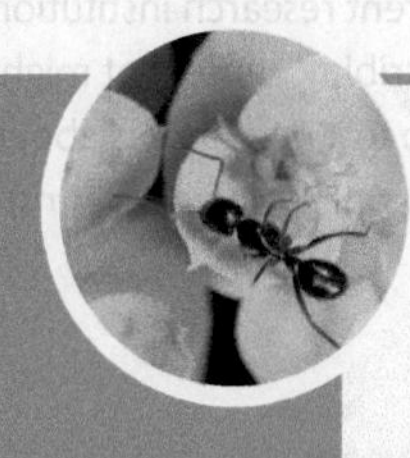

# Distance: tracing the origins and evolution of the domesticated apple

**RELATED TECHBOXES**

**TB 5.2:** High-throughput sequencing

**TB 11.1:** Distance methods

**RELATED CASE STUDIES**

**CS 3.1:** SNPs

**CS 11.1:** Networks

Velasco, R., Zharkikh, A., Affourtit, J., Dhingra, A., Cestaro, A., Kalyanaraman, A., Fontana, P., Bhatnagar, S. K., Troggio, M., Pruss, D., Salvi, S., Pindo, M., Baldi, P., Castelletti, S., Cavaiuolo, M., Coppola, G., Costa, F., Cova, V., Dal Ri, A., Goremykin, V., Komjanc, M., Longhi, S., Magnago, P., Malacarne, G., Malnoy, M., Micheletti, D., Moretto, M., Perazzolli, M., Si-Ammour, A., Vezzulli, S., Zini, E., Eldredge, G., Fitzgerald, L. M., Gutin, N., Lanchbury, J., Macalma, T., Mitchell, J. T., Reid, J., Wardell, B., Kodira, C., Chen, Z., Desany, B., Niazi, F., Palmer, M., Koepke, T., Jiwan, D., Schaeffer, S., Krishnan, V., Wu, C., Chu, V. T., King, S. T., Vick, J., Tao, Q., Mraz, A., Stormo, A., Stormo, K., Bogden, R., Ederle, D., Stella, A., Vecchietti, A., Kater, M. M., Masiero, S., Lasserre, P., Lespinasse, Y., Allan, A. C., Bus, V., Chagne, D., Crowhurst, R. N., Gleave, A. P., Lavezzo, E., Fawcett, J. A., Proost, S., Rouze, P., Sterck, L., Toppo, S., Lazzari, B., Hellens, R. P., Durel, C-E., Gutin, A., Bumgarner, R. E., Gardiner, S. E., Skolnick, M., Egholm, M., Van de Peer, Y., Salamini, F., Viola, R. (2010) The genome of the domesticated apple (*Malus× domestica* Borkh.). *Nature Genetics*, Volume 42(10), pages 833–839.

> *Apples are known to have been gathered in the Neolithic and Bronze Age in the Near East and Europe, and all archaeological findings indicate a fruit size compatible with those of the wild* M. sylvestris, *a species bearing small astringent and acidulate fruits. Sweet apples corresponding to extant domestic apples appeared in the Near East around 4,000 years ago, at the time when the grafting technology used to propagate the highly heterozygous and self-incompatible apple was becoming available. From the Middle East, the domesticated apple passed to the Greeks and Romans, who spread fruit cultivation across Europe*

**Keywords**
network, whole genome, gene duplication, genome assembly, hybridization, Linnean nomenclature, synteny, distance, Hamming, domestication

**Background**

The apple has been an important crop plant for many human cultures, with desirable varieties largely propagated by grafting[1] (**Figure 11.64**). The wild relatives of apples are many and widespread, so hybridization between lineages may have left a complicated signature in the genome. Which wild species contributed to the development of the domestic apple (*Malus × domestica* Borkh.)? In case you were wondering, the '×' in the species name is a marker of a hybrid taxon—the hybrid is given a specific epithet (in this case, *domestica*) but the name doesn't reveal the source of the parent taxa other than the genus (intergeneric hybrids have the cross before the genus name). 'Borkh.' is the authority, which refers to the person who first formalized this name in a taxonomic publication (in this case, the botanist Moritz Balthasar Borkhausen, who described the domestic apple in his 1803 manual on forest trees).

**Aim**

A large collaborative team of researchers from 18 different research institutions contributed to the genome sequencing and analysis. As well as describing genes that might be involved in the formation of the distinctive fruit, the team aimed to test hypotheses about the origin of the domestic apple: was it derived from the European crabapple (*Malus sylvestris*) or wild apple from Central Asia (*Malus sieversii*)?

**Figure 11.64** Apples are pome fruits: the fruit develops from the receptacle at the base of the flower. This means that the fruit is made from parent tree tissue, not grown from the fertilized ovule. Because of this, apples do not 'breed true'—growing an apple tree from the seed of a particularly delicious apple won't necessarily give you a good crop, because that delicious apple was the result of the genome of the tree it grew on, not the genome of the seed it contained. Some apple varieties are produced from seedlings generated by crossing different varieties together. For example, this apple tree is a Pink Lady, otherwise known as Cripp's Pink, an Australian variety developed in the 1970s by John Cripps by crossing a Golden Delicious with a Lady William. Some new apple varieties are identified as 'sports', branches on existing trees in which a somatic mutation occurred during growth that made the fruit growing from that branch noticeably different from the rest of the tree. The mutant variety can then be propagated vegetatively. For example, Ruby Pink apples are all descended from a single limb found growing on an otherwise normal Pink Lady tree in an orchard. Ruby Pinks have been registered as a new variety because the fruit is a different shape, colour and texture, and it matures later than standard Pink Ladies. There are a large variety of apples available, and effort is being made to keep 'heritage' varieties alive. One of my favourite heritage apple varieties is the quirky Peasgood Nonsuch, which originated from a single ancestral apple seed planted in a pot by Mrs Peasgood in 1858 in Lincolnshire, UK.

Photographs by Lindell Bromham.

**Methods**

They used a combination of Sanger sequencing and high-throughput sequencing, using genome maps produced by breeding experiments to anchor the genome assembly. Genome assembly for the apple is complicated by presence of repetitive elements, multiple copies of genes due to ancestral polyploidy and high levels of heterozygosity (variation in their sequence between homologous chromosomes), so they could only assemble around three quarters of the genome. They used iterative assembly, grouping contigs into metacontigs, ultimately constructing linkage groups for each of the 17 chromosomes. They used a variety of methods for gene prediction, then blasted the putative amino acid sequences to find homologs in other species. They estimated genetic distances between three species—apple, peach, and pear—based on alignments of EST data. Homologous genes were aligned and synonymous changes were used to estimate the age of duplication events. Domestic apples may have had genetic input from a variety of wild *Malus* species, so the history is unlikely to conform to a simple bifurcating evolutionary tree[2]. To examine relationships between apple lineages, they sequenced 23 genes from a range of cultivars and wild species, then constructed a network based on uncorrected (Hamming) distances.

**Results**

The genome analysis suggested a relatively high number of genes, with over 57,000 putative genes. They estimated that over 40 per cent of the genome is made up of transposable elements. Comparing the chromosomes to each other revealed areas of collinearity, which may be the result of genome duplications, however, many chromosomes also contained sections with no obvious matches elsewhere in the genome. Peaks in the synonymous distances between chromosomal segments were interpreted as representing the different ages of multiple rounds of genome duplication. The genetic distance network shows that the sequences from the domestic apple varieties are more similar to those from *M. sieversii* than *M. sylvestris* (**Figure 11.65**).

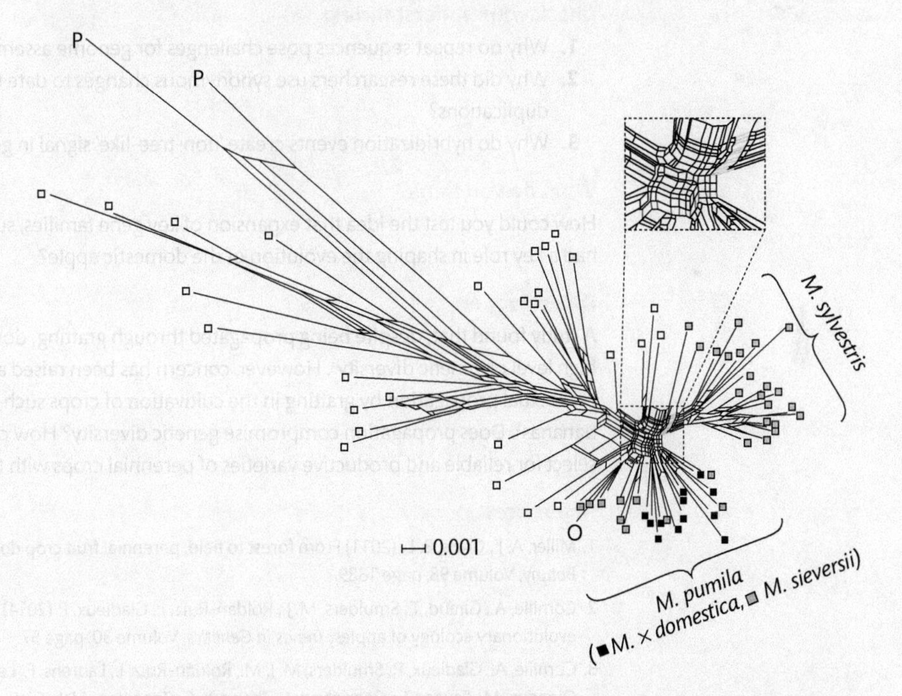


**Figure 11.65** Network of domesticated apple varieties (*M.* × *domestica*) and wild *Malus* species. This network is a splits graph based on observed differences between sequences from an alignment of 23 different genes. The red colour highlights splits that support a group containing domestic apples (*M.* × *domestica*), *M.* × *asiatica* (a Chinese cultivated apple), and two wild apples from Central Asia (*M. sieversii, M. orientalis*). Splits grouping European crab apples, *M. sylvestris*, are highlighted in green.

Reprinted by permission from Macmillan Publishers Ltd: *Nature Genetics*, Volume 42, pages 833–839 (2010), copyright 2010.

### Conclusions

The authors suggest that the domestic apple is derived primarily from the Central Asian wild apple. The genome of the domestic apple bears the signature of whole genome duplication events at various stages in its evolutionary history. The estimated dates of the duplications vary from 30 to 65 million years ago, depending on the calibration used. They found that, in comparison to other crop plants with genome sequences, apples had larger numbers of genes in several key gene families, including the transcription factor genes in the *MADS-box* family, which have been implicated in flower and fruit development.

### Limitations

While a series of whole genome duplications provides a plausible hypothesis for the patterns in the apple genome, the uncertainty in the patterns and dates of these events, and the areas without obvious duplicates in the genome, could indicate that the patterns of synteny were generated by many smaller duplication events that each involved only part of the genome. Dating the origin of genome duplications relies on making assumptions about the rate of molecular evolution (see Chapter 14). While these results suggest that the Central Asian wild apple is the ancestor of the domestic apple, the non-tree-like signal in the network allows for genetic input from the European crab apple[3].

### Further work

Identifying gene families that are expanded in apples does not, in itself, provide evidence that these genes were critical in the formation of key characteristics. However, they provide an interesting starting point for investigation. Investigation of the population genetics of both domesticated and wild apples may allow comparison of two alternative explanations of the reticulate pattern in the distance network, hybridization and incomplete lineage sorting[4] (differing gene histories due to polymorphism in the ancestral population).

### Check your understanding

1. Why do repeat sequences pose challenges for genome assembly?
2. Why did these researchers use synonymous changes to date the origin of genome duplications?
3. Why do hybridization events create 'non-tree-like' signal in genomic data?

### What do you think?

How could you test the idea that expansion of key gene families, such as the *MADS-box* genes, had a key role in shaping the evolution of the domestic apple?

### Delve deeper

A study found that, despite being propagated through grafting, domesticated apples retain a high level of genetic diversity[3]. However, concern has been raised about the wisdom of reliance on asexual propagation by grafting in the cultivation of crops such as apples, grapes and bananas[5]. Does propagation compromise genetic diversity? How could you balance the need to select for reliable and productive varieties of perennial crops with the maintenance of variation?

### References


1. Miller, A. J., Gross, B. L. (2011) From forest to field: perennial fruit crop domestication. *American Journal of Botany*, Volume 98, page 1389.
2. Cornille, A., Giraud, T., Smulders, M. J., Roldán-Ruiz, I., Gladieux, P. (2014) The domestication and evolutionary ecology of apples. *Trends in Genetics*, Volume 30, page 57.
3. Cornille, A., Gladieux, P., Smulders, M. J. M., Roldán-Ruiz, I., Laurens, F., Le Cam, B., Nersesyan, A., Clavel, J., Olonova, M., Feugey, L., Gabrielyan, I., Zhang, X-G., Tenaillon, M. I., Giraud, T. (2012) New insight into the history of domesticated apple: secondary contribution of the European wild apple to the genome of cultivated varieties. *PLoS Genetics*, Volume 8: e1002703.
4. Harrison, N., Harrison, R. J. (2011) On the evolutionary history of the domesticated apple. *Nature Genetics*, Volume 43, page 1043.
5. Myles, S. (2013) Improving fruit and wine: what does genomics have to offer? *Trends in Genetics*, Volume 29, page 190.

# Hypotheses

## *Seeing the wood for the trees*

*"Out of the same facts of anatomy and development men of equal ability and repute have brought the most opposite conclusions. . . . Facts of the same kind will take us no further. The issue turns not on the facts but on the assumptions. . . . If facts of the old kind will not help, let us seek facts of a new kind"*

Bateson, W. (1894) *Materials for the study of variation: treated with especial regard to discontinuity in the origin of species.* Macmillan and Co.

## What this chapter is about

Phylogenies map paths of descent. But there are many possible phylogenies that could explain how we ended up with the DNA sequences we have observed. Phylogenetic analysis is a way of generating and comparing different hypotheses for the evolution of a set of sequences, so requires a criterion for choosing the most plausible history. Most methods compare the plausibility of different phylogenies, using a model of molecular evolution which states the relative probability of different kinds of sequence changes. The strength of support for different possible phylogenies can be compared by asking whether slightly different data, or differences in the analysis, would have produced the same result. Phylogenies can also be used to test hypotheses about evolution by asking what patterns we would expect to see if the hypothesis was true.

### Key concepts

- Phylogenies represent possible evolutionary histories
- Hypotheses concerning the evolutionary past and processes can be tested in a statistical framework

# Comparing trees

*Picturing history*

Phylogenies themselves have a long history of descent. The first diagrams that map lines of descent and divergence of lineages predate evolutionary theory. For example, in 1766 the French naturalist Antoine-Nicolas Duchesne included a diagram in his treatise on the natural history of strawberries that showed his ideas about the lines of descent from the wild strawberry to nine different cultivated varieties. Jean-Baptiste Lamarck is sometimes credited with creating the first evolutionary trees, because he depicted the descent of animal groups with lines connecting ancestral stock to their descendants, including some bifurcations where a single ancestral lineage splits to give rise to two descendants.

The development of evolutionary biology in the early twentieth century generated a need for diagrams that illustrated the evolutionary history of living species, connecting fossil taxa to their living descendants, and illustrating changes in form and function over time. **Figure 12.1** provides a charming example, showing the relationship between past and present elephant species, drawn by the palaeontologist Henry Fairfield Osborn and published in 1934. Many of these hand-drawn phylogenies included information on the relative diversity of lineages throughout geological time, as well as mapping significant changes in form or distribution of species. These phylogenies represent a distillation of their author's knowledge and opinions. Through their detailed

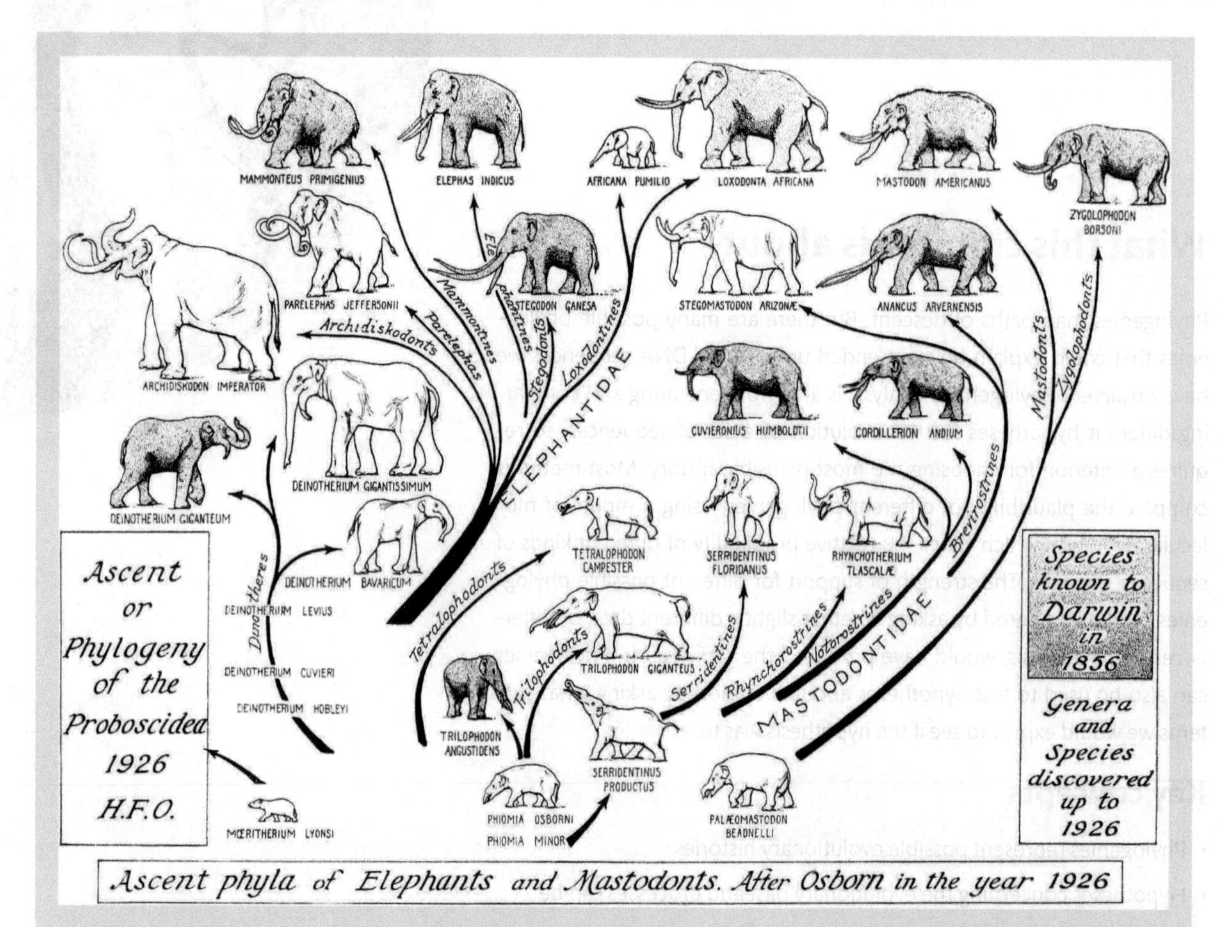


**Figure 12.1** The evolution of elephants according to Henry Fairfield Osborn, renowned fossil hunter and a founding director of the vertebrate palaeontology department at the American Museum of Natural History. This phylogeny is from a 1934 paper in which Osborn used the adaptive radiation of elephants, increasing in diversity over time, to support his notions of aristogenesis, or the unfolding of evolutionary potential over time.

investigation of a particular group of interest, naturalists formed ideas about the way traits had changed over time, as the species had evolved and adapted. A phylogeny could encapsulate a wide range of disparate information on time periods, environments, morphological traits and relationships together in one easily digestible diagram, a visual representation of a complex evolutionary history.

Sadly, these endearing hand-drawn phylogenies are now practically extinct. Almost all phylogenies, whether based on molecular, morphological, behavioural or other data, are now generated using computational algorithms that infer phylogeny according to a defined set of rules, typically within a statistical framework that weighs alternative explanations of the data and produces the most well-supported tree according to some optimization strategy. What is gained is objectivity and repeatability: given the same data and the same methods and parameters, anyone would get the same answer. What is lost, though, is the personal, quirky touches, and the feeling of connection to the wisdom of an expert. Modern phylogenies may be more defensible and, for want of a better word, more sciency, but they are also in some ways just a bit duller.

While 'desirograms' like **Figure 12.1** may be subjective, based on personal intuition rather than the results of an objective algorithm, there is one very important way in which they are the same as all the phylogenies produced today by the latest whizz-bang computer programs. These trees are all hypotheses about evolutionary descent. Each tree contains a possible history, one proposed explanation for the pattern of descent that could connect and explain the diversity we have observed. Any particular phylogeny may be disproved by new data or new methods or a re-evaluation of assumptions. In fact, few phylogenies stand immutable. If you happen to be reading this book in 2030, for historical interest only, you may find that few of the phylogenies represented in this book have stood the test of time unchallenged (except, perhaps, the ones I made up in the last chapter). In this chapter, we are going to focus on the concept of the phylogeny as a hypothesis, as a statement about the most plausible path and mode of descent, given a set of assumptions about how evolution proceeds.

## Possible histories

In Chapter 11, we discussed distance trees and networks, where patterns of similarity are used to cluster sequences into nested groups. Since descent with modification is expected to leave a hierarchy of similarities, branching diagrams based on similarities can be used to propose evolutionary relationships.

But many people who produce molecular phylogenies intend to do something more than describe patterns of similarity in the data. Phylogenies are usually constructed in order to work out the evolutionary history of a set of taxa. When we construct phylogenies, we are not simply describing the end points of the process (which sequences are most similar?), we are aiming to reconstruct the underlying process itself (what series of changes occurred to produce these sequences?). This is essentially what we did in **Figures 11.27** to **11.33**: we interpreted each split in the alignment as a substitution event that had occurred in one particular lineage, then was inherited by its descendant lineages. We mapped the inferred substitution events onto particular lineages (edges) in the phylogeny (**Figure 12.2**).

**Figure 12.2**

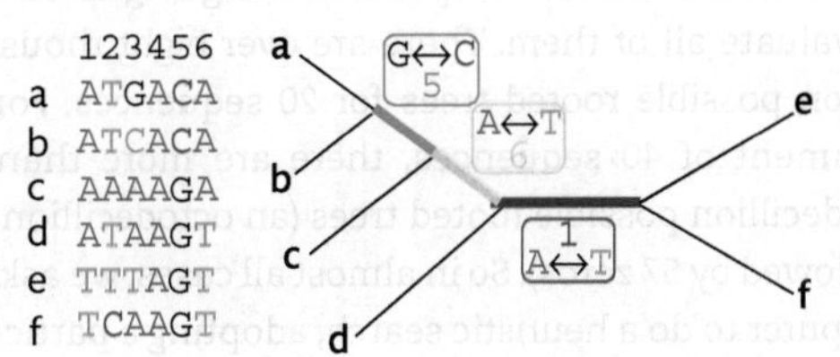


This fundamental difference—process instead of pattern—requires a different approach to solving phylogenies. For distance methods, there is (usually) a single tree that best describes the pattern of similarities. In contrast, if we are aiming to create a tree that maps a hypothetical series of changes that took place to create the sequences, then we can imagine many different possible evolutionary paths that could all lead to the pattern of sequences we see today. So, to describe our DNA sequences using a phylogeny that maps evolutionary change, we are going to need a way of creating and evaluating different trees.

Remember that we don't want to have to do all the work ourselves, we want a computer to do it. The only way a computer can solve this kind of problem is if we give it a set of instructions. We tell the computer to generate one possible phylogeny that describes the evolution of our sequences, then use some kind of rule to score how plausible that phylogeny is as an explanation of our data, then generate a different phylogeny, score that one for plausibility, and keep generating and scoring trees until the computer can report the tree that got the highest plausibility score. This sounds pretty straightforward, but we need to tell the computer how to do both parts of the process: generate a phylogeny, and score it for plausibility.

If this is sounding familiar, you may recall that we described a similar process for telling a computer how to find the best alignment for a set of sequences (Chapter 10): generate an alignment, score it, change it slightly, score it again, and keep going until you have found the alignment with the best score. For two sequences, each 300 nucleotides long, there are over $10^{100}$ possible alignments, so any useable computer program needs to have an efficient strategy for searching for the alignment that best describes the data (**TechBox 10.2**). Similarly, for any given alignment of sequences, there are a very large number of possible phylogenies that could explain how they evolved. For the six species in **Figure 12.2** (*a* to *f*), there are 945 possible rooted trees, or 105 possible unrooted trees. For small numbers of taxa, a computer can do an exhaustive search, scoring every possible tree. But for most datasets, there are so many possible trees that even the fastest computer is not going to be able to evaluate all of them. There are over eight thousand trillion possible rooted trees for 20 sequences. For an alignment of 40 sequences, there are more than an octodecillion possible rooted trees (an octodecillion is a 1 followed by 57 zeros). So in almost all cases, we ask the computer to do a heuristic search, adopting a particular way of generating and testing alternative trees that is likely to find the best explanation for our data without having to try every single possible tree.

***Rooted and unrooted trees are explained in Chapter 11***

One possible search strategy can be summarized like this: start with a distance-based tree, cut the tree at a randomly chosen edge, then reattach that edge somewhere else in the tree. Now you have a slightly different tree than you started with, and you can ask if it is better than the original, using whatever your chosen method of evaluating how well the tree explains the data. If the rearranged tree scores better than the starting tree, keep it and use it as the basis of more rearrangements, if it scores lower than the original tree, abandon it and go back to the first tree. Keep repeating until you find that no further rearrangements can improve the score of your tree, which you then accept as the best tree for this data. This approach is referred to as 'hill climbing'—because you only take steps in an upward direction (to a tree with a better score), never downward (to a tree with a lesser score: **TechBox 12.1**). An alternative approach would be to weight the decision whether to move to the new tree by the difference in the score between the new tree and the old tree: the better the score, the more likely it is you will adopt the new tree, but there is always some chance you will stay on the old tree (**TechBox 13.2**). You can also conduct more than one search, by starting your series of possible trees again from a different beginning, or you could occasionally make a big change that generates a radically different tree.

Because of the impossibly large number of alternative trees, your search method will never test every single possible tree, so you can't guarantee that one of the untested trees wouldn't have had a higher score. But in practice these algorithms work very well. However, this searching process—modifying a tree, scoring it, comparing it, modifying it again—is time-consuming. The more taxa you have in your tree, the more possible trees there are and the longer it will take to find the best one.

## Evaluating alternative trees

> *A wise man, therefore, proportions his belief to the evidence . . . All probability, then, supposes an opposition of experiments and observations, where the one side is found to overbalance the other and to produce a degree of evidence, proportioned to the superiority.*
>
> Hume, D. (1748) "Of Miracles" in *An Enquiry Concerning Human Understanding.*

How are we going to score alternative trees in order to decide which is the best? The tree shown in **Figure 12.2** provides a reasonable explanation of the pattern of substitutions seen in the alignment. But it is not the only possible explanation of that pattern of substitutions. There are other possible trees that could also give rise to the same set of sequences. To illustrate this, I generated an alternative tree by randomly pulling off some branches and reconnecting them elsewhere. I disconnected *f* from its position next to *e* and I reconnected it next to *b*, then I disconnected *c* and stuck it next to *a*, and so on (**Figure 12.3**).

**Figure 12.3**

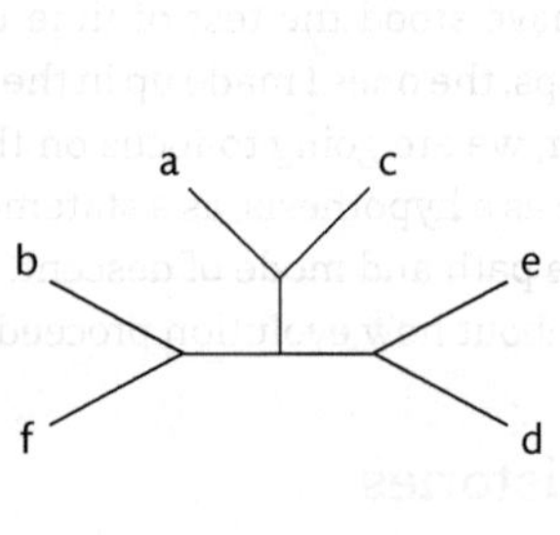

This is an alternative evolutionary history for our sequences. In other words, the first tree was one way that our present-day sequences could have been generated, and this is another way. To explain how this possible history gave rise to the sequences we have in our alignment, we need to map substitution events onto the tree.

So, to explain the pattern of nucleotides we see in site 1, we need to show how *a, b, c* and *d* all ended up with an A, but *e* and *f* ended up with a T. On this tree, *e* and *f* aren't grouped together, so we can't say that the reason that they both have a T is that they both inherited it from a common ancestor (unless it was then subsequently lost three separate times, in the *b*, *d* and *a-c* lineages). Instead, we might say that *e* and *f* both have a T in position 1 because they both independently acquired a substitution from A to T (**Figure 12.4**).

**Figure 12.4**

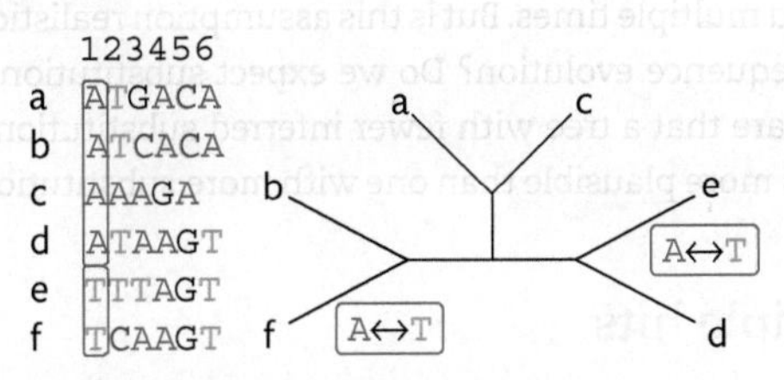


This tree requires two A to T substitutions to explain site 1, whereas the previous tree (**Figure 12.2**) required only one. We can map the other substitutions on in the same way. For example, *a, b, c* and *d* are not grouped together, so we can't simply say that the ancestor to all four populations underwent a substitution from T to A, so we have to infer multiple substitutions from T to A (**Figure 12.5**). Similarly, to explain the C at site 5 in both *a* and *b* we are going to need to believe either that it was gained independently in both *a* and *b* or that it was lost in both *c* and *f*.

**Figure 12.5**

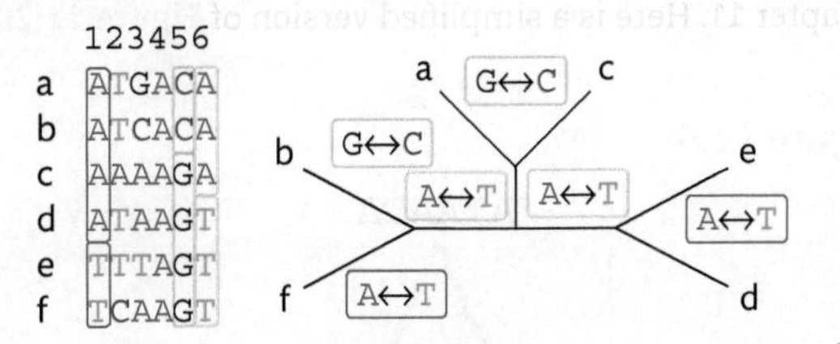


We now have a way of scoring and comparing our alternative trees. We can map substitutions onto our alternative trees, then ask 'how many substitutions does each tree imply happened during the history of these sequences?' We are not going to bother mapping on the shared ancestral characters (found in all species) or the unique derived characters (found in only one species) because these don't split the data into groups, so will be the same in all possible trees. We are only going to map the shared derived characters, which define groups of sequences.

***For explanation of shared ancestral characters, unique derived characters and shared derived characters, see Chapter 11***

Here is a comparison of the number of substitutions we need to map on to our two alternative evolutionary hypotheses, Tree 1 (**Figure 12.2**) and Tree 2 (**Figure 12.3**), in order to explain the sequence data we observe (**Figure 12.6**).

**Figure 12.6**

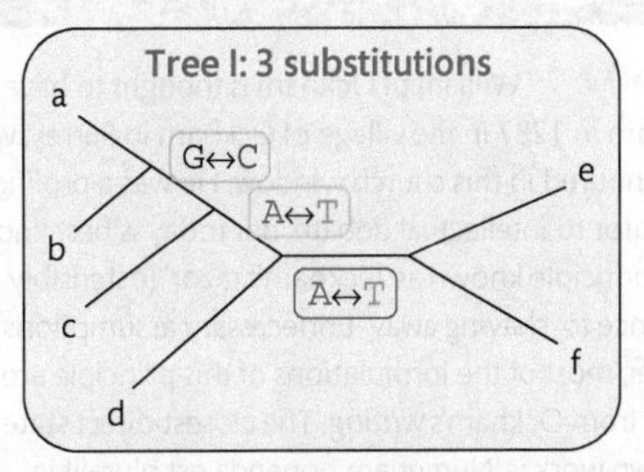


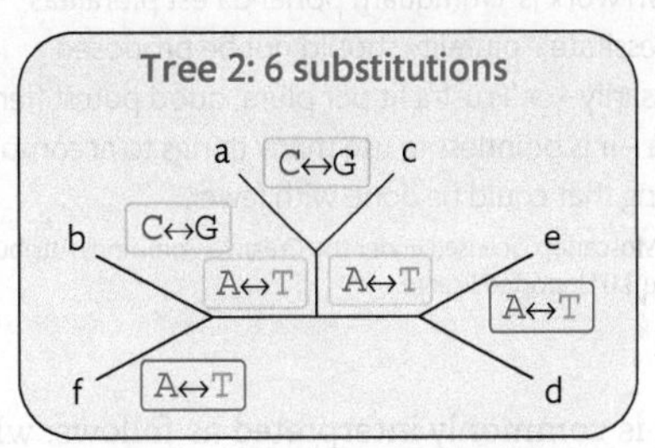


Tree 2 requires that we assume more substitutions events than Tree 1. Which one of these trees do you feel gives the most plausible explanation of the history of these sequences? Given no other information, you might decide that Tree 1 is the most plausible, because it requires us to assume fewer independent evolutionary changes. This is how most phylogenetic methods work: in order to explain our observations (DNA sequences) we generate different possible hypotheses (trees) and we compare their plausibility, and pick the one that seems to provide the most likely explanation. But the criteria for choosing the most plausible tree will depend on the method you use.

### *Parsimony*

In this example in **Figure 12.6**, we employed the principle of parsimony to decide which tree is more believable. We chose the explanation that required us to infer the fewest number of evolutionary events (in this case, an 'event' is a substitution of one base for another). A more general form of parsimony is often referred to as Ockham's razor, regarded by many as a key principle of scientific enquiry. Whatever William of Ockham's original intention was (**Figure 12.7**), the principle that bears

**Figure 12.7** William of Ockham is thought to have been born in 1287 in the village of Ockham in Surrey, where he is honoured in this church window. He was a prolific contributor to intellectual debate, but today is best known for the principle known as 'Ockham's razor' (ostensibly as a reference to 'shaving away' unnecessary assumptions). However, most of the formulations of this principle are not actually from Ockham's writing. The closest direct statement in his own work is 'Numquam ponenda est pluralitas sine necessitate'—plurality should not be proposed unnecessarily—or 'Frustra fit per plura, quod potest fieri per pauciora'—it is pointless to use many things to accomplish something that could be done with fewer.



his name is commonly interpreted as follows: when assessing competing explanations for a phenomenon, go with the explanation that requires the fewest number of ad hoc assumptions, until such time as it is proven false. Ad hoc assumptions are the additional things we have to believe in order to fit a hypothesis to our observations: for example, additional events that must have happened for our hypothesis to be true, or unobserved processes that must be operating to make that hypothesis plausible. We want to minimize the number of ad hoc assumptions because they are things that we have to invent without direct supporting evidence, and they have no other role than to prop up a particular hypothesis. The fewer events or processes we have to take on faith, without direct evidence, the easier it is to believe a particular hypothesis.

It's important to recognize that the principle of parsimony, or Ockham's Razor, does not tell us which explanation is more likely to be true. Instead, it is a guide for our behaviour when comparing alternative explanations for a particular observation: go with the hypothesis that requires the fewest ad hoc assumptions until it is proved false. This prevents us from inventing fancy underlying phenomena when we could just as easily explain what we have observed using mechanisms we already know about.

When we use the principle of parsimony to decide between alternative trees, we are stating that the tree with the fewest inferred substitutions is always the most plausible. Parsimony is one of the most common ways of estimating phylogenies from non-molecular data (**Figure 11.12**). The central assumption of parsimony is that evolutionary change is rare enough that when we observe two species that share exactly the same character, the most likely explanation is that they both inherited it from a common ancestor, rather than each evolved it independently. This may be true for many morphological datasets, particularly for complex traits that seem unlikely to have evolved multiple times. But is this assumption realistic for DNA sequence evolution? Do we expect substitutions to be so rare that a tree with fewer inferred substitutions is always more plausible than one with more substitutions?

## Multiple hits

The differences that we can observe between aligned sequences may not be all of the substitutions that have occurred in these sequences. In Chapter 10, we discussed one of the special features of molecular evolution that makes it, in some ways, more difficult to interpret than morphological evolution: DNA only has four character states. If an A changes to a G then back to an A again, we are not going to be able to tell that we have a different A than the one we started with. An A is an A. And if the A changes to a G then a C, all we will have is the start point (we began with A) and the end point (we now see a C), and we won't know that there was a G in between.

Consider the case of the diverging onychophorans from Chapter 11. Here is a simplified version of **Figure 11.20**.

**Figure 12.8**

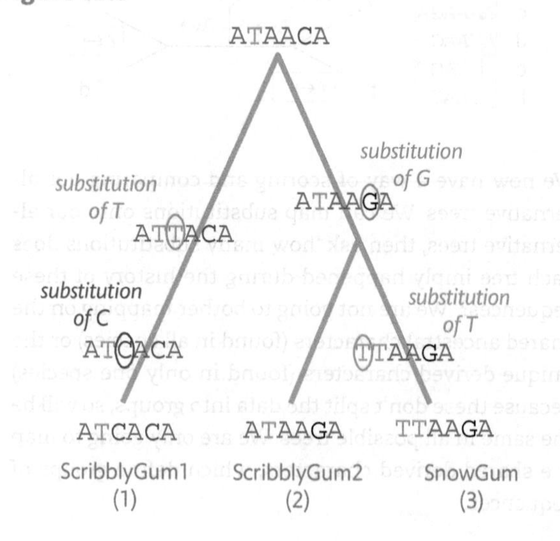

We can draw this phylogeny as an unrooted tree and count the number of substitutions that occurred.

**Figure 12.9**

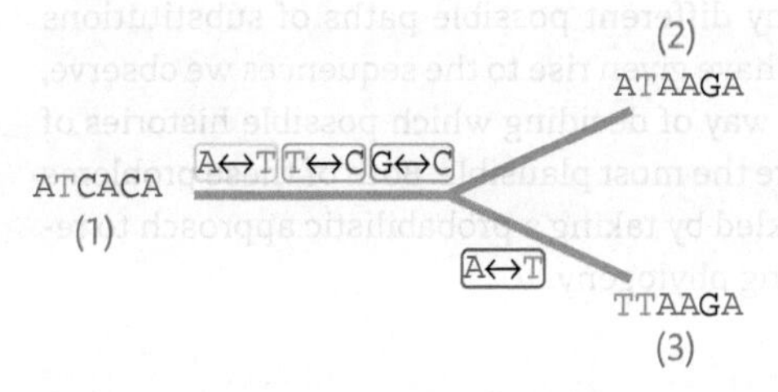


We can see that there have been three substitution events between sequence (1) and sequence (2), and four substitutions between sequence (1) and sequence (3). But if we were trying to reconstruct this phylogeny using only the sequences at the tips, we would miss one of these substitutions. Try constructing a distance matrix from these sequences (recording the absolute number of differences between sequences rather than proportions: figure 12.10).

**Figure 12.10**

| | (1) ATCACA | (2) ATAAGA | (3) TTAAGA |
|---|---|---|---|
| (1) ATCACA | 0 | 2 | 3 |
| (2) ATAAGA | | 0 | 1 |
| (3) TTAAGA | | | 0 |

Our distance matrix suggests there have been only two substitutions between sequences (1) and (2), but we know there have really been three substitutions—you can count them by following the path from sequence (1) to sequence (2) on the unrooted tree (**Figure 12.9**). Similarly, our distance matrix records three substitutions between (1) and (3), but there have actually been four. How did we miss one? The same site changed more than once: the third site in the sequences changed from A to T to C. But all we have left to show for this chain of events is the end product, which is a sequence with a C. We would have the same end product whether there had been one change (from A to C) or six changes (from A to T to C to A to G to T to C). Substitutions that occur at the same site, overwriting previous changes, are referred to as 'multiple hits'.

*We return to the subject of multiple hits in Chapter 13*

Parsimony specifically ignores multiple hits. If two sequences differ by one nucleotide, then parsimony states that the best explanation is that there has been a single substitution between them. But remember that the most parsimonious solution is not necessarily the true explanation. For molecular data, we recognize that there might be extra substitutions that have been covered up by multiple hits. So, how can we account for changes that we cannot directly observe? We need some way of saying 'I observe this pattern, but I believe that there may have been more events that I can't directly observe'. Obviously, we could let our imaginations run wild and put in lots of hypothetical changes. For example, we could have mapped the following changes onto our tree (**Figure 12.11**).

**Figure 12.11**

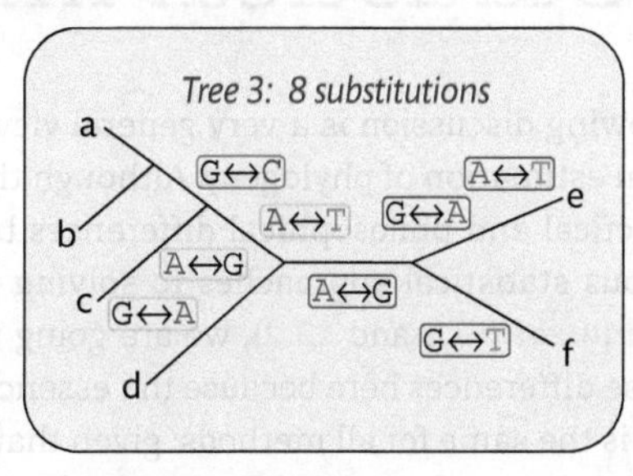


This tree has exactly the same branching order as Tree 1 (**Figure 12.6**), but it involves more substitutions, because many of the substitutions are multiple hits. In this tree, *e* and *f* both have a T at site 1, not because they inherited it from a common ancestor, but because they both acquired it independently (**Figure 12.12**).

**Figure 12.12**

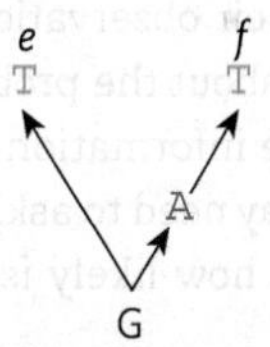


And although *a*, *b* and *c* all have an A at position 5, this position underwent two substitution events on the lineage leading to c (**Figure 12.13**).

**Figure 12.13**

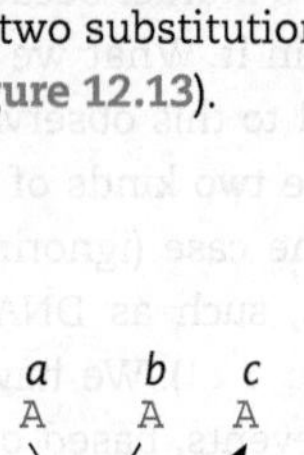

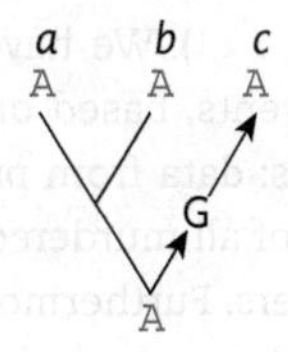

Tree 3 with eight substitutions may seem intuitively less likely than Tree 1 that only requires three substitutions. But Tree 3 is not impossible. Molecular data is such that we can well imagine that some nucleotides have undergone many substitution events even if they end up looking similar. If all nucleotide changes are equally likely, then if a particular site changes more than once, there is a one in four chance that it will return to its original base and be therefore undetectable.

So we have struck two problems in our quest to make an evolutionary tree that is a map of the substitution events that have occurred. One is that multiple hits erase past substitutions, so we need a way of inferring how many unseen changes are likely to have occurred. And the other problem is that, given that we can construct many different possible paths of substitutions that could have given rise to the sequences we observe, we need a way of deciding which possible histories of the data are the most plausible. Both of these problems can be tackled by taking a probabilistic approach to reconstructing phylogeny.

# Statistical inference of phylogeny

The following discussion is a very general view of the statistical estimation of phylogeny. Although there are both practical and philosophical differences between the various statistical approaches to solving phylogeny (**TechBoxes 12.1** and **13.2**), we are going to gloss over those differences here because the essence of the problem is the same for all methods: given that we observe a certain set of sequences in the present day, how can we use what we know about molecular evolution to help us decide which evolutionary tree provides the most plausible version of the evolutionary history of the data?

We employ this kind of probability-based thinking in many ordinary situations. A classic example is the court case, where the judge or members of the jury are asked to decide which is the most likely scenario for past events given their observations of the outcomes, plus a set of beliefs about the probability of certain occurrences, and some information specific to this case. For example, they may need to ask: 'given that a woman has been murdered, how likely is it that her husband killed her?'

The dead woman is an observation. It is a known fact that there has been a murder because we have a corpse with stab wounds in it. What we don't know is what series of events led to this observation of a murdered corpse. But we have two kinds of information we can bring to bear on the case (ignoring, for the moment, other observations, such as DNA samples from the crime scene: **Figure 5.21**). We have our experience of the probability of events, based on our knowledge of how the world works: data from previous cases might suggest that a third of all murdered women have been killed by their partners. Furthermore, when we have a murdered woman who was previously abused by her partner, then the chance that she was killed by her partner is much higher (possibly as high as 80 per cent). And we have some relevant information about this particular situation: police recorded a previous assault on this particular woman by her husband.

We can come up with many hypotheses for how we ended up with the observed situation. For example, the defence may present a hypothesis that the woman was killed by an intruder, while the prosecution may present a hypothesis that she was killed by her husband. We combine our knowledge of the world (murdered women have often been killed by their husbands, particularly if they have been abused) with information about this case (this wife was abused by her husband) to help us to judge which of these hypotheses is more likely. But, based on this alone, we cannot say which hypothesis is true. Indeed, we cannot rule out that the woman was killed by aliens conducting an exobiological experiment, but we can say that given what we know about the world and the particulars of this situation, it does not seem as plausible an explanation (**Figure 12.14**).

We can apply the same kind of approach to solving phylogeny. We have gone out into the world and sampled some DNA sequences. Barring laboratory errors, these sequences are 'facts'. They are observations about the world. They are, in short, what we know to be true. What we don't know is the exact series of events that gave rise to those DNA sequences. But we can come up with many different hypothetical paths of substitutions that produced the sequences we observe. We want to find the tree that gives us the most plausible explanation for our observations. To do this we need some kind of statement of belief about how sequences evolve: this is our model of molecular evolution.

Figure 12.14

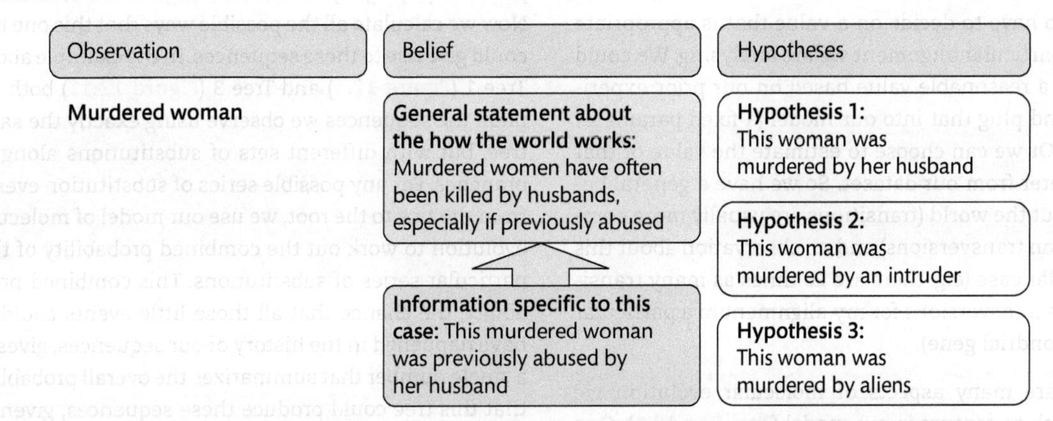


## Models of molecular evolution

> *Attempts at reconstructing evolutionary trees using computers are leading to a clarification of our basic ideas as to how it should be done. It has become particularly clear that any attempt at producing an evolutionary tree must be based on a specific model, for only then can proper statistical procedures be adopted and only then are the assumptions implicit in the method clear for all to see . . . As the methods and computer programmes develop from the prototypes of today, it will become possible to handle the vast amount of information in the published data . . .*
>
> A. F. Edwards (1966) *New Scientist*, Volume 19, pages 438–440.

The model is a statement of belief about the process of molecular evolution. When we weighed up competing explanations in our hypothetical murder case (**Figure 12.14**), we separated our beliefs into the general statements about the world and specific statements about this particular case. We do the same thing for molecular data (**Figure 12.15**). The general statements are our model. The specific information that tailors the model to a given situation are the parameter values for that model.

For example, one common aspect of models of molecular evolution is the transition/transversion ratio. Most people who have compared DNA sequences will have noticed that, in general, there are far more transition substitutions (changes from one pyrimidine (C or T) to another, or from one purine (A or G) to another) than transversions (changes from a pyrimidine to a purine or vice versa). So our model could contain the general statement that transitions have a higher probability of occurring than transversions. If we were comparing two possible phylogenies for a given set of sequences, and one phylogeny required us to infer 6 transitions whereas the other required us to infer 6 transversions, then we would generally consider the transition tree to be more probable than the transversion tree.

We can include this knowledge in our model of molecular evolution by having a different substitution rate for transitions and transversions (**TechBox 13.1**). These two rates are now parameters of our model. But what value should these parameters have? Exactly how much more common are transitions than transversions? The degree to which transitions outnumber transversions varies between different sequences. For example, mitochondrial sequences tend to have much higher ratios of transitions to transversions than nuclear sequences. So if we include a parameter

Figure 12.15

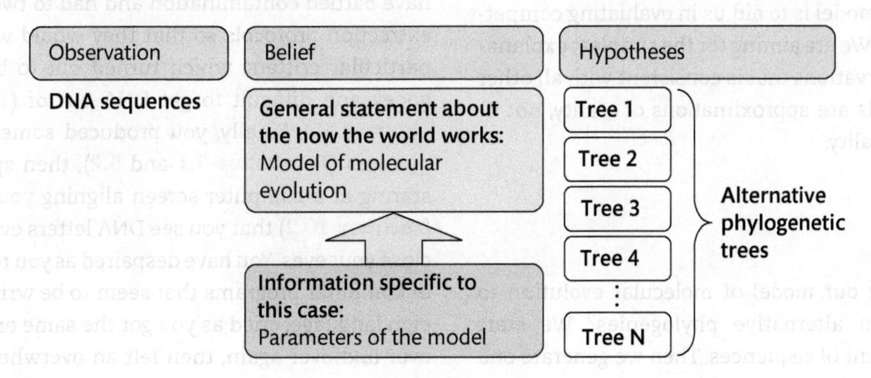

for transition/transversion ratio in our model, we are going to have to decide on a value that is appropriate to the particular alignment we are analysing. We could choose a reasonable value based on our prior experience, and plug that into our model (a fixed parameter value). Or we can choose to estimate the value of that parameter from our dataset. So we have a general belief about the world (transitions are usually more common than transversions) and an observation about this particular case (e.g. there are 20 times as many transitions as transversions for my alignment of a particular mitochondrial gene).

There are many aspects of molecular evolution we may wish to capture in our model (TechBox 13.1). One important observation is that, for most functional sequences, some sites undergo many more substitutions than other sites. If we ignore this fact, then we will underestimate the number of substitutions that have occurred between the sequences we observe. We may also need to take account of differences in base composition. Some genomes have a greater proportion of Gs and Cs than As and Ts. In a GC-rich genome, we expect changes from any nucleotide to G or C to be more common than those to A or T. So a phylogeny with lots of changes from G to C would be more probable than one with the same number of changes from G to A.

***Substitution models are discussed in more detail in Chapter 13***

Molecular evolution is complex. We don't expect two sequences to evolve in exactly the same manner, because there will always be different factors affecting patterns of substitution: fluctuations in population size, responses to changing environment, interactions between alleles, and so on. Clearly our model of evolution can only ever capture a very, very small part of this complexity. But we should not fall into the trap of thinking that a model is an attempt to fully describe the process of sequence evolution. Not only is that impossible, it wouldn't even be useful. Remember that the sole purpose of our model is to aid us in evaluating competing hypotheses. We are aiming for the simplest explanation of our observations that is consistent with all other evidence. Models are approximations of reality, not illustrations of reality.

## Likelihood

How do we use our model of molecular evolution to choose between alternative phylogenies? We start with an alignment of sequences. Then we generate one possible phylogeny with these sequences at the tips. Now we calculate all the possible ways that this one tree could give rise to these sequences. In the example above, Tree 1 (Figure 12.6) and Tree 3 (Figure 12.11) both explain the sequences we observe using exactly the same tree, but with different sets of substitutions along its branches. For any possible series of substitution events, from the tips to the root, we use our model of molecular evolution to work out the combined probability of that particular series of substitutions. This combined probability, the chance that all those little events could all have happened in the history of our sequences, gives us a single number that summarizes the overall probability that this tree could produce these sequences, given all the possible patterns of substitutions that could have occurred along the tree (TechBox 12.1).

Now we generate another possible tree and repeat the process. In fact, we have to do it all over again for thousands of alternative trees. Then when we have done so, we will have created a set of possible trees, each of which has a likelihood score. At this point we could simply say 'give me the tree with the highest likelihood' and consider that the best possible phylogeny for this data, given this method and these assumptions. But the next-best tree might have a likelihood that is only slightly lower than our best tree. And if that is the case, we have to accept that if our data had been slightly different, or we had used a slightly different model, we might have got a different answer. So we need to ask ourselves: if two hypotheses are only slightly different in some measure of plausibility, is that really enough to reject one and accept the other?

### *How good is my tree?*

You have spent six months in the lab, grinding up your favourite little creatures and extracting their DNA (alas, it is still the case that most DNA extraction techniques involve the death of the unfortunate DNA donor, but the ethical biologist should always try to think of a non-destructive way of getting their samples). You have battled contamination and had to tweak the DNA extraction protocols so that they would work on your particular critters, which turned out to be heinously gooey and difficult to get DNA out of (TechBox 2.2, Figure 2.13). Finally, you produced some useable sequences (TechBoxes 5.1 and 5.2), then spent so long staring at a computer screen aligning your sequences (TechBox 10.2) that you see DNA letters every time you close your eyes. You have despaired as you read manuals of computer programs that seem to be written in a foreign language, cried as you got the same error message over and over again, then felt an overwhelming sense of gratitude as someone on a user group posts a helpful reply to your desperate plea for assistance. Eventually you get the program running and you wait with hopeful trepidation for an answer. Finally the program pings and you have a tree! Yippee, you have the answer! You jump up and down and shout, print out the tree and wave it at passers-by and immediately head out to celebrate, although none of your non-biological friends can understand why you are so elated.

At the risk of raining on your parade, at this point I have to remind you that your tree is not 'the answer'. It is 'the working hypothesis'. Remember, the sequences are the facts (if you have faith in your lab skills and the veracity of the sequencing procedures). The tree is a hypothesis that explains those facts. It may be a very good hypothesis, and it may well be true. But you don't know that. All you know is that given your method and the assumptions you had to make to apply that method, you have come up with this tree as the most likely explanation of your data.

However, you might look carefully at your tree and realize that things are not entirely as you expected. Perhaps one sequence is coming out in a strange place on the tree (e.g. **Case Study 10.2**). Or you might find that your tree disagrees with a previously published phylogeny in several important respects (e.g. **Case Study 2.2**). At this point there are four options: one, defend your tree as the 'true tree' by force of will (irrational as it sounds, this is a surprisingly common option); two, pretend that your tree is fine by ignoring the silly bits and concentrating on the bits you like (another very popular option); three, gather more data and see if it changes your picture (always a good option if available); and four, try to find out just how convincing your tree is for this dataset (in fact this is not so much an option as highly recommended behaviour).

## Testing phylogenetic hypotheses

There is a common misconception that you cannot conduct experiments in evolutionary biology because you cannot directly witness past events or rerun evolutionary history. But we don't need to go back in time to conduct experiments in evolutionary biology. An experiment is a test of competing hypotheses against observations. You start with a hypothesis and you use it to make predictions about what you would expect to see if the hypothesis were true.

Take, for example, the origin of the human immunodeficiency virus, HIV. It is clear that HIV is related to simian immunodeficiency viruses (SIV) found in other primate species, but how did it get into humans? One hypothesis was that HIV was spread to humans in the late 1950s through a contaminated oral polio vaccine that had been manufactured using chimp kidneys sourced from Kisangani (then known as Stanleyville) in the Democratic Republic of Congo. This hypothesis, which came to be known as the OPV-AIDS hypothesis, had relatively little support in scientific communities, but it was championed by some researchers. The OPV-AIDS hypothesis stirred heated controversy, with detractors frustrated that it was given any credence whatsoever and supporters claiming a conspiracy by the mainstream scientific community to deny them a voice. Because it undermined confidence in vaccination programmes, it was very important to test the OPV hypothesis to determine if there was any evidence for the claim that contaminated vaccines caused a transfer of a virus from lab animals to humans (**Figure 12.16**).

The OPV-AIDS hypothesis leads to predictions about the topology (branching order) of a phylogeny containing SIV and HIV sequences. Firstly, SIV sequences from the

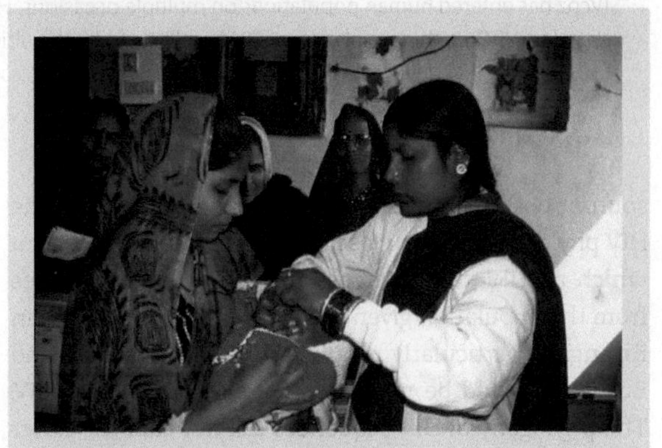

**Figure 12.16** The development of the orally administered polio vaccine was a breakthrough that offered the World Health Organization the chance to make polio the second communicable disease to be eradicated by vaccination (after smallpox). But vaccination programmes rely on public confidence. For example, in 2003 a regional government of Nigeria stopped the polio vaccination programme after rumours circulated that the vaccines were infected with HIV and contained covert anti-fertility drugs. Although vaccination resumed in 2004, the halt in the programme caused an outbreak that spread through Nigeria and to at least five other countries. The eradication of polio has thus been delayed, but will hopefully be achieved in the near future. South-East Asia, including India, was finally declared polio-free in 2014. You can follow the progress of polio eradication online (and make a donation to help with the effort needed to tackle the last remaining areas where polio is endemic).

Photograph: Chris Zahniser, B.S.N., R.N., M.P.H.

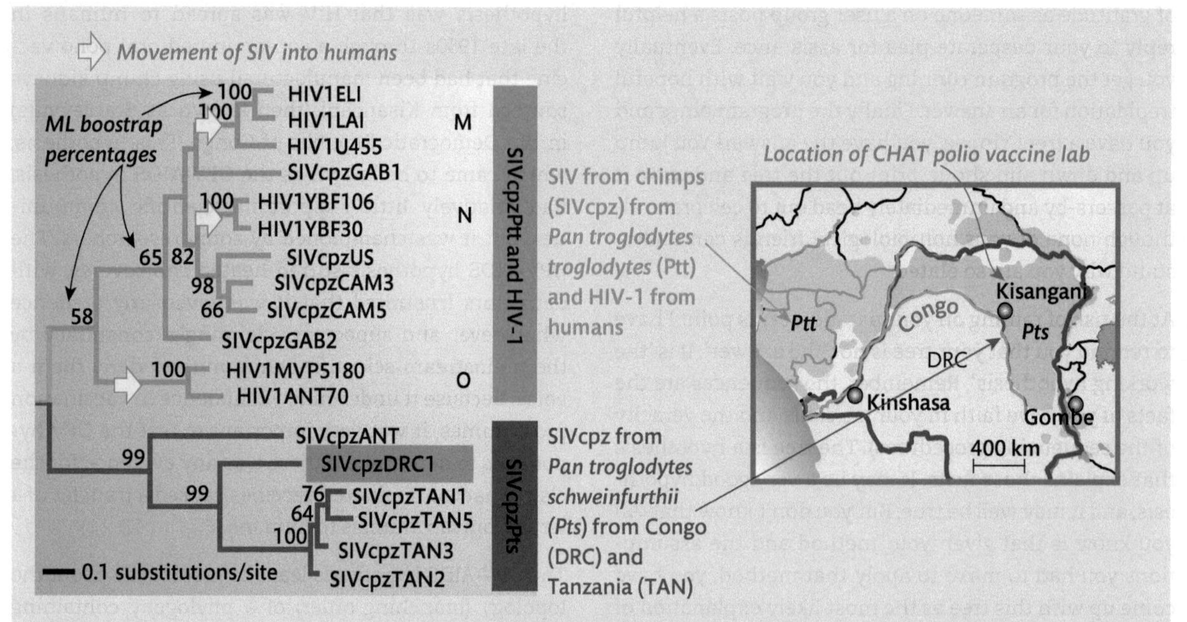


**Figure 12.17** Maximum likelihood phylogeny of sequences of two genes (*gag* and *nef*) from a range of human and chimpanzee (*Pan troglodytes*) immunodeficiency viruses (HIV and SIV). HIV1 is nested with the clade of chimpanzee SIV (SIVcpz) sequences, demonstrating that HIV1 is derived from SIVcpz. Furthermore, this phylogeny supports previous studies that show that SIVcpz has entered human populations on multiple occasions. However, the SIVcpz sequence from Kisangani is apparently distantly related to HIV1 so this population of SIV is unlikely to be the ultimate source of HIV1.



source pool of chimpanzees should be at the base of the HIV phylogeny (that is, close to the ancestral lineage from which all HIV strains arose). Secondly, HIV sequences from the population given the polio vaccine produced in Kisangani, particularly people in in Congo, Rwanda and Burundi, should be more closely related to chimp SIV. The phylogeny of HIV sequences did not confirm either of these predictions. It shows that HIV-1 sequences (the main pandemic strain, found worldwide) group with many different chimp SIV sequences, suggesting that HIV has moved from chimps into humans on multiple occasions (**Figure 12.17**). It also shows the chimp SIV from the region surrounding the vaccine laboratory, which had been suggested to be the source of any chimp tissue used in the lab, groups with sequences from infected chimps from Tanzania, not with HIV sequences. These groupings have strong statistical support in the data (**TechBox 12.2**). This suggests that while HIV-1 did come from chimps, it did so on numerous occasions, with no evidence to suggest that the Kisangani polio vaccine lab was the primary source of zoonotic transfer of SIV to humans.

Molecular phylogenies are just one source of historical information. The best way to test the reliability of a phylogenetic hypothesis is to compare your molecular phylogeny to independent sources of evolutionary information, and see if they all tell the same story. The researchers who tested the OPV-HIV hypothesis using molecular phylogenies considered their results in light of other lines of evidence, such as the failure to find evidence of HIV or chimp DNA in archived polio vaccine stocks. In other cases, independent evidence might come from palaeontology, biogeography, comparative morphology, development biology, and so on. If your tree is at odds with all of these lines of evidence, then you might want to take a hard look at your data and methods in order to ask why your answer is different. But often molecular phylogenies are applied to cases where no other lines of evidence are available, so this option may not be possible. Even in the absence of any independent evidence you can ask just how strongly your sequence data support your phylogenetic hypothesis.

## Phylogenetic support

An important part of carrying out any experiment is to consider the uncertainty in your observations. There are many reasons why your phylogeny may not reflect the true history of those sequences. What if your tree is wrong? You can ask how robust your result is to the assumptions you have made in your analysis. For example, you could try a different method or change the model you employ and see if you get the same result. Did you make decisions during the alignment of your sequences that could influence your result? Maybe you could try cutting out the ambiguously aligned sites and see if you get the same answer. Do different parts of the alignment all support the same phylogeny?

In addition to exploring the robustness of your conclusion, you could use some formal statistical techniques for assessing how well your tree is supported by your data. For example, bootstrapping your data allows you to ask 'if my data were slightly different, would I still get the same tree?' (**TechBox 12.2**). This test involves resampling sites from your alignment to generate alternative datasets that are slightly different from the original. If resampling the data results in a totally different tree, this tells you that your alignment provides only marginal support for the top-scoring tree: had you sequenced different sites, or had a slightly longer or shorter alignment, you might have got a different tree. In this case, you can't be as confident that you have found the true biological history of your taxa. But if, no matter how you dice it up, your data always return the same tree, then you can say that the phylogenetic signal in the data all points to the same story.

Alternatively, instead of asking 'is my tree the right tree?' you could ask 'can I safely reject alternative trees?' Many phylogenetic methods produce a distribution of trees that are all nearly equivalent in their plausibility (**TechBoxes 12.1** and **13.2**). This allows you to test your conclusions against a range of alternative phylogenetic solutions. This can be translated into formal statistical tests of support for particular phylogenetic hypotheses (**Figure 12.17**).

Consider, if you will, legless lizards. Loss of limbs, and the development of a long, snake-like body, is surprisingly common in lizards: seven different families of lizards include limbless forms (**Figure 12.18**). How many different times has the legless form evolved? Taxa with strongly modified morphology can be difficult to place in a phylogeny, so molecular phylogenies can help work out where such strange beasts fall in their family trees. One particular study asked how two worm-like genera of legless lizards, *Dibamus* and *Feylinia*, were related to the rest of the skink family. Easy, you might think, stick the sequences in an alignment, run the phylogeny program, and look at the tree. But, as is often the case in molecular phylogenetics, these researchers found that when they changed the assumptions of the analysis, or used a different model of molecular evolution, or analysed different sequences, they got different phylogenies. There was not a single phylogeny for these sequences, but many alternative stories that could be told, depending on what assumptions you were prepared to make about the evolution of the sequences. So did the researchers give up and go home? No. Did they just pick the tree they liked best and pretend that they hadn't seen the alternative phylogenies? No. They asked what parts of the history of these taxa they could be certain of, no matter what assumptions were made during the analysis.

Even though they couldn't say for sure which alternative solution was the correct phylogeny, they could consider all the alternative phylogenies and ask whether some conclusions were supported in all the different possible trees. All the best-scoring trees grouped one of the legless lizards, *Feylinia*, in a clade of skinks with reduced limbs. So *Feylinia* evolved within the skinks, as an extreme example of a short-legged lineage. But

**Figure 12.18** Is it a snake? Is it a worm? No, it's a lizard! These two legless lizards, (a) *Dibamus* and (b) *Lialis*, represent independent origins of the snake-like form. There have probably been over 20 origins of limblessness in lizards (including the highly successful snake lineage).

for the other legless lizard, *Dibamus*, they couldn't be sure exactly where they should go in the tree. However, they could make a statement about where they didn't go: none of the best-scoring trees had *Dibamus* placed inside the skink family. So, despite the lack of clear phylogenetic resolution on the placement of these two taxa, they could confidently say that they represented two separate evolutionary origins of snake-like bodies, that each arose independently from different lineages of limbed lizards. We will meet this kind of hypothesis testing again—'I don't know exactly what the answer is but I know what it's not'—when we look at estimating dates from molecular data, in Chapter 14.

However, you will sometimes find that your DNA sequences do not clearly distinguish between alternative hypotheses. This may be because there have been too few substitutions to produce sufficient splits in the alignment to resolve all the groups (low signal). Or it may be that there have been too many substitutions, so that the historical record has been largely overwritten (high noise). But, as we discovered in Chapter 11, failure to find a single tree to represent your data may not be a problem of either low signal or high noise (**TechBox 11.2**). It could be that your data carries the signal of the true evolutionary history of your data, but that history is not a simple story of populations splitting again and again to give a clear hierarchy of similarities. When lines of descent do not follow a simple hierarchical branching pattern, then we say that evolution is not wholly tree-like. It is important to keep this in mind if you find that your sequences do not unambiguously point to one phylogenetic tree.

### *Alternative histories*

One of the advantages of molecular data as a source of phylogenetic information is that we can, roughly speaking, consider each nucleotide site to be an independent recorder of history. This is an important basis of the statistical inference of phylogeny, since virtually all methods in statistics assume that datapoints are independent of each other, so that each site independently records its own evolutionary history. This independence has an important corollary: if sites have different histories, then their individual records of history will not be the same. Remember that when you make a phylogeny of molecular data you are tracing the history of the DNA sequences themselves. You probably hope that the phylogeny you construct from DNA sequences reflects the evolutionary history of the organisms you got the DNA from. But in most organisms, different parts of the genome have different evolutionary histories, because the genome is not inherited as a single unit. Sexual reproduction and recombination blends alleles with different histories into the same genome. For example, your mitochondrial genes were inherited from your mother. If you have a Y-chromosome, then you got that from your father. If your mother and father are unrelated, then your mitochondrial sequences have a different history from your Y-chromosome sequences. The separate histories of mitochondrial and Y-chromosome sequences can be exploited by researchers wishing to trace human movement (see **Case Study 4.2**).

But even sequences on the same chromosome can have different histories. You inherited half of your alleles from your mother, and she inherited half of her alleles from her father, who inherited half of his alleles from his mother, and so on. So different loci in your genome will have different ancestries. Imagine you compared two alleles at a particular locus in your genome with those at the same locus in your cousin's genome. It happens that, at this locus, you and your cousin both have a copy of an allele that originated in your paternal grandfather. These two copies of this allele—your copy and your cousin's copy—have a recent common ancestor, only two generations ago. But the second allele at this locus differs between you and your cousin, because your allele was inherited from your mother (who is descended from Tasmanian aborigines) and your cousin has inherited an allele from her mother (who is descended from Alaskan Inuit: **Figure 12.19**). These two alleles have a much older common ancestor, over fourteen hundred generations ago. So these two alleles at a single locus in your genome have entirely different histories. More broadly, different parts of a genome can tell quite different stories (**Case Study 10.2**).

Many techniques in population genetics, and increasingly in phylogenetics, make use of the fact that alleles can have different histories. For example, the individuals in a small population will tend to be more closely related, so alleles are likely to have more recent common ancestors. Alternatively, a population with lots of migrants will contain alleles from different populations, so these alleles are likely to have more distant common ancestors. The variation in the trees inferred from different sequences can be useful. But it might create a mess if you try to infer a single phylogeny from sequences with different histories, because there won't be a single tree that describes the path of inheritance of all the sequences. Instead, some splits in the data will support one tree, and some will support a different tree.

Conflicting histories can be most clearly seen in phylogenies of recombinant genomes. For example, the

**Figure 12.19** Nowadluk Ootenna, an Inupiat woman from Alaska, photographed in 1907 wearing a fur-lined parka. Inupiat is one of the Inuit language groups, which are distributed across the Arctic in Greenland, Canada and the USA. Incidentally, the word 'anorak' is derived from an Inuit language, whereas 'parka' comes from the Nenets language, so the use of parka and anorak in English is a case of horizontal transmission in the evolution of languages.

human immunodeficiency virus (HIV) can undergo recombination between strains and this can complicate attempts to reconstruct the evolutionary history of HIV. What's more, it seems that HIV can recombine with the related Simian Immunodeficiency Viruses (SIV). This seems to be how the N-strain of HIV-1 originated. In contrast to the globally distributed M-strain of HIV-1, which infects tens of millions of people worldwide, the N-strain of HIV-1 is very rare, with only a handful of known cases, most of which were identified in Cameroon. Phylogenies based on genes from one end of the genome group N-type HIV-1 with the global M-type HIV, but phylogenies based on genes from the other end of the genome group N-type HIV with chimp SIV (**Figure 12.20**). A plausible explanation is that the genome of N-type HIV is recombinant, formed when a single individual was infected with both HIV-1 and SIV, and that these two different virus genomes recombined to create a new virus.

Recombination occurs most commonly between closely related genomes, such as members of the same species, or similar viral strains. But some lineages swap DNA with more distant relatives. Movement of DNA between the genomes of unrelated individuals is commonly referred to as horizontal gene transfer. This has great practical implications. Many bacteria seem to be able to swap genes and this can lead to the rapid spread of antibiotic resistance across different types of bacteria. Similarly, there is a concern that hybridization between plant lineages could provide a conduit for herbicide-resistance alleles from domestic plants to spread into wild species. Horizontal gene transfer also creates problems for phylogenetics because it creates a genome with a mosaic of histories, leading to contradictory phylogenies based on different parts of the genome (**Case Study 7.2**). Some biologists suggest that the movement of sequences from one lineage to another in the early stages of the evolution of the biological kingdoms may have been so common that evolutionary history may be better represented with an interconnected network (see Chapter 11). Consequently, there is currently a vigorous debate about whether the tree of life is really a tree at all. And so we come, somewhat surprisingly, full circle back to hand-drawn desirograms that represent a best guess at the rather complex evolutionary history of a set of species (**Figure 12.21**).

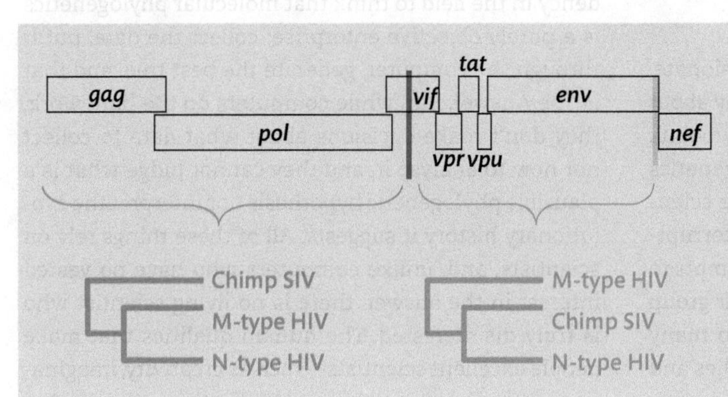


**Figure 12.20** Origins of rare N-type HIV virus: A phylogeny of the genes from the 5' end of the genome groups the N genome with global M-type HIV-1, but a phylogeny of genes from the 3' end of the genome groups N-type HIV with SIV from chimpanzees from Cameroon. This result suggests that N-type HIV is recombinant: the 5' end appears to be derived from an M-type HIV-1 strain, but the 3' end is derived from a chimpanzee virus.
Adapted from Roques et al. (2004).

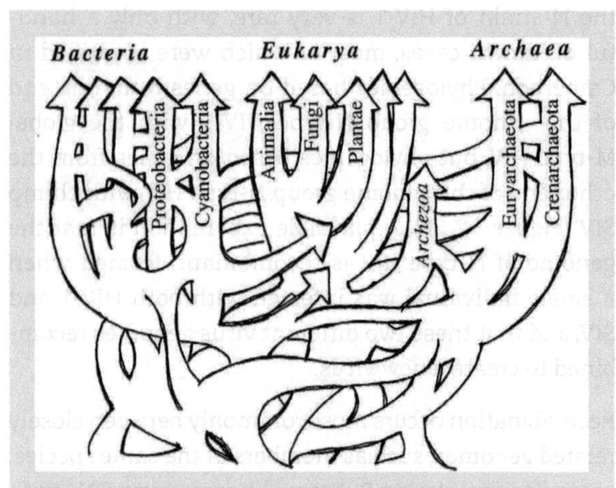


**Figure 12.21** Ford Doolittle's tree of life published in 1999, showing horizontal transfer of genetic material between the kingdoms. While the majority of phylogenies are computer generated, the occasional hand-drawn masterpiece still makes it into print.

From Doolittle, W. F. (1999) *Science*, Volume 284(5423), pages 2124–2128. Reprinted with permission from AAAS.

## Which phylogenetic method should I choose?

> "*I have steadily endeavoured to keep my mind free so as to give up any hypothesis, however much beloved (and I cannot resist forming one on every subject), as soon as facts are shown to be opposed to it. Indeed, I have had no choice but to act in this manner, for with the exception of the Coral Reefs, I cannot remember a single first-formed hypothesis which had not after a time to be given up or greatly modified.... On the other hand, I am not very sceptical,–a frame of mind which I believe to be injurious to the progress of science. A good deal of scepticism in a scientific man is advisable to avoid much loss of time, but I have met with not a few men, who, I feel sure, have often thus been deterred from experiment or observations, which would have proved directly or indirectly serviceable.*"
>
> Charles Darwin (1887) in *The life and letters of Charles Darwin*, including an autobiographical chapter. Volume 1, ed. F. Darwin; London: John Murray.

The field of phylogenetics is surprisingly passionate. Many people working in this area feel very deeply about their favourite hypothesis and alternative viewpoints can make their blood boil. I have been to phylogenetics conferences marred by heated exchanges where scientists have stood up and shouted at each other, interrupting presentations on the results of the latest attempts to uncover the evolutionary tree of some particular group of organisms. The literature gives testament to many long-running feuds, with series of published replies and counter-replies that can run for years. These feuds may concern competing solutions to the same phylogenetic problem, but some of the most heated conflicts in the past have concerned not the phylogenies themselves but the methods used to derive them.

One of my first experiences in scientific research was to work for a 'pheneticist' (who used clustering algorithms to produce phylogenies), who spoke in the most vehement terms about the despicable 'cladists' (who used parsimony to produce phylogenies). I then moved universities and was taught by cladists, who were smugly assured that they had displaced the old-fashioned and patently misguided pheneticists. At my next university, researchers looked down on cladists from a great height, sure that their chosen phylogenetic method (maximum likelihood) was the most superior approach. Not long after that, Bayesian methods were top of the tree and some researchers would scoff at anyone who dared to use any other method. All of this in two decades since I started doing scientific research. The moral of the story is that when someone tells you that their phylogenetic method is the ultimate solution, try holding your breath and waiting for the next method, it won't be far away.

The history of phylogenetics could be interpreted as a move from the subjective toward the objective. Subjective hand-drawn phylogenies have been largely replaced by algorithmic computer-generated phylogenies. For molecular data, methods that score and rank trees according to a distance or parsimony have been largely replaced by methods that place phylogeny estimation within a statistical framework that makes explicit assumptions about molecular evolution and weighs the relative levels of support for different phylogenetic hypotheses. The more sophisticated these methods get, the more computationally intensive they become. But, although the methods may be objective, the scientists that use the methods are not.

The reason I am pointing this out is that there is a tendency in the field to think that molecular phylogenetics is a purely objective enterprise: collect the data, put it through the computer, generate the best tree, and that is The Answer. But, while computers do the hard work, they don't make decisions about what data to collect nor how to analyse it, and they cannot judge what is a plausible phylogenetic hypothesis nor interpret the evolutionary history it suggests. All of these things rely on scientists, and, unlike computers who have no vested interest in the answer, there is no living scientist who is truly disinterested. The human qualities that make people excellent scientists—such as creativity, imagination, enthusiasm, passion and dedication—also tend to make people get rather attached to certain ideas and ways of doing things.

However, while recognizing that it's our humanness that makes science possible and fun, we must be wary of becoming too attached to favourite methods or beloved hypotheses. You will tend to notice, as you wander through the scientific literature, that papers published by a particular research team will always produce evidence in favour of Hypothesis X, while a different research team, investigating the same phenomenon, will always produce studies in support of the alternative Hypothesis Z. What's going on? If each research team is conducting fair experiments in good faith, why are they always coming up with evidence opposing each other? Shouldn't it frequently be the case that Team A conducts an analysis that shows that actually Team B were right all along? Why is proving yourself wrong so uncommon in the scientific literature?

We don't need to accuse anyone of malevolence or dishonesty in order to recognize that people have a tendency to favour the evidence that suggests their favourite hypothesis is indeed correct, and they will tend to downplay the analyses that speak against it. If running your analysis with a particular method or assumption gives you a lovely-looking tree that fits your hypothesis just perfectly, you might be inclined to focus on that one rather than on all the other analyses you did where the answer was not so attractive. You will probably also find yourself coming up with a whole list of reasons why that method really was the best way to ask the question, and the other less attractive trees were actually the result of inferior approaches.

I don't think scientists should pretend to be disinterested robots. But remember that the aim of the game is to discover the truth and if the story you have been telling yourself about your data turns out not to be true, then the sooner you find out that it is wrong, the better. There is nothing good to be gained from fooling yourself or others. So be honest in your appraisal of the story in your data and always give serious consideration to the alternative explanations. Personally, when I feel my humanness getting the better of my objectivity, I remind myself of Darwin's words: 'I have steadily endeavoured to keep my mind free so as to give up any hypothesis, however much beloved ... as soon as facts are shown to be opposed to it.'

# Conclusion

A phylogeny is a story that explains how we got the sequences we see today. Of course, some stories are true and some aren't. You can compare the plausibility of different phylogenetic explanations for a particular alignment of sequences by using a model of molecular evolution that tells you which kinds of changes are more likely to have happened. When you do this, you may find that there are many alternative phylogenies that are all as believable as each other. You should explore how well your data supports a particular phylogeny: would you have got a different answer if you had done things slightly differently or does the story always come out the same no matter how you tell it? Inferring the evolutionary history of DNA sequences may reveal complicated narratives. Different parts of your data may be telling different stories because they have different histories, due to the mixing and diverging of alleles in populations, recombination between individual genomes, or transfer of genes from one species to another.

Phylogenies can also be used to test hypotheses about evolutionary past and processes. We can ask what phylogenetic pattern we would expect to find if a particular hypothesis was true. In many cases, you may find that you can't find one particular tree that provides a perfect explanation of your data, but you may find that you can rule out particular hypotheses as being very unlikely to have produced the sequences that you have. When weighing up alternative hypotheses, we should constantly remind ourselves of the principle of Ockham's Razor: don't invent fancy explanations when a simple one will do the job, but reject the simple explanation when it is shown to be inadequate.

## Taking stock: what have we learned so far and where are we going next?

We have seen how genetic variation arises continuously as mutations change individual genomes, and that some of these genetic variants rise in frequency by selection or drift until they replace all other variants at that locus in that population. Thus, by a process of substitution, populations constantly acquire genetic differences that distinguish them from populations that they do not exchange alleles with. Eventually, populations become genetically distinct so that they cannot interbreed. They continue to acquire substitutions, and to divide and diverge, leaving a nested hierarchy of changes to the genome. When we compare DNA sequences from different lineages, we can use the patterns of shared substitutions to reconstruct the history of population divergence. But there are always many possible stories we could tell to explain how we ended up with the sequences we observe, so phylogenetic inference is a process of comparing the plausibility of different hypotheses regarding the evolutionary history that gave rise to the DNA sequences we have.

Chapters 13 and 14 emphasize how phylogenetic analysis is a matter of comparing the plausibility of alternative explanations of evolutionary history, given what we know about the way genomes evolve. Very often, what we know about molecular evolution is that it is almost always trickier to decipher than it first appears. We can illustrate these points by considering the way that the rate of DNA change varies across the genome, over time, and between lineages. Once considered simple and clock-like, we now recognize that DNA sequence evolution is acted on by such a great variety of factors that we can expect rich and complex patterns to be the norm, not the exception. In Chapter 13, we will see how the more we look into variation in rate of molecular evolution, the more fiendishly complex it appears. Then in Chapter 14, we will use the estimation of dates of divergence to illustrate that, while our inference methods cannot hope to precisely capture all this complexity, molecular data is a useful source of historical information as long as we are aware of the extent to which different assumptions about evolutionary processes lead us to very different answers. We must explore the bounds of our confidence with openness and honesty.

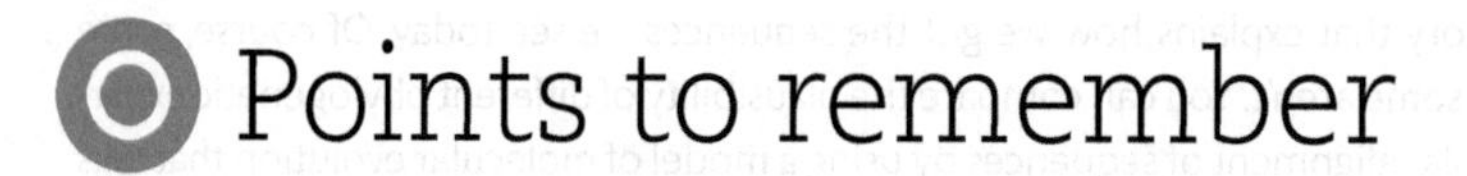

# Points to remember

### Comparing trees

- Phylogenies represent hypotheses about the evolutionary history of a set of sequences.
- For any set of sequences, there are a large number of possible phylogenies that could explain their evolution.
- Phylogenetic analyses generate alternative trees then score the likelihood that each tree could have given rise to the observed sequences.
- Parsimony is the principle that the tree with the fewest inferred substitutions is the most plausible, but this is rarely a good description of the evolution of DNA sequences.
- Multiple substitutions that occur at the same site erase previous changes, so models of evolution need to be able to infer substitutions that cannot be directly observed.
- A model of molecular evolution is a statement about the relatively probability of different kinds of changes used to evaluate the plausibility of different possible evolutionary histories.

### Statistical inference of phylogeny

- Phylogenetic analyses can produce many alternative phylogenies for any given dataset: the plausibility of these alternative hypotheses needs to be evaluated given the signal in the data, and in light of evidence from other sources.

- Statistical comparison of phylogenetic hypotheses asks whether we would expect to get a particular pattern of data if a given hypothesis were true.
- Phylogenetic analysis may lead not to the unambiguous identification of a single plausible hypothesis, but the rejection of alternative hypotheses as being unlikely to have given rise to the observed sequences.
- The mosaic nature of genomes, generated by sexual reproduction, recombination, hybridization and horizontal gene transfer, can lead to different sequences supporting different phylogenies.

# Ideas for discussion

**12.1** In what situations will distance methods, parsimony and likelihood-based methods produce similar phylogenies? When will they tend to differ?

**12.2** Most phylogenetic studies involve many different analyses, but typically present only one (or few) trees. How should a scientist decide which of the alternative outcomes to present? Or should they present all of them, even if there were hundreds? How might the alternative phylogenies be usefully reported?

**12.3** Supporters of the OPV-AIDS hypothesis (that HIV spread to humans via contaminated polio vaccine) have suggested that if any other hypothesis was subject to the same level of scrutiny, it too would be found to be problematic. How well does a hypothesis need to fit the data to be an acceptable explanation? Or does it just need to be better than the alternatives?

# Examples used in this chapter


Brandley, M. C., Schmitz, A., Reeder, T. W. (2005) Partitioned Bayesian analyses, partition choice, and the phylogenetic relationships of scincid lizards. *Systematic Biology*, Volume 54, page 373.

Doolittle, W. F. (1999) Phylogenetic classification and the universal tree. *Science*, Volume 284, page 2124.

Jegede, A. S. (2007) What led to the Nigerian boycott of the polio vaccination campaign? *PLoS Medicine*, Volume 4, page e73.

Poinar, H., Kuch, M., Pääbo, S. (2001) Molecular analyses of oral polio vaccine samples. *Science*, Volume 292, page 743.

Roques, P., Robertson, D. L., Souquière, S., Apetrei, C., Nerrienet, E., Barré-Sinoussi, F., Müller-Trutwin, M., Simon, F. (2004) Phylogenetic characteristics of three new HIV-1 N strains and implications for the origin of group N. *AIDS*, Volume 18, page 1371.

Worobey, M., Santiago, M. L., Keele, B. F., Ndjango, J. B., Joy, J. B., Labama, B. L., Dhed'A, B. D., Rambaut, A., Sharp, P. M., and Shaw, G. M. (2004) Origin of AIDS: Contaminated polio vaccine theory refuted. *Nature*, Volume 428, page 820.

HEROES OF THE GENETIC REVOLUTION

12

# Hélène Morlon

*"Tremendous progress has been made in the last decade in both the development of phylogenetic models for understanding diversification and the integration of phylogenetic biology with ecology. The biggest advance in this ongoing integration may yet have to come, and will rest on our ability to embrace the use of diversification models in community ecology, the science of interaction networks and conservation biology."*

Morlon, H. (2014) Phylogenetic approaches for studying diversification. *Ecology Letters*, Volume 17, page 508.

**EXAMPLE PUBLICATIONS**

Morlon, H., Chuyong, G., Condit, R., Hubbell, S., Kenfack, D., Thomas, D., Valencia, R., Green, J. L. (2008) A general framework for the distance-decay of similarity in ecological communities. *Ecology Letters*, Volume 11, page 904.

Morlon, H., Parsons, T. L., Plotkin, J. B. (2011) Reconciling molecular phylogenies with the fossil record. *Proceedings of the National Academy of Sciences USA*, Volume 108, page 16327.

**Figure 12.22** Hélène Morlon: while rock climbing influenced her choice of PhD, she is now into kitesurfing.

Photograph by Tamatoa Gillot.

It is a curious fact that, while choice of PhD topic is influenced by many factors including chance, opportunity, interest and location, most researchers in biology seem to continue to work on the same topic as their PhD thesis for much of their careers. It's a kind of scientific niche conservatism. Not so for Hélène Morlon, who has traversed a path from maths, to microbiology, to macroevolution and macroecology (**Figure 12.22**). Perhaps because of this varied background, her work is distinguished by its focus on developing new ways to analyse phylogenetic patterns of diversity in light of ecology, biogeography and palaeontology. Morlon is primarily concerned not with how we estimate phylogeny from molecular data, but how we can use those phylogenies to understand the processes that drive patterns of biodiversity over space and time.

After an undergraduate degree in maths at the École Normale Supérieure in Paris, Morlon undertook a Masters project in theoretical evolutionary biology. This project used the theory of adaptive dynamics to model sexual selection and explore how the evolution of female's preferences can lead to exaggerated male ornamentation. After this, she was sure she wanted to study ecology, but wasn't sure what field. So, naturally, she picked the topic of her PhD research based on the proximity of the lab to good rock climbing sites, studying environmental biology in south-eastern France. Her first scientific publications are on the uptake of selenite in a single-celled alga, *Chlamydomonas*. This work led her to a postdoctoral position in Jessica Green's research group at the University of California, where she worked on factors determining diversity in bacterial communities, developing new theoretical approaches to modelling how the decay of similarity of species assemblages over distance is influenced by population distribution over the landscape. She then moved to Joshua Plotkin's research group at the University of Pennsylvania, combining theoretical models with analytical methods and empirical data to look at patterns of biodiversity over macroevolutionary time. She developed new phylogenetic analysis methods that could infer loss of diversity over time, applying them to whale molecular phylogeny and fossil data to show that while some clades were actively radiating, others were in decline.

After a short postdoc at the University of California Berkeley, Morlon joined the CNRS (French National Centre for Scientific Research), building a research group that develops new phylogenetic approaches to macroecology and macroevolution. Her recent work has spanned ecological and evolutionary processes, including community assembly (the processes that govern the addition of species to a community by speciation and migration), broadscale geographic patterns in biodiversity (such as the higher species richness in the tropics) and the process of speciation (inferring the processes underlying adaptive radiations). She has written and released software for analysing phylogenies, including contributing to the popular R package Picante, and, like most academics, she makes her software freely available for anyone to use.

Bioinformatics and phylogenetics have provided a rich field for exploration by researchers with a mathematical background, but sometimes mathematical approaches to phylogenetic analyses can lack biological realism. Morlon's work is distinguished not only by new analytical approaches based on theoretical models, but with a solid grounding in empirical ecology and evolutionary biology that gives them a distinctly 'real world' feel.

TECHBOX **12.1**

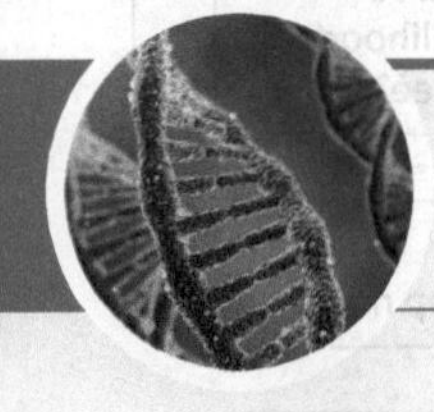

# Maximum likelihood

**RELATED TECHBOXES**

**TB 12.2:** Bootstrap

**TB 13.2:** Bayesian phylogenetics

**RELATED CASE STUDIES**

**CS 12.1:** Epidemiology

**CS 12.2:** Prediction

**Summary**

Alternative phylogenies can be compared by considering the probability that a tree gave rise to the sequences you have, given a set of assumptions about sequence evolution.

**Keywords**

model, parameters, optimization, likelihood ratio test, hill-climbing, -lnL

R. A. Fisher was a key figure in the development of mathematical approaches to evolutionary genetics (**Figure 7.16**). In fact, Fisher's contributions to statistics were as great as his contributions to biology, and statisticians are sometimes surprised to hear Fisher referred to as a geneticist. Amongst his many achievements, Fisher was responsible for the development of the maximum likelihood (ML) approach, which was later adapted to estimating phylogenies from DNA sequences by Joe Felsentein (**Hero 11**).

When we estimate phylogeny from DNA sequences, we start with a set of sequences observed in the present day and ask what is the most probable series of substitutions that happened in the past to produce these sequences. To do this, we need a model of molecular evolution that states the probability of different kinds of substitution events occurring. The parameters that will be included in the model are decided before the analysis commences, but the optimum values of these parameters for this particular dataset can be either fixed before the analysis begins or estimated during the procedure. The aim of a maximum likelihood analysis is to find the set of parameter values that maximizes the probability of the data, given the model you have applied. In phylogenetics, the most important parameter that we vary is the tree itself: we try different topologies (branching orders) and edge lengths (branch lengths) until we find the tree that is the most likely explanation for the sequence data we observe. The model may also allow different kinds of substitutions to have different probabilities of occurring (**TechBox 13.1**). Since the value of these parameters will influence the likelihood score we calculate for any given tree, we can also optimize these parameters by finding the values that give us the tree with the highest likelihood.

The procedure followed during a maximum likelihood analysis is something like this (**Figure 12.23**):

**Figure 12.23**

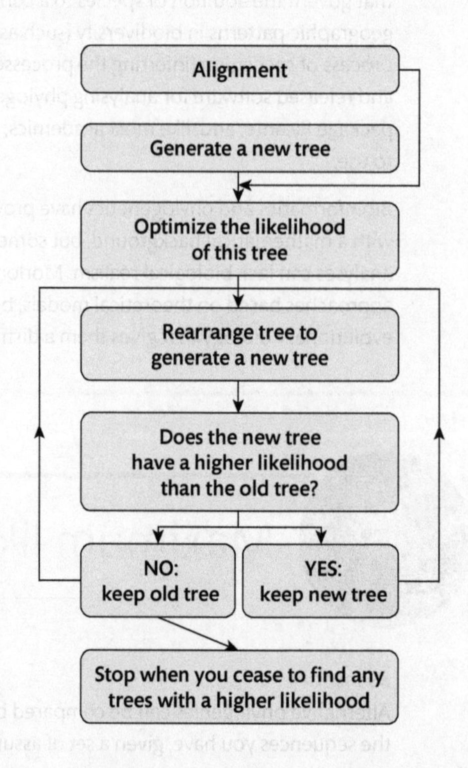


1. **Alignment:** Remember that this is the most critical step in any phylogenetic analysis of DNA sequences (see **TechBox 10.2**). The outcome of the analysis depends wholly on whether the sequences are accurately aligned such that homologous sites can be compared. Your phylogeny will be nonsense if you are not comparing homologous sites.

2. **Starting tree:** This could be a randomly selected tree, but a common way to generate a starting tree is to construct a distance phylogeny from the alignment (**TechBox 11.1**).

3. **Optimize likelihood:** Once you have a tree that you wish to estimate the likelihood of, the observed data is mapped onto the tips of that tree, then the parameter values of the model are varied until the values that maximize the likelihood of that tree are found. We can break down this procedure to estimate the likelihood of a given tree as follows:

(i) **For each site:** For each column (site) in the alignment, the nucleotide states are mapped onto the tips of the trees. These are the observed end states of a process, but there are many possible sets of substitutions along that tree that could have generated that particular pattern of nucleotides at the tips of the tree. Using the substitution probabilities given in the model, the site likelihood is calculated by summing the probabilities of each of the proposed substitution events, starting at the tips of the tree and working down through the shared nodes until you reach the root.

(ii) **Over all sites:** On the assumption that each site evolves independently, the probability of this tree producing the whole alignment is the product of all of the site likelihoods. This gives one likelihood score for this particular tree given this particular model and a particular set of parameter values. Now it's time to fiddle the parameter values to find their optimum value for this tree.

(iii) **Over all parameter values:** The value of a free parameter is changed slightly, and the likelihood of the tree recalculated (as in steps (i) and (ii)). Remember that the edge lengths of the tree are free parameters, so the change may be to slightly shorten or lengthen one of the branches. Or it may be a change to one of the substitution probability parameters, such as the transition/transversion ratio. Iterate, varying parameter values and calculating the likelihood over and over again. Those parameter values that give the highest likelihood are retained.

4. **Rearrange the tree** to generate a new tree. For example, you might cut the tree at one edge and re-join it somewhere else.

5. **Compare the new tree to the old tree:** Calculate the likelihood of the new tree using the procedure of step 3 above. If the new tree has a higher likelihood than the old tree, keep it in memory. If the new tree has a lower likelihood than the old tree, forget it, and keep the old tree in memory. So at any given point, the tree you keep in memory is the best one you have found so far.

6. **Rearrange the tree again** and repeat steps 3 to 5. Keep going until you never find a new tree with a higher likelihood than the one you hold in memory. At this point, accept the one in memory as the best tree for this data given the assumptions of the model.

### Likelihood scores

The likelihood of a given phylogeny is the probability of the data given the model. In other words, it is the probability that you would have ended up with these sequences if this was the real tree, and if sequences evolved in the way you assume they do. Being a probability, the likelihood is a number between 0 and 1. But, in general, this is not the number that your phylogenetics program will report to you. Remember that to get the likelihood score for a dataset given a particular tree, you had to multiply the probabilities across all the sites in the sequence. Because each of these site probabilities is a number between 0 and 1, the product of likelihoods across all sites in the alignment is going to be a very small number, so most tree scores are reported as the natural log (ln), which is $\log_e$ of the likelihood value (where *e* is Euler's number, approximately 2.71). Taking the log of a number between 0 and 1 gives you a negative value. That is why the tree score you get is often labelled '–lnL'. So the likelihood score of a phylogeny will typically be a largish number with a minus sign in front of it. Because log scores are negative, you need to remember that the score closer to zero is better. In other words, a tree with an lnL score of –1300.50 is a more believable explanation of your data, given the model, than a tree with a score of –1450.70.

### Advantages of maximum likelihood

The reliance on an explicit model of molecular evolution is a strength of ML because it allows ML to be adapted to different situations and datasets. It also makes it quite clear that the results you get are conditional on the model you use, hopefully prompting you to test the robustness of your conclusion with different models. One way of doing this is to use a Likelihood Ratio Test, which compares the likelihood of the data given two different models, one of which has one (or more) extra parameter. The simpler model is rejected only if the more parameter-rich model is found to be significantly better fit to the data. In fact, likelihood gives a formal statistical basis for testing all aspects of the phylogeny estimation, including comparing specific phylogenetic hypotheses (**TechBox 12.2**).

### Disadvantages of maximum likelihood

The iterative procedure underlying ML estimation can take a long time. The more sequences you have, the more nodes in the tree over which you must calculate site likelihoods. The longer your alignment, the more site likelihoods you must calculate. The more free parameters in your model, the more times you must recalculate the joint likelihood for each tree topology for each parameter value. And since the edge lengths of the tree are also free parameters, the more sequences you have the greater the number of edges in your tree that must be varied in length during your likelihood optimization. So ML estimation of phylogenies very easily turns into a computational marathon. The 'hill-climbing' heuristic search method means that you only ever move 'upwards' to a tree with a better likelihood, so it is possible to miss out on the best tree by getting stuck on a local optimum that cannot be improved in a single step. Maximum likelihood is a robust method that has been shown to perform well under many circumstances. However, there are also situations in which maximum likelihood is consistently misleading, so as with all phylogenetic estimation methods, caution should always be exercised when interpreting results. Remember that even if you are using a program that outputs the single maximum likelihood tree (the one with the best likelihood score), the best tree may only have a slightly better score than the next-best trees, so you should explore ways of considering the distribution of solutions around the maximum.

TECHBOX 12.2

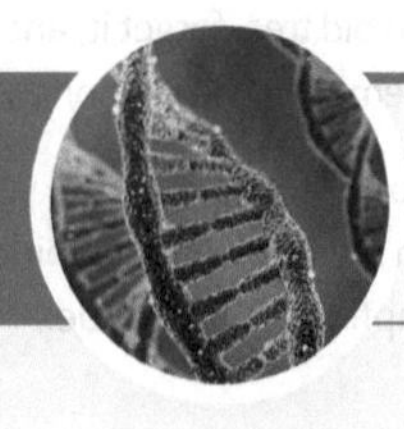

# Bootstrap

**RELATED TECHBOXES**

TB 11.2: Phylogenetic networks

TB 12.1: Maximum likelihood

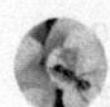

**RELATED CASE STUDIES**

CS 12.1: Epidemiology

CS 12.2: Prediction

**Summary**

Resampling your data provides a way of asking if you would get the same phylogeny if the data were slightly different.

**Keywords**

phylogenetic support, parametric bootstrap, alignment, simulated data, likelihood, statistical significance, resampling, hypothesis testing, null distribution, whole genome

It is a well-known fact that you cannot lift yourself off the ground simply by pulling on your own shoes (feel free to test this hypothesis for yourself right now). The bootstrap statistical technique performs the statistical equivalent of picking yourself up by the bootstraps (**Figure 12.24**) by testing the signal in the data using only the data itself (thus it shares a common etymology with 'rebooting' a computer. A computer 'boots up' because it has to use one of its own programs to start all of its other programs). The bootstrap was introduced to phylogenetics by Joe Felsenstein (**Hero 11**) as a way of testing the strength of phylogenetic signal in an alignment[1]. It takes a single alignment, then randomly selects columns to include in a replicate alignment which is similar, but not identical, to the original. Although you end up with the same number of sites in the alignment, any given site from the original alignment may be included once, or twice, or many times, or not at all. You can repeat this sampling process again and again to create a large number of replicate datasets, all based on the original dataset but not exactly the same.

Now you estimate a phylogeny from each of your replicate datasets. If you always get the same phylogeny, no matter how many times you resample your data, then you can confidently say that, given your chosen phylogenetic method, these data point unambiguously to a particular tree. But what if some of your replicates give rise to different topologies? Say you made 100 replicate datasets, and 99 of them grouped *a* with *b*, but one tree out of the hundred grouped *a* with *c*. Then you would say that the *a*–*b* group had 99 per cent bootstrap support, and you might feel fairly comfortable that this was the best hypothesis for your data. But what if 55 of the replicate alignments grouped *a* with *b*, and 45 of them grouped *a* with *c*? Now how confident are you that your data supports the *a*–*b* grouping? A low bootstrap percentage tells you that if

**Figure 12.24** Baron Munchausen is said to have lifted himself out of the sea by pulling on his own bootstraps. Bootstraps are not shoelaces, but the loops at the top of the boot to help you pull them on, as in the iconic Australian footwear, Blundstone Boots (affectionately referred to as Blunnies).

Photograph: Ted Phelps.

you had slightly different data you might get a different tree. Clearly, then, we would have less confidence in a node with low bootstrap value (even though it may be correct).

Just as there are conventions about what makes an experimental result statistically significant (usually a probability of your result occurring by chance of less than five per cent), so there is a convention about what level of bootstrap support is acceptable (usually greater than 95 per cent). However, it is up to you to decide what level of support you find plausible. The bootstrap is just a tool for helping you judge how well your data support a hypothesis, it does not tell you whether your hypothesis is likely to be true or not. In fact, it would be circular reasoning to generate a hypothesis from a particular dataset, then test it using the very same dataset. Instead, the bootstrap is testing how strongly your dataset supports that particular phylogeny, given the particular method you are using. If there is a bias in your data or your method, then you might find you have very strong support for the wrong tree. Caution must be exercised when considering bootstrap values generated for very large datasets, such as whole genome sequences. Because the resampled datasets are so large, the variation between them is relatively small, generating very high bootstrap percentages on virtually all nodes, even if there is a substantial amount of conflict in the data. For example, a study that estimated phylogeny of yeasts based on whole genome sequences returned a tree with 100 per cent bootstrap support on every node, even though trees estimated from each of the separate genes all disagreed with each other[2].

Also remember that there are two possible causes of low bootstrap values: low signal or high noise (**TechBox 11.2**). If there have been very few substitutions in your sequences, then there may be few informative splits (see Chapter 11). Resampling the data may miss the small number of substitutions that reveal the history of the sequences, so some of the replicate datasets could support an alternative tree. Low signal may be ameliorated by collecting more data to increase the number of informative sites. Alternatively, there could be lots of informative splits in the data, but they do not all support the same tree. High noise occurs when multiple hits overwrite the signal, or when parallel substitutions are acquired independently in different lineages, or when evolution is not tree-like (e.g. hybridization). In this case, there is no point adding more of the same sort of data, because it just adds more mess (though getting different data might be helpful—sequencing a different gene, for example).

### Parametric bootstrapping

The standard bootstrap is effectively asking 'if my sequences were slightly different, would I still get the same tree?' The parametric bootstrap turns this around to ask 'what is the chance that a different tree could have produced my sequences?'[3] This is useful if you want to ask whether your best tree is significantly better than an alternative hypothesis. Say you conduct a phylogenetic analysis and the maximum likelihood tree supports Hypothesis A. But can you safely reject Hypothesis B as an explanation of your data? Is it possible that, if Hypothesis B is actually true, you could, by chance, have ended up with the data you have? To test this, you can generate replicate datasets on the tree that supports hypothesis B, by using a substitution model to evolve simulated sequences along a tree. Programs that produce simulated data usually start with a randomly generated sequence at the root of the tree, then move up the tree, asking at each node what the probability is that each site will have undergone a substitution. Substitutions are accumulated until the tips are reached, giving the final sequences. This process can be repeated to produce hundreds of simulated datasets, each one of which represents a possible outcome of DNA sequences evolving along this tree, given a particular substitution model. You can then estimate a phylogeny for each of the simulated datasets and see how often you recover either of your two alternative phylogenies. You can also estimate the likelihood of each dataset given each of your alternative phylogenies and compare it to the likelihood of the true dataset on each of these phylogenies. This gives you a distribution of likelihood differences for these two alternative trees from the simulated data. Now you can measure the difference in likelihood of your observed dataset given your two phylogenetic hypotheses to ask 'if my alternative hypothesis is really true, what is the chance that I could have got my observed likelihood difference by chance?' (The term 'parametric' refers to using this null distribution to test for significance of a likelihood difference.)

If this is a bit hard to follow, then perhaps it's best if we consider an example involving starfish (on the assumption that adding starfish is bound to improve nearly any situation: **Figure 12.25**).

**Figure 12.25** How cool are these starfish?

Molecular phylogenies have shaken up the classification of animals, as some taxa that were previously grouped on the basis of particular similarities have been shown to be only distantly related and other very disparate animals have been brought together as surprisingly close relatives. For example, many people had assumed that the rather unprepossessing marine worms called hemichordates were related to the chordate lineage, because they have classic chordate characteristics, such as a head-end with a pharynx, a back-end with a muscular tail, and a dorsal nerve chord running between the two. But molecular phylogenies often placed hemichordates on the lineage leading to the echinoderms, which have very un-chordate-like body-plans, with radial symmetry, a circular nervous system, no head, no tail, and no brain

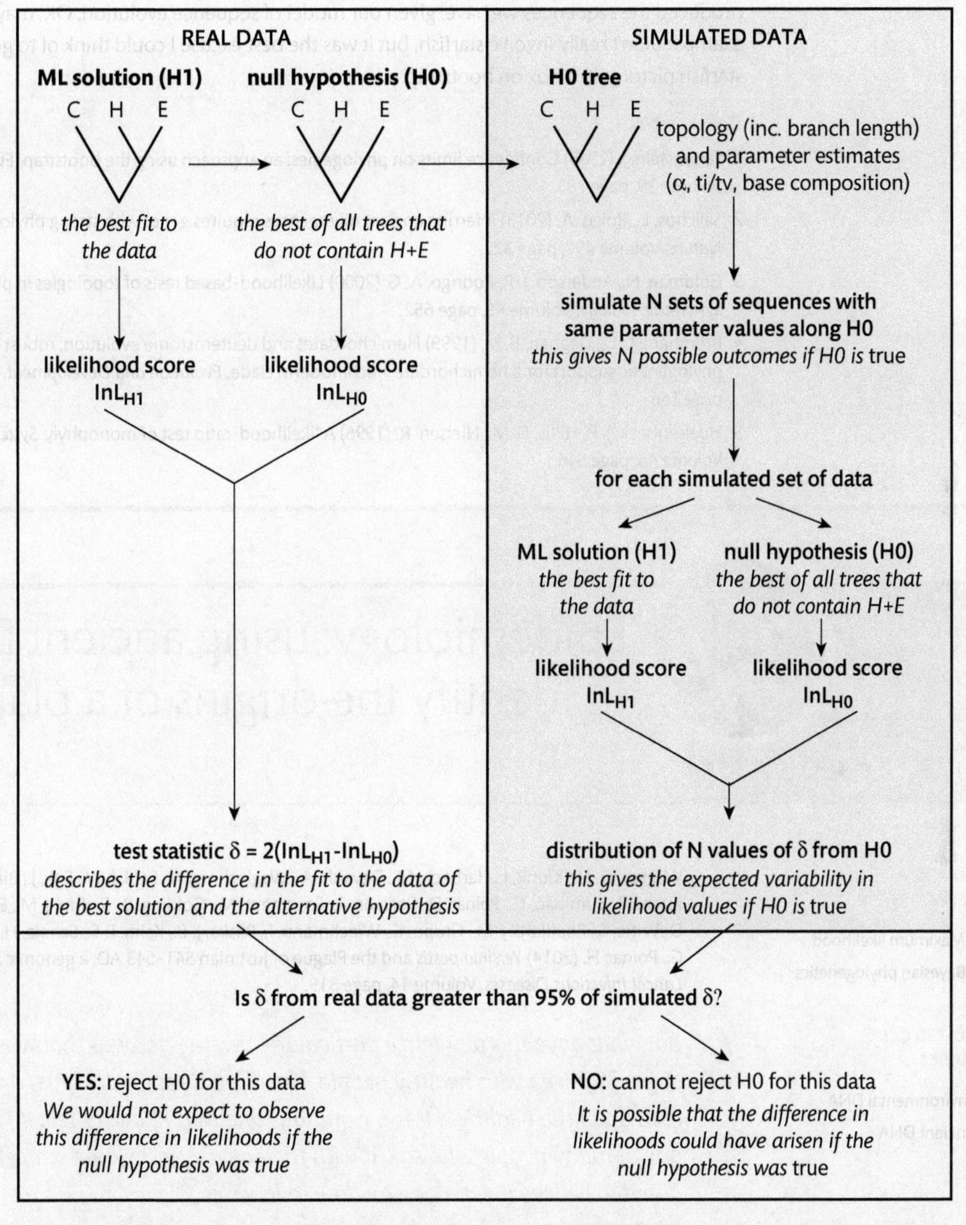


**Figure 12.26** Parametric bootstrap is one way of comparing phylogenetic hypotheses. Here, the parametric bootstrap is illustrated by comparing the best-fit phylogenetic tree that groups hemichordates (H) with echinoderms (E) to the exclusion of chordates (C) to an alternative hypothesis. In this example, the best tree (H1) is compared to any next-best tree that doesn't contain the H + E grouping, but often the comparison will be made between two defined alternative phylogenetic hypotheses.

(e.g. starfish). But, possibly because of the deep divergence times of hundreds of millions of years, the phylogenetic support for placing hemichordates on the echinoderm lineage was often poor. So just how well did the molecular data support the hemichordate + echinoderm grouping? Is it possible that even if hemichordates are actually more closely related to chordates that you could, just by chance, get a maximum likelihood phylogeny with the hemichordate + echinoderm grouping? To test how likely this is, we need to ask whether we could get the observed difference in likelihood between the two phylogenetic hypotheses even if the alternative hypothesis was true[4]. By simulating data along the alternative tree (H0), you can ask how often you would expect to get a tree supporting H1 even if H0 was true (**Figure 12.26**)[5]. In this particular case, we can reject the hemichordate + chordate grouping as unlikely to have produced the sequences we have, given our model of sequence evolution. OK, maybe that example didn't really involve starfish, but it was the best excuse I could think of to get some starfish pictures in a box on bootstrap.

### References


1. Felsenstein, J. (1985) Confidence limits on phylogenies: an approach using the bootstrap. *Evolution*, Volume 39, page 783.
2. Salichos, L., Rokas, A. (2013) Inferring ancient divergences requires genes with strong phylogenetic signals. *Nature*, Volume 497, page 327.
3. Goldman, N., Anderson, J. P., Rodrigo, A. G. (2000) Likelihood-based tests of topologies in phylogenies. *Systematic Biology*, Volume 49, page 652.
4. Bromham, L. D., Degnan, B. M. (1999) Hemichordates and deuterostome evolution: robust molecular phylogenetic support for a hemichordate + echinoderm clade. *Evolution and Development*, Volume 1, page 166.
5. Huelsenbeck, J. P., Hillis, D. M., Nielsen, R. (1996) A likelihood-ratio test of monophyly. *Systematic Biology*, Volume 45, page 546.


CASE STUDY 12.1

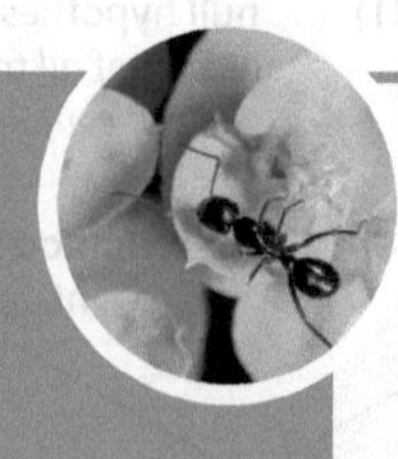

# Epidemiology: using ancient DNA to identify the origins of a plague

RELATED TECHBOXES

TB 12.1: Maximum likelihood

TB 13.2: Bayesian phylogenetics

RELATED CASE STUDIES

CS 2.1: Environmental DNA

CS 2.2: Ancient DNA

Wagner, D. M., Klunk, J., Harbeck, M., Devault, A., Waglechner, N., Sahl, J. W., Enk, J., Birdsell, D. N., Kuch, M., Lumibao, C., Poinar, D., Pearson, T., Fourment, M., Golding, B., Riehm, J. M., Earn, D. J. D., DeWitte, S., Rouillard, J-M., Grupe, G., Wiechmann, I., Bliska, J. B., Keim, P. S., Scholz, H. C., Holmes, E. C., Poinar, H. (2014) *Yersinia pestis* and the Plague of Justinian 541–543 AD: a genomic analysis. *The Lancet Infectious Diseases*, Volume 14, page 319.

> *But what gave this pestilence particularly severe force was that whenever the diseased mixed with healthy people, like a fire through dry grass or oil it would rush upon the healthy.... It is a wondrous tale that I have to tell: if I were not one of many people who saw it with their own eyes, I would scarcely have dared to believe it, let alone to write it down.*

Boccacio, G. (1471) *The Decameron*, Translated from Italian by Richard Hooker

**Keywords**

maximum likelihood, bootstrap, calibration, hypothesis testing, ancient DNA, pandemic, phylogeny, bacteria, infectious disease

### Background

The Black Death, one of the world's worst infectious disease pandemics, caused the death of around half the population of Europe in the fourteenth century. The Black Death has been attributed to the bacterium *Yersinia pestis*, which causes both bubonic and pneumonic plague. The Black Death was the second of three great pandemics of plague. During the Third Pandemic of plague, starting in the mid-nineteenth century, researchers identified *Y. pestis* as the causative agent, and made the link between rats as carriers of plague and fleas that acted as vectors for the bacteria to infect humans. Analyses of DNA samples confirmed the presence of *Y. pestis* in victims of an earlier plague which spread through the Eastern Roman (Byzantine) Empire between the years 541 and 543. This First Pandemic, possibly the first large-scale outbreak of plague, is named for the emperor of the time, Justinian I, who caught the disease but survived.

### Aims

Do the three plague pandemics—First (Justinian), Second (Black Death), and Third (nineteenth century)—represent three separate outbreaks of *Y. pestis*? Or did the plague persist in human populations throughout this period, with the number of infections occasionally rising in frequency to pandemic levels?

### Methods

DNA was extracted from teeth from two individuals from a mass grave in a cemetery in Bavaria, Germany. DNA degrades over time, so target DNA will be in small amounts in old samples. They enriched the target DNA in solution by using 'bait libraries'[1], generated from modern genome sequences of both *Y. pestis* and human mitochondrial DNA. Sequences from the two Justinian *Y. pestis* samples were aligned against 130 contemporary *Y. pestis* strains, as well as a *Y. pestis* genome from the Black Death (Second Pandemic)[2] dated to 1348, and a maximum likelihood phylogeny of the sequences estimated. They also used Bayesian analysis to examine the effect of different calibration strategies on the phylogeny estimation, comparing analyses where the sequences were isochronous (samples all assumed to be taken at the same time) to 'tip dated' sequences (where the ages of the ancient DNA sequences are included).

### Results

Radiocarbon dating of the teeth placed the samples in the time of Justinian, though with wide margin of error (up to 100 years). The human mitochondrial sequences were consistent with samples from people of European or Middle Eastern extraction. The two Justinian samples form a closely related pair at the end of a long, well-supported branch (with 100 per cent bootstrap support), that was distinct from the strains of *Y. pestis* associated with the later Second and Third pandemics (**Figure 12.27**). In contrast, the Black Death samples are at the base of a clade that led to Third Pandemic and contemporary human strains. To test how strongly supported this grouping is, they compared the likelihood of the highest likelihood tree to a tree where the Justinian sequences were the direct ancestor of the later plagues, concluding that they could reject the grouping of the three pandemics as significantly poorer fit to the data.

### Conclusions

Because the two Justinian plague sequences cluster together, separate from the later plague pandemic sequences, researchers conclude that the Justinian plague represented a novel cross-species infection from rodents to humans. They suggest that none of the later pandemics, or contemporary *Yersinia pestis* strains, are descendants of the Justinian plague. In contrast, the Second Pandemic strain may have persisted in rodent reservoirs, eventually giving rise to the Third Pandemic and contemporary infections. The authors interpret their study in light of increasing geographic reach of *Y. pestis* in humans over time, with each subsequent pandemic having a wider reach and greater persistence (**Figure 12.28**).

### Limitations

This analysis shows no clear relationship between time and accumulation of genetic change: the older sampled sequences do not have shorter branch lengths than the more recent samples.

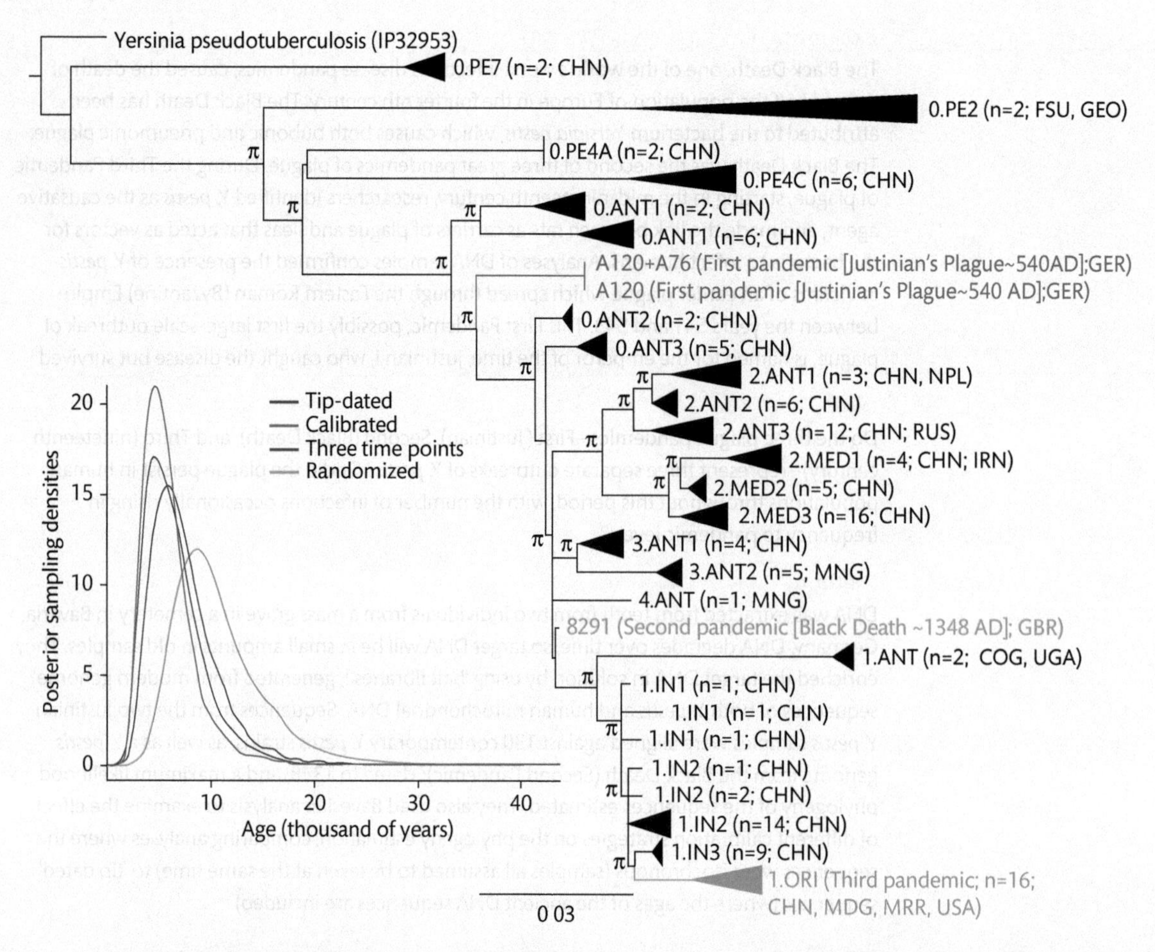


**Figure 12.27** Phylogeny of the plague bacterium (*Yersinia pestis*). Each of the pandemics comes out in a different place in the tree (First = red, Second = green, Third = blue), clustered amongst modern samples (in black). The phylogeny shown was generated using maximum likelihood, but they also conducted Bayesian analyses that varied with how many samples were dated: all samples dated (Tip-dated), first two (Calibrated) or all three (Three Timepoints) epidemics dated and all others arbitrarily dated at year 2000, and for comparison a Randomized calibration where the dates of the first and second pandemic strains were swapped. The countries from which the contemporary samples were taken are: CHN = China; FSU = Former Soviet Union; GEO = Georgia; GER = Germany; NPL = Nepal; RUS = Russia; IRN = Iran; MNG = Mongolia; GBR = Great Britain; COG = Republic of Congo; UGA = Uganda; MDG = Madagascar; MRR = Myanmar.



Therefore branch lengths in this phylogeny cannot be interpreted as indicative of evolutionary time. It has been suggested by other researchers that the rate of molecular evolution in *Y. pestis* is highly variable over time[3]. Not only does this make attempts to date plague evolution using molecular data unreliable, it will also affect estimates of phylogeny through its impact on branch length. There are only two First Pandemic plague sequences, derived from individuals that were buried in the same cemetery, so there may have been more diversity in the First Pandemic plague strains than is captured in this phylogeny. In this sense, comparing the maximum likelihood phylogeny to one where the two Justinian sequences from Bavaria are at the base of the later pandemics may represent an unfair comparison, because more First Pandemic samples might have broken up the long branch and made alternative phylogenetic hypotheses more plausible.

### Further work

Justinian plague taken from different geographic areas would allow a more robust test of the hypothesis that the Justinian plague lineages did not persist. Similarly, showing that the three

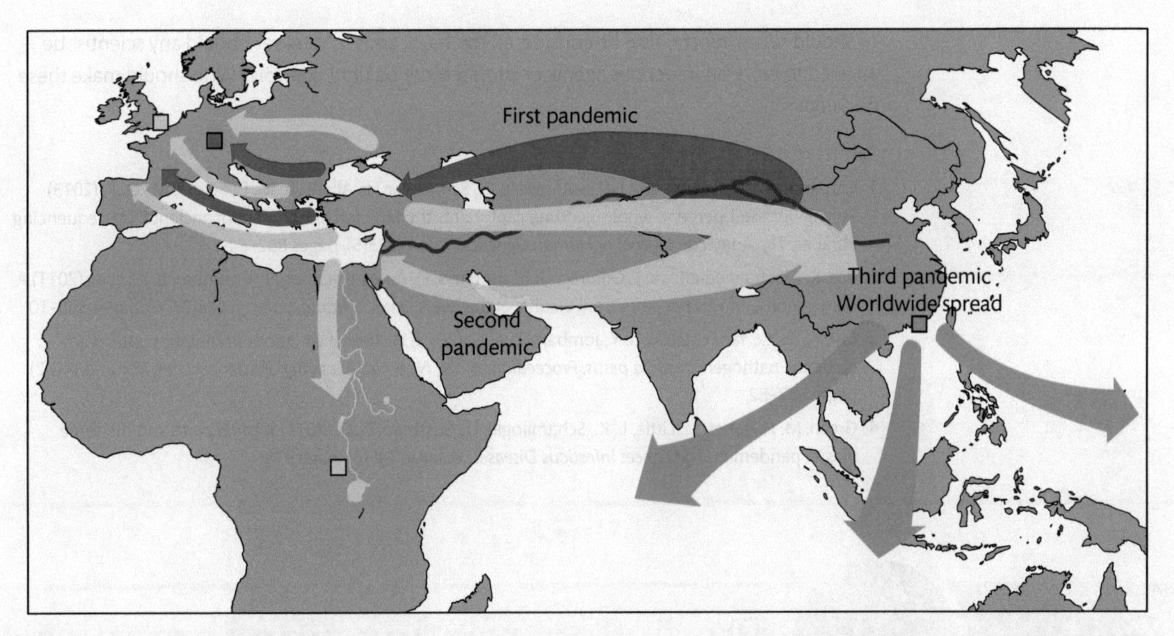


Figure 12.28 Hypothesis for the history of the three plague pandemics based on phylogenetic analysis of *Yersinia pestis* genome sequences. In the same way that a phylogenetic tree is one possible explanation of how the observed sequences came to be, this diagram represents one possible historical explanation for how we got the inferred phylogeny (**Figure 12.27**). The First (Justinian) Pandemic is said to have arrived in Constantinople aboard grain ships from Egypt, but is assumed to have originated in China then moved along trade routes. Similarly, the Second (Black Death) Pandemic is considered to have arisen in China by zoonotic transfer from rodents to humans, then travelled along the Silk Road to Europe, and from there to Africa. This may have established plague as an endemic infection in rodent populations in many places, which served as a reservoir for the emergence of the Third Pandemic and a continuing source of infection today. What parts of this analysis are well supported by the data, and what aspects are speculation? How might you test how well the data fits this particular hypothesis for the origins of the three plague pandemics?



pandemics occur on different parts of the tree doesn't prove beyond doubt that they represent three independent zoonoses from rodents to humans[4]. If it were possible to obtain plague sequences from the times between the current ancient DNA sequences they might 'fill in the gaps' between the pandemic lineages, however this currently does not seem to be a very likely prospect.

### Check your understanding

1. Do any of the sampled Justinian plague sequences represent the ancestors of modern *Yersinia pestis*?
2. Why did researchers reject the hypothesis that plague persisted in human populations between the three pandemics?
3. How does the phylogeny in **Figure 12.26** support the hypothesis presented in **Figure 12.27**?

### What do you think?

Plague still causes deaths, with thousands of new infections every year. Could DNA analysis help to understand the dynamics of *Yersinia pestis* outbreaks and how it emerges in humans? Could such analyses contribute to containment or prevention?

### Delve deeper

Ancient DNA sequencing of infectious agents from past pandemics has been criticized on the grounds that it could lead to a re-emergence of diseases through accidental release of bacterial or virus DNA. How should such threats be judged? Publication of genome sequences from dangerous pathogens has also raised concerns about the possibility that the information could be used to develop bioweapons. Should all results of scientific investigations be made public, or should some information be considered too hazardous to release? Should any scientist be allowed to work on infectious agents or should there be tight controls? Who should make these decisions?


### References

1. Carpenter, M. L., Buenrostro, J. D., Valdiosera, C., Schroeder, H., Allentoft, M. E., Sikora, M., et al. (2013) Pulling out the 1 percent: whole-genome capture for the targeted enrichment of ancient DNA sequencing libraries. *The American Journal of Human Genetics*, Volume 93(5), pages 852–64.
2. Bos, K. I., Schuenemann, V. J., Golding, G. B., Burbano, H. A., Waglechner, N., Coombes, B. K., et al. (2011) A draft genome of *Yersinia pestis* from victims of the Black Death. *Nature*, Volume 478(7370), pages 506–10.
3. Cui, Y., Yu, C., Yan, Y., Li, D., Li, Y., Jombart, T., et al. (2013) Historical variations in mutation rate in an epidemic pathogen, *Yersinia pestis*. *Proceedings of the National Academy of Sciences* USA, Volume 110(2), pages 577–82.
4. Green, M. H., Jones, L., Little, L. K., Schamiloglu, U., Sussman, G. D. (2014) *Yersinia pestis* and the three plague pandemics. *The Lancet Infectious Diseases*, Volume 14(10), page 918.


CASE STUDY 12.2

# Prediction: relating ethnobotanical resources across different cultures

**RELATED TECHBOXES**

TB 3.2: Biobanking

TB 12.1: Maximum likelihood

**RELATED CASE STUDIES**

CS 1.2: Forensics

CS 11.2: Distance

Saslis-Lagoudakis, C. H., Savolainen, V., Williamson, E. M., Forest, F., Wagstaff, S. J., Baral, S. R., Watson, M. F., Pendry, C. A., Hawkins, J. A. (2012) Phylogenies reveal predictive power of traditional medicine in bioprospecting. *Proceedings of the National Academy of Sciences USA*, Volume 109, page 15835

> *". . . we propose a more sophisticated framework of identifying plants with high medicinal potential based on traditional medicine, combining two criteria: phylogenetic signal and cross-cultural agreement. We have shown that lineages fulfilling these two criteria are significantly richer in plants with demonstrable bioactivity than a random sample."*

### Keywords

bioprospecting, phylogenetic signal, tip shuffle, mean phylogenetic distance (MPD), null distribution, clustering, herbarium

### Background

Ethnobotany—the study of plant use across human cultures—has been a source of valuable information for the development of pharmaceuticals. Many thousands of different plants are used in traditional medicine around the world. To make screening for potential pharmaceuticals more efficient it would be good to have a way of predicting which traditional remedies are more likely to be efficacious. If unrelated cultures use the same (or similar) plants for the same purposes, this might suggest that they have independently identified useful botanical resources.

### Aim

Researchers collated lists of medicinal plants from three different regions (the Cape of South Africa, New Zealand, and Nepal: **Figure 12.29**), then used phylogenetic analysis to ask whether the plants traditionally used in each of the three regions were a random sample of their local flora, or whether there were any significant similarities across regions in their choices of medicinal plants.

**Figure 12.29** Native vegetation within agricultural landscapes can serve as sources of ethnobotanical remedies. In Nepal, collecting and selling traditional plant-based remedies contributes to many rural household incomes.

Photographs by Barry Bromham.

### Methods

Researchers used sequences of the *rbcL* chloroplast gene, obtained from sequencing herbarium material and collating published sequences, to estimate maximum likelihood phylogenies for each of the three regional floras (which ranged in size between 4,000 and 9,000 species), including one representative species for about 80 per cent of the genera found in each of the floras. They also combined the separate regional datasets to estimate a genus-level tree that included all three floras. They identified species used in 13 categories of medical conditions (for example, for treating eye conditions, gastrointestinal complaints, or fertility). Medicinal use was recorded for between 3 and 14 per cent of the flora of each region. They estimated the 'phylogenetic signal'—the degree to which medicinal plants are clustered in related lineages—using the mean phylogenetic distance (MPD), which is the mean path distance between medicinal plants on each phylogeny. This is calculated by summing the branch lengths connecting all possible pairs of genera containing medicinal plants. If medicinal plants tend to be from related lineages, then the paths connecting them on the phylogeny will be, on average, shorter than a sample of random pairs of genera from across the phylogeny. So to assess the significance of the observed MPD value, they compared it to a null distribution, produced by randomly distributing the tip states across the phylogeny 1,000 times (a procedure often referred to as a tip-shuffle). This null distribution provides the range of values of MPD expected to occur by chance if medicinal plants were randomly distributed on the phylogeny. Then they compared the phylogenetic distribution of medicinal taxa across all three regions by calculating the mean path distance of each genus containing medicinal plants from one region to all medicinal taxa in the other. If people from different regions use related lineages, the average path lengths connecting them will be lower than expected from a random sample.

### Results

Several categories of medicinal plant use had significant phylogenetic signal: that is, genera containing medicinal species in the category were more closely related to each other than would be expected if medicinal plants were randomly distributed between genera. MPDs between medicinal floras of the different regions were significantly smaller than expected by chance, suggesting that plants from some clades are more likely to be included in traditional medicines across the three different regions. Clades identified as containing a significantly higher number of medicinal plants were more likely to occur in more than one of the cultures than if each population selected plants entirely independently of the others. The researchers also compared the ethnobotanical lists to genera from which known pharmaceutical compounds have been derived, and found that lineages containing plants used in traditional medicine also contained more taxa identified as producing pharmaceutical substances than expected by chance.

### Conclusions

The non-random distribution of clades with larger-than-expected numbers of medicinal plants in the cross-cultural analysis suggests that the three different cultures are targeting more of the same plant lineages than would be expected by chance. They argue that phylogenetic signal indicates that selection of medicinal plants is not just based on mythical properties or placebo effects, but is influenced by inherited traits of the lineages that influence the bioactivity of plant material, a conclusion also supported by the overlap between the medicinal plants and those that have been identified as containing pharmaceuticals.

### Limitations

In this study, phylogenetic clustering of ethnobotanical resources is taken as a sign that people are independently selecting similar useful compounds from plants. The same clustered pattern could also be generated if, instead of the chosen species being particularly efficacious, people were avoiding particular lineages, for example not using harmful plants (which might cluster on the phylogeny if they share the same phytochemicals or physical defences) or not utilizing plants that were rare or difficult to access (given that related lineages often tend to be found in the same areas or habitats). The significance of clustering can be influenced by the definition of use categories: for example a similar study on South African ethnobotany that used different use categories found little evidence of phylogenetic signal[1].

### Further work

In any hypothesis testing, it is important to avoid ascertainment bias, where data collection is targeted to those observations that support the hypothesis. Screening plant species that are not used in traditional medicine for bioactive compounds would provide a more robust test of the hypothesis that traditional medicines have been chosen for their high levels of active ingredients. Similarly, an analysis that looks for shared features of the clades with few ethnobotanical resources might reveal collection biases (such as geographic spread, rarity, or growth habit). Future tests would ideally take into account differences in species richness (do genera containing more species tend to contain more medicinal plants?), distribution (do genera found closest to areas of human occupation get incorporated into traditional uses?) and rarity (are rare plants more or less likely to be used than common plants?).

### Check your understanding

1. Why did the researchers compare medicinal plant use in three different cultures?
2. Why is average path length between pairs of genera used as a measure of targeted plant selection in ethnobotanical uses?
3. Why did they calculate the distance between randomly chosen genera?

### What do you think?

What other fields would benefit from phyloprediction, where the relatives of identified lineages can be investigated to see if they have particular properties? How would you design the analysis?

### Delve deeper

This study suggests that choosing closely related lineages to lineages with known ethnobotanical or pharmaceutical properties will maximize chances of success in bioprospecting. An opposite strategy, that of selecting sets of species that maximize the phylogenetic distance between lineages, has been suggested as a means of making conservation decisions: that we can use phylogenetic placement to focus conservation effort on regions or sets of taxa that maximize the amount of differences between them, thus optimizing conservation strategies to preserve the highest disparity. Do you think that phylogenies can be used to optimize the preservation of biological disparity? And is this a worthy basis for conservation strategy?

### Reference

1. Yessoufou, K., Daru, B. H., Muasya, A. M. (2014) Phylogenetic exploration of commonly used medicinal plants in South Africa. *Molecular Ecology Resources*, Volume 15, page 405.

# Rates

13

## *Tempo and mode*

*"How fast, as a matter of fact, do animals evolve in nature? That is the fundamental observational problem of tempo in evolution. It is the first question that the geneticist asks the paleontologist"*

**Simpson, G. G. (1944)** ***Tempo and Mode in Evolution.*** **Columbia University Press.**

## What this chapter is about

Genomes constantly accumulate changes, but the rate of change can vary between lineages. Anything that increases the number of mutations, such as copying the DNA more often or being exposed to more mutagens, could increase the rate of molecular evolution. Similarly, anything that influences the number of mutations that go to fixation, such as population size and patterns of selection, can also affect rates of molecular evolution. The upshot of this is that many species characteristics, such as physiology, life history, ecology and habitat, can influence the average rate of molecular evolution. We can study the kind of species traits that influence the rate of genomic change by comparing estimates of substitution rate, using statistical analyses that allow for the patterns of descent created by the evolutionary process.

### Key concepts

- Rate of molecular evolution is influenced by the mutation rate and substitution rate, both of which can vary across the genome, over time and between lineages
- Branch lengths on a phylogenetic tree, representing estimates of amount of evolutionary change, can be estimated using a substitution model to predict how many substitutions have occurred given the observed differences between sequences

# Rate of evolutionary change

## *An odd fish*

In December 1938, Marjorie Courtenay-Latimer, the curator of the East London Museum in South Africa, spotted an unusual fish in the day's catch of a fishing trawler: 'I picked away the layers of slime to reveal the most beautiful fish I had ever seen. It was five foot long, a pale mauvy blue with faint flecks of whitish spots; it had an iridescent silver-blue-green sheen all over. It was covered in hard scales, and it had four limb-like fins and a strange little puppy-dog tail. It was such a beautiful fish—more like a big china ornament—but I didn't know what it was' (Figure 13.1). The trawler captain had likewise never seen one before. Courtenay-Latimer took it back to the museum (much to the chagrin of the taxi driver who did not like the look of the sixty kilogram fish). The strange blue fish was not to be found in any of the books she consulted. When she examined it, it became clear to her that it belonged to a group of fish that were known only from fossils. Her gut feeling was confirmed by a fish taxonomist who identified the strange blue fish as a coelacanth and named it after her: *Latimeria*.

Coelacanths belong to one of the oldest lineages of bony fish. Their fossil record extends from around 400 million years (Myr) ago until around 80 Myr ago. But coelacanths disappear from the fossil record toward the end of the Cretaceous, which is the period in which the last dinosaur fossils are found. Finding a live coelacanth was therefore almost as unexpected as finding a live dinosaur. Not only had the coelacanth emerged from the sea 80 million years after its apparent extinction, it appeared not to have changed much in all that time. *Latimeria* is similar in morphology to fossil coelacanths, despite the tens of million years that separate it from its extinct relatives. The coelacanth is often held up as an extreme example of evolutionary stasis. While the world's climate had gone in and out of ice ages, while the mammals and birds evolved out from under the feet of dinosaurs to dominate many terrestrial, aerial and aquatic niches, while human populations expanded and wreaked massive changes on global ecosystems, the coelacanth had apparently stayed much the same.

In 1997, a coelacanth was spotted in an Indonesian fish market by two Americans on their honeymoon (one of whom happened to be a marine biologist). The original specimen was apparently sold and eaten, but after nearly a year of questioning fishermen and offering a reward, another coelacanth was caught alive (Figure 13.2). Over ten thousand kilometres from the African population, the announcement of the Indonesian coelacanth was nearly as surprising as Courtenay-Latimer's original discovery. The Indonesian coelacanth was virtually identical to the African ones, differing mainly in the colouring, not blue with flecks of white as described by Courtenay-Latimer, but brown with flecks of gold. However when DNA sequences from the African and Indonesian coelacanths were compared, they were found to differ at around four per cent of sites. This amount of genetic divergence would be expected from lineages that had been separated for millions of years. The two populations of coelacanth continued to accumulate molecular change despite the low rate of morphological evolution.

**Figure 13.1** The African coelacanth (*Latimeria chalumnae*) can weigh as much as a person and live as long. They have a jelly-filled rostral organ on the front of the head which might be used for electro-detection of prey (and may explain their odd habit of standing on their heads: are they scanning the sea floor for tasty morsels?).

**Figure 13.2** The first live specimen of the recently discovered Indonesian coelacanth, swimming with Arnaz Mehta Erdmann, who happened to spot a coelacanth in a fish market while on honeymoon in Sulawesi.

Image courtesy of Arnaz Mehta.

## Variation in the rate of evolution

Darwin proposed that evolutionary change is continuous, but he didn't claim it would occur at the same rate in all lineages, or uniformly across all periods of evolutionary history. Lineages can differ dramatically in their pace of morphological change and diversification. In the time that it has taken the two living coelacanth species to do practically nothing (morphologically speaking), the Hawaiian honeycreepers have gone bananas (morphologically speaking). From an initial ancestral species that arrived in the newly formed archipelago less than 10 million years ago, the honeycreeper lineage in Hawaii has given rise to more than 50 different species with a wide variety of colours, shapes and ways of life (**Figure 13.3**). Some of these species are endemic to islands that are less than one million years old, suggesting a very rapid rate of speciation and divergence.

George Gaylord Simpson (**Figure 13.4**) was a palaeontologist who was fascinated by such variation in the rate of evolution of different biological groups. Why did some lineages, like the honeycreepers, produce so many different forms, while others, such as the coelacanth, produced so few? Simpson coined the term 'tempo and mode of evolution' to encapsulate the way the pace (tempo) and type (mode) of evolutionary change can vary between lineages, over different periods in evolutionary history, or in different places. Simpson was interested in the way that palaeontology and genetics could be combined to shed light on the rates and mechanisms of evolutionary change.

To compare the tempo and mode across lineages or periods or places, you need to be able to estimate rates of evolutionary change. To measure the rate of evolution, you need some measure of the amount of

**Figure 13.3** There are more than 50 species of Hawaiian honeycreepers (Drepanidinae), all descended from a single colonizing species which reached Hawaii less than 10 million years ago. More than half of the known honeycreeper species are now extinct, and 19 more species are listed as threatened or vulnerable to extinction, including the (a) 'i'iwi (*Vestiaria coccinea:* vulnerable to extinction due to habitat loss through land clearance and introduced diseases); (b) Palila (*Loxioides bailleui:* previously near extinct due to habitat destruction, now persisting in a small population); (c) 'Akeke'e (*Loxops caeruleirostris*: critically endangered due to the effects of invasive plants and animals and (d) 'Akikiki (*Oreomystis bairdi*: critically endandered due to avian malaria and introduced rats).

**Figure 13.4** George Gaylord Simpson was one of the key figures shaping the neo-Darwinian synthesis, in which population-based mechanisms were used to explain the evolution of biodiversity. Simpson interpreted patterns in the fossil record in light of Darwinian principles, and had a particular interest in the way that the rate of evolutionary change varied over time and between different lineages.



evolutionary change accumulated, and you need to know the period of evolutionary time over which the change occurred. Simpson's timescale came from the fossil record, allowing him to compare rates of change in taxa with a continuous fossil record. For example, he showed that the rate of change in the dimensions of horse teeth had accelerated in some geological periods, and slowed down in others. He also demonstrated that the rate of speciation had been several times faster in the horse lineage than it had been in ammonites.

In this chapter, we focus on the way that the rate of molecular evolution can vary between lineages. First, we are going to need to work out how to measure rates of DNA sequence evolution. Next, we will think about the kind of factors that can result in different rates in different lineages. Then we need to examine the particular challenges associated with comparing rates of molecular evolution between species, before we can follow G. G. Simpson's example and ask what factors influence the tempo and mode of evolution at the genomic level.

## Estimating branch lengths

The first step we have to take to study the rates of molecular change is to estimate the amount of molecular change between two species. Branch lengths of a molecular phylogeny usually represent the number of substitutions estimated to have occurred along each lineage, so the length of each branch is the result of both a particular rate of substitution and the amount of time the lineage has had to accumulate substitutions. If you have a long branch in a phylogeny, you know that this sequence has accumulated many substitutions, but you don't know if that's because it has been diverging for a long time, or because it has a fast substitution rate, or both. If you know the rate, you can convert a branch length to time. If you know the evolutionary time, you can convert a branch length to a rate.

***A branch can also be called an edge, see Figure 11.22***

Estimating branch length might seem like a pretty easy thing to do: once you have an alignment (see Chapter 10), you can just count the number of differences between species. A count of observable differences, sometimes referred to as a Hamming distance, is a measure of similarity between sequences. But is it a fair measure of amount of molecular evolution? As we saw in Chapter 12, the problem with using a count of the number of places where sequences are is that it misses substitutions that have been overwritten by subsequent changes (Figure 12.9). If a substitution occurs on top of a previous substitution, it effectively erases the earlier change. The longer two sequences evolve independently, the more substitutions they will accumulate, and the more substitutions they accumulate, the greater the chance that substitutions will occur in the same sites as previous changes. If an A changes to a T and then to a G, then all we can see is the G, and we have no way of directly observing the bases that came before it. Multiple hits represent a loss of evolutionary signal, because they erase past evolutionary changes.

***The problem of multiple hits was introduced in Chapter 12***

Consider the following example, where we will follow a single site in a sequence as it undergoes substitutions in two descendant lineages (Figure 13.5).

**Figure 13.5**

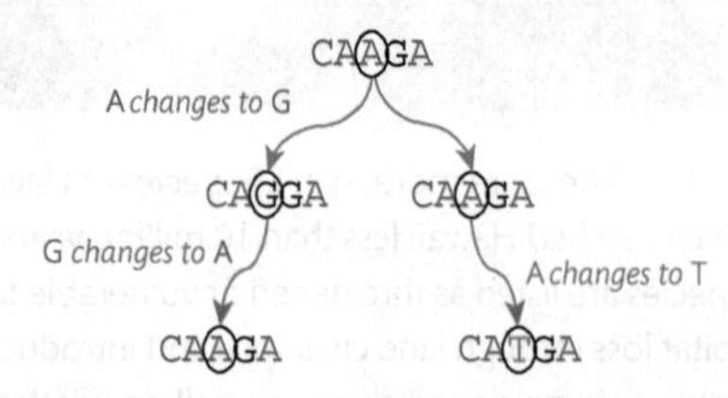


We can see that, since their last common ancestor, there have been three substitutions at this site, two in

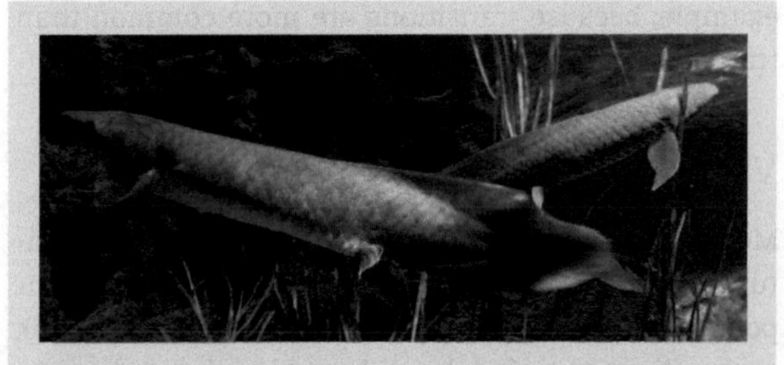

**Figure 13.6** The Australian lungfish (*Neoceradotus forsteri*, also known as the Queensland lungfish) is an extreme case of an 'EDGE' species: Evolutionarily Distinct, Globally Endangered. Fossil evidence suggests this species closely resembles species that lived hundreds of millions of years ago, which suggests that it has persisted since the Cretaceous with relatively little change. This makes it one of the oldest known vertebrate species. Even more interesting, as one of the few surviving members of an ancient lineage of fish, the Queensland lungfish plays an important role in reconstructing the evolution of the vertebrates. However, this species only occurs naturally in two rivers: the Burnett River, which was dammed in 2005, and the Mary River, which the Queensland Government proposed to dam in 2006. The Mary River dam proposal was defeated in 2009, largely on environmental grounds, including consideration of the vulnerability of the unique Queensland lungfish. Thanks to any readers of the first edition of this book who answered my call to sign petitions to stop the Mary River dam—your help may be needed again in future, so please stay vigilant.
Photograph by Lindell Bromham.

one lineage, one in the other. But when we compare the sequences of the descendants, all we can see is a single difference: where one sequence has an A, the other has a T. We have missed two thirds of the changes that have happened, because they have been overwritten by subsequent changes. If we are interested in using the number of substitutions between lineages as a measure of evolutionary time, then clearly we are going to need a way of estimating how many substitutions we miss due to multiple hits.

Since every new substitution that occurs has some chance of overwriting a previous substitution, the problem of multiple hits gets worse the more substitutions occur. Look at these mitochondrial DNA sequences from the *cytochrome b* gene of an Indonesian coelacanth (*Latimeria menadoensis:* **Figure 13.2**), African coelacanth (*Latimeria chalumnae:* **figure 13.1**) and Australian lungfish (*Neoceratodus forsteri:* **Figure 13.6**).

**Figure 13.7**

```
Latimeria_menadoensis   AACATCCGAAAGACACACCCACTAATTAAA
Latimeria_chalumnae     AACATCCGAAAGACACACCCGCTAATTAAA
Neoceratodus_forsteri   AATATCCGAAAAACACACCCGCTCCTAAAG
```

The two species of coelacanth have probably been evolving separately for at least 10 million years, and possibly a lot longer. In this small sequence, there are two nucleotides that differ between these species (**Figure 13.7**). The coelacanths last shared a common ancestor with the lungfish at least 400 million years ago. Yet the coelacanths differ from the lungfish at only eight nucleotides in this sequence. The lungfish-coelacanth split is at least ten times older than the coelacanth-coelacanth split, but it has only four times as many substitutions. What is going on?

The most likely explanation is that substitutions have continued to accumulate throughout the long history of the lungfish and coelacanth lineages, but that many of these substitutions overwrote previous changes. Since we can only observe the end points of the process, we cannot directly observe any past substitutions that have been erased. When the same sites change again and again, sites become saturated with changes, and the hierarchical patterns of similarity that reveal signal of history are lost. Saturation obscures evolutionary history, making it difficult for us to read the story in DNA.

The suspicion that the surprisingly low number of differences between these sequences is due to multiple hits is supported by an inspection of the pattern of substitutions in these sequences. If you look at the alignment in **Figure 13.7** you will see that nearly all of the substitutions occur in the third position in the alignment.

**Figure 13.8**

| | 123 | 123 | 123 | 123 | 123 | 123 | 123 | 123 | 123 | 123 | 123 |
|---|---|---|---|---|---|---|---|---|---|---|---|
| Latimeria_menadoensis | AAC | ATC | CGA | AAG | ACA | CAC | CCA | CTA | ATT | AAA | ATC |
| Latimeria_chalumnae | AAC | ATC | CGA | AAG | ACA | CAC | CCG | CTA | ATT | AAA | ATT |
| Neoceratodus_forsteri | AAT | ATC | CGA | AAA | ACA | CAC | CCG | CTC | CTA | AAG | ATT |

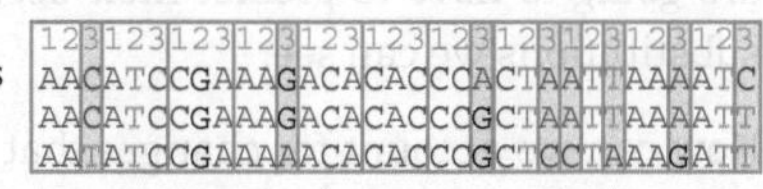

As you will recall from Chapter 6, protein-coding sequences such as this one are read in threes (codons: **Figure 13.8**). There are more different codons than there are amino acids to be coded for, so most amino acids can be coded for by more than one three-base codon. If you look closely at the genetic code (**TechBox 6.1**), you can see that changes in the third codon position often do not change the meaning of the codon. So although the two coelacanths have AAC as the first codon while the lungfish has AAT, both of these codons specify the amino acid asparagine. Similarly, in the last codon in this short alignment, ATC (Indonesian coelacanth) and ATT (Comoran coelacanth, lungfish) both code for isoleucine. Check the rest of the changes on your own using the genetic code table in **Figure 6.18**. Are they all 'silent' (synonymous) changes that alter the nucleotide sequence but not the protein specified by it? You will notice there is one change in the first codon position, where the lungfish has CTA (leucine) and the colecanths have ATT (isoleucine). These two amino acids, leucine and isoleucine, are structurally and functionally very similar, so we can guess that making this swap might not make a big difference to protein function. Amino acid changes that swap between functionally similar amino acids are likely to happen more often than those that put in a completely different kind of amino acid (e.g. swapping a positively charged amino acid for a negatively charged one: Figure 6.18). The point to take on board here is that some sites have more freedom to change than others, and those sites will tend to collect a lot of changes and become saturated more rapidly.

*Synonymous (silent) and nonsynonymous (replacement) substitutions are explained in* TechBox 7.2

The practical result of saturation is that genetic distance does not always increase as a linear function of time. If you compare two sequences and find they differ at four per cent of sites, you cannot assume that they diverged twice as long ago as two sequences that differ at two percent of sites. If we wish to use genetic distance to estimate evolutionary time, we are going to need to use the observable differences (Hamming distance) to estimate the number of substitutions that have really occurred between the two sequences. By definition, we can't observe these overwritten changes directly, so we are going to have to predict their occurrence from the substitutions we can see.

How can we account for changes that we can't directly observe? We can use a model of molecular evolution, which states the probabilities of different types of substitutions, to predict how many additional changes have occurred and been subsequently overwritten (TechBox 13.1). Such a model allows you to say 'if I see this many changes, how many am I likely to have missed?' For example, if you observed only two substitutions across an alignment of 1,000 nucleotide sites, you might conclude that, since there have been relatively few substitutions, the chances that one of them occurred on top of a previous substitution is pretty low. But if you find that 450 out of 1,000 sites vary between sequences in your alignment, then you would expect that the chance of one of those substitutions overwriting a previous one is pretty high. Models of substitution probabilities provide a way of guessing how many changes you are likely to have missed due to multiple hits. Substitution models need to account for the different probabilities of different kinds of changes. For example, because transitions are more common than transversions, they will tend to saturate more rapidly.

## Variation in rates across sites

Models also need to account for different rates of substitution between sites in the genome. We can see the importance of variation in rates across sites when we look at the alignment of coelacanth and lungfish sequences (Figure 13.8). Knowing that some sites accumulate multiple hits at a faster rate than other sites has important implications for estimating amount of genetic change between two sequences. Imagine we observe 200 differences between some aligned sequences of 1,000 bases, then we would consider that roughly a fifth of sites had undergone a substitution. But if we noticed that all of the 200 substitutions were in the third codon position, then we could see that the majority of the third codon positions had changed, and therefore we would expect that a large proportion of these overwrote previous changes, and that the sequences are more saturated than they might first appear. Similarly, if you observe only ten differences between two long sequences, you might assume a shallow divergence. But if you look closely at the sequences and find that, rather than being scattered across the sequence, all ten substitutions are in a particular region of the gene, you might reassess your conclusion—perhaps there are only ten differences because only ten sites are free to change without destroying the function of the gene product. In this case, these ten sites might have been changing back and forth for aeons, erasing past history with every new substitution.

*Chapter 8 explains why less important regions of a sequence have a greater rate of change*

Substitution rates vary dramatically across the genome. Some variation may be due to different mutation and recombination rates across the genome. But much of the variation will be due to differential action of selection. In Chapter 8, we saw that the *nef* gene in HIV has a fast rate of substitution in sites associated with evading the host immune system (Figure 8.19). And in Chapter 2 we saw that an active site in RNA polymerase II gene has changed so little over hundreds of millions of years that it is recognizably similar in humans, rats, flies, bread mould, rice and bacteria (Figure 2.2). These are two extreme examples—one so fast it changes within a year, the other so slow it doesn't change for a billion years. But we expect variation in rates between sites to be widespread, because substitution rates will vary between coding and non-coding regions, between the stems and loops of RNA-coding genes, between the introns and exons of protein-coding genes, and between the first, second and third codon positions within a gene. These differences in rates between sites may be accounted for by a model based on known elements of a sequence—for example, defining intron and exon boundaries or codon positions—or can be estimated from the data (**TechBoxes 12.1** and **13.2**).

*Chapters 7 and 8 explore the influence of selection on rates of change*

# Comparing rates

*The basis of our knowledge is not in chemical entities and laws, but is in occasions of experience, which when systematically organised and observed can be rated as experiments*

Waddington, C. H. (1974) 'How much is evolution affected by chance and necessity?' in *Beyond Chance and Necessity* ed. John Lewis. Garnstone Press.

We have seen how the rate of molecular evolution varies across sites in the genome. Now we will consider how rate of molecular evolution can vary systematically between species. Given all that we have learned about molecular evolution thus far, how it is shaped by mutation, selection and population structure, we should not be the slightest bit surprised to find that the rate of molecular evolution also varies between lineages. This is because factors that affect both the rate at which mutations are generated, and the proportion of these mutations that become substitutions, can vary between species.

Let's start by thinking about the ways that a species' average mutation rate might be influenced by its biology. Broadly speaking, there are two ways that heritable changes can be made to the DNA sequence in the genome: imperfectly repaired DNA damage (Chapter 3) and mistakes made in copying DNA (Chapter 4). Both of these sources of mutation are influenced by a multitude of factors that can differ between species. We can begin by asking whether some species might suffer more DNA damage than others.

## Mutation rate variation

Given that many mutagens come from the environment, such as UV light, radiation, and toxic chemicals, could some species suffer more exposure to mutagens than others? Scientists in the 1950s developing the use of irradiation as a way of sterilizing food were surprised to find that there was one bacterium that could survive high doses of radiation (and therefore go on to spoil the irradiated food). Subsequent research has shown that *Deinococcus radiodurans*, otherwise known as Conan the Bacterium, can survive doses of radiation 1,000 times the lethal dose for a human, due to its amazing ability to heal a broken genome (**Figure 13.9**). All bacteria have repair pathways that can repair X-ray induced double-strand breaks in DNA, but the repair systems in *Deinococcus* are so remarkably efficient that even after its genome is smashed to pieces it takes only three or four hours to completely rebuild it, with no apparent cost in survival and viability. Why would a species evolve the ability to withstand massive, artificially induced doses of radiation? This superpower is likely to be a by-product of the ability to survive other genome-smashing experiences, particularly desiccation.

*Deinococcus*' remarkable DNA repair is an extreme example (though some other dessication-resistant species, such as bdelloid rotifers, have evolved similar genome-rebuilding abilities: **Figure 9.11**). But it does suggest that living in DNA-damaging environments could possibly lead some species to evolve better DNA repair systems. For example, bacteria living in high altitude lakes in the Andes are exposed to high levels of mutagenic UV-B rays, and it has been suggested that they have evolved very high levels of efficiency of repair of UV-B related DNA damage. So, paradoxically, while we might expect species living in mutagenic environments to have higher mutation rates, the opposite may be true. In at least some cases, species in mutagenic environments evolve very efficient DNA repair in order to survive, so they do not suffer higher rates of mutation rate due to DNA damage.

One type of mutagen that no living organism can avoid is reactive oxygen species (ROS), which are by-products of metabolism. When cells make their energy molecules (ATP) in the mitochondria, some electrons leak out of the electron transport chain and form oxygen radicals, particularly superoxide radicals ($O_2\bullet^-$). 'Radical' refers to molecules that have at least one unpaired electron, and they generally have high chemical reactivity. So when these guys are bouncing around inside a cell, they have a tendency to inflict damage on macromolecules, including DNA. We might expect this damage to be worst in the mitochondrial powerhouses of the cell

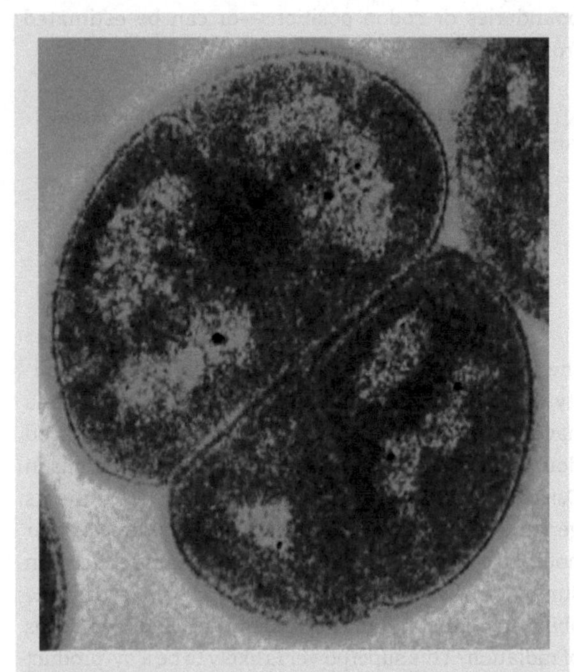

Figure 13.9 An unlikely superhero. Species able to tolerate conditions that would annihilate most organisms are referred to as extremophiles. They don't get much more extreme than the bacterium *Deinococcus radiodurans*, which can survive cold, dessication, and ionizing radiation one thousand times greater than the lethal dose for humans. Consequently, species of *Deinococcus* can be found in a range of challenging environments, including radioactively contaminated sites, hot springs, hypersaline soil, desert sands and Antarctic rocks. Here you can see *Deinococcus* as a tetrad of four cells, each of which carries multiple copies of the genome. These multiple copies provide a template for reassembling the genome by double-strand break repair after it has been smashed to smithereens by radiation, dessication or any number of really bad things. *Deinococcus* also has enhanced resistance to reactive oxygen species (ROS) that can damage DNA. Interestingly, asexual bdelloid rotifers seem to have evolved some of the same strategies for genome repair following dessication, including carrying four copies of the genome to use as templates for reconstruction following damage.

Figure 13.10 My metabolic rate is higher than yours. In mammals, mass-specific metabolic rate scales with body size, so that a gram of tissue in a small mammal, like this pygmy possum, burns more fuel than a gram of tissue in a big mammal, like you.



where the ATP is produced. Indeed, mutation rates are higher in the mitochondrial genome than in the nuclear genome in animals (though not in plants). The degree of metabolic damage sustained by the mitochondrial genome has been proposed as a key factor shaping life expectancy.

Metabolic rate in animals is largely predictable from the type and size of organism. In general, a gram of tissue in a small animal burns more fuel per unit time than a gram of tissue in a big animal, so the smaller the animal, the more ROS it will produce per unit time (Figure 13.10). Body size is also a good predictor of variation in rate of molecular evolution between species, in mammals, birds and other reptiles, and even in plants. On the whole, bigger-bodied species have slower rates of molecular evolution. Some people have suggested that these two observations—smaller animals have both higher metabolic rates and faster rates of molecular evolution—can be taken as evidence that metabolism drives molecular evolution.

There are two problems with this conclusion. The first is that lots of other species traits also scale with size. In mammals and birds, for example, smaller-bodied species not only have higher mass-specific metabolic rates than the larger ones, they also tend to have shorter lifespans, more rapid turnover of generations, and higher reproductive outputs. So all of these traits that scale with body size naturally also scale with rate of molecular evolution. When body size increases, generation time tends to go up (therefore fewer DNA replications per unit time), metabolic rate tends to go down (therefore fewer free oxygen radicals produced), and population size tends to go down (so fewer nearly neutral substitutions fixed: see TechBox 8.2). So how can we tell which species traits are really causing the relationship, and which ones are indirectly linked through correlation with other traits?

The second problem is that, because species are the products of evolutionary descent, we should expect closely related species to be similar in many different respects (see Chapter 9). Two species of mice are likely to be more similar in metabolic rate, generation time, fecundity, and DNA repair mechanisms, than either is to a more distantly related species, such as a dugong or a fruitbat. Evolution creates hierarchies of similarity, so degree of relatedness very often predicts degree of similarity in a whole range of species traits. Simply plotting one trait against another can imply that they are causally linked when they may be only indirectly correlated. This leads to a problem you will have no doubt encountered before: correlation does not prove causality. For example, a recent study showed that, when you compare data from different countries, linguistic diversity is significantly correlated with the number of traffic accidents (**Figure 13.11**). Why? Is it due to complications of reading multi-lingual road signs? Does speaking many languages affect spatial ability or co-ordination? It seems more likely that there are a chain of other variables that link these two together. For example, larger countries with bigger populations might have more language groups, more roads and more cars. The same study also found that the amount of chocolate consumed per capita predicts the number of serial killers in a country, an interesting observation to add to the much-hyped finding that chocolate consumption is linked to the number of Nobel prize winners.

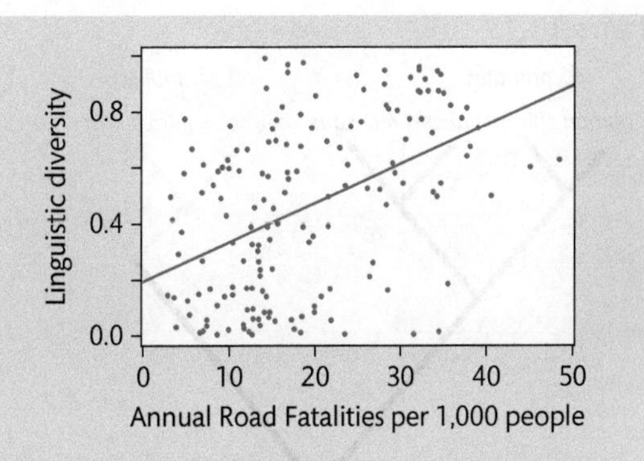


**Figure 13.11** Correlation does not prove causality: Countries with greater linguistic diversity have more fatal traffic accidents per capita.

Image courtesy of Roberts, S. and Winters, J. (2013) *PLoS ONE*, Volume 8(8):e70902, licensed under the Creative Commons Attribution 2.5 license.

## Accounting for descent

> *We should not invoke biological principles where statistics suffices.*

George C. Williams (1966) *Adaptation and Natural Selection: A critique of some current evolutionary thought*. Princeton University Press.

Before we can move ahead with our examination of the factors that govern rate of molecular evolution in different species, we are going to have to solve these two problems: many species traits scale together and related species tend to be more similar in many ways. Both of these problems can create spurious relationships in our data which we might erroneously interpret as revealing evolutionary mechanisms when they only reflect an indirect association between traits. We can minimize the effect of both of these problems by using a phylogenetic comparative approach. We could make comparisons between related species and say 'does the one with the higher metabolic rate tend to have the higher rate of molecular evolution?' If we collect data from many different traits then we can ask a slightly more nuanced question, such as 'Does the species with the higher metabolic rate tend to have a higher rate of molecular evolution, even when we take into account the difference in body size between them?'

Let's consider a hypothetical example. Imagine that we want to test the hypothesis that body size drives rate of molecular evolution in mammals. We have measurements of body size from eight species, for which we also have estimates of the rate of molecular evolution. We plot size against rate and get a nice linear relationship (**Figure 13.12**).

**Figure 13.12**

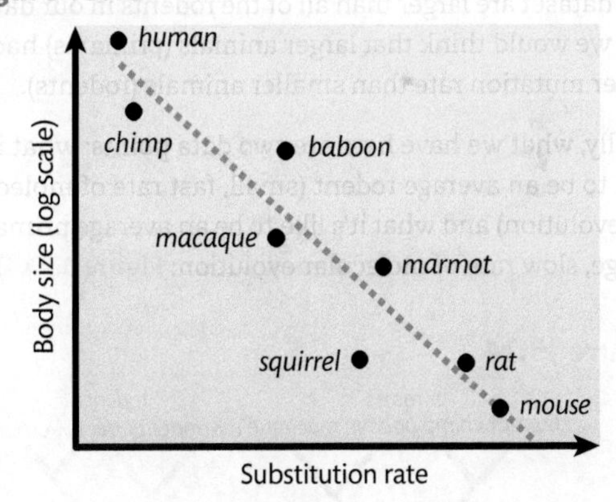


This shows that size is correlated with substitution rate across these species: species with smaller body sizes also tend to have higher rates of molecular evolution. But we have to be careful how we interpret the meaning of this association. We can consider these datapoints in an evolutionary context, by plotting them on a phylogeny (**Figure 13.13**).

**Figure 13.13**

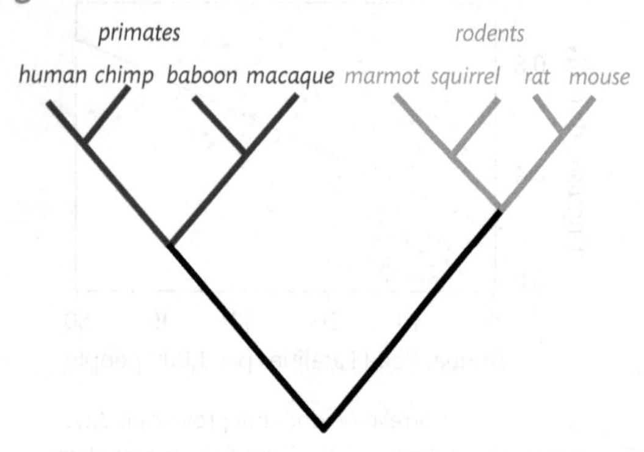


All the primate species are more closely related to each other than they are to any of the rodent species, and vice versa. This matters because the primate species have all inherited characters from their common ancestor, so the similarity between them could be due to descent, rather than because they have each independently evolved similar traits in response to their needs. Similarity by descent is important because it can create apparent trends in the data. In this case we don't know whether body size is causally related to the rate of molecular evolution, or it just happens to be that all our primate species are bigger than all our rodent species, and rodents happen to have a faster rate of molecular evolution. For example, primates have more efficient excision repair than rodents (see Figure 3.2). If the ancestral primate evolved improved DNA repair, then all its descendants might also have lower mutation rates, quite independently of any change in body size. But because all four primates in our dataset are larger than all of the rodents in our dataset, we would think that larger animals (primates) had a lower mutation rate than smaller animals (rodents).

Really, what we have here are two data points: what it's like to be an average rodent (small, fast rate of molecular evolution) and what it's like to be an average primate (large, slow rate of molecular evolution: Figure 13.14).

**Figure 13.14**

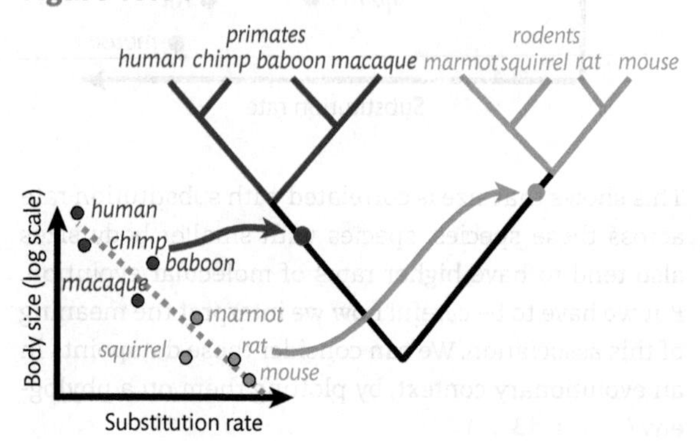


Because so many things vary between rodents and primates—size, life history, ecology, DNA repair mechanisms, and so on—we can't tell what is really driving the differences between them in molecular rate and which traits are just coming along for the ride. What if having nails instead of claws decreases your rate of molecular evolution? We would get the same pattern as we have now. If we want to tease apart the effects of phylogenetic patterns (related species will tend to be similar by descent), then we need to take the degree of relatedness into account. There are a couple of different ways of doing this. The simplest is to compare pairs of related species.

Contemplate, for a moment, the differences between a rat and a mouse. They last shared a common ancestor around 12 million years ago, so any differences between them must have evolved since they diverged. Rats are bigger than mice, so since their common ancestor, either mice got smaller or rats got bigger, or both. Rats have a slower rate of molecular evolution than mice, so since their last common ancestor, either the rats have slowed down, or the mice have sped up, or both. On its own, this doesn't tell us very much. But if we can get a lot of different comparisons, where two related species have evolved different body sizes, then we start to build a picture of whether the large one tends to have a slower rate of molecular evolution. In this example, we have four such observations (Figure 13.15).

**Figure 13.15**

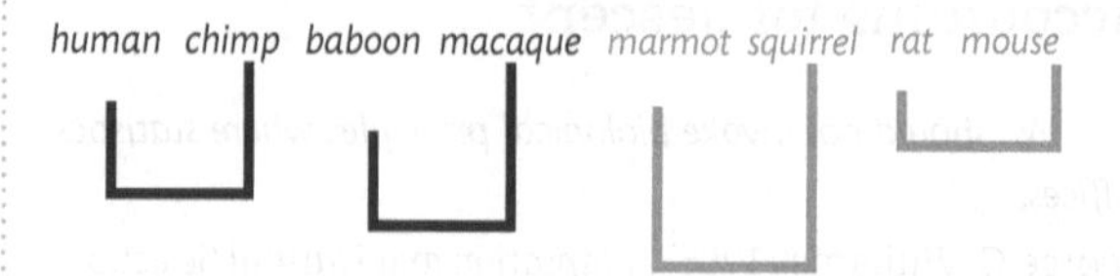


Each of these pairs of species is like an independent test of the relationship between size and rates. Humans and chimps shared a common ancestor, possibly between five and 10 million years ago, so they both started with the same size and rate of molecular evolution. Any difference in rates and size between humans and chimps has evolved since their shared ancestor and this difference has evolved completely independently of whatever was happening in the rat-mouse lineage as they diverged. Rat-mouse and human-chimp are two different tests of the relationship between size and rates. If we collect many different independent comparisons, each of which shares a more recent common ancestor with each other than they do with any other species in the dataset, then we can be sure that the differences between each pair evolved independently from the differences between any other pair (Case Study 13.1).

We have solved the problem of non-independence of related species. Now we can ask: when we look at pairs of lineages, is the one with the smaller body size usually also the one with the faster rate? If we could collect enough comparisons then we could do a statistical test of the relationship between rates and body size, and we might also be able to include more data to control for the co-variation between species traits, for example, between body size, metabolic rate, and generation time. A good rule of thumb for selecting comparisons that represent independent datapoints to include in your analysis is that the path connecting the two members of a pair on your phylogeny should not cross the path connecting any other pair (there are alternative methods that use statistical analyses to correct for the amount of shared history between comparisons, though for various reasons these are tricky to apply to the study of substitution rates).

As it turns out, for many datasets you can plot metabolic rate against rate of molecular evolution and get a significant correlation. But often when you re-do the analysis, taking into account the fact that related species will tend to resemble each other not only in metabolic rate but also in many other aspects of their biology, you find that the link between metabolic rate and rate of molecular evolution can be entirely explained by the fact that metabolic rate scales with size, longevity, generation time, and fecundity. Now the question becomes: what have body size, longevity, generation time, and fecundity got to do with rate of molecular evolution?

## Generation time effect

The generation time is the average amount of time it takes for an individual of a given species to reproduce—you can think of this as the time taken for an embryo to grow into a reproductive adult that produces another embryo. Mice can go through 50 generations in the time that it takes a human to grow to maturity. Reproduction naturally involves copying the whole genome to make gametes. We might therefore expect that the mouse's genome has gone through many, many more replications per year than the human. So the mouse genome, copied over and over again, has had the chance to accumulate many more DNA replication errors in any given period of time. It's not surprising, then, that the shorter the generation time, the higher the mutation rate. Generation time has been shown to correlate with rate of molecular evolution in animals, and small, annual plants tend to have faster rates of molecular evolution than large perennial plants (**Case Study 4.1**).

*DNA replication errors are covered in Chapter 4*

Generation time is not the only species trait that influences the number of times the genome is copied. For example, we saw in Chapter 4 that males and females can have markedly different mutation rates per generation, because the production of sperm usually involves many more cell divisions than the production of eggs. By the time a sperm is ready to join with an egg to create an embryo and kickstart a new generation, its genome will have accumulated many more copy errors than the genome in the egg (**Figure 4.1**). Bird species that have more intense sexual selection have higher rates of molecular evolution. This may be because males in strongly sexually selected species must produce more sperm, and the more sperm is produced, the more cell divisions the germline undergoes, so the more copy errors accrue per generation.

*Male-biased mutation is explained in Chapter 4*

The DNA copy error effect on rates of molecular evolution is a plausible explanation of many patterns of lineage-specific rates of molecular evolution, but it raises an important question. We have already considered how species might evolve higher rates of DNA repair if they live in mutagenic environments. Shouldn't we also expect species that go through more genome copies per year to evolve better DNA copy fidelity to avoid the build-up of harmful replication errors (see **Case Study 3.2**)?

### Fine-tuning mutation rates

Given the cost of deleterious mutation, you might consider that all organisms should evolve toward the minimum possible mutation rate. But this is evidently not the case: natural populations contain alleles that have positive or negative effects on the mutation rate, so there is variation that would allow selection to drive the mutation rate lower. This suggests that the average mutation rates in a population are not at the lowest level they could be. Why would natural selection fail to reduce the mutation rate when deleterious mutation exacts a high cost on fitness?

*Causes and costs of mutation are discussed in Chapter 3*

Mutation rates are the outcome of a balancing act between the cost of mutations that disrupt essential information in an organized genome, and the cost of making and running the barrage of repair mechanisms needed to remove copy errors and DNA damage. This balancing act is most clearly observed in viruses and bacteria. Experiments in which bacteria must adapt to a changing environment can favour 'mutator' strains that have reduced DNA repair, thus a higher rate of mutation. Mutator strains can get ahead because they have a higher rate of production of advantageous mutations. The mutator allele goes to fixation by hitchhiking on the successful mutations it produces. But, as we saw in Chapter 3, very few mutations are beneficial, most are expected to be deleterious. While raised mutation rate leads to occasional wins, it comes at the cost of accumulating more deleterious mutations. The individual with a raised mutation rate may produce more defective offspring who are less able to pass on the allele for higher mutation rate to their offspring, leading to a reduction in the frequency of the allele in the population.

*Hitchiking is explained in Chapter 7*

The balancing act may be more subtle in multicellular organisms, but they must nonetheless balance repair against mutation. Compare a large-bodied mammal, like an elephant, to a small-bodied creature, like a mouse. The large animal must make more cells to build its body, so it goes through more somatic (body cell) cell divisions in a lifetime and probably more germ cell divisions per generation as well. Each cell division carries a risk of DNA copy errors when the genome is replicated. A large animal typically takes longer to reach maturity—an elephant must grow for a decade before it can reproduce, while a mouse can start having babies when it is only a month or two old. Larger mammals also tend to produce fewer offspring and invest more resources in each. A female elephant will probably have fewer than ten babies in a lifetime, whereas a mouse might have fifty or more (Figure 13.16). Thinking about this from a population genetic point of view, a deleterious mutation that causes the loss of one elephant child represents a ten per cent reduction in her mother's reproductive output, whereas the loss of one mouse child might only be a two per cent reduction in her mother's reproductive output. So a big animal has more opportunity for harmful mutations to occur and the impact on fitness of a lethal mutation in the body or germline is relatively higher. Not surprisingly, there is evidence that mutation rates in mammals scale negatively with body size, longevity and generation time (that is, large, long-lived, late-maturing species have lower mutation rates) and positively with fecundity (the more offspring, the higher the mutation rate: Figure 13.17, Hero 13).

While observations are consistent with a role for natural selection in shaping mutation rates, much work remains to be done before we can confidently speak of causes rather than simply describing the correlations. Note that another common correlate of large size, long life, and low fecundity is small effective population size. There will typically be far more individuals in a population of mouse-sized animals than there will be

**Figure 13.16** Precious baby: species that have fewer offspring per parent tend to invest more in parental care. Pregnancy in elephants lasts nearly two years and the baby might not be weaned until it is three years old (when the baby's growing tusks start to make feeding a bit awkward).

**Figure 13.17** Rockfish (*Sebastes*) vary greatly in their longevity; while members of some species of *Sebastes* rarely live beyond ten years old, some rockfish species are known to contain individuals that are over 200 years old. Like mammals, longer-lived rockfish have lower mitochondrial mutation rates. But this can't be due to reduction in DNA replication frequency because, unlike mammals, in rockfish fecundity increases with age. So the longer a rockfish lives, the more eggs she produces and, presumably, the more times her germline DNA is copied. Rockfish might provide some evidence that selection acts to reduce mitochondrial mutation rates in longer-lived species.

Photograph by Chad King (SIMoN/MBNMS).

in a population of elephant-sized animals. Remember that selection is likely to be less efficient in species with small population sizes (**TechBox 8.2**). In a small population, mutations conferring slight improvements in DNA repair will be lost more readily, and mutations causing small drops in repair efficiency can go to fixation by chance. So it has been suggested that large-bodied multicellular organisms might actually end up with higher mutation rates per site per replication due to the limits on the power of selection to keep mutation rates low (**Case Study 3.2**).

## Lineage-specific rates

We have already seen how different parts of the genome can have different substitution rates. Sites under positive selection can have fast substitution rates (**Figure 8.19**). Conserved sites under negative selection can have slow rates (**Figure 2.2**). Sites can vary in the ability to change without harm to the organism, for example third codon positions tend to evolve faster than first and second codon positions (**Figure 13.8**). In addition to variation in the rate of molecular evolution between different parts of the genome, there are consistent differences in rate of molecular evolution between species. Substitution rate varies across the HIV genome, but, on average, sites in the HIV genome evolve about a million times faster than sites in the human genome. Even more closely related organisms can have consistently different rates of molecular evolution. A gene in a mouse evolves, on average, several times faster than the equivalent gene in a human.

Obviously, the action of selection will vary markedly between species, as they respond and adapt to different environments. But the number of sites under positive selection in the genome is typically very small. Much more of the genome is under negative selection, but the impact of negative selection on substitution rates will vary over the genome and is likely to affect genes differently in different species. So while selection is an important force shaping substitution rates, it is expected to have varying effects across the genome and is unlikely to explain consistent, genome-wide differences in rate of molecular evolution between species. But there is an important property that can vary dramatically between species, and over time and space, which influences the effectiveness of selection operating on all parts of the genome.

In Chapter 8, we saw that population size influences the rate at which mutations go to fixation. This is because effective population size governs the balance between the effectiveness of selection at promoting advantageous mutations and removing deleterious mutations and the disruptive influence of genetic drift, which can result in mutations being lost (even despite their advantage) or going to fixation (even despite their deleterious effects). If the effective population size is reduced fewer slightly positive mutations will go to fixation by selection, but more slightly deleterious mutations will be fixed by drift. Given that advantageous mutations are probably quite rare, they will usually be vastly outnumbered by slightly deleterious mutations, so the net effect will be an overall rise in substitutions in small populations.

This effect of population size on the outcomes of selection and drift can lead to all sorts of patterns in lineage-specific rates of molecular evolution. Most obviously, if one species is confined to a smaller area than its sister species, it is likely to have a smaller population, and therefore a higher substitution rate. Species confined to islands tend to have higher rate of fixation of mutations by drift than their wider-ranging mainland relatives. Bacteria normally have very large population sizes, but endosymbiotic bacteria live in small groups sequestered inside their host's body (**Case Study 10.1**). These bacteria have much smaller effective population sizes, and higher substitution rates, than their free-living relatives. Keep in mind that anything that influences the average effective population size—such as area available, climatic variation, sex ratio, mating patterns, inbreeding—is expected to influence the rate of substitution, right across the genome.

## Comparing substitution rates

There are many different features of species that can influence their rate of molecular evolution. If all we have is an alignment of DNA sequences from different species, how can we tell how fast they have been evolving? To compare the tempo and mode of molecular evolution across lineages or periods or places you need to be able to estimate rates of evolutionary change. To measure the rate of evolution you need to estimate the amount of evolutionary change accumulated and you need to know the period of evolutionary time over which the change occurred. If you know both the amount of time and the amount of change then you can calculate an 'absolute rate', which describes the average number of changes per unit time (for example, so-many substitutions per site per million years). Estimating an absolute rate of change for your sequences will only be practical if you know how much time it took for the observed differences to accumulate—that is, if you are able to calibrate your rate estimates with some known dates of divergence (**TechBox 14.1**). For example, the oldest known fossils from both the mouse and rat lineages are around 11 Myr old, so we can assume that there has been at least 11 Myr worth of differences accumulated in each lineage since they split. The rate of change in HIV has been estimated by comparing genomes from viruses sampled over several decades. But reliable calibrations are, unfortunately, not available for the majority of lineages. This means that for many of our favourite species, we cannot hope to calculate the absolute rate of DNA change. But we can still learn an awful lot about differences in rates, and the factors that shape the tempo and mode of molecular evolution, by comparing relative rates of change between lineages.

## *Relative rates tests*

There are a lot of different tests for detecting variation in rate of molecular evolution, but it is worth considering one of the simplest, because it illustrates how we can infer relative differences in rates even when we can't estimate absolute rates. And unlike most statistical tests for rate variation, we can work this one out with a pencil (grab one now before we start, I promise this is not difficult). The relative rates test uses the estimated number of changes between sequences in an alignment to compare the amount of change in a pair of lineages since they last shared a common ancestor. Except in some rare cases of experimental evolution or fast-evolving pathogens, we don't have the sequence from the shared ancestor, so we need to be able to use only the sequences from living species to work out how many changes have occurred along each lineage.

Imagine that we want to investigate the tempo and mode of molecular evolution in raspberries (*Rubus* subgenus *Idaeobatus*), and we are particularly interested in the native Hawaiian raspberries *Rubus hawaiensis* (common name 'Ākala: **Figure 13.18**). We want to know if substitution rate in the island endemic 'Ākala differs from other raspberry lineages, such as its close relative, the salmonberry *Rubus spectabilis*. For convenience, let's say we get sequences that look just like this (**Figure 13.19**):

**Figure 13.19**

```
Rubus_hawaiensis    CAAAATCGAACCCACATCCCAGGTACCCTTACACCCTTTAA
Rubus_spectabilis   CAAAATCGAACCCACATTCAAGGTAGCCTTATATCATTTAA
```

**Figure 13.18** Berry diverse: the genus *Rubus* contains hundreds of species including raspberries and blackberries. The salmonberry (*Rubus spectabilis*, (a)), native to North America, is thought to be the closest relative of one of the native Hawaiian raspberries, the 'Ākala (*Rubus hawaiensis*, (b)). The other native raspberry, the 'Ākalakala (*Rubus macraei*), is considered to have arisen from an independent dispersal from another species. Conservation of native raspberries in Hawaii is complicated by the environmental threat posed by invasive species, including *Rubus niveus*, (**Figure 13.21**). Attempts to use biological control agents against the invasive species may endanger the local endemic raspberries. Should the native raspberries be put under threat to enact a conservation programme that may benefit many endemic threatened species?

For these sequences, there are six differences between the ʻĀkala and the salmonberry. From this information alone, we cannot tell whether these lineages have the same or different substitution rate, because, although we can see which positions in the alignment have undergone a substitution since these lineages last shared a common ancestor, we don't know on which lineages the substitutions happened. So the history of substitutions in these lineages could look like any of these (or other) reconstructions (**Figure 13.20**).

**Figure 13.20**

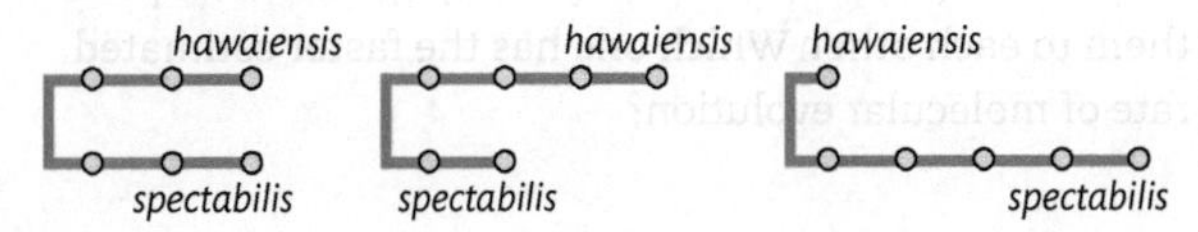


We can solve this problem by comparing the observed number of substitutions between each of these sequences and another more distantly related lineage, referred to as an outgroup. In this case we are going to include the Ceylon raspberry, *Rubus niveus* (**Figure 13.21**). The ingroup species (*hawaiensis* and *spectabilis*) share a more recent common ancestor than either does with the outgroup (*niveus*).

We want to use what we can measure directly (the differences between the three sequences) to estimate what we can't measure directly (the changes between each of the ingroup species and their last common ancestor). When we measure the distance from either ingroup to the outgroup, the contribution of substitutions on the lineage between the outgroup and the last common ancestor of the ingroup is the same for both species (**Figure 13.22**). Any difference in the distance from each ingroup species to the outgroup is due to substitutions that have accumulated in each of the ingroup lineages since their last common ancestor.

**Figure 3.21** The Mysore raspberry (or hill raspberry, *Rubus niveus*) is grown for its edible fruit, but has become invasive in many parts of the world including Hawaii and the Galapagos islands.



**Figure 13.22**

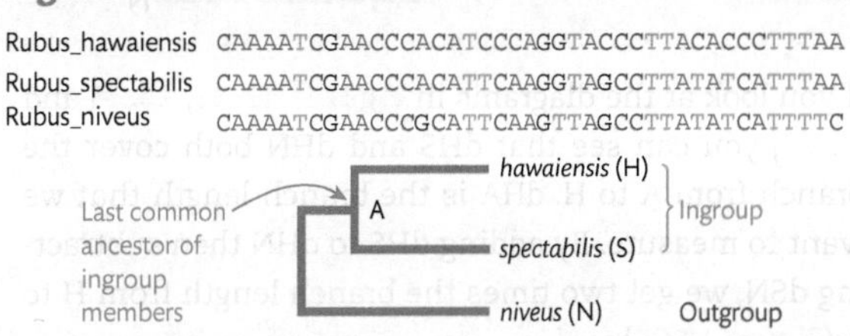


We want to be able to compare the length of the branch that connects A to H (we will call this distance dHA) to the length of the branch connecting A to S (dAS: **Figure 13.23**). To calculate dHA, first we measure the distance between both ingroup species (dHS), which in this case is six differences.

**Figure 13.23**

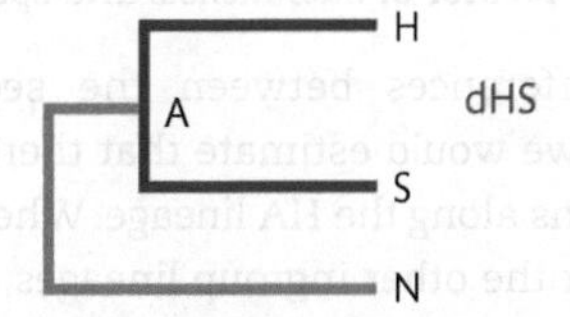


Then we measure the distance between H and the outgroup (dHN, ten differences: **Figure 13.24**):

**Figure 13.24**

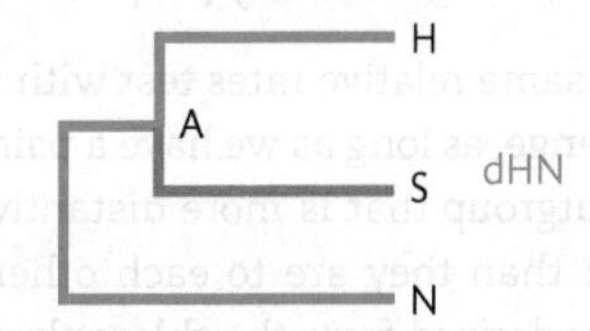


And then we measure the distance between N and the outgroup (dSN, four differences: **Figure 13.25**):

**Figure 13.25**

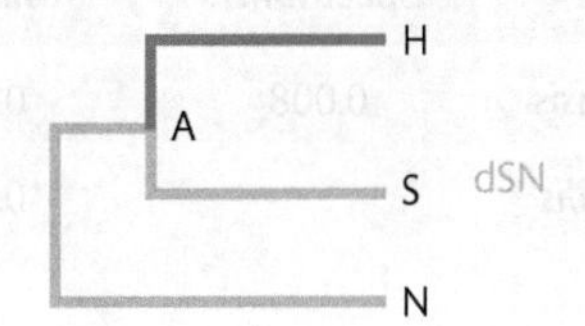

Figure 13.26

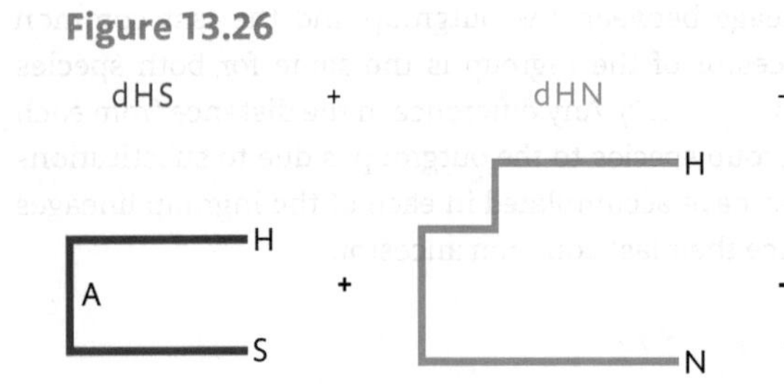


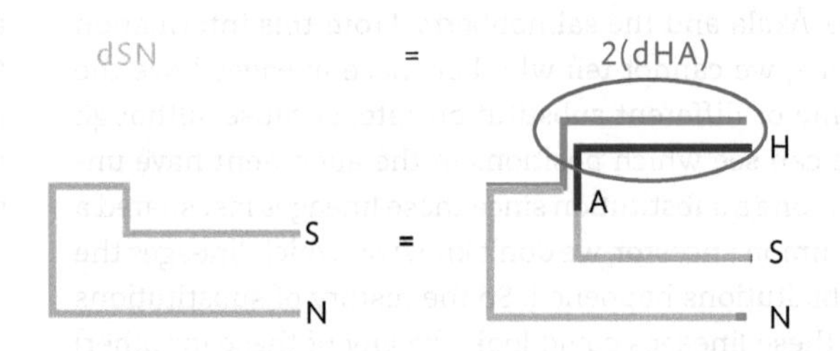


If you look at the diagrams in Figures 13.23, 13.24 and 13.25, you can see that dHS and dHN both cover the branch from A to H. dHA is the branch length that we want to measure. By adding dHS to dHN then subtracting dSN, we get two times the branch length from H to A (Figure 13.26).

So now all we need to do is divide that number by two to get the branch length from *hawaiiensis* to the ancestor of *hawaiiensis* and *spectabilis*. Easy! We have estimated the branch length we can't measure directly (from an extant species to its ancestor) using measures of branch length between the three living species we have sequences for. Now, I want you to work out, using exactly the same logic, how to measure the distance from *spectabilis* to the common ancestor of *hawaiiensis* and *spectabilis*.

Using the differences between the sequences in Figure 13.22, we would estimate that there have been six substitutions along the HA lineage. When we repeat the process for the other ingroup lineages, adding dHS to dNS, subtracting dHS and dividing by two, we get a branch length of zero. We conclude that all six of the differences between *hawaiensis* and *spectabilis* occurred on the *hawaiensis* lineage. This might be telling us that the Hawaiian 'Ākala (*R. hawaiensis*) has a much faster substitution rate than the salmonberry (*R. spectabilis*).

We can do the same relative rates test with any kind of measure of change, as long as we have a pair of ingroup taxa and an outgroup that is more distantly related to either of them than they are to each other. Here is a distance matrix derived from the chloroplast *ndhF* gene for these raspberries (Figure 13.27):

Figure 13.27

| | *R. spectabilis* | *R. niveus* |
|---|---|---|
| *R. hawaiensis* | 0.008 | 0.020 |
| *R. spectabilis* | | 0.012 |

Now you get to use your pencil again to work out what the distance from R. *hawaiensis* to the shared ancestor is (dHA = (dHS + dHN–dSN)/2) is, then work out what the distance from R. *spectabilis* to the shared common ancestor is (dSA = (dHS + dSN–dHN)/2), then compare them to each other. Which one has the faster estimated rate of molecular evolution?

## Testing for rate variation

Much as it was refreshing to be able to do this exercise without using any electricity (unless you cheated and used a calculator), there are two ways in which this relative rates test we have performed is too simple. Firstly, just counting the observable differences between these sequences risks underestimating the number of substitutions that have actually occurred, as past substitutions may have been erased by multiple hits. If we fail to account for multiple hits, we may underestimate the number of substitutions that have occurred in either lineage, potentially masking the rate variation we are trying to detect. Secondly, even if the substitution rate is constant, there can be variation in the absolute number of substitutions along branches of the same age due to the stochasticity of the substitution process. In other words, a difference in branch length does not necessarily imply a different underlying substitution rate. To judge whether any observed difference in branch length is more than we would expect from chance alone, we need to place our relative rates test within a statistical framework.

So we will need to estimate the branch lengths using a decent model of molecular evolution that allows for biases in the rates of different kinds of nucleotide substitutions, for variation in rates across sites, and for the sloppiness of the substitution process (TechBox 13.1). Then we need some way of testing whether these patterns of substitutions could be explained by the same underlying substitution rate, or whether we need to infer variation in rate between the lineages. We might contrast a model in which our three raspberry lineages all have the same rate to one in which the 'Ākala lineage has a different rate to the mainland and outgroup lineages and ask if the model with rate differences has a significantly higher fit than the one where they all have the same rate. Whenever we compare models, we need to be sure that we haven't just improved the fit by adding any random parameter to soak up some of the random variance in substitution rate, so we will need to use a formal statistical test to ask whether adding extra rate categories really is a more reasonable description of the data, such as a likelihood ratio test (see **TechBox 12.1**) or the Akaike information criterion (AIC ).

As with any statistical test, the power of all of these tests to detect rate variation depends on the data. Substitution rate estimates are particularly tricky because, statistically speaking, they don't behave like many other biological traits, like body size or longevity. Substitution rates may look like normal, continuously varying biological traits, but they are actually based on count data, because the number of differences between sequences must start at zero (at the common ancestor), then changes accrue over time. Imagine that lineage A has a 20 per cent faster substitution rate than its sister lineage B. They both start with the same ancestral sequence and at any point in time, each sequence has a chance of acquiring a substitution. Perhaps the first substitution just happens to occur in B: at this point a measure of relative rates would suggest B has a higher substitution rate than A. Then a substitution occurs in A: now it looks like they both have the same relative rates. Then another substitution occurs in A: now it looks like A is twice as fast as B. The point is that when there are few substitutions, estimates of rates will be unreliable. Once we have left A and B to evolve separately for 10 million years, the rates start to settle down: A has acquired 120 substitutions for every 100 acquired by B. But now we start to get into the territory of multiple hits, and we have the problem of trying to reconstruct number of substitutions from saturated sequences. Lineages with a fast rate of molecular evolution will saturate faster than their relatives with a slower rate of evolution. So with rate estimation we have trouble at the shallow end—too few substitutions lead to inaccurate rate estimates—and trouble at the deep end—too many substitutions erase the signal of history and processes (**Hero 13**).

Lack of power to detect significant rate differences is a problem if you want to use these tests to prove that rates are constant in your sequences. If you fail to detect rate variation, you don't know for sure whether it is because all the sequences are evolving at the same rate, or because you did not have the power to detect rate variation. But lack of power is less of a problem if you are conducting a comparative study across many different sequences in order to examine patterns in the tempo and mode of molecular evolution (e.g. **Case Study 13.2**). In this case, lack of power can work against you and stop you from seeing the rate variation shaping your data, but if you do see a consistent pattern you might feel that there is real signal coming from your sequences. For example, we might hypothesize that the Hawaiian raspberry has a faster rate of molecular evolution because it lives on an island so has a smaller population size than its mainland relatives. Using only this one comparison between the 'Ākala and the salmonberry can't tell us which aspect of the biology of *Rubus hawaiensis* is influencing its substitution rate, or even if the faster rate is due to an idiosyncratic event in the history of this lineage. But if we were to gather data from many different island endemic lineages, and compare each of them to their mainland relatives, and we found that, significantly more often than expected by chance, the island lineages had a faster rate of substitution, then we could start to believe we have seen a general pattern. This pattern, in turn, may help us predict cases when rates will vary between lineages, and might ultimately lead to more biologically realistic models of rate change (**TechBox 13.2**).

# Conclusion

"The fact that we expect our theories to have exceptions makes it hard to test them . . . It makes me envious of my colleagues in molecular biology. They can usually settle their problems by experiment: I seem to live with mine. Of course, my problems are more interesting."

John Maynard Smith (1990) 'Taking a chance on evolution', *New York Review of Books*: 14 June 1990.

So many factors affect the rate of molecular evolution that it is usually impossible to say why one particular species has a faster or slower rate than its relatives. Analysis of the African coelacanth genome suggests that protein-coding genes evolve slower than in other fish. Why? Is it because they have a small population (less than 400 individuals have been recorded to date)? Because they have long generation times and few offspring so have fewer DNA replications per unit time? Because they live in the cool depths of the ocean and maintain relatively low metabolic rates? Because they are large and long-lived so invest in better DNA repair to ensure they can maintain themselves into healthy old age? Because they live in caves deep underwater and can't be bothered evolving? We cannot draw conclusions from single examples, but by comparing rates across many different species, allowing for phylogenetic patterns of relatedness, we can start to detect and understand general patterns.

The story of the coelacanth illustrates another very important point: the genome continues to record evolutionary history even when the fossil record stops and morphology is in stasis. The converse is also true: when morphological evolution accelerates, the genome continues, by and large, to steadily accumulate changes. Like the coelacanth, the honeycreeper genomes have continued to diverge at a steady rate (**Figure 13.3**). In fact, if you plot the amount of genetic divergence between honeycreeper species against the age of the island they come from, you get a surprisingly linear relationship. This means that if we compared the DNA from a species of honeycreeper to its relatives, we could make a decent estimate of how long it had been since it shared a common ancestor with its closest known relative.

The relationship between time elapsed and genetic difference accumulated gives us an amazingly useful tool to use in evolutionary biology. For example, if we could use molecular data to estimate the age of the last common ancestor of the two living coelacanth species, we might solve the mystery of how two nearly identical species came to be found in small unconnected populations living on opposite sides of the globe. Did they separate with the breakup of the Gondwanan land mass? Is the African population formed from recent immigrants who rode the currents from Indonesia? The problem is that, until we actually know how old the coelacanth species are, we can't estimate the rate of molecular evolution. So at this stage, we don't know for sure if the two coelacanth species have an ancient ancestor and a slow rate of molecular evolution, or a recent ancestor and a fast rate of molecular evolution (or anything in between). This conundrum illustrates a frustrating problem that we will have to confront in the next chapter on molecular dating.

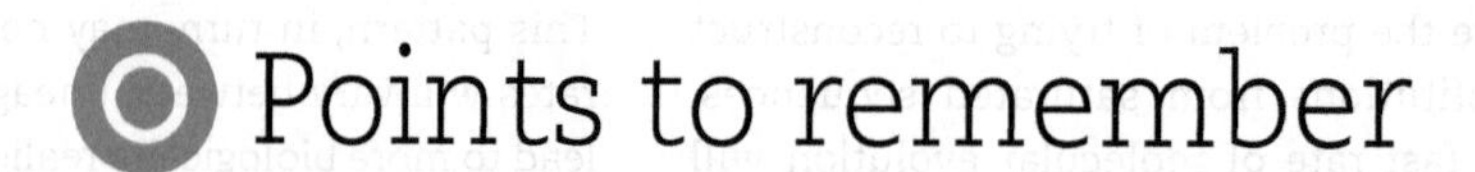

## Points to remember

### Rate of evolutionary change

- Genomes continuously accumulate change but the rate of change varies across the genome, and between lineages, and over time.
- To estimate rate of change, you need to be able to estimate the number of substitutions that have occurred, including those obscured by multiple hits.
- Multiple hits, when substitutions occur on previous substitutions, prevent us from making a direct count of the number of substitutions that have occurred.
- A model of molecular evolution that describes the probability of different kinds of substitutions allows prediction of the number of substitutions that have occurred given the observed difference between sequences.

### Comparing rates

- To uncover patterns in rates of molecular evolution we need to control for covariation of species traits and similarity due to descent.

- One way of controlling for similarity by descent is to make many comparisons between pairs of species that do not overlap on a phylogeny, because any differences between each pair of species have evolved since their last common ancestor, independently of any other such pair.
- Mutation rate could be affected by exposure to mutagens, although this may be compensated by increased efficiency of DNA repair.
- The more times DNA is copied, the more opportunity for mutations to occur through replication errors.
- Selection can act to balance the cost of mutation against the cost of repair and this balance might vary between lineages.
- Population size can influence genome-wide substitution rates by altering the balance between the effectiveness of selection and fixation by chance.
- Lineages with smaller effective population sizes tend to have higher substitution rates due to the increased rate of fixation of slightly deleterious mutations.
- A relative rates test uses measures of amount of genetic change between a pair of lineages and a more distantly related outgroup to determine which of the pair has accrued more change since their last common ancestor.

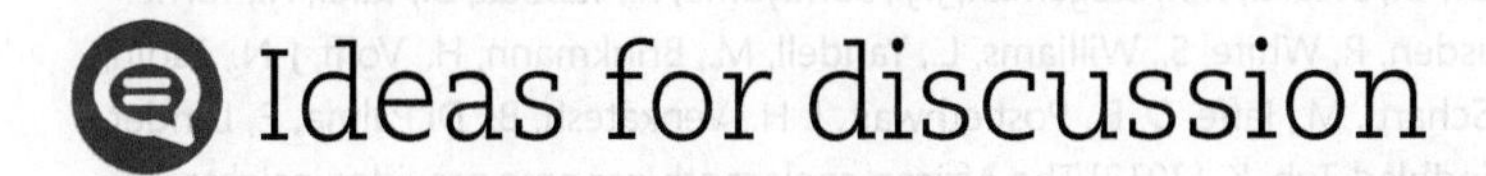

## Ideas for discussion

**13.1** Under what circumstances, if any, will natural selection favour increases in mutation rate?

**13.2** Is there any way of recovering historical signal from saturated sites? Should they be deleted from alignments before analysis?

**13.3** How would you test the hypothesis that mitochondrial mutation rate limits longevity?

## Sequences used in this chapter

Table 13.1: *Cytochrome b* sequences used in Figure 13.7

| Description from GenBank entry | Accession |
|---|---|
| *Latimeria menadoensis* mitochondrial DNA, complete genome | AP006858 |
| *Latimeria chalumnae* mitochondrial DNA, complete sequence, isolate: Comoro_1 | AP012198 |
| *Neoceratodus forsteri* mitochondrion, complete genome | AF302933 |

# Examples used in this chapter

Albarracin, V., Pathak, G., Douki, T., Cadet, J., Borsarelli, C., Gurtner, W., Farias, M. (2012) Extremophilic *Acinetobacter* strains from high-altitude lakes in Argentinean Puna: remarkable UV-B resistance and efficient DNA damage repair. *Origins of Life and Evolution of Biospheres*, Volume 42, page 201.

Amemiya, C. T., Alfoldi, J., Lee, A. P., Fan, S., Philippe, H., MacCallum, I., Braasch, I., Manousaki, T., Schneider, I., Rohner, N., Organ, C., Chalopin, D., Smith, J. J., Robinson, M., Dorrington, R. A., Gerdol, M., Aken, B., Biscotti, M. A., Barucca, M., Baurain, D., Berlin, A. M., Blatch, G. L., Buonocore, F., Burmester, T., Campbell, M. S., Canapa, A., Cannon, J. P., Christoffels, A., De Moro, G., Edkins, A. L., Fan, L., Fausto, A. M., Feiner, N., Forconi, M., Gamieldien, J., Gnerre, S., Gnirke, A., Goldstone, J. V., Haerty, W., Hahn, M. E., Hesse, U., Hoffmann, S., Johnson, J., Karchner, S. I., Kuraku, S., Lara, M., Levin, J. Z., Litman, G. W., Mauceli, E., Miyake, T., Mueller, M. G., Nelson, D. R., Nitsche, A., Olmo, E., Ota, T., Pallavicini, A., Panji, S, Picone, B., Ponting, C. P., Prohaska, S. J., Przybylski, D., Saha, N. R., Ravi, V., Ribeiro, F. J., Sauka-Spengler, T., Scapigliati, G., Searle, S. M. J., Sharpe, T., Simakov, O., Stadler, P. F., Stegeman, J. J., Sumiyama, K., Tabbaa, D., Tafer, H., Turner-Maier, J., van Heusden, P., White, S., Williams, L., Yandell, M., Brinkmann, H., Volff, J-N., Tabin, C. J., Shubin, N., Schartl, M., Jaffe, D. B., Postlethwait, J. H., Venkatesh, B., Di Palma, F., Lander, E. S., Meyer, A., Lindblad-Toh, K. (2013) The African coelacanth genome provides insights into tetrapod evolution. *Nature*, Volume 496, page 311.

Bartosch-Harlid, A., Berlin, S., Smith, N. G. C., ller, A. P., Ellegren, H. (2003) Life history and the male mutation bias. *Evolution*, Volume 57, page 2398.

Fleischer, R. C., McIntosh, C. E., Tarr, C. L. (1998) Evolution on a volcanic conveyor belt: using phylogeographic reconstructions and K-Ar based ages of the Hawaiian islands to estimate molecular evolutionary rates. *Molecular Ecology*, Volume 7, page 533.

Gladyshev E, Meselson M (2008) Extreme resistance of bdelloid rotifers to ionizing radiation. *Proceedings of the National Academy of Sciences USA*, Volume 105, page 5139.

Howarth, D. G., Gardner, D. E., Morden, C. W. (1997) Phylogeny of *Rubus* subgenus *Idaeobatus* (Rosaceae) and its implications toward colonization of the Hawaiian Islands. *Systematic Botany*, Volume 22, page 433.

Hua, X., Cowman, P. F., Warren, D., Bromham, L. (2015) Testing the link between longevity and mutation rates in rockfish using Poisson regression of substitution rates. *Molecular Biology and Evolution*, Volume 32, page 2633.

Roberts, S., Winters, J. (2013) Linguistic diversity and traffic accidents: Lessons from statistical studies of cultural traits. *PLoS ONE*, Volume 8, page e 70902.

Slade, D., Radman, M. (2011) Oxidative stress resistance in *Deinococcus radiodurans*. *Microbiology and Molecular Biology Reviews*, Volume 75, page 133.

Sniegowski, P., Gerrish, P. J., Lenski, R. E. (1997) Evolution of high mutation rates in experimental populations of *E.coli*. *Nature*, Volume 387, page 703.

Woolfit, M., Bromham, L. (2005) Population size and molecular evolution on islands. *Proceedings of the Royal Society B: Biological Sciences*, Volume 272, page 2277.

HEROES OF THE GENETIC REVOLUTION

13

# Xia Hua (華夏)

*"In a nutshell, Xia is simply brilliant. At our lab meetings, Xia routinely comes up with exciting new ideas that could easily launch new papers and new research programs. She is also a modest, kind, and generous person, and a real pleasure to have in the lab (she is almost always very, very cheerful and enthusiastic)."*

John Wiens (2012) personal communication.

**EXAMPLE PUBLICATIONS**

Hua, X., Wiens, J. J. (2010) Latitudinal variation in speciation mechanisms in frogs. *Evolution*, Volume 64(2), pages 429–43.

Hua, X., Wiens, J. J. (2013) How does climate influence speciation? *American Naturalist*, Volume 182(1), pages 1–12.

**Figure 13.28** Xia Hua explains her new method for studying the correlates of variation in substitution rate during one of our lab meetings. This new method, which we call xPGLS (Xia's Phylogenetic General Least Squares regression), was first used to describe the relationship between mitochondrial mutation rates and longevity in rockfish (see **Figure 13.17**).

I love working with Xia Hua. Typically, when there is a discussion in the lab meeting on how we are going to analyse a particular dataset and all the complicating factors we have to account for and the tangle of parameters of interest, Xia will listen quietly and intently to all the discussions, then say 'why don't you just analyse it like this?' and come up with an elegant method that no-one else had thought of (**Figure 13.28**).

Hua always wanted to be a scientist and she studied biology at Fudan University in Shanghai. She undertook undergraduate research in herpetology, and, having read a paper on frog diversity patterns by John Wiens, she repeated the analysis on her own data and sent him the results. Wiens was impressed by Hua's research, and offered her a PhD place at Stony Brook University in USA. She had initially wanted to investigate sexual selection in some unusual polymorphic frogs, but unable to gain support for this field and lab-focused project, she instead turned to theoretical work, which can be done with little more than a decent computer and your own brain (especially if you happen to have a brain like Xia's). She taught herself mathematics and statistics by reading scientific papers, then took over teaching the graduate biostatistics course that had been taught by the renowned biostatisticians Robert Sokal and James Rohlf (both of whom made her feel at home, Sokal because of shared language, thanks to his PhD in Shanghai, and Rohlf through a shared love of dim sum). Her thesis took several different approaches to understanding speciation, including biogeographical niche modelling in frog species using GIS, theoretical modelling of factors affecting speciation, and the role of climate in driving species formation. After her PhD, she was offered a job in industry modelling the evolution of resistance in GM crops, but happily Hua decided to follow the academic path all the way to my research group in Australia.

Hua's work is characterized not only by its inventive uses of mathematical and statistical approaches, but also in the way it combines understanding of biological processes across many different levels of organization, from the genome to the population to the community to global patterns in biodiversity. She has a reliable instinct for the biological plausibility of mathematical or statistical explanations and a healthy scepticism for the amount of signal that can be derived from often messy biological data. Her enthusiasm is infectious. When she sees a new way of answering an interesting question, she is prone to jumping up and down with excitement, quite literally. She is so full of good ideas that she gets involved in a large number of different projects, so in the past few years she has developed new analytical methods for investigating macroevolutionary patterns from phylogenies, differences in substitution rates between lineages (**Figure 13.17**), drivers of diversification (**Case Study 13.2**) and the mechanisms of language change. But she has bigger issues in her sights, hoping to develop models of evolution that unite change at the molecular level with speciation at the population level and diversification at the lineage level. Watch this space.

# TECHBOX 13.1

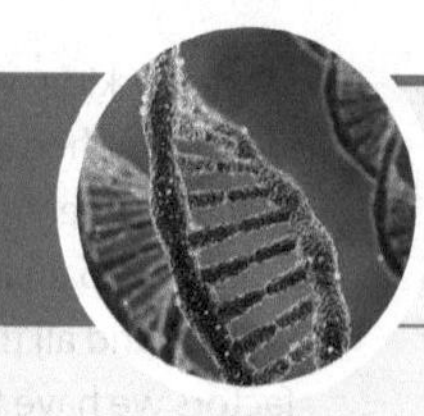

# Substitution models

**RELATED TECHBOXES**

**TB 6.1:** Genetic code

**TB 10.2:** Multiple sequence alignment

**TB 12.1:** Maximum likelihood

**RELATED CASE STUDIES**

**CS 13.1:** Rates

**CS 13.2:** Diversification


## Summary

To estimate changes we can't directly observe, we need a statement of the likelihood of different kinds of substitutions.



## Keywords

rates across sites, transition, transversion, base composition, likelihood ratio, model selection, substitution rates, branch length, GC bias, JC69, HKY85, K80, F81, general time-reversible (GTR)


To measure the genetic divergence between DNA (or protein) sequences, we need to estimate the number of substitutions that have occurred since they were both copied from a shared common ancestor. One way to estimate number of substitutions is to simply count the number of differences between the sequences. But this count will fail to include multiple hits, because we can't directly observe substitutions that occur in the same sites as previous substitutions. Since we cannot directly observe overwritten changes, we must use a model to predict how many changes we have missed. Our model reflects the probability of different kinds of substitutions occurring. The values we give the parameters of the model are often tailored to a particular dataset. We may choose a value based on some prior knowledge, for example if we happen to know the typical transition bias for our particular sequence. Or we may estimate these values directly from the data, either from a preliminary analysis of the data (e.g. estimating base frequencies from the alignment), or as part of the branch length optimization (e.g. finding the values of parameters that optimize the likelihood of a given phylogeny: see **TechBoxes 12.1** and **13.2**).

## Nucleotide substitution models

The simplest model considers all nucleotide changes equally likely. So, if you have an A in a particular spot in a sequence, and it undergoes a substitution, it is just as likely to change to a G or a T or a C. This means we need only one rate parameter, which is the chance that any given base will change to any other base. Nucleotide substitution models are typically referred to by the first letter of the authors' surnames plus the year of publication: this one is the Jukes-Cantor model (JC69: **Figure 13.29**).

**Figure 13.29**

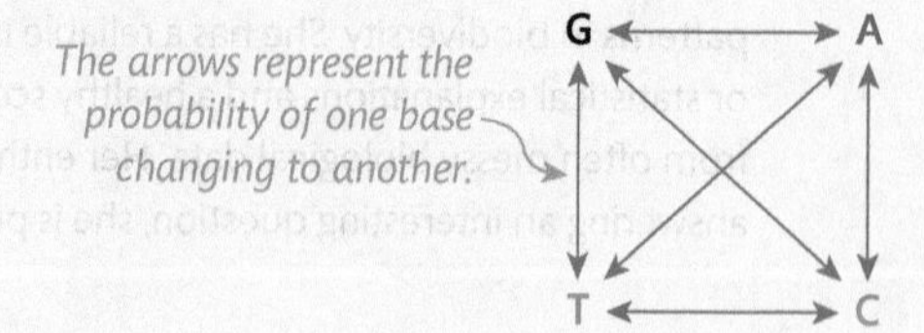

But for most sequences, not all base substitutions are equally frequent. Typically, transitions are more common than transversions. We can allow for this by specifying one rate for changes that swap a purine for another purine, or a pyrimidine for another pyrimidine (transitions) and one for changes that swap a purine for a pyrimidine or a pyrimidine for a purine (transversions). Now we have a model with two rate parameters, which is attributed to Motoo Kimura (**Figure 8.2**), so is called the K80 model. We are going to represent the two rate categories with different coloured arrows, so we assume that all the red transitions occur with one rate, and all the green transitions with another (**Figure 13.30**).

**Figure 13.30**

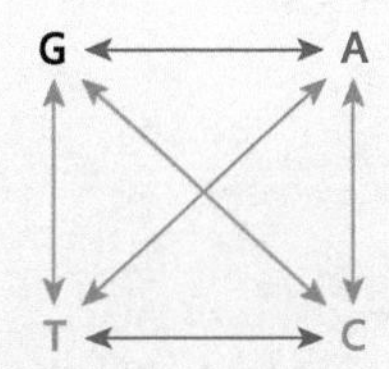


By assuming that all kinds of transitions are equally likely, we are making another unspoken assumption, that the bases all occur in roughly equal amounts. In reality, most genomes have a tendency to have more of some kinds of nucleotides than others, a phenomenon known as base composition bias. Chargaff's rules (see Chapter 2) tell us that there should be equal amounts of the paired nucleotides, so we should expect the percentage of Gs in a genome to match the amount of Cs, and the percentage of As to match the percentage of Ts. But genomes vary in the amount of Gs and Cs compared to As and Ts. For example, wheat has 56 per cent Gs and Cs, but peas have only 42 per cent GC[1]. If the sequences have a strong GC bias, we should expect changes to G or C to be more frequent than changes to A or T. If we allow base compositions to vary, then we have more parameters to estimate from the data: we need to estimate the transition rate, the transversion rate, and the proportions of As, Cs, Ts and Gs (**Figure 13.31**). Models that include these parameters include the Hasegawa-Kishino-Yano (HKY85) and Felsenstein (F81: **Hero 11**) models.

**Figure 13.31**

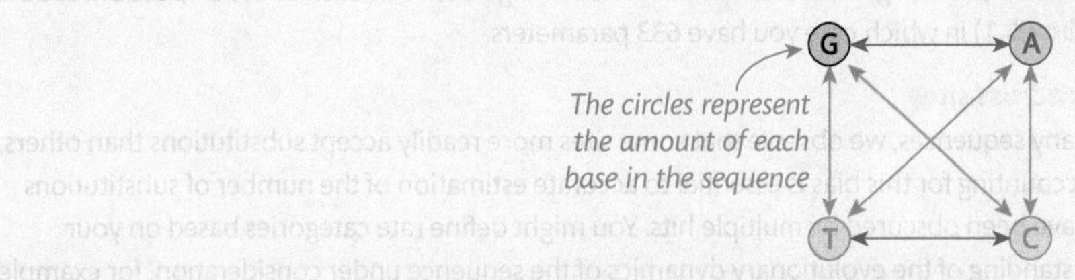


We might even have separate parameters for the different 'flavours' of transitions by allowing one rate for pyrimidine transitions and one for purine transitions, in which case we have an extra parameter to estimate in the Tamura-Nei (TN93) model (**Figure 13.32**).

**Figure 13.32**

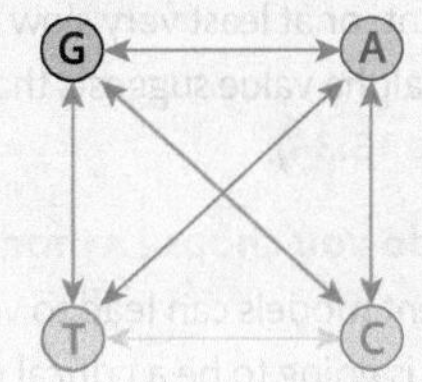

Or we can go the whole hog and give every kind of substitution its own rate parameter, plus the parameters for the amount of each kind of base (**Figure 13.33**). This is the most general possible model, as it makes no prior assumptions about what kind of changes are more likely than others. When we get to this point we cease to name the model after its parents, and call it the 'Generalized Time-Reversible' Model (GTR: though if you were sentimental you could call it T86, after Simon Tavaré).

**Figure 13.33**

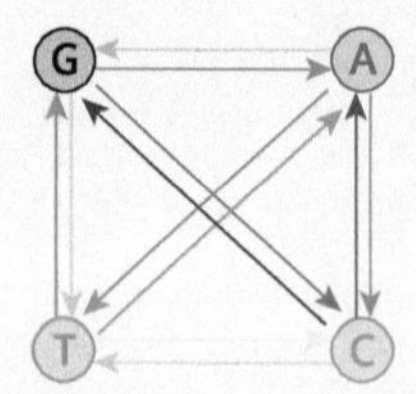


### Amino acid substitution models

There are only four nucleotides, but there are twenty common amino acids, so protein sequences have a lot more possible substitutions than DNA sequences. But not all amino acid substitutions are equally likely. If you are analysing protein-coding sequences you should take these differential amino acid substitution probabilities into account. Broadly speaking, there are two ways to do this. Firstly, you could base your model on observations of which amino acid changes are more common. There are a number of empirically determined matrices of amino acid substitution probabilities, created by analysing large numbers of protein alignments. For example, the PAM matrix (Point Accepted Mutation, but sometimes also referred to as the Dayhoff matrix: **Hero 10**) was based on the observed substitutions in 71 sets of related protein families, and the WAG (Whelan and Goldman) matrix was estimated from 3905 sequences from 82 protein families. However, these average substitution probabilities may not exactly match the evolutionary dynamics of your particular sequence. Alternatively, you could estimate the amino acid substitution probabilities directly from your own data. For example, you can extend the GTR model to have a parameter for each kind of amino acid substitution, plus a parameter for the proportion of each of the 20 amino acids—but then you have 209 parameters to estimate. Or you can go wild and have a rate of change between each of the 64 possible codons (**TechBox 6.1**) in which case you have 633 parameters.

### Rates across sites

For many sequences, we observe that some sites more readily accept substitutions than others, and accounting for this bias is essential to accurate estimation of the number of substitutions that have been obscured by multiple hits. You might define rate categories based on your understanding of the evolutionary dynamics of the sequence under consideration, for example allowing first and second codon positions to have a different rate than third codon positions for a protein coding gene, or putting stem sites in a different rate category than loop sites for an rRNA gene. Or you could define a general model of rate variation, and use the estimation procedure to find the pattern of rates across sites that gives the highest likelihood. Many models of molecular evolution use a gamma distribution, for which a shape parameter (called alpha, $\alpha$) changes the distribution of rates across sites. A low alpha value suggests that most sites are invariant, or at least very slow to change, but some sites have an intermediate or fast rate, but a high alpha value suggests that the majority of sites fall within an intermediate rate category (**Figure 13.34**).

### How do you choose a model?

Different models can lead to very different estimates of branch length. Clearly, choosing the best model is going to be a critical part of branch length estimation. If you don't know for certain what the particular dynamics in your sequence are, then it might seem the best approach

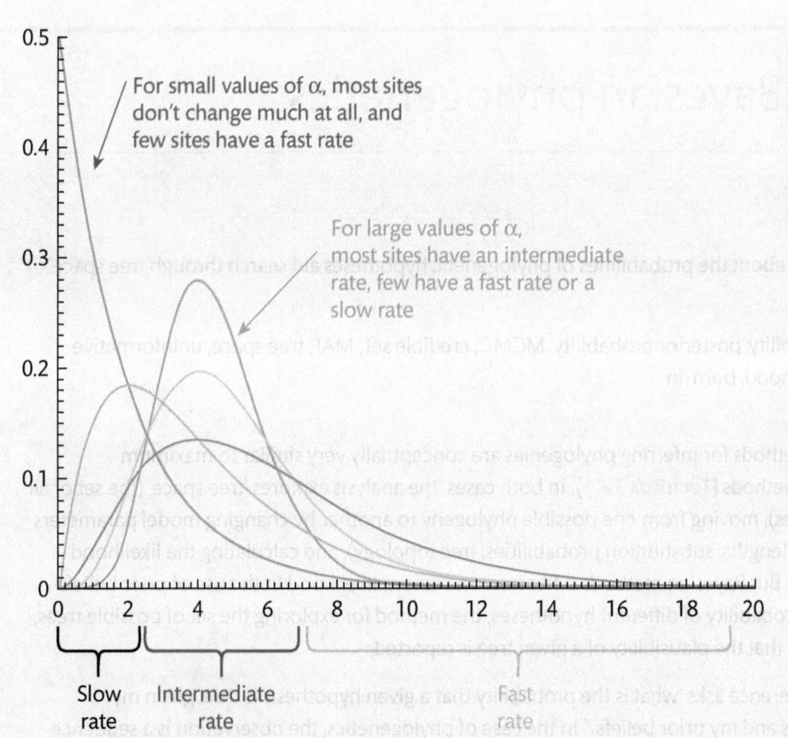


**Figure 13.34** Variation in rates across sites in an alignment is often described using the gamma distribution, α, which describes how many sites are expected to fall into each rate category. The shape of the distribution is determined by both shape and scale parameters, but these two parameters are often set equal in substitution rate models.

to choose the most complex model that makes the fewest a priori assumptions about the evolutionary process. But, perhaps counter-intuitively, models with more free parameters can in some cases lead to greater errors due to 'overparameterization': the inclusion of so many free parameters that even random variance can be accounted for and nearly any possible answer can be entertained. In other words, an overparameterized model has low explanatory, or predictive, power. It might just be shouting out the noise in the data, not telling a coherent story. On a practical level, the number of free parameters to be estimated increases the computation time. So the aim is to choose the simplest adequate model, the one that does the maximum amount of explanatory work with the least amount of fuss (recall Ockham's razor: **Figure 12.7**).

You may pick a favourite model a priori, using your feel for the molecular evolutionary patterns of the sequences under consideration. Or you may use a formal statistical test to compare the 'fit' of different models to your data, selecting the one that best describes the patterns in the data with the fewest number of parameters. Many model tests ask whether adding a parameter provides a significant increase in the likelihood of the observed data, beyond the automatic increase in fit that we expect by adding any extra free parameters. You can also partition your data, so that you fit different models to different parts of the dataset: again this could be done a priori, for example giving different genes or different codon positions different models because you recognize they might be evolving in different ways. Or the partitioning could be done analytically, where you test different partitioning schemes and pick the one that provides the best fit to your data. Whether you choose your model by instinct or statistics, you would do well to test the robustness of your assumptions by seeing if your conclusions hold when you analyse your data with a different model.

## Reference

1. Kawabe, A., Miyashita, N. T. (2003) Patterns of codon usage bias in three dicot and four monocot plant species. *Genes & Genetic Systems*, Volume 78, page 343.

TECHBOX 13.2

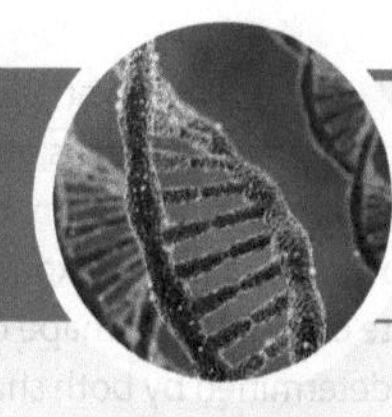

# Bayesian phylogenetics

RELATED TECHBOXES

TB 12.1: Maximum likelihood

TB 12.2: Bootstrap

RELATED CASE STUDIES

CS 12.1: Epidemiology

CS 14.2: Dates

**Summary**

Prior beliefs about the probabilities of phylogenetic hypotheses aid search through tree space.

**Keywords**

prior probability, posterior probability, MCMC, credible set, MAP, tree space, uninformative priors, likelihood, burn-in

Bayesian methods for inferring phylogenies are conceptually very similar to maximum likelihood methods (**TechBox 12.1**). In both cases, the analysis explores 'tree space' (the set of all possible trees), moving from one possible phylogeny to another by changing model parameters (e.g. branch lengths, substitution probabilities, tree topology), and calculating the likelihood of each tree. But Bayesian methods differ from maximum likelihood in the use of prior beliefs about the probability of different hypotheses, the method for exploring the set of possible trees, and the way that the plausibility of a given tree is reported.

Bayesian inference asks 'what is the probability that a given hypothesis is true, given my observations and my prior beliefs?' In the case of phylogenetics, the observation is a sequence alignment, the hypothesis is a particular phylogenetic tree, and the estimate of probability is based on a model of the evolutionary process. While maximum likelihood (ML) incorporates prior beliefs about the way the world works in the form of a model of sequence evolution, Bayesian inference goes a step further than this, assigning prior probabilities to alternative hypotheses before we have seen the data. Prior probabilities reflect the chance that any hypothesis is true, regardless of the data. The use of prior beliefs is both the best thing and the worst thing about Bayesian inference. If we do know, before we start, which hypotheses are more likely to be true (or, conversely, which are very implausible) then we can use this information to dramatically reduce the amount of time needed to search and evaluate different hypotheses. But, given that our prior beliefs will influence the outcome of the analysis, we need to be sure that we do not choose inappropriate priors that will lead our analysis astray.

Actually, in phylogenetics, we usually don't have reliable prior knowledge of which hypotheses are more probable than others before we start the analysis. Therefore, phylogenetic applications of Bayesian inference tend to use uninformative priors for many variables. Uninformative priors are distributions of prior probabilities that do not bias the outcome towards a particular hypothesis, so that the posterior probability simply reflects the likelihood. Some parameter values are given 'flat priors' that state that all values (within certain bounds) are equally likely. But not all attempts to set uninformative priors use flat distributions. For example, the prior distribution of edge lengths (branches) on each tree may be set using a particular model of lineage speciation and extinction, such that large departures from the expected tree structure will be penalized and end up with a lower posterior probability. In this case, before we even start, we are saying that we don't know what the tree is but we are pretty sure it is more likely to have an even distribution of nodes, rather than having all the bifurcations at the base or the tips of the tree. It's up to you to think carefully about all these priors before you start your analysis, and consider whether they are reasonable statements to make about your particular dataset.

Whatever probability distribution you choose, if your aim is to have uninformative priors that do not bias your result, then it is critical to test whether the priors you set are influencing the outcome of the analysis. One way to test this is to sample from the prior distribution and compare it to an equivalent sample from the posterior distribution: this is like running your analysis again without the data. If the two are the same it suggests that the answer you get is being determined by priors and not influenced by the data you are analysing. If you get the same answer whether or not you include the sequence data, then the data has not given you any extra information than you had to start with, suggesting either that there isn't a very clear story in the data, or that the priors have overwhelmed the data. That ought to give you pause for thought.

There are a number of different Bayesian methods for phylogenetic inference, but the basic approach is something like this (**Figure 13.35**):

**Figure 13.35**

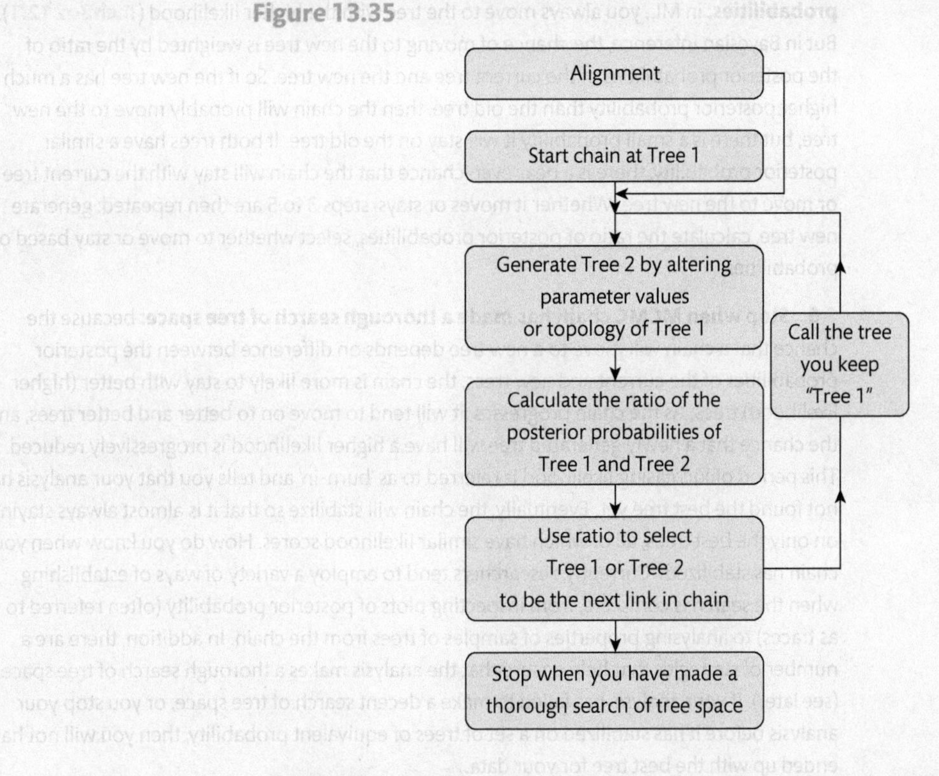


1. **Alignment:** as with all methods, if your alignment is not reliable, neither is your phylogeny (**TechBox 10.2**).

2. **Starting tree:** commonly a randomly chosen tree to ensure independence of runs (chains).

3. **Generate a new tree:** Bayesian methods make use of a procedure called Markov Chain Monte Carlo (MCMC) to explore the set of all possible trees. A Markov Chain describes the movement of a system through a series of states: at each moment the chain may move to a new state or stay in its current state, and the movement is not influenced by past states (the chain has no 'memory' of where it has been). Monte Carlo refers to random sampling of numbers (the name is a reference to a famous casino, highlighting the role of chance). So an MCMC algorithm takes a random walk through tree space. In practice it does this by randomly altering one or more parameters of the current tree to produce a slightly different tree[1]. The chain might then move to the new tree, or it might stay on the current tree. The chance of it moving depends on the ratio of the posterior probabilities of the two trees.

4. **Calculate the ratio of posterior probabilities of the two trees:** the prior probability is the probability of a hypothesis being true before the data is taken into account. All variables—such as tree topology, branch lengths, and parameters of the substitution model—must have prior probabilities, even if they are uniform (i.e. no value is more probable than any other). The posterior probability is the probability of a hypothesis being true after you have considered

it in light of your data. The posterior probability is arrived at by calculating the likelihood of a tree (TechBox 12.1), multiplying it by the prior probability of that tree, then dividing by the probability of the data. It would be difficult, if not impossible, to calculate the probability of the data, so estimating the posterior probability of a single tree is generally not possible. But calculating the ratio of posterior probabilities neatly allows the unknowns to be cancelled out, making the calculation tractable.

5. **Movement of chain to new tree is conditioned by the ratio of posterior probabilities:** in ML, you always move to the tree with the higher likelihood (TechBox 12.1). But in Bayesian inference, the chance of moving to the new tree is weighted by the ratio of the posterior probabilities of the current tree and the new tree. So if the new tree has a much higher posterior probability than the old tree, then the chain will probably move to the new tree, but there is a small probability it will stay on the old tree. If both trees have a similar posterior probability, there is a near even chance that the chain will stay with the current tree or move to the new tree. Whether it moves or stays, steps 3 to 5 are then repeated: generate new tree, calculate the ratio of posterior probabilities, select whether to move or stay based on probabilities.

6. **Stop when MCMC chain has made a thorough search of tree space:** because the chance that a chain will move to a new tree depends on difference between the posterior probabilities of the current and new trees, the chain is more likely to stay with better (higher likelihood) trees. As the chain progresses, it will tend to move on to better and better trees, and the chance that a newly generated tree will have a higher likelihood is progressively reduced. This period of increasing likelihood is referred to as 'burn-in' and tells you that your analysis has not found the best tree yet. Eventually, the chain will stabilize so that it is almost always staying on only the best trees, all of which have similar likelihood scores. How do you know when your chain has stabilized? Currently, researchers tend to employ a variety of ways of establishing when the search is complete, from inspecting plots of posterior probability (often referred to as traces) to analysing properties of samples of trees from the chain. In addition, there are a number of strategies that help ensure that the analysis makes a thorough search of tree space (see later). If your analysis has failed to make a decent search of tree space, or you stop your analysis before it has stabilized on a set of trees of equivalent probability, then you will not have ended up with the best tree for your data.

As with maximum likelihood, you can choose the best tree (referred to as the maximum *a posteriori* (MAP) tree), or you can report a 'credible set' of trees by starting at the MAP tree and progressively adding the next-best trees to the set until you have some specified cumulative probability (usually 95 per cent)[2]. Because the MCMC is less likely to move away from a tree with a high posterior probability, the chain spends more time on better trees. So, once you have discarded trees from the 'burn-in' period before the search reached the highest posterior probability trees, you can indicate the support for each node in the tree by randomly sampling trees from the posterior distribution and asking what percentage of trees in that sample contain a particular node. This is conceptually similar to the bootstrap (TechBox 12.2), but much faster as it does not involve calculating the likelihood of replicate datasets.

The posterior sample of trees also gives you a way of encapsulating phylogenetic uncertainty, by allowing you to consider a sample of trees that are plausible alternative phylogenetic hypotheses to explain your data. You can then repeat whatever phylogenetic analysis you are doing on alternative trees, which is a way of asking 'if my tree were slightly different would I get the same answer?' However, the MCMC chain adds a complication to the interpretation of a credible set of trees. The chain wanders from one tree to the next, so two trees close to each other in the chain will tend to be more similar, because they will be separated by fewer changes than trees sampled from disparate parts of the chain. If you sample trees from the same part of the chain, then you will tend to end up with a very similar set of trees. If you want a fair representation of all the trees of near-equivalent plausibility, then you need to randomly sample at different parts of the chain.

### Advantages of Bayesian inference

For the sake of brevity, we will ignore the more fundamental (and rather vigorous) debates on whether a Bayesian or Maximum Likelihood approach is more logically defensible. Most phylogeneticists who choose Bayesian methods do so for their practical advantages: many of the most popular phylogenetics packages are based on Bayesian inference, they offer a relatively fast way to estimate trees under rate-variable ('relaxed clock') models (see **TechBox 14.1**), they provide a convenient way of accounting for phylogenetic uncertainty by allowing you to sample many equivalent phylogenetic hypotheses from the posterior distribution, and they generally report measures of support for all nodes[3]. In common with all such support measures, these are not indicators of how 'true' your tree is, but how strongly your data supports your conclusions given the assumptions you made in your analysis. In the case of Bayesian inference, remember that you start with a prior belief about the probability of different outcomes, so the posterior probabilities reflect these prior beliefs updated from what you have learned from the data given the model used[4]. To be honest, a lot of people prefer Bayesian support values to other measures of support, such as the bootstrap (**TechBox 12.2**), because they have a likeable tendency to give you higher numbers, making you feel happier about your tree.

### Disadvantages of Bayesian inference

The increase in speed over maximum likelihood (ML) methods is a consequence of sampling fewer trees. After all, a Bayesian method must still calculate the likelihood of each tree considered, just as an ML method does. But unlike ML methods, the Bayesian chain does not attempt to calculate the likelihood of the nearby trees, nor for all possible branch lengths per tree. Instead, it draws random samples of trees (each with a particular topology and set of branch lengths). So the performance of Bayesian methods compared to ML must depend on how well tree space is sampled. An MCMC chain can get stuck on local optima just as ML can: strategies can be adopted for jumping off local peaks to explore different parts of tree space, or for running multiple chains from different starting points and seeing if they converge on the same set of trees. Similarly, the percentage values on nodes are reliable only if they are based on a random sample of trees from the posterior. In reality, these samples are often correlated with each other, biasing the posterior probabilities. Never forget you can get very high support values for a completely misleading phylogeny (see **TechBox 12.2**). As with all methods discussed in this book, be an intelligent user, not a mindless button-pusher. Don't be fooled into thinking that a more sophisticated method can be relied upon to give you the right answer every time. 'Sophisticated' often means 'more assumptions', even if the nature of these assumptions is not always obvious to the end user. Remember that the phylogeny you get is contingent on the decisions you made in your analysis, find out what the prior beliefs and the models you are using mean, and always keep your assumptions in the front of your mind when interpreting the results.

### References

1. Huelsenbeck, J. P., Ronquist, F., Nielsen, R., Bollback, J. P. (2001) Bayesian inference of phylogeny and its impact on evolutionary biology. *Science*, Volume 294, pages 2310–4.
2. Huelsenbeck, J. P. (2002) Potential applications and pitfalls of Bayesian inference of phylogeny. *Systematic Biology*, Volume 51 (5), Pages 673–88.
3. Holder, M., Lewis, P. O. (2003) Phylogeny estimation: traditional and Bayesian approaches. *Nature Reviews Genetics*, Volume 4 (4), Pages 275–84.
4. Simmons, M. P., Pickett, K. M., Miya, M. (2004) How meaningful are Bayesian support values? *Molecular Biology and Evolution*, Volume 21 (1), Pages 188–99.

CASE STUDY 13.1

# Rates: flightless insects have faster substitution rates

RELATED TECHBOXES

TB 8.2: Population size

TB 12.1: Maximum likelihood

RELATED CASE STUDIES

CS 3.2: Mutation rate

CS 13.2: Diversification

Mitterboeck, T. F., Adamowicz, S. J. (2013) Flight loss linked to faster molecular evolution in insects. *Proceedings of the Royal Society B: Biological Sciences*, Volume 280, page 20131128.

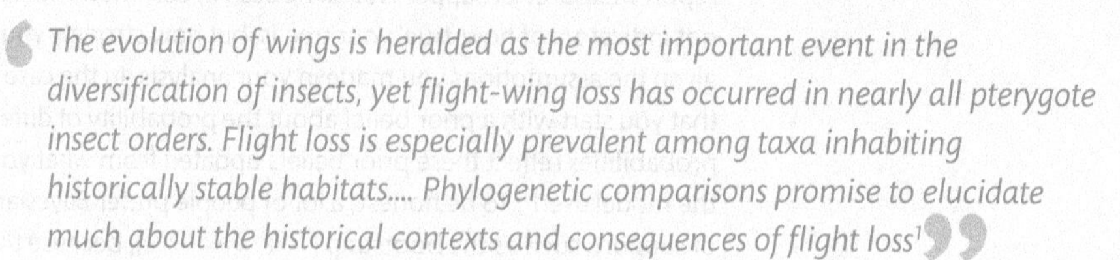

*"The evolution of wings is heralded as the most important event in the diversification of insects, yet flight-wing loss has occurred in nearly all pterygote insect orders. Flight loss is especially prevalent among taxa inhabiting historically stable habitats.... Phylogenetic comparisons promise to elucidate much about the historical contexts and consequences of flight loss[1]"*

## Keywords

effective population size (*Ne*), synonymous, nonsynonymous, d*N*/d*S*, sister pairs, selection, drift, metabolism

## Background

Although flight seems to have been a key factor in the evolutionary radiation of the insects, many lineages have subsequently evolved flightless forms (**Figure 13.36**). Flightlessness is likely to result in changes in metabolism, and reduction in population size and dispersal ability, all factors that have been proposed as important drivers of tempo and mode of molecular evolution.

## Aim

Because flightlessness has evolved thousands of times independently in many different insect groups, it presents an ideal case for a comparative test of the effect of flightlessness on rates of

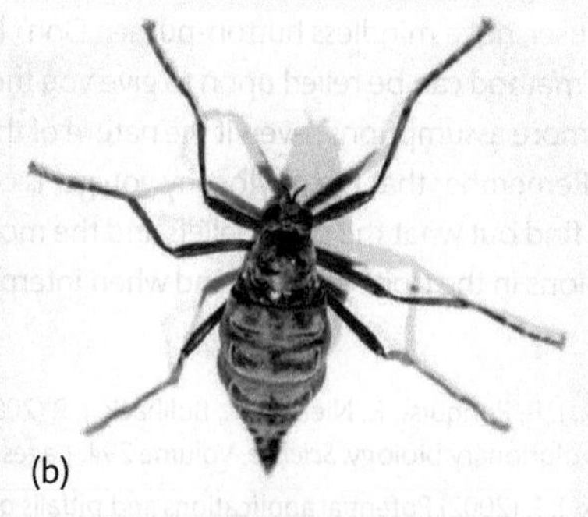

**Figure 13.36** What do you call a flightless species of fly? A walk, of course. Flightless species occur in many different dipteran lineages. (a) Some flightless flies, such as snow flies (*Chionea*) retain their halteres (the stubby second set of wings you can see on house flies), possibly as sense organs. Snow flies might have lost their wings due to the difficulties of generating energy for flight in sub-zero temperatures. Thanks to their own internal antifreeze, snow flies walk over the snow. (b) In some species, only one species is wingless, as in the soldier fly (*Boreoides subulatus*) where females are flightless but males have wings. (c) Other flightless flies, like this stilt-legged fly, *Badisis ambulans*, have lost both wings and halteres and barely look like flies at all. Larvae of the ant-like *Badisis ambulans* develop inside the pitchers of the Albany Pitcher Plant (*Cephalotus follicularis*), a threatened plant species found only in the biodiversity hotspot of south-western Western Australia. Presumably if the plant succumbs to extinction (partly because of collecting pressure for horticultural specimens), then the flightless stilt-legged fly will too.

molecular evolution. Each independent origin of a flightless lineage is like running a separate experiment on the effect of flight loss on molecular rates.

### Methods

The researchers selected published phylogenies of insect groups that contained both flight-capable and flightless species, but they did not include examples where flightlessness was connected to another transition that could affect rates, such as island taxa or parasites. Eight phylogenies contained multiple transitions to flightlessness (between 2 and 13 inferred origins) and seven phylogenies had a single transition, providing a total of 49 independent origins of flightless lineages. They used two methods of analysis. For the whole-tree method, lineages in the phylogeny were designated either flight-capable or flightless (deeper lineages were assigned to a background rate to avoid biasing the flight-capable rate estimate), then the average substitution rate for each category was estimated using maximum likelihood. For the sister clades analysis, they selected 40 clades where a flightless lineage could be compared to its flight-capable sister lineage (though some of these were subsequently rejected due to too few substitutions between the sister lineages). They estimated the synonymous (d*S*) and non-synonymous (d*N*) substitution rates and the ratio of d*N*/d*S* for each tree or sister clade. They used relative rates tests to compare rates between flight-capable and flightless sister clades, then used a two-tailed binomial test to ask whether the flightless lineages were either slower or faster than flighted lineages more often than expected by chance.

### Results

For all whole-tree analysis, the flightless lineages had higher d*N*/d*S* than the flight-capable in the mitochondrial genes, but there was no significant pattern for overall rates. For the mitochondrial sister clades, three quarters of the flightless lineages had higher d*N*/d*S* and higher d*N* than the flight-capable lineages, but there was no significant difference in d*S*. The overall rates were higher in the flightless lineages for the mitochondrial genes, but not for nuclear. Nuclear genes did not show consistently higher d*N*/d*S* or overall substitution rates in any of the analyses.

### Conclusions

There is a consistent and significant trend to higher d*N*/d*S* in mitochondrial genes in the diverse flightless insect lineages considered in this study, when compared to their flight-capable sister lineages. Since reduction in population size should affect both nuclear and mitochondrial genomes, the authors attribute this pattern to reduced metabolic activity in flightless insects leading to a relaxation on selection in mitochondrial genes, increasing the number of nonsynonymous changes that are fixed by drift rather than removed by selection.

### Limitations

The lack of a clear pattern for nuclear sequences may be because they are less affected by flight loss than the mitochondrial genes, but it may also be due to lower power (fewer substitutions analysed). The sister clade analyses represent phylogenetically independent instances of the origin of flightlessness, but some of the whole-tree analyses include separate analyses of different genes from the same taxa. In this case, the data cannot be pooled in the same statistical analysis, because the same history is being sampled again and again, so there are only six independent datapoints for the mitochondrial sequences. Within each insect group examined, inference of the origins of the flightless lineages was based on molecular phylogenies, which has some potential for circularity: if flightlessness influences the rates and patterns of substitution, it could possibly affect the placement of taxa in the tree.

### Further work

This study includes species in which only females are flightless, as well as species where both sexes are flightless, but there was no notable difference in the results between the two categories of flightlessness. This seems curious, given that female-only flight loss might have different effects on population size, dispersal, and selection on metabolic genes than those in which all individuals are flightless. Does this suggest that the reduction in metabolic costs in females is driving the trend? Or that some other feature of these species influences rates, that is correlated to but not directly caused by loss of flight in one or both sexes?

### Check your understanding

1. Why were wingless parasitic insects excluded from this study?
2. Why did they only use protein-coding sequences for this study? Why not rRNA or non-coding sequences?
3. Why did flightless lineages have relatively more nonsynonymous substitutions, but not more synonymous substitutions, than their flight-capable relatives?

### What do you think?

Can you think of any confounding factors in this study that might generate a non-causal correlation between flightlessness and rates? If so, how would you test whether those other factors were actually driving the observed relationship?

### Delve deeper

The authors interpret the lack of relationship between synonymous substitution rate and flightlessness as evidence against the hypothesis that metabolic rate drives mitochondrial mutation rates. Is this a fair conclusion? What evidence has been used to support this hypothesis? If you were to design a comparative study to test the influence of metabolic rate on rate of molecular evolution, which lineages would you consider most informative, and what data would you need?

### Reference

1. Wagner, D. L., Liebherr, J. K. (1992) Flightlessness in insects. *Trends in Ecology & Evolution*, Volume 7, page 216.

CASE STUDY 13.2

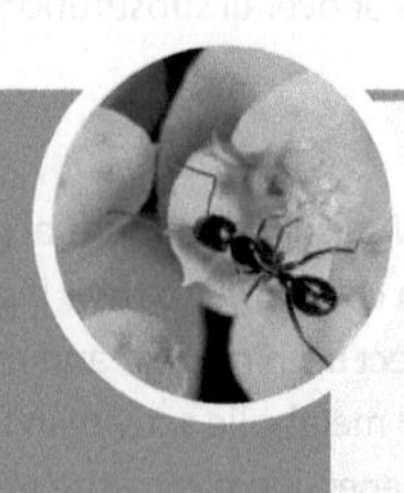

## Diversification: mutation rates are linked to species richness in plants

RELATED TECHBOXES

TB 8.1: Neutral theory

TB 8.2: Population size

RELATED CASE STUDIES

CS 3.2: Mutation rate

CS 4.1: Replication

Bromham, L., Hua, X., Lanfear, R., Cowman, P. F. (2015) Exploring the relationships between mutation rates, life history, genome size, environment and species richness in flowering plants. *American Naturalist*, Volume 185, page 507.

> *Evolution is a process of Variation and Heredity. The older writers, though they had some vague idea that it must be so, did not study Variation and Heredity. Darwin did, and so begat not a theory, but a science*[1]

### Keywords

biodiversity, species richness, speciation, extinction, substitution rate, synonymous, nonsynonymous

### Background

A growing number of studies have shown that lineages with higher diversification rates tend to have higher rates of molecular evolution[2-4]. But there has been relatively little progress in explaining the cause of this observed correlation. Is it possible to untangle the many possible contributing factors, such as life history, climate, population size, selection, speciation and mutation, and uncover the mechanisms underlying the link?

### Aim

We set out to contrast three possible explanations for the observed link between rates of molecular evolution and net diversification: (1) faster molecular rates increase the speed of diversification; (2) higher speciation rates drive faster rates of molecular evolution; (3) diversification and molecular evolution are indirectly linked because they are both correlated with another factor.

### Methods

If you compare two lineages that are each other's closest relatives, then a difference in number of species between these two sister clades may reflect a difference in the net rate of diversification (the net result of addition of lineages by speciation and removal by extinction: **Figure 13.37**). Similarly, any difference in branch lengths between the two sister clades could represent a difference in the number of substitutions each has acquired since their common ancestor. This study built on a previous study that found that plant families with greater numbers of extant species had faster rates of molecular evolution than their sister families with fewer species[5] (**Case Study 4.1**). We collected DNA sequences from mitochondrial, chloroplast and nuclear genomes for pairs of sister families, along with extant number of species per family, and family averages for genome size, plant height, temperature, UV exposure and latitude. Diagnostic tests were used to identify and exclude family pairs deemed too shallow for reliable rate estimation[6]. The differences between sister pairs were analysed in several ways. Many of the variables scale closely together—for example, temperature and UV tend to increase with decreasing latitude—so several forms of regression analysis, designed to allow for this collinearity, were used. Then we used path analysis to evaluate the support for the alternative hypotheses. Path analysis tests predefined causal models against the patterns of variance in the data[7].

### Results

Height was a strong predictor of mitochondrial and chloroplast synonymous substitution rates and $dN/dS$, and nuclear rRNA substitution rates. Genome size was negatively related to

**Figure 13.37** Why do some plant families have so many species while others contain only a few? The (a) Asteraceae (daisy family) and the (b) Calyceraceae (which are so unexciting they don't even have a common name) are sister families that last shared a common ancestor over 80 million years ago. Since then, the Asteraceae lineage has produced over 23,000 recognized species while, from the same starting point, the Calyceraceae lineage has produced fewer than 200 species. Clearly, there are a great many factors that will contribute to differences in diversification rate. Could mutation rate be one of these factors? Could increasing the supply of variation speed the rate of diversification? Raised mutation rate could increase the standing variation available for adaptation to new niches or changing environments, but it seems more plausible that increasing the supply of mutation speeds the rate at which genomes from different populations become incompatible due to genetic conflicts.

synonymous substitution rates and dN/dS in the nuclear protein-coding gene. Synonymous substitution rates were positively associated with species richness for the chloroplast and nuclear protein-coding genes. Latitude was negatively associated with nonsynonymous substitution rates in chloroplast genes. Curiously, some analyses suggested that higher UV was associated with lower mitochondrial substitution rates, and temperature was negatively associated with species richness, but these results were not consistent across all analyses.

### Conclusions

The consistent pattern that emerged from these analyses was that the family average values for height and genome size correlated with synonymous substitution rates in the organelle and nuclear genomes respectively, that synonymous substitution rates were correlated across all three genomes, and within each genome, nonsynonymous rates were correlated with synonymous rates (Figure 13.38). There was also weaker evidence that species richness was positively associated with substitution rates. We concluded that the most plausible explanation for these patterns was that plant life history shapes mutation rates, which drive greater rates of substitution across all three genomes, which increase the speed at which the genomes of diverging populations become incompatible, reducing the time to speciation, and therefore increasing the speciation rate.

### Limitations

While the best fitting model supports an influence of rate of molecular evolution on diversification rate, the alternative explanation that differences in diversification rate drive substitution rate variation cannot be rejected. Instead, we relied on an argument from plausibility: it is difficult to think of a convincing mechanism whereby diversification rate could directly influence mutation rates (as opposed to increasing substitution rates through selection or population subdivision). Similarly, although the results give no indication of reduction in population size or changing action of selection accompanying speciation (because there is no link between dN/dS and species richness), absence of evidence is not evidence of absence. Furthermore, since species richness is the net outcome of both speciation and extinction, it is possible that lineages with a higher mutation rate have lower extinction rates due to greater levels of standing variation allowing adaptation to changing environments. More generally, it is questionable how meaningful it is to test family average environmental variables such as UV or temperature: species within widespread families may experience very different environmental conditions.

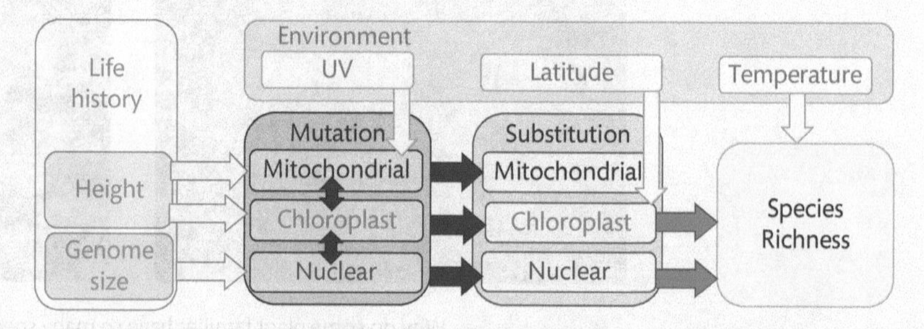


**Figure 13.38** An attempt to make sense of it all. Flow charts that try to capture complex networks of interactions are aptly referred to as 'horrendograms'. Remember that diagrams such as this are often attempts to simplify complex and messy statistical results to bring out a clear narrative. Is this a brave attempt at synthesis, or a misleading whitewash? For example, not all correlations are shown in this figure, or it would be covered with a confusing tangle of arrows. Not every alignment analysed supported each link in this chain, for example the mitochondrial sequences did not show evidence of a link between species richness and rates of molecular evolution. The links between environmental factors and rates are indicated even though they were not strongly supported by the analysis. On the one hand, attempts to simplify results into a neat narrative risk distorting the results, but on the other hand, statistics without interpretation are sterile, and scientists have an obligation to provide plausible explanations for their results. What do you think?

### Further work

It is perplexing that the chloroplast genes show consistent links between species richness and synonymous substitution rates, a result also seen for the nuclear protein-coding gene in some analyses, but this relationship is not supported from the analysis of mitochondrial genes. Does this indicate that there is something special about the relationship between chloroplast genome evolution and diversification? Or simply a difference in power between the three datasets? Or is the chloroplast result a statistical anomaly (though it has been reported in studies of other plant lineages too[8])?

### Check your understanding

1. Why did this study use pairs of families? Wouldn't it be easier to look for a relationship between family size and molecular rates by conducting a correlation analysis across all families?
2. Is species richness of a family a reliable way of measuring the speciation rate?
3. Why did the authors conclude that the relationship is driven by difference in mutation rate, not population size or selection?

### What do you think?

Studies that include many different biological variables can report associations that may not be meaningful, but have occurred due to the inclusion of many free parameters, which can end up describing random variance. In this study, the relationship between height and rates was strong and consistent, the association between species richness and rates was weak but consistent and in line with previous findings, but the associations between environmental factors and rates were weak and inconsistent and not in the direction suggested by previous studies. How should you decide which relationships to report and explain, and which to ignore? How do you balance the risk of over-explaining a random result against ignoring a weak but important signal? Should scientists stick to objective reporting of the patterns, or should they provide subjective interpretations in light of their own opinions on the underlying processes, deciding which results are important and which can be ignored as irrelevant to the story?

### Delve deeper

It has been suggested that differential mutation rates underlie one of the most prominent patterns in biodiversity: that species richness increases towards the equator. The suggested links are that higher energy environments drive faster mutation rates, either directly or indirectly through increased growth rates, and that this flows through to lineage diversification. What evidence supports this hypothesis? How would you test it, controlling for covariation between environmental factors, geographic distribution and species traits?

### References

1. Bateson, W. (1910) 'Heredity and variation in modern lights'. In *Darwin and Modern Science*, ed. Seward, A. C. Cambridge University Press.
2. Lanfear, R., Ho, S. Y. W., Love, D., Bromham, L. (2010) Mutation rate influences diversification rate in birds. *Proceedings of the National Academy of Sciences USA*, Volume 107, page 20423.
3. Webster, A. J., Payne, R. J. H., Pagel, M. (2003) Molecular phylogenies link rates of evolution and speciation. *Science*, Volume 301, page 478.
4. Barraclough, T. G., Savolainen, V. (2001) Evolutionary rates and species diversity in flowering plants. *Evolution*, Volume 55, page 677.
5. Lanfear, R., Ho, S. Y. W., Davies, T. J., Moles, A. T. A. L., Swenson, N. G., Warman, L., Zanne, A. E., Allen, A. P. (2013) Taller plants have lower rates of molecular evolution: the rate of mitosis hypothesis. *Nature Communications*, Volume 4, page 1879.
6. Welch, J. J., Waxman, D. (2008) Calculating independent contrasts for the comparative study of substitution rates. *Journal of theoretical Biology*, Volume 251, page 667.
7. Shipley, B. (2002) *Cause and correlation in biology: a user's guide to path analysis, structural equations and causal inference*. Cambridge University Press.
8. Duchene, D., Bromham, L. (2013) Rates of molecular evolution and diversification in plants: chloroplast substitution rates correlate with species richness in the Proteaceae. *BMC Evolutionary Biology*, Volume 13, page 65.

# 14 Dates

## *Telling the time*

*"Nature's system is not that of preserving a perfect history of her past creation, but only a broken and imperfect one"*

Charles Lyell (1856) Scientific Journal No. II. In *Scientific journals on the species question*, ed. Wilson, L. G. (1970) Yale University Press.

## What this chapter is about

The longer two lineages have been separated, the more differences we expect to see when we compare their genomes. Therefore, comparisons between DNA sequences can help us estimate when lineages diverged from each other. To estimate evolutionary time from DNA sequences we need to know the rate of molecular evolution. However, the rate of molecular evolution varies between lineages. This complicates the estimation of time from DNA sequences, but if we can express the uncertainty in date estimates with honest confidence intervals, we can use molecular date estimates to test hypotheses in evolutionary biology.

## Key concepts

- Molecular date estimates rely on assumptions about rates of molecular evolution
- Confidence limits on date estimates can be used to test evolutionary hypotheses

# Confidence

## *When animals exploded*

The fossil record of the animal kingdom starts with a bang. During the early- to mid-Cambrian period (which runs from 541 to 485 million years ago), the fossil record explodes with animal diversity. Many fundamentally different kinds of animals all appear more or less simultaneously in the fossil record. We see the first arthropods with jointed legs, complex eyes, and a segmented body covered in a hard cuticle (**Figure 14.1**). There are molluscs, some protecting their soft bodies with shells, spikes or scales. There are echinoderms with pentameral (five-pointed) symmetry. There are polychaete worms whose delicate hairy bodies are preserved in astoundingly fine detail. And there are a great variety of legged worms, some soft-bodied like today's velvet worms (**Figure 11.13**), others arrayed with formidable plates or spikes along their backs.

Many of these complex Cambrian animals, with their diverse ways of life and different bodily organizations, can be recognized as members of modern animal phyla. A phylum is the highest level of animal taxonomy, containing large collections of species that all share a fundamental similarity of construction. For example, arthropods are characterized by having a hard jointed exoskeleton, distinct body sections and paired limbs (think of a spider or an ant: **Figure 14.2**). Annelids are characterized by elongated segmented bodies with a mouth at one end and an anus at the other (think of an earthworm). Echinoderms have pentameral symmetry, no head or eyes, and a water vascular system that performs both as a circulatory system (instead of blood) and a hydraulic system (moving tube feet and tentacles).

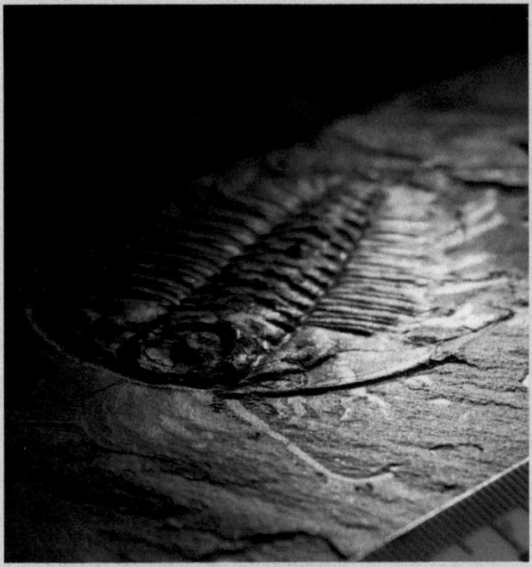

**Figure 14.1** The Cambrian fossil record includes exquisitely preserved soft-bodied creatures and records some fine details of anatomical structures, such as the antennae on this *Olenoides* trilobite.

Image courtesy of Smith609 at en.wikipedia, licensed under the Creative Commons Attribution Share Alike 3.0 Unported license.

These fundamentally different body organizations must have all ultimately originated from a single ancient animal ancestor. But the fossil record does not record a continuous series of animal forms linking the modern phyla to this ancestor, with lineages initially resembling each other then gradually becoming as distinct as the phyla we recognize today. The startling thing about the Cambrian is that clearly identifiable members of almost all of the readily fossilizable animal phyla, such as arthropods, molluscs, and echinoderms, all appear in the fossil record during a period of little more than 10 million years (**Figure 14.3**: not surprisingly, at least a third of all animal phyla do not appear in the Cambrian because they have no fossil record at all).

There are fossils of earlier creatures, which might represent the ancestors of the metazoans. Animal-like fossils from before the Cambrian are named for the Ediacaran range in South Australia, where they were discovered by the geologist Reg Sprigg in 1946. Late in the day, when the low angle of the sun caused even minor impressions in the rock to cast tell-tale shadows, Sprigg saw the faint shapes of jellyfish-like creatures pressed into the rust-coloured sandstone. Since then, fossils of Ediacaran species have been found all over the world. They form a diverse collection of relatively simple multicellular organisms. They have no bones, no teeth, no legs, no eyes, no hard armour. The Ediacarans have been described as 'flat earthers', pudding-like creatures living a simple life, lounging around on the seafloor (**Figure 14.4**).

But at the end of the Precambrian, the 'Garden of Ediacara' disappears, and the animal fossil record undergoes a striking transformation. Small animals with shells appear. Some of these shells seem to have boreholes in them, suggesting the presence of specialized predators. Tracks show that animals were moving, and burrows suggest they began to colonize the sediment instead of just sitting on the ocean floor. By the mid-Cambrian, the diversity of animal forms display a range

**Figure 14.2** Animal phyla, such as (a) Arthropoda, (b) Annelid, and (c) Echinodermata are often considered to represent fundamental differences in body plan.

(c) Reproduced courtesy of the National Estuarine Research Reserve Centre.

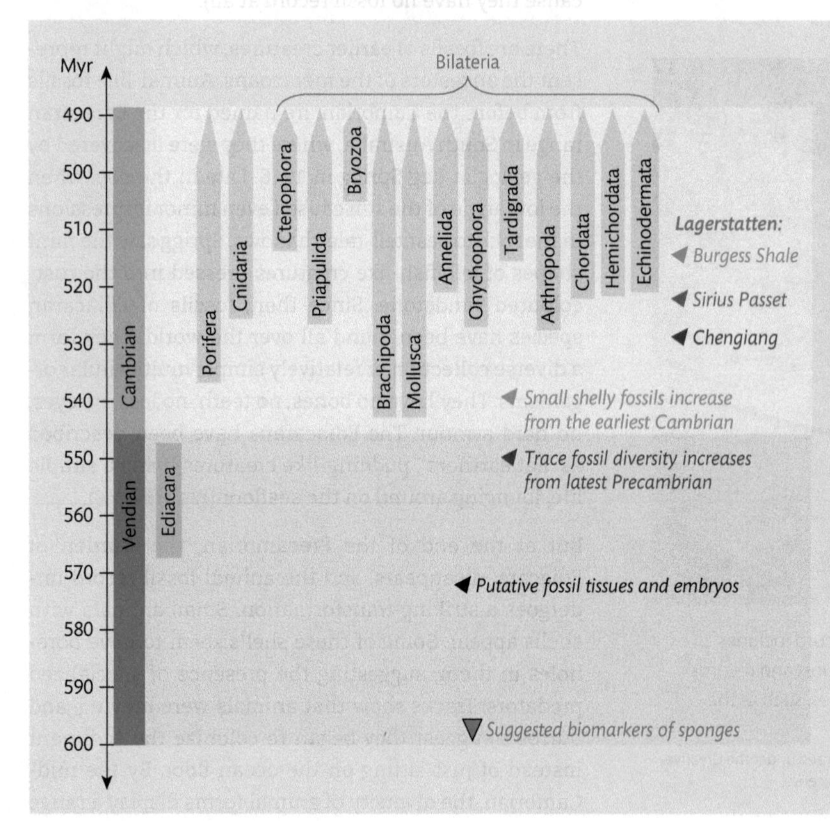


**Figure 14.3** The known fossil ranges of around half of the bilaterian phyla begin in the early- to mid-Cambrian period. Note that the dates shown in this figure are approximate only, it is not intended to give precise ages for fossil beds or first appearances of phyla. There is much debate about exactly when the earliest members of some phyla occur. For example, while some palaeontologists consider that various Ediacaran taxa are early cnidarians, others say the first unambiguous cnidarian fossil is Cambrian age. Similarly, biomarkers (diagnostic chemical signatures) of sponges have been reported from sediments at least 630 Myr old, and putative sponge spicules and body fossils have been described from Ediacaran assemblages, but the veracity of all sponge fossils older than the Cambrian boundary have been called into question.

**Figure 14.4** The Ediacaran fauna is known from locations all over the world from the latest Precambrian period (579–542 Myr ago). Since their discovery, Ediacaran fossils such as (a) *Spriggina*, (b) *Dickinsonia*, and (c) *Cyclomedusa* have variously been interpreted as ancestral animals, or modern metazoans, or a sister group to modern animals, a completely independent experiment in multicellularity, or even lichens or giant unicells. Debate continues as to whether the fauna contains early members of bilaterian lineages.



of ways of life—burrowing, grazing, swimming, eating, and being eaten. These new ways of life were characterized by more sophisticated morphological accessories such as appendages, sense organs and armour.

One of the aspects of the Cambrian record of animals that makes it so fascinating is the extraordinary level of fine detail in some of the Cambrian period fossils, which are preserved in fine-grained shales that reveal even the soft-parts of the animals. These fossil beds with exceptional preservation (termed lagerstätten) give us an astounding snapshot of animal life half a billion years ago. The fossils are of such high quality that in some cases it has been possible to examine the internal organs of ancient chordates, look through the crystalline eyes of a trilobite, describe the last meal of a priapulid worm, and explore the colours generated by refraction from the surface of polychaete body hairs.

The transition from simple soft-bodied creatures to complex modern animals is not in itself surprising. Clearly the first multicellular animals must have been much simpler and less diverse than their descendants. What is surprising is the apparent speed of the transition. If it is true that the different animal body plans evolved within 10 million years, then the pace of change is unlike anything seen in the history of life before or since the Cambrian period. Consider the remarkable evolutionary radiation of the Hawaiian honeycreepers (**Figure 13.3**): in 10 million years they have undergone an explosive radiation that produced over fifty different species with a variety of colours, beak shapes, and ways of life. But, impressive as this is, the honeycreepers are all still birds, sharing the same basic body plan. Most lineages produce rather less morphological change in 10 million years, or, in the case of the coelacanths, hardly any change at all (**Figures 13.1** and **13.2**). In the same period of time as it has taken honeycreepers to change their beak shapes and coelacanths to do virtually nothing, fundamentally different kinds of animal construction and physiology are said to have evolved.

The synchronous appearance of many animal phyla has led some biologists to doubt that this diversification could have been achieved by the gradual accumulation of many mutations, each of relatively small effect, by selection and drift. Instead, it has been proposed that the rapidity of evolution in the Cambrian is due to special evolutionary mechanisms, such as the evolution of key developmental genes or gene networks, or a consequence of particular conditions present in the Cambrian but never before or since. No wonder that the Cambrian explosion has been called the greatest biological mystery of all times.

## Fossilization: a chance at immortality

The fossil record is a rich and bountiful source of past history, providing the primary source of temporal information in evolutionary biology. Biologists from all walks of life rely on fossil information not only to give a timescale for evolutionary change, but also as a unique source of information on past forms (**TechBox 14.1**). Just as you should think critically about the molecular data you analyse, it's important to think carefully about what palaeontological information represents, and the assumptions that you make when interpreting fossil data in terms of patterns of change and evolutionary processes.

When contemplating the fossil record, remember that fossilization is exceedingly rare. Most organisms that ever lived rotted away to nothing when they died. Only the occasional lucky corpse encountered conditions that preserved its form after death. In fact, most species have little or no fossil record. For example, the Platyhelminthes (flatworms) are a diverse animal phylum. This lineage is over half a billion years old and probably contains over ten thousand living species. If I was to make a ballpark guess at how many flatworms have ever lived, I would think it would be at least 12 trillion individuals. Yet only one of these 12 trillion individuals was lucky enough to have achieved palaeontological glory, preserved in a piece of Eocene amber (though there are also flatworm eggs reported from coprolites—fossilized faeces—from sharks and dinosaurs). Despite their antiquity, abundance, and diversity, flatworms are just the sort of small, soft-bodied creatures that tend not to leave many informative fossils.

A lineage will only be represented in the fossil record if some of its members happen to be captured in layers of sediment that remain undisturbed for long enough to turn into stone, and the stone survives the perturbations of geology long enough to be uncovered by a lucky palaeontologist, and the fossil contains features that allow it to be identified as a member of that lineage. So organisms that are unlikely to land in sediment (e.g. desert-dwelling shrubs), or unlikely to survive burial intact (e.g. soft squishy worms), or unlikely to remain in undisturbed rock (e.g. birds on a volcanic island), or in a place underexplored by palaeontologists (e.g. a small mammal from the New Guinea highlands), or have few distinctive physical features (e.g. many bacteria) are a lot less likely to have an informative fossil record. What is the chance of a species of mycoplasma being preserved undamaged in sedimentary rock then being discovered by someone who can recognize it as a fossil and not just a spot on a rock?

In other words, the fossil record does not provide a complete or continuous record of evolutionary history. Instead, the fossil record is patchy. It's patchy in time (some periods are not recorded), in space (some areas are not recorded), and in its biological coverage (some lineages are not recorded). This does not in any way lessen the importance of fossils, which provide the primary source of information on evolutionary history. But it does mean that it is not always simple to interpret gaps in the fossil record. We have seen that gaps in the fossil record can be substantial. Some, such as the half-billion year near-absence of flatworms, may be unsurprising due to poor fossilizability. Others, such as the 80 million year gap in the coelacanth record, are harder to explain (see Chapter 13). If a lineage is missing from the fossil record, can we be sure that the species was absent from that place and time, or could it be the species was present but failed to leave identifiable fossils? When can we consider that absence of evidence is convincing evidence of absence?

## Alternative hypotheses

> *Consequently, if the theory be true, it is indisputable that before the lowest Cambrian stratum was deposited, long periods elapsed, as long as, or probably far longer than, the whole interval from the Cambrian age to the present day; and that during these vast periods the world swarmed with living creatures*

Darwin, C. (1872) *On the origin of species by means of natural selection: or the preservation of favoured races in the struggle for life.* 6th edn. John Murray, London

Darwin was sure that while the Cambrian explosion represented the start of the fossil record of animal evolution, it did not mark the start of the lineages themselves. Imagine turning on a video camera halfway through a party. Someone watching the tape the next day might guess that people arrived one by one in a relatively sober state, even though the recording suggests they miraculously appeared all at once in a state of combined hilarity. In Darwin's day, the fossil record was so clearly an imperfect and discontinuous record of evolutionary change that he could make a plausible and well-reasoned argument that the absence of earlier fossils did not indicate an absence of the animals themselves before the Cambrian.

As more fossils were collected, many gaps in the fossil record were filled in. The palaeontological record of animal life continues to expand and amaze. Yet, if anything, the increase in the resolution and detail of the animal fossil record has served to emphasize the discordance between the Precambrian and Cambrian periods. If members of modern animal phyla, with eyes, shells, armour or jointed legs, evolved before the Cambrian, then it is difficult to explain why there is no clear evidence of their Precambrian existence being preserved alongside the soft-bodied Ediacarans. There has been an ongoing debate about the relationship of Ediacaran fossil taxa to modern metazoans, with some people regarding them as ancestral forms that precede the origin of modern metazoans, others recognizing them as early members of modern phyla, or as a side branch from the early metazoan lineage, or even an entirely independent 'experiment' in multicellularity. Whatever the true affinities of the Ediacarans are, there is no doubt that the fossil record suggests a dramatic transition in form and function from the simple, soft-bodied Ediacarans to the sophisticated fauna of the Cambrian.

In Chapter 12, we talked about comparing the plausibility of different evolutionary hypotheses. But in explaining the Cambrian explosion of animal diversity, we seem to have two equally implausible explanations. One is that the modern animal phyla gradually diverged and diversified during the Precambrian but somehow managed to avoid leaving any undeniable evidence of their existence in the Ediacaran period. The other is that the Cambrian fossil record bears witness to the origin of fundamentally different animal body plans in the time normally taken for species to diverge slightly in shape and colour. How are we to weigh up these alternative hypotheses? It would be nice if we could call upon a reliable independent witness to help us weigh up these competing explanations.

## Divergence

All genomes carry a record of their evolutionary history. We have seen how the patterns of substitutions in DNA allow evolutionary relationships to be uncovered (Chapters 11 and 12). Now we will explore how we can use DNA data to estimate the timing of evolutionary events. Since molecular change accumulates continuously, we expect that two lineages that have a recent common ancestor will have fewer differences between their genomes than either has with a more distantly related lineage. More specifically, if we can estimate how many genetic changes have occurred since two lineages split, and if we know how fast such changes accumulate in these lineages, then we can use the amount of genetic change to predict when those two lineages diverged from each other.

Inferring evolutionary time from molecular changes has always been a controversial topic, although these days papers with molecular date estimates are so common that it's really only cases where the molecular dates are in disagreement with accepted wisdom that stir up heated debate. Rather than focus this chapter on sensible examples of molecular dating that nicely illustrate just how easy the whole thing is, we are going to torture ourselves by looking at the most difficult example I can think of. Dating the origin and diversification of the animal kingdom is probably the most controversial application of molecular dating that has ever been. But because estimating molecular dates for the diversification of the animal kingdom has been such a difficult task, it has provided an important testing ground for improving molecular dating techniques. This makes it a good illustration of not only the advantages of molecular dating, but also the complications and problems. Of course, the Cambrian explosion is not just an informative case study in molecular dating, it is also a critical test case for our understanding of tempo and mode of evolution. Many people have interpreted the Cambrian animal fossil record as the signature of a period of astonishingly high rates of change, and even as proof that unusual mechanisms of evolutionary change operate only at special times in life's history. We need to work out whether this is true or not if we want to fully understand evolution.

As we saw in Chapter 12, when there are conflicting explanations for an evolutionary event, we need to use those hypotheses to make predictions, then test the predictions against observations. Explanations for the Cambrian explosion span a range of complex ideas incorporating phenomena from ecology, biochemistry, geology, and so on. But to simplify this example, let's contrast just two simple alternative hypotheses for the origins of the animal phyla. One, which we will call the Cambrian explosion hypothesis, is that all of the animal phyla diverged during, or just before, the Cambrian period, implying a very high rate of morphological evolution in the earliest part of the animal diversification. The other, which we will refer to as the Precambrian slow-burn hypothesis, is that the diversification of the animal lineages only appears to be sudden because an earlier period of more gradual diversification has not been detected in the fossil record (**Figure 14.5**).

Clearly there are many different types of observations that could be used to test these hypotheses, using information from the geological and palaeontological records, from systematics and biogeography, from genetics and developmental biology. But, of course, this chapter

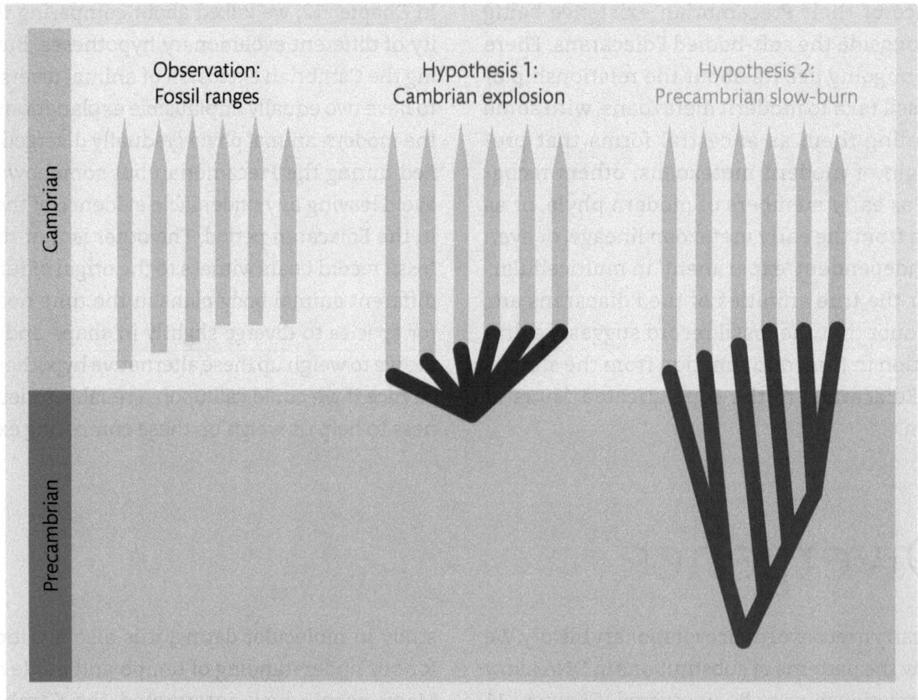


**Figure 14.5** Contrasting hypotheses: here we imagine two distinct explanations for the origin of animal phyla, but clearly there are many possible ways to tell this story.

is about estimating evolutionary time from molecular sequence data, so that is what we are going to focus on here. We want to know if the amount of genetic divergence between animal phyla is compatible with a divergence of lineages in the early Cambrian, or much earlier.

## Divergence dates

> *Mutations happen, and such changes gather in the genetic code like bad memories stocking a guilty conscience: the effect is cumulative. These accreted mutations can provide a kind of clock, which can be reckoned in terms of millions of years if the right part of the genome is examined. There are 'fast' clocks and 'slow' clocks, and to try and look back into the Precambrian we need almost the slowest clocks of all, located in parts of the genome that are enormously conservative. We need to look for the genetic Collective Unconscious shared by all animals*

Fortey, R. A. (2000) *Trilobite! Eyewitess to evolution*. Harper Collins

We saw in Chapter 11 that the longer two populations have been separated, the more substitutions we expect to see between their DNA sequences. If we knew how many substitutions accumulated every million years then we could convert measures of substitutions to evolutionary time. Estimation of evolutionary age of lineages from molecular data is commonly referred to as molecular dating. The phrase 'molecular clock' is less helpful because it can be interpreted as implying that DNA change ticks in a regular, clock-like manner, at the same rate in all lineages. As we will have already seen in Chapter 13, uniformity of rates is usually not a very good description of molecular evolution.

To contrast the Cambrian explosion and Precambrian slow-burn hypotheses (**Figure 14.5**), we want to know if the diversification of the animal phyla occurred during the Cambrian period or sometime before it. Obviously, we can't compare all the genomes that have descended from the animal radiation, but we don't need to, because the genome of any living species of animal carries the traces of its origins. If I compare homologous sequences from two living species from different animal phyla, then their last common ancestor should date to a point just before those phyla diverged. If the Cambrian explosion hypothesis is true, then all of the bilaterian phyla originated during one short evolutionary radiation, in the early- to mid-Cambrian (most tests of the Cambrian explosion focus on the divergence times of bilaterian lineages rather than the origins of sponges

**Figure 14.6** Say hello to your relatives. Sponges (phylum Porifera) represent one of the basal divergences of the metazoan phylogeny. 'Basal' does not imply they are poor cousins—both Porifera and Cnidaria (corals, jellyfish and allies) are among the top ten most diverse animal phyla. 'Early branching' does not mean primitive—sponges have as much time to evolve as you have since you last shared a common ancestry with them (whenever that was). Just as with any lineage, the characteristic features of modern sponges may not have been present in the earliest representatives of the lineage, but may have evolved since. While many have assumed sponges are represented in Precambrian deposits, none of the putative sponge fossils or biomarkers have been accepted by all researchers. This is odd, given that sponge spicules, small structural elements containing silica or carbonates, are just the sort of material that tends to fossilize well. Taking the record at face value, sponges show much the same pattern as other animal phyla: a puzzling absence in the Precambrian followed by a flowering of diversity in the Cambrian period.



and cnidarians: **Figure 14.6**). In this case, the date of divergence between any two sequences from different animal phyla should occur in roughly the same time, presumably in (or just before) the early Cambrian (**Figure 14.7**).

But if the Precambrian slow-burn hypothesis is true, then the divergence date between any two animal phyla should be long before the early Cambrian. So estimating the date of divergence of any of the pairs of species shown in **Figure 14.7** (or any of millions of possible pairs of species from different phyla) provides a test of the Cambrian explosion hypothesis.

To estimate the divergence dates between phyla, we can select suitable DNA sequences and compare them between representatives of different animal phyla. Clearly, we will need to target a fundamental and important sequence, so that it is present in the genomes of all animals, and evolves slowly enough that it is recognizably homologous between the species being compared. For example, we could align the methionine adenosyltransferase gene from some molluscs (perhaps a clam and a mussel) and some arthropods (say, two species of damselfly—a bluet and a spreadwing: **Figure 14.8**). Here is a tiny section of that alignment (**Figure 14.9**).

**Figure 14.7**

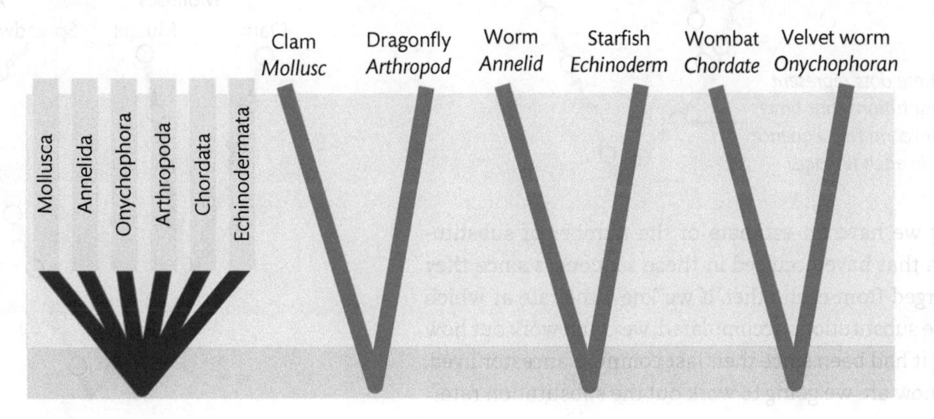

**Figure 14.8** Two types of damselfly, (a) a bluet and (b) a spreadwing.

**Figure 14.9**

| | | Species | Sequence |
|---|---|---|---|
| | Mussel | Mytilus edulis | ATCACAGTATTCAGCTACGGAACATCGA |
| | Clam | Nucula proxima | ATCACGGTTTTCAGCTATGGCACCTCTG |
| Damselflies | | Lestes congener | ATCACCGTTTTTGACTATGGCACATCAA |
| Damselflies | | Enallagma aspersum | ATTACGGTTTTTGACTATGGGACATCAA |

*Chapter 10 explains the importance of alignment for evolutionary analysis of DNA sequences*

Then we estimate the number of substitutions that have occurred between these sequences. Suppose that when we reconstruct the phylogeny, using whatever is our favourite method, we happen to get the following branch lengths (**Figure 14.10**).

**Figure 14.10**

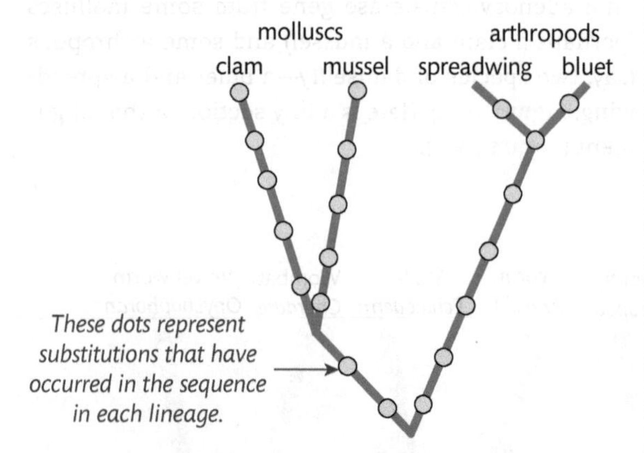


Now we have an estimate of the number of substitutions that have occurred in these sequences since they diverged from each other. If we knew the rate at which these substitutions accumulated, we could work out how long it had been since their last common ancestor lived. But how are we going to work out the substitution rate?

We need some known time points that will allow us to calculate the number of substitutions that occur per million years. Happily, clams and mussels are just the sort of creatures that fossilize well. They live in or near sediments, they have a hard outer casing that is resistant to stress and decay, and this outer casing contains a wealth of features that are used to distinguish different species. Not surprisingly, these lineages have an informative fossil record, which suggests that the oldest known member of either lineage is at least 485 Myr old. The split between the two lineages must have occurred sometime before this first fossil (**Figure 14.11**). So now we have a minimum age of the last common ancestor of the clam and mussel (**TechBox 14.1**).

**Figure 14.11**

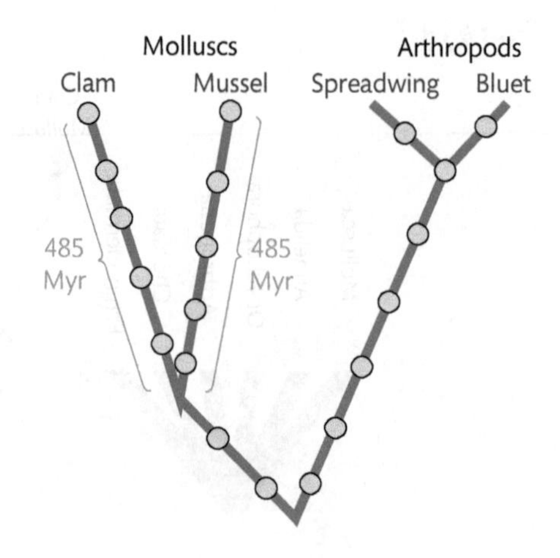

In this case, we can see that we have ten substitutions occurring between the clam and mussel sequences. Since each lineage has had at least 485 Myr to evolve these differences, there has been approximately one substitution every 100 Myr in this sequence in these lineages.

So now we have a rate of substitution, calculated from a known date of divergence, that we can use to estimate our unknown divergence dates. We want to know the date of the split between the two phyla, so we need to know how many substitutions have occurred between the arthropods and molluscs. Counting from an arthropod to a mollusc on this tree, there are 14 substitutions. If we expect approximately one substitution every 100 Myr, then we would guess that it took a total of 1,400 Myr to accumulate these substitutions. The differences between two species include changes that have accumulated in both lineages since their common ancestor, so it represents the sum of the two paths from the present back to the ancestor. If rates of substitution are roughly the same across the whole tree, then the information we have in this example suggests that these phyla have been evolving separately for around 700 Myr (**Figure 14.12**).

**Figure 14.12**

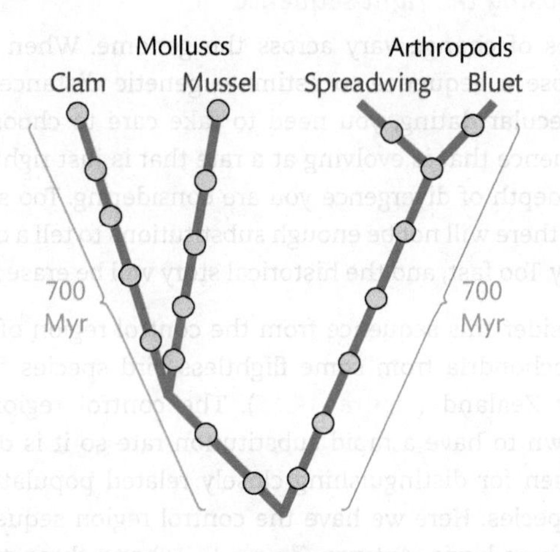


Well, that was easy, wasn't it?

In practice, estimating molecular dates is not nearly as straightforward as this example. That's because both of the things we need to know—the number of substitutions that have occurred and the rate at which they accumulate—are fairly tricky to estimate. First, let's consider the things that make estimating number of substitutions difficult, and then we will move onto the thorny problem of estimating substitution rates.

## Estimating number of substitutions

The earliest molecular dating studies simply lined up the sequences and counted the differences between them, but as we saw in Chapter 13, this approach is likely to underestimate the number of substitutions that have occurred because multiple hits will erase the historical signal of previous substitutions. In the previous chapter, we discussed the difficulties associated with accurately estimating the number of substitutions that have occurred along a lineage, using a model of substitution to predict substitutions that we can't directly observe. Choosing a model of molecular evolution may not seem to be the most exciting part of a molecular dating analysis. But if you are analysing molecular data because you actually want to know what happened in the past, then you need to be aware of the impact that model choice has on the answers you are getting (**TechBox 13.1**).

Consider, for example, a study that used alignments of seven different proteins, taken from a range of modern animals, to estimate dates of divergence for major bilaterian animal lineages. When they assumed that all sites evolved at the same rate, they got a date estimate of 573 millions of years ago. Since this date estimate corresponds with early Ediacaran type fossils, this result was considered to support the contention that the Ediacaran fauna contains the earliest members of the animal kingdom. But when the authors used exactly the same data but a different assumption about patterns of evolution—that rates could vary between sites—they got a very different answer. By allowing rates to vary between sites, they estimated the bilaterian phyla diverged at 656 Myr ago, long before any clear fossil evidence of their existence (**Figure 14.13**).

**Figure 14.13**

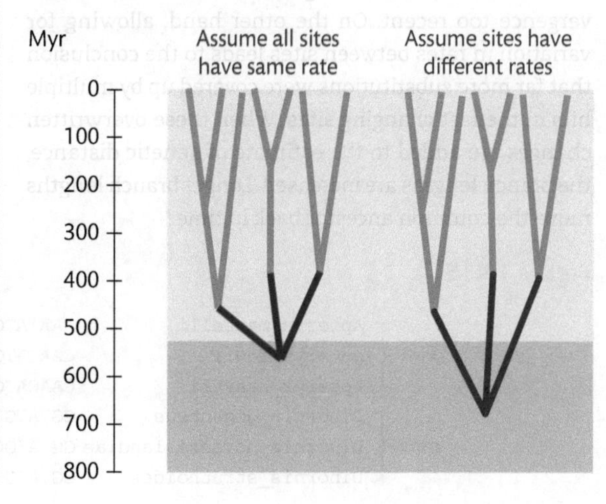

Why did allowing for different substitution rates across the sequence change the date estimate? We can see why when we look at the sequences used to produce these date estimates. **Figure 14.14** shows one short section of the amino acid sequence of one of the genes used in the study (methionine adenosyltransferase). In this small part of one alignment, which includes only half of the species used in the analysis, you can clearly see that some sites change more readily than others. Many of the amino acids have been preserved through over half a billion years of evolution, suggesting that they are strongly conserved by selection. Most of the differences between these sequences occur in only half of the sites (**Figure 14.14**).

**Figure 14.14**

| | |
|---|---|
| *hydrozoan* | AGDQGLMFGYATDETDELMPLTCVYAHKLSQKLSDLRRNGDMG |
| *flatworm* | AGDQGLMFGYATDETEECMPLTVVLAHGLNYRLSECRRREVLP |
| *bristleworm* | AGDQGLMFGYATDETEECMPLTVVLAHKLNAKIAELRRKGVLA |
| *sea urchin* | AGDQGLMFGYATDETEELMPLTVVLAHGMNKRLAACRRDGTMA |
| *anemone* | AGDQGLMFGYASDETEECMPLTCVLAHALTQKLAEIRH-GGDR |
| *sand dollar* | AGDQGLMFGYASDETEECMPLTVVLAHALTVKMAEIRH-GGDR |
| *hemichordate* | AGDQGLMFGYATDETEECMPLTVVLSHALNKKLAENRRNGTMP |
| *mussel* | AGDQGLMFGYATDETEECMPLTVVLAHKLNAKIAELRRKGVLA |
| *clam* | AGDQGLMFGYATDETEECMPLTIVLSHQLNAKIGTLRRNGTFP |
| *bluet* | AGDQGLMFGYATDETEECMPLTVVLAHRLNQRIAQLRRTGVFW |
| *spreadwing* | AGDQGLMFGYATDETEECMPLTVVLAHKLNQRIAQLRRNGVFW |

*invariant sites* — *slow changing sites* — *fast changing sites*

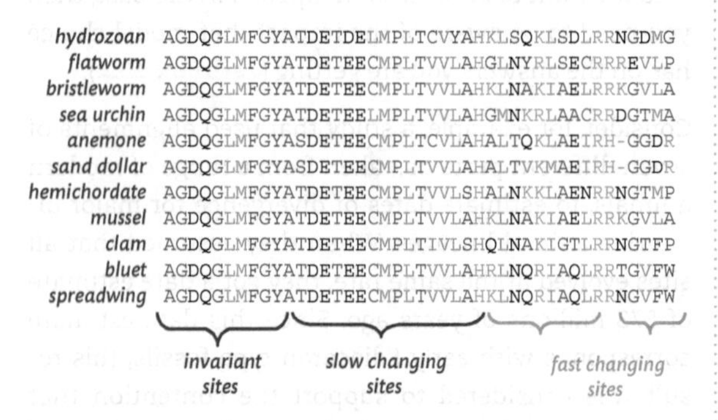

Sites that more readily accept substitutions are, *ipso facto*, more likely to undergo multiple hits than conserved sites. Therefore the average rate of change across all sites is not an accurate predictor of the percentage of sites that will have accumulated multiple hits. If we calculated an average rate across all sites, we would underestimate the number of substitutions that had occurred in the rapidly changing sites. And if we underestimate the number of substitutions that have occurred, then we will underestimate the time taken to accumulate these substitutions, making our estimated dates of divergence too recent. On the other hand, allowing for variation in rates between sites leads to the conclusion that far more substitutions were covered up by multiple hits in the fast changing sites. When these overwritten changes are added to the estimate of genetic distance, the branch lengths are increased. Longer branch lengths move the common ancestor back in time.

The accuracy with which you estimate genetic divergence depends critically on the model of molecular evolution (**TechBox 13.1**). It is your model that allows you to correct for multiple hits and account for variation in rate of change between sites in a sequence. If your model does not adequately capture the patterns of substitution in your sequences, then you may fail to allow for many substitutions that have occurred and been overwritten, or you may falsely infer many more substitutions than have actually occurred. The problem of multiple hits will be exacerbated where relatively few sites in a sequence are able to change. If the free-to-vary sites change again and again, then they will lose historical signal, but this may not be obvious from a consideration of the number of substitutions across the whole alignment, because the sequences will differ only at a small number of variable sites.

How do you know which model is giving you the right answer? Unless you already have reliable information on divergence dates from another source, you don't. The best you can do is give serious consideration to choosing an appropriate model for the sequences you are analysing, and test what effect model choice is having on the answer you are getting, to give you a realistic appreciation of your margins of error.

## Choosing the right sequence

Rates of change vary across the genome. When you choose a sequence to estimate genetic distance for molecular dating, you need to take care to choose a sequence that is evolving at a rate that is just right for the depth of divergence you are considering. Too slow, and there will not be enough substitutions to tell a clear story. Too fast, and the historical story will be erased.

Consider this sequence from the control region of the mitochondria from some flightless bird species from New Zealand (**Figure 14.15**). The control region is known to have a rapid substitution rate so it is often chosen for distinguishing closely related populations or species. Here we have the control region sequence for three kiwis (*Apteryx*: **Figure 14.16**) and three moas (*Dinornis*: see **Case Study 2.2**). We want to use this alignment to get a picture of the patterns of divergence between all six species.

**Figure 14.15**

| | | |
|---|---|---|
| kiwi | Apteryx_mantelli | CAATATGACTAGCTTCAGGCCCATTCATTCCCCGCGCACTACC |
| | Apteryx_rowii | CAACATGACTAGCTTCAGGATCATTCATTCCCCGCGCACTACC |
| | Apteryx_haastii | CAACATGACTAGCTTCAAGACCATTCATTCCCCGCGCACTACC |
| moa | Dinornis_giganteus | GGTATGCGCTAGCTTCAGGAACCCTAAGTCCATATGTCATGCC |
| | Dinornis_novaezealandiae | GGTATGCGCTAGCTTCAGGAACCCTAAGTCCATATGTCATGCC |
| | Dinornis_struthoides | GGTATGCGCTAGCTTCAGGAACCCTAAGTCCATATGTCATGCC |

**Figure 14.16** Kiwis, found only in New Zealand, are flightless birds related to emus and ostriches.
Reproduced courtesy of Phil Brown Photography.

This particular alignment includes substitutions that distinguish the three kiwi sequences from each other. In this short sequence, there are differences between the kiwi sequences at nearly 15 per cent of sites. You can imagine that if you took a long enough sequence you would get a nice sample of informative substitutions for estimating the divergence time between these three kiwi species. But when we look at the same sequence in three moa species, there are no differences between these sequences. This particular alignment would be no good for estimating the divergence between these moa species, because there are not enough substitutions. Conversely, this sequence would not be useful for estimating the relationships between kiwis and moas because there are too many substitutions between them. Half of all the nucleotides in this sequence differ between kiwis and moas, suggesting that the sequence has accumulated so many substitutions since the divergence between these two lineages that many of the sites will be saturated, and will have lost historical signal. This example illustrates that rather than there being 'good' and 'bad' sequences, we have to choose the sequence to suit the problem. This one sequence might have an ideal rate of change for investigating kiwi divergences, but too slow for investigating moa divergences, and too fast for investigating the split between kiwis and moas.

Sequence selection is not just a matter of choosing which gene you want to use. It may also involve choosing which sites you wish to compare within a gene. Here are sequences from four different animal phyla, for a gene frequently used to estimate the timing of the metazoan radiation (**Figure 14.17**). These sequences are part of the 18S ribosomal RNA gene in Onychophora (velvet worms: **Figure 11.13**), Rotifera (wheel animalcules: **Figure 9.11**), Nematoda (roundworms: **Figure 14.18**) and arthropods (here represented by a mosquito: **Figure 5.9**).

Across a single short sequence in the alignment, you can see evidence of variation in rate of substitution. In the first half of the alignment, there are substitutions in around 20 per cent of sites. But in the second part of the alignment, over 90 per cent of sites have substitutions. In this saturated region, we can no longer detect similarities that reveal the shared ancestry of these sequences. If we are interested in the date of divergence between these sequences, then the saturated sites are no help to us, because the number of differences between these sequences is no longer a good indicator of time since divergence. Since saturated sites hold no useful information for us, there is no point including them in our analysis. Indeed, including saturated sites could mislead our analysis. Therefore, it is prudent to exclude saturated regions of an alignment from any analysis aiming to estimate branch lengths that reflect evolutionary time.

*Chapter 10 explains the importance of alignment to establishing homology*

**Figure 14.17**

```
Nematoda    GCCCTAGTTCTAACCGTAAACTATGCCAATTAGCGATCTGCTGGTGTTAATCATGACCCAGCAGG
Arthropoda  GCCCTAGTTCTAACCGTAAACGATGCCAATTAGCAATTGGCAGACGCTACTACATTCCCTGCTCT
Onychophora GCCCTAGTTCTGACCGTAAACGATGCCACCTGACGATCCGCCAGGGTTACTACCATGACTCGGCG
Rotifera    GTCCTAGTTCCAACCATAAACGATGCCAACTAAGCATTAGCTGCCGTTAATTAAACACACAGCTA
```

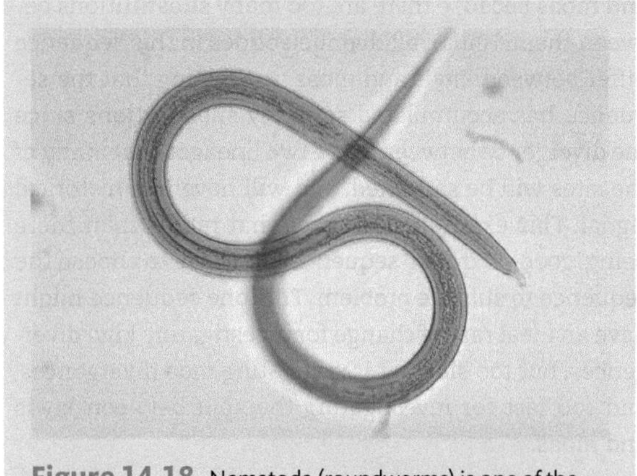

**Figure 14.18** Nematoda (roundworms) is one of the most diverse animal phyla with over a quarter of a million recognized species, at least half of which are parasitic. Nematode larvae often lack key diagnostic features of adult flatworms and can be difficult for a non-expert to identify. DNA barcoding databases can help to take the guesswork out of diagnosing nematode infections.
Photograph: CDC/Dr Mae Melvin.

Another way to improve estimates of the divergence date between lineages is to take the longest possible sequences that you can. Whenever there is variation in the quantity we want to estimate, we will have more confidence in our estimates if we have based them on a large sample size. If you wish to measure body size in a population of penguins, and you measure only two individuals, you cannot be sure that your measurement accurately reflects the population average. If you happened to sample two larger-than-average penguins, then your estimate will be higher than the true population average. In fact, the chance of the average size of your two-penguin sample being the same as the overall average for the population is quite low. But now if you measure sixty penguins, your sample estimate is more likely to be close to the true population average. In a large sample, the occasional large individual is unlikely to derail your estimate, because it will probably be balanced by the occasional small one.

We meet a similar problem when we use samples of substitutions to estimate the time frame of genetic divergence. In most cases, the sequence you choose to analyse is just a tiny fraction of the genome, but you are hoping that the substitutions you observe are a representative sample of the differences between the genomes. If you choose a sequence that has few substitutions over the time period you are considering, then you could end up with an estimate much higher or lower than the real degree of divergence between these sequences. In **Figure 14.15**, the kiwi sequences differ from each other by only one or two substitutions. This in itself is not enough to estimate the depth of divergence between these species. But if you took a sequence fifty times longer than this, then you might get fifty times as many substitutions, and the combined effect of all of these differences might be enough to give a clear picture of the patterns of divergence between them. Longer sequences are the main route to larger sample sizes, and larger sample sizes are often the key to being able to develop and test hypotheses. However, bigger is not always better: caution must also be exercised when interpreting the results from very large datasets, such as whole genome sequences, because not all parts of the dataset may be telling the same story. In this case, some patterns in the data can overwhelm any alternative signal, potentially giving you a great deal of confidence in a misleading conclusion.

## Accuracy and precision

We have seen that too many substitutions can lead to inaccurate estimates of genetic distance due to the problem of multiple hits. Now let's consider the opposite problem: if there are too few changes between our sequences it will decrease the precision of our estimates of the rate of molecular evolution. Not surprisingly, estimates of substitution rates will be best when you have a lot of substitution events to consider.

Here is a hands-on example. Feel your pulse and count the number of heart beats that occur in two seconds. Now I want you to calculate your heart rate, expressed as number of beats per minute. To estimate the number of heart beats per minute, all you have to do is take the number of beats you felt in two seconds then multiply it by 30. The first time I tried this I measured two beats in two seconds. So my first estimate is that my heart rate is 60 beats per minute. I took another 2-second measurement and this time I got three beats in two seconds, so my second estimate is 90 beats per minute. That's a 50 per cent measurement error on a fairly regular heartbeat 'clock'. Now try using a bigger sample of heart beats to estimate your heart rate, by measuring the number of heart beats in 30 seconds. The first time I got 34 beats in 30 seconds, the next time 35. These measurements result in estimates of 68 and 70 beats per minute respectively. That's less than 3 per cent measurement error. So although I was measuring the same quantity (heart rate), and the absolute difference in the measurements was the same in both cases (samples differed by 1 heart beat), the estimate based on a small sample of beats was less precise (had a greater margin of error) than the estimate based on a larger sample of beats.

The same principle applies to estimating the substitution rate. When there are few substitutions between two sequences, every new substitution makes a relatively large difference to the rate estimate. This would not be so bad if molecular change accumulated evenly. But the problem of estimating substitution rate from a small number of observable substitutions is made more difficult by the fact that, rather than ticking regularly like a stopwatch, the molecular 'clock' has an irregular tick rate. In other words, the molecular clock is a sloppy clock.

## *Sloppy clocks*

A sloppy clock is one that does not tick regularly and evenly, like a metronome, but has an erratic tick rate, governed by chance events. Geochronological clocks, based on radioactive decay, are sloppy clocks. For example, potassium-argon (K-Ar) dating utilizes the radioactive decay of potassium isotopes ($^{40}K$) to argon isotopes ($^{40}Ar$) in a rock sample. Argon gas can escape from molten rock, but when the rock solidifies, any $^{40}Ar$ produced by decay of $^{40}K$ will be trapped in the rock sample. As more of the potassium in the rock decays to argon, the ratio of $^{40}K$ to $^{40}Ar$ will drop. At any given point in time, every atom of potassium has the same chance of decaying to argon, but it is impossible to say exactly when any given atom will release an electron and decay. While we cannot say precisely how many atoms will have decayed in a given time period and we certainly can't predict which atoms have decayed and which have not, we can estimate an average rate of decay and use it to convert measures of isotope frequencies to estimates of geological time.

Similarly, although we can describe an average rate of molecular change, and we can estimate the probability that a particular site will change in a given time period, we cannot say exactly which nucleotides in a DNA sequence will change and when. This is because the process of substitution is influenced by chance at many different stages: which mutations arise, when they arise, whether these mutations are lost from the population or go to fixation, whether the populations with particular substitutions persist or are wiped out. It is therefore not surprising that molecular change does not 'tick' regularly like a metronome. Instead, the interval between substitutions varies.

To explore the implications of using a sloppy clock to estimate time, compare the distribution of changes along these two paths (**Figure 14.19**).

**Figure 14.19**

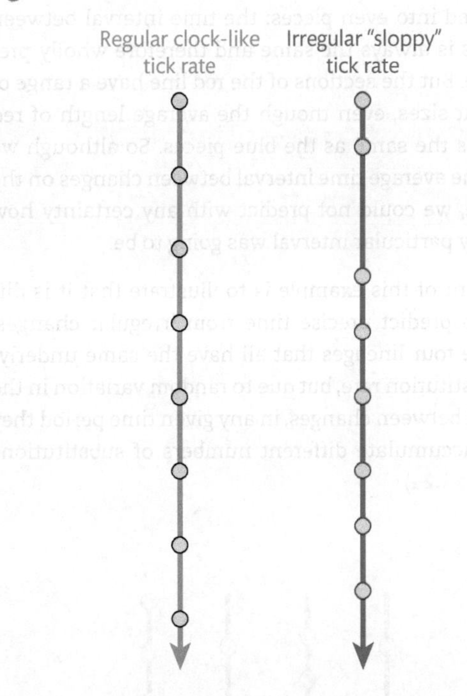


Both of these series of changes have the same overall rate of change (eight changes each in the same time period). So the average length of time between each change is the same in both cases. The difference is that in the blue line, the interval between changes is always the same, but the interval between changes varies on the red line. This is easier to see if we chop the lines up, cutting them at the point a change occurs (**Figure 14.20**). Each piece represents the interval between two changes. If we pile up the pieces we can compare the distribution of time intervals between changes.

**Figure 14.20**

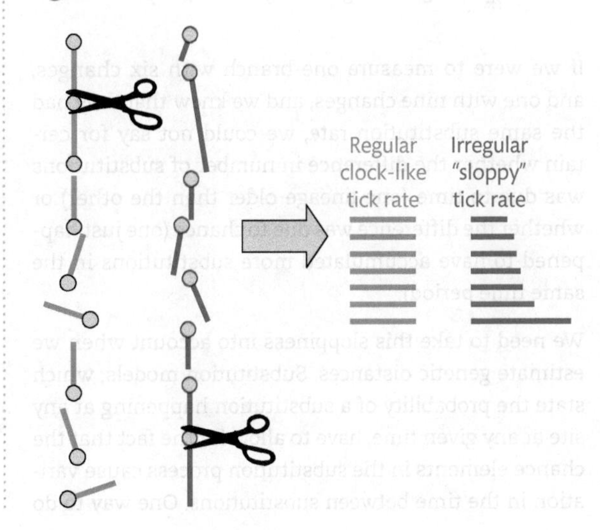

You can see that the blue line, with the regular tick rate, is divided into even pieces: the time interval between changes is always the same and therefore wholly predictable. But the sections of the red line have a range of different sizes, even though the average length of red pieces is the same as the blue pieces. So although we know the average time interval between changes on the red line, we could not predict with any certainty how long any particular interval was going to be.

The point of this example is to illustrate that it is difficult to predict precise time from irregular changes. Imagine four lineages that all have the same underlying substitution rate, but due to random variation in the interval between changes, in any given time period they might accumulate different numbers of substitutions (Figure 14.21).

**Figure 14.21**

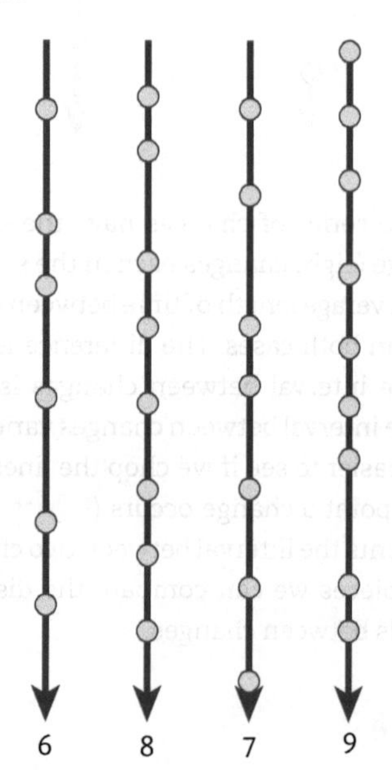


If we were to measure one branch with six changes, and one with nine changes, and we knew that they had the same substitution rate, we could not say for certain whether the difference in number of substitutions was due to time (one lineage older than the other) or whether the difference was due to chance (one just happened to have accumulated more substitutions in the same time period).

We need to take this sloppiness into account when we estimate genetic distances. Substitution models, which state the probability of a substitution happening at any site at any given time, have to allow for the fact that the chance elements in the substitution process cause variation in the time between substitutions. One way to do this is to presume that, although the exact substitution intervals are unknown, they are likely to follow a particular distribution of interval lengths. This distribution might be such that the majority of substitutions occur at an interval that is not very different from the average interval length. Intervals only slightly longer or shorter happen quite frequently, but intervals much longer or shorter occur occasionally.

Imagine cutting the lines in Figure 14.21, piling them up as we did before, then counting the numbers of intervals of each length. To make it easier to see, I have coloured the intervals according to their length (Figure 14.22).

**Figure 14.22**

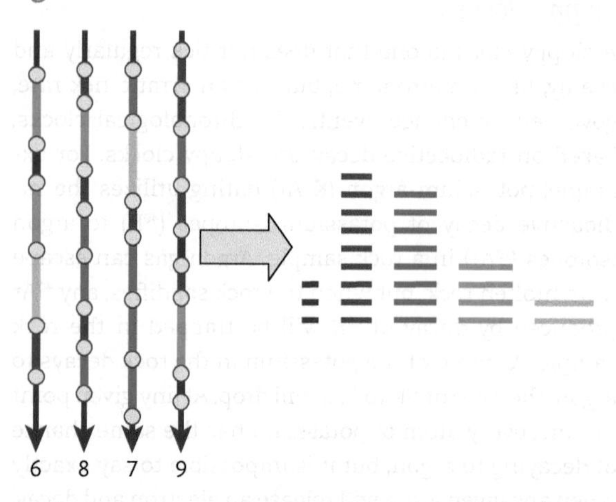


We could draw this as a distribution of interval lengths (Figure 14.23).

**Figure 14.23**

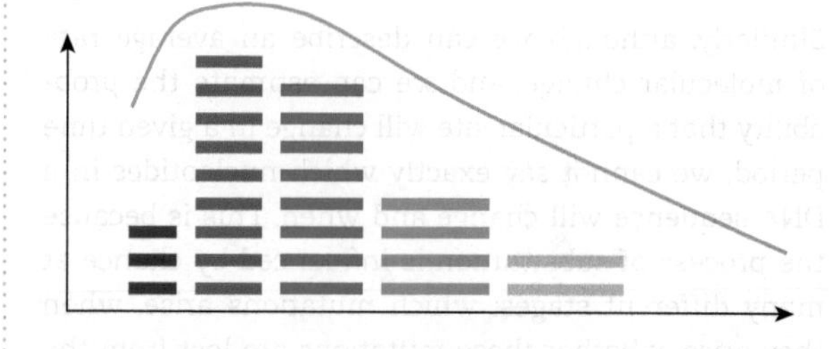

If we had a large enough sample of known interval lengths, then we might be able to describe a distribution that was typical of substitution patterns for our data. With this distribution, we can predict the scatter of substitution events in time along a lineage. We can also use this distribution to estimate the probability that a certain number of substitutions happened in a particular period of time (TechBox 13.1).

## Dating with confidence

If the molecular clock is so sloppy, then how can anyone pretend they can estimate dates of divergence from sequence data? The sloppiness of a clock does not negate its usefulness. You might hope that this Christmas someone gives you a super accurate desk clock synchronized by satellite with an atomic time server. But in the meantime, you have to rely on the cheap watch you got last Christmas. Let's say that your cheap watch ticks on average once a second, but the ticks are not always exactly one second apart. In a ten-second interval, you might get ten ticks, or maybe eight ticks, or sometimes twelve. If you sat there for an hour and recorded the number of ticks per minute, then not many minutes would have exactly 60. But the average ticks per minute, taken over that whole hour, would be around 60. This clock would be imprecise, but it might be reasonably accurate. If the clock said it was 10.30, then, give or take 5 minutes, it might be more or less right. And, in the end, if you don't know what the time is, then even an imprecise clock tells you something useful.

If we know (or can guess) the distribution of interval times between substitutions, then we can predict the likely range of time periods that could have produced the number of substitutions we observe. If we observe six substitutions, and we know the distribution of intervals between substitutions follows the distribution given in **Figure 14.23**, then how long did it take those six substitutions to accumulate? Let's call the shortest interval length one million years duration, and the longest 5 million years. I drew six intervals at random from the distribution, using a random number generator (actually, I used the last column of numbers in the phonebook as a handy source of pseudorandom numbers, but that's good enough for our purposes). Here are the first five branch lengths I produced by drawing at random from the distribution of substitution intervals (**Figure 14.24**). These branches have lengths of 20, 12, 17, 10, and 19 Myr respectively.

**Figure 14.24**

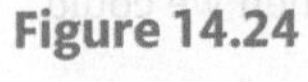

Of course it would be possible to get a very short path (6 Myr) or a very long path (30 Myr), but these would be pretty unlikely to occur (**Figure 14.25**).

**Figure 14.25**

If we continue to draw random sets of six substitution intervals, we could build up a distribution of the possible lengths of time it would take to accumulate six substitutions. To speed this process up, I got a friend to write a little computer program that did the random sampling for me. **Figure 14.26** shows the distribution of 100,000 randomly sampled branch lengths.

**Figure 14.26**

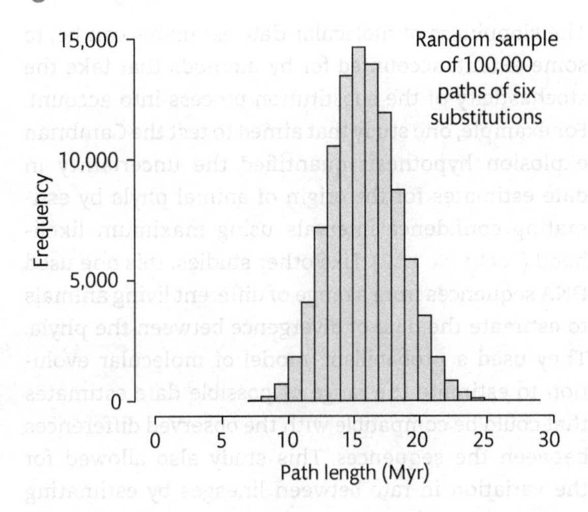


You can see that most of the branch lengths are clustered around the mean value of 16 Myr. Although the random sampling occasionally produces a very short branch (the minimum here is 7) or a very long branch (the maximum is 29 Myr), 95 per cent of the estimates lie between 11 and 22 Myr. So if we observed a lineage with six substitutions, we could not say exactly how old it was, but we could say that, given what we know about this substitution rate for these sequences, the lineage was likely to be somewhere between 11 and 22 Myr old. In this hypothetical case, we based our distribution on observed substitution intervals. But in a normal situation we will not have that information. So most methods assume that substitution intervals vary according to some pre-defined distribution, the dimensions of which can be tailored to the data under consideration (see **TechBox 13.1**).

### *A sloppy clock is better than no clock*

The sloppiness of the substitution process prevents us from saying exactly how long it took to produce the observed number of substitutions. But we can use a distribution of possible path lengths to describe how likely the value is to fall between any given range of path lengths. In other words, we can define a confidence set—the range of values that we are quite confident the true estimate falls within. This is akin to the cheap watch example above, where we said the time was 10.30, give or take 5 minutes. We don't know for sure whether it's 10.25 or 10.35 or somewhere in between, but we are pretty certain it isn't lunchtime yet. Which brings us to an important point: even if you don't know the precise date of divergence, with confidence intervals you can reject certain values as being highly unlikely. In other words, even a sloppy clock can be used to test hypotheses about divergence times.

The sloppiness of molecular date estimates can be, to some extent, accounted for by methods that take the stochasticity of the substitution process into account. For example, one study that aimed to test the Cambrian explosion hypothesis quantified the uncertainty in date estimates for the origin of animal phyla by estimating confidence intervals using maximum likelihood (**TechBox 12.1**). Like other studies, this one used DNA sequences from a range of different living animals to estimate the date of divergence between the phyla. They used a probabilistic model of molecular evolution to estimate the range of possible date estimates that could be compatible with the observed differences between the sequences. This study also allowed for the variation in rate between lineages by estimating separate substitution rates for different phyla using a range of calibration dates (**TechBox 14.1**). The results reflected the range of possible divergence times that could have produced the observed DNA sequence data, given an assumed pattern of sloppiness of the clock and allowing for variation in substitution rate between lineages. The confidence intervals were embarrassingly large (**Figure 14.27**).

These estimates are horribly imprecise. The confidence intervals go so far back into the past that they cannot reject the possibility that the animal kingdom originated billions of years ago. At this point you might throw up your hands and say that this just proves that molecular dating is a hopeless cause. But even imprecise estimates can be informative if they are accurate—that is, if the confidence intervals, however wide, contain the true value. In this case, even though the confidence intervals span over a billion years, they do not include the early Cambrian or latest Precambrian. These estimates lack the precision that would allow us to say exactly when the animal kingdom diversified. But, if these estimates are accurate, then they can tell us when the animal kingdom didn't diversify: the results obtained using these data, methods and models of evolution suggest that these lineages must have diverged long before the first undisputed fossil evidence of metazoans. Of course, if the results are inaccurate, due to problems

**Figure 14.27**

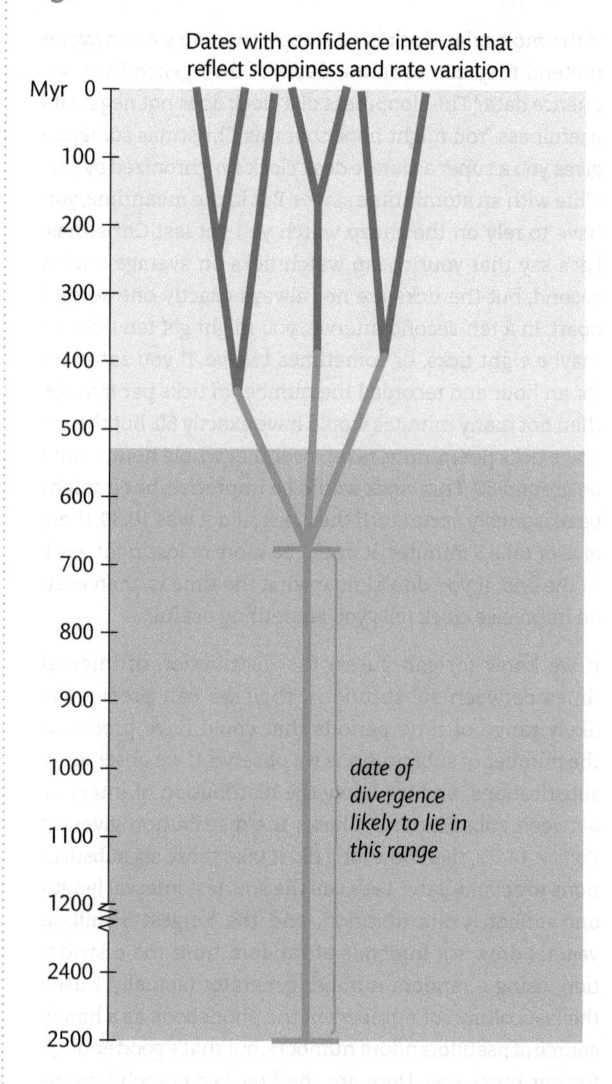


with the data, methods or assumptions, then we could be led astray.

Molecular dates are certainly imprecise, but that doesn't necessarily make them unusable. Far more important is the question of whether molecular dates are inaccurate, in which case the confidence intervals, however wide or narrow, may fail to contain the true date of divergence. So far we have considered the imprecision arising from the sloppiness of the tick rate. But the pattern of substitutions is more than sloppy. As we saw in Chapter 13, substitution rates can also vary consistently between lineages, and this lineage-specific variation in rate can generate inaccurate molecular date estimates.

# Molecular dating

*...ignorance more frequently begets confidence than does knowledge: it is those who know little, and not those who know much, who so positively assert that this or that problem will never be solved by science*

Darwin, C. (1871) *The descent of man, and selection in relation to sex*. John Murray

The earliest molecular dating studies were applied to sequences for which the rate of substitution seemed to have been constant in many lineages over a long time. Typically, these studies either just assumed rate constancy, or supported the claim for uniform rates by plotting observed sequence differences between lineages of known divergence times. If we know that the rate of molecular evolution has been the same in all lineages, then we only need a single calibration rate with which to date the rest of the tree. To estimate the date of the animal radiation, some researchers have sought to construct a database of sequences that evolve at a slow and steady rate, rejecting any genes that show significant rate variation between lineages. When the rate of molecular evolution in these apparently clock-like sequences is calculated using known dates of divergence, the molecular date estimates obtained for the radiation of the animal phyla are very variable, but all much older than the Cambrian (ranging from 630 Myr to 1,200 Myr: **Figure 14.28**).

**Figure 14.28**

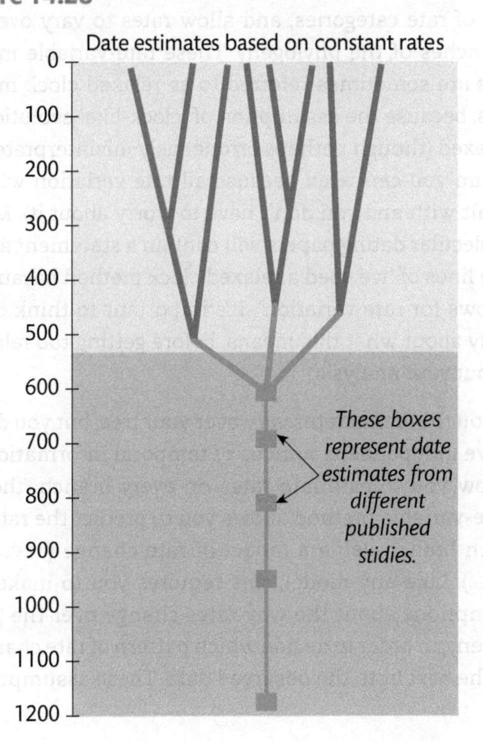


The problem with this approach is that the tests used to detect and reject sequences that vary in rate between lineages sometimes have relatively low power to detect rate variation. This means that when they are applied to sequences with relatively few substitutions—such as short sequences with slow rates of divergence—then the tests may fail to reject all of the rate-variable alignments. If rate-variable sequences are included in a molecular dating study that relies on the assumption of rate constancy, then the calibration rate will not be an accurate reflection of the amount of time taken to produce the observed substitutions, so the date estimates could be inaccurate.

In most cases, we cannot rely on finding truly clock-like sequences, so we will need to find a way of allowing for rate variation in our date estimation procedures. One way of doing this is to use multiple calibration dates to estimate substitution rates separately for different parts of our phylogeny. If we want to estimate lineage-specific rates directly from our alignments, then we will need at least one calibration for every different rate we estimate. This may be tractable if rate changes are relatively rare, for example if each phylum tends to have a characteristic rate of molecular evolution. But we have seen that there are many factors that influence rate even between closely related species (e.g. **Case Study 13.2**). In fact, significant rate variation exists at all levels of the animal kingdom, between species, between families, between phyla.

***Variation in substitution rate between lineages is discussed in Chapter 13***

This brings us to two important conclusions. Firstly, constant rates of substitution may be a feature of some datasets, but clock-like molecular evolution should be regarded as the exception, rather than the rule. There are so many factors that influence the rate of substitutions that we should expect rate variation to occur in many, if not most, datasets. Secondly, if rates can vary consistently between species, then we should expect rates to evolve along lineages. This is hardly surprising given all that we have learned about the influence of species traits on the rate of molecular evolution (Chapter 13). If we compare two species of snake and we find that one has a shorter generation length than the other, then we do not expect that the difference in generation time evolved the instant that the two ancestral snake populations became reproductively isolated. Instead, we might assume that the two populations became gradually more distinct in generation time. We might also expect that substitution rates could have evolved in concert with generation time in these species: as the generation length became shorter in one species, its rate of substitution might have increased. If rates of molecular evolution evolve along phylogenies, just as other species traits do, then rates could be in a continuous state of change over the tree.

If substitution rates evolve along phylogenies, then we would not be surprised to find that every branch in our phylogeny has a different rate of molecular evolution. If we wanted to estimate these rates directly from our sequence data, then we would need a calibration for every branch of the tree. This would be a marvellous way to proceed, but there are few groups that have such a well-known evolutionary history (and, of course, if we already know when all the lineages diverged then we don't really need molecular dates for that group). If we don't have calibrations for every branch in the tree, then we cannot estimate rates directly. The only way to allow for frequent rate change, if we cannot measure it directly, is to predict changes in substitution rate using a model of rate change.

## Modelling rate change

We have already seen how substitution models can be used to predict the occurrence of substitutions that cannot be directly observed (**TechBox 13.1**). The same principle can be applied to estimating changes in substitution rate itself. Most molecular dating studies now use methods that allow rates to vary over the tree, finding the pattern of substitution rates that maximizes some measure of the plausibility of the solution. If the rates along the tree can be predicted, then one or more calibrations can be used to convert the branch lengths into measures of evolutionary time.

This approach sounds simple, but in reality, a large number of different assumptions must be made in order to decide between different possible solutions. If we allow rates to change on any branch of the phylogeny, then when we estimate a certain number of substitutions between two lineages, we have to decide whether to reconstruct that branch length with a slow rate (therefore a deep divergence) or a fast rate (therefore a shallow divergence) or anywhere in between (**Figure 14.29**).

**Figure 14.29**

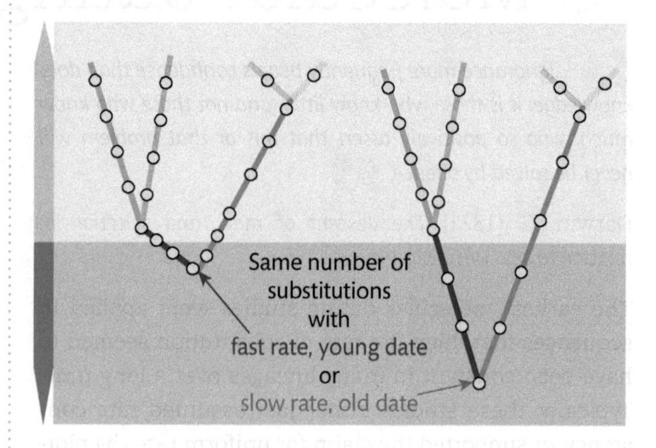


All molecular dating methods must make assumptions about the way substitution rates behave over phylogenies. Some methods, referred to as strict clock methods, start with the assumption that average substitution rates are much the same in every lineage at all time periods (**TechBox 14.2**). An alternative is to accept that different parts of a phylogenetic tree have different rates, but that some branches will share a common rate, based on their place in the phylogeny or due to the presence of some rate-influencing trait. Methods that seek to assign branches of a phylogeny to a number of defined rate categories are often referred to as local-clock methods. Or you can say that you don't expect your tree to conform to either a uniform rate or a predictable set of rate categories, and allow rates to vary over all branches of the phylogeny. These rate-variable methods are sometimes referred to as relaxed clock methods, because the assumption of 'clock-like' evolution is relaxed (though perhaps erroneously misinterpreted to mean you can relax because all rate variation will be dealt with and you don't have to worry about it). Many molecular dating papers will contain a statement along the lines of 'we used a relaxed clock method because it allows for rate variation'. It's important to think carefully about what this means, before getting too relaxed about your analysis.

If you think that rates vary over your tree, but you don't have independent sources of temporal information to allow you to estimate rates on every branch, then a rate-variable method allows you to predict the rate on each branch using a model of rate change (**TechBox 14.2**). Like any model, this requires you to make assumptions about the way rates change over the phylogeny, in order to decide which pattern of rate changes is the best fit to the observed data. These assumptions include some kind of function that describes what patterns of changes in rate are most likely, or the distribution of substitution rates that you expect to find (**TechBox 14.2**). But there are also a great many other less obvious assumptions concerning the underlying evolutionary processes that created these sequences. You need these statements to decide which kinds of solutions are more believable, otherwise you cannot solve the problem illustrated in **Figure 14.29**: how to choose between describing your data using a fast rate and a recent date, or a slow rate and an older date. Molecular dating methods that allow changes in rate are very complex, although the casual user of molecular dating software may be largely unaware of all the statistical machinery that is being employed to produce their date estimates.

For example, one of the ways that rate-variable methods weigh the plausibility of different solutions is to have a model of speciation and extinction of lineages that lets you compare different tree shapes (**TechBox 13.2**). When you apply one of these models, then you are effectively saying, before you even start your analysis, that you don't think a solution that involves some extremely long branches and many extremely short branches is as believable as one that has an even distribution of branch lengths, and that a tree where all the nodes (branch points) are crowded into one part of the tree is not as likely as one in which the nodes are distributed throughout the phylogeny. It's important to think about this when you are interpreting the result you get. If you are testing the hypothesis that all of the animal phyla originated in the early Cambrian, do you really want to start by saying that you expect that rate of diversification is even over time?

## Challenging assumptions

Many rate-variable methods also start with an assumption that taxon sampling is random, so that any one of the descendants of the lineage under consideration is just as likely to be included in the tree as any of the others. In reality, this is rarely the case. If we were to conduct random sampling on the metazoan tree, then at least a third of our sampled species would be arthropods (which would be mostly beetles), and most of the rest would be molluscs and chordates. But this does not reflect the way that most people construct datasets for dating the origins of animal phyla. Instead, most molecular dating studies aim to include representatives of as many phyla as possible, even though some phyla have orders of magnitude more species than other phyla. In this case, the assumptions of the method may be violated before you even start the analysis.

The assumptions inherent in rate-variable methods can make a big difference to the outcome. For example, researchers applied a rate-variable method to the same data that was used to produce the estimates in **Figure 14.27** and came up with a very different answer. Whereas the local clock method produced estimates greater than 680 Myr ago (with massive confidence intervals), the relaxed clock method on the same data produced date estimates between 498 and 616 Myr (with much narrower confidence limits). When faced with the same data, why did the local clock method reconstruct the branch lengths as a slow rate and deep date, whereas the relaxed clock method reconstructed the branch lengths as a fast rate and recent date? The difference in date estimates can only be coming from the assumptions made in the analyses, because it's not coming from the data.

Exploration of the results of the relaxed clock study showed that the young date estimates rested critically on the assumption of a constant rate of lineage speciation and extinction throughout the phylogeny, and also on the assumption that the sequences used were a random sample of the tips of the phylogeny. Both of these assumptions were violated for this particular dataset due to the way pairs of species were selected to represent particular calibrated nodes in the animal phylogeny. Highly diverse clades, such as Tetrapoda (21,000 extant species), were represented by the same number of tips in the tree as less diverse lineages, such as Dipnoi (lungfish, six extant species). The fact that these species were not a random sample, but chosen to define particular deep splits in the animal phylogeny, means that the actual distribution of nodes within the phylogeny will be very different from the expectation that was described in the priors (see **TechBox 14.2**). This might not matter if we could be sure that the signal from the data could overwhelm that from the priors, but in this particular case, investigation demonstrated it was the assumptions made in the analysis that were telling the story, not the data.

The large number of different parameters included in these models makes it possible to come up with many alternative solutions by varying the assumptions made. So it is not surprising that when different research groups have applied rate-variable molecular dating methods to estimating the timing of the divergence of the animal phyla, they have come up with a range of estimates. Of course, these studies have also used different sequences, species and calibrations, which may also add to the wide variation in the date estimates, which range from just before the Cambrian to a billion years ago (**Figure 14.30**).

**Figure 14.30**

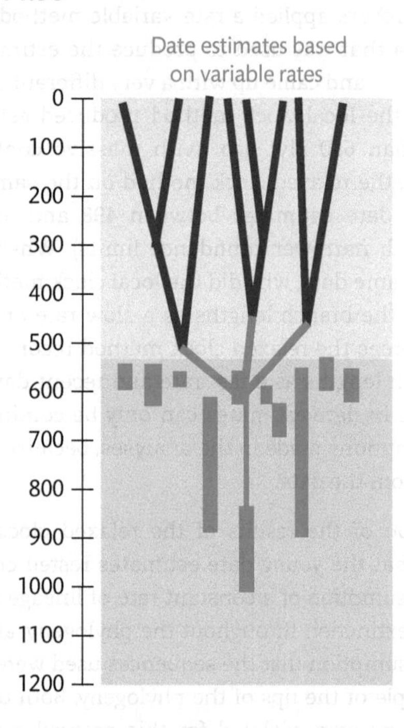


Many of the more recent rate-variable date estimates are distributed between 630 and 540 million years ago. On the assumption that the Ediacarans represent early members of lineages leading to the modern metazoan phyla, this clustering of recent estimates gives some researchers confidence that we are converging on the truth. However, if you look carefully at the analyses used to produce these date estimates, you will find that many of these studies start with an assumption that the dates are unlikely to be older than the Ediacaran period, so older dates are rejected as implausible.

Rate-variable molecular dating methods often won't work without bounds being set on the allowable range of dates, because otherwise almost any answer could be entertained from older than the earth to the day before yesterday. Because of this, many studies set bounds on the root (deepest divergence in the tree), so the researcher must decide the lowest possible date estimate before they begin the analysis. This means that while the upper limit (minimum ages) of phylum divergences are set by the fossils found in the Cambrian, the lower limit (maximum ages) of phylum divergences are bounded by the researchers' prior belief. For example, one molecular dating study used a range of calibrations in a Bayesian analysis (**TechBox 13.2**), including a prior belief that the date for the metazoan radiation was not older than 635 Myr. Not surprisingly, their molecular date estimates suggested a metazoan radiation occurring around 635 Myr ago. In such cases, you must ask yourself whether you are asking the data a question or telling it the answer.

Given that molecular date estimates vary so much with the data, methods or assumptions employed, how can we tell which estimates are right and which are wrong? Since we don't know which model is right, the best approach is to explore the robustness of our results to different assumptions, to compare the estimates to other lines of evidence, and to consider the implications of the dates in light of what we know about evolution at the molecular and organismal levels. Keep the assumptions you make in the analysis in the front of your mind when considering the results, and be careful to recognize when prior beliefs are dictating the outcomes, such that the analysis ends up just telling us what we want to hear. Most importantly, we must recognize the sources of error in molecular date estimates and use these to describe a range of possible date estimates, rather than aiming to produce a single magic number.

The difference patterns of rate variation inferred from each of these models serves to highlight that you should not think of these methods as simply reading patterns of rate variation from the data. Instead, these methods are working out what parameters they need to make a given model fit the data. Remember that 'best fit' under any given model does not necessarily mean 'good fit'. It might be an absolutely lousy description of the real history of your data, and it might be giving you completely misleading results. But if all you do is put your data in the machine and hit the 'Go!' button, you will never know. As with all analyses discussed in this book, you need to stop and think how much you care about finding out the truth. If all you care about is publishing your work, then use whichever method is currently the most fashionable. If you really want to know what happened, then explore your data and work out just what your analyses are really telling you, and humbly acknowledge the limits to your ability to make certain statements about the historical events that created your data.

## *Why did the Cambrian explode?*

> *The Cambrian, so beautifully documented . . ., was clearly a special time when animal life exploded into the wide range of forms that we associate with the modern faunas of today. So what does it mean to find molecular estimates of animal . . . divergences that are so much older? That is the challenge, both to paleontologists and phylogeneticists*

Levinton, J. (2001) *Genetics, Paleontology, and Macroevolution*, 2nd edn. Cambridge University Press

At the present time, we cannot use molecular data to put precise date estimates on the divergence of the animal phyla. But this does not make molecular information useless. After all, we do not reject the fossil record when faced with evidence that it is less than perfect, such as the 80 Myr hiatus in the coelacanth record, or the near absence of flatworm fossils. Instead, palaeontologists work to derive as much information as possible from the palaeontological record, despite its obvious sloppiness.

We know that there is information about the timing of animal diversification in the molecular record. We can see this in the short alignment shown in **Figure 14.9**. Fossil evidence suggests that the two damselfly lineages have been separated for around 120 Myr while the clam and mussel last shared a common ancestor at least 485 Myr ago. Sure enough, there are far fewer differences between the more recently diverged damselflies that there are between the clam and the mussel. On the whole, we can reasonably expect that sequence alignments should be able to provide us with information about divergence times. But, as we have seen, molecular date estimates are usually imprecise and potentially inaccurate, so we must exercise caution when interpreting molecular evidence for the timescale for evolution, just as we exercise caution when interpreting tempo and mode of evolution from the fossil record.

This is not the place to review the debate about the nature, causes and consequences of the metazoan radiation, discussions of which fill greater volumes than this one. But we can summarize the evidence from molecular data as follows: when we compare DNA sequences from living members of different animal phyla, we find there are more substitutions between them than we would expect from only half a billion years of evolution, based on what we can observe concerning the range of substitution rates in modern animal lineages. There are several possible explanations for this observation. One is that phylum-level lineages diverged long before the Cambrian, although we have no unambiguous evidence of their earliest history. To support this hypothesis, we would have to think of a good reason why early members of lineages leading to modern phyla have not been found, or not recognized, from the Precambrian geological record. An alternative explanation is that rates of molecular evolution were so much faster in the earliest part of the animal radiation that molecular dates based on post-Cambrian substitution rates greatly overestimate the origin of these lineages. To believe this hypothesis, we would need to come up with a good reason why all early animal lineages would have had systematically faster rates of molecular evolution than all of their descendants (**Figure 14.31**).

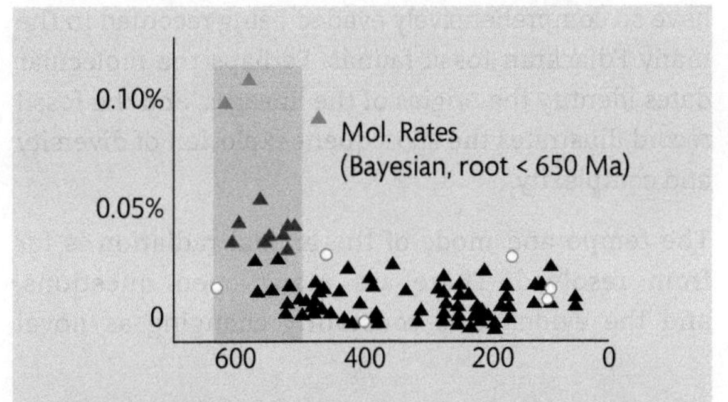


**Figure 14.31** Fast and young: this graph shows the rates estimates produced by a Bayesian 'relaxed clock' analysis of arthropod sequences, with a prior belief that the root cannot be older than 650 Ma (Mega-annum, or million years ago). This forced the analysis to fit very high rates to the basal lineages in the arthropod phylogeny, so that the earliest arthropods were estimated to have substitution rates five times greater than any post-Cambrian arthropod lineage. What could make rates run so much faster in the Cambrian? When they relaxed the root age constraint, they estimate a root age of around 940 Ma, with substitution rates in the range of subsequent lineages. What could explain such a deep divergence despite lack of any recognizable arthropod fossils in the Precambrian? How can we decide between these two alternative descriptions for the same dataset—surprisingly old dates and unsurprising rates, or surprisingly fast rates and unsurprising dates?

Reprinted from Lee, M. S. Y. et al. (2013) *Current Biology*, Volume 23(19), pages 1889–1895. Copyright © 2013, with permission from Elsevier.

Sometimes, when there are two extreme hypotheses, the answer is eventually found to lie somewhere in between. Perhaps that is how different ideas about the tempo and mode of the diversification of the animals will be resolved. Some palaeontologists feel that a range of evidence points to an origin of animal lineages between 700 and 600 million years ago, based on putative fossilized embryos, possible animal traces, and suggestive biomarkers—plus this may coincide with the rise in multicellular forms of fungi and plants. While molecular estimates cover a bewilderingly large range of possible phylum divergence dates, the majority seem to centre around this period, but as we have seen, this may be more a result of shared biases in analysis rather than indicating a convergence on the true date. Maybe, as we uncover more fossil evidence and learn more about molecular evolution, we might find that the fossil and molecular evidence is converging. However, even if the roots of the animal radiation lie deep in the

Precambrian, there seems little doubt that there was a remarkable transformation across the Cambrian boundary. It is difficult to see how metazoan ancestors possessed of shells, carapaces, teeth or spikes could have so comprehensively evaded being recorded in the many Ediacaran fossil faunas. Perhaps the molecular dates identify the origins of the lineages, and the fossil record illustrates the subsequent explosion of diversity and complexity.

The tempo and mode of the animal radiation is far from resolved. There are many open questions, and the evidence is constantly changing as novel discoveries are made, new data analysed, and old observations re-evaluated. Molecular data may play some role in resolving these issues, but it is unlikely to provide all the answers. Hypothesis testing in evolutionary biology usually requires the combination of as many different lines of evidence, and the application of as many alternative modes of analysis, as possible. For myself, I remain truly perplexed by both the fossil record and the molecular record of animal evolution, but I am looking forward to seeing what emerges in the years ahead.

# Conclusion

There is temporal information in DNA: the longer genomes evolve separately the more different they will be. We cannot predict evolutionary time from molecular data with unswerving accuracy or deadly precision. However, in science we are accustomed to the idea of presenting estimates with confidence intervals that reflect our certainty that the true value lies within a range of estimates. For molecular dates, we should expect the confidence intervals to be quite wide, because there are so many things that can influence the number of substitutions that accumulate in a lineage over time, not only the sloppiness of the substitution process, but also the variation in rates between lineages. Given that we rarely know what the underlying rates are, our confidence limits must include our uncertainty in the patterns of rate evolution, as well as the possible effect on date estimates of the various assumptions inherent in our models. Sometimes our confidence intervals might be formally expressed as a distribution of values derived from statistical analysis, other times they may represent the upper and lower estimates obtained by varying the assumptions on which the date estimates are based.

The fossil record remains the primary source of information on the evolutionary past. But palaeontological dates are not available for all taxa. The majority of lineages have no fossil record, so we must find other ways to investigate the tempo and mode of evolution. Furthermore, the fossil record can give a misleading impression of the tempo and mode of evolution in some cases, so it would be handy to have an independent timescale to contrast it to and identify areas of discrepancy that require further investigation. We are liable to be led astray if we expect either the palaeontological or molecular record to always give us perfect point estimates of divergence dates. Instead, the most robust approach to investigating the tempo and mode of evolution is to combine the strengths of both the palaeontological and molecular records, along with as many other sources of information as possible, and to constantly examine our findings in light of the assumptions we have made in our analyses. Improving our understanding of the tempo and mode of molecular evolution will help us to unlock the wealth of information on evolutionary past and processes that is written in the genome.

# Points to remember

## Confidence

- The simultaneous first appearance of many animal phyla has been interpreted as rate of evolution greater than any other known lineage or period.
- Fossilization is rare and biased toward certain kinds of organisms, time periods and locations.
- While the Cambrian explosion of animal diversity has been attributed to lack of early animal fossils, this explanation is difficult to reconcile with the rich record of soft-bodied Ediacarans.
- Molecular data can be used to contrast the Cambrian explosion hypothesis (all phyla originate in or just before the Cambrian period) to the Precambrian slow burn (phyla originate long before the Cambrian).
- Molecular change is a 'sloppy clock' because the interval between substitutions varies, making time estimates imprecise.
- Given a distribution of intervals between substitutions, we can estimate the range of time intervals that could have produced the observed number of substitutions.
- If date estimates are imprecise (wide confidence limits) but accurate (confidence limits contain the true value), they can be used to reject dates as incompatible with the data, given the assumptions of the analysis.

## Molecular dating

- The longer two lineages evolve separately, the more differences will accumulate between their genomes, so the amount of genetic differences between species should reflect the time elapsed since they last shared a common ancestor.
- Estimating divergence dates requires a way of inferring the number of substitutions that have occurred and the rate at which they accumulated.
- Failing to account for biases in substitutions across sites or between types of changes can bias date estimates.
- Molecular dating requires selecting sequences with appropriate rates of change to give enough substitutions to be informative without losing signal through multiple hits.
- Strict clock studies rely on the assumption that the average rate of substitutions is the same in all lineages under consideration.
- Local clocks assign each branch in the phylogeny to a rate category, based on a priori definition, relationships, or through model fitting.
- Relaxed clock methods allow all branches to have their own rates, using a model of rate change to select the best-fitting pattern, or drawing rates from an assumed distribution.
- Rate-variable methods rely on a large number of assumptions about the evolutionary process, which may include prior beliefs about divergence dates.

# Ideas for discussion

**14.1** Could you design a molecular dating analysis to estimate the age of the common ancestor of all living species?

**14.2** Could early metazoans have had a much higher rate of molecular evolution than any contemporary species? How could you test this hypothesis?

**14.3** In cases where we have no other temporal evidence to compare them to, can we trust molecular dates?

## Sequences used in this chapter

**Table 14.1: Methionine adenosyltransferase sequences used in Figure 14.9**

| Description from GenBank entry | Accession |
|---|---|
| *Obelia sp.* KJP-2004 methionine adenosyltransferase mRNA, partial cds. | AY580240 |
| *Stylochus sp.* KJP-2004 methionine adenosyltransferase mRNA, partial cds. | AY580254 |
| *Chaetopterus sp.* KJP-2000 methionine adenosyltransferase mRNA, partial cds. | AY580185 |
| *Metridium senile* methionine adenosyltransferase mRNA, partial cds. | AY580247 |
| *Strongylocentrotus purpuratus* methionine adenosyltransferase mRNA, partial cds. | AY580282 |
| *Dendraster excentricus* methionine adenosyltransferase mRNA, partial cds. | AY580198 |
| *Saccoglossus kowalevskii* methionine adenosyltransferase mRNA, partial cds. | AY580278 |
| *Mytilus edulis* methionine adenosyltransferase mRNA, partial cds. | AY580273 |
| *Nucula proxima* methionine adenosyltransferase mRNA, partial cds. | AY580233 |
| *Lestes congener* methionine adenosyltransferase mRNA, partial cds. | AY580226 |
| *Enallagma aspersum* methionine adenosyltransferase mRNA, partial cds. | AY580212 |

**Table 14.2: Mitochondrial control region sequences used in Figure 14.15**

| Description from GenBank entry | Accession |
|---|---|
| *Apteryx mantelli* mitochondrion, partial genome. | AY016010 |
| *Apteryx australis rowii* haplotype Okarito 5 ATPase subunit 8 and ATPase subunit 6 genes, complete cds; mitochondrial genes for mitochondrial products | AY150600 |
| *Apteryx haastii* mitochondrion, complete genome | NC_002782 |
| *Dinornis giganteus* mitochondrion, complete genome | NC_002672 |
| *Dinornis novaezealandiae* specimen-voucher CM_Av30497 mitochondrial control region, partial sequence | AY299875 |
| *Dinornis struthoides* specimen-voucher CM_Av8872 mitochondrial control region, partial sequence | AY299874 |

**Table 14.3: 18S ribosomal RNA sequences used in Figure 14.17**

| Description from GenBank entry | Accession |
|---|---|
| *Stilbonema majum* 18S rRNA gene | Y16922 |
| *Anopheles maculatus* 18S ribosomal RNA gene, complete sequence | AF440198 |
| *Euperipatoides rowelli* voucher NZAC03006004 18S ribosomal RNA gene, partial sequence | GQ911186 |
| *Brachionus patulus* 18S ribosomal RNA gene, complete sequence | AF154568 |

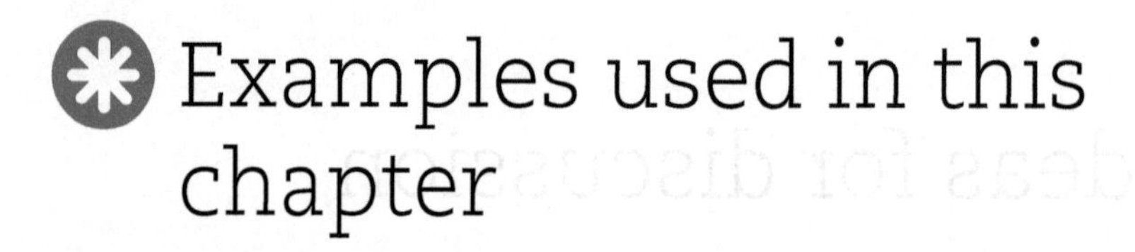

## Examples used in this chapter

This list includes the sources of date estimates used in figures.

Antcliffe, J. B., Callow, R. H., Brasier, M. D. (2014) Giving the early fossil record of sponges a squeeze. *Biological Reviews*, Volume 89, page 972.

Aris-Brosou, S., Yang, Z. (2002) Effects of models of rate evolution on estimation of divergence dates with special reference to the metazoan 18S ribosomal RNA phylogeny. *Systematic Biology*, Volume 51, page 703.

Aris-Brosou, S., Yang, Z. (2003) Bayesian models of episodic evolution support a late precambrian explosive diversification of the metazoa. *Molecular Biology and Evolution*, Volume 20, page 1947.

Ayala, F. J., Rzhetsky, A., Ayala, F. J. (1998) Origin of the metazoan phyla: molecular clocks confirm palaeontological estimates. *Proceedings of the National Academy of Sciences USA*, Volume 95, page 606.

Bromham, L., Rambaut, A., Fortey, R., Cooper, A., Penny, D. (1998) Testing the Cambrian explosion hypothesis by using a molecular dating technique. *Proceedings of the National Academy of Sciences USA*, Volume 95, page 12386.

Bromham, L., Rambaut, A., Hendy, M. D., Penny, D. (2000) The power of relative rates tests depends on the data. *Journal of Molecular Evolution*, Volume 50, page 296.

Blair, J. E., Hedges, S. B. (2005) Molecular phylogeny and divergence times of deuterostome animals. *Molecular Biology and Evolution*, Volume 22, page 2275.

Douzery, E. J. P., Snell, E.A., Bapteste, E., Delsuc, F., Philippe, H. (2004) The timing of eukaryotic evolution: Does a relaxed molecular clock reconcile proteins and fossils? *Proceedings of the National Academy of Sciences USA*, Volume 101, page 15386.

Edgecombe, G. D., Giribet, G., Dunn, C. W., Hejnol, A., Kristensen, R. M., Neves, R. C., Rouse, G. W., Worsaae, K., Sorensen, M. V. (2011) Higher-level metazoan relationships: recent progress and remaining questions. *Organisms Diversity & Evolution*, Volume 11, page 151.

Gu, X. (1998) Early metazoan divergence was about 830 million years ago. *Journal of Molecular Evolution*, Volume 47, page 369.

Lee, M. S. Y., Soubrier, J., Edgecombe, G. D. (2013) Rates of phenotypic and genomic evolution during the Cambrian explosion. *Current Biology*, Volume 23 (19), page 1889.

Lynch, M. (1999) The age and relationships of the major animal phyla. *Evolution*, Volume 53, pages 319–325.

Peterson, K. J., Butterfield, N.J. (2005) Origin of the Eumetazoa: Testing ecological predictions of molecular clocks against the Proterozoic fossil record. *Proceedings of the National Academy of Sciences USA*, Volume 102, page 9547.

Peterson, K. J., Lyons, J. B., Nowak, K. S., Takacs, C. M., Wargo, M. J., McPeek, M. A. (2004) Estimating metazoan divergence times with a molecular clock. *Proceedings of the National Academy of Sciences USA*, Volume 101, page 6536.

Peterson, K. J., Cotton, J. A., Gehling, J. G., Pisani, D. (2008) The Ediacaran emergence of bilaterians: congruence between the genetic and the geological fossil records. *Philosophical Transactions of the Royal Society B: Biological Sciences*, Volume 363, page 1435.

Rehm, P., Borner, J., Meusemann, K., von Reumont, B. M., Simon, S., Hadrys, H., Misof, B., Burmester, T. (2011) Dating the arthropod tree based on large-scale transcriptome data. *Molecular Phylogenetics and Evolution*, Volume 61, page 880.

Rota-Stabelli, O., Daley, A. C., Pisani, D. (2013) Molecular timetrees reveal a Cambrian colonization of land and a new scenario for ecdysozoan evolution. *Current Biology*, Volume 23, page 392.

Sperling, E. A., Pisani, D., Peterson, K. J. (2011) Molecular paleobiological insights into the origin of the Brachiopoda. *Evolution & Development*, Volume 13, page 290.

Wang, D. Y-C., Kumar, S., Hedges, S. B. (1999) Divergence time estimates for the early history of animal phyla and the origin of plants, animals and fungi. *Proceedings of the Royal Society of London B.*, Volume 266, page 163.

HEROES OF THE GENETIC REVOLUTION

14

# Andrew Rambaut

**Figure 14.32** Andrew Rambaut and his son Hamish on Port Meadow in Oxford: a fine demonstration of genetic inheritance.

Photograph: Jo Kelly.

**EXAMPLE PUBLICATIONS**

Rambaut, A., Posada, D., Crandall, K. A., Holmes, E. C. (2004) The causes and consequences of HIV evolution. *Nature Reviews Genetics*, Volume 5, pages 52–61.

Hedge, J., Lycett, S. J., Rambaut, A. (2013) Real-time characterization of the molecular epidemiology of an influenza pandemic. *Biology Letters*, Volume 9, page 20130331.

Andrew Rambaut's introduction to computational biology came about when he won a competition to create a 'biomorph' using code supplied with Richard Dawkin's 1986 book *The Blind Watchmaker*. This led him to study zoology at Edinburgh, then to a doctorate at the University of Oxford (**Figure 14.32**). He joined Paul Harvey's evolutionary biology group just as they were beginning to exploit the power of DNA sequence analysis to uncover patterns of evolution. Rambaut played a key role in developing software for detecting general evolutionary patterns from phylogenies, including programs for making statistically independent comparisons in evolutionary biology and for detecting changes in diversification rate of lineages over time. He also wrote a widely used program for simulating the evolution of DNA sequences along phylogenies, and a really nifty alignment editor. While these methods are applicable to inferring evolutionary patterns from the phylogeny of any group of organisms, Rambaut's research has generally focused on the application of these methods to understanding emerging diseases, such as HIV (human immunodeficiency virus), dengue fever, SARS (severe acute respiratory syndrome), Hepatitis C and West Nile viruses.

Rambaut has been at the forefront of many new methods in molecular dating. He developed one of the first variable-rate maximum likelihood dating methods, and applied it to estimating the timing of the radiation of animal phyla. He also devised a dating method that uses non-contemporaneous sample dates of sequences to estimate rates of molecular evolution along virus phylogenies. He co-developed the phylogenetic inference package, BEAST, that implements a range of techniques in molecular dating within a Bayesian framework. This was one of the first molecular dating methods that allowed users to specify various models of rate change along the phylogeny, and to treat calibration dates as prior probability distributions rather than point estimates.

Rambaut's programs begin as tools developed for his own research, but he makes his programs freely available for anyone who wishes to use them. This is a fine example of the way that academia should ideally operate, with the products of research and development being shared to speed the advancement of the field, and to allow any results to be tested independently by other scientists. More recently, he has been involved in the analysis of viral sequences during major outbreaks of new diseases, and he posts the results of analyses in real time, to provide an up-to-date source of information for the public, press and medical professionals. His epidemiology website not only keeps the research results flowing to the wider world for current outbreaks such as ebola (2014), Middle East respiratory syndrome (MERS-CoV, 2013), and avian flu (H7N9, 2013), but also includes the development of new ways of analysing and reporting epidemiological data, such as real-time maps of outbreaks.

Rambaut's programs are used in many of the studies reported in this book (even counting all the papers cited in the case studies that include him as a co-author will underestimate his impact because a lot of the other papers use his methods). His basic approach to programming seems to be to start with the acronym, think of a name that matches the acronym, then design the icon and the pull-down menus, and only then begin writing the code to make the thing work. Consequently, his programs (unlike so many academic software packages) usually have catchy names, are nice to look at and easy to operate (though his enthusiasm rarely extends to writing a manual).

Despite an amazing track record of producing useful programs and interesting papers, Rambaut is sufficiently modest that he was somewhat perturbed by being described here as a 'hero', and asked me to rename this box 'useful people of the genetic revolution'. He also wants me to point out that he is a friend of mine so, like all of the heroes in this book, my selection can hardly be described as unbiased. I have been fortunate enough to work with Andrew and to take part in the rush of excitement that comes when he seizes on an idea for a novel method, or gets his hands on new data to explore, or his attention is brought to a questionable hypothesis to dismember. He is a constant innovator, so if you are analysing DNA sequences you would do well to keep an eye on his website for new programs that do neat stuff. In the words of one blogger following the MERS-CoV outbreak: 'if you aren't following Andrew Rambaut, you should be'.

TECHBOX 14.1

# Calibration

**RELATED TECHBOXES**

**TB 13.1:** Substitution models

**TB 14.2:** Molecular dating

**RELATED CASE STUDIES**

**CS 14.1:** Calibration

**CS 14.2:** Dates

**Summary**

To estimate dates of divergence from molecular data, you must know the rate of change, which can be calculated from changes accumulated over a known time period.

**Keywords**

fossil, ancient DNA, biomarker, island, divergence dates, maximum and minimum bounds, substitution rates, crown group, stem lineage

Given that rates of molecular evolution can vary between lineages (see Chapter 13), substitution rates should, wherever possible, be estimated independently for different lineages rather than assumed to be the same as any other lineage. If we can estimate the number of substitutions that have occurred since two sequences were copied from a common ancestor (**TechBox 13.1**), we can convert that to an absolute rate of change if we know how much time has elapsed since the sequences we are comparing last shared a common ancestor.

Calibration is the use of any independent sources of information to define the amount of time taken to accumulate the observed substitutions on a phylogeny. The rate might be calibrated on a well-established rate for a related group of lineages (e.g. **Case Study 14.2**), or it might be estimated from the data by calculating the number of substitutions that have occurred in a known time period (**Case Study 14.1**). Both methods ultimately rely upon an independent source of temporal information, such as palaeontological specimens (e.g. the first recorded fossil of a particular lineage), biogeographic events (e.g. the splitting of two continents), ecological events (e.g. the co-evolution of parasites with their hosts), or even the ages of the DNA sequences themselves (e.g. sample dates of viruses). On the whole, it is best to avoid using molecular dates derived from other studies to calibrate a molecular clock, as it introduces a worrying circularity to the date estimation process and has the potential to magnify systematic errors. Predefined calibration rate estimates derived from a different lineage, such as two per cent per million years, might give you a ballpark feeling for the amount of divergence between your sequences (**Figure 14.33**), but given the degree to which lineages can vary in rate of molecular evolution, there are few situations in which you can use a predefined rate with any degree of confidence.

**Figure 14.33** How slow am I? Many studies of mitochondrial divergence use an assumed rate of change of two per cent per million years, but turtles have a much slower rate of molecular evolution than the mammals on which this rate was based. Various studies on turtles have suggested that 0.5 per cent per million years is more appropriate for turtle mitochondrial genes. Using this rate, researchers estimated that the Sonoran desert tortoise (*Gopherus agassizii*) has been separated from its sister species for around 5 million years. This is a decent first guess, and in the absence of any other information tells us something useful. But actually we don't know if the mitochondrial genes in this tortoise really do evolve at an average rate of 0.5 per cent per million year. There are many peculiar features of the biology of this tortoise that might impact on its substitution rate, such as their slow life history: they don't reach sexual maturity until around 15 years of age, lay relatively few eggs per year, but can live to 60 years or more. While the basic calibration rate gives us a good starting point, a more conservative approach would be to express this as a range of reasonable hypotheses: for example to consider the implied range of date estimates obtained by assuming that the rate is the fastest that has been reported for turtles, or by assuming it is the slowest.

If the rate of molecular evolution is uniform over all the lineages we wish to date, then we would need only one good calibration to estimate the rate of change. Many early molecular clock studies focused on finding a single reliable fossil calibration. But if the rate of molecular evolution varies, then we are going to need as many calibrations as possible. Just as we cannot expect molecular dates to be precise point estimates, we should not expect our calibrations to be without error either. So we will need to estimate confidence limits around the calibrations, and somehow use those confidence limits to inform our date estimation. These confidence limits may reflect the measurement error inherent in the calibration date. For example, geological clocks are sloppy clocks, like molecular clocks, so geochronological dates are usually reported with confidence intervals. But the error on calibration dates should reflect not just the confidence in the age estimate of the calibrating fossil, but also the confidence with which that fossil can be assigned to a branching point (node) in the phylogeny (**Figure 11.22**).

To estimate the rate of molecular evolution, we need to know the date at which two or more lineages diverged from each other. In other words, calibrations are nodes in a phylogeny with a known age. But most calibrations do not mark the exact point in time when two sequences split from a single ancestor to give rise to two genetically isolated lineages. Imagine we want to estimate the rate of molecular evolution in two lineages, A and B (**Figure 14.34**). The earliest known fossil of either lineage is an extinct species, let's call it A4, which has all the defining features of the taxa in lineage A. We might calculate the rate of change in these lineages by assuming that the age of this fossil (which we will call T1) is a good approximation of the date of the split between lineages A and B.

**Figure 14.34**

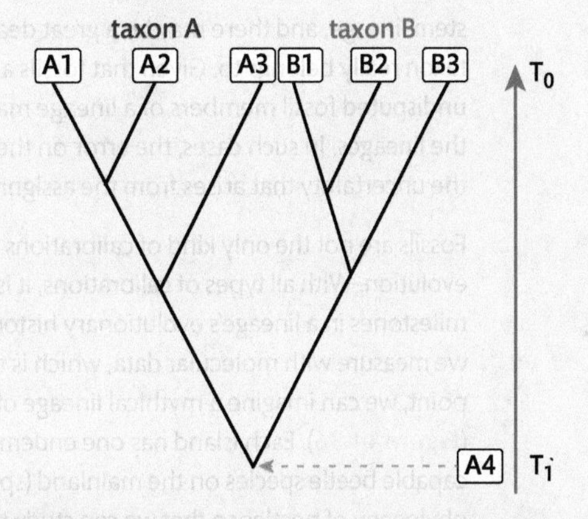


But does the first identifiable member of a taxon really mark the split between the lineages? The molecular divergence between lineages, marked by a node in the phylogeny, occurs around the time the populations become reproductively isolated. But at this point, the populations will still be genetically nearly identical. Recognizable differences between lineages are likely to evolve sometime after that split. Furthermore, the defining characteristics of a taxon may not evolve all at once, but may accrue in a lineage over time. This means that when we use a fossil taxon to estimate the rate of molecular evolution, we need to think about how closely the age of the fossil matches the split between lineages.

In this case, we know that fossil A4 is in the crown group of lineage A, because it has all the defining features of taxon A. But we don't know what the interval of time is between the divergence of lineages A and B and the development of the crown characters of taxon A (**Figure 14.35**).

**Figure 14.35**

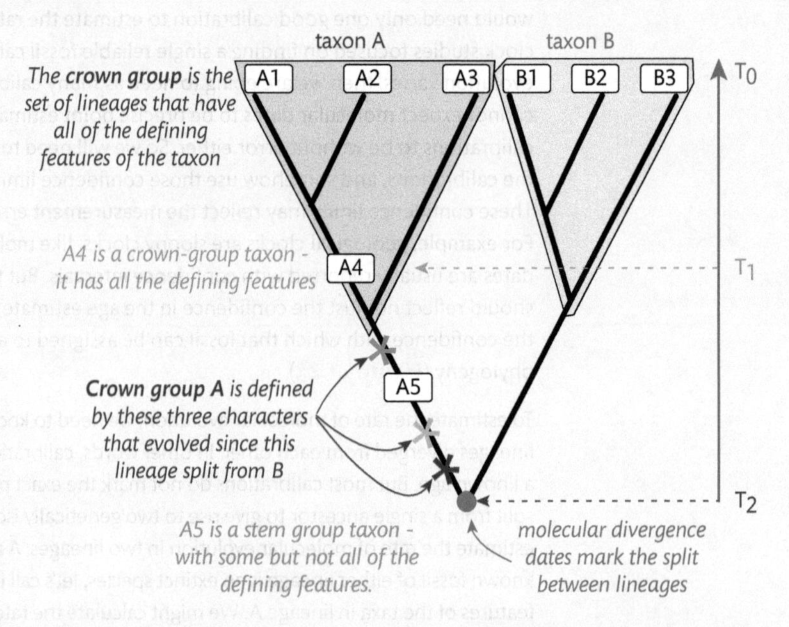


We might also have another fossil species, A5, that has some but not all of the defining characteristics of taxon A, in which case we might assign A5 to the stem lineage of A. In some cases, particularly for taxa with a continuous fossil record, the oldest stem group fossil might be readily recognizable, reliably dated, and very close to the point of genetic divergence. But in other cases, it may be difficult to assign early members of the lineage to the correct stem lineage, and there may be a great deal of debate about which lineage a particular fossil taxon really belongs to. Given that fossils are, for most biological groups, very rare, the earliest undisputed fossil members of a lineage may substantially post-date the genetic split between the lineages. In such cases, the error on the age of the calibration itself may be overwhelmed by the uncertainty that arises from the assignment of the calibration to a point on a phylogeny.

Fossils are not the only kind of calibrations that can be used to estimate rates of molecular evolution. With all types of calibrations, it is important to remember that they represent different milestones in a lineage's evolutionary history, and most do not coincide exactly with the event we measure with molecular data, which is the genetic isolation of populations. To illustrate this point, we can imagine a mythical lineage of flightless beetles endemic to a particular island chain (**Figure 14.36**). Each island has one endemic beetle (species A to F), which are related to a flight-capable beetle species on the mainland (species M). We want to be able to date the molecular phylogeny of beetles so that we can study the tempo and mode of the island radiation.

**Figure 14.36**

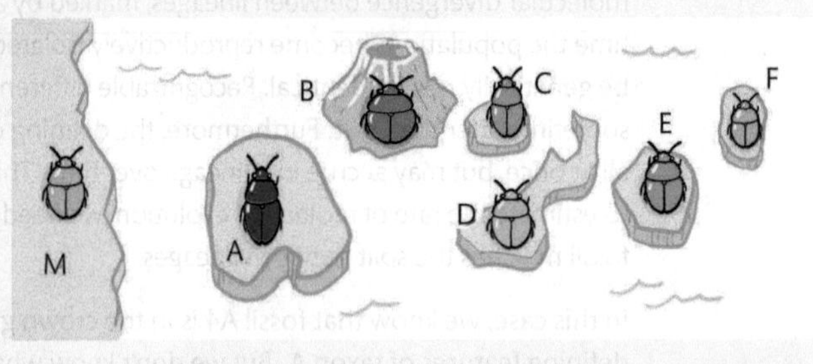


We are fortunate, in this imagined case study, to have a wealth of temporal information. We know that the island chain formed in a volcanic event approximately 10 million years (Myr)

ago. The first known fossil of the flightless lineage is preserved in the ashbed on island A from a volcano that erupted on island B 2 million years ago. There is ancient DNA from a subfossil preserved in a bog on island F, carbon dated to 50,000 years before present. There is also a biochemical calibration available from island C: sediments dated to 2.0±0.8 Myr contain a particular biomarker for the presence of this type of beetle. Combining all these observations, you might draw the calibrations on the phylogeny as follows (**Figure 14.37**).

**Figure 14.37**

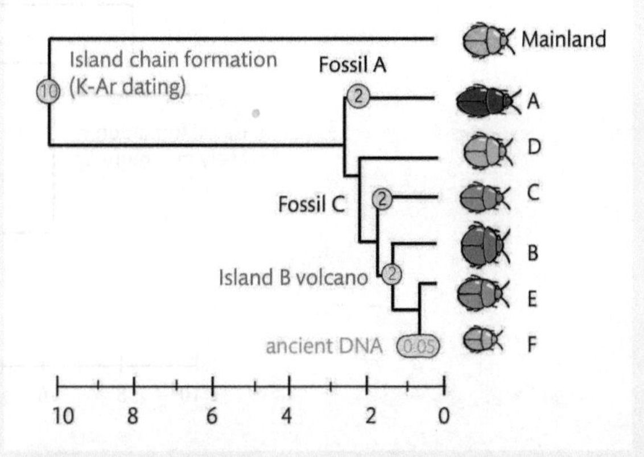


We might conclude, based on this information, that all of the island endemic beetles species date from around 2 million years ago, except for the outermost island which was only colonized recently. Thus we might propose that the radiation of these flightless beetles was driven primarily by rapid recolonization of islands after the recent volcano. But representing these calibrations as point estimates would be an oversimplification of the temporal information we have. Instead, we should consider how this information informs confidence ranges for divergence events on our phylogeny.

The island beetle lineage is unlikely to have split from the mainland population before the formation of the island, but it could have colonized any time after. In fact, the nearest mainland relative is capable of flight and is a regular visitor to the island, so the populations could have interbred even after the colonization of the island. At some point the island form became so distinct that it could no longer interbreed with the mainland immigrants and, from this point on, gene flow ceased and the island population began to accumulate unique substitutions. So the island formation date provides information on the maximum age of the lineage, but it does not tell us when the genetic split between the island and mainland lineages actually occurred.

The fossil of the flightless form sets a minimum age on the flightless endemic clade: we know the flightless form was fully developed by the time this fossil was preserved, but it could have evolved any time before that point. So the split between the mainland and island beetles was probably after island formation at 10 Myr and before the first distinct fossil of the island lineage at 2 Myr.

The volcano on island B would probably have wiped out any beetles on island B at the time, though we can't be sure they didn't manage to persist in some sheltered cave somewhere. So we are fairly sure that species B is younger than 2 Myr, but we don't know how much younger, and we can't rule out that it is older. The biomarker for the beetle lineage tells us that there were beetles on island C sometime between 2.8 and 1.2 Myr ago, but it does not tell us which beetle lineage it was. As it happens, the biomarkers are the last trace of a population of species D that lived on island C until it was wiped out by the volcano 2 Myr ago. So we have placed this information on the wrong node on the phylogeny. It tells us nothing about the age of species C (which, as it happens, colonized Island C 1 Myr after the eruption). And finally, the ancient DNA date from the subfossil on island F represents the age of the sequence itself, not the origin of the lineage.

So we could draw a rather different picture of the diversification of these flightless beetles (**Figure 14.38**):

**Figure 14.38**

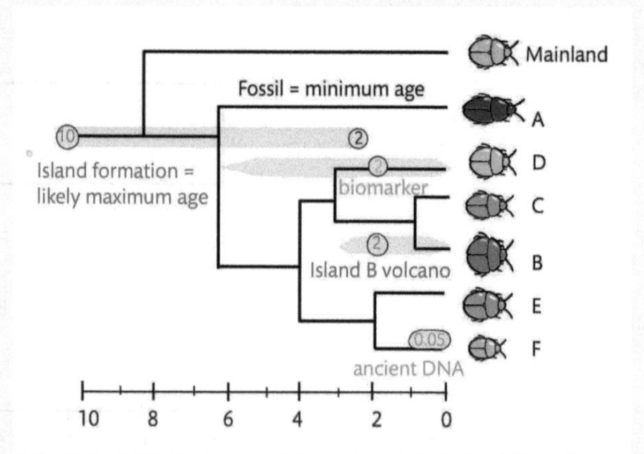


This phylogeny suggests a very different tempo and mode of island beetle evolution than the first one. Instead of a burst of diversification around 2 Myr ago, there is a steady divergence into different lineages throughout the history of the island lineage. This alternative phylogeny does not support the hypothesis that the diversification of the flightless beetles was primarily driven by radiation into empty niches after the volcano on island B.

If you are wondering what on earth these mythical beetles can teach you, let me summarize. Some calibrations, like island dates, are likely to represent maximum bounds on divergence dates (the oldest possible date of origin). Some calibrations, like fossils or biomarkers, tell you only that a lineage was present at a particular time, not when it originated, so provide probable minimum bounds on divergence dates (the youngest possible date of origin). Some calibrations, like ancient DNA, represent the age of the sequence itself, not the date of divergence from a common ancestor. And all calibrations should be considered in light of the confidence with which we can assign nodes in the phylogeny to particular ranges of dates.

Calibrations rarely provide precise point estimates on the splits in a molecular phylogeny. Think about exactly what any given piece of information is telling you about the biological lineages they represent, and try to incorporate any uncertainty into molecular date estimation, either by repeating the analysis with different possible calibration dates, or by incorporating confidence limits on calibrations into the dating procedure. New molecular dating methods allow calibrations to be included with probability distributions of error around them, so that you can quantify how sure you are that the real divergence occurs on any particular date. Always try to explore what the implications for your results would be if your calibrations are inaccurate—can you try your analysis with only a subset of the calibrations? Or with different error distributions? How much is the assumed age of the root influencing the answers you get? As with most things in phylogenetic analysis, there are no magic bullets, no perfect solutions. Instead, in common with all the techniques we have discussed in this book, you should gather as much information as possible, consider that information in light of its margins of error, and weigh different sources of information up against each other.

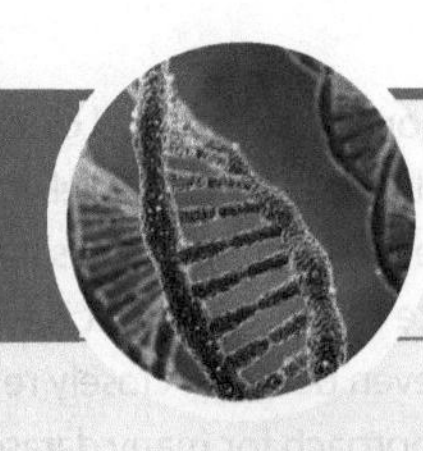

# TECHBOX 14.2 Molecular dating

RELATED TECHBOXES

TB 13.2: Bayesian phylogenetics

TB 14.1: Calibration

RELATED CASE STUDIES

CS 14.1: Calibration

CS 14.2: Dates

**Summary**

All molecular dating methods make assumptions about rate variation which determine the dates estimated.

**Keywords**

strict clock, local clocks, relaxed clocks, rate-variable dating, calibration, confidence

Molecular dating uses measures of difference between sequences to estimate the date of divergence of biological lineages. In order to estimate time from molecular sequence data, you need to be able to make assumptions about the rate of change in those sequences, and you need some way of calibrating the rate of molecular change against an absolute timescale (**TechBox 14.1**). There are a wide range of dating methods, all of which employ different assumptions about substitution rates. In common with the rest of this book, I won't discuss details of specific methods or programs—there are far too many and more are being invented all the time—but instead will discuss some of the broad issues that must be addressed in these kinds of analyses. We will briefly consider three broad approaches, but there are methods that do not fit neatly into these categories.

**Strict clocks** are the simplest model of substitution rate evolution, resting on the assumption that all of the sequences in your alignment have the same underlying rate of substitution. If this is the case, then we only need one calibration rate to date the whole phylogeny. Because it has high predictive power, this is the most powerful and practical molecular clock model: if rates are constant over the phylogeny, and we know the date of one node (branching point) in the phylogeny, then we can estimate the age of all other nodes. A constant substitution rate may provide a fair description of molecular evolution of some datasets (**Case Study 14.2**), but we should expect that the rate of substitution will vary between lineages for most phylogenetic datasets. If you apply an analysis that assumes constant rates to sequences that do not share the same average rate of change, the date estimates could be seriously misleading. One approach to using strict clocks is to test alignments for evidence of rate variation, reject any datasets that show significant variation, and assume that all the rest have constant rates. There are two problems with this approach. Firstly, many of the tests used have low power to detect and exclude rate variable data, so you might end up including sequences with undetected rate variation, which could bias the date estimates[1]. Secondly, the detect-and-exclude approach will not be practical when rate variation is common, as there will be few if any suitable datasets to use for strict clock dating[2].

**Local clocks** allow lineages within a phylogeny to have different substitution rates by assigning the edges (branches) in the phylogeny to different rate categories (for definition of edge see **Figure 11.22**). The rate categories might be defined by the availability of calibrations, or some other form of prior information (for example, free-living versus parasitic lineages), or defined by phylogenetic relationships (e.g. all lineages in a defined clade have the same rate), or lineages may be assigned to rate categories using an optimization approach. Once rate categories are defined, you can estimate a rate for each category based on available calibrations or using prior beliefs about rate distributions. By limiting the number of rate categories, local clocks make rate-variable dating statistically tractable without having to make as many of the assumptions as the relaxed clock methods. If you have sufficient calibrations to estimate rates across the phylogeny, and you are confident that changes in substitution rate are reasonably rare so that they occur mainly between lineages defined by different calibrations, then the local clock model provides a useful method for inferring node dates. But as with the strict clock model, if the assumption of rate constancy within rate categories is violated, then the date estimates may be inaccurate. Given that many different factors can influence the rate of substitution, it is reasonable to assume that the rate of molecular evolution can vary even between closely related lineages, which will invalidate the application of a local clock approach for many datasets.

**Relaxed clocks** are thus called because the requirement for rates to be constant has been relaxed, although they have also had the effect of making people feel a lot more relaxed about using molecular dating even though we know that rate variations are common. In these methods, every branch (edge) in the phylogeny can have its own rate. It would be impossible to directly estimate a substitution rate for every edge unless you had an accurate calibration for every node (and if you did, you wouldn't need molecular date estimates). So these methods use a model of molecular evolution to predict changes in rate along the phylogeny, or draw rates from an assumed distribution of substitution rates. Since there is little or no independent evidence of rate changes being employed by these methods, the accuracy of prediction of rate changes will depend critically on how well the model describes the data.

Ideally, we would develop a model of rate change based on a solid understanding of the dynamics of molecular evolution in the lineages under consideration. But, in most cases, we do not have a clear idea of what is driving rate variation, nor how much rate variation to expect. Even if we did know what determined rate variation in a particular group, we might require a lot of extra biological information to predict patterns of rate change—like generation time or population size—which may not be available for all lineages in the phylogeny (especially the internal nodes representing extinct taxa). So relaxed clock methods use statistically tractable but biologically arbitrary models of evolution. For example, most Bayesian relaxed clock methods rely on a 'tree prior' which specifies the prior probability of different tree topologies (e.g. a Yule prior which assumes a constant speciation rate); a 'rate prior' which determines the distribution of rate categories; and a 'date prior' that describes the relative heights of nodes in the tree given the calibrating information. All of these priors can make a big difference on the outcome of a dating analysis[3, 4], and yet most users give them relatively little attention.

### Different models give different date estimates

*Given the simplicity of most models, it is possible that model selection in modern systematics is analogous to an overweight man shopping in the petites department of a women's clothing store. A particular garment might fit the portly man best, but this does not imply a good overall fit*

Gatesy, J. (2007) A tenth crucial question regarding model use in phylogenetics. *Trends in Ecology & Evolution*, Volume 22, page 509

It is important to bear in mind that molecular dating methods are not 'reading' rate variation from the data, they are fitting the best solution given the assumptions of the method. The best fit may be a terrible fit, and it may be giving you completely wrong answers. One way to illustrate this is to consider the different solutions you will get using different methods or assumptions.

For most dating studies, we have little idea about the underlying pattern of rate variation, but in some cases we do have prior information on rate variable. Researchers wished to date the origin of a particular parasitic plant clade[5]. But parasitic plants generally have faster substitution rates than their free-living relatives[6]. They tried three different rate variable methods: two local clock methods and a relaxed clock method. All of these methods assigned higher rates to the parasitic clade, but the pattern of rate variation varied greatly between the different methods, resulting in a wide range of date estimates (**Figure 14.39**). These patterns of rate change can't all be correctly reflecting the true history of the sequences, but we don't have any way of knowing

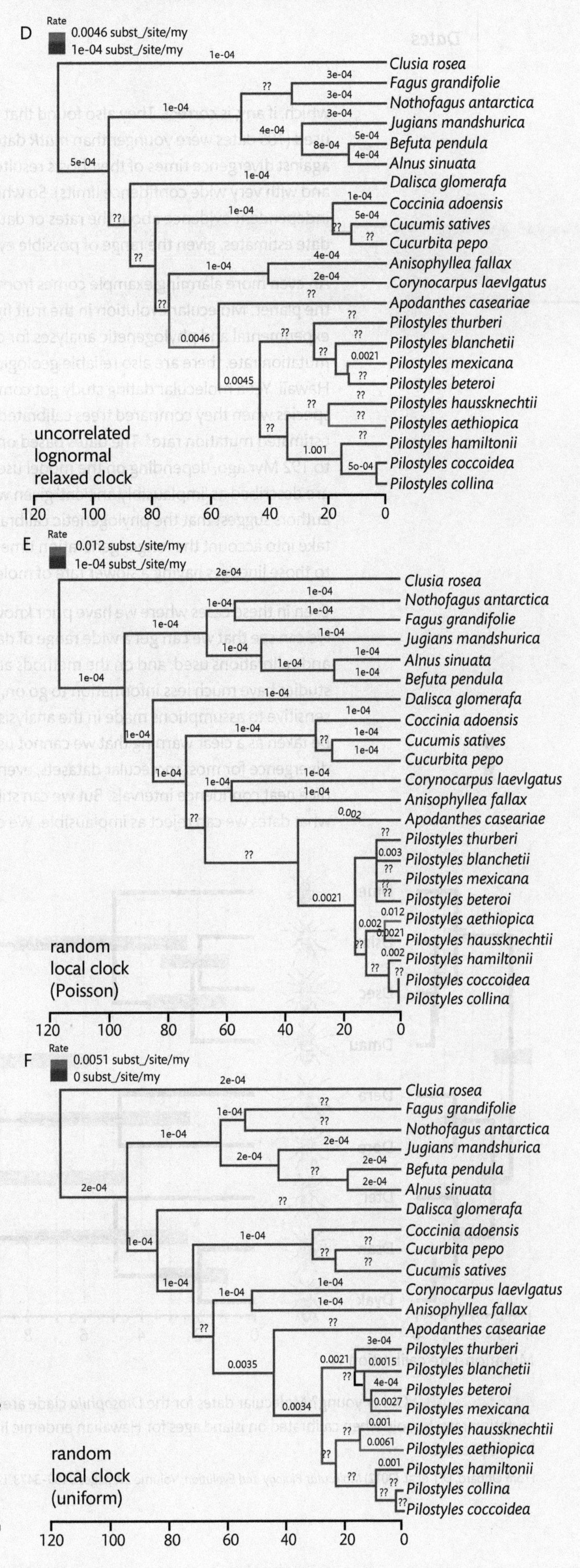


**Figure 14.39** Three different molecular dating methods applied to the same dataset of 18S rRNA sequences from plants in the Curcubitales, including parasitic plants from the genera *Pilostyles* and *Apodanthes*. The scale bar is in tens of millions of years before present. The coloured branches indicate the inferred substitution rates, with red and dark red the highest rates. The Poisson random clock model assigns the fastest rates to the tips, so the date of the base of the *Pilostyles* clade is very young, whereas the relaxed clock assigned faster rates to the base of the clade, giving much older dates. The uniform random clock model is somewhere in between. Is it possible to decide which solution is the most reasonable? Or should we aim to test all possible rate-variable methods to give a wide range of possible estimates?

Redrawn from Bellot, S. and Renner, S. S. (2014) *Molecular Phylogenetics and Evolution*, Volume 80, pages 1–10, with permission from Elsevier.

which, if any, is correct. They also found that the dates varied depending on which gene they used (*18S* dates were younger than *matR* dates) and the choice of calibrations (calibrating against divergence times of their hosts resulted in much older estimates than fossil calibrations, and with very wide confidence limits). So which analysis gives the right answers? Without independent evidence about the rates or dates, the best we can do is show the range of possible date estimates, given the range of possible evolutionary scenarios.

An even more alarming example comes from one of the most well-studied organisms on the planet. Molecular evolution in the fruit fly genus *Drosophila* has been the subject of experimental and phylogenetic analyses for decades. Not only are there good estimates of mutation rate, there are also reliable geological calibrations for island endemic lineages in Hawaii. Yet a molecular dating study got completely different dates of divergence for *Drosophila* species when they compared trees calibrated with the island dates with those based on the estimated mutation rate[7]. The dates based on phylogenetic calibrations varied from 26 Myr ago to 192 Myr ago, depending on the model used to estimate dates (**Figure 14.40**). The older dates are described as 'implausibly ancient' given what is known about the history of *Drosophila*. The authors suggest that the phylogenetic calibrations may be misleading in this case by failing to take into account the longer generation times of Hawaiian *Drosophila*, which might contribute to those lineages having a slower rate of molecular evolution than other *Drosophila* lineages.

Even in these cases where we have prior knowledge about rate variation or reliable calibrations, we can see that we can get a wide range of date estimates depending on the sequence data and calibrations used, and on the methods and parameters of the models. Most phylogenetic studies have much less information to go on, but we can assume that the results are just as sensitive to assumptions made in the analysis as in these well-studied examples. This should be taken as a clear warning that we cannot usually expect to be able to precisely infer dates of divergence for most molecular datasets, even if the methods we use give us nifty estimates with nice neat confidence intervals. But we can still test hypotheses using molecular data by asking what dates we can reject as implausible. We can ask: what assumptions would we have to make

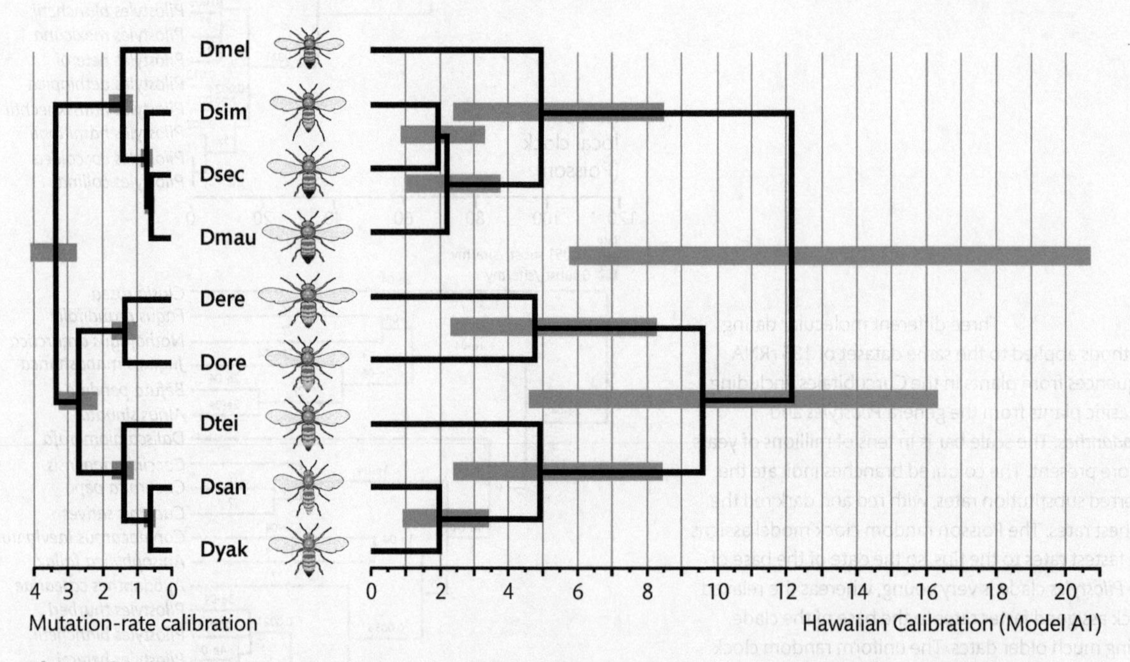


**Figure 14.40** Old or young? Molecular dates for the *Drosophila* clade are young when calibrated on the established mutation rate, but old when calibrated on island ages for Hawaiian endemic lineages. The scale bars are in millions of years before present.

From Obbard, D. J. et al. (2012) *Molecular Biology and Evolution*, Volume 29, pages 3459–3473, by permission of Oxford University Press.

to have date estimates that encompass a proposed date of divergence? Do these assumptions provide a plausible description of this dataset? What range of values fall within believable parameter space, and what values fall outside that space?

When reading molecular dating studies, you need to scrutinize the methods section carefully, and look for evidence that they have examined the robustness of their findings to the assumptions made in the analysis. How did the authors choose and define their calibration points, particularly the date of the root node? Would they have got the same answer if they had used an alternative set of calibrations (for example, how different would the estimates be if one of the calibrations was removed from the analysis, or if the bounds on the root date were changed)? Did they try different parameter values or models, and if they did, how much did the results change?

In statistical estimation procedures there is a trade-off between adding parameters that capture important and consistent patterns in the data, and over-fitting a complex model and producing inaccurate or imprecise results. Poor model fitting can lead to high confidence in the wrong results. Narrow confidence bounds on your date estimates do not necessarily mean you are getting closer to the truth, instead it could be a reflection that your prior beliefs are dominating your results, giving your data relatively little voice[8]. Given that we rarely understand all of the forces acting on substitution rates for a given dataset, we have no way of checking whether our models are providing useful explanations of our data. As with all techniques outlined in this book, the best approach is to explore your data: test the influence of assumptions, compare to multiple lines of evidence, and consider all estimates in light of the confidence you have in your data and methods.

## References


1. Bromham, L. D., Rambaut, A., Hendy, M. D., Penny, D. (2000) The power of relative rates tests depends on the data. *Journal of Molecular Evolution*, Volume 50, page 296.
2. Welch, J. J., Bromham, L. (2005) Molecular dating when rates vary. *Trends in Ecology and Evolution*, Volume 20, page 320.
3. Dos Reis, M., Zhu, T., Yang, Z. (2014) The impact of the rate prior on Bayesian estimation of divergence times with multiple loci. *Systematic Biology*, Volume 63, page 555.
4. Warnock, R.C., Parham, J. F., Joyce, W. G., Lyson, T. R., Donoghue, P. C. (2015) Calibration uncertainty in molecular dating analyses: there is no substitute for the prior evaluation of time priors. *Proceedings of the Royal Society B: Biological Sciences*, Volume 282, page 20141013.
5. Bellot, S., Renner, S. S. (2014) Exploring new dating approaches for parasites: The worldwide Apodanthaceae (Cucurbitales) as an example. *Molecular Phylogenetics and Evolution*, Volume 80, page 1.
6. Bromham, L., Cowman, P. F., Lanfear, R. (2013) Parasitic plants have increased rates of molecular evolution across all three genomes. *BMC Evolutionary Biology*, Volume 13, page 126.
7. Obbard, D. J., Maclennan, J., Kim, K-W., Rambaut, A., O'Grady, P. M., Jiggins, F. M. (2012) Estimating divergence dates and substitution rates in the *Drosophila* phylogeny. *Molecular Biology and Evolution*, Volume 29, page 3459.
8. Dos Reis, M., Yang, Z. (2013) The unbearable uncertainty of Bayesian divergence time estimation. *Journal of Systematics and Evolution*, Volume 51, page 30.

CASE STUDY

**14.1**

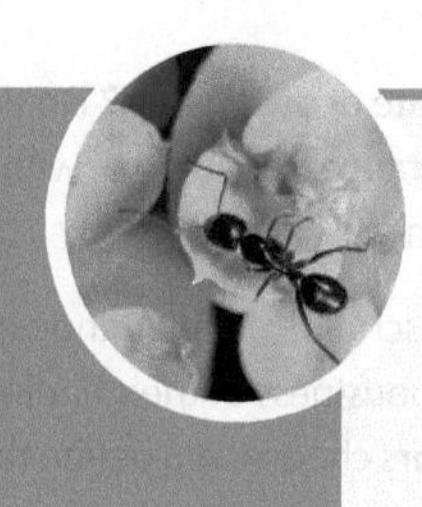

# Calibration: did kauri survive the Oligocene drowning of New Zealand?

**RELATED TECHBOXES**

TB 13.2: Bayesian phylogenetics

TB 14.1: Calibration

**RELATED CASE STUDIES**

CS 2.2: Ancient DNA

CS 13.2: Diversification

Biffin, E., Hill, R. S., Lowe, A. J. (2010) Did kauri (*Agathis*: Araucariaceae) really survive the Oligocene drowning of New Zealand? Systematic Biology 59: 594.

*The Eocene-Oligocene boundary marks half time for New Zealand. From a slow start and, so far as we know, a relatively quiet first half, everything was about to change. The Oligocene (33–24 Mya) is most notable for being the low point of New Zealand's slow submergence beneath the waves. A rise in global sea level occurring at the time was something our low-lying, waterlogged subcontinent could not withstand*[1]

**Keywords**

molecular dating, relaxed clock, confidence, crown, stem, fossil, dispersal, biogeography

## Background

If you ever want to start a fight at a New Zealand ecology and evolution conference, try mentioning 'Oligocene drowning'. Then stand back. New Zealand has an interesting biogeographic history: it began as a part of the Gondwanan supercontinent, then split from Australia between 83 and 75 million years ago (Mya) to form a separate, geologically active chain of islands. It has usually been assumed that some of the native flora and fauna are relicts of Gondwana, while others have settled the archipelago since. But the hypothesis that the entire archipelago was submerged when sea levels rose in the Oligocene leads to the prediction that all New Zealand (NZ) terrestrial lineages must have colonized the land after it re-emerged from the waves. This would mean that charismatic NZ taxa such the large flightless moas (**Case Study 2.2**) must have originated from colonists that dispersed over the Tasman Sea.

## Aim

Molecular dating might be able to distinguish two hypotheses for the origin of endemic NZ taxa: Gondwanan relics that split from their ancestral stock over 70 Mya, or colonists that arrived after the Oligocene seas retreated after 23 Mya. It had previously been claimed that the long branch connecting Kauri (*Agathis australis*: **Figure 14.41**) to the rest of the Southern Hemisphere conifer family Araucariaceae was evidence for its pre-Oligocene isolation in NZ[2], and thus it must have survived the drowning period. In this paper, the authors challenge that conclusion by re-examining how the phylogenetic inference of Kauri's history in NZ is influenced by assumptions made in the analysis.

## Methods

The topology (branching pattern) of the Araucariaceae phylogeny could support either an origin of Kauri as a relict of Gondwana (in which case the age of the split between kauri and the other members of the family would date to the breakup of the Australia-NZ land mass) or a post-drowning colonist (in which case the split will be post-Oligocene in age). So the interpretation of the phylogeny rests entirely on how you convert the genetic distance between Kauri and its relatives to time. DNA sequences from the chloroplast genes *rbcL* and *matK* from 24 species were aligned and a phylogeny estimated using a rate-variable Bayesian method, under two different calibration scenarios (**Figure 14.42**). In the crown group calibration scenario, fossil

**Figure 14.41** This Kauri tree is so impressive it has its own name: Te Matua Ngahere (Father of the forest). It is reckoned to be the widest kauri alive, the second tallest, and is probably over a thousand years old.



evidence was considered to provide age constraints on the crown group of the Araucariaceae and for splits within the family, with a soft age constraint on the upper limits of node ages (meaning that older ages of the node becomes less plausible the further you get from the specified calibration date). In the stem group calibration scenario, the fossil evidence was considered to date the origin of the stem group of the Araucariaceae, and a different calibration was given to the split between *Agathis* and its sister genus *Wollemi*. The two scenarios also differ in the number of calibrations used: four for the crown group scenario and two for the stem group scenario. In both dating scenarios, the maximum age was set at the appearance of conifers in the fossil record at 320 Mya.

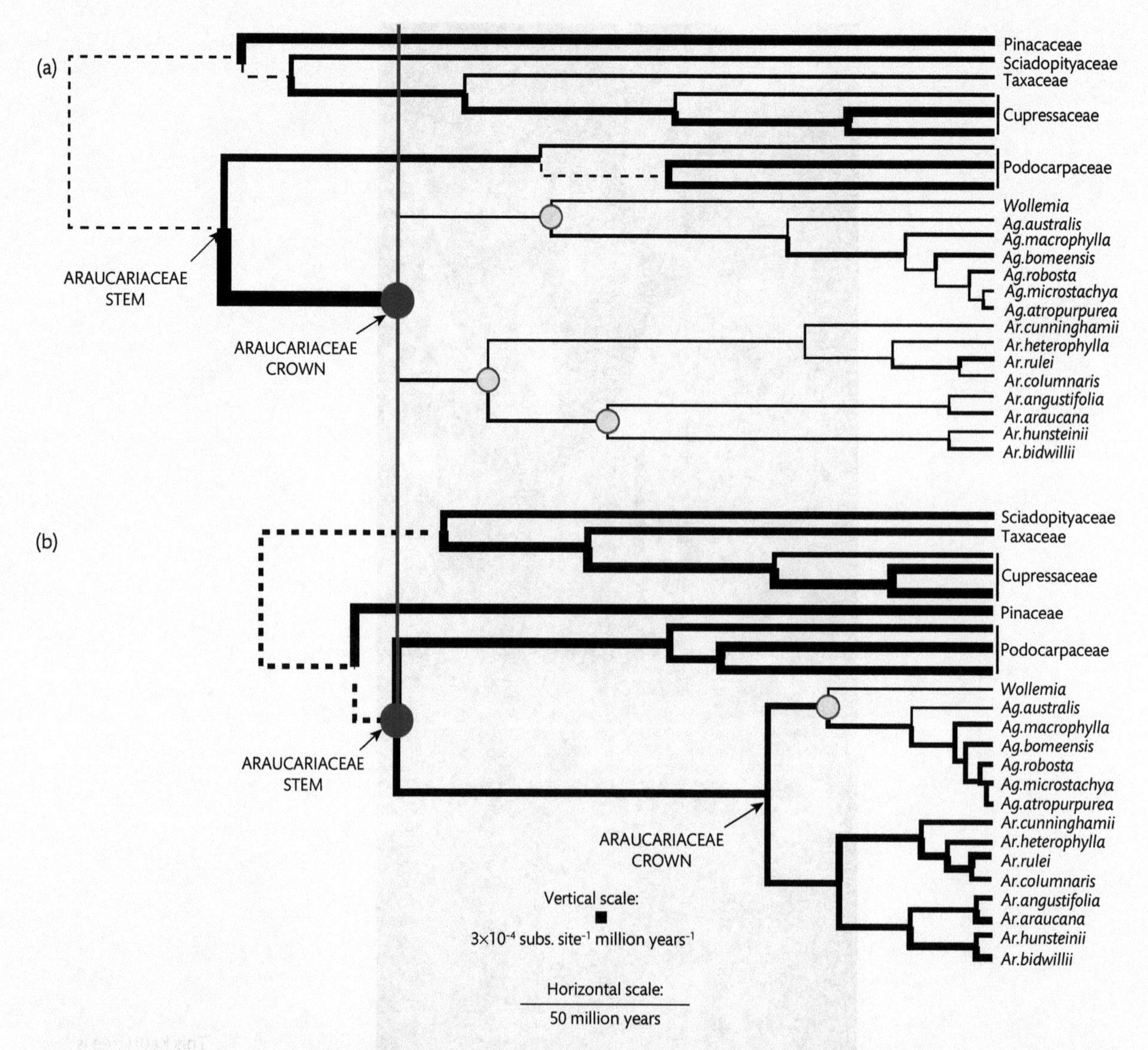


**Figure 14.42** Drowning or waving? Two alternative solutions for molecular dates for the Araucariaceae, including the endemic New Zealand kauri (labelled *Ag. australis*). The calibration scenarios differ in: the placement of Mesozoic fossils (red circle) at the base of either the crown group (a) or the stem group (b) of the Araucariaceae; which fossils are considered reliable evidence of the origin of the *Agathis* lineage (two different orange circles); and inclusion of two additional calibration points (yellow circles). The wider the black edges on the phylogeny, the faster the rate on that branch. You can see that, for the same sequences and topology, the relaxed clock method used provides very different solutions under the two calibration scenarios, giving the Araucariaceae clade either slow rates with an old date (a) or faster rates and a younger date (b).

Figure from Biffin, E., Hill, R. S., Lowe, A. J. (2010) *Systematic Biology*, Volume 59(5), pages 594–602, by permission of Oxford University Press.

### Results

Both analyses suggest substantial rate variation between lineages, but because the stem calibration scenario implies a more recent origin of the clade, it requires faster rates of molecular evolution to explain the data (**Figure 14.42**). Because they infer different patterns of rates across the tree, the two calibration scenarios gave entirely different date estimates. While the stem group scenario gave an estimated date of the node separating kauri from its neighbour within the Oligocene period, the confidence limits on dates from the crown group scenario span from the Oligocene to the Cretaceous, which allows Kauri to be either a relict or a recent arrival.

### Conclusions

This study highlights that the results obtained in molecular dating studies rest critically on the assumptions made in the analysis. The pattern of rates inferred is that which best fits the model of evolution assumed, so changing the model or its parameters changes the inferred pattern of rates, and therefore the divergence date estimates. Interpreting particular fossils as marking the age of the stem group or the crown group made tens of millions of years difference to the date estimates, and determined whether the analysis supported entirely different evolutionary scenarios for the origin of an endemic taxon (Gondwanan origin vs post-Oligocene colonist).

### Limitations

The two dating scenarios differed not only in the interpretation of fossil calibrations but also in their number, which could have influenced the date estimates obtained. This study only contrasted calibration scenarios and did not test the influence of other assumptions in the analysis on the date estimates obtained.

### Further work

Many of the molecular dates used to support Gondwana origins of NZ endemics are from invertebrates such as velvet worms, mite harvestmen and centipedes, and in some cases molecular dates suggest these divergences are even older than the Australia-NZ split[3]. If these taxa are as old as the dates suggest, then there are two possible explanations: either they persisted in NZ throughout the Oligocene, implying that there was sufficient land area to support many endemic taxa, or they recolonized post-drowning NZ despite having apparently little obvious capacity for long-range dispersal. An alternative explanation is that the NZ taxa have much slower rates of molecular evolution, which would push date estimates back in time. It would be interesting to look for any evidence of systematic rate variation that might be biasing date estimates in these cases.

### Check your understanding

1. Why can the same Mesozoic fossil date (red circles in **Figure 14.42**) be placed on different parts of the tree?
2. What is the difference between a minimum and maximum calibration? Give an example of each from this study.
3. Given that the same method has been applied to the same dataset, why do the two trees in **Figure 14.42** have different patterns of rate variation?

### What do you think?

What criteria should be used for selecting calibration dates for molecular dating analyses? Should you prefer a few reliable dates over a larger number of less certain dates? Should you prefer some kinds of fossils over others, for example using leaf fossils rather than pollen[4]? Can molecular systematicists be relied upon to choose calibration points for their phylogenies, or can these decisions only be reliably made by palaeontologists?

### Delve deeper

Rate-variable dating methods rely on assumptions about the evolutionary processes that produced the sequence data being analysed. Some of these assumptions describe the diversification process of the tree, usually assuming a constant rate of speciation over time, and either no extinction or constant extinction rates. But if New Zealand did get submerged in the Oligocene, partially or totally, we might expect a high extinction rate at that time. This would skew the distribution of nodes in the tree away from the expected pattern under a uniform process, and could result in date estimates that are too young[5]. What could be the consequences of a mismatch between tree priors (see **TechBox 14.2**) and evolutionary history of a group? How would you test whether these priors are biasing date estimates? Is it possible to come up with a bias-free prior on diversification rates?

### References

1. Gibbs, G. W. (2006) *Ghosts of Gondwana: the history of life in New Zealand*. Craig Potton Publishing.
2. Knapp, M., Mudaliar, R., Havell, D., Wagstaff, S. J., Lockhart, P. J. (2007) The drowning of New Zealand and the problem of *Agathis*. *Systematic Biology*, Volume 56, page 862.
3. Giribet, G., Boyer, S. L. (2010) 'Moa's Ark' or 'Goodbye Gondwana': is the origin of New Zealand's terrestrial invertebrate fauna ancient, recent or both? *Invertebrate Systematics*, Volume 24, page 1.
4. Sauquet, H., Ho, S. Y. W., Gandolfo, M. A., Jordan, G. J., Wilf, P., Cantrill, D. J., Bayly, M. J., Bromham, L., Brown, G. K., Carpenter, R. J., Lee, D. M., Murphy, D. J., Sniderman, J. M. K., Udovicic, F. (2012) Testing the impact of calibration on molecular divergence times using a fossil-rich group: the case of *Nothofagus* (Fagales). *Systematic Biology*, Volume 61, page 289.
5. Sharma, P. P., Wheeler, W. C. (2013) Revenant clades in historical biogeography: the geology of New Zealand predisposes endemic clades to root age shifts. *Journal of Biogeography*, Volume 40, page 1609.

CASE STUDY **14.2**

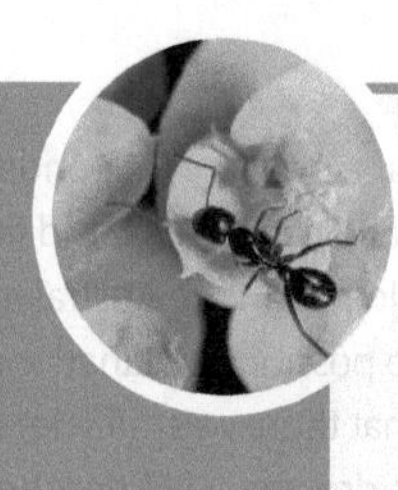

# Dates: using phylogenies to trace the source of disease outbreaks

**RELATED TECHBOXES**

**TB 14.1:** Calibration

**TB 14.2:** Molecular dating

**RELATED CASE STUDIES**

**CS 5.2:** Genome

**CS 12.1:** Epidemiology

de Oliveira, T., Pybus, O. G., Rambaut, A., Salemi, M., Cassol, S., Ciccozzi, M., Rezza, G., Gattinara, G. C., D'Arrigo, R., Amicosante, M., Perrin, L., Colizzi, V., Perno, C. F., Benghazi Study Group (2006) Molecular Epidemiology: HIV-1 and HCV sequences from Libyan outbreak. *Nature*, Volume 444, page 836.

> *The tragedy for the nurses is finished; now starts the tragedy for the children*

Vittorio Colizzi, quoted in *The New York Times*, 29 July 2007

### Keywords

molecular dating, relaxed clock, confidence intervals

### Background

In March 1998, six new foreign staff members arrived at the Al-Fateh Hospital in Benghazi, Libya. Two months later, the hospital noted its first case of HIV-1 infection. Within six months of the arrival of the foreign staff, 111 children who had attended the hospital were found to be HIV-1 positive. An investigation by the World Health Organization (WHO) eventually revealed that 418 children had been infected by HIV-1 in the Al-Fateh Hospital outbreak and many of these children were also infected with hepatitis C (HCV). At least 60 of these children have since died. A number of medical workers from the hospital were accused of involvement in the deliberate infection of the children. Six foreign medical workers were detained in prison in 1999, then sentenced to death in 2004. The death sentence was upheld in a retrial in 2007, though subsequently commuted to life imprisonment. Lawyers from an international charity who were representing the medical workers appealed to international experts to conduct an independent assessment of the evidence. This study was a response to that appeal.

### Aim

Following the discovery of the Al-Fateh Hospital (AFH) outbreak, just over half of the infected children were sent to hospitals in Europe for assessment and treatment. The median age of these children at the time of diagnosis was only three and a half years old. Most of these children were asymptomatic, but some had begun to develop symptoms of AIDS. This international team of researchers had access to samples of HIV-1 from 44 of the children who were treated in Europe. If all of the children were infected by the foreign medical workers using a common stock of HIV, then you would expect that the divergence between their DNA sequences would be consistent with the genomes having a common ancestor in 1998. These researchers used molecular phylogenetics and molecular dating to test this hypothesis.

### Methods

HIV-1 RNA was extracted from plasma samples, amplified and the *gag* gene sequenced. These sequences were blasted against GenBank (**TechBox 1.1**) and against the Los Alamos HIV sequence database, to identify the most similar sequences. This procedure resulted in a reference database of 56 HIV-1 *gag* sequences from Africa and Europe. The AFH and reference sequences were aligned automatically then adjusted manually (**TechBox 10.2**). The researchers tested which nucleotide substitution model best fitted the data (**TechBox 13.1**), then used that model to estimate a phylogeny of the sequences using both maximum likelihood (**TechBox 12.1**) and Bayesian methods (**TechBox 13.2**). To estimate a substitution rate for this alignment, the researchers collated a reference set of the same genomic region from 48 HIV-1 sequences of known sample dates, spanning two decades. HIV-1 evolves so rapidly that sequences sampled in different years will be measurably different, so these sample dates could be used to estimate the substitution rates. They used both a strict clock and a Bayesian rate-variable method (**TechBox 14.2**) to estimate rates of substitution: these estimates were similar in both methods, and did not vary greatly over the tree for the Bayesian analysis. The estimated substitution rate was used to infer the age of the AFH cluster of sequences under a number of different epidemiological models (e.g. whether the population size of viruses stayed the same or grew exponentially over time) and different substitution models (e.g. allowing rates to vary across all sites, or specifying a codon model of variation in rates between sites).

### Results

The AFH HIV-1 sequences all formed a distinct clade on the phylogeny, suggesting that the children had all been infected by a single strain of HIV-1. However, under nearly all of the models used, the date of origin of this clade of AFH sequences was estimated to be before the arrival of the foreign medical workers in March 1998. Furthermore, the confidence intervals on these estimates did not contain March 1998, so the probability that they could have originated from the foreign medical workers was considered to be effectively zero. Similar results were obtained for the hepatitis C (HCV) sequences (**Figure 14.43**).

### Conclusions

These results suggest that the Al-Fateh children were all infected with the same strain of HIV-1, but that this strain originated before the foreign medical workers arrived at the hospital. Those children for whom medical information was available had all had invasive medical procedures at the hospital and several different international reports cite poor hygiene practices at the hospital as the probable cause of the infections. The high incidence of co-infection with HCV was also put forward as evidence of nosocomial (hospital-based) infection due to poor hygiene. The six accused medical workers were released in 2007 and allowed to return to their countries of origin, though their release was primarily the result of negotiations rather than due to persuasive scientific evidence.

### Limitations

The sequences from the AFH outbreak were compared to sequences from an international database, rather than to local controls (sequences of the virus taken from the same population[1]). The molecular date estimates rely on the assumption that the rate of molecular evolution in the AFH HIV sequences is the same as those from the reference database, and that substitution rates in HCV will be similar to those from a previously studied nosocomial transmission history in Ireland. If the children had been deliberately infected with different strains of HIV, but those infecting strains were all sampled from the sample population, then it would be possible to get

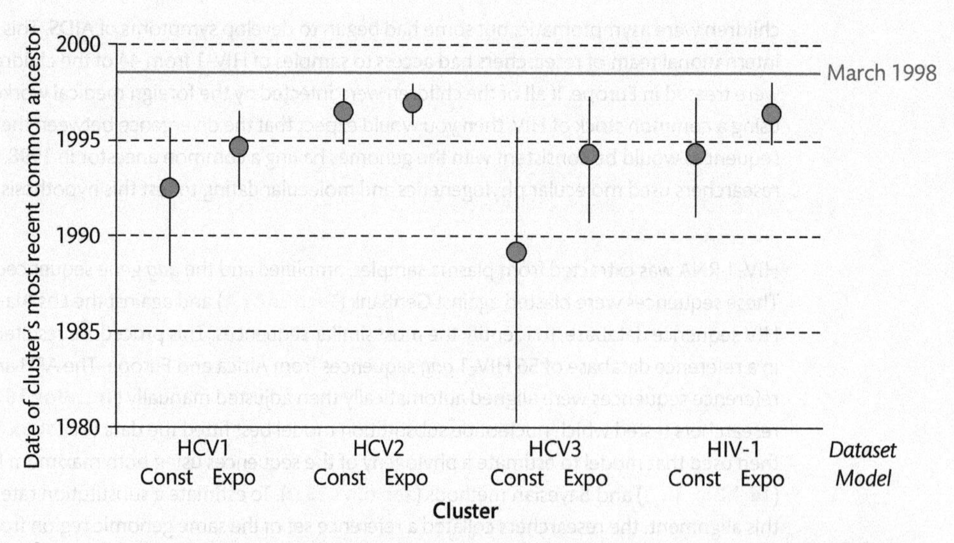


**Figure 14.43** Estimated dates of the most recent common ancestor for each cluster of hepatitis (HCV) and HIV-1 sequences taken from children infected at the Al-Fateh Hospital, derived under two different models of evolution: Const (which assumes constant population size) and Expo (which assumes exponential growth). The vertical lines represent confidence intervals.



a clade of related sequences whose date preceded the time of infection—but this explanation is much less parsimonious than the conclusion that they were all infected from the same source of contamination prior to March 1998.

### Future work

This analysis included two different epidemiological models in their phylogenetic analysis. It has since been shown that substitution rates in HIV differ within individuals, between individuals in a transmission chain, and between strains, which might in future be incorporated into more sophisticated models of rate change to estimate dates of HIV transmission[2]. The AFH cohort has led to identification of several putative cases of child-to-mother HIV transmission through breastfeeding[3]. The prevalence of such an infection route is currently little understood and needs further investigation[4].

### Check your understanding

1. What led these researchers to reject the hypothesis that the foreign workers had infected the AFH children?
2. How did they calibrate their molecular date estimates for the AFH outbreak?
3. Why were estimates from the strict clock and relaxed clock methods both used? What conclusions did the researchers reach by comparing both methods?

### What do you think?

What further analyses could be done to check the reliability of these molecular dates? What recommendations would you give to researchers investigating the source of disease outbreaks using phylogenetic analysis?

### Delve deeper

DNA sequence analysis is being increasingly used as legal evidence in cases of alleged deliberate infection or negligence[1, 5]. Just as the legal use of DNA fingerprinting techniques has been confused by inappropriately confident statistics, so the use of molecular phylogenetic analysis in court cases will require careful attention to, and clear communication of, the confidence limits on molecular phylogenies and date estimates. Will it be possible to give judges and juries a critical understanding of molecular date estimates? What guidelines should be put in place to determine reliability of molecular date estimates in a legal setting? Given that all of the bioinformatics tools are available to anyone, should provision of such evidence be restricted to particular people? If so, how would expertise in this area be recognized?

### References


1. Vandamme, A-M., Pybus, O. G. (2013) Viral phylogeny in court: the unusual case of the Valencian anesthetist. *BMC Biology*, Volume 11, page 83.
2. Vrancken, B., Rambaut, A., Suchard, M. A., Drummond, A., Baele, G., Derdelinckx, I., Van Wijngaerden, E., Vandamme, A-M., Van Laethem, K., Lemey, P. (2014) The genealogical population dynamics of HIV-1 in a large transmission chain: bridging within and among host evolutionary rates. *PLoS Computational Biology*, Volume 10, page e 1003505.
3. Visco-Comandini, U., Cappiello, G., Liuzzi, G., Tozzi, V., Anzidei, G., Abbate, I., Amendola, A., Bordi, L., Budabbus, M. A., Eljhawi, O. A. (2002) Monophyletic HIV type 1 CRF02-AG in a nosocomial outbreak in Benghazi, Libya. *AIDS Research and Human Retroviruses*, Volume 18, page 727.
4. Little, K. M, Kilmarx, P. H., Taylor, A. W., Rose, C. E., Rivadeneira, E. D., Nesheim, S. R. (2012) A review of evidence for transmission of HIV from children to breastfeeding women and implications for prevention. *The Pediatric Infectious Disease Journal*, Volume 31, page 938.
5. Ou, C. Y., Ciesielski, C. A., Myers, G., Bandea, C. I., Luo, C. C., Korber, B. T. M., Mullins, J. I., Schochetman, G., Berkelman, R. L., Economou, A. N., Witte, J. J., Furman, L. J., Satten, G. A., Macinnes, K. A., Curran, J. W., Jaffe, H. W. (1992) Molecular epidemiology of HIV transmission in a dental practice. *Science*, Volume 256, page 1165.

# You are a scientist

## *What do I do now?*

*"A mind which has once imbibed a taste for scientific enquiry, and has learnt the habit of applying its principles readily to the cases which occur, has within itself an inexhaustible source of pure and exciting contemplations . . . he walks in the midst of wonders"*

Herschel, J. F. W. (1851) *Preliminary discourse on the study of natural philosophy*. Longman, Brown, Green & Longman.

## What this chapter is about

You have learned how to collect data, either by extracting and sequencing DNA from biological samples, or by searching public databases. You have learned how to align the sequences, estimate the amount of evolutionary change between them, and use that information to construct phylogenetic trees and estimate evolutionary timescales. More importantly, you have learned to think about DNA sequence in an evolutionary context, comparing sequences and looking for patterns that reveal descent with modification by selection or drift. Now it is time to put all that you have learned into practice.

### Key concepts

- Being a scientist is about what you do, rather than your qualifications and employment
- You can do real scientific research and make a valuable contribution to our shared knowledge

It is surprising, given the amount of time and paper devoted to exploring the issue, that there is no agreed definition of what science actually is. Many attempts have been made to define 'the scientific method', or to describe the motivations and behaviour of scientists in terms of philosophical principles or social rules. Perhaps we face the same problems defining science as we do defining species (Chapter 9). If different individuals or groups do science in different ways, our definition may need to be flexible, suited to the purpose and occasion for which we wish to apply it. I am not cavalier enough to wish to wade into this learned debate. All I wish to say is that being a scientist is about the way you do things. It is not dependent on your qualifications or employment. And that is particularly true of the ideas and techniques discussed in this book. The point of this little coda is to say that, whoever you are, dear reader, you should now feel happy to go out and start applying the basic principles you have learned in this book.

There are several important stages in most scientific investigations. First, you need to think of an issue that puzzles or interests you, or a problem that needs solving, or competing hypotheses that need to be resolved. Then, you need to devise some way of shedding light on that problem or idea. And then you need to communicate what you have learned to other people. You can do all of this. What's more, the field of molecular evolution makes all of these stages particularly accessible to you, because there is much that can be done without access to research budgets or specialist equipment.

One of the great principles of science is sharing. When a scientist reports research they have done, they should provide everything that an informed reader needs to verify those claims. In this way, science is fundamentally different to many other activities, such as commerce, where success may depend on keeping secrets, or the arts, where the outputs may be specific to a particular artist and not able to be reproduced by other practitioners. Obviously, as in any other field of human endeavour, there are selfish scientists who will not share their data, there are ego-driven scientists who wish to prevent others from making progress in the same field, and sadly there are, very rarely, dishonest scientists who cheat and make up their data or manipulate their results. There are also cases where sharing data, methods or materials may not be in the public interest (for example, genetic manipulation of dangerous pathogens). But on the whole, one of the marvellous things about molecular evolutionary biology is that almost all scientists working in the field make their data and methods freely available to anyone who wants to use them.

GenBank is a paragon of sharing (**TechBox 1.1**). Virtually every DNA sequence ever produced is submitted to this global database, and is then available to anyone in the world with access to an internet connection. This is a beautiful example of returning the fruits of scientific labours to the people: research dollars poured into DNA sequencing by governments and private enterprise produce a resource to which we all have equal access. New data is being generated at a phenomenal rate, greatly outstripping the rate at which it is being analysed. There are many fascinating questions in evolutionary biology that can be answered simply by playing around with sequences available on GenBank.

Not only is the data freely available, but so are many of the computer programs used to analyse those data. Some software for molecular genetic analysis is produced by companies that aim to make a profit. But all of the programs I use in my research are produced by academics in universities or research organizations who make their programs freely available to anyone who wants to download them. In addition, there are publicly funded server-based applications that not only provide the software but also run the analysis for you. This is handy, because it means that you do not have to own a computer with sufficient capacity for analysis, as long as you can access an internet connection. Again, the provision of software and servers for molecular genetic analysis is an admirable case of research funded by governments returning a resource to anyone in the community who wishes to use it.

Scientific research is worth little if the results are not communicated to anyone else. This is why most researchers publish the results of their work in one form or another, usually in a peer-reviewed scientific journal. If you are interested in a particular area, the first thing you should do is read about what others have been doing in this field. Review articles, which summarize the state of play in a particular field, are often the most useful place to start. Unfortunately, scientific papers can be difficult to read. If you find your head swimming with jargon, or you feel like crying with frustration at your inability to understand the methods in a particular paper, take heart in the fact that many scientists feel the same. We all struggle at some time to understand a paper, or to stop our eyes from closing while perusing some less than riveting details. But even if you just read the titles and abstracts of papers, you will begin to get a feel for the current issues and the approaches people are taking to address them (an abstract is a short summary of the aims and results of the research, at the beginning of the paper).

You can use public search engines such as PubMed or GoogleScholar to search the scientific literature for papers on a particular topic, or by specific authors. While many journals require a subscription or a pay-per-view fee to access the whole paper, you can usually read at least the abstract for free, and increasing numbers of scientific journals are making their contents freely available online. You will be able to access a wider range of journals if you are a member of a library with paid subscriptions to scientific journals. Try visiting the webpages of scientists working in the field, as they may have summaries of their research or copies of their papers to download. You can even write to scientists and ask for a reprint (copy) of their paper if you cannot get the paper by any other means.

The public availability of scientific literature, molecular sequence data, analytical software, and servers for molecular genetic analysis means that, unlike many fields of scientific endeavour, research in molecular evolution is technically open to anyone. Evolutionary biology is, it must be admitted, predominantly in the hands of the professionals—academics employed at universities, or scientists working in research organizations. Science is most easily practised within professional communities, partly because of the availability of resources such as laboratories and libraries, but more importantly because of the availability of like minds to discuss ideas with. But evolutionary biology was founded on the investigations of passionate and dedicated amateurs. And the post-genomic age makes participation in evolutionary research accessible to anyone determined enough to apply themselves to it.

This book will not give you everything you need to undertake research in molecular evolution. But I hope it gives you a starting point. Most importantly, I hope it stirs your interest, stokes your enthusiasm and fires up your courage. Now the thing to do is get out there and get your hands dirty: think of a question and work out how you could answer it, contrast hypotheses and think of how you would test them, delve into the data and see what you find. Do not fuss about whether your question is too big or too small, as long as it is a question you find interesting, and you think you can see a way of shining some light upon it. The only way to start is begin with something, anything, and see where it leads you. Good luck. I hope you have fun.

# Glossary

Since this book is intended to be accessible to entry-level students with a basic grounding in biology, this glossary provides definitions of some words and phrases that may be familiar to some readers but mysterious to others. It is not an exhaustive dictionary of biological terms used in the book, but contains explanations of a number of key concepts that can be confusing or ambiguous. I have bundled together related terms to make meanings clearer, so many terms in the glossary are cross-referenced to other entries. This cross-referencing doesn't imply equivalence. For example you may look up 'zygote' and be referred to 'gamete': these words mean very different things but are explained in the same paragraph. I have not included terms that occur only in one place in the text, and are defined at first mention (for example, 'affine gap penalty' is used only in **TechBox 10.2**, where its meaning is explained), so if you don't find a word in the glossary, try looking in the index for the page on which it is first mentioned. Finally, a word of warning: as with any language, the meaning of scientific words can vary between users. The purpose of this glossary is to explain terms as they are used in this book. You may find some words included here that are used in a different sense by other authors.

**3'** (three-prime) and **5'** (five-prime) refer to the two different ends of a polynucleotide strand. The 5' end terminates with a phosphate molecule which is attached to the 5' carbon of the sugar molecule. The 3' end terminates with a sugar, the third carbon atom of which has a free hydroxyl group which forms a phosphodiester bond with the phosphate of a newly added nucleotide. This is why DNA synthesis always proceeds in the 5' to 3' direction—new bases can be added to the 3' end (carbon with free hydroxyl group), not to the 5' end (phosphate). See **TechBox 2.1**.

**454 sequencing**: see pyrosequencing

**Acquired characteristics** are modifications to phenotype that occur during an individual's lifetime. Some early evolutionary thinkers believed that acquired characteristics of the parent could be inherited by its offspring. Weismann (**Hero 2**) argued convincingly that acquired characteristics could not be inherited by an individual's offspring, so did not contribute to descent with modification (evolution). See germline.

**Active site**: see enzyme

**Adaptation** can refer to the process whereby a trait has evolved, or to the end product of this process. An adaptation is a feature of an organism that suits it to a particular way of life, often described as increasing the 'fit of an organism to its environment'. The recognition of adaptations does not depend upon the mechanism of their generation: for example a clear treatise on adaptation was given by William Paley in 1802 (i.e. long before Darwin's theory of evolution was published) where he described how living organisms are notable for having features that are 'formed . . . for the purpose which we find it actually to answer'. However, natural selection is the only known mechanism for generating adaptations.

**Adenine (A)** is one of the four DNA bases that make up the 'alphabet' of the genetic code. See base.

**Afrotheria**: see placental

**AIDS**: see human immunodeficiency virus (HIV)

**Alignment** is the process of arranging homologous sequences so that comparisons can be made between the character states (nucleotides or amino acids) that have descended from a shared ancestral character (see **TechBox 10.2**).

**Allele** refers to one of two or more alternative versions of a heritable trait. The term allele predates molecular genetics, and can be applied to alternative forms of any heritable trait that are transmitted independently and segregate randomly. For examples, alleles can be the different variations on the possible genetic sequence at a particular locus that are present in the same population. Since diploid individuals carry two copies of every locus, they may be heterozygous (carry two different versions—alleles—of the same locus, one on each chromosome) or homozygous (two copies of the same allele). Alleles at closely linked loci will tend to be inherited together: sets of linked alleles form a haplotype.

**Alpha globin**: see haemoglobin

**Amino acids** are the building blocks of proteins. There are 20 common amino acids, which vary from each other in the functional groups attached to the central carbon atom. The sequence of bases in a protein-coding gene specifies an exact sequence of amino acids when it is translated see **TechBox 6.1**.

**Analogy**: see homology

**Ancient DNA (aDNA)** is, in the broad sense, any DNA sample derived from non-fresh material. Examples mentioned in this text include DNA extracted from museum specimens (pinned flies, preserved thylacines), frozen samples (woolly mammoths) and subfossils (moa bones). DNA decays over time, and currently there are no reliable DNA sequences from samples over a million years of age. Because the amount and quality of DNA in a sample reduces with age, ancient DNA extraction and amplification relies on sterile techniques to prevent the sample DNA being overwhelmed by contaminating DNA from other sources.

**Anemia** (anaemia) is a disease that results from insufficient oxygen being transported in the blood. This may have a genetic cause (for example, a haemoglobinopathy) or an environmental cause (for example, iron deficiency).

**Annotation** is the addition of information about a DNA sequence to its entry in a database, for example the gene name, chromosomal location, or regulatory and coding regions of the sequence. Large-scale sequencing projects tend to use automated annotation to identify sequence features by similarity to a reference database of sequences of known function.

**Anthrax** is an infectious disease of livestock and humans, caused by the bacterium *Bacillus anthracis*. It is transmitted by spores which can enter the body through the lungs (causing pulmonary anthrax), the digestive system (causing gastrointestinal anthrax) or the skin (causing cutaneous anthrax).

**Antimutator**: see mutator

**Arthropoda** is a phylum of animals, including spiders, insects and crustaceans, characterized by segmented bodies, paired jointed limbs and an exoskeleton (a hard covering over the body) made of chitin.

**Artificial selection**: see selection

**Artiodactyl**: see placental

**Assembly**: see high-throughput sequencing.

**Association studies** aim to identify genetic markers that are significantly associated with a particular trait, such that individuals with that marker have a higher chance of having that trait than individuals without the marker. Association studies may target a particular informative pedigree, or may be applied to samples from a population (for example, a study based on a biobank). A genome-wide association study (GWAS) uses a sample of people to compare phenotypic outcomes (for example, particular diseases or desirable agricultural outcomes) to SNP alleles, then uses statistical analyses to identify significant associations between alleles and outcomes (that is, if individuals carrying a particular SNP or haplotype have a greater chance of having the phenotypic outcome than expected by chance). See **TechBox 3.2**.

**Autapomorphy** is a trait unique to one lineage from a group of related lineages being compared in a systematic analysis. Because it is not shared with other related lineages, an autapomorphy cannot be used to determine phylogenetic relationships between lineages, but it can be used to estimate rates of change along a single lineage.

**BACs** (bacterial artificial chromosomes) are circular DNA molecules into which can be inserted a DNA sequence from another organism (say, several hundred kilobases of the human genome), which can then be replicated in bacterial cells.

**Bacteriophage** (also known as phage) are viruses that infect bacteria. Because phage have simple genomes and are amenable to laboratory experiments, phage genetics has played an important role in the development of molecular genetics (e.g. see **Figure 2.5**)

***Bacillus***: see anthrax

**Barnacles** are arthropods in the class Cirripeda: underneath the calcareous plates that form their 'shell', most barnacles have typical aspects of the arthropod body plan like segmented legs and antennae. The larvae are free-swimming then settle on a surface, but the adult form is immobile, cemented to the surface.

**Bases** are small biomolecules that make the four 'letters' of the DNA 'alphabet': adenine (A), cytosine (C), guanine (G) and thymine (T). RNA molecules have a slightly different alphabet, using uracil (U) instead of T. A and G are purines (double-ring bases) and T and C are pyrimidines (single-ring bases). A nucleotide consists of a base attached to a sugar molecule which is attached to a phosphate. DNA is a polynucleotide because it is made of series of nucleotides joined together by phosphodiester bonds which link the sugar of one nucleotide to the phosphate of the next nucleotide in the chain. The bases of one polynucleotide can form hydrogen bonds with the bases of another, linking two nucleotide strands. Watson-Crick pairing rules state that A pairs with T, and G pairs with C (in RNA molecules U pairs with A as T does in DNA). In DNA, it is these bonds between bases that bind the two strands of the double helix together. In RNA, base pairing can bind one part of the polynucleotide to another to generate secondary structure.

**Base composition bias** refers to uneven representation of the four DNA bases, A, C, T and G. Chargaff's rule states that the amount of each base should match its pair (G = C, A = T), but the relative amounts of GC may vary with respect to AT. For example, bacterial genomes may vary in base content from 25 per cent GC (and therefore 75 per cent AT) to 75 per cent GC (and 25 per cent AT). Base composition can also vary across a single genome. Base composition bias can influence molecular phylogenetic inference because of its impact on substitution probabilities (see **TechBox 13.1**).

**Bayesian inference** is a form of statistical inference, which begins with a set of prior beliefs (prior probabilities), then uses the observed data to modify those beliefs in the face of the evidence (posterior probabilities). See **TechBox 13.2**.

**Beta globin**: see haemoglobin

**Bilaterian** can refer broadly to any animal having bilateral symmetry, such that if you cut it down the plane of symmetry you would get two similar halves (imagine slicing a cat in half from between the ears to the tail, and you would get two similar cat-halves). But in this book, the term bilaterian is used to refer to a clade of animals that includes all members of the superphyla Lophotrochozoa (including molluscs and annelids), Ecdysozoa (including arthropods and nematodes) and Deuterostomes (including echinoderms and chordates). The bilaterian clade excludes the earlier-branching animal lineages of Cnidaria (jellyfish and their kin) and Porifera (sponges).

**Biodiversity** is a contraction of the phrase 'biological diversity', and is intended to capture the variation in numbers and kinds of organisms. Often, biodiversity is measured by summing the total number of species found in a given area, however some biodiversity measures are more sophisticated, for example including the distribution of organisms among different types, or capturing the total amount of evolutionary divergence as represented by phylogenetic branch length.

**Biogeography** is the study of the geographical distribution of organisms.

**Bioinformatics** is, broadly speaking, the application of computational and statistical techniques to the analysis of biological data, although common usage of the word applies more narrowly to statistical analysis aimed at detecting patterns in large collections of molecular sequence data.

**Biostratigraphy**: see geochronology

**BLAST** is a bioinformatic method for searching a database of sequences for the closest match to a query sequence. See **TechBox 1.2**.

**Body plan** refers to a basic level of physical organization in animals. Body plan is sometimes interpreted as fundamental differences between deep lineages, such as phyla.

**Branch**: see edge

**Calibration** can be used as a verb in molecular phylogenetics, as the process of converting some measure of genetic distance or branch length to time using a known date of divergence, or to refer to the date of divergence used in such an analysis: see **TechBox 14.1**.

**Carnivorans** are mammals from the order Carnivora, which includes cats, dogs, bears, weasels and seals. Not to be confused with the term 'carnivore' which refers to any meat-eating animal (or plant).

**Chain termination sequencing method**: see Sanger method, **TechBox 5.1**

**Chance**: see random

**Character** is a broad term meaning any discrete heritable trait possessed by an organism, that may vary independently of other such characters. Each character has a number of different character states (possible variants of that character). Systematics is based on the identification of informative character states that define sets of related lineages: see homology.

**Chargaff's rules** describe the relative amounts of the four bases in DNA molecules. Chargaff's first rule states that the amount of A equals the amount of T, the amount of G equals the amount of C. This observation was important for the discovery of Watson-Crick pairing. Chargaff's lesser-known second rule is that the relative amounts of the four bases differ between species, a phenomenon now often referred to as base composition bias. See bases.

**Chiasma** (plural: chiasmata): see crossing-over

**Chloroplasts** are a type of plastid, which are organelles found in photosynthesising cells of plants and algae. Chloroplasts capture the energy from light and use it to drive a series of reactions that result in chemical energy storage. Chloroplasts, like mitochondria, were originally derived from free-living bacteria, and retain a small circular genome. Chloroplast DNA sequences are frequently used in plant systematics.

**Chordata** is a phylum of animals characterized by having a 'head and tail' body plan, with a notocord (a supporting rod down the back), a dorsal nerve cord, a pharynx (feeding structure on the head e.g. gills in fish) and a muscularized post-anal tail. You are a chordate, so you can check these features on your own body, except that, in humans, the tail is only present in the early embryo, and the notocord is replaced during development.

**Chromosomes** are cellular structures containing the genome, consisting of nucleic acids and supporting proteins (such as histones). Some small genomes are circular, such as those of bacteria, mitochondria and chloroplasts (see **Figure 5.4** for an example). Larger genomes tend to be divided into a number of linear packages (see **Figure 5.7b** for an example)—we refer to each separate package, inherited as a unit, as a chromosome. Diploid cells carry two copies of the genome, on pairs of homologous chromosomes that each contain the same genes (but may have different alleles). Before cell division, each chromosome replicates to produce two sister chromatids, held together at a region called the centromere, which is important for proper segregation of chromatids into daughter cells at cell division (see **Figure 5.6**).

**Clade** represents a group of related lineages united by descent from a common ancestral lineage. Phylogenies generally describe sets of nested clades, each defined by sharing a common ancestral node (see **Figure 11.22**). Clades, defined by the set of all lineages descended from a specific node in a phylogeny, may be used as the basis of taxonomic grouping under the cladistic framework for systematics.

**Cloning** has several broad meanings in molecular genetics. Cloning may refer to any method for producing copies of a particular sequence, for example by inserting the sequence into a cultured cell (e.g. see BACs). Cloning may also refer to the production of a colony of cells derived from a single cell, or to the production of offspring that are genetically identical (or near identical) to a single parent. Naturally occurring clones may be produced by asexual reproduction or when a single embryo splits to form identical twins. Artificial clones may be produced by transferring the nucleus of an adult cell into an anucleate embryo. See **TechBox 10.1**.

**Coding region** typically refers to a DNA sequence that codes for an amino acid sequence—that is, the exons of a protein-coding gene. However, the term is sometimes used more broadly to refer to all of the transcribed DNA in the genome, whether it codes for proteins or RNA molecules.

**Codon** is a three-base sequence that specifies a particular amino acid (or a stop codon, which signals the end of a peptide). When a messenger RNA transcript is translated into a protein sequence at the ribosome, each codon is matched to the complementary recognition sequence on the correct transfer RNA molecule, which joins the corresponding amino acid to the growing peptide chain (**Figure 2.8**). See translation.

**Confidence intervals** (CIs) refer to a range of likely estimates for a parameter, reflecting the precision of the parameter estimate. If a study reports the 95 per cent confidence intervals around an estimate, they are stating that there is a 95 per cent chance that the true value falls within the stated range, given the assumptions of the analysis. Note that the true value may fall outside the CIs if the assumptions of the analysis are incorrect, or if the model is not a good description of the evolution of those sequences.

**Copy errors** are changes in the DNA sequence that arise from imperfect replication of a nucleotide strand. Every time the genome is copied there is a small but finite chance that the sequence in the copy will not be exactly the same as in the parent strand. The more times DNA is copied the more errors will accumulate. See Chapter 4.

**Crick, Francis** (1916–2004): one of the co-discovers of the structure of DNA, who also made significant contributions to understanding how the genome specifies phenotype, for example predicting and decoding aspects of the genetic code, and formalizing the central dogma. see **Hero 6**.

**Crossing-over** is exchange of genetic material between **homologous chromosomes,** resulting in **recombination** (disruption of linkage) between genetic markers (see **Figure 5.6**). Unequal crossing over, which can occur when repeat sequences become misaligned, results in the net loss of sequences from one chromatid and the net gain of sequences in the other (**Figure 5.10**). The crossover points are referred to as chiasmata (singular: chiasma).

**Cryptic species** are those that do not show any detectable phenotypic differences from one or more other species, despite evidence of consistent and sustained reproductive isolation from other such populations. Cryptic species are often detected by observing a significant level of genetic divergence from otherwise similar populations of a particular organism. See **Case Study 9.2**.

**Cryptozoology** is the study of elusive, and probably illusive, organisms, for which there is no conclusive proof. See **Case Study 1.1**.

**Cytosine (C)** is one of the four DNA bases that make up the 'alphabet' of the genetic code: see bases.

**Darwin, Charles** (1809–1882), though he was not the first to describe evolution, essentially created the science of evolutionary biology through his rigorous investigation of the evidence for the transformation of species over time, and by providing a plausible mechanism for evolutionary change (descent with modification by natural selection).

**Degenerate** refers to redundancy of information. The genetic code is said to be degenerate because many different codons can specify the same amino acid (a non-degenerate code would have exactly one codon for each amino acid: See **TechBox 6.1**). **Degenerate primers** are a set of primers with slightly different sequences, used when the exact target sequence is unknown (a non-degenerate primer would bind to one specific sequence only: **TechBox 6.2**).

**Desirogram** isn't actually a proper scientific term. I got it from Ainsley Seago, who created many of the illustrations for this book. It refers to a phylogeny that is a manifestation of someone's opinion about the evolution of a group, so it's a subjective statement of belief rather than an objective result of a phylogenetic analysis. But really, when you stop and think about it, most published phylogenetic trees are 'desirograms' because the tree that makes it into print is the one that the researchers believe tells the right story, even though alternative stories could be supported by different analysis or different data.

**Dicot** is a plant from one of two major groups of flowering plants. The dicots include most flowering trees, daisies, strawberries and so forth. The other major group is the monocots which includes the grasses and palms.

**Dideoxy sequencing method**: see Sanger method (**TechBox 5.1**)

**Diploid** cells have two copies of each chromosome. For example, most human cells are diploid, carrying two copies of each of the 22 autosomal chromosomes (each copy containing the same genes but potentially carrying different alleles) and two sex chromosomes (two Xs or an X and a Y). Human germ cells (sperm and eggs) are haploid, carrying one copy of each of the autosomes and one sex chromosome.

**Divergence date** is the point in time when a single interbreeding population became separated into two separate lineages. The divergence date marks the age of the **last common ancestor** of a clade, but usually predates the evolution of distinct characteristics that define members of the descendant lineages. In phylogenetic analysis, the divergence date is marked by a node (branching point) where two lineages split. In some cases the divergence date between lineages may represent a discrete event when two populations separated, but often divergence will happen gradually over a protracted period so there is no precise date at which the lineages become entirely separate.

**Diversification** is the generation of many distinct lineages from an original ancestral lineage. More formally, the net diversification rate of a clade is defined as the outcome of gain in lineages over a time period as a result of the addition of lineages by speciation and the loss of lineages by extinction.

**DNA amplification** is the production of many copies of a DNA sequence. The polymerase chain reaction (PCR) is a common means of amplifying a DNA sequence. See **TechBox 4.2**.

**DNA barcoding** typically refers to the practice of assigning biological samples to species by comparing a DNA sequence from the sample to a curated taxonomic database (see **TechBox 9.2**). DNA barcoding can also refer more broadly to the conviction that the base sequence at a single locus (for example, a mitochondrial gene sequence) may be able to act as a universal means of species identification. DNA barcoding may also be used not only for species identification but also for species discovery (assigning undescribed specimens to new species) or species delimitation (deciding how many species a set of samples should be divided into).

**DNA fingerprinting** refers to any method for genotyping individuals, revealing a unique combination of alleles carried by an individual that can be used to distinguish them from all other individuals in the population. DNA fingerprints were originally based on restriction length fragment polymorphism (RFLPs), but now more commonly based on a set of microsatellite loci, or SNPs at defined polymorphic loci (see Chapter 3).

**DNA hybridization** refers to the technique of combining DNA from multiple sources, heating to separate the double-stranded helices into single strands, then cooling so that the single strands reanneal into double strands. When a helix forms between DNA strands from different sources, the strength of binding between the strands is proportional to the number of complementary bases that match between the sequences. The amount of heat required to separate the hybridized strands will be proportional to the number of matched bases between the strands, so the melting temperature reflects the similarity between DNA sequences from different sources (see Chapter 5). In this way, DNA hybridization can be used as a measure of the amount of difference between the two genomes (**TechBox 11.1**).

**dNTP** (deoxyribonucleotide triphosphate) is the form of nucleotide used in DNA synthesis, so dNTPs must be added to DNA amplification reactions such as the polymerase chain reaction (PCR).

**Dominant** alleles create the same phenotypic effect whether present in the homozygous or heterozygous state, so it only takes one copy of the allele to generate the trait.

**Downstream** refers to any sequences on the 3' end of the sequence of interest—so for example the polyadenylation signal is usually downstream of the coding sequence of a gene, found after the stop codon. **Upstream** refers to sequences on the 5' end of a sequence of interest—for example, transcription initiation sequences are usually found before the start codon of a gene. See Chapter 6.

**Drift** is a change in allele frequencies in a population from one generation to the next due to incomplete sampling of the alleles in the parent generation. Drift refers specifically to random fluctuations in allele frequencies across generations (not changes in frequency due to selection, migration or mutation). The effect of drift on allele frequencies is most pronounced for neutral alleles, or in populations with small effective population size.

***Drosophila*** is a genus of fruit flies commonly used in genetic experiments, as they can be reared in large numbers in the laboratory and have short generation times (**Figure 3.6**). *Drosophila melanogaster* was developed as a model genetic organism by Thomas Hunt Morgan in the early part of the twentieth century, then adopted by other geneticists such as Theodosius Dobzhansky (**Figure 3.9**). Early genetic studies were aided by the very large (polytene) chromosomes in the salivary glands of fruit flies, marked with banding patterns that allowed the inheritance of alleles to be tracked, making inheritance of chromosomal regions observable under the microscope.

**Echidna**: see monotreme

**Echinoderms** are members of the animal phylum Echinodermata, including starfish, sea cucumbers, sea urchins, crinoids and brittlestars.

**Echolocation** is the detection of objects using reflected sound. Animals such as whales, bats, and some birds (e.g. cave swiftlets) emit sounds then use the echoes of these sounds from solid objects to build a picture of their surroundings.

**Edge** refers to the line connecting two nodes in a phylogeny (including the lines connecting an internal node to a tip). This term is derived from mathematical graph theory, but is used by biologists as a synonym for 'branch' (**Figure 11.22**).

**Effective population size ($N_e$)** broadly represents the number of parents that contribute alleles to the next generation. This number is generally much smaller than the total number of breeding individuals in the population, because generally not all individuals in a population successfully reproduce, and even those that do may not pass all of their alleles on to viable offspring. See TechBox 8.2

**Empirical** refers to observation or measurement: an empirical estimate is derived from the data. **Theoretical** refers to predictions made from consideration of hypotheses; theoretical estimates are derived from the predictions of a model.

**Endangered species** are species that are considered to be threatened with extinction. Endangered is also a defined category under the IUCN red list, which is a scheme for classifying species according to the perceived likelihood of extinction, drawing on information from the global population size of the species, extent of the species distribution, and the rate of decline in distribution and population. There are three recognized levels of extinction threat under the IUCN classification: vulnerable, endangered and critically endangered.

**Endemic species** are found only in the area under consideration and nowhere else. Therefore endemism is considered with reference to a specific area.

**Endonuclease** is an enzyme that cuts a DNA or RNA molecule in the middle of the strand by breaking the phosphodiester bonds holding adjacent nucleotides together. An **exonuclease** removes nucleotides from the end of a nucleotide strand.

**Enzymes** are proteins that perform a specific catalytic role, changing the form of some substrate by actively making or breaking chemical bonds. The active site of an enzyme refers to the part of the peptide that binds to a substrate and performs some catalytic function. Proteins with enzymatic functions usually have a name ending in 'ase', for example DNA polymerase catalyses the formation of a polymer of nucleotides to make a DNA polynucleotide.

**Epitope** is the part of a molecule recognized by the immune system, which can form antibodies specific to that epitope.

***Escherichia coli (E. coli)*** is a species of bacterium found in the human digestive system (and also present in the environment). Like *Drosophila*, *E.coli* is a model organism in genetics research (Figure 3.5).

**ESTs (expressed sequence tag)** are produced by sequencing the messenger RNA content from a cell, so represent the sequences being transcribed by that cell. Genes can be identified by comparing sequences from ESTs to genomic data or by hybridizing ESTs to chromosomes to determine the location of the gene. The messenger RNA content of cells is increasingly being used in high-throughput sequencing as a way of capturing a large number of exonic sequences.

***Eucalyptus*** (gum trees) is a speciose genus of trees, found in Oceania, forming a predominant part of many Australian ecosystems (Figures 4.35, 9.3 and 9.10). Due to their rapid growth rates, eucalypts are a common timber species around the world, though they have become invasive species in many places. The smell of eucalyptus oil often makes Australians feel nostalgic.

**Eugenics** in the broad sense refers to the active intervention in reproduction to achieve a change in the frequency of heritable traits within a particular population, usually applied only to human genetics not to other forms of artificial selection. However, some people apply the word eugenics only to the selection of desirable traits whereas others include programmes aimed at reducing the incidence of heritable disease.

**Eukaryote** cells have their genome contained within a distinct organelle called the nucleus. Most eukaryotes also have mitochondria. Animals, plants, fungi and protists are eukaryotes. The genome of prokaryotes, the archaebacteria and bacteria, is not enclosed in a nucleus.

**Eusocial** species live in social groups that show a high degree of co-operation amongst individuals, with reproductive division of labour (a small number of individuals produce offspring which the rest of the colony help to raise) and often specialization to different tasks (food gathering, defence etc.) Examples of eusocial animals include naked mole rats, honey bees and termites (Figure 10.7).

**Evolutionary tree**: see phylogeny

**Exhaustive search** considers all possible states: in phylogenetics, an exhaustive search evaluates every possible phylogeny for a set of taxa. A **heuristic search** uses some strategy to consider a sub-set of all possible states, with the aim of finding the optimum state without having to evaluate every possibility; in phylogenetics, a heuristic search makes a partial exploration of tree space, using a search strategy directed to finding trees that provide the best explanation of the data (e.g. with a higher likelihood or posterior probability).

**Exons** are the parts of protein-coding genes that specify an amino acid sequence. **Introns** are regions of protein-coding genes, which may contain regulatory elements but are excised from the messenger RNA transcript before translation.

**Exonuclease**: see endonuclease

**Expression**: see gene expression

**Extinction** occurs when the last member of a species, or other distinct taxon, dies. Extinction represents the loss of a unique lineage (or a unique set of alleles).

**Fisher, Ronald Aylmer** (1890–1962) was a key figure in the development of population genetics, as well as establishing many key statistical techniques such as maximum likelihood (and whenever you use 0.05 as a cut-off for statistical significance, think of Fisher: **Figure 7.16**). Fisher demonstrated that even alleles with a very small advantage over other alleles in the population could go to fixation by selection in a large randomly-mating population. Fisher was a committed eugenicist, believing that selective breeding was an important tool in improving human health and prosperity.

**Fitness** can be defined in many ways (see **TechBox 7.1**), but usually reflects the average reproductive success of different types of individuals, genotypes or alleles in a population. In this book, we consider that the fitness of an allele (or genotype) is reflection of its relative selective advantage or disadvantage relative to the other alleles in the population, and thus influences the chance of that allele going to fixation.

**Fixation** of an allele occurs when the frequency of that allele in a given population reaches 1, so that all members of the population carry the same allele at a particular locus and there is no polymorphism at that locus. See also substitution.

**Flanking sequence** refers to the DNA sequence either side of the locus of interest.

**Flo** (2013–), the most laudable chicken I have ever had the pleasure to meet, a fine layer and a good egg (**Figure 6.20b**). Officially given the title The Brave and Resourceful Flo after being the sole survivor of a fox attack, living on her wits in the bush until her humans returned.

**Frameshift** mutations result from the insertion or deletion of bases from a protein-coding sequence in multiples other than three, which disrupts the way that the codons are read from the subsequent (downstream) sequence, changing the amino acid sequence specified by the rest of the sequence. See also indel.

**Gamete** is a haploid reproductive cell, such as sperm or eggs. In diploid organisms, gametes are produced by meiosis. Gametes from two parents fuse to form a diploid zygote, which develops into an embryo.

**Gap**: see indel

**Gene** is a surprisingly slippery term, sometimes used in the abstract sense to indicate an independent heritable trait, sometimes used to describe a DNA sequence with particular features that allow it to be transcribed (see Chapter 6), sometimes to indicate a variant of a heritable trait (allele), and sometimes as a short-hand for any locus in the genome. Mendel was the first to describe the action of genes—discrete inherited units of heritable information—but the term 'gene' was introduced in 1909 by Wilhelm Johanssen (who also introduced the words 'genotype' and 'phenotype' in order to distinguish the heritable component of variation from that caused by the environment). Genes began to take on a physical reality when in 1910, the great fruit fly geneticist Thomas Hunt Morgan showed that particular genes were located on specific chromosomes, however the nature of the gene was unknown. In the 1940s, genes were defined as the units of hereditary information, each of which produces a particular protein. With the discovery of the genetic code, a gene could be recognized as a particular DNA sequence that carries the information needed to produce a gene product, such as a peptide or RNA molecule. 'Gene' appears to be somewhat like 'species', of which Charles Darwin said 'No one definition has as yet satisfied all naturalists; yet every naturalist knows vaguely what he means when he speaks of a species'. Thomas Hunt Morgan declared, in his Nobel prize acceptance speech in 1933: 'At the level at which the genetic experiments lie it does not make the slightest difference whether the gene is a hypothetical unit, or whether the gene is a material particle'.

**Gene expression** is the process whereby the information in the genome is converted into biological structures and processes. Typically this involves the recognition of regulatory elements by the transcription machinery, which makes an RNA transcript of the gene. The RNA may form a functional product, or it may act as a template for the construction of a peptide.

**Gene family** is a set of related genes, generated by duplication from an ancestral sequence. Related copies of the same gene within the same genome are referred to as paralogs, to distinguish them from orthologs (related genes in different lineages). In other words, paralogs are produced by gene duplication, orthologs are produced by speciation (lineage divergence). For example, the two alpha globin genes in the human genome, *alpha globin 1* and *2*, are paralogs, but *human alpha 1* and *chimp alpha 1* are orthologs.

**Gene flow** describes the movement of alleles from one population to another, typically by migration of individuals, movement of gametes, or through hybridization where populations meet.

**Gene pool**: see population

**Gene regulation** is the control of gene expression, so that the gene product is produced in appropriate amounts when and where it is needed, in response to external or internal signals.

**Generation time effect** refers to the prediction that lineages that have a shorter generation turnover time (that is, the time it takes for an embryo to become an adult and produce another embryo) should have a higher mutation rate, because their genomes are copied more often per unit time and therefore are expected to accumulate more DNA copy errors. The generation time effect has been observed for DNA sequences in many vertebrates, and has also been proposed for plants.

**Genetic code** is the set of 64 possible three-base codons that correspond to 20 amino acids (see **TechBox 6.1**).

The universal genetic code is not actually universal, but is the code used in the majority of genomes. However, there are many known variations on the code, differing from the universal code by one or several codons. The code is said to be degenerate because multiple codons specify the same amino acid, and these synonymous codons are effectively interchangeable without altering the protein made from the sequence. See **TechBox 2.3**.

**Genetic drift**: see drift

**Genetic marker** is any detectable allele that allows a certain piece of DNA to be identified. Markers can be used to track the inheritance of linked alleles (haplotype).

**Genetically isolated**: see population

**Genome** refers to all of the DNA inherited as a coherent set. The human nuclear genome is arranged on 23 different chromosomes. A diploid cell contains two copies of the nuclear genome (46 chromosomes total); haploid gametes contain only one copy of the nuclear genome (23 chromosomes). Most human cells also contain multiple copies of the mitochondrial genome, a circular DNA molecule replicated within the mitochondria, and transmitted to the next generation in the cytoplasm of the egg cell. Most plant cells also contain a chloroplast genome.

**Genome-wide association study (GWAS)** is the analysis of a large set of genetic markers, distributed throughout the genome, to detect any markers that are significantly associated with a particular phenotypic outcome (found in more people with that trait than expected by random sampling alone).

**Genomic system** is short-hand for the core hereditary system of nucleic acids plus proteins common to all life on earth. I must admit that I made this term up while I was writing this book, so it is unlikely to appear anywhere else.

**Genotype** can refer to the total genetic information carried by an organism (as distinct from its phenotype); the specific alleles carried by an individual at a particular locus; or to the process of determining the alleles carried by an individual (see DNA fingerprinting).

**Genus** is a group of related species, defined within a taxonomic hierarchy. In binomial nomenclature, genus is the first part of a species formal name.

**Geochronology** is the discipline in earth sciences concerned with determining the absolute or relative ages of geological strata (layers of rock, minerals or sediments with defined age ranges). The most common dating techniques make use of the decay of radioactive elements such as uranium. Many rock types cannot be dated directly (such as sedimentary rocks) so their age is inferred by measuring the age of underlying and overlaying strata, or by the presence of fossils of characteristic species (a process known as biostratigraphy).

**Germline** cells can pass genetic information to future generations. Soma (body) cells do not copy their genomes to the next generation. In most animals, the germline is set aside early in development as the gamete-producing cells; all other body (soma) cells die when the individual dies. The situation is less clear cut in many plants, where the gametes can be formed from body tissues, and parts of the adult body can reproduce vegetatively (grow a new individual from a piece of the parent's body). See **Case Study 4.1**.

**Googol** is $10^{100}$, or ten duotrigintillion, or a one followed by one hundred zeros.

**Gradualism** refers to a model based on the cumulative effect of many small changes. Darwinian evolutionary theory proposes that most evolutionary change is gradual, achieved by a series of substitutions of alleles each of relatively small (or no) phenotypic effect. Gradualism does not refer to speed of change (which may vary over time or between lineages) and does not discount periods of rapid change.

**Guanine** (G) is one of the four DNA bases that make up the 'alphabet' of the genetic code: see bases.

**Guthrie test** (heel prick test) is the practice of taking a blood sample from newborn babies by pricking the baby's heel and blotting the blood onto a card. The blood samples are then analysed for evidence of a range of metabolic disorders, including phenylketonuria. In most cases, the Guthrie card is stored with the baby's and mother's names, and the place and date of birth. Stored Guthrie cards represent a biobank, and have been used in medical research, in court cases, and to identify victims of a terrorist attack.

**Haeckel, Ernst** (1834–1919) was a doctor, scientist and artist who used comparative development and anatomy to illuminate evolution. See Chapter 11.

**Haemoglobin** is a blood protein, made of two alpha globin chains and two beta globin chains, that transports oxygen around the body. **Haemoglobinopathies** are disorders arising from mutations in the various globin genes, for example thalassemia which results from underproduction of one of the globin chains.

**Haldane, John Burdon Sanderson** (known as JBS: 1892–1964) was a biochemist, physiologist and evolutionary biologist, one of the key contributors to the formation of the neo-Darwinian synthesis. Haldane was also a great popularizer of science, publishing many readable and entertaining books and essays, and, as a committed socialist, he wrote a regular column for *The Daily Worker*. His larger-than-life personality and outrageous behaviour generated countless stories and legends, retold with relish (and probably improved) by his student John Maynard Smith. See **Hero 7**.

**Hamilton, William Donald (Bill)** (1936–2000) was a key figure in the development of the 'gene-centred' view of evolution, developing ideas of kin selection to explain the evolution of eusociality, and evolutionary arms races to explain the evolution of sex as a means of generating genetic variability (particularly as an adaptation to parasitism).

**Hamming distance** is a term from information theory that describes the number of positions at which two strings of symbols differ from each other. For two aligned nucleotide or amino acid sequences, the Hamming distance is the observed number of differences between them see p-distance.

**Haploid**: see diploid

**Haplotype** is a set of linked alleles that are usually inherited together as a unit.

**Helicase** is an enzyme that separates the two strands of a DNA helix so it can be replicated.

**Heterozygote** is a diploid organism that has two different alleles at the locus of interest, whereas a homozygote carries two copies of the same allele. The heterozygosity of a population is the proportion of heterozygous at a particular locus, so is a measure of the genetic variability of a population. Severe reduction in effective population size is likely to lead to inbreeding (mating between relatives) which is expected to reduce the heterozygosity of a population, because an increasing proportion of the population carry alleles copied from the same recent ancestor. Inbreeding will thus ultimately reduce the number of polymorphic loci.

**Heuristic search**: see exhaustive search

**Higher taxa**: see taxon

**High-throughput sequencing (HTS)** is a catch-all term for a large (and growing) number of techniques that produce large amounts of sequence data through parallelized reactions: see **TechBox 5.2**. High throughput sequencing typically results in large numbers of short sequence fragments (reads) which must then be combined into longer sequences using bioinformatics analysis (assembly). In this book, I use 'high-throughput sequencing' rather than the common phrase 'next generation sequencing' (next-gen, or NGS) because it seems to have a broader generality. Contrast to Sanger sequencing **(TechBox 5.1)**.

**Homolog**: see gene duplication

**Homology** refers to similarity by descent. If you wish to uncover evolutionary relationships, it is important to distinguish homologies (which are evolutionary signal) from analogies (which represent noise, because they do not reflect descent). Homologous traits are similar because they were copied from the same ancestral trait. Analogous traits are superficially similar traits that have been arrived at independently in different lineages (i.e. same ends from different starts). Some analogies are the result of convergent evolution, whereby similar selection pressures promote the evolution of the same (or similar) features in different species: for example, many different insect lineages have evolved the same amino acid changes that confer resistance to insecticides (see **Case Study 7.1**). Other analogies may result from chance, for example a particular site in a DNA sequence may happen to acquire an adenine at the same position in two different lineages. Sequence alignment is the process of arranging DNA or protein sequences so that homologous sites, originally copied from the same position in an ancestral gene, can be compared in order to uncover evolutionary patterns.

**Homozygous, Homozygosity**: see heterozygote

**Horizontal gene transfer (HGT)** is the movement of DNA between individuals other than the transmission of the genome from parent to offspring. Examples include the bacteria taking up genes from the genomes of distantly related lineages, the transfer of genes from the mitochondrion to the nucleus, the formation of recombinant virus genomes, or the movement of DNA from parasite to host (see **Case Study 10.2**). Horizontal gene transfer produces non-tree-like signal in sequence data (see Chapter 11).

**Human immunodeficiency virus (HIV)** is a retrovirus that infects cells of the immune system, reducing the effectiveness of the immune response and causing acquired immune deficiency syndrome (AIDS) which increases vulnerability to infections and particular cancers.

**Huntington's disease** is a neuromuscular disorder caused by inheritance of a single dominant allele. The *Huntington disease* gene *HD* contains a trinucleotide repeat region which causes disease if it contains more than a threshold number of repeats.

**Huxley, Thomas Henry** (1825–1895) was a doctor and scientist, who used comparative anatomy to investigate and illustrate evolution. Huxley is famous for his vociferous promotion of Darwin and evolutionary theory, for which he earned the epithet Darwin's Bulldog, but he was also a tireless champion of scientific research and education.

**Hybridization** is used in two senses in this book. Firstly, it refer to the production of a hybrid individual who has inherited alleles from two distinct populations (including between populations which are considered to be separate species: **Figure 9.10**). Secondly, in molecular genetic terms, the word hybridization refers to joining together nucleotide strands from different sources by complementary base pairing.

**Hypothesis** is a proposed explanation for observed data, which can then be tested by experiments or observations. In some cases, the relative level of support for different hypotheses is contrasted. In other cases, the emphasis may be on falsifying a particular null hypothesis by showing that the observed data could not have been produced if the null hypothesis was true, given certain assumptions about the evolutionary process and data sampling. The null hypothesis is a statistical tool used to generate patterns of data expected if the process of interest does not operate: if the null hypothesis cannot be rejected, then the pattern in the data could have been produced without the process of interest operating.

***In vitro*** means a process carried out in the laboratory, not within a living organism (Latin for 'in glass', presumably referring to test tubes and such like). *In vivo* ('in a living thing') is a process occurring in a living cell, or within an organism.

**Inbreeding** results from mating between relatives, either by non-random mate selection, or random mating within a consistently small population. Inbreeding is expected to result in loss of genetic variability and increasing homozygosity, which may increase the incidence of heritable recessive diseases.

**Indel** stands for 'insertion or deletion', where one or more nucleotides are added or taken away from a sequence. Indels are represented by gaps in an alignment.

**Information** is a very tricky word to define, but in this book we will consider two senses in which the genome might be said to contain information. Firstly, parts of the genome contain information in the form of instructions for making RNA and protein molecules that have specific function. This information is transferable from the genome to the cytoplasm, and between one generation and the next, through complementary base pairing. Secondly, the genome contains information in the sense that related sequences are likely to be more similar, so the genome contains a record of its own evolutionary history. If you want to have a very long argument with a philosopher of biology, then trying to define biological information is a good place to start.

**Informative pedigree**: see pedigree

**Interbreeding population**: see population

**Intron** is a DNA sequence within a protein-coding gene that does not code for an amino acid sequence. Introns are transcribed but then removed from the processed messenger RNA transcript (see **Figure 6.12**). Some introns are removed by an enzyme complex called the spliceosome, but some introns are self-splicing, containing sequences that allow them to fold and excise from the mRNA. Most intron sequences have a rapid substitution rate, compared to the exons, and are sometimes used to estimate the neutral rate of substitution, however it is important to remember that introns can contain functional sequences that will be under selection.

**Invariant sites** are columns in an alignment where all sequences have the same nucleotide. Note that invariant is with respect to a particular collection of sequences and does not imply the site could not change in some other sequence.

**IUCN red list**: see endangered species

**Junk DNA** has been used to describe DNA sequences in the genome that have no apparent function in building or maintaining the organism. This term is best avoided due to the potential for misinterpretation and the tendency to make people very cross. The term is not used much these days as it has become apparent that some of the non-genic DNA in the genome codes for other things, such as regulatory elements or small RNA molecules, and much of it is derived from viruses or transposable elements, so whether it is considered functionless or not might depend on who it is functioning in aid of. There is a lively debate about how much of the intergenic DNA in the genome (all of the sequences between genes) makes any contribution to phenotype, and how much is essentially neutral. See also non-coding DNA.

**Kingdom** is one of the highest taxonomic categories. Various numbers of kingdoms have been recognized, though most current schemes include six kingdoms: animals, plants, fungi, protists (single-celled eukaryotes), bacteria and archaebacteria.

**Label**, in molecular genetics, generally refers to a chemical added to a molecule to allow its detection, for example a radioactive label (in traditional Sanger sequencing) or a fluorescent label (in some forms of high-throughput sequencing such as pyrosequencing).

**Lagerstätte** are fine sedimentary deposits that record an unusually high degree of fossil detail and diversity.

**Leading strand** is the newly synthesized strand of DNA that is being extended toward the replication fork, in a continuous strand. The **lagging strand** is the newly synthesized DNA strand on the other side of the helix that must run away from the replication fork, creating a series of short strands (Okazaki fragments) that must be joined together to make a continuous DNA sequence. See **TechBox 4.1**.

**Likelihood**: see maximum likelihood

**Likelihood ratio test** compares the likelihood of observing a particular dataset under two (or more) alternative hypotheses. See **TechBox 12.1**.

**Lineage** is a line of descent, drawn as an edge (branch) on a phylogeny, linking a series of populations all descended from a common ancestor.

**Linkage** relates to the joint inheritance of alleles that are physically connected together on a chromosome. Linked alleles may form identifiable haplotypes.

**Locus** (plural: loci) is a broad term meaning any location in the genome. Locus is often used to refer to an independent heritable unit of genetic diversity, such as a polymorphic site in the sequence, a microsatellite region, or a gene.

**Lysis** is the rupture of the cell membrane to spill the cell's contents.

**Macroevolution**: see microevolution

**Malaria** is a disease caused by infection by the single-celled protist *Plasmodium*, which infects over 200 million people and kills at least half a million people annually (particularly young children: **Figure 7.15**). Malaria is characterized by fever, headache and vomiting, and is transmitted by the bite of an *Anopheles* mosquito. There are at least five different species of *Plasmodium* that can cause malaria, and around thirty different species of *Anopheles* that transmit the infection (**Figure 5.9**).

**Marker**: see genetic marker

**Marsupials** are one of three major lineages of mammals (the others are monotremes and placentals), including possums, kangaroos and koalas, characterized by development of young in a pouch on the outside of the body.

**Maximum likelihood** is a statistical technique for comparing the plausibility of different hypotheses. In phylogenetics, maximum likelihood is used to compare the probability that a given phylogeny would have produced the observed sequence alignment, given a particular model of molecular evolution.

**Maynard Smith, John** (known as JMS: 1920–2004): I asked Kim Sterelny (a philosopher of biology) to write a definition of JMS and this is what he wrote: 'One of the most brilliant, and certainly the most sane, of the great UK twentieth-century biologists, who largely built modern evolutionary biology. His most distinctive contribution was to incorporate game theory within evolutionary biology' (**Figure 7.5**).

**Meiosis**: see mitosis

**Melt** refers to heating a DNA sample to break the hydrogen bonds that form between the complementary bases on each strand, resulting in single strands of DNA rather than double helices. The melting temperature (temperature needed to convert of the majority of the double-stranded DNA in the sample to single strands) will depend on a number of factors, including the GC content of the DNA: since GC pairs have three hydrogen bonds but AT pairs only two, it takes more energy to melt GC-rich DNA.

**Mendel, Gregor** (1822–1884) revealed the particulate nature of genetic inheritance by conducting breeding experiments on pea plants, so was the first quantitative geneticist. The significance of this work became apparent only after Mendel's death, when it became the basis of the neo-Darwinian synthesis. Incidentally, Mendel apparently suffered from severe exam anxiety, which you may find comforting next time you are facing an exam.

**Messenger RNA** (mRNA) is an RNA polynucleotide made as a complementary copy of a gene by base pairing. It is produced in the nucleus, and then processed and transported to the cytoplasm where it is translated into amino acid sequence on the ribosome. See Chapter 6.

**Methylation** is the addition of a methyl group to a protein or nucleotide chain. DNA is methylated at cytosines; in mammalian genomes, the majority of CpG sites (where a C is next to a G) are methylated. Methylation plays a role in gene expression (e.g. gene silencing), and the methylation state of genes may be inherited (see **Case Study 6.2**).

**Microarray** is a solid surface (usually a chip) onto which are fixed a series of polynucleotides. When a sample of labelled DNA or RNA is added to the microarray, it will hybridize to any complementary sequences on the chip, and the label will reveal the position of the matching sequence on the array, thus allowing the identity of the sample sequence to be determined.

**Microevolution** is descent with modification by changes in the frequency of heritable variants in a population over generations. Mechanisms of microevolution include selection and drift. **Macroevolution** refers to evolutionary patterns that can be observed by comparing different lineages. Most biologists consider that macroevolutionary phenomena arise from microevolutionary processes; that is, that microevolution and macroevolution are different views of the same underlying process. For example, speciation is a microevolutionary process involving genetic change within populations, but patterns of species richness are observed comparing the relative numbers of species produced by different lineages. Taking a macroevolutionary perspective involves asking whether there are any interesting evolutionary patterns observable only when considering patterns in species and other higher taxa.

**Microsatellites** are loci in the genome where the same short nucleotide sequence (typically less than ten bases long) is repeated multiple times. The number of repeats has a high mutability, so microsatellite loci are often useful as genetic markers to detect within-population variability, and for genotyping individuals.

**Microsporidia** are unicellular intracellular parasitic eukaryotes, once thought to be one of the oldest lineages of eukaryotes on account of the simplicity of their genome and phenotype, but now considered to be highly simplified fungi.

**Mismatch** is a base pair other than those specified by Watson-Crick pairing rules. When the incorrect bases are opposite each other in a DNA helix, they cannot bond properly. **Mismatch repair** detects these incorrect base pairs, and excises one or more bases on one of the strands, and replaces them with the correct nucleotide sequence.

**Mitochondria** (singular: mitochondrion) are energy-generating organelles in eukaryotic cells, originally derived from a symbiotic prokaryotic cell, and retaining a small circular genome with a small number of functional genes. In animals, mitochondrial DNA is usually, though not exclusively, maternally inherited.

**Mitosis** is the normal process of cell division, whereby each daughter cell gets an identical (or nearly identical) copy of the genome of the parent cell. Meiosis is a special form of cell division associated with the production of gametes in sexually reproducing organisms. In meiosis, a diploid cell undergoes two rounds of cell division, one in which each daughter cell receives one copy of each chromosome, then each divides again to produce four haploid gametes. Unlike mitosis, the products of meiosis are genetically non-identical.

**Moa** were flightless birds native to New Zealand, from the ratite family which includes kiwis, emus and ostriches. Moa went extinct not long after the arrival of humans in New Zealand, less than one thousand years ago.

**Mobile genetic elements** are any DNA sequences capable of moving from one part of the genome to another, also known as transposable elements.

**Model** is an abstract description of the behaviour of a system, which can be used to interpret observed patterns or predict future occurrences.

**Molecular clock** has two meanings. Firstly, it may refer to the assumption (or observation) of constant rates of molecular evolution in all lineages under consideration. Secondly, it may refer to the method of molecular dating, where the amount of divergence between sequences is used to estimate the age of the last common ancestor of those sequences. These two concepts are not equivalent: the molecular clock assumption underlies various analytical techniques (not just estimating dates), and molecular dating does not necessarily involve assuming constant rates of molecular evolution. See **TechBox 14.2**.

**Monocot**: see dicot

**Monotremes** are a lineage of egg-laying mammals including the platypus and echidnas found in Australia and Papua New Guinea (**Figure 10.3**). Monotremes can hunt for insects using electrodetection, using their soft beaks to pick up the electromagnetic radiation given off by moving animals. Platypuses are one of the few mammal species that are poisonous, as male platypuses are able to inject venom using spurs on their hindlimbs.

**Morgan, Thomas Hunt** (1866–1945) was a pioneering geneticist who established the use of the fruit fly, *Drosophila*, in genetical research, and used them to demonstrate that genes, localized to chromosomal regions, were the agent of heritability.

**Morphology** typically refers to the observable physical characteristics of an organism, though sometimes the term is used synonymously with phenotype.

**Multiple hits** occur when two or more nucleotide substitution events have occurred at the same site in a DNA sequence, such that only the last substitution is directly observable. Multiple hits obscure the true number of substitutions that have occurred in a sequence. See saturation.

**Mutagen** is any agent capable of causing mutation.

**Mutation** is a permanent, heritable change to the information in the genome. Most of the mutations we discuss in this book are point mutations that change the base sequence of DNA, but mutations can also alter DNA structure (e.g. chromosomal rearrangements).

**Mutator** and antimutator refer to heritable variation in mutation rate among members of a population. A mutator has a higher-than-average mutation rate, antimutators have a lower-than-average mutation rate.

**Natural selection**: see selection

**$N_e$**: see effective population size

**Nearly neutral theory** is an enrichment of the neutral theory to encompass a continuum of mutation effects. In particular, the nearly neutral theory predicts that the fixation probabilities of 'nearly neutral' mutations, which are slightly advantageous or slightly deleterious, can contribute significantly to patterns of molecular evolution. See **Hero 8, TechBox 8.2**.

**Negative selection**: see selection

**Neo-Darwinian synthesis** (also referred to as the Modern Synthesis) describes the body of ideas developed primarily in the first half of the twentieth century linking the evolutionary theories of Darwin and his contemporaries to Mendelian genetics, demonstrating the power of natural selection to drive change in allele frequencies, thus vindicating the hypothesis of gradualism. Synthesis refers here to the bringing together of previously disparate fields of study to an agreed position that they were all studying aspects of the same fundamental evolutionary process. Key players in the development of the neo-Darwinian synthesis include Fisher, Haldane and Wright who forged a mathematical framework for population genetics, combining the effects of selection and drift on allele frequencies. The neo-Darwinian synthesis was extended by researchers such as Theodosius Dobzhansky (**Figure 3.9**), who provided lab-based evidence of mutation and substitution in *Drosophila*, and G. G. Simpson (**Figure 13.4**), who demonstrated the compatibility of Darwinian mechanisms with observed patterns in the fossil record.

***Neurospora crassa*** is a haploid fungus that, like *Drosophila* and *E. coli*, is a classic model organism for research in genetics. It can be grown on simple medium in the laboratory, and produces arrays of sexual spores contained within a body that allows the products of segregation to be easily detected.

**Neutral alleles** are variations of a trait that are functionally equivalent, such that none has a selective advantage over the others. Since the frequency of a neutral allele is not influenced by selection, it will fluctuate due to random events (see drift). It is important to note that neutral alleles are not necessarily phenotypically silent: observable phenotypic variants may be selectively equivalent. **Neutral sites** are positions in the genome where all nucleotides states are selectively equivalent: since changing the nucleotide will not affect the function of the sequence, the substitution rate at those sites should reflect the mutation rate. **Neutral theory** suggests that a significant proportion of observed substitutions are fixed by drift, not by selection (see **TechBox 8.1**).

**Next generation sequencing** (NGS): see high-throughput sequencing

**Nodes** are terminal points of branches (edges) in a phylogeny, such that each branch is defined by two nodes (the beginning and the end of the edge). Internal nodes are the branching points where a lineages splits to give rise to two or more descendant lineages (**Figure 11.22**). The tips of a phylogeny, which in molecular phylogenies usually represent the sequences from the alignment, are also nodes. The root node is the primary branching event at the base of a rooted phylogeny.

**Noise**: see signal

**Non-coding** is a confusing term, because sometimes it is used to refer to any DNA sequence that does not code for a protein (including RNA genes), sometimes it is used to represent all non-transcribed DNA sequences (including regulatory elements), and sometimes it means apparently functionless DNA (in the past referred to as 'junk DNA' though the term is less widely used now).

**Nuclear DNA**: see nucleus

**Nucleotides** are the basic units of nucleic acids. Each nucleotide consists of one base (A, C, T or G) bound to a sugar molecule which is bound to a phosphate molecule (**TechBox 2.1**).

**Nucleus** is a membrane-bound organelle in eukaryote cells that contains the nuclear DNA. The nuclear DNA is the primary genome of a eukaryotic cell, typically packaged into multiple chromosomes.

**Null hypothesis** (or null model): see hypothesis

**Okazaki fragments**: see leading strand

**Oligonucleotide** is a short nucleotide strand. The term is typically used to refer to artificially created polynucleotides.

**Onychophora** is a phylum of adorable little caterpillar-like animals, otherwise known as velvet worms or peripatus (**Figure 11.13**).

**Open reading frame** (ORF) is a DNA sequence that does not contain a stop codon so could be translated into a continuous amino acid sequence without interruption. Note that this refers to the potential to code for a peptide, it does not necessarily imply that a given sequence is indeed transcribed and translated. See Chapter 6.

**Organelles** are structural subunits of cells that have specific functions, for example mitochondria and chloroplasts.

**Orthologs** are homologous copies of a sequence produced by the divergence of lineages: see gene family.

***Oryza sativa*** is the scientific name for the most commonly cultivated species of rice.

**Overparameterization (overfitting)** is the inclusion of too many parameters in a model, which may increase the fit of the model to the data, but decreases the explanatory or predictive power of the model by making the model explain one specific dataset (potentially describing noise rather than signal) rather than provide a general explanation of a pattern or process.

**Paralogs** are copies of genes produced by gene duplication: see gene family.

**Parsimony**, as applied to phylogenetic reconstruction, is the principle that the most reasonable phylogeny is the one that requires the inference of the smallest number of evolutionary changes.

**Pauling, Linus** (1901–1994) pioneered the study of the structure and evolution of proteins, and won two Nobel prizes, one for chemistry and one for peace.

**p-distance** is the Hamming distance divided by the number of sites compared (also called uncorrected p-distance).

**Pedigree** is a family history, usually drawn as a branching diagram with male family members represented as squares and females as circles. In genetics, pedigrees are used to uncover patterns of inheritance, for example to identify carriers of an allele associated with a particular trait. An **informative pedigree** is one in which the known phenotypes and family relationships provide sufficient information to allow the genetic basis of a trait to be identified.

**Peptides** are chains of amino acids, held together by peptide bonds which are covalent bonds between the carboxyl group of one amino acid and the amino group of the next amino acid. This reaction is referred to as a condensation reaction because formation of the bond results in the release of a water molecule ($H_2O$). Proteins are formed of single or multiple peptides, which usually adopt a particular three-dimensional structure determined by the amino acid sequence.

**Phage**: see bacteriophage

**Phenetics**, as applied to phylogeny reconstruction, is the principle that the relationships between lineages can be determined by analysis of measures of similarity.

You could have a hearty debate about whether distance methods (**TechBox 11.1**) are phenetic or not, if you are so inclined.

**Phenotype** generally refers to the observable properties of an organism, particularly its morphology and behaviour. However, phenotype is sometimes used more broadly to encapsulate all of the results of gene expression, including development and metabolism, in order to provide a contrast to genotype, which refers only to the genetic information contained in the genome. Phenotype often refers to a 'partial phenotype', the expression of a trait or traits of interest, rather than to the entire morphology.

**Phenylketonuria** (PKU) is a heritable metabolic disorder caused by a recessive allele, homozygotes for which cannot metabolize phenylalanine, which can then build up in the nervous system and brain causing onset of mental retardation during childhood.

**Phosphates** are one of the three units in a polynucleotide strand: see nucleotide (**TechBox 2.1**).

**Phosphodiester bonds** join nucleotides to make a polynucleotide chain, through a covalent bond between 5' phosphate group of one nucleotide to the 3' hydroxyl group of the sugar of the next nucleotide, catalysed by a polymerase enzyme (**TechBox 2.1**).

**Phylogeny** is a representation of the evolutionary history of biological lineages as a nested series of branching events, each marking the divergence of two or more lineages from a common ancestral lineage. **Phylogenetics** is the discipline of inferring phylogenies—there are many different phylogenetic techniques including those based on maximum likelihood and Bayesian inference. See Chapters 11 and 12.

**Phylum** (plural: phyla) is a taxonomic category: kingdoms are divided into phyla.

**Placental** mammals, more formally known as Eutheria, are one of three major lineages of mammals (the others being monotremes and marsupials). The placentals have been split into three major clades, largely based on molecular phylogenies: the Afrotheria (including the elephants, hyraxes, dugongs, aardvarks and others); Laurasiatheria (including artiodactyls, cetaceans, carnivorans, horses, rhinos and others); and Euarchontoglires (including rodents, bunnies, primates, bats and others).

**Plankton** refers to the microscopic organisms, including animals, plants and algae, that float in surface waters.

**Platypus**: see monotreme

**Polymerase** is an enzyme that catalyses the formation of phosphodiester bonds that bind nucleotides together in a DNA or RNA molecule.

**Polymerase chain reaction (PCR)** is a laboratory method for amplifying DNA, in which DNA is heated to separate the double helices, cooled so primers bind to specific sequences in the single-stranded DNA, then heated with polymerase which makes complementary strands starting from the primer sequences. The cycle is repeated many times, resulting in exponential increase in the number of copies of the amplified sequence (**Figure 4.33**).

**Polymorphism** occurs when there is variation within a population for a given trait. At the molecular genetic level, polymorphism is the presence of two or more alleles for a given locus in a population.

**Polynucleotide**: a biopolymer consisting of a chain of *nucleotides* bound together by *phosphodiester bonds*. See bases.

**Population**, in evolutionary genetics, represents an interbreeding set of individuals capable of combining genetic material to produce offspring. Populations may be kept distinct from all other such populations by reproductive isolation mechanisms that prevent the formation of hybrid offspring with parents from different populations. These isolating mechanisms may prevent potential parents coming together (for example, distance between populations or behavioural differences in mating rituals), or may prevent the development or reproduction of offspring resulting from hybrid crosses (such as genetic incompatibility). Note that a population may have 'fuzzy borders' if there are low levels of genetic exchange between connected populations, through occasional hybridization or rare migration events (see **Figure 9.10**). Some populations maintain genetic distinctness even in the face of ongoing genetic exchange with other populations (see **Case Study 9.1**).

**Positive selection**: see selection

**Posterior probability, Prior probability**: see Bayesian inference

**Priors** are a special form of assumption used in Bayesian phylogenetic analyses: see **TechBox 13.2**.

**Prokaryotes** (bacteria and archaebacteria) are single-celled organisms without a nucleus or mitochondria: see eukaryote.

**Promoter** is a sequence located near a gene to which RNA polymerase binds to begin transcription of the gene. See Chapter 6.

**Proofreading** is an exonuclease function possessed by some polymerase enzymes that removes incorrect nucleotides from the end of the newly synthesized DNA strand (see **Figure 4.13**).

**Protein**: see peptide

**Protein-coding** sequences are parts of the genome that can be transcribed and translated into an amino acid sequence.

**Pseudogene** is a non-functional version of a gene, produced by gene duplication or by inactivating mutations within a previously function gene. Because they are not subject to selection, pseudogenes are assumed to accumulate substitutions at the neutral rate.

**Purines** are the double-ring bases, adenine and guanine.

**Pyrimidines** are the single-ring bases, thymine, cytosine and uracil.

**Pyrosequencing** is a high-throughput sequencing method, one of the first common alternatives to traditional Sanger sequencing. Rather than producing a series of fragments, each ending at a particular nucleotide in the sequence, pyrosequencing uses light reactions during DNA synthesis to report the identity of each nucleotide added to the growing strand. 454 sequencing is a particular parallel pyrosequencing technology that allows very large amounts of sequencing to be done in a short time period. 454 technology was used for whole genome sequencing, and producing partial genome sequences from ancient DNA samples such as Neanderthal and woolly mammoth, however the 454 platform has been discontinued.

**Quagga** is a subspecies of zebra that went extinct when the last individual died in captivity in an Amsterdam zoo in 1883. See Chapter 10.

**Random** processes are undirected with respect to a particular outcome. Therefore it is important to define 'random with respect to what?' (see Chapter 5). A process that is random with respect to particular outcomes (e.g. mutations are random with respect to fitness) may be biased towards certain other outcomes (e.g. mutations are more likely to occur in certain places in the genome). In this book, 'random' is used synonymously with stochastic and chance.

**Read**: see high-throughput sequencing

**Recessive** alleles affect phenotype only when in the homozygous state, not when heterozygous. Recessive diseases are only evident when the carrier has a 'double-dose', carrying a copy of the disease allele on both homologous chromosomes.

**Recombination** is the exchange of genetic material between chromosomes or between genomes: see crossing-over. Recombinant DNA has DNA from two different sources, thus breaking up normally linked genes or alleles.

**Redundancy**: see degenerate

**Reduction/division**: see meiosis

**Reinforcement** is selection for reproductive isolation mechanisms that reduce the incidence of hybridization between members of different populations.

**Relaxed clock** may refer in the broad sense to any phylogenetic method that does not rely upon the assumption of a uniform constant rate of change in all lineages (an assumption sometimes referred to as a molecular clock). In practice, relaxed clock is usually used to describe the set of methods that allow each branch in the phylogeny to have a different rate, either by drawing those rates from a distribution of possible rates (uncorrelated) or by 'evolving' rates along the phylogeny according to a defined model of rate change (correlated). See **TechBox 14.2**.

**Repeat sequences** contain multiple tandem (side-by-side) copies of the same nucleotide sequence, which are prone to change in repeat number. Microsatellites loci contain short nucleotide repeats: see microsatellite.

**Replacement** substitutions change the amino acid sequence of the resulting protein.

**Restriction enzyme** is an endonuclease that recognizes a specific nucleotide sequence and cuts the DNA strand wherever that sequence occurs. Restriction fragment length polymorphism (RFLP) is variation between genomes in a population in the number or location of a particular restriction sequence, so that when DNA is digested with a specific restriction enzyme, individuals will be characterized by different fragment numbers and lengths which can be separated on a gel: see DNA fingerprinting.

**Retroviruses** replicate by using reverse transcriptase to make a DNA copy of their RNA genome, and the DNA copy is then inserted into the host's genome. The inserted retroviral genome, referred to as a **provirus**, contains regulatory elements that co-opt the host's transcription machinery which makes RNA copies of the viral genome. These RNA viral genomes can act as transcripts for the production of viral proteins, or as genomes to be packaged into new virus particles. Endogenous retroviruses (ERVs) are viral genomes (or the remains thereof) embedded in the host genome and inherited by offspring when the genome is replicated. Some ERVs are gained by infection, others may replicate within the genome, and others are incapable of replicating. See **Case Study 5.2**.

**Reverse transcriptase** is a polymerase enzyme that makes a complementary DNA strand from an RNA template. Retroviruses use reverse transcriptase to make the provirus which is inserted into the host genome: see retrovirus.

**Ribose**: see sugar

**Ribosomal RNA** (rRNA) is a component of the ribosome, the cellular organelle responsible for translating the information in a messenger RNA transcript to the amino acid sequence of a peptide. Transcription of rRNA genes produces an RNA polynucleotide which then adopts a secondary structure by complementary base pairing between specific parts of the sequence. Some rRNA genes are commonly used in phylogenetics, including the mitochondrial-encoded 12S rRNA and the nuclear-encoded 18S rRNA (also referred to as small subunit RNA, or SSU) and 28S rRNA (also called the large subunit RNA or LSU). The 'S' refers to the size of the molecule produced by the gene, measured in Svedburgs, which reflects the rate of sedimentation under centrifugation—the larger the S value the bigger the particle. rRNA genes have different patterns of molecular evolution than protein-coding sequences, for example indels are more common, and rates of change may vary between stem and loop regions of the sequence.

**Ribosome**: see ribosomal RNA

**RNA polymerase** is an enzyme that makes an complementary RNA strand to match a DNA template strand.

**Sanger method** is the traditional standard approach to DNA sequencing, where DNA is replicated in the presence of modified nucleotides that halt synthesis when incorporated, producing fragments of different lengths which can be separated on a gradient: see **TechBox 1.2**.

**Saturation** is the loss of evolutionary information in a sequence through multiple hits which erase the signal of descent.

**Scientific literature** describes research articles published in 'scholarly' (professional science) journals, rather than in the popular press. Articles in the scientific literature should fully describe the research so that, technically speaking, any suitably qualified and equipped person could repeat the research. Articles published in the scientific literature should have been through peer review, where the article is first sent to a number of independent scientists who evaluate the article (usually anonymously) in order to detect any flaws in the research.

**Seago, Ainsley** (1981–) is an entomologist, illustrator, and cartoonist, almost certainly the only person ever to have created both a molecular phylogeny of ladybird beetles and a pirate ferret called Cecilia. Look for her cartoons about science, bugs and life in general online.

**Segregation** is the division of the diploid genome into haploid gametes in meiosis. Segregation refers more specifically to the separation of alternative alleles for one or more loci into gametes such that they can be inherited separately. Some geneticists refer to polymorphic alleles as 'segregating in the population' (or segregating in a particular pedigree).

**Selection** (natural selection) is change in the frequency of an allele as a result of its influence on its own chances of being passed to the next generation (see Chapter 7). Alleles under **positive selection** have an increased chance of being included in the next generation relative to other alleles in the population. In a large population, positive selection will usually result in the fixation of an allele so that it replaces all other variants in the population (though this will not always be the outcome of positive selection: see Chapters 7 and 8). Alleles under **negative selection** have a decreased chance of being included in the next generation: any alleles that cause reduction in their carriers' chances of reproduction will tend to reduce in frequency until lost from the population. Because negative selection results in the removal of alleles from the population, it commonly results in the conservation of sequences. **Artificial selection** refers to cases of selection resulting from intervention by humans, usually restricted to active intervention through selective breeding, but occasionally also used to describe any selective response to human activity, whether intended or not.

**Selective sweep** occurs when selection on a specific locus affects the frequency of alleles at neighbouring (linked) loci. Alleles in loci that are linked to a locus under positive selection may be swept to fixation along with the selected allele, regardless of their own effects on fitness (see **Figures 7.10** to **7.13**). Selective sweeps result in the reduction in nucleotide diversity at loci linked to a trait under strong positive selection, detected through the overrepresentation of particular haplotypes in a population.

**Signal** is patterns in the data created by the process of interest. Noise is random variation resulting from stochastic processes, not from the process of interest. The signal-to-noise ratio of a dataset reflects how clearly those data reflect the process of interest, therefore whether the data can be used to discriminate between alternative hypotheses.

**Silent** changes alter the nucleotide sequence of the genome but do not have a noticeable effect on phenotype. For example, synonymous changes to the nucleotide sequence of a protein-coding gene do not change the amino acid sequence of the resulting protein, so they are phenotypically silent. Note that 'silent' is not exactly the same as 'neutral'. Silent changes may have a fitness cost or benefit (for example, synonymous codon bias suggests that there may be an efficiency gain in using certain codons), and non-silent changes may be effectively neutral (for example, a nonsynonymous change may have a noticeable effect on phenotype that has no implications for fitness).

**Simpson, George Gaylord** (1902–1984) was a palaeontologist who contributed to the codification of the neo-Darwinian synthesis by showing how palaeontological patterns could be interpreted in light of Darwinian gradualism, natural selection and population genetics (**Figure 13.4**).

**Simulated data** is data produced artificially, rather than observed from natural systems. In phylogenetics, data is simulated according to a particular model of molecular evolution (e.g. a substitution model used to 'evolve' a set of sequences from a single ancestral sequence). The simulated data may be used to test the effectiveness of methods of phylogeny reconstruction, or to generate an expected pattern of substitutions to use as a null hypothesis against which another hypothesis can be tested.

**Single nucleotide polymorphism (SNP)** is a locus in the genome where the nucleotide sequence differs between members of a population. See **TechBox 3.1**.

**Soma**: see germline

**Species** is one of the lowest levels in the taxonomic hierarchy, given a scientific name consisting of *Genus species* (e.g. *Homo sapiens*). The precise definition of species is a matter of debate, and there are many alternative species concepts that set out criteria for delineating species (see Chapter 9). An undescribed species is a set of organisms that probably form a distinct species but have not been given a formal taxonomic description with a scientific species name (e.g. **Case Study 9.2**). It is likely that the majority of living species have not yet been described.

**Spencer, Herbert** (1820–1903), the originator of the phrase 'Survival of the fittest' to describe natural selection (a phrase adopted by Darwin in later editions of the *Origin of Species*), used evolutionary principles to inform models of social change.

**Substitution** is the loss of all alternative alleles in a population but one, so that all members of the population carry the same allele at that locus. Substitution should not be confused with mutation, which changes the genetic information in a single individual. A mutation may become a substitution if it increases in frequency over generations until it reaches fixation.

**Sugar**, in the broad sense, refers to a class of short-chain carbohydrate molecules. Ribose is a pentose (five-carbon) sugar, with four carbons arranged in a ring with oxygen, with four hydroxyl (OH) groups. Deoxyribose is the version of this sugar found in DNA, in which one of the hydroxyl groups is replaced with a hydrogen.

**Synonymous codon**: see codon

**Synonymous substitution rate** is an estimate of the rate at which substitutions occur that change the nucleotide sequence but not the resulting protein-coding sequence. On the assumption that synonymous changes are effectively neutral, the synonymous substitution rate is often interpreted as reflecting the underlying mutation rate of the sequence.

**Systematics** is the study of the relationships of living things, nowadays commonly conducted in a phylogenetic framework.

**Taxonomy** is the practice (and the output) of classifying biological diversity. Taxonomy divides organisms into defined units, known as taxa (singular: taxon). A taxon could refer to any division of organisms into groups, whether species, sub-species, phyla, and so on. The phrase 'higher taxon' generally refers to taxonomic levels above species (or above the level referred to). See Chapter 9.

**Theoretical**: see empirical

**Thylacine** (Tasmanian tiger) refers to a lineage of marsupial predators, originally distributed throughout Australia and Papua New Guinea. Thylacines disappeared from the Australian mainland before European settlement (probably more than a thousand years ago, possible due to competition with introduced dingos), but survived on the island of Tasmania until the first half of the twentieth century (**Figures 9.2** and **10.16**).

**Thymine (T)** is one of the four DNA bases that make up the 'alphabet' of the genetic code: see bases. A **thymine dimer** (also called thymidine dimer) forms when two adjacent thymines bond together, rather than pairing with the A on the opposite strand (see **Figure 3.2**).

**Topology** is the branching order of a phylogenetic tree. Trees that contain the same taxa but differ in the way those taxa are connected together are said to differ in topology. Topology is sometimes used to distinguish the branching order from information about branch lengths: two phylogenies that depict the same relationships between taxa but show different amounts of change along lineages have the same topology but differ in branch length (see **Figure 11.39**).

**Trait** is generally used synonymously with character to indicate a discretely heritable aspect of phenotype.

**Transcription** is the production of a complementary RNA strand from a DNA sequence in the genome. This RNA transcript can move from the nucleus to the cytoplasm, where it may be directly involved in cellular processes (for example, ribosomal RNAs and transfer RNAs) or translated into a protein sequence at the ribosome (messenger RNA).

**Transfer RNA** (tRNA) is a small RNA molecule, often depicted as having a clover-leaf secondary structure, that carries amino acids to the ribosome for protein synthesis. Each type of tRNA carries a particular amino acid and has an anticodon that can pair to the complementary codon in messenger RNA, ensuring that the messenger RNA is accurately translated into a specific amino acid sequence (**Figures 2.8, 6.4, 6.15**).

**Transition** is a mutation (or substitution) involving the exchange of one pyrimidine for another, or one purine for another. A transversion changes a purine to a pyrimidine or vice versa. In most sequences, transitions are more common than transversions, so the transition-transversion ratio is an important component of many substitution models.

**Translation** is the conversion of information from the nucleotide sequence of a messenger RNA molecule into the amino acid sequence of a peptide, which occurs on the ribosome.

**Transposable elements**: see mobile genetic elements

**Tree-like** data can be represented by a branching diagram, where lineages can split into two descendant lineages. Non-tree-like data does not follow a simple hierarchical branching pattern. See Chapter 11.

**Uncorrected distance**: see Hamming distance

**Unequal crossing over**: see crossing over

**Upstream**: see downstream

**Uracil**: see bases

**Velvet worm**: see onychophoran

**Vulnerable species**: see endangered species

**Wallace, Alfred Russel** (1823–1913) was a British naturalist who independently discovered the principle of natural selection, and who made a major contribution to establishing the field of biogeography.

**Watson-Crick pairing**: see base

**Weismann, Friedrich Leopold August** (1834–1914) placed Darwin's evolutionary theory within the framework of heritability and development: see **Hero 2**.

**Wright, Sewall** (1889–1988) was one of the three key figures in the development of population genetic theory in the first half of the twentieth century (the others being Haldane and Fisher), promoting the importance of drift in determining allele frequencies and embedding the concept of the adaptive landscape in evolutionary biology (**Figure 7.5**).

**Zygote**: see gamete

# Index


1,000 genomes project 22, 66–7, 70

## A

accession number, GenBank 20, 21
acquired characteristics 34–5, 487
*ad hoc* assumptions 374
adaptation 73, 268, 273, 300–1, 487
  convergent 6, 7, 298
  whales 2, 5–6
adenine (A) 48, 487
advanced sleep phase syndrome (ASPS) 68
affine gap penalty 317
Afrotheria 7, 8
*Agathis australis* (Kauri tree) 476–9
AIDS *see* human immunodeficiency virus (HIV)
Al-Fateh Hospital (AFH) HIV outbreak, Libya 480–3
ALEX protein 194–6
aliens 2, 10, 62, 185, 376
alignment 23–4, 302–9, 487
  automated 306, 308, 317–18
  checking 319–20
  gaps 303–4, 317–18
  as hypothesis 320
  manual 305–6, 308–9, 318–19
  multiple sequence alignment 317–20
  scoring 306–8
alleles 67–8, 135, 487
  alternative histories 382
  common 85–6
  divergence and 267–8
  dominant 84, 218, 491
  genetic drift 232–3, 236, 250
  neutral 208–9, 220–1, 233–4, 290, 499
  recessive 212, 239, 260–1, 500
  *see also* fixation
alpha globin genes 207, 241
  knock-out mutations 207–8
alpine marmot 144
alternative reading frames 195–6
alternative splicing 177–8, 194
amino acids 40, 186–8, 487
  substitution models 426
  *see also* codons
amplification *see* DNA amplification
anaemia 207–8, 488
  sickle cell 211
analogous traits 300–1
ancient DNA 223–4, 355, 469–70, 487–8
  analysis 53, 57–60, 396–400
annealing temperature 133–4, 189–191
annelids 439, 440
annotation 20–1, 488
*Anopheles funestus see* mosquitoes
ant, carpenter 321–4
anthrax 111–13, 488
  bioterrorist attack 113
antibiotic resistance 225–8, 383
antibody production 136
aphids 348
  nicotine tolerance 161–3
*Apis mellifera* (honeybees) 301
apple, origins and evolution 365–8
Approximately Unbiased (AU) test 326
*Apteryx* (kiwi) 448–9
Archezoa 302
Arctic vegetation changes 54–6
arthropods 439, 440, 459, 488
  body plan 9, 10
artificial selection 230, 502
  *see also* selection
asexual reproduction 92, 255–8, 275, 410
  species definition 271
association studies 84, 488
  genome-wide association study (GWAS) 88–9, 494
autapomorphies 304–5, 342, 488
automated alignment 306, 308, 317–18
  gap penalty 317–18
  substitution score 317
autoradiograph 19
Avery, Oswald 37

## B

*Bacillus* bacteria 111
  *B. anthracis* 111–13, 274
  hierarchies of similarity 111–13
BACs (bacterial artificial chromosomes) 488
background selection 254
bacteria, endosymbiotic 321–4, 362–4
bacteriophage 37, 488
barcoding *see* DNA barcoding
barnacles 9, 72, 199, 488
base composition bias 425, 488
base pairing 49–51, 98
bases 47–8, 488
Bateson, William 167, 369
Bayesian methods 428–31, 488
bees
  honey (*Apis mellifera*) 301
  stingless (*Melipona*) 301
beetles 199, 291–3
Benzer, Seymour 168–9, 170
beta-galactosidase 171, 173–4
  *see also lacZ* gene
beta globin genes 175–6, 244
  malaria resistance 209, 210, 211
beta thalassemia 208, 211
bilaterian 444, 488–9
binomial nomenclature 263–4
biobanking 84–7
  ethical issues 85–7
biodiversity 54–6, 489
  molecular evolution 434–7
  surveys 291–3
  unified neutral theory of 243
biofuel production 88–90
biogeography 489
  New Zealand study 476–9
bioinformatics 157, 489
biological classification *see* classification
biological species concept 270–1
  *see also* species
bioluminescent endosymbiotic bacteria 362–4
bioprospecting 402
bioterrorism 112, 113
birds of paradise 237
Black Death 397–9
BLAST (Basic Local Alignment Search Tool) 23–5, 192, 285, 489
*Blochmannia chromaiodes* (bacterial endosymbiont) 321–4
blowfly, insecticide resistance 222–4
body plan 9–10, 489
body size 410–13
bootstrap 326, 381, 392–6
  parametric bootstrap 394–6
*Bos gaurus* (gaur) 313
bottleneck 254, 363
branch length estimation 406–8, 418
bucardo 313
*Buchnera* (bacteria) 348
butterflies 177, 269

## C

C-value paradox 134, 148
cacao 287
calibration 465–70, 489
Cambrian explosion 439–41, 443–5, 458–60
*Camponotus* (carpenter ants) 321–4
*Canis* 263
  *C. latrans* (coyote) 272
  *C. rufus* (red wolf) 263, 272
  *C. simensis* (Ethiopian wolf) 263
capillary electrophoresis 142
*Capra pyrenaica pyrenaica* (bucardo) 313
carnivorans 489
carnivorous plants 43
carotenoid synthesis 348
cat 298–9
cell division 134
  meiosis 135–6
  mitosis 498

cell lysis 52
Central Dogma 41
centrifuge 52
*Cepaea nemoralis* (snail) 203
*Ceratotherium simum* (white rhinoceros) 66, 67
chain termination sequencing *see* Sanger sequencing
chance, role of 234–6
  *see also* genetic drift; neutral theory
character 489
Chargaff, Erwin 38
Chargaff's rules 38, 425, 489
Chase, Martha 37
Chaucer, Geoffrey 107–8
chiasma 136
chickens 192, 230–1
Chilean blob mystery 2–3, 10, 25–8
chimpanzee, HIV origin 380
chloroplast 52, 125, 135, 285, 435–7, 489
chocolate 287
Chordata 9–10, 334, 394–5, 441, 489
chromosomes 36–7, 134–7, 489
  crossing-over 136, 139
  homologous pairs 135
  inversion 138
  rearrangement 138
ciliates 91–3
CITES *see* Convention on International Trade in Endangered Species (CITES)
clade 340–1, 489
clams 446–7, 459
classification 264–6
  hierarchical arrangement 266, 331
cloning 489–90
  by sequencing 314–15
  endangered species 313–14
  extinct species 314
clustering 354–6
Cnidaria 444–5
code *see* genetic code
coding region 490
CODIS (Combined DNA Index System) 143
codons 490
  start codons 171, 187
  stop codons 171, 172, 187
  *see also* genetic code
codon usage bias 187, 220, 240
coelacanths 404, 407–8
colicin 206
column 52, 81, 157
common eider 97
complementary base pairing 49–51, 98
complexity 242–4
  genome size 133–4, 148, 243
  non-adaptive increases 244
confidence intervals 453–4, 481, 490
consensus network 363, 364
consensus sequence 67
conservation 28, 272
  cloning and 313–14
  inbreeding concerns 239, 258–61
  *see also* endangered species
contamination 22, 25–6, 106
  GenBank submissions 21–2
continuity principle 35, 42
control region 57–8
Convention on International Trade in Endangered Species (CITES) 29
convergent adaptation 6, 7, 298
coprolite (fossilised faeces) 55–6, 59, 442
copy errors 100, 490
  error rates 96
  hierarchy of similarity 106–7, 109–13
  mismatch repair 101–2
  molecular evolutionary rate 413
  *see also* mutation
  proofreading 100–1
CoT curve 134
Courtenay-Latimer, Marjorie 404
coyote (*Canis latrans*) 272
Crick, Francis 38–9, 185, 490
crime *see* forensic applications
crossing-over 136, 490
  unequal 139
Crow, James 78–9
cryopreservation 316
cryptic species 274, 490
cryptozoology 27–8, 490
Cundick, Mary 212
*Cuscuta* (dodder) 327
cytosine (C) 47, 490
cytotoxic T lymphocytes (CTLs) 242

**D**

Darwin, Charles 15, 34–5, 197–9, 201, 229, 262, 295, 329, 490
  barnacle studies 199
  *The Descent of Man* 198–9
  *The Origin of Species* 198, 202, 229, 262, 295, 330
*Dasyurus see* quoll
databases *see* DNA databases
DataDryad 20
dating *see* molecular dating
Dawkins, Richard 335
Dayhoff, Margaret Oakley 312–13
de-extinction 313–16
  cloning 314–15
  genome editing 315
  quagga 296–7, 311
  thylacine 309
deCODE database, Iceland 84–5
degenerate 185, 186, 490
  degenerate primers 191
*Deinococcus radiodurans* (Conan the bacterium) 409, 410
denature
  DNA 104, 105, 121–3
  proteins 105
deoxyribonculeic acid *see* DNA
deoxyribose 48
descent with modification 200
desirogram 370–1 490
diabetes 84–5
*Diaea* (spider) 201
Diazinon insecticide resistance 223–4
*Dibamus* (lizard) 381–2
*Diceros bicornis* (black rhinoceros) 67
dicot 265–6, 490
dideoxy sequencing *see* Sanger sequencing
*Dinornis see* moas
dinosaur phylogeny 337
diploidy 67, 135, 490
directed mutation 74
disease *see* epidemiology; specific diseases
dispersal ability 280, 432–3
distance methods 345–7, 354–8
directed mutation 74
divergence 267–9, 443–54, 491
  dates 444–7, 490–1
  mosquitoes 138
  reinforcement 288–90, 501
diversification 330–1
  influence of mutation rate 434–7
DNA
  discovery of 37
  excess, costs of 243–4
  heating effects 104
  sampling 42
  structure *see* DNA structure
  ubiquity of 42
  X-ray diffraction image 38
  *see also* extraction of DNA; mitochondrial DNA; repetitive DNA
DNA amplification 102–6, 121–3, 491
  contamination 106
  *in vitro* replication 103
  polymerase chain reaction (PCR) 103–6
DNA barcoding 274, 284–6, 305, 491
  database 285
  species discovery 286
  species identification 284–6
  weevil diversity analysis 291–3
DNA damage 62–5
  *see also* mutation
DNA databases
  barcode databases 285
  deCODE database, Iceland 84–5
  ethical issues 86–7, 143–4
  forensic databases 86–7
  genetic screening 212
  *see also* GenBank
DNA fingerprinting 142–5, 491
DNA hybridization 134, 355, 356, 491
DNA library 159
DNA polymerase 99, 100, 105, 119–20
  DNA amplification 123
  Taq polymerase 105–6, 123

*see also* polymerase chain reaction (PCR)
DNA profiling 143–4
DNA purification 52
DNA repair 71–3
  *Deinococcus* bacteria 409
  mismatch repair 101–2
  pathways 63–4
  proofreading 72–3, 100–1
DNA replication 96–102, 118–20
  copying by base pairing 98
  error rates 96
  proofreading 100–1
  replication bubbles 99, 118
  replication fork 99, 119–20
  replication origins 98, 119
  template reproduction 97–8
  *see also* copy errors
DNA sequencing 19
  high-throughput sequencing 70–1, 157–60, 318–19, 323–4, 495
  Sanger sequencing 19, 153–6, 502
  shotgun sequencing 19
  SNP detection 80–1
  whole genome sequencing 20, 71
DNA structure 38–9, 47–51
  base pairing 49–51
  bases 47–8
  helix 51
  phosphate 48
  sugars 48
DNA transcription 39–40, 170–3, 503
dN/dS *see* nonsynonymous to synonymous substitution ratio
Dobzhansky, Theodosius 61, 74
dodder 327
dog 298–9, 304
dominant alleles 84, 218, 491
double helix 49, 51
*doublesex (dsx)* gene 177
Down's syndrome 138
downstream 171, 175, 491
Drake's rule 91
drift *see* genetic drift
*Drosophila* (fruit flies) 62–3, 70, 172, 177, 202, 240, 474, 491
  *D. melanogaster* 33, 68–70, 109
  *D. subobscura* 202
dunnart 298–9

## E

echidna 298
echinoderms 9, 334, 394–6, 439, 440, 491
echolocation 5, 491
EcoTILLING 82
*Ectopistes migratorius* (passenger pigeon) 296
ectoplasmic gliding 335
EDGE species (Evolutionarily Distinct, Globally Endangered) 407
edges (branches) 340–1, 472, 491
Ediacarans 439, 441, 447, 458, 460
effective population size ($N_e$) 83, 91–3, 149, 238, 242–5, 253–5 324, 414–5, 432, 492
eider 97
eletrodetection (in predators) 298
electrophoresis 142
elephants, 370, 414
Ellis-van Creveld (EvC) syndrome 239
empirical 492
ENCODE project 182
endangered species 492
  cloning 313–14
  cryopreservation 316
  inbreeding 258–61
  trade in, forensic applications 28–30, 144–5
endemic species 492
  New Zealand studies 57–60, 258–61, 476–9
endogenous retrovirus (ERV) 146, 164–6, 359–60
endonuclease 101, 123, 142, 492
endosymbiotic bacteria 91, 321–4, 362–4
Entrez search engine 20
environment 205
  genetic 206–8
  influence on mutation rate 409–10
  environmental niche modeling 281
environmental change 204–5
  evolutionary rescue 225–8
enzymes 33, 63–4, 98–9, 141–2, 158–9, 492
  lactase enyme *see lacZ* gene
  esterase enzyme 222–4
  RuBISCo (ribulose 1,5-bisphosphate carboxylase/oxygenase) 266
epidemiology 164, 396–400
  tracing the source of disease outbreaks 480–3
epistasis 225
epitope 242, 492
*Equus quagga see* quagga
error correction *see* copy errors
*Escherichia coli* 69, 171, 492
  antibiotic resistance development 225–8
  colicin production 206
  *see also lacZ* gene
ESTs *see* expressed sequence tags
ethanol 25–6, 28
ethical issues
  biobanking 85–7
  DNA databases 86–7, 143–4
  DNA sampling 42, 85–6, 143, 378
  eugenics 212–13
ethnobotany 400–2
*Eucalyptus* (gum trees) 273, 492
  *E. largiflorens* × *E. gracilis* 273
  *E. pauciflora* 338–41
  *E. regnans* (mountain ash) 264–6
  *E. rossii* 338–41
eugenics 212–13, 492
eukaryote 492
*Euperipatoides rowelli* (Onychophora) 338
*Euprymna scolopes* (bobtail squid) 362–4
eusociality 238, 301, 492
E-value 23
evolutionarily significant units (ESUs) 279, 281–2
evolutionary convergence 6–7, 278, 298–300
evolutionary rescue 225–8
evolutionary trees *see* phylogeny
exhaustive search 23, 272, 492
exons 175, 176–7, 192, 194–5, 492
exploratory probes 191
expressed sequence tags (ESTs) 492
expression *see* gene expression
extinction 67, 225–7, 233, 239, 255–8, 272, 296, 309, 313–5, 492
  mammoths 54–6
  quagga 296–7
  thylacine 264, 296, 309
  gastric brooding frog 315
  *see also* de-extinction
extraction of DNA 42–3, 52–4
extremophile 105, 410

## F

faeces
  fresh 27, 42, 55
  fossilized 55–6, 59, 442
Felsenstein, Joseph 353–4, 392
Felsenstein Zone 353
*Feylinia* (lizard) 381–2
fingerprinting *see* DNA fingerprinting
Fisher, Ronald Aylmer 203, 212, 389, 492–3
fitness 205–13, 217–19, 493
  definitions of 217–18
  hybrids 268–9
  inbreeding effect 238–9, 258–61
fixation 200–1, 204, 208–9, 219, 220–1, 233–5, 238–9, 242, 250, 254, 414–5, 493
flanking sequences 141–2, 493
flightless insects 432–4
flower colour evolution 251, 288–300
forensic applications
  forensic databases 86, 86–7
  illegal rhino horn products 66
  whale meat surveillance 28–30
  timber trade 144
formaldehyde 25–6
fossil record 442–3
  Cambrian fauna 439–41
  fossilization 441–2
four-fold degenerate 186–7
foxtail millet 148
Fragile X syndrome 141
frameshift mutation 65, 176, 493
Franklin, Rosalind 38, 117
frequency-dependent selection 205–6
Fred 74

frog
  Darwin's mouth brooding 315
  gastric brooding 315
frozen zoo 316
fruit flies *see Drosophila*
fugu 176

## G

G-protein-coupled receptors (GPCR) 191
Galapagos finches 204–5
Galton, Francis 35, 169
gamete 96–7, 135–6, 236, 313–4, 493
gaur (*Bos gaurus*) 313
GenBank 19–23, 26, 266, 485
  accession number 20
  barcode submissions 285
  errors and contamination 21–2
  sequence information and annotation 20–1, 22
gene conversion 136
gene expression 170–1, 493
  regulation 173–5, 181–2, 493
  *see also* transcription; translation
gene families 191–3, 206–7, 493
gene flow 260, 273, 279, 282–3, 289–90, 493
general mixed Yule clusters (GMYC) 291–3
generation time effect 124–6, 410–1, 413–15, 455–6, 474, 493
genes 168–75, 493
  discovery of 36–7
  diversity 79
  duplication 139, 161–2, 192, 240–1
  exons 175, 176–7, 492
  identification of 82–3, 178–81
  introns 175–6, 177–8, 496
  names 171–3
  number of 132–3
  overlapping 194–6
  RNA genes 179
  structure 170–1
  *see also* gene expression
genetic code 186–9, 493–4
  alternative codes 187, 188
  degeneracy 186–7
  multiple proteins from a single gene 194–6
  redundancy 186, 240 *see also* degenerate
  start codons 171, 187
  stop codons 171, 172, 187
genetic drift 231–3, 234–6, 249–52, 267, 491
  neutrality 233–4
  population size 236–9
genetic load 230–2
genetic marker 68, 79–83, 89–90, 137, 143, 281–2, 494
genetic resource banks 313–14
genetic screening 212
genome 35, 39, 494
  assembly 321, 323, 366–7
  duplication 137
  history 106–7
  replication 134–6
  size *see* genome size
  viral 37
  whole genome sequencing 20, 71
genome size 65, 66, 91, 132–8, 170
  complexity 133–4, 148, 243
  costs of excess DNA 243–4
genome-wide association study (GWAS) 88–9, 494
genomic system 32–3, 40, 110–11, 494
genotype 39, 47, 57, 65, 79, 142–3, 168, 170, 494
  genotyping individuals 258–60
genus 263, 494
geochronology 451, 467, 494
*Geospiza fortis* (Galapagos finch) 204
germline 35, 41–2, 46–7, 91–2, 96, 124–5, 164–5, 413–4, 494
giant ground sloth 53, 201
Gibbon Ape Leukemia Virus (GALV) 166, 359
Gillespie, Rosemary 278
globin genes 206–8, 300
  alpha globin 207–8
  beta globin 175–6, 208
  duplication 241
  mutations 206–8
*GNAS* gene 194–6
googol 318, 494
*Gopherus agassizii* (Sonoran desert tortoise) 466
gradualism 202, 494
Great Chain of Being 132
guanine (G) 48, 494
gum trees *see* Eucalyptus
Guthrie (heel-prick) test 86, 494

## H

Haeckel, Ernst 332, 335, 494
haemoglobin 206, 207, 494
  *see also* globin genes
haemoglobinopathies 206–8
Haldane, John Burdon Sanderson 199, 203, 211, 216–17, 231, 494
Hamilton, William Donald 212, 494–5
Hamming distance 291, 367, 406, 408, 495
haploid 134–6, 490
haplotype 82–7, 220, 495
  haplotype block 82, 137, 222
HapMap project 82, 83
helicase 118–9, 495
helicase-dependent amplification 123
hemichordates 395–6
herbarium collections 42, 54–6, 401
herbicide resistance 383
heredity, principles of 33–5
Hershey, Alfred 37
Hershey–Chase experiment 37–8
heteroduplex 81–2
heterozygosity, loss of 239
heterozygote 194–5, 211, 218, 268, 495
  heterozygote advantage 211, 217
heuristic search 23, 372, 492
hierarchy of similarities 106–13
  *Bacillus* bacteria 111–13
  history 110–13
  *see also* systematic hierarchy
high-throughput sequencing 70–1, 89, 157–60, 323–4, 495
  alignment 318–19
  pyrosequencing 500
hihi (stitchbird) 259–60
*Hippopotamus amphibius* 6
histones 300
hitchhiking 209–10, 254, 414
HIV *see* human immunodeficiency virus (HIV)
homology 23, 296–302, 495
  alignment 302
  DNA analysis 301–2
  homologous sites 304–5
  similarity by descent 300
homozygosity 239, 254, 258, 495
honeybees 301
honeycreepers 405
hops 89
horizontal gene transfer (HGT) 189, 325–7, 348, 359, 383, 495
  between parasites and hosts 146–7, 325–7
horse phylogeny 354–5
host shift 161–3
hotspots
  biodiversity 273, 432
  indel 322
  mutational 73
  recombinational 82, 136–7
housekeeping genes 256–7, 268, 300, 363–4
Hua, Xia 414, 423–4, 434–7
Hubbell, Stephen 243
human evolution 209–13
human genome project 181
human immunodeficiency virus (HIV) 495
  Al-Fateh Hospital (AFH) outbreak, Libya 480–3
  cytotoxic T lymphocyte epitope 242
  mutation/substitution rate 204, 242, 408, 415
  origins 379–80, 383
  recombination 383
*human Period 2 (hPer2)* gene 68
hunting 28–9
Huntington's disease 84, 140–1, 212, 495
  gene structure 176
Huxley, Thomas Henry 199, 202, 495
hybrid incompatibility 268–9

hybrids 24, 137–8, 268–9, 271–3, 288–9, 314
hybrid speciation 137, 273, 327, 348–9, 365–7
species definition 271–3
hybridization 273, 348, 365–7, 495
SNP detection 81
*see also* DNA hybridization
hypothesis 320, 374, 376–7, 379, 382, 384–5, 495–6

**I**

ice age 4–5, 54–6
Icelanders 84–5
identity 23
illegal logging of tropical hardwoods 144
immigration disputes 142, 144
immunological distance 354–5
*in vitro* 496
inbreeding 238–9, 496
coefficient 258–61
endangered species 258–61
fitness impact 238–9, 258–61
indels *see* insertions and deletions (indels)
individual identification
wildlife forensics 29
microsatellite 142–4
SNP 65–7
information 496
informative pedigree 84
informed consent 85–6
insecticide
Bt (*Bacillus thuringiensis* toxin) 111
makes caterpillars floppy (MCF) 173
nicotinoids 161–2
organophosphates 222–4
resistance 161–2, 211, 222–4
insertions and deletions (indels) 241, 303–4, 306–7, 496
evolutionary significance 321
identification of in endosymbiont genomes 321–4
introns 175–6, 177–8, 496
alternative splicing 177–8, 194
invariant sites 426, 448, 496
*Ipomoea purpurea* (morning glory) 251
isothermal 123
isozymes 234

**J**

Jeffreys, Alec 142
Johannsen, Wilhelm 36, 168, 169
jumping genes *see* transposable elements
junk DNA 496

**K**

Kauri tree 476–9
Kimura, Motoo 231, 248, 250
kingdom 109, 264–5, 302, 334, 348, 384, 496
animal kingdom *see* Metazoa
kite (*Milvus*) 272
kiwis 448–9
knock-out mutations 146, 172, 179
alpha globin 207–8
Koala Retrovirus (KoRV) 164–6, 359–10
Koch, Robert 112

**L**

label 81, 123, 154, 159, 190, 496
fluorescent 81, 156
radioactive labelling DNA 37–8, 156
lactase enyme, *see lacZ* gene
*lacZ* gene 171
expression regulation 173–4, 181
repressor protein 173–4
lagerstätten 440–1, 496
landscape genetics 279–82
last glacial maximum (LGM) 55, 56
*Latimeria* (coelacanth)
*L. chalumnae* (African coelacanth) 404
*L. menadoensis* (Indonesian coelacanth) 407
leading strand 99, 103, 120, 496
lek 237, 253
Lento plot 359
leopard (*Panthera pardus*) 269
leukemia viruses 166, 359–61
*Lialis* (lizard) 381
library creation 159
likelihood 378–9
*see also* maximum likelihood (ML) approach
likelihood ratio test 391, 496
lineage 496
lineage-specific rates 413, 415–6
LINEs (long interspersed elements) 146
linkage 137, 208–9, 216, 254, 496
linkage disequilibrium 80
linkage map 89–90, 367
Linnaeus, Carolus 264
lion (*Panthera leo*) 269
*Liriopsis pygmaea* (parasitic isopod) 72
lizards
legless 381–2
side-blotched 205
local clocks 471–2, 473
locus 67, 141, 496
repeat (microsatellite) locus 141, 143
transposon locus 148
barcoding locus 284–5
long interspersed elements *see* LINEs
long terminal repeats (LTRs) 146, 147
*Lucilia* (blowfly) 222–4
lungfish
Australian (*Neoceratodus forsteri*) 143, 407–8
marbled (*Protopterus aethiopicus*) 65, 66
Lyell, Charles 202, 438
lysis 42, 53, 496

**M**

McClintock, Barbara 145, 149, 152–3
MacDonald-Kreitman (MK) test 221
macroevolution 245, 497
maize 147, 148
*makes caterpillars floppy (mcf)* gene 173
malaria 138–9, 211, 497
anti-malarial drug resistance 219
avian malaria 405
beta globin alleles and malaria resistance 209–11
male-driven evolution 96–7
*Malus* (apple) 365–8
Mammals 297–9, 410
evolutionary history 7
phylogeny 8, 148
mutation rate 91, 96–7
rate of molecular evolution 410, 414
mammoths 42, 55–6, 314–15, 370
manual sequence alignment 318–19
Markov Chain Monte Carlo (MCMC) procedure 429–31
*Marmota marmota* (alpine marmot) 144
marsupial mole 298, 299, 305
marsupials 264, 297–300, 497
mate choice 144, 198, 268
in conservation 260
effective population size 253
*see also* lek; reinforcement
matriphagy (mother-eating) 201
maximum likelihood (ML) approach 257, 291, 326, 363, 389–92, 497
likelihood scores 391
procedure 390–1
Maynard Smith, John 15, 46, 78, 96, 203, 205, 206, 232, 419, 497
mean phylogenetic distance (MPD) 401
medicinal plants 400–2
megafauna 54–6
meiosis 134–6, 139 *see also* gamete
*Melipona* (stingless bees) 301
melt (separate DNA strands) 103–4, 118–19, 133, 497
in DNA amplification 121–3
melting temperature 81–2, 121, 190–1, 497
as a measure of similarity 355
Mendel, Gregor 36, 67, 168, 497
messenger RNA (mRNA) 39–41, 170–1, 497
transcriptome 159, 174–5
translation 40, 170–3, 194, 503
metabolic rate 410–11
Metazoa 439, 443, 445 *see also* Cambrian explosion
methylation 73, 102, 182, 194, 497
microarray 81, 497
microevolution 245, 497
microsatellites 139, 141–3, 161–3, 497
divergence analysis 288–300
individual identification 142–4
species identification 29
population history 130, 288–9
family relationships 144–5, 259–61

microsporidia 302, 497
Miescher, Friedrich 37
migration 127–30
*Milvus migrans* (black kite) 272
*Milvus milvus* (red kite, Cape Verde kite) 272
mimicry 177, 270
minimum evolution (ME) method 357
minisatellites 141
mismatch 497
  alignment scores 309, 317, 320
  SNP detection 81
  microsatellite mutation 140
  repair 101–2
mitochondria 4, 128, 130, 301–2, 409–10, 497–8
mitochondrial DNA 25–9
  genetic code 187, 188
  genome 98, 119, 325
  horizontal gene transfer 325–7
  maternal inheritance 4, 128–30, 382
  mutation rates 410
  nuclear copies 58, 189, 285, 319
mitosis 498
MK (McDonald-Kreitman) tests 221
moas 57–60, 448–9, 498
  phylogeny 58–9
mobile genetic elements 498
  *see also* transposable elements
mocker swallowtail butterfly (*Papilio dardanus*) 270
model 498
  maximum likelihood analysis 389–91
  substitution models 424–6, 452
  rate (clock) models 456–7, 471–74
  sequence evolution 192, 253, 255, 257, 336, 376–8, 408
  sequence alignment 318
  *see also* null model
modifier loci 208
modern synthesis 74, 217, 245,
  *see also* neo-Darwinian
mole
  European 298–9, 304–5
  marsupial 298–9, 304, 305
molecular clocks 498
  local clocks 471–2, 473
  relaxed clocks 456, 457, 459, 472, 473, 501
  sloppy clocks 451–4, 467
  strict clocks 471
molecular dating 58–60, 444–7, 455–60, 471–5
  calibration 465–70
  rate change modelling 456–7
  substitution rate estimation 447–53
  *see also* molecular clocks
molecular evolution
  body size relationship 411–13
  generation time effect 413–15, 493
  lineage-specific rates 413, 415–6
  models 377–8
  rate variation 406, 408–19
  species richness 434–7
  variation across sites 408–9, 426
  *see also* mutation; substitution
Molecular Operational Taxonomic Unit (MOTU) 55, 56, 286
molecular phylogenetics 10–11, 335–40
molecular taxonomy 274–5
Monera 335
monotremes 297, 298, 498
monogamy 144
Moreau de Maupertuis, Pierre-Louis 34
Morgan, Thomas Hunt 152, 168, 498
Moritz, Craig 281
Morlon, Hélène 388–9
morphology 5, 498
  morphological vs molecular change 59–60, 250, 270, 275, 298, 301–2
  phylogeny 298, 301–2, 336–7
  species identification 270, 275, 285
morphospecies 272–3, 291, 292
mosquitoes 138–9, 211
  evolutionary divergence 138
moss piglets 181
mountain ash (*Eucalyptus regnans*) 126, 264, 265
Muller, H. J. 212–13
  Muller's Ratchet 256
multiple hits 374–6, 406–7, 498
multiple sequence alignment 317–20
museum specimens 57–8
mussels 446–7, 459
mutagens 63, 409, 498
  reactive oxygen species (ROS) 409–10
  ultraviolet (UV) light 63–5
mutation 62, 498
  accumulation 91–2
  advantageous 72–3, 221
  deleterious 71, 221, 257
  directed mutation 74
  frameshift mutations 65, 176, 493
  knock-out 146, 179
  linkage effects 208–9
  nearly neutral 243–4, 251, 498
  neutral 220, 221, 233–6, 249–50
  nonsense mutation 176
  point mutations 62–3
  randomness 73–4
  rate *see* mutation rate
  synonymous and nonsynonymous 220–1, 252, 256–7
  transition 73–4, 503
  *see also* DNA repair; substitution
mutation rate 68–73, 409–11, 413–15
  environmental influences 409–10
  generation time effect 413–15, 493
  laboratory study 91–3
  metabolic rate 410
  mitochondrial DNA 410
  species richness 434–7
  variation 73, 409–11
  *see also* mutation; substitution rate
mutational hotspots 73
mutator 72–3, 414, 498
*Myzus persicae* (aphids) 161–3


## N


National Center for Biotechnology Information (NCBI) 19
natural selection 34, 201–2, 502
  *see also* selection
nearly neutral theory 243–4, 251, 498
Nee, Sean 242
negative selection 202, 220, 239–40, 241, 502
neighbour-joining 354
nematodes 54–6, 450
  bacterial symbionts 173
  *Pratylenchus coffeae* (smallest animal genome) 66
$N_e$ *see* effective population size
neo-Darwinian 46, 74, 217, 245, 406, 498
  *see also* modern synthesis
*Neoceratodus forsteri* (Australian lungfish) 143, 407–8
nested hierarchy *see* hierarchy of similarities
networks 349–50, 358–61, 367
  consensus network 363, 364
*Neurospora crassa* 33, 109, 498
neutral alleles 208–9, 220–1, 233–4, 290, 499
neutral theory (molecular evolution) 230–6, 249–52
neutral theory of biodiversity 243
New Zealand
  biogeographic history 476–9
  endemic species studies 57–60, 258–61, 476–9
Newfoundland dogs 238
next generation sequencing (NGS) *see* high throughput sequencing (HTS)
nicotine tolerance, aphids 161–3
nodes 335, 340, 341, 499
nomenclature 263–4
  binomial system 263–4
non-coding DNA 182, 243–4, 499
  non-coding SNP 68
nonsense mutation 176
nonsynonymous mutations 195, 220–1
nonsynonymous to synonymous substitution ratio (dN/dS) 221, 252, 256–7
*Notiomystis cincta* (hihi/stitchbird) 259–60
nuclease inactivation 52

nuclein 37
nucleotides 48–9, 121, 499
homologous 304–5
substitution models 424–6
nucleus 4, 39–40, 313, 322, 499
null model 243
null distribution 401
null hypothesis 343, 495

## O

Oceania settlement models 127–30
Ockham's razor 373–4, 427
Ohta, Tomoko 248–9, 251
Okazaki fragments 120, 496
oligonucleotide 121, 189, 499
onychophorans 337–41, 374–5, 499
operon 171
open reading frame (ORF) 178, 191–2, 499
OPV-AIDS hypothesis 379–80
orchid 270, 271, 348, 349
origin of replication 98, 119
organelles 322, 499
organelle DNA 52, 58, 189, 286–5, 319
*see also* mitochondria; chloroplast
orthologs 499
*Oryza sativa* (rice) 32, 499
Osborn, Henry Fairfield 370
overparameterization 427, 499

## P

p-distance 291, 499 *see also* Hamming distance
palindromic sequences 174
palynological *see* pollen
PAM (Point Accepted Mutation) matrix 312, 317, 426
pangenesis 34–5
*Panicum virgatum* (switchgrass) 88–90
*Panthera*
*P. leo* (lion) 269
*P. pardus* (leopard) 269
*Papilio dardanus* (mocker swallowtail) 270
paralogs 319, 361, 499
*Paramecium tetraurelia* 91–3
parametric bootstrap 394–6
parasites 72, 145–6, 301–2
parasitic plants 325–7, 472–3
parataxonomist 292–3
parsimony 373–4, 375, 499
passenger pigeon 296
paternity assessment 142, 144
Pauling, Linus 37, 499
*Pax6* gene 172
PCR 28–30
pedigree 84, 164, 499
informative pedigree 84, 488
Penny, David 127–30, 149, 334
Peptide 40, 171, 499 *see also* protein
penetrance 84
phasmid 255–7
phenetics 384, 499–500
phenotype 39, 47, 65, 168–70, 500
phenylalanine hydroxylase gene (*PAH*) 205
phenylketonuria (PKU) 205, 212, 213, 500
*Phlox* 288–300
phosphate 48–51, 500
phosphodiester bonds 48–50, 119, 500
*Photorabdus luminescens* 173
phylogenetic species concept 271–2
phylogeny 5, 57–8, 332–5, 500
alternative histories 382–4
conflicting data 347–9, 359–61
dinosaurs 337
distance methods 345–7, 354–8
horses 354–5
mammals 8
method selection 384–5
minimum evolution 357
molecular 10–11, 335–40
morphological 298, 301–2, 336–7
multiple hits 374–6, 406–7, 498
networks 349–50, 358–61
non-tree-like evolution 348–9
reading trees 343–4
reconstruction 340–4
splits 342–3
statistical inference 376–85
tree comparisons 370–6
tree evaluation 378–9
uncertainty 380–2
phylogeography 279–83
phylotranscriptomics 174
phylum (plural: phyla), 9, 265, 439, 500
*Physeter catadon see* sperm whales
pipeline 80, 157
placenta 146–7
placental mammals 297, 500
plague pandemics 396–400
plankton 5, 500
plants
carnivorous 43
crop plants 82, 88–6, 148, 316, 368
diversity 54–6
herbarium collections 54–6, 316, 400–2
medicinal 400–2
parasitic plants 325–7, 472–3
substitution rate 124–6, 413, 434–7
taxonomy 265–6
*see also* specific plants
*Plasmodium see also* malaria
*Platycercus flaveolus* (rosella) 271
platypus 298
ploidy 134, 136
*see also* diploid; haploid; polyploidy
point mutations 62–3, 498
polio vaccine 379
pollen 54–6
polymerase 118–23, 500
*see also* DNA polymerase; RNA polymerase
polymerase chain reaction (PCR) 103–6, 121–3, 500
primers 103–4
Taq polymerase 105–6
temperature cycling 104, 123
polymorphism 66, 205–6, 221–2, 500
restriction fragment length polymorphism (RFLP) 141–2, 491
insertion polymorphism 148
*see also* single nucleotide polymorphisms (SNPs)
polyploidy 134–5, 136, 349, 367
speciation and 137, 349
population 500
population size 236–9, 252–5
effective population size 93, 238, 253, 492
estimation 143, 254–5
genetic drift 236–9, 250
Porifera (sponges) 444–5, 489
*Porphyra purpurea* (red algae) 135
positive selection 192–3, 202–3, 220, 239, 502
substitution rates 241–2
Precambrian slow-burn hypothesis 443–5
prestin 5
primase 99, 103, 104, 120
primers 99, 120, 189–91
degenerate 191
design of 190–1
DNA amplification 103–4, 121, 123
prions 41
probes 123, 158–9, 164–5, 189–91
promoter 145, 162, 171, 173–4, 194, 500
proofreading 72–3, 100–1, 500
efficiency 100–1
protein-coding sequence 20, 171, 175–6, 220–1, 242, 319–20, 500
protein 18, 24, 36, 39–40, 110, 170–1, 186–91
*see also* translation; peptide; protein-coding gene
proteinase 52, 58
*Protopterus aethiopicus* (marbled lungfish) 65, 66
pseudogenes 139, 180, 192–3, 220–1, 240–1, 322–4, 500
pure-bred line 36, 168–9, 238
purine 47–9
pyrimethamine resistance 219
pyrimidine 47, 49
pyrosequencing 501

## Q

quagga 24, 355, 500
de-extinction 296–7
extinction 296
quoll (*Dasyurus*) 298, 299, 300, 304

## R

radiation damage
resistance 409

radiation damage (*continued*)
UV radiation 63–5
*see also* DNA repair
*Rafflesia* 325–7
Rambaut, Andrew 464–5
random 500
mutation 71, 73–4
random matches in DNA 24, 143–4, 190, 319
random primers 104, 123
random sampling 200, 239, 401, 430–1, 453
random walk 236, 429
*see also* drift
raspberries 416–18
rate of molecular evolution 409–19
body size relationship 411–13
branch length estimation 406–8
environmental influences 409–11
generation time effect 413–15, 493
lineage-specific rates 413, 415–16
variation 405–6
rate (clock) models 456–7, 471–4
*see also* mutation rate; substitution rate
Ray, John 331
reactive oxygen species (ROS) 409–10
reading fram 194–6
reanneal 81, 133–4
reasons to be cheerful 1, 2, 3
recessive alleles 212, 239, 260–1, 500
recombination 136–7, 383, 490, 500
hotspots 82, 136–7
V(D)J recombination 136, 146
reduced representation sequencing 80, 89, 158
reduction-division *see* meiosis
redundancy 186, 240
regulatory sequences 146, 162, 181–2, 288–90
identification of 181–2
reinforcement 288–90, 501
relatedness
population genetics 258–60
taxonomy 264, 266, 275, 297
phylogenetic comparative methods 412
relaxed clocks 456, 457, 459, 472, 473, 501
repair pathways *see* DNA repair
repeat sequences 139–45, 501
comparing 141–5
Fragile X syndrome 141
Huntington's disease 141
tandem repeats 139, 141
terminal repeats 145–6
*see also* microsatellite; VNTR; STR
repetitive DNA 132–4
origins of 134
*see also* repeat sequences
replacement substitutions 241, 501
replication *see* DNA replication
replication bubbles 99, 118
replication fork 99, 119–20
replication origins *see* origin of replication
reproductive technology 313–14
restriction enzymes 141–2, 501
*see also* RFLP
retroviruses 146, 501
endogenous (ERV) 146, 164–6
*see also* human immunodeficiency virus (HIV); Koala Retrovirus (KoRV)
reverse transcriptase 41, 164–6, 501
restriction enzymes 89, 141–2, 148, 158, 164
RFLP (restriction fragment length polymorphism) 141–2, 491
*Rheobatrachus* (gastric brooding frog) 315
rhinoceros
black (*Diceros bicornis*) 67
white (*Ceratotherium simum*) 66, 67
woolly 55
*Rhinoderma darwinii* (Darwin's mouth brooding frog) 315
ribonculeic acid *see* RNA
ribose 39, 48
ribosomal RNA (rRNA) 179, 307, 501–2
rRNA genes 179, 449, 501
ribosomes 39–40, 174, 179
rice 32
rifampicin resistance 225–8
RNA 39
messenger RNA (mRNA) 39, 497
ribosomal RNA (rRNA) 179, 307, 501–2
sequence alignment 306–8
structure influence on gene sequence 179–81
transfer RNA (tRNA) 39–40, 180, 503
RNA polymerase 32–3, 120, 171, 300, 502
sequence conservation 107–10, 239–40
synthesis 39–40
rockfish 414
rosella (*Platycercus flaveolus*) 271
rotifers 275, 409
Rowell, Dave 338
RuBISCo (ribulose 1,5-bisphosphate carboxylase/oxygenase) 266, rBCL gene 266, 285
*Rubus* (raspberries) 416–18


## S


*Sacculina* (barnacle) 72
sampling of DNA 42, 84–6
ethical considerations 42, 85–6, 143, 378
Sanger, Fred 18–19
Sanger sequencing 19, 153–6, 502
satellite DNA 134, 139–40
*see also* microsatellites
saturation 407–8, 449, 502
*scala naturae* 132
scientific literature 485–6, 502
scientific research 485
dissemination of results 485–6
Seago, Ainsley 490, 502
*Sebastes* (rockfish) 414
segregation 36, 82, 502
selection 92–3, 267–8, 502
artificial 502
background selection 254
costs of 231
detection of 220–2
evolutionary rate relationship 415
frequency-dependent 205–6
negative 202, 220, 239–40, 241, 502
positive 202–3, 220, 239, 241–2, 502
power of 202–5
costs of 231
mathematical theory 203
sexual selection 413
selective coefficient 218, 243, 251, 253
selective sweeps 209–10, 217, 219, 221–2, 289,502
sequence evolution 175–82
RNA polymerase II beta 107–10
*see also* molecular evolution; mutation
sequencing *see* DNA sequencing
sex chromosomes 67, 97, 382,
sexually-deceptive flowers 271
sexual dimorphism 59, 97, 433
sexual reproduction 96–7, 136–7, 255–8, 268, 326, 382
short interspersed elements *see* SINEs
short tandem repeats (STR) 139–42, 234
*see also* repeat sequences
shotgun sequencing 19
shrew 298–9
sickle cell anaemia 211
signal 308–9, 349–50, 358–61, 363, 502
silent changes 502
silica 52
simian immunodeficiency viruses (SIV) 379–80
similarity 345–50
*see also* hierarchy of similarities
Simpson, George Gaylord 403, 405–6, 502
simulated data 394–6, 502
SINEs (short interspersed elements) 146, 148
single molecule DNA sequencing 53, 157
single nucleotide polymorphisms (SNPs) 65–8, 79–83, 503
analysis of 82–3, 142
crop development application 88–90
detection of 79–82
individual identification 65–7
tracking inherited traits 67–8
six-fingered dwarfism *see* Ellis-van Creveld (EvC) syndrome

sloppy clocks 451–4, 467
sloth poo 53
small populations *see* population size
*Somateria mollissima* (eider duck) 97
somatic nuclear cell transfer 314–15
*sonic hedgehog* (SHH) gene 173
SOS response 63–5
speciation 134, 267–8
  chromosomal inversions 136
  polyploid 137
  rate 434–6, 472
  reinforcement 288–90, 501
  sympatric 163
species 269–75, 503
  biological species concept 270–1
  classification *see* taxonomy
  cryptic species 274
  definition 272–4
  discovery 286
  DNA barcoding 274–5, 284–6
  identification 272, 284–6
  morphospecies 291, 292
  phylogenetic species concept 271–2
spectral plot 359
Spencer, Herbert 503
sperm 96–7
sperm whales 2–5, 10
spiders
  matriphagy 201
  radiation 278
spliceosome 176, 179, 496
splicing
  alternative 177–8, 194
  trans splicing 177
sponges 440, 444–5
sponge pudding 107
spontaneous generation 34
Sprigg, Reg 439
squid, bobtail 362–4
standing variation 204–5, 222–4, 239, 435–6
starfish 9, 394
start codons 171, 187–8, 194
stingless bees 301
stitchbird (hihi) 259–60
stop codons 171, 172, 178, 186–7, 192, 195
  alignment 241, 319
strand displacement amplification 123
strict clocks 456, 471, 481
structure of DNA *see* DNA structure
substitution 200, 204, 233, 267, 406–8, 503
  models 424–7
  estimation 447–53
  patterns of 239–44
  relationships 337–40
  selection effects 220, 241–2
  variation across sites 408–9, 426
  synonymous versus nonsynonymous mutations 220–1, 252, 256–7
  *see also* molecular evolution
substitution rate 415–19
  comparing 415–18
  flightless insect study 432–4
  lineage-specific rates 413, 415–16
  neutral substitution 240–1, 250–1
  rate change modelling 456–7
  relative rate tests 416–18
  variation 406–9, 415–19, 418–19
  *see also* molecular evolution; substitution
sugar 503
  DNA structure 38–9, 47–51, 120
  metabolism 171, 174
  taste 192–3
Svalbard Global Seed Vault 316
Swift, Jonathan 145
switchgrass crop development 88–90
symbiont 362, 365
synapomorphies 305, 342
syncytin gene 146, 147
synonymous mutations 220–1, 503
  nonsynonymous to synonymous substitution ratio 221, 252, 256–7
  selection 244
systematic hierarchy 297–300
  classification 266, 331
systematics 503

# T

*Takifugu rubripes* (fugu) 176
tandem repeats 139, 141
  *see also* repeat sequences
Taq polymerase 105–6
tardigrades 181
target amplification 158
targeted enrichment 189
Tasmanian tiger *see* thylacine
Tasmanian West Coast Monster 2, 26
taste receptor genes 191–3
TATA box 181, 182
taxon (*plural:* taxa) 340
taxonomy 263–6, 272, 503
  classification 264–6, 331
  DNA taxonomy 274–5
  relatedness 266
Tay-Sachs disease 212
temperature cycling 104, 123
terminal repeats 145–6
termination sequences 174
*Theobroma cacao* (cacao) 287
thermal cycling 123
*Thermophilus aquaticus* 105
thylacine (*Thylacinus cynocephalus*) 264, 303–4, 503
  extinction 296, 309
  homology 297–300
  resurrection 309, 314–15
thymine (T) 47, 503
  repair pathway 63–5
  UV damage 63
TILLING (Targeting Induced Local Lesions In Genomes) 81–2
*Timema* (walking stick insect) 256–7
topoisomerases 119
topology 379, 392, 395, 428, 476, 503
tortoise, Sonoran desert 466
trait 503
trans splicing 177
transcription 39–40, 170–3, 503
transcriptome 159, 174–5
transfer RNA (tRNA) 39–40, 185, 503
  structure 180
  tRNA synthetase 110–11
transgenic 162–3
transition 73–4, 424–5, 503
transition/transversion ratio 377–8, 425
translation 39–41, 170–3, 188, 503
transposable elements 145–9
  endogenous retroviruses (ERV) 136, 146–7, 164–6,
  LINEs (long interspersed elements) 146
  SINEs (short interspersed elements) 146, 148
  terminal repeats 145–6
transposition 146–7
  genomic record of 147–9
tree of life 332–6
  non-tree-like evolution 348–9
  reading trees 343–4
  rooted and unrooted trees 343–4, 359
  *see also* phylogeny
tree-like data 128
trees, evolutionary rate 124–6
*Trigonopterus* (weevils) 291–3
Turkmen people 108
twins 79, 147, 313–14, 490
Type 2 diabetes 84–5

# U

UK Biobank 85
ultraviolet light (UV) damage 63
  DNA repair pathway 63–5
ultrametric 356–7
uncorrected p-distance 291
unified neutral theory of biodiversity 243
unwind 118–21
*Uta stansburiana* (side-blotched lizard) 205

# V

*Vampyrella* 335
variable number of tandem repeats (VNTRs) 141
variation 198–205
  genetic load 230–2
  human populations 210–12
  standing variation 204–5
V(D)J recombination 136, 146
velvet worms (Onychophora) 337–8, 439, 479, 499
venomous mammals 298
*Vibrio fischeri* (bioluminescent symbiotic bacterium) 362–4
voucher specimens 285, 291

## W

walking stick insect 256–7
Wallace, Alfred Russel 213, 504
Watson, James 38–9
weevil diversity, New Guinea 291–3
Weismann, Friedrich Leopold August 35, 46–7, 504
Wet Tropics region, Northern Australia 281–2
whales 5
  DNA use to trace whale meat 28–30
  evolutionary history 5–9
  *see also* sperm whales
Whippo Hypothesis 6
whole genome sequencing 19, 20, 71, 80, 366, 393
wildlife forensics *see* forensic applications
wobble 180
wolf
  coyote hybridization 272
  red (*Canis rufus*) 263, 272
  Tasmanian *see* thylacine
woolly mammoth *see* mammoths
Wright, Sewall 203, 504

## X

XLαs protein 194–6

## Y

Y-chromosome 128, 130, 138
Yangtze river dolphin (baiji, *Lipotes vexillifer*) 315
*Yersinia pestis* (plague bacterium) 397–9

## Z

zebroid 24
zeta globin gene 241